ISBN 978-1-5278-0212-4
PIBN 10902531

This book is a reproduction of an important historical work. Forgotten Books uses
state-of-the-art technology to digitally reconstruct the work, preserving the original format
whilst repairing imperfections present in the aged copy. In rare cases, an imperfection in
the original, such as a blemish or missing page, may be replicated in our edition. We do,
however, repair the vast majority of imperfections successfully; any imperfections that
remain are intentionally left to preserve the state of such historical works.

1 MONTH OF
FREE
READING

at
www.ForgottenBooks.com

SUBTROPICAL ENTOMOLOGY

WALTER EBELING
ASSOCIATE PROFESSOR OF ENTOMOLOGY
AND
ENTOMOLOGIST IN THE EXPERIMENT STATION
UNIVERSITY OF CALIFORNIA, LOS ANGELES

PUBLISHED BY
LITHOTYPE PROCESS CO.
123 FOLSOM STREET
SAN FRANCISCO, CALIF., U.S.A.

Dedicated to

PROFESSOR HENRY J. QUAYLE,

for over forty years a pioneer in
research on the biology and
control of subtropical
fruit pests.

PREFACE

This book is the outgrowth of teaching experience in a course called "Insects Affecting Subtropical Fruit Plants" required of students majoring in Subtropical Horticulture in the College of Agriculture, University of California, Los Angeles. The students qualifying for this course have had a good educational background in the physical and biological sciences, but the majority have had no introductory course in general entomology. The primary consideration in the preparation of this volume has been to adapt it to the needs and educational background of the students in the aforementioned course. Thus the subject matter and its arrangement are not in all chapters necessarily suited for the widest possible popular usage of the text.

A brief orientation on the scope and importance of the science of entomology, on morphology, metamorphosis, and classification, and on the recognition of the orders of insects, is first presented as a preparation for further study. Then follows a section on entomological organization and legislation. Cultural and mechanical control measures are only briefly discussed, for they are little utilized in subtropical entomology. A discussion of the chemistry, toxicology and formulations of the insecticides, as well as the equipment used in their application, form a prominent part of this volume. Oil spray and fumigation are especially emphasized because they loom relatively large in the citrus pest control picture. Next follows a concise treatment of the theoretical basis of biological control and the rôle of beneficial insects in the control of subtropical fruit pests.

The above subject matter comprises 18 of 30 lectures presented in the course. The question may arise, at this point, as to whether the title of the course and the text will be justified in the next 12 lectures. It will be found, however, that the background so far presented has made possible an efficient and easily understandable presentation of the subject matter pertaining more specifically to biology and control of subtropical fruit pests, which is to follow.

Since the course is designed primarily as a supplement to the major in Subtropical Horticulture, the pests discussed in the last ten chapters of this text are grouped according to the varieties of subtropical fruits which they attack, rather than according to the more common method of grouping according to insect orders, in the sequence of their evolutionary development, without any relationship to their hosts. For a given crop, however, the pests are discussed in the order of their phylogenetic relationships, beginning usually with the lower phyla (nematodes and snails), the mites, and then the insects in the ascending evolutionary status of the orders.

Citrus fruit pests have been given by far the greatest attention. The control of citrus insects and mites generally constitutes, from an economic standpoint, the major cultural operation in a citrus orchard in California, and apparently in the majority of citrus-growing regions, and it is certainly the most complex and the most subject to change. Continual changes in pest control methods are brought about in part by the resistance which some insects develop to existing insecticides, which makes necessary the development of new control measures. Likewise the increasingly greater numbers of new compounds of insecticidal value, resulting from the unceasing expansion of the nation's huge chemical industry, result in frequent changes in the pest control program as old insecticides give way to new. Improvements in pest control equipment have a similar effect. An exception to this general tendency may be noted in the case of the armored scale insects, in the control of which oil spray and fumigation have to date not been superseded by new insecticides.

Changes in the pest control program may also be caused by a number of other factors which are responsible for changes in the relative importance of the various species of pests. These factors will later be discussed in detail.

SUBTROPICAL ENTOMOLOGY

Special emphasis is given to pests and pest control problems of California, but the subtropical fruit pests of other states are listed, and a discussion of the economic importance and control of the more important of these pests is briefly presented. In Chapter 19, a series of maps indicate the location and extent of the citrus regions of the world. Each map is accompanied by a list of citrus pests, arranged in the order of their importance, for the various citrus-growing regions indicated on the map.

Although in this text main emphasis is placed on the citrus pests of California, the greater part of the information contained therein is applicable throughout the citrus regions of the world. World commerce has made the majority of citrus pests cosmopolitan, within the limits of their ecological requirements. This is especially true of regions connected by direct trade routes, as, for example, California and Australia. In nearly all citrus-growing regions of the world, the introduced species are far more serious pests than the indigenous species.

The most widespread of the citrus scale insects is the purple scale, Lepidosaphes beckii (Newm.), and control measures found to be successful against this insect in one region can usually be applied, perhaps with modifications depending on economic and social conditions, in others. The California red scale, Aonidiella aurantii (Mask), the yellow scale, Aonidiella citrina (Coq.), the Florida red scale, Chrysomphalus aonidum (L.), the dictyospermum scale, Chrysomphalus dictyospermi (Morg.), and the rufous scale, Selanaspidus articulatus (Morg.) are so similar in structure, life history, habits, and response to the usual insecticides, that experience gained in one region against one or more of these pests is of value in combating the other species in widely separated regions. The same may be said for the soft scale insects and for the nematodes, snails, mites, thrips, aphids, bugs, whiteflies, moths, beetles and fruit flies attacking citrus.

Common pest control problems also exist throughout the subtropical regions of the world with respect to some of the pests attacking grapes, walnuts, avocados, olives, figs, dates, and other subtropical fruits discussed in this volume.

Teaching experience has shown that appropriate illustrations, especially photographs, greatly aid in effective presentation of a course in entomology. Likewise, illustrations supply information that is the most easily assimilated and the most effectively retained by students, professional entomologists, pest control operators, and fruit growers. In this volume the photographs of the subtropical fruit pests are for the most part original, and are considered to be an important auxiliary to the text.

It is hoped that this volume may be revised often enough to keep the subject matter reasonably up to date. Consequently, the writer shall be grateful for any suggestions as to possible improvements at the times of reprinting or revision, in the way of more effective organization or presentation as well as greater clarity or readability of any or all sections of the text. Likewise the writer will be glad to have his attention called to any errors that may occur in the text, as well as such omissions as are not justifiable in view of the scope and purpose of the book.

The wide range of subject matter encompassed by this text, with much of which the writer is relatively unfamiliar, indicates the extent of his indebtedness to many individuals and many sources of information, too many to enumerate. The pioneer work of Professor H. J. Quayle on the subtropical fruit pests and their control is contained in many publications extending over a period of four decades and including his widely read book, Insects of Citrus and Other Subtropical Fruits. I wish to express my indebtedness for this great source of original information.

The Pest Control Bureau of the California Fruit Growers Exchange, which until January 1, 1949 was under the direction of Mr. R. S. Woglum, has rendered the citrus industry a great service through the monthly distribution of the "Pest Control Circular". The frequent reference to the Circular in this book is a measure of my indebtedness to the entomologists of the Exchange Pest Control Bureau.

PREFACE

I have been particularly fortunate in having obtained the cooperation of many colleagues in the University of California who have been so generous and so helpful in reviewing and criticizing the chapters of the text dealing with their particular specialties and fields of research. Among these are A. D. Borden, A. M. Boyce, G. E. Carman, P. H. DeBach, E. O. Essig, S. E. Flanders, F. A. Gunther, D. L. Lindgren, E. G. Linsley, R. L. Metcalf, H. S. Smith, and E. M. Stafford. Others within the University have also been helpful in many ways.

Mr. H. J. Ryan, County Agricultural Commissioner for Los Angeles County, read and criticized the chapter on "Entomological Legislation and Organization". Federal, state and county agencies as well as commercial companies have contributed much information and many photographs, for which credit is given at the point of insertion. The preparation of the maps and the listing of citrus pests, in the order of their importance, in the various citrus-growing areas of the world, involved a voluminous correspondence with many government agricultural officials in nearly all the citrus-growing countries of the world. Without their generous and courteous cooperation, the greater part of the information contained in Chapter 19 could not have been obtained.

Mr. Roy J. Pence, Principal Laboratory Technician in the Division of Entomology at the University of California, Los Angeles, was responsible for the great majority of the original photographs appearing in this volume. His ingenuity and skill both in the preparation and the photography of insect material merit special commendation. In the course of this work, Mr. Pence has made several worthy scientific contributions in the field of photomicrography which have been published in scientific journals.

Walter Ebeling

University of California,
Los Angeles 24, California,
April, 1949.

ERRATA

Page
viii . . . . Line 21
538 . . . . Line 20
            Selanaspidus articulatus should be _Selenaspidus articulatus_.
690 - 162a._____1916· Life history and feeding records of a series of
             California Coccinellidae. Univ. Calif. Publ. Ent. 1:251-99.
                 (this carries a cleared name -- Clausen, C. P.)

TABLE OF CONTENTS

CHAPTER I
ORIGIN AND SCOPE OF APPLIED ENTOMOLOGY

The term "entomology" is derived from the Greek _entom_, meaning "cut in." The term "insect", derived from the Latin, also means "cut in." These terms both refer to the conspicuous segmentation which gives insects an appearance of having been "cut in" at various places throughout the length and breadth of their anatomy.

Aristotle, who proposed the term _entomon_ for insects, and who devised an alary system of classification in 384 B.C., might be considered as the "father" of Entomology, although Hippocrates, the "father of medicine", had written much about insects a half a century earlier. The old Greek civilization that nurtured Hippocrates and Aristotle was soon to fall, its demise being attributable in part to the mosquito-borne malaria introduced by invading Persian armies.

After the cultural blackout of the middle ages, interest in the scientific study of insects was revived. As long as 300 to 400 years ago remarkable studies on the morphology and life history of insects were being made. Illustrations were made that have remained unexcelled in accuracy and quality of workmanship. Amateur entomologists had collected assiduously, and taxonomy as a science had developed to a remarkable degree. On the other hand applied entomology had been limited to occasional feeble sorties against the insect enemy with such questionable weapons as sage, leek, wormwood, pepper, salt, coffee grounds, ashes, lye, vinegar, soot, urine, and animal dung. Many amusing concoctions widely used as late as the nineteenth century are cited in Lodeman's "The Spraying of Plants" (1896). The tendency was to use concoctions offensive to the human senses or having their origin in ancient and medieval superstitions.

The approximate rate of "progress" in applied entomology over a period of about 18 centuries is indicated by two quotations

(1) Pliny 77 A.D.: "Ants are a great pest to trees; they are kept away however, by smearing the trunk with red earth and tar; if a fish, too, is hung up in the vicinity of the tree, these insects will collect in that one spot."
(2) United States Commissioner of Agriculture, 1865, under remedies and preventives for the apple-worm and the curculio: "Strong-smelling herbs, such as tansy and elder-leaves and blossoms or other nauseous matters not agreeable to the olfactory nerves of the insect are hung among the branches, in hopes the insect will give them a wide berth." (773)

Man's passive resistance to the inroads of his insect enemies throughout the many centuries of historic record can hardly be ascribed to the absence of urgent need for action. The insects have always exacted a heavy toll from his available food supply. In the many instances of insect plagues, the luckless inhabitants of wide areas were often reduced to starvation. There is a familiar ring in the words of the prophet Joel as he bewailed the woes of the ancient Hebrews in these poignant words: "That which the palmer-worm hath left hath the locust eaten; and that which the locust hath left hath the canker-worm eaten; and that which the canker-worm hath left hath the caterpillar eaten."

Probably mankind has suffered most, however, from the insects which carry disease organisms. Besides exacting their appalling toll of death and suffering, insect vectors of disease had been, until World War II, more decisive than armies, guns, and generals in affecting the course of war, and had thus been decisive factors in the course of human history.

It may be argued that the causal relation of insects and disease was for the most part unknown until recent times, but man has certainly always been aware that insects were robbing him of a large part of his food supply. The question remains to be answered then, as to why men of learning, leisure, and unquestioned ability, have for centuries examined, dissected, pinned, described and bestowed Latin names upon insects without any concerted effort to combat them.

The disproportionate rate of development of taxonomic as compared to applied entomology would make fascinating subject material for a sociologist, for it involves a basic social phenomenon of fundamental importance in man's social and economic development. Until comparatively recent times the distinction between social classes was very rigorously observed, and chief among the distinguishing features of these classes was their manner of employment. The "upper classes" carefully avoided productive labor or any activity which could be interpreted as being economically useful, this type of activity being strictly limited to slaves and serfs or, later, to the disdained "working classes". From the time of the Greek philosophers to the present, a degree of conspicuous leisure, and more especially an exemption from contact with such industrial processes as serve the immediate everyday purposes of human life, have always been recognized as perquisites of the life of an upper class gentleman.

Since until comparatively recent times only the upper classes had the opportunity to become educated, the absence of such a socially debasing activity as is implied in the term "applied entomology" among educated people is readily understood. A century or more ago the collection and classification of insects could be engaged in by gentlemen of high social standing only because such activity carried with it no vulgar connotation of usefulness, the economic utilization of taxonomy having fortunately not been anticipated at that time. Taxonomy, therefore, was considered a hobby without a utilitarian taint, which could be engaged in by gentlemen pursuing "honorable" professions, or who, by virtue of their pecuniary circumstances, could be classified as "gentlemen of leisure."

L. O. Howard, for 33 years the Head of the U. S. Bureau of Entomology, was, in 1916, the first economic entomologist to be elected to the National Academy of Sciences. He wrote in his History that he had a suspicion that he might have failed to gain this distinction if it had been thoroughly understood by the members of the Academy that he was "so pronouncedly utilitarian" in his work and his views.

The relation of the above discussion to our present day problems in applied entomology becomes apparent when we reflect that it is precisely in those parts of the world where the traditional ideas concerning the social degradation connected with labor and economically useful mental or physical activity has to the greatest extent disappeared, that progress in applied entomology is the most rapid. On the other hand, there are many countries where one must call on a social inferior to empty a beaker in laboratory investigations, and where the actual handling of spraying or dusting equipment would forever relegate an investigator to a lower social status. In such countries there exists a lack of understanding and even a disdain for mechanical matters, a serious handicap in modern investigations in the field of applied entomology. Needless to say, in countries where this spirit still predominates, progress in applied sciences is practically nil.

In the United States, all social classes have had a chance to obtain higher education and technical training, and the results of this social amalgamation among educated people can not fail to be noticed by many foreign visitors who visit a typical American entomological research laboratory. The economic slant of research is here eagerly sought rather than begrudgingly accepted. In part, of course, this change in attitude is also caused by the exigencies of industrial efficiency and can to some extent be noted in all highly industrialized nations.

Even more indicative of the new spirit in scientific research is the work of the field entomologist. It has become respectable for him to work in the field in khaki clothes, carrying on the physical tasks of insecticide application, inspection, and the painstaking and often monotonous and routine determination of the results of insecticide treatments. As a result of his personal field experience, the entomologist is better able to demonstrate to others the proper uses of insecticide materials and equipment. He obtains valuable information from his personal contact with growers, pest control operators and field workers, especially with regard to the development of new equipment and new techniques of application. He gains new ideas for fields of further research from his intimate association with practical problems. His usually higher salary, as compared to those who are engaged in non-economic phases of entomology, has broken down the last psychological barriers so long erected about the term "applied."

Indicating the recent origin of the science of applied entomology is the following paragraph from L. O. Howard's A History of Applied Entomology, quoted from The Practical Entomologist, and written in 1865, only 84 years ago:

"The agricultural journals have from year to year, presented in their columns, various recipes, as preventative of the attacks, or destructive to the life, of the "curculio", the "apple-moth", the "squash-bug", etc. The proposed decoctions and washes we are well satisfied, in the majority of instances, are as useless in application as they are ridiculous in composition, and if the work of destroying insects is to be accomplished satisfactorily, we feel confident that it will have to be the result of no chemical preparations, but of simple means, directed by a knowledge of the history and habits of the depredators."

The great L. O. Howard, who probably influenced the course of American economic entomology far more than any other man, showed that he himself at the time of the writing of his History (1930) had not become entirely emancipated from the feeling expressed in the above paragraph when he wrote that "the trained entomologist remembers all the time that his main idea must be to find some means of controlling insects that will obviate the use of expensive chemical and mechanical measures such as spraying and dusting." For better or for worse, a very large number of economic entomologists at present do not "remember all the time" the above advice, for they spend all their time synthesizing new insecticides or testing myriads of new chemical compounds for their possible insecticidal uses. As a consequence of this tendency of recent years, it was possible for a prominent American entomologist to propose, in the presidential address at the fifty-eighth annual meeting of the American Association of Economic Entomologists in 1946, that entomologists should begin thinking in terms of nation-wide eradication of age-old cosmopolitan pests of man and his domesticated animals, such as the house fly, the horn fly, cattle grubs, cattle lice, screwworm, Argentine ant, and other species (543).

It must be admitted that this systematic and purposeful exploitation of the world's gigantic chemical industry has met with spectacular successes. In World War II, in the disease-ridden Pacific theatre, entire islands were freed in a few hours of what could in all seriousness be called "The Insect Menace" a decade ago, by the application of DDT dust by plane. And DDT itself, rather than being an isolated miracle, was merely the harbinger of a rapid succession of chlorinated hydrocarbon derivatives of high-insecticidal and acaricidal value, such as the DDT analogs, benzene hexachloride, chlordan, Compounds 118 and 497, and chlorinated camphenes, comprising a truly formidable array of insecticides. Still more recently certain organic phosphates (tetraethyl pyrophosphate, parathion) have exhibited spectacular insecticidal effect, and the potentialities of man's chemical warfare against insects in the immediate future stagger the imagination.

The recognition of the dependence of the physical and biological properties of insecticides upon molecular structure has resulted in more or less of a standardization of the various ways in which toxicological problems are being approached. These have been summarized by Martin (592) as follows: ". . . straightforwardly as by the synthetic route proceeding from the simple molecule to the more elaborate and determining the effects of substitution and addition on toxicity; cunningly, as in the statistical method in which the degree of uniformity with which the individuals of a given population respond to the toxicant and its derivatives is made to provide information on the mechanism of toxic action; aggressively, as in the analytical method in which the structure of the toxicant is modified and dissected to expose the groupings responsible for toxicity. Finally there are the physiological methods, a labyrinth of ways which start from a guess. If, for example, toxicity is the result of an inhibition of a specific enzymatic activity, tests in vitro may be possible using the isolated enzyme system as test subject. Today a popular hypothesis is based on the similarities in structure between toxicant and an essential metabolite permitting the toxicant to displace the latter with consequent interference with metabolism."

A still further refinement in the chemical approach is suggested by recent research at the University of California Citrus Experiment Station, which indicates that through molecular rearrangement, insecticides may be made to retain their effectiveness against some species of insects while losing their toxicity to other species. A selective molecular rearrangement to make possible the synthesis of insecticides toxic to injurious species and non-injurious to entomophagous and other beneficial insect species is indicated by these investigations. "Selective insecticides" having little or no effect on beneficial insects have been successfully used in Great Britain, especially in the control of aphids (742, 743).

Plant hormones and growth-regulating substances are now widely used as weed killers. Allen (6) suggests a possible similarity between the phytotoxic effect of these compounds and similar symptoms noted after insect feeding and suggests further that plant growth regulating substances "may be a new tool in the hands of the entomologist in explaining the nature of toxin activity following insect feeding." Allen discusses the possibility of chemicals having both insecticidal effect and the ability to induce favorable growth changes in plants and thereby suppressing certain adverse effects resulting from the insect feeding. In this connection should be mentioned the promising results obtained at the Citrus Experiment Station in adding a growth-regulating substance, 2,4-D, to spray oils to suppress certain adverse effects sometimes accompanying an oil spray, such as leaf drop and fruit drop (848, 849).

Mechanical advances in equipment for the application of insecticides have kept pace with the advances in the field of insecticide chemistry (figs. 1 to 4) and each year brings substantial new gains in this field of endeavor.

Fig. 2. Citrus spraying in Florida in 1885. From Hubbard's "Insects Affecting the Orange."

Fig. 1. Application equipment used 50 years ago. Left, heath whisk; right, an "improved" model connected by a hose to a knapsack reservoir. From Glasgow (359) after Lodeman.

It is breaking with venerable tradition among entomological writers, even as exemplified in the latest textbooks, to suggest that man has a better than even chance for survival against admittedly very formidable insect enemies. Yet the writer ventures the opinion that in his post-DDT thinking

Fig. 3 (left). Present day citrus spraying in California. Courtesy R. S. Woglum.

Fig. 4 (right). The helicopter and the aerosol, representing two modern contributions to pest control. Courtesy Bell Aircraft Corp.

and planning, man will come to consider applied entomology as the maintenance of a constant vigilance against a potentially dangerous foe, rather than, as in the past, an attempt to share the food resources of the world on a more or less equitable basis with a resourceful and respected enemy.

Spectacular successes with new insecticides do not decrease the public interest in applied entomology, but on the contrary "entomology" and "entomologist" become household words, public support for entomological research is increased, and student enrollment in entomology courses is increased. Regardless of how effective an insecticide may be, a still more effective and less expensive insecticide or combination of insecticides will be sought to supplant it. More efficient methods will be developed for the application of insecticidal sprays, dusts, aerosols, and fumigants. New insect and mite species are continually swelling the list of injurious pests. They must be identified, their life histories must be carefully studied, and control investigations must be undertaken. Increasing commerce and increasing speed of transportation between distant parts of the world, as exemplified by the transport and passenger plane, increase the rate of introduction of new injurious species despite the utmost vigilance of quarantine officials.

Some insects in the course of a variable number of years become resistant to insecticides which formerly kept them in check, and new measures must be developed to control these species. This has an important bearing on economic entomology and will later be discussed in considerable detail.

All new insecticides must be carefully tested for years as to their possible phyticidal effect and their effect on warm blooded animals, as well as in regard to their insecticidal effectiveness. In this connection the amount and chemical nature of the residues left by the various insecticidal preparations must be carefully determined, usually by standardized and readily reproduceable methods. Sometimes an adverse effect of an insecticide is not apparent until, after repeated trials, a combination of circumstances is encountered, such as among the various climatic factors or soil conditions, or certain conditions pertaining to the physiology of the plants, under which the adverse effects are magnified. Sometimes an accumulative adverse effect of an insecticide on plants or warm blooded animals is noted only after years of experimental or semi-commercial trials. The Federal Food and Drug Administration is the final arbiter as to the usability of new insecticides on food crops, and this agency has been admirably vigilant in affording the protection to public health for which it was organized.

The biologies of insects must continue to be investigated, for regardless of the insecticidal effectiveness of a new compound, its most efficient use depends on its application at periods and under conditions in which the pest is most vulnerable, and this, depending on the species, may be at any one of its stages from egg to adult, and may be at one or more of the seasons of the year. All this information must then be considered in relation to the requirements of the plant or animal to be protected. Often in the application of insecticides to plants, it is found that the least damage to the plant or its fruit is caused when the insecticide is applied at a certain period of the year, and this determines the time of application rather than the relative vulnerability of the various stages in the life history of the pest species to be treated.

Finally, all the possibilities of biological control, the combating of injurious insects by means of their natural enemies, usually entomophagous insects, must be fully investigated. Biological control is the least expensive and often the most effective means of control of noxious insect species and has also been successfully employed in the control of noxious weeds. This field of investigation alone will engage many well-trained entomologists for an indefinite period.

Entomologists have recently been suggesting the possibility that insecticides may be developed, or possibly have already been developed, which will prove to be so highly effective that the populations of injurious pest species may be reduced to extremely small numbers, resulting in the extinction, at least in large areas, of their natural enemies by means of starvation if not by direct insecticidal action. It is speculated that this disruption of the natural balance between the injurious species and their parasites and predators might result in a worse situation than was encountered before the use of the so-called "super-insecticides". It is further speculated that such a development might result in a great modification in the use of the "super-insecticides", or even in the abandonment of their use in some instances.

Although such a state of affairs might in some instances result, it does not follow that a defeatist attitude is justified. It would appear that in such an event (1) a continuous series of new compounds might be synthesized to combat such species as may not be effectively controlled by the existing insecticides while at the same time having been freed of their natural enemies, and that (2) entomophagous insects may be restocked, in orchards or regions in which they have been seriously depleted, from insectaries designed for efficient mass production of those species of parasites or predators which have been eradicated by the action of the "super-insecticides", if such a procedure should prove to be a necessary adjunct to the general pest control program.

The greatest apprehension is felt concerning the treatment of large, continuous areas, such as would often be the case in the treatment of forest insects. It is speculated that, the predators and parasites having been exterminated by the insecticide or by starvation, the injurious species might quickly increase to greater numbers than it had ever reached under natural conditions, and that yearly treatment might become necessary in areas where formerly treatment was rarely, if ever, practiced. Likewise the effect of widespread eradication of insects on fish and game; and the effect of an eradication of bees and other pollinators on fruit and other crops, might lead to serious consequences.

There is no doubt that the high degree of effectiveness of recently synthesized insecticides has, paradoxically, posed new problems for entomologists which will make demands as never before on a coordination of all phases of entomological research, biological as well as chemical. Even before the advent of the war-borne "super-insecticides", Dr. E. M. Patch voiced the opinion that at sometime in the future, entomologists will be as much concerned with the protection of insects as they now are with their destruction (680).

ORIGIN AND SCOPE OF APPLIED ENTOMOLOGY

Insects comprise about 70 per cent of the described species of animals. Some entomologists estimate that millions of species of insects remain as yet undescribed and unnamed. The economic importance of insects is proportionate to their numbers. It is quite generally agreed by those who have estimated the annual losses to agricultural crops occasioned by insects, that this loss amounts to about 10 per cent of the value of the crops. In recent years the value of these crops in the United States has been between 15 and 20 billion dollars per year. In the protection of the crops, well over a half a billion dollars worth of insecticides and fungicides are annually utilized, and certainly the outlay for equipment for the application of these materials must be even greater. In a single California county, the cost of pest control in 1946 was $3,612,000. Crop loss from insect and disease damage was estimated to be at least an equal amount, making a total loss from pests of well over $7,000,000 for the year.[1] To really gain an insight into the potential importance of insects in man's economy, however, one should reflect on the losses which would occur if the control measures now used should be discontinued for even a single year.

No doubt without his unremitting fight against insects, man could exist on this planet in only a small fraction of his present numbers. Yet even in the United States, where the use of insecticides is probably more highly developed than anywhere else in the world, less than half of our crops are being treated each year. The potentialities for economic advantages from the more efficient and intensive, as well as more extensive use of insecticides and fungicides, are thus seen to be very great, and offer to man a challenge which, no doubt, he will meet.

The above estimate of insect depredations does not include the losses incurred by insects affecting human health as carriers of disease. Nor does it indicate the importance of insects from a positive standpoint, that is, the benefit that is derived from them. The monetary value of insect-pollinated crops alone is no doubt greater than the losses sustained from injurious insect species. The Division of Bee Culture of the U. S. Bureau of Entomology and Plant Quarantine lists 53 fruit and seed crops which depend upon honeybees for pollination or yield more abundantly when bees are plentiful (25).

Insects produce products valuable to man, such as silk, honey, beeswax, and shellac; they serve as food for other animals, such as fish, game birds, fur-bearing animals, hogs and domestic fowl, and, in some parts of the world, as food for the human being also; they are of great importance to man as parasites and predators of injurious insects; they serve as scavengers; and even their use in scientific investigations, as in the use of fruit flies in genetics investigations, is of value that can hardly be estimated in terms of dollars and cents.

---

[1] From the Annual Report of the Agricultural Commissioner, County of Los Angeles, for the year ending June 30, 1946.

CHAPTER II
MORPHOLOGY AND METAMORPHOSIS

THE CLASS INSECTA

The insects comprise the largest and the most highly developed group in the phylum Arthropoda, a phylum which includes, besides the insects, the crustaceans, millipeds, symphylids, centipedes, scorpions, spiders, mites and other forms. The insects may be distinguished from other arthropods by a number of structural characteristics. The segmented body is segregated into three regions, the head, thorax, and abdomen. Insects possess 1 pair of antennae, 3 pairs of legs, and the majority have 2 pair of wings, although some have 1 pair and others none. Respiration takes place by means of a system of tubes called <u>tracheae</u> which divide and subdivide until they end in the delicate <u>tracheoles</u> which reach every organ, tissue and cell of the body.

DIRECTIONAL ORIENTATION

Certain terms are used in biology to designate the various regions of an organism. The following terms will be found to be useful: <u>anterior</u>, at or toward the front; <u>posterior</u>, at or toward the hindermost part; <u>dorsal</u>, of or belonging to the upper surface; <u>ventral</u>, of or pertaining to the under surface of the abdomen; <u>proximal</u>, the part of the appendage nearest the body; <u>distal</u>, the part of the appendage farthest from the body; <u>caudal</u>, pertaining to the posterior or anal extremity; also, many such words as lateral, marginal, etc., which are self-explanatory. In description one may use adjectives formed from such words as thorax, abdomen and pleuron, as, for example, thoracic ganglia, abdominal appendages, and pleural regions.

THE EXOSKELETON

The skeleton of an insect is external (fig. 5). It forms a tube-like support for and protection of the internal organs. This outer shell is light, but very strong. This is demonstrated by the tremendous impact that an insect can withstand without injury, as, for example, when striking a solid object in flight. The hard material on the exterior comprises a thin covering of the entire insect known as the <u>cuticula</u>. It is secreted by a single continuous layer of living cells called the <u>hypodermis</u> which is supported on a non-cellular <u>basement membrane</u>. These three layers make up the insect cuticle. Until recently it has been assumed that the polysacharride <u>chitin</u> is the main constituent of the cuticle of insects and other arthropods, and one often finds such terms as "chitinous exoskeleton", in fairly recent literature. Richards (736) points out the error of this generalization. He states that chitin may in a few cases comprise as much as 60% of the cuticle, but it may also be present in amounts as small as 1-2%. Richards states that anthropod cuticle may be thought of as a continuous protein layer which usually contains chitin within and waxes on the outer surface. Fraenkel and Rudall (341) conclude that "the peculiar and characteristic pattern of insect cuticles is due to an intimate association of chitin and protein chains."

Fig. 5. Comparison of skeleton of a mammal and an insect. A, bony endoskeleton of a cat; B, chitinous exoskeleton of a bee. From Metcalf and Flint's FUNDAMENTALS OF INSECT LIFE.

The body wall does not decompose, but remains in its original form after the insect dies, which makes possible the preservation of insects for hundreds of years with their original life-like appearance. Thus the mounting and preservation of insects is a comparatively easy task.

MORPHOLOGY AND METAMORPHOSIS

The exoskeleton of insects cannot be readily dissolved away by strong acids, alkalis and solvents, but fortunately for man's attempts to control insects by means of contact poisons, the exoskeleton can be penetrated by solvents, surface active materials, vapors, and gases. Likewise the spiracles, which are the openings for the tracheae, form a channel of entry for liquids of low surface tension, vapors and gases.

An examination of figure 5 will show that insects are divided by constrictions into ring-like pieces called segments. Connecting the segments are thin, flexible portions of the cuticula called conjunctivae. These conjunctivae may be folded into the body of the insect, allowing for expansion between segments much in the manner of an expanding accordion. Other constrictions or infoldings of the cuticula called sutures may be seen running in various directions on the surface of the exoskeleton. The areas bounded by sutures are called sclerites (see fig. 6). The

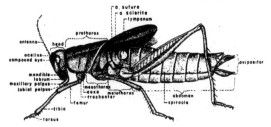

Fig. 6. One of the lubber grasshoppers, Rhomalea microptera. This species has very short wings.

shape and position of the sclerites are often used in the description of species in some insect groups. Likewise the hairs, spines, setae, ridges, and other excrescences are used in taxonomy. These may be merely projections of the cuticula or they may be associated with the underlying body structures, such as the sense organs.

The dorsal or upper face of each segment is called the tergum, the ventral or lower face is called the sternum, and each lateral face is called a pleuron. Each of these faces is made up of one or more sclerites.

MOLTING

An insect can increase in size until the cuticula restricts further expansion. The cuticula must then burst, apparently after having been softened by substances secreted by special molting-fluid glands or the hypodermal cells in general. The old weakened cuticula is split by internal pressure and muscular action, the split occurring along some line of least resistance which varies in its location among the various insect species. The shedding of the cuticula or "skin" of an insect is called molting (fig. 7). It appears to be a difficult and laborious process and a critical period in the life of an insect. The old cuticula cast off by an insect is called the "cast skin" or exuviae.

Fig. 7. An aphid which has just completed the process of molting, leaving the cast skin behind. X 18. Photo by Roy J. Pence.

THE HEAD

The head appears to comprise a single segment, but it is believed, from embryological observations, to be a fusion of five or six segments possessed by the primitive insect progenitors. The arrangement of the skeletal features of a grasshopper is fairly well

9

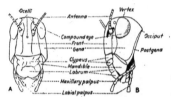

Fig. 8. Sclerites and appendages of the head of a grasshopper. From Metcalf and Flint's FUNDAMENTALS OF INSECT LIFE.

defined, and the grasshopper will, therefore, be used as a generalized type. As seen from figure 8, the principal external regions of the head are designated as vertex, front or frons, gena, clypeus, and labrum. The eyes, antennae and mouthparts are borne on the head.

Eyes:- An adult insect usually possesses both compound and simple eyes. The compound eyes are peculiar to the Insecta. They are composed of many small hexagonal panes, fitted closely together, which are called facets. Each facet is the exposed face of an independent lens and an independent eye unit called an ommatidium. Depending upon the species, there may be from 50 to 30,000 ommatidia in a single compound eye.

Matheson (595) removed the outer portion of the surface of the compound eye of a dragonfly, cleaned away all the cellular structures under the facets, and mounted the remaining structure dry with the exterior surface directed downward on a slide. Through the facets he took a photograph of a butterfly. Each facet produced a complete picture. We still cannot know, however, what an insect actually sees or comprehends. An insect is extremely sensitive to movement in its environment, and it may be surmised that if each image is reproduced hundreds or thousands of times in the compound eye, the sensitivity of the insect to movement will be greatly increased.

Photomicrographic material from a dragonfly's compound eye was somewhat differently prepared by Roy J. Pence. Approximately the outer third of the ommatidium, consisting of the outer cuticula, the cornea, the crystalline cone cells, corneaginous cells and iris pigment cells, was removed and soaked overnight in a clearing and preserving solution of lactic acid, alcohol and distilled water. This softened and separated the pigment from the crystalline structure so that the pigment-bearing portions could be easily removed. The photomicrograph was made through the remaining eye structures soaked in the solution. Figure 9 shows photomicrographs of objects both near and at a considerable distance.

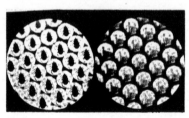 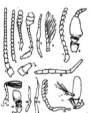

Fig. 9 (left). Photomicrograph of a tick (left) and a building 700 feet distant (right) through a small portion of the compound eye of a dragonfly. Photo by Roy J. Pence.

Fig. 10 (right). Various forms of antennae. After Matheson (595). Courtesy Comstock Publishing Company.

Besides the two compound eyes, an adult insect usually has three simple eyes or ocelli (fig. 8) but may have two, one, or none, depending on the species. Simple eyes have only a single facet. The larvae of some orders of insects, such as the flies, beetles, butterflies, and ants, never have compound eyes. They may have from one to six or more simple eyes on each side of the head. The nymphs of such insects as the bugs, grasshoppers and dragonflies may have compound eyes.

The eyes of insects cannot be focused, and they have no protection equivalent to the eyelids and eyelashes of the vertebrates. Protection is afforded by the same cuticula that covers the entire insect body, and which continues in an unbroken sheet over the eyes. Over the eye surface, however, the cuticula is thin and transparent, and admits light.

Antennae:- The antennae of insects are extremely variable in form (fig. 10). Various names, for the most part self-explanatory, have been more or less generally accepted to designate the different types of antennae: bristle-like, filiform (thread-like), moniliform (bead-like), serrate (saw-like), pectinate (comb-like), clavate (club-shaped), capitate (knobbed), lamellate (bearing many plates), and plumose (feather-like). Many forms of the above types, as well as miscellaneous unclassified forms of antennae, may be found among the insects.

Antennae may be used as tactile organs, or may possess organs for smelling or hearing. In general the antennae are used as sensory aids in locomotion, for locating food and finding mates, and it appears that the ants use their antennae to communicate with others of their species.

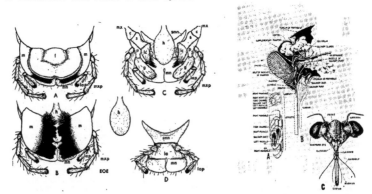

Fig. 11. Biting and chewing mouthparts of a grasshopper. A, dorsal or front aspect; B, same with labrum removed; C, same with labrum and mandibles removed; D, labium; l, labrum or upper lip; m, mandibles or jaws; mx, maxillae or second pair of jaws; c, cardo; s, stipes; lc, lacinia; p, palpifer; g, galea; mxp, maxillary palpi; h, hypopharynx or tongue; la, labium or lower lip; lap, labial palpi; mn, mentum; smn, submentum. Fr. E. O. Essig: COLLEGE ENTOMOLOGY. Copyright, 1942 by The Macmillan Company & used with their permission.

Fig. 12 (right). Piercing-sucking mouthparts. A, cross section of stylets; B, sagittal section of head of a cicada; C, front view of head of a cicada. From Metcalf and Flint's FUNDAMENTALS OF INSECT LIFE.

Mouthparts:- Insects are divided into those having chewing (mandibulate) mouthparts (fig. 11) and those having sucking (haustellate) mouthparts (fig. 12). The chewing mouthparts are the more primitive. The mouthparts of the grasshopper may be used as a fairly generalized type of the chewing mouthparts. The upper lip or labrum and lower lip, or labium oppose each other in a vertical plane. The labrum is a sclerite of the head, but the labium is a true mouthparts appendage. The actual biting and chewing appendages are the mandibles and maxillae. These operate in a horizontal plane, that is, from side to side. The mandibles are the first pair of "jaws" and can be seen upon removing the labrum and labium. They are strong, heavily chitinized structures possessing teeth and grinding surfaces. The mandibles are hinged on to the clypeus and the genae. They are operated by two sets of muscles, one closing them against each other and the other pulling them apart.

The second pair of "jaws" are the maxillae. As can be seen in figure 11, a maxilla is made up of a number of parts. The part modified for the grinding of the food is called the lacinia. It will be noted that the maxillae bear segmented sensory palps. These bear tactile hairs and probably also organs of smell and taste.

The lower lip or labium performs a function similar to the lower lip of vertebrates. In addition it bears a pair of sensory palps whose function is similar to those of the maxillary palps. As might be surmised from the paired palps, the labium is a fusion of primitively separate appendages, the "second maxillae".

The epipharynx, which in the case of the grasshoppers forms the inner face of the labrum, is a sensory organ believed to contain end-organs of taste. It is of interest principally because in some of the sucking insects it becomes modified into important structures.

The hypopharynx, forming the floor of the mouth cavity, is a tongue-like organ. The salivary duct opens at its posterior margin, where the hypopharynx joins the labium.

The chewing mouthparts are the most primitive type, from which the various sucking types have been derived by means of various interesting modifications. The sucking mouthparts are thus said to be homologous with the chewing mouthparts, even though they may differ greatly both in appearance and function. The following types of sucking mouthparts among adult insects are distinguished by Metcalf and Flint (599).

1. The "rasping-sucking" type of mouthparts is possessed by the thrips. It is considered by some authorities to be intermediate in structure between the chewing and piercing-sucking types. The mouthparts are somewhat asymmetrical, the right mandible being vestigial (see fig. 512). The left mandible and the maxillae can be protruded through a circular opening at the conical apex of the head. To free the plant juices, a thrips thrusts its head downward and backward in the manner of a pick-ax, the stylets piercing the leaf or fruit cuticle. The plant sap is then withdrawn through the mouth cone, which, when closely pressed to the plant surface, probably can provide considerable suction. Barnhart (63) reached the conclusion that the citrus thrips does not scrape the leaf cuticle, as many believe, but merely pierces and sucks. Judging from histological studies of cross sections of infested avocado foliage, it would appear that the same can be said for the greenhouse thrips (see p. 647). The term "rasping" with reference to the mouthparts of thrips appears to be inappropriate, although it is well established in the literature.

2. The piercing-sucking type of mouthparts is characterized by a tubular, usually jointed beak, which encloses the stylets (fig. 12) which are modifications of the mandibles and maxillae of the chewing insects, sometimes supplemented or replaced by the labrum-epipharynx and hypopharynx. Metcalf and Flint (599) recognize several subtypes of the piercing-sucking mouthparts, namely, the "bug or hemipterous sub-type", the "louse or anoplurous sub-type", and the "common biting-fly or dipterous sub-type" (including in the latter mosquitoes and horse flies), the "special biting fly or muscid sub-type" (including the stable fly, horn fly, and tsetse fly), and the "flea or siphonapterous sub-type".

3. The sponging type of mouthparts is illustrated by the house fly (fig. 13). In the house fly, the end of the fleshy, elbowed and retractile labium is specialized into a large sponge-like organ, the labella. The labella contains many furrows or channels, called pseudotracheae, through which liquid material is drawn by capillarity. From the pseudotracheae the liquid food is drawn up through the food

12

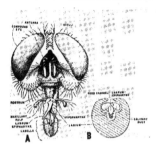

Fig. 13. Sponging type of mouthparts possessed by the house fly. From Metcalf and Flint's FUNDAMENTALS OF IN-SECT LIFE.

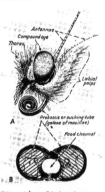

Fig. 14. Siphoning type of mouthparts possessed by the Lepidoptera. After Metcalf and Flint's FUNDAMENTALS OF IN-SECT LIFE.

Fig. 15. Mouthparts of a wasp, Vespa sp., showing the mandibles, (M) as well as the lapping tongue, (T). Photo by Roy J. Pence.

channel into the oesophogus. It is not possible for insects with this type of mouthparts (non-bloodsucking Muscidae, Syrphidae, and many other Diptera) to pierce the skin.

4. In the siphoning type, possessed by the adult Lepidoptera (moths and butterflies) the proboscis is derived almost entirely from the maxillae of the more primitive mandibulate insect. The two elongated maxillae fit together to form a tube which, in the resting position, is coiled and held close to the underside of the head, (fig. 14). The labial palpi are frequently large and conspicuous. Except for a few species, notably the fruit-piercing moths of some tropical countries, moths and butterflies can not pierce the animal or plant epidermis. They feed by uncoiling the proboscis and sucking exposed liquids, such as nectar, through its full length. Müller (629) illustrated a proboscis of a sphingid moth from Brazil found to be eleven inches long and called attention to the fact that Darwin had predicted that such proboscces must exist in order to pollinate certain orchids with nectaries of this length.

5. In the chewing-lapping type of mouthparts (fig. 15), illustrated by the bees and wasps, the mandibulate origin of the labrum and mandibles is readily recognizable. The mandibles may be modified to serve as tools for a variety of tasks performed by the various species possessing such mouthparts. The maxillae and labium are elongated, forming a long tongue which is of great importance in feeding.

The mouthparts of immature insects tend to be even more variable than those of the adults. The type of mouthparts possessed by the adult insect does not necessarily indicate the type possessed by the young of the same species. In the ma-

13

SUBTROPICAL ENTOMOLOGY

jority of the lower orders of insects (from an evolutionary standpoint) the immature insects, because of their incomplete metamorphosis, have mouthparts much the
same in appearance and function as those of the adults. The principal orders of
this type are the Protura, Thysanura, Collembola, Orthoptera, Isoptera, Anoplura,
Thysanoptera, and Hemiptera. In the Odonata, Trichoptera, Hymenoptera, and Diptera,
the mouthparts of the immature insects usually differ greatly from those of the
adults. In the Lepidoptera, the mouthparts of the larvae are always different from
those of the adults, the former having mandibulate and the latter haustellate
(siphoning) mouthparts. The larvae may have in their mouths the spinning organs of
the silk or salivary glands. The cocoon or attachments for the pupa are formed of
silk threads spun by the mouth of the larva. Adult Neuroptera and Coleoptera have
biting mouthparts, and the immature life stages have mouthparts similar to those of
the adults except that in some species (ant lions, lacewing fly larvae, and diving
beetles) the mandibles are grooved to make possible the sucking up of the blood of
the insect's prey after the body has been pierced.

THE THORAX

The second of the three body regions, called the thorax, is composed of three[1]
segments: the prothorax, the mesothorax, and the metathorax (fig. 6). Each segment bears a pair of legs, and the mesothorax and metathorax bear the wings.
Likewise the latter two segments usually each bear a pair of spiracles in the case
of adult insects. The locomotory appendages of adult insects are thus confined to
the thorax, but among the immature insects, locomotion may be aided by two to eight
pairs of fleshy, unjointed prolegs on the abdominal segments (Lepidoptera and
Hymenoptera) or locomotory appendages may be absent (Diptera, Coleoptera and
Hymenoptera).

Legs:- Three pairs of jointed legs are practically always present in adult insects and generally present in immature stages. Practically all adult animals
having six legs are insects,[2] and the name of the class, Hexapoda, is based on this
fact.

The five parts of the legs are held together by joints and are operated by internal muscles. The five parts, beginning next to the thorax, are called coxa,
trochanter, femur, tibia, and tarsus (fig. 6). These terms should be remembered
because they are so often found in entomological literature, especially in taxonomic literature. The trochanter in some hymenopterous species has two segments
and the tarsus may have from one to six segments.

Usually there are two claws on the tarsus and there are frequently pads (pulvilli) beneath the claws which offer greater purchase against smooth objects. The
ability of the pulvilli to cling to smooth surfaces is usually enhanced by many
small hairs, and these sometimes, as in the case of the house fly, exude a sticky
substance. Between the claws there may be another type of pad or lobe called the
arolium. Among the thrips (Thysanoptera) the claws are very small, or they may be
entirely absent, and the tarsus terminates in a large bladder-like arolium, which
has considerable suction force to enable the thrips to cling to objects with little
or no aid from the claws. Among the scale insects, sucking lice, and some biting
lice, there is only one claw on the tarsus.

A great variation can be seen among the legs of different insect species, depending on the mode of life to which the insect has become adapted. A rapidly
running insect like the tiger beetle (Cicindelidae) will have long, slender legs,
while an insect that leaps, such as a grasshopper, flea, or flea beetle (Haltica)
will have enlarged femora or coxae, at least on the hind legs. The legs of insects that dig into soil or wood are short and stout, and may be armed with spurs.
The fore legs may be specialized for the grasping of prey, as in the case of the
praying mantis and ambush bugs. Aquatic insects have legs adapted for swimming,

[1]In the Hymenoptera, the first abdominal segment (propodeum) is incorporated in the thorax, so
the thorax is functionally (though not morphologically) four-segmented.

[2]Some adult mites of the Podapolipalidae and Trichadenidae also have only three pairs of legs.

14

and these lack claws. Arboreal insects, such as the Cerambycids, may have the tarsi greatly expanded and hairy and the claws and pulvilli are strongly developed to afford the maximum grip.

Wings:- Insects are the only invertebrates with wings and the only animals with two pairs of wings. This, like the three pairs of legs, is another feature which can be used to readily separate most insects from the many small invertebrate species with which they might be confused. Only adult insects have wings, and some groups (proturans, silverfish and springtails) appear never to have had wings in the course of their evolutionary history, while others (lice, fleas, females of the mealybugs and scales, the lepidopterous family Psychidae, and some aphids and ants) have no wings, but had winged ancestral forms. The absence of wings in this latter group of insects does not indicate a primitive evolutionary status, but on the contrary, suggests a high degree of morphological specialization for their particular mode of existence

Typically there are two pairs of wings: one pair on the mesothorax and the other pair on the metathorax. The latter pair in the case of the true flies (Diptera) is replaced by vestigial appendages called haltere and is believed by some to function as balancing organs. In the male coccids the hind pair of wings is replaced by two spines, one in each side, which hook into the weakly developed pair of wings of the mesothorax, giving them added rigidity. Among the Coleoptera the first pair of wings (elytra, singular elytron) are horny and shield-like. When in flight, most beetles hold their fore wings aloft to each side, like, the wings of a glider, depending on the membranous second pair of wings for propulsion. The second pair of wings may be longer than the first pair, but are folded away under the elytra when the beetles are at rest.

The color of the wings of the lepidoptera depends on the color and arrangement of myriads of "scales" attached shingle-like to the wings of these insects (fig.16).

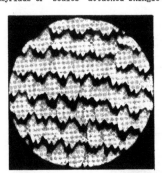 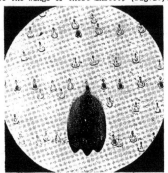

Fig. 16 (left). The shingle-like "scales" of a butterfly wing. X 56. Photo by Roy J. Pence.

Fig. 17 (right). A "scale" of butterfly wing with its pedicel fit into a "socket" in the wing membrane. The other sockets have the scale removed. X 140. Photo by Roy J. Pence.

The shingled arrangement of these scales strengthens an otherwise membranous structure that might prove inadequate for rapid flight. Often each scale is strengthened by minute longitudinal corrugations (fig. 17). These diffract light rays, resulting in the brilliant color and iridescence usually noted in butterflies. Each scale has a pedicel at its base which serves as an anchor point. The pedicel is enlarged in the middle and forms the "ball" which fits, in a locked position, into the "socket" cavity in the cuticula of the wing membrane (fig. 16).

15

The musculature of the thorax is highly developed to suit the needs of the great muscular activity involved in flight. This is most conspicuous among the Diptera and Aculeate Hymenoptera, and these insects are among the most active of the insect fliers. The wings of flies and bees move so rapidly that they become invisible.

If the wing pads of a nymph of the lower winged insects or of the chrysalis of Lepidoptera are examined, it will be noted that each wing is simply a double layered hollow sac folded out from the body wall, in the lumen of which are nerves, tracheae, and body fluid. The double-layered nature of the wings may also be easily demonstrated by placing small adult beetles in a damp location until the wings have absorbed enough water to become bloated, thus revealing their sac-like construction (fig. 18).

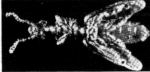

Fig. 18. Saw-toothed grain beetle which had been placed overnight in a moist chamber. Note that the second pair of wings are bloated with water. X 21. Photo by Roy J. Pence.

The tracheae may determine the future pattern of the wing venation but among the higher insects the veins may be formed before the permanent tracheae. The upper and lower walls of the wing sac become fused and appear to be a single membrane, usually thin and transparent. The pattern of the veins of the wings and the size, shape and location of the cells which comprise the areas between the longitudinal and the cross veins, are usually constant and are consequently much used in the description of species, genera and other categories. The veins and cells have universally accepted names (fig. 19) which are often encountered in taxonomic literature. Wing venation is one of the morphologic characters often used to establish the course of evolutionary development and phylogenetic relationship of insect groups.

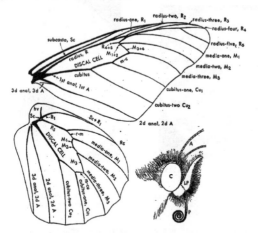

Fig. 19. Wing venation of the monarch butterfly, Danaus menippe (Hübner), and head of same. A, antenna; C, eye; P, proboscis; L. P. labial palpus. Fr. E. O. Essig: COLLEGE ENTOMOLOGY. Copyright, 1942 by The Macmillan Company & used with their permission.

16

THE ABDOMEN

The segments of the third region of the body, called the abdomen, are somewhat similar in appearance (fig. 6). Each segment possesses two sclerites, the dorsal sclerite being the tergum and the ventral sclerite the sternum. These sclerites are joined by the thin, membranous conjunctivae, similar to the way in which the abdominal segments are joined. The conjunctivae permit expansion of the abdomen, and during egg laying the abdomen of the female may be distended to fantastic proportions, as in the case of the termites.

Adult insects are usually without abdominal appendages except for those used in reproduction, called genitalia (fig. 6). The genitalia will later be discussed under the heading "The Reproductive System". Cerci and caudal filaments are borne on the eleventh segment of some insect species. If eleven abdominal segments can be distinguished, the spiracles normally occur on the first to the eighth segments. In many species of insects a fusion of abdominal segments takes place posteriorly, and it may be impossible to recognize more than five or six segments. It will be noted that in some grasshoppers (Acrididae) the auditory organ, called the tympanum is located on the first abdominal segment (fig. 6).

Larvae of some insects may have fleshy, unjointed projections of the abdomen, known as prolegs (see fig. 545) which are used as legs and are a considerable aid to locomotion for some of the more elongated forms. It is of value to remember that the larvae of Lepidoptera never have more than five prolegs, while the larvae of certain Hymenoptera, the sawflies (Tenthredinidae) may have from 6 to 8 pairs.

INTERNAL ANATOMY

The grasshopper will again serve as a generalized type of insect, this time to illustrate the internal anatomy of the Insecta (fig. 20). It will be recalled that the exoskeleton was likened to a tube serving as an outer skeleton. The alimentary tract can be thought of as a tube within this outer tube, and extending from the mouth to the anal opening. This analogy is more striking in the case of the larvae of the higher insects. Between the two tubes is the body cavity, sometimes called hemocoele.

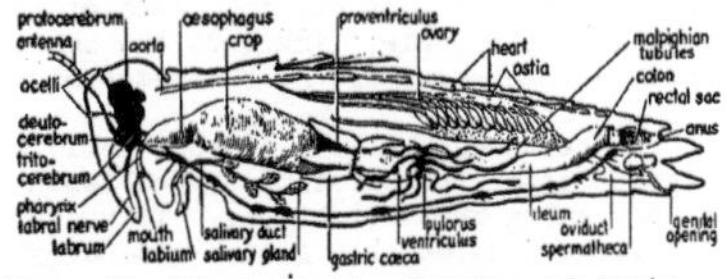

Fig. 20. Diagrammatic drawing of the internal systems of a grasshopper. Fr. E. O. Essig: COLLEGE ENTOMOLOGY. Copyright, 1942 by The Macmillan Company and used with their permission.

Alimentary Canal:- The alimentary canal (fig. 20) is divided into three regions: the fore-intestine, the mid-intestine, and the hind-intestine. The fore-and mid-intestines are connected by the cardiac valve and the mid and hind-intestines are connected by the pyloric valve. The fore and hind-intestines are divided into several more or less distinguishable parts, which are well illustrated by the grasshopper (fig. 20). The parts of the fore-intestine of the grasshopper are the pharynx (into which the mouth opens), the esophagus, the crop, and the gizzard. The latter occurs in insects which eat hard substances, and it appears to have a grinding and straining function like the gizzard of birds. The crop, which is a food reservoir, is absent in some insects. The pharynx is particularly well developed in the sucking insects.

The mid-intestine is not differentiated into regions, but a variable number of evaginations of variable size and shape and in various locations arise from the mid-intestine. These are called gastric caeca, and in these the digestive enzymes are produced.

In the hind-intestine of the grasshopper (fig. 20), the small intestine, (ileum), the large intestine (colon), and the rectum may be distinguished. The Malpighian tubules discharge into the anterior end of the hind-intestine. They are variable in number and in length, but the point at which they are attached to the

17

alimentary canal can always be used as a means of establishing the point of junc-- ture between the mid-intestine and hind-intestine. The function of the Malpighian tubules is primarily excretory; they remove from the blood stream the waste prod- ucts of katabolism and empty them into the hind-intestine. The excretory products, unlike those discharged from the kidneys of higher animals, are semisolid and insoluble.

The length and complexity of the alimentary tract of an insect species de- pends on the kind of food consumed, the more long and complex systems being found among the herbivorous insects, while those of the entomophagus species and the in- sects feeding on concentrated food are relatively short and simple.

Circulatory System:- The "heart" of an insect, stated in simplest terms, is a dorsal tube (fig. 20) closed at the posterior end and opened at the anterior end, and with small valve-like openings (ostia) located at regular intervals along its sides. Blood enters the ostia on dilation of the heart, but the ostia close when the heart is contracted, and their valve-like arrangement prevents the blood from leaving. Since the contractions start at the posterior end of the heart and pass forward, the blood is pushed anteriorly, passing through the aorta, a constricted portion of the anterior portion of the heart. The blood first bathes the brain, and from there it passes posteriorly through the ventral portion of the body, bath- ing all the internal organs and passing into the appendages. In contact with the alimentary canal, the blood picks up by osmosis the products of digestion. While the blood is bathing the Malpighian tubules, the waste products which have not been removed by respiration are taken up by the excretory system. The insect blood has no red blood corpuscles and plays no part in respiration.

Nervous System:- In common with all arthropods, the nervous system of the in- sects is ventrally located (fig. 20). The primitively paired ganglia of each seg- ment are usually fused together, and so, also, are the double cords connecting the ganglia. Nerves pass out from each of the ganglia. In the head, above the oesoph- agus, lies the brain, linked to the smaller suboesophageal ganglion by paired con- nectives which pass around the oesophagus.

Since so much of the nervous activity of insects is on a simple receptor- effector basis, the brain does not serve as important a coordinating function as it does among the vertebrates. If the brain is removed, an insect may continue to live for a considerable period and carry on many activities of a non-purposeful nature.

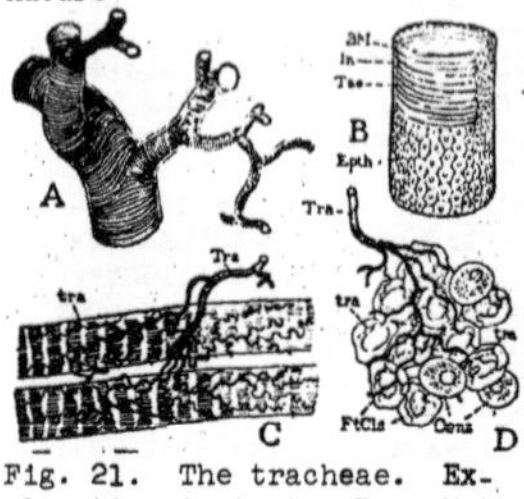

Respiratory System:- The location of the spir- acles (fig. 6) has already been mentioned in the dis- cussion of the thorax and abdomen. The spiracles are the openings of the air tubes, known as tracheae, which divide and subdivide as they penetrate all por- tions of the body. The tracheae finally divide into still smaller tubes, usually less than one micron in diameter, called tracheoles. These reach every cell in the body. In fig. 21 the tracheoles are shown in a muscle fibre (C) and a piece of fat body (D). The tracheoles (tra) are seen to lie on fat cells (Ft. cls.) but not on the oenocytes (Oens). The cells ob- tain oxygen and give off the waste gases of metabol- ism by diffusion through the walls of the tracheoles.

Fig. 21. The tracheae. Ex- planation in text. From Met- calf and Flint's FUNDAMEN- TALS OF INSECT LIFE.

By microscopic examination of insect tissue, the presence of the tracheae can always be determined by the fact that they are lined with cuticula in fine, spirally arranged taenidia (fig. 21, tae) which prevent the collapse of the deli- cate tracheal tubes.

Many insects as, for example, parasitic insects spending their larval instars in the bodies of their hosts, respire through their skin, even though they may have a tracheal system. Some insects may live for long periods without oxygen. Wiggles-

18

worth (948) reports an oestrid larva having remained alive 17 days immersed in oil. California red scale (<u>Aonidiella</u> <u>aurantii</u> Mask) have remained alive as long as 40 days after having been placed on glass slides, covered with refined spray oil, then sealed in with glass cover slips. The majority survived under these conditions for 10 days or more. If the cover slips were not placed on the scales after they were bathed in oil, they died much more rapidly; none survived longer than 10 days. Partial access to oxygen seemed to hasten death. In oxygen-free atmosphere red scales lived as long as 7 days. (265)

<u>Reproductive System</u>:- As stated before, the external reproductive organs of insects are usually found toward the tip of the abdomen. Sometimes the male and female genitalia are so similar that it is difficult to differentiate them. Secondary sexual characters may differentiate the sexes, as for example, the shape of the eyes of some Diptera. The shapes of the bodies of the two sexes may be quite different, and the males are usually smaller than the females. Often the females are wingless and the males are winged, although in some, such as the fig wasp (<u>Blastophaga</u>), the opposite is true.

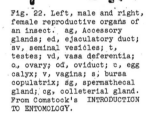

Fig. 22. Left, male and right, female reproductive organs of an insect. ag, Accessory glands; ed, ejaculatory duct; sv, seminal vesicles; t, testes; vd, vasa deferentia; o, ovary; od, oviduct; c, egg calyx; v, vagina; s, bursa copulatrix; sg, spermathecal gland; cg, colleterial gland. From Comstock's INTRODUCTION TO ENTOMOLOGY.

The female reproductive organs (fig. 22, right) consist primarily of a pair of <u>ovaries</u>. Each ovary normally consists of several <u>ovarioles</u> within which the eggs are developed. From the ovaries issue a pair of <u>oviducts</u> which usually unite and lead to the <u>vagina</u>. There is usually a sac-like pouch called the <u>seminal receptacle</u> or <u>spermatheca</u>, which stores sperm. When the eggs are being laid, some sperm may be forced out upon them. Since some insects mate only once, the spermatheca performs an important function. The <u>accessory glands</u> or <u>colleterial glands</u> secrete the "cement" for fastening eggs together, waterproofing them, attaching them to objects, or forming capsules to enclose a number of eggs. The external female genitalia consist of an <u>ovipositor</u> with which the female can thrust eggs into the ground, into plant tissues, or into the bodies of animals, or the ovipositor may be modified into a sting and be associated with poison glands.

The number of eggs that can be laid by an insect may vary from one to many thousand. The queen honeybee may lay 2,000 to 3,000 eggs per day for weeks at a time, and the termite queen can lay millions of eggs.

All insects reproduce by means of eggs, but not all insects lay eggs. Those which lay eggs are said to be <u>oviparous</u>. Their eggs may be fertilized as they pass down the vagina by sperm stored in the spermatheca. If they are not fertilized, reproduction is said to be <u>parthenogenetic</u>. Nourishment within the egg is by means of the yolk, as in the case of the birds. Unlike the birds, however, the insects do not incubate the eggs and, in fact, usually do not care for the eggs in any way whatsoever after they are laid. They instinctively lay the eggs on or near the food which will be consumed by the hatched young.

Among many species the eggs hatch before they are deposited, the insect thus giving birth to active young. Such insects are said to be <u>viviparous</u>. Among such species also, the production of young may be by parthenogenesis as, for example, among the aphids. The term <u>ovoviparous</u> is often found in the literature. The term refers to the type of reproduction in which eggs with well-developed shells hatch within the body of the female and active young are then produced. The most recent tendency is to consider that the term ovoviparous is superfluous, because all insects lay eggs, whether these hatch within the body or, as is more often the case, they are laid by the female and hatch outside (353, p. 117; 595, p. 118).

The males of ants, bees and social wasps are produced from unfertilized eggs and the females and workers from fertilized eggs. The males of some parasitic Hymenoptera and aleyrodids are also produced from unfertilized eggs.

Curious means of increasing the efficiency of reproduction may be found among insects. The storage of sperm in a spermatheca and reproduction by parthenogenesis have already been mentioned. In a method of reproduction known as polyembryony, a few to as many as 1500 individuals may be produced from a single egg, which splits at an early stage of its development into a number of embryos. The resulting individuals are, of course, all of the same sex. This type of development is found most often among the parasitic Hymenoptera. Oviposition among these insects is not easily accomplished, especially in active insects and when ants are present. The production of a large number of young per oviposition is thus an obvious advantage.

Among some insects, chiefly in the dipterous family Cecidomyiidae, there occurs an unusual type of parthenogenesis known as paedogenesis. In this type of reproduction the larvae, or in rare instances the pupae, give birth to active young, a few to as many as several hundred individuals may be produced by this method. Several generations of larvae may be produced before pupation takes place and male and female adults are again formed.

The rate of reproduction can be increased by the deposition of young in an advanced stage of development. This occurs chiefly among several families of the parasitic Diptera. Full grown larvae or puparia are deposited by these insects.

In a diagrammatic sketch of the male and female reproductive organs, the structural analogy can be seen (fig. 22). Comparable to the ovaries are the male testes leading to the vasa deferentia which unite to form the ejaculatory duct. Accessory glands secrete the fluid with which the sperms are mixed. The external genitalia consist of the genital claspers and the penis.

The external genitalia, especially of the male, are used extensively in the description of species because they often show more constant and more distinct differences than any other structural characters to be found on the insect.

MUSCULATURE

Insects have a large number of muscles. The external skeleton allows for a large area of attachment for muscles compared to the bony internal skeleton of vertebrates. The muscles of insects are soft, but strong, and they are cross striated, alternate light and dark bands crossing the fibers. The muscles are yellowish or colorless. In man there are said to be 696 muscles, but Lyonet found about 4000 in the caterpillar of the goat moth Cossus ligniperda.

The muscles of insects contract and relax with phenomenal rapidity, as illustrated by the great speed of wing vibrations. Insect muscles are also very strong in relation to the insect's body weight. Insects are able to lift from 15 to 25 times their body weight.

INSECTS WITHOUT METAMORPHOSIS

Metamorphosis means change in form. The insects of the three orders lowest in the scale of evolution, the Protura, Thysanura, and Collembola, do not go through a change in form after birth, any more than do the vertebrates. A young silverfish or springtail, for example, appears so much like the adult that one would immediately infer that the newly-hatched young and the adult are of the same species. The young insect changes in size, but very little in shape. It will be noted that the three orders in which no metamorphosis occurs comprise the subclass Apterygota, the insects which have no wings and are descended from wingless ancestry.

INSECTS WITH GRADUAL METAMORPHOSIS

Since no insect has visible wings immediately after birth, any insect which develops wings must undergo some degree of metamorphosis during its life history.

20

Therefore metamorphosis may be found among all the insects of the subclass Pterygota. These are insects that have wings in the adult stage or, if wings are absent, they were lost during the course of evolution from winged ancestry. Grasshoppers, earwigs, termites, lice, thrips, bugs, aphids, and scales are representative of the insects with gradual metamorphosis. Since the immature stages, called
nymphs, have about the same kind of habitat and food as the adults, little internal
change during the life history of these insects is necessary. The most conspicuous
change other than increase in size, during the development of insects with gradual
metamorphosis, is the difference in the appearance of the externally developed wing
buds (fig. 23).

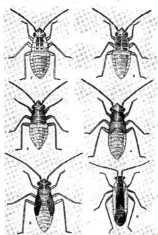

Fig. 24. Metamorphosis
of a dragonfly. 1, naiad
under water; 2, 3, & 4,
naiad out of water and
adult emerging; 5, newly
emerged adult. After
Matheson (595). Courtesy
Comstock Publishing Co.

INSECTS WITH INCOMPLETE
METAMORPHOSIS

Fig. 23. Metamorphosis of the bug
Lopidea robiniae. 1 to 5, nymphal
instars; 6, adult. After M. D.
Leonard.

Three orders of aquatic insects,
the Ephemeroptera (mayflies), Plecoptera (stoneflies), and Odonata
(dragonflies), are included by some
authors in a group having incomplete metamorphosis. Unlike the previous group,
these insects, during their life history, undergo a radical change in habitat
(fig. 24). The stages corresponding to the nymphs of the insects with gradual
metamorphosis, called naiads, are aquatic, while the adults are aerial. Obviously
the morphological changes consequent upon a change from aquatic to terrestrial
existence must of necessity be greater than those which accompany the changes occurring among insects with gradual metamorphosis, which undergo no change in habitat or food habits.

According to Matheson (595), the most distinctive characteristics of insects
with incomplete metamorphosis are the following: "(1) The nymphs and adults live
in entirely different habitats. (2) The nymphs possess many modifications, as
tracheal gills, legs for clinging, clambering, or burrowing, bodies for swimming,
and mouthparts for taking food in the water. (3) Special adaptions are required
for the adults to escape from the last nymphal skin. (4) The adults are aerial and
in the case of dragonflies and damselflies require new methods of capturing food or,
in the case of mayflies and most stoneflies, take no food, or very little, but in-
flate the alimentary canal with air to aid in flight."

INSECTS WITH COMPLETE METAMORPHOSIS

The higher orders of insects, including the four which have the greatest number of described species, the Coleoptera (beetles), Lepidoptera (butterflies, moths), Hymenoptera (bees, ants), and Diptera (flies, mosquitos) undergo what is known as complete or complex metamorphosis (figs. 25 & 368). The young of the insects with complete metamorphosis, which are known as larvae, are so different in their habitat, their food, and their appearance from the adult, that no one unfamiliar with their life history would suspect them of being merely different stages in the development of the same individual. What is there, for example, about the habitat, food habits, and appearance of a codling moth larva, the "worm" of the wormy apple, that would lead an uninformed person to suspect that it is the immature stage of the codling moth? For such radical transformations, the insects with complete metamorphosis have a life stage not possessed by any of the previously discussed groups, which is called the pupal stage. The pupal stage is a quiescent stage, there usually being no movement or feeding. The larva usually pupates in some place where the pupa will be hidden, for the pupa, being motionless, is also defenseless. A protection for the pupa often is woven or in some other way prepared by the larva (fig. 26). All activity in the pupal stage is concentrated on the complex, chemical changes associated with the breaking down (histolysis) and building up (histogenesis) of tissues which must take place before the insect can emerge from its pupal skin transformed for its radically new life habits. In a moth, for example, sexual organs appear, chewing mouthparts are transformed to coiled siphoning mouthparts, wings appear, compound eyes and antennae are formed, and in general an insect whose simple, cylindrical bodily form was previously specialized for feeding and storing energy, is transformed into a highly active and sensitive organism, specialized for flying about, finding a mate, and locating a place for oviposition. Obviously, means of locomotion, sense organs, and reproductive organs which the larvae did not require are now essential to the adults. One of the advantages of complete metamorphosis is that the larvae are specialized in habits and form for feeding, growing, and storing energy and the adults are specialized for mating, reproducing, and effecting a dispersal of the species. In addition, the pupae may provide a means of passing through seasons of adverse weather conditions or periods of food shortage.

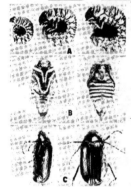

Fig. 25. Complete metamorphosis. Life stages of the June beetle Phyllophaga fusca. A, three larval instars; B, pupae; C, adults. X 1. Photo by Roy J. Pence.

Fig. 26. A portion of an avocado leaf folded and webbed together for protection of the omnivorous looper during pupation. X 0.6. Photo by Roy Pence.

It will be noted, in the comparison of insects with complete metamorphosis with those having gradual metamorphosis, that the latter, as stated before, develop their wing buds internally in the larval instars. They become external in the pupal stage, but hidden under the pupal sheathe. They are first seen in the adult stage.

Neither the pupa nor the adult ever grows. The layman often supposes that a small house fly, for example, is a fly which has not yet completed its growth. Differences in sizes of adults, however, merely indicate differences in the amount of growth attained by the larvae, and this, in turn, is largely dependent upon availability of food.

MORPHOLOGY AND METAMORPHOSIS

The change of form from larva to adult never fails to excite the wonder and imagination of all people; the development of mammals, including man, appears so simple and uneventful in comparison. In fairness to the latter, however, it should be pointed out that their greatest changes in form occur during embryonic development; changes which, in fact, recapitulate the greater part of their long and eventful phylogenetic development. The morphologic changes of the higher insects are the more spectacular because they occur during the post-embryonic development of the animal.

THE TERMS "INSTAR" AND STAGE"

Now that some insight has been gained of the meaning of metamorphosis, the distinction between the terms instar and stage can be more readily understood. An instar is the period between molts, and can only refer to immature insects with simple metamorphosis, or the nymphs, naiads, or larvae of other insects. It can refer only to one stage of any insect, the stage during which growth takes place. Each instar is heavier than the previous instar. A stage is a period of great changes in structure and appearance without change in weight. A grasshopper has three stages: egg, nymph, and adult. A moth has four stages: egg, larva, pupa, and adult.

DIFFERENCES BETWEEN NYMPHS AND LARVAE

Since the naiads of aquatic insects with incomplete metamorphosis are of comparatively little interest to the majority of entomologists, a tabulation will be made only of the factors which differentiate nymphs and larvae.

Nymph	Larva
1. Appearance similar to that of adult.	1. Appearance radically different from that of adult, typically cylindrical.
2. Habitat and food same as adult.	2. Habitat and food often greatly different than that of adult.
3. Successive nymphal instars become progressively more like the adult in appearance.	3. Last larval instar appears no more like the adult than the first.
4. No inactive stage between nymph and adult.	4. An inactive pupal stage for transformation from larva to adult.
5. Wing buds develop externally.	5. Wing buds develop internally.
6. Has few, if any, structures different than those of adult.	6. Has temporary structures, such as prolegs, not retained by adult.
7. Mouthparts same type as those of adult.	7. Mouthparts often different than those of adult, (e.g. caterpillars, chewing; moths, sucking).
8. Has compound eyes if compound eyes are possessed by adult.	8. Never has compound eyes.

DIAPAUSE

Common to all forms of life, and apparently as necessary as any other life function, is the phenomenon of rest or diapause, involving various forms of cessation of activity. It may take the form of a temporary cessation of development or a temporary cessation of activity or both.

Diapause most often occurs in the egg or pupal stages, although it may occur in other stages of development. It may be caused by changes in temperature, moisture, food, water or oxygen or may be imposed by such internal factors as heredity, enzymes, or hormones. Some believe that diapause may be brought about by autointoxication or by the accumulation of some chemical, somewhat like muscular fatigue in higher animals.

An extensive study was made of the factors influencing the diapause of the oriental fruit moth, Grapholitha molesta (Busck), which normally enters diapause in the pupal stage in the late summer and fall. It was found that whether or not the oriental fruit moth goes into a diapause is determined by, (1) the number of hours of light per day during the larval feeding period and, (2) the temperature during the larval feeding and prepupal periods. (237)

The diapause phenomenon is of interest in connection with biological control, since it affects the utilization of entomophagous insects. If the introduced predator or parasite undergoes a diapause that is not synchronized with that of the host species, its establishment may be impossible. (328)

CHAPTER III
CLASSIFICATION. THE ORDERS OF INSECTS

BINOMINAL NOMENCLATURE

It has been shown that the members of the class Insecta have in common a number of characteristics which definitely separate them from other classes of the phylum Arthropoda: the centipedes, spiders, crabs, etc. (fig. 27). Likewise the

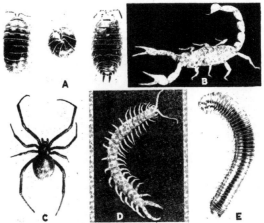

Fig. 27. Some arthropods which are not insects: A, pill-bug and sowbug (Crustacea); B, scorpion and C, black widow spider, Latrodectus mactans (Arachnida); D, a centipede (Chilopoda); E, a milliped (Diplopoda).

Photo by Roy J. Pence.

class Insecta can be divided into a number of orders, each of which includes insects with certain common characteristics that indicate more or less of a relationship. For example, any casual observer would recognize the points of difference between grasshoppers (order Orthoptera) with their four wings of uniform width (the front pair generally thickened), chewing mouthparts, and gradual metamorphosis, as compared with the flies (order Diptera) with their single pair of wings, sponging or piercing mouthparts, fused thorax, and complete metamorphosis. The orders can usually be readily distinguished by relatively inexperienced students of entomology.

The orders are divided into families, the families into genera (sing. genus), the genera into species (sing. the same) and the species sometimes are divided into subspecies or varieties. The classification of man in comparison to that of a well-known American insect, the Rocky Mountain grasshopper, Melanoplus mexicanus spretus (Walsh), will illustrate the significance and the relative magnitude of the various categories universally accepted in both plant and animal classification. It will be found, however, that only a relatively few species of insects have been divided into taxonomically distinct subspecies.

25

	Man	Grasshopper
Kingdom	Animal	Animal
Phylum	Chordata	Arthropoda
Class	Mammalia	Hexapoda
Order	Primates	Orthoptera
Family	Hominidae	Locustidae
Genus	Homo	Melanoplus
Species	sapiens	mexicanus
Subspecies	Caucasian	spretus

In addition to the above categories, further divisions have been used to facilitate the classification of certain groups whose large size or structural peculiarities make further division desirable. Orders may be divided into suborders. Some authors, for example, believe that the order Hemiptera should be divided into the suborders Heteroptera (true bugs) and Homoptera (leafhoppers, whiteflies, aphids, mealybugs, and scales), but others prefer to regard each of the two suborders as orders (Hemiptera and Homoptera, respectively). A number of families may be grouped into a superfamily, or a family may also be divided into a number of subfamilies. A subfamily may in turn be divided into tribes, each consisting of a number of genera.

From the International Rules of Zoological Nomenclature, the following rules pertaining to the proper designation of the aforementioned groups appear (152):

Article 2. The scientific designation of animals is uninominal for sub-genera and all higher groups, binominal for species, and trinominal for sub-species.

Article 3. The scientific names of animals must be words which are either Latin or Latinized, or considered and treated as such in case they are not of classic origin.

Article 4. The name of a family is formed by adding the ending idae; the name of a sub-family by adding inae, to the stem of the name of its type genus.

Supplemental rules of nomenclature are found in the Banks and Caudell Code, which was written to supplement the International Rules of Zoological Nomenclature as adopted at the Fifth International Zoological Congress at Berlin in 1901 (307):

Section 110. The name of a superfamily shall be formed by replacing the idae of one of the included families with oidea.

Section 111. The name of the tribe (a prime division under the subfamily) shall be based on that of an included genus, and shall end in ini. One of the tribes under a subfamily shall be based on the same genus as that of the tribe of which it is a part.

Again quoting from the International Rules of Zoological Nomenclature:

Article 8. A generic name must consist of a single word, simple or compound, written with a capital initial letter, and employed as a substantive in the nominative singular.

Article 13. While specific substantive names derived from names of persons may be written with a capital initial letter, all other specific names are to be written with a small initial letter. Examples: Rhizostoma Cuvieri or Rh. cuvieri, Francolinus Lucani or F. lucani, Hypoderma Diana or H. diana, Laophonte Mohammed or L. mohammed, Oestrus ovis, Corvus corax.

Article 14. Specific names are:

(a) Adjectives, which must agree grammatically with the generic name. Example: Felis marmorata.

26

(b) Substantives in the nominative in apposition with the generic name. Exmple: Felis leo.

(c) Substantives in the genitive. Examples: rosae, galliae, sancti-pauli, anctae-helenae.

The layman is inclined to think of scientific names as a pedantic gesture, or f having perhaps some vague academic significance of small concern to anyone beides the taxonomist. Actually it is only the scientific names that are of uniersal significance, meaning the same thing the world over. The common names of nsects are variable, even within a relatively limited area, and may mean nothing o people in other areas within the same country, to say nothing of those in other arts of the world. Likewise, common names may give little or no indication of he taxonomic relationship of the insects to one another. Consider, for example, he terms "citrus whitefly" and "blackfly". These terms are worse than meaningless, or they are actually misleading. They refer to insects that are not even flies. he corresponding scientific names, Dialeurodes citri and Aleurocanthus woglumi, ay at once indicate to the entomologist what kind of insects these are. If not, e can quickly find in any entomological library that they belong to the family leyrodidae, and are somewhat related not to flies, but to the aphids and scale insects.

Always the first step in investigating an insect problem is to establish the identity (scientific name) of the insect in question. If the insect is not known to the investigator, or if he is not absolutely certain of its identity, he may send specimens to the United States National Museum for identification, or if he knows to what group (family or genus) the insect belongs, he may send his specimens directly to some taxonomist who is a specialist in that particular group of insects. When he is certain of the scientific name of the insect, the investigator proceeds to examine the entomological literature for information on this insect. No matter in what country the literature originated, the insect will nearly always have the same scientific name, for insects are named according to universally accepted standards of nomenclature.

The literature may reveal to the investigator much that is already known about the insect, such as its economic importance, geographic distribution, host plants, life history, seasonal history, habits, tropisms and ecological relationships, climatic or other physical factors favorable or unfavorable to its abundance, natural enemies, and previous investigations on control measures. The obtaining of all this wealth of information is predicated on the correct identification of the insect species and the ability to use the world's entomological literature on the basis of a standardized system of nomenclature. Without a universally accepted system of classification and naming of insects, no other phase of entomological investigation could be systematically and efficiently pursued.

Before the time of Linnaeus, a great Swedish naturalist of the 18th century, the name of an insect was a short Latin description. One can imagine how cluttered with names entomological literature would be if such a system of nomenclature had been retained. Linnaeus proposed that a plant or animal should carry a name to be compounded of the names proposed for the genus and the species of the insect in question. These were always Latin names, for Latin was the universal language among scholars at the time of Linnaeus. Latin had an additional value for use in "scientific names" in that it became a dead language and its form and usage are now stereotyped, to be used with unchanged meaning by people of all languages and in all parts of the world. One hears much nowadays about the desirability of a universal language. A universal language has been used for centuries by biologists in the naming of plants and animals, and if this had not been the case, the chaotic condition of biological literature would be difficult to imagine. Certainly natural science would have been greatly retarded. It should be noted that chemists are also in world-wide agreement as to chemical nomenclature, having adopted a universal system at a conference held at Geneva, Switzerland in 1892.

Linnaeus divided the insects into seven orders: Coleoptera, Hemiptera, Lepidoptera, Neuroptera, Hymenoptera, Diptera, and Aptera. He described and named a

surprisingly large number of species, as a perusal of general entomological literature will show. Since Linnaeus perfected the binominal system of classification,
and used it in his famous general treatise on animals, called Systema Naturae, the
tenth and last edition of this publication, which appeared in 1758, is universally
accepted as the starting point in binominal nomenclature. Any specific name proposed before the date of publication of the tenth edition of Systema Naturae is
not considered as valid. In all the naming of species subsequent to this date,
however, the rules of priority are strictly followed. If a species is described
and named and subsequent research reveals that someone else had already described
and named the same insect, the later name becomes invalid.

In writing a scientific name, the generic and specific names are underscored
once. The scientific names of plants and animals are thus found to be underscored
in correctly written manuscripts. When published, these names are supposed to be
written in italics. The generic name is written first and the first letter is
capitalized. The specific name follows, and always begins with a small letter, except that when derived from the name of a person it may be written with a capital
initial letter. In entomological literature, however, such words are nearly always written with a small initial letter. For complete information, the author's
name follows the name of the species, but it is not underscored or written in
italics. It will be found that in most entomological literature the author's name
is abbreviated. The modern tendency, however, is to write out the author's name
completely. There are now too many names to abbreviate without duplication or confusion. Likewise, beginning students and those not familiar with the insect group
in question are not apt to know the meaning of the majority of the abbreviations.

The house fly may be taken as an example to illustrate the correct form of
binominal nomenclature. It is one of the many species which was named by Linnaeus.
The scientific name of the house fly is Musca domestica Linnaeus (or Musca domestica
Linn. or L.). One knows from the scientific name that Linnaeus described the house
fly as the species domestica and that he placed it in the genus Musca. Let us take
another example, the codling moth of apples, pears, and walnuts, another well-
known cosmopolitan insect pest. The scientific name of this insect is Carpocapsa
pomonella (Linnaeus). The brackets around "Linnaeus" show that he named the codling moth pomonella but that he placed it in some other genus.[1] Later some specialist in this particular group of the Lepidoptera decided that the codling moth
showed a greater taxonomic relationship to insects which belong in the genus
Carpocapsa, a genus which had not yet been recognized in Linnaeus' time.

It has been seen from the various categories of classification listed on
page 26 that the phylum is the most inclusive of the subdivisions of the animal
kingdom. In the Arthropoda the various classes share these characteristics in
common: bilateral symmetry, segmented bodies, paired jointed appendages, chitinous exoskeleton, dorsal blood vessel and ventral nervous system. Some of these
characteristics are possessed by classes of animals far down in the scale of evolution. For example, the members of the phylum Annulata, to which the earthworms belong, possess bilateral symmetry, segmented bodies, a chitinous protective cuticle,
a dorsal blood vessel and a double ventral nerve-cord segmented into a series of
ganglia. However, they do not possess jointed appendages. No other phylum of
animals possesses all of the characteristics listed above. Within the phylum
Arthropoda, however, there are some points of difference which relegate different
groups of animals to different classes. The class Insecta, for example, is the
only group of Arthropods which possesses three body regions, three pairs of legs,
and wings.

When an insect taxonomist (systematic entomologist) is given a large group of
insects to separate into their respective orders, families, genera and species, he
first attempts to sort out the species he recognizes from previous experience.
Many he will not recognize as to species, but he may know to what genus they belong.
Others he may be able to identify as far as the family to which they belong; the
remainder he may be able to group into their respective orders, but it is possible

---

[1]Like so many of the smaller moths, this species Linnaeus placed in the genus Tinea.

CLASSIFICATION. THE ORDERS OF INSECTS

that even an experienced taxonomist may not be familiar with some of the more obscure orders. For aid in tracing down the relationships of the insects of whose
identification he is not certain, the taxonomist will use "keys" by means of which
the insects can be traced through their orders, families, genera, and to their
species. Keys have been prepared for all of the above categories of classification. They are based on the work of a large number of specialists in the various
groups of insects and are, of course, subject to revision as new information is
gained regarding the systematic relationships of the various insect groups. The
use of a key, as will be seen, depends on some knowledge of the structural characteristics of the group of animals (or plants) for which the key has been worked
out. Likewise a familiarity with the terminology used in the classification of
the group in question is also essential.

The following simple and admittedly incomplete key was designed for the present course, in which, because of the vast field that it is attempted to cover in a
single semester, time is not available to key down all the orders or to take into
consideration all the variations even within the orders represented in the key.
The key is usable, however, for the specimens, representing common examples of the
principal insect orders, which will be distributed for identification in the laboratory. In this course no special attempt is made to acquaint the student with
insect families. The recently-published book, The Insect Guide, by Ralph B. Swain,
has very brief and generalized descriptions of the principal insect families. These
are accompanied by illustrations, mostly in water color, of one or more of the most
representative or most typical species of the family. It would appear to the
writer that this book, and possibly a number of others of not too technical a
nature, would admirably serve the needs of those who do not plan to specialize in
entomology, yet wish to familiarize themselves with the more common families of
insects.

SIMPLIFIED KEY TO THE PRINCIPAL ORDERS OF INSECTS[1]

A. Wings usually present.

B. Mouthparts formed for biting.

C. Tip of abdomen with paired, forcep-like
appendages (Earwigs)................................DERM AP'TERA

CC. Tip of abdomen without paired, forcep-like appendages.

D. Fore wings membranous or leathery, not horny in texture.

E. Fore wings leathery, hind wings folded fan-
like (Grasshoppers, crickets, roaches)......ORTH OP'TERA

EE. Fore wings membranous, hind wings not folded fan-like.

F. Abdomen usually constricted at base,
often with a sting or specialized
ovipositor (Ants, bees, wasps).........HYMEN OP'TERA

FF. Abdomen not constricted at base, without
sting or specialized ovipositor.

G. Antennae short, inconspicuous.

H. Hind wings smaller than fore
wings; abdomen with several
thread-like filaments
(Mayflies)...................EPHEM ER OP'TERA

[1]This key was found among the lecture notes of the late Dr. Ralph H. Smith.

29

HH. Hind wings not smaller than
fore wings; abdomen without
long, thread-like filaments
(Dragonflies, damselflies)...O DON'ATA

GG. Antennae long, conspicuous.

    H. Tarsi with less than five segments.

        I. Hind wings broader than
fore wings, folded in
repose (Stoneflies).....PLEC OP'TERA

    HH. Tarsi five-segmented.

        I. Wings naked or very slightly hairy.

            J. Mouthparts prolonged,
beak-like (Scorpion-
flies)............MEC OP'TERA

            JJ. Mouthparts not pro-
longed or beak-like
(Lacewings, ant-
lions)............NEUR OP'TERA

        II. Wings densely clothed with
hair (Caddisflies)......TRICH OP'TERA

    DD. Fore wings horny, shield-like, forming cover for
hind wings (Beetles).........................COLE OP'TERA

BB. Mouthparts formed for sucking.

    C. Wings densely clothed with scales; mouthparts coiled
below head (Butterflies, moths).....................LEPID OP'TERA

    CC. Wings not clothed with scales; mouthparts not coiled below head.

        D. Mouthparts sunken into head capsule, asymmetrical
(Thrips)...................................THYSAN OP'TERA

        DD. Mouthparts external in repose, symmetrical.

            E. Wings usually four; mouthparts forming a jointed beak.

                F. Basal portion of wings thickened,
leathery, or horny; distal portion
membranous (True bugs)...............HEM IP'TERA

                FF. Wings membranous throughout (Aphids,
scale insects, leafhoppers, cicadas,
psyllids)...........................HOM OP'TERA

            EE. Wings two; mouthparts stylet-like or sponge-
like, not forming a jointed beak (Flies)....DIP' TERA

AA. Wings absent.

    B. Mouthparts well developed.

        C. Body flattened dorso-ventrally; prothorax small,
ring-like (Lice).....................................ANO PLUR'A

CLASSIFICATION. THE ORDERS OF INSECTS

    CC. Body flattened laterally; prothorax large,
        hood-like (Fleas).....................................SIPHON AP'TERA

  BB. Mouthparts vestigial.

      C. Abdomen ten or eleven segmented, terminated in forceps
        or long filaments (Bristletails, silverfish).........THYSA NU'RA

      CC. Abdomen with not more than six segments, often
        terminated in a springing apparatus (Springtails).....COLLEM'BO LA

SYNOPSIS OF THE ORDERS OF INSECTS

Sub-Class Apterygota (Ap'terygo'ta). Primitively wingless insects without metamorphosis.

Protura (Protu'ra). Proturans

METAMORPHOSIS: Slight or wanting.

WINGS: None.

MOUTHPARTS: Piercing-sucking.

MISCELLANEOUS: Very small, slender, white insects. Systematic position unsettled.
May belong to a separate class, the Myrientomata. Abdomen 12-segmented and
has a pair of appendages on first 3 segments; tracheae may be present or absent; without visible antennae; blind.
Only about 45 species of proturans have been recorded since they were first
discovered by Silvestri in 1907. They are principally found in leaf mold.
Proturans walk only on the second and third pair of legs and hold the first
pair in front of and above the head as tactile organs (fig. 28).

Fig. 28. Two proturan species, showing the natural as-
pects of the fore legs, which function as antennae. Fr.
E. O. Essig: COLLEGE ENTOMOLOGY. Copyright, 1942 by The
Macmillan Company & used with their permission.

Thysanura (Thy'sanu'ra). Silverfish, bristletails, firebrats.

METAMORPHOSIS: Slight or wanting.

WINGS: None.

MOUTHPARTS: Chewing.

MISCELLANEOUS: Small; elongated, flattened, naked or scaly bodies; abdomen with
10 or 11 segments, with variable number of caudal appendages and in some in-

31

stances with rudimentary legs on abdominal segments; cerci usually long and filiform, or short with few segments, or represented by a pair of forceps; antennae long and filiform; compound eyes present in some species; degenerate in others, wanting in some. This order of universally distributed insects is represented by the silverfish (fig. 29), which are common household pests. They are injurious to book bindings, wall paper, and similar articles containing starch or glue size. They may be poisoned by means of sodium fluoride baits or trapped in jars which are taped on the outside to enable the insects to climb in, but which, because of their smooth inner surface, prevent them from escaping.

Fig. 29. A silverfish. Photo by Roy J. Pence.

Collembola (Collem'bola). Springtails, snowfleas.

METAMORPHOSIS: Slight or wanting.

WINGS: None.

MOUTHPARTS: Chewing mouthparts sunken into the head.

MISCELLANEOUS: Very small to minute; compound eyes; eyes are grouped, ocelli-like. Antenna short, rarely have more than 5 or 6 segments; 6 abdominal segments, the 4th usually bearing a pair of appendages (the springing organs (furcula)), and the 3rd bearing a pair of appendages which act as a catch to hold the furcula when it is at rest; tracheal system usually absent; Malpighian tubules absent.

The collembola (fig. 30) are sometimes very abundant, and some species are injurious to plants. They may be found in mushroom beds, maple sap buckets, and on the surface and in the soil of fields and woodlands wherever there is shade, but they can be found in greatest abundance under boards or logs, where they find both darkness and moisture. They travel about at night. The chief characteristic of their habits for which they are known is their ability to spring into the air by means of their furcula, which is doubled under the abdomen and held in place by the "catch", then released like a spring.

Fig. 30. Collembola or "springtails". In this figure, the forcula is in place in some specimens and extended in others. X 10. Photo by Roy J. Pence.

Sub-Class II. Pterygota (Pter'ygo'ta) Winged insects, with some secondarily wingless.

Division A. Exopterygota (Exopter'ygo'ta) Heterometabolous insects - incomplete metamorphosis.

Orthoptera (Orthop'tera) Locusts, grasshoppers, crickets, cockroaches, walkingsticks, mantids.

METAMORPHOSIS: Incomplete.

WINGS: Four, sometimes greatly reduced or wanting. When present, fore wings thickened and modified into somewhat hardened tegmina, though with distinct venation; hind wings broad, fan-like, membranous and folded longitudinally when at rest.

MOUTHPARTS: Chewing.

MISCELLANEOUS: For this brief and elementary treatment of orders the conventional concept of the Orthoptera is retained, although there are good grounds for the splitting of the Orthoptera into a number of orders. Essig, in his College Entomology, includes under Orthoptera the locusts, grasshoppers, katydids, crickets, and mole crickets, and the remaining insects of the old order Orthoptera he places in the orders, Grylloblattodea (grylloblattids), Blattaria (cockroaches), Phasmida (walkingsticks), and Mantodea (mantids).

Among the Orthoptera are found some of the largest of all insects. A grasshopper in Venezuela is as long as 6-1/2 inches and walkingsticks attain a length of 10 inches in Africa. While the fore wings (tegmina) of most Orthoptera are very drab, the hind wings are often brightly colored. The grasshoppers, katydids and crickets are "singing" insects, the males being able to make sounds of more or less musical quality by means of stridulating organs. The sounds are made by means of definite structures of various types in the legs and wings; the legs may be scraped against the hardened veins of the fore wings or, in other species, the wings may be rubbed against each other. Some grasshoppers can produce sound only in flight by rubbing the costa of the hind wings against the under surface of the thickened veins of the tegmina. The "ears" possessed by some Orthoptera have already been alluded to. These may occur on the abdomen (fig. 6) or on the legs. The locusts (family Acrididae) have been, since times of earliest historic record, the cause of widespread agricultural losses and even occasional famine. The migratory locust, Locusta migratoria, occurs over vast areas of the Eastern Hemisphere, normally contained in its breeding grounds, but occasionally, according to Uvarov, crowding results in the production of a migratory phase. These sweep over vast areas in incredible numbers, often destroying practically all vegetation. The Rocky Mountain locust, Melanoplus mexicanus spretus, has done great damage in the Great Plains region of the United States and Canada. It is believed that a general lessening of the severity of its attacks is caused by cultivation in its usual breeding grounds, which results in the breaking up of the egg cases (oothecae) in the soil. Grasshoppers and locusts may be destroyed by poison bran mash (see p. 383) or by means of some of the newer synthetic organic insecticides (chlordan, benzene hexachloride).

Cockroaches (fig. 31) are predominantly tropical and subtropical insects, but some species have been disseminated into temperate regions, being able to survive in homes, bakeries, restaurants and similar protected places where food is available. They are very objectionable because they frequent dark, filthy places, have an objectionable odor, and may be carriers of disease. They may be controlled by distributing in appropriate places the commercial cockroach powders.

Other Orthoptera are of interest because of their interesting forms and habits. The praying mantis, so-called because of the manner in which it holds its enlarged forelegs, uses these forelegs for capturing and holding its prey. The walkingsticks (fig. 32) closely resemble the twigs or leaves of their environment, and afford us some of the most striking examples of protective mimicry.

Fig. 31 (above). Oriental cockroach, Blatta orientalis. Female with ootheca. X 1.0. Photo by Roy J. Pence.

Fig. 32 (at right). Arizona walkingstick, Diapheromera arizonensis, a striking example of protective mimicry. X 0.50. Photo by Roy J. Pence.

33

SUBTROPICAL ENTOMOLOGY

Isoptera (Isop'tera). Termites

METAMORPHOSIS: Incomplete.

WINGS: Four, when present, with both pairs of similar size and shape, laid flat on back when at rest, and capable of being shed by means of basal fractures.

MOUTHPARTS: Chewing.

MISCELLANEOUS: Social and polymorphic species living in colonies composed of winged and wingless reproductive forms together with numerous apterous, sterile soldiers and workers. Abdomen wide where it joins the thorax, and ending in a pair of cerci.

The termites, sometimes called "white ants", can sometimes be seen in large numbers in logs in the forests, in wood lying in contact with the ground, or in timbers in buildings. The subterranean species of the family Rhinotermitidae, are very destructive either to wooden structures that are in contact with the ground or which can be reached by means of tubes built by the termites out of soil and body secretions. Often building timbers are so badly infested with termites that, while the exterior may appear to be sound, the interior may be a mass of tunnels and frass. Often the termites may not be seen until in the spring when great swarms of the winged forms may appear and give warning of previously unsuspected termite damage.

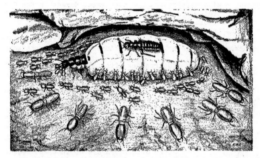

Fig. 33. A sketch drawing of a tropical termite colony showing greatly enlarged queen with king on her abdomen and with many workers feeding and grooming her. Soldiers with enlarged heads and jaws are guarding the workers. Smaller soldiers may be seen among the workers, acting as traffic police. After Matheson (595). Courtesy Comstock Publishing Company.

The termite colony is an example of regimentation par excellence. The colony consists of a number of castes. Only a few of the eggs develop into individuals that can reproduce, the remainder devoting themselves to the care of the "kings" and "queens" and their myriads of offspring (fig. 33). The wings of the reproductive individuals are used for only a single flight, after which they are broken off near the base and mating then takes place. Some of the eggs develop into castes known as soldiers and workers, which are never winged. The soldiers, the defenders of the colony, have massive heads with greatly enlarged mandibles.

The ability of termites to live on cellulose was a mystery until it was found that their digestive tracts contain protozoa possessing enzymes which break down the cellulose into products termites can assimilate. If these protozoa are removed, the termites eventually die of starvation.

34

Prevention of attack by subterranean termites depends primarily on proper construction, which reduces the possibility of contact between the soil and the timbers. However, the soil around the foundations of a building can be freed of termites by soaking it with sodium arsenite solution. Likewise DDT has been successfully used in combating termites.

Dermaptera (Dermap'tera). Earwigs.

METAMORPHOSIS: Incomplete.

WINGS: Usually four; fore wings modified into very short, leathery tegmina without veins; hind wings semicircular, membranous, with veins highly modified and disposed radially and, when at rest, folded both lengthwise and crosswise; wingless forms common.

MOUTHPARTS: Chewing.

MISCELLANEOUS: Caudal end of body bears a pair of unjointed appendages, the cerci, which resemble forceps.

The earwigs, of which there are only about a dozen species in North America, are easily recognized by their short, leathery fore wings and the large forceps-like cerci at the end of the abdomen. The European earwig, Forficula auricularia, (fig. 34), is a widely distributed species which was first reported in Newport, Rhode Island, in 1901 and in Portland, Oregon, in 1909. It breeds in garbage dumps, manure piles, lawn cuttings, leaves, and other kinds of debris, and has but one generation a year. In this country the European earwig will feed on dead or live animal matter, but also on many kinds of plants, and is therefore a pest. It may also become a household pest.

Embioptera (Embiop'tera). Embiids

METAMORPHOSIS: Incomplete in male, absent in female.

WINGS: Males usually winged, possessing 4 wings which are elongate, membranous, and with reduced venation.

MOUTHPARTS: Chewing.

Fig. 34. An earwig, Forficula auricularia. X 2.5. Photo by Roy J. Pence.

MISCELLANEOUS: Solitary or gregarious insects living in silken tunnels; females larviform; cerci 2-segmented; usually asymmetrical in male. The embiids (fig. 35) comprise a small order of less than 150 species. They live in silken nests or galleries under stones and other objects, the silk being spun by glands located on the swollen first segments of the tarsi of the front legs. Their webbed galleries are the most conspicuous indication of their presence. These insects shun the light and leave their tunnels only at night. Their food is thought to be wholly vegetable, chiefly dead and decayed.

Corrodentia (Corroden'tia). Psocids, book lice, bark lice.

METAMORPHOSIS: Incomplete.

WINGS: When present, 2 pairs; membranous, fore pair the larger and with extensive pterostigmata; when at rest, held roof-like over abdomen.

MOUTHPARTS: Chewing.

Fig. 35. Embia californica Banks. Apterous female X 3.5. Photo by Roy Pence.

35

MISCELLANEOUS: Small or minute in size; relatively long antennae; prothorax small, cerci absent; tarsi 2-or 3-segmented.

One group of psocids, all formerly included in the family Atropidae, includes the book lice, which are found in unused books (occasionally in the books of college students!) and in dark, damp buildings. The book lice have no wings or ocelli and have more than 13 segments in the antennae. Other psocids live on the bark of trees and in foliage. In southern California psocids can frequently be seen on the older leaves of citrus trees, especially if they contain "sooty mold" fungus, and are sometimes mistaken for aphids. Molds and fungi are used as food by these species, and none are known to be of economic importance. Psocids sometimes become a household nuisance by overrunning cereals and materials of a starchy nature. They may sometimes be found in alarming numbers throughout the house. Note the striking resemblance of the psocids (fig. 36) to the Mallophaga (fig. 39).

Thysanoptera (Thy'sanop'tera). Thrips

METAMORPHOSIS: Incomplete, but deviates from the usual type of incomplete metamorphosis in that the last two nymphal stages (prepupa and pupa) are quiescent.

WINGS: Four narrow membranous wings, nearly veinless, fringed with hairs; some species wingless.

MOUTHPARTS: Rasping-sucking, asymmetrical, there being only one mandible.

MISCELLANEOUS: Tarsi 1-or 2-segmented, terminating in a protrusible bladder.

Fig. 36. A psocid. X 56. Note close resemblance to the Mallophaga (fig.39). Photo by Roy Pence.

The suborder Terebrantia includes those species in which the abdomen of the female terminates in a saw-like ovipositor for inserting her kidney-shaped (reniform) eggs into the epidermis of leaves or fruit. These species are generally quite active, the greenhouse thrips (fig. 37) being a conspicuous exception, and include the most important pests. Males may be absent, but even when present, reproduction may occur by parthenogenesis.

The suborder Tubulifera includes large, inactive species, in which the female has no ovipositor, the eggs being laid on the plant surface. The last segment of the abdomen of both sexes is tubular, as contrasted to the cone-shaped terminal abdominal segment of the female Terebrantia or the bluntly rounded terminal segments of the males. The Tubulifera may be found in the flower heads of the Compositae, under bark, in galls, in moss and turf, under leaves, etc., where they may be predaceous on small insects and mites or may feed on plant exudations or on dead and decayed vegetable material. A few species, however, are injurious to cultivated plants.

The order Thysanoptera contains two species injurious to subtropical fruits in California: the citrus thrips and the avocado greenhouse thrips. The South African orange thrips is an important pest of citrus in South Africa.

Fig. 37. Greenhouse thrips, Heliothrips haemorrhoidalis. X 56. Photo by Roy J. Pence.

Ephemeroptera (Ephemerop'tera). Mayflies

METAMORPHOSIS: Incomplete.

WINGS: Usually 4, membranous, hind wings smaller than fore wings and sometimes wanting, triangular in outline, generally gauzy, with many longitudinal cross veins, and folded vertically over back when at rest.

36

MOUTHPARTS: Those of adults vestigial or absent, those of the nymphal forms (naiads) chewing.

MISCELLANEOUS: Antennae very short; adults with 2 or 3 very long, slender, many-segmented "tails". Naiads aquatic; with paired tracheal gills on sides and back of abdomen, which also has 2 or 3 slender "tails".

The delicate gauzy-winged, long-tailed adults of the mayflies may be found in large numbers around lights, or near their breeding grounds. At the right season one may observe these mating flights or "dances" near the breeding grounds. The swarm, consisting only of males, rises and falls until a female enters. She is then seized by a male and the pair depart. The "dance" of the males continues, with the females being seized as soon as they enter the swarm. They lay their eggs soon after mating. (595)

The order name is derived from the fact that the mayfly adult lives one of the briefest existences known in the winged state among insects, being with some species a day or two, with some a couple of days longer, but with some even less. Since the adults rarely if ever consume food, their intestinal tract is modified so as to contain large amounts of air for buoyancy. It no longer functions as a digestive tract. (639)

Odonata (Odona'ta). Dragonflies, damselflies.

METAMORPHOSIS: Incomplete.

WINGS: Four membranous, elongate, finely net-veined wings of about equal size.

MOUTHPARTS: Chewing. In naiads the labium of the mouthparts modified into a pre-hensile organ used in capturing prey, and folding like a mask over the face when not in use.

MISCELLANEOUS: Antennae very small, compound eyes large; abdomen long and slender; naiads aquatic, respire by means of rectal or caudal gills. Adults and naiads predaceous. The Odonata are divided into the Anisoptera (dragonflies) and the Zygoptera (damselflies). These two well-defined suborders (fig. 38) differ as follows:[1]

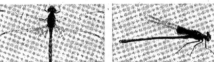

Fig. 38. Dragonfly (left) and damselfly (right), showing difference in position of wings when at rest. Photo by Roy J. Pence.

Dragonflies	Damselflies
"Hind wings broader at base, not folded but held in a horizontal position at sides of body when at rest. Strong flyers.	"Two pairs of wings of same size and shape, narrow at base; folded back over the abdomen or up over the back, like those of a butterfly, when at rest. Feeble flyers.
"Eyes do not project from the side of the head.	"Eyes projecting, constricted at the base.
"Eggs laid on the water or on aquatic plants.	"Eggs thrust into the stems of aquatic plants.

[1]From FUNDAMENTALS OF INSECT LIFE by Metcalf & Flint, 1932. Courtesy of McGraw-Hill Book Co.

"Nymphs breathe through tracheal gills inside of the rectum, and the forcible ejection of water from the anus propels the nymphs forward."

"Nymphs breathe by three, leaf-like, tracheal gills, projecting from the end of the abdomen."

Both the nymphs (naiads) and adults of the Odonata are predaceous. The Odonata are probably of some benefit as predators on horseflies, gnats and mosquitos, and certainly as food for fish. However, the naiads are said to devour fingerlings, and certain species are intermediate hosts of flukes which are parasitic in poultry and waterfowl.

Plecoptera (Plecop'tera). Stoneflies

METAMORPHOSIS: Incomplete.

WINGS: Four membranous wings, held flat over back in repose; in the majority of genera the hind wings much larger than fore wings and folded in plaits on abdomen when at rest; venation variously modified.

MOUTHPARTS: Chewing, though weakly developed and frequently vestigial in adult.

MISCELLANEOUS: Antennae long, filiform with 25 to 100 segments; abdomen of adult bears cerci that are usually long and many-segmented; immature forms (naiads) aquatic, campodeiform and with tuffed tracheal gills commonly present, though variably located.

The drab-colored stoneflies may be seen near swift streams with stony bottoms, resting on stones, trees, and bushes, or flying for short distances over the water. The naiads may be found under stones or trash, in streams or along their margins. Their cast skins are a common sight on stones and shrubs projecting over a stream or lake shore. They are of value to man as food for fishes.

Anoplura (An'oplu'ra). Biting lice and sucking lice

METAMORPHOSIS: Incomplete or none.

WINGS: None.

MOUTHPARTS: Greatly modified; adapted for chewing in suborder Mallophaga, for piercing-sucking in suborder Siphunculata.

MISCELLANEOUS: These are small, flattened, wingless insects living as external parasites of birds and mammals; legs short, with tarsi adapted for clinging to the host, antennae short, 3-to 5-segmented; eyes reduced or absent.

The order Anoplura may be divided into two suborders, the Mallophaga (biting lice) (fig. 39), and the Siphunculata (sucking lice) (fig. 40). Some authors, however, regard, the Mallophaga as a distinct order (287). Both occur continuously on the bodies of warm-blooded animals. They glue their eggs to the hairs or the feathers of their hosts. The biting lice subsist on bits of hair or feathers, skin scales, or the dried blood from scabs, and some species can obtain blood by puncturing the bases of the young feathers. The

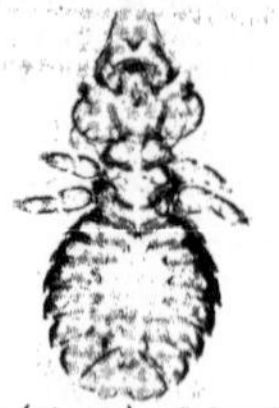

Fig. 39 (above). Suborder Mallophaga. A "biting louse". X 56. Photo by Roy J. Pence.

Fig. 40 (at right). Human head louse, Pediculus humanus humanus L. X 20. Photo, Pence.

38

great majority of biting lice occur on birds, but three families are restricted to mammals, attacking, among other hosts, cattle, horses, sheep, goats, dogs, and cats. Nearly all species infesting mammals have only one tarsal claw which, with modification of the end of the tibia, forms a grasping organ for holding to the hairs of its host. The species on birds have two tarsi. As compared to the sucking lice, the biting lice are of little economic importance.

The suborder Siphunculata, constituting the order Anoplura of certain authors, are all parasitic on mammals and their mouthparts are modified for sucking, but are completely withdrawn into the head when not in use. The tarsi are 1-segmented and terminate in a single claw much in the manner of the biting lice occurring on mammals. The sucking lice can be distinguished from the biting lice not only by the difference in mouthparts, but also because the head of the sucking lice is narrower, and more pointed. The head and body lice of man, Pediculus humanus var. humanus and P. humanus var. corporis, the cooties of World War I fame, are the carriers of typhus and trench fever, and have caused fearful losses of human life, especially in times of war and pestilence. In World War II, the timely development of DDT, a remarkably effective insecticide against lice, reduced casualties attributable to louse-borne diseases to a very low number. Besides the head and body lice and the crab louse of humans, different species of lice attack cattle, oxen, horses, dogs, hogs, and goats.

Hemiptera (Hemip'tera). Bugs, leafhoppers, aphids, mealybugs, scales

METAMORPHOSIS: Incomplete, except in some highly specialized forms.

WINGS: Four, or secondarily wingless (2 wings in male Coccoidea).

MOUTHPARTS: Piercing-sucking, with palpi lacking. The labium is modified into the form of a dorsally grooved sheath which is usually jointed, within which lie the needle-like stylets, the mandibles and maxillae.

MISCELLANEOUS: Antennae, with rare exceptions, are 2-to 10-segmented; tarsi 1-to 3-segmented, with 1 or 2 claws, with or without arolia or empodia; abdomen with few to 10 segments, with cerci absent.

Suborder Heteroptera (Het'erop'tera)

This suborder is often given order status and comprises a reasonably well-defined group of insects characterized primarily by the fore wings (hemelytra) being thickened in the basal half, while the apical half is thin and membranous (fig. 41, B). The wings lie flat in the back when at rest, and the membranous apical halves of the wings overlap. The tarsi are usually 3-segmented. The beak is attached well forward near the front end of the head.

The aquatic bugs of this suborder have short antennae concealed in grooves on the under side or back of the head. The bugs most likely to catch the popular fancy are the giant water bugs or electric light bugs (Belastomatidae), obtaining the latter common name from the fact that they are frequently found around electric lights. These are large species, feeding on insects, tadpoles, and small fish. The females of two genera lay their eggs on the backs of the males, secreting on them a waterproof glue to which the eggs are attached (fig. 41, A). The eggs are carried about by the males until they hatch.

The terrestrial Heteroptera have conspicuous antennae which are longer than, and extend out in front of, the head. They are of great economic importance. The predaceous species include the semi-aquatic water striders, the assassin bugs, bedbugs, damsel bugs, ambush bugs, stink bugs, and leaf bugs (mirids and capsids).
The plant-infesting bugs include the stink bugs, leaf bugs, lace bugs, chinch bugs and squash bugs. Among the stink bugs is the harlequin bug, Murgantia histrionica Hahn, a very destructive pest of cruciferous plants in the Southern States. The tarnished plant bug, Lygus pratensis Linnaeus, is a representative of the family Miridae. Best known of the Lygaeidae is the very destructive chinch

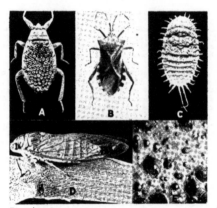

Fig. 41. Some Hemiptera. A, a male water bug
with eggs of female on its back, X 1.00; B,
a coreid bug, X 2; C, Baker's mealybug, X 11;
D, a cicada *in situ*, slightly enlarged; E,
California red scale *in situ*, X 5. Photo by
Roy J. Pence.

bug, *Blissus leucopterus* Say, which annually causes millions of dollars of damage
to corn and wheat crops. Among the Coreidae, the well-known squash bug, *Anasa
tristis* De Geer, is a serious pest of cucurbitaceous plants.

Suborder Homoptera (Homop'tera)

The suborder (or order) Homoptera is characterized by the wings, when present,
being membranous throughout (fig. 41, D), or sometimes of a somewhat leathery texture throughout, but never with the "hemelytra" which characterize the Heteroptera.
The hind pair of wings are shorter and wider than the front pair. When the wings
are at rest they usually are held roof-shaped over the abdomen. All Homoptera are
plant feeders and the excrement of many species is practically pure sugar. It is
known as honeydew and supports the growth of the familiar sooty mold fungus which
blackens the foliage and fruit of plants upon which the honeydew-producing species
appear.

Prominent among the Homoptera are the aphids or "plant lice", which are
economically important on the majority of agricultural plant crops. Their bodies
may be naked or may be covered with a cottony or mealy wax, like those of the
wooly apple aphis and mealy plum aphis. The mealybugs (fig. 41, C) are characterized by nearly always having this cottony or mealy waxy covering over their bodies.

The scale insects derive their name from the "scale" which the members of the
family Diaspididae secrete to form a protective "armor" covering the body completely, at least from above (fig. 41, E). The adult male scale insects (Coccidae) have
only one pair of wings and a complete metamorphosis. The female scales lose their
eyes, wings and legs in the first molt. The beak of the Homoptera is attached near
the back of the head and often appears to arise from between the front coxae.

In the majority of the citrus-growing regions of the world, the Homoptera are
more important as pests than all other species combined, the scale insects usually
being the worst offenders. Injurious scale insects also attack the olive, avocado,
and fig. The aphids and whiteflies (Aleyrodidae) also are important citrus pests

40

the world over. The grape leafhopper, <u>Erythroneura</u> <u>elegantula</u> Osburn, is an important pest of grapes.

Division B. Endopterygota (Endopter'ygo'ta). Holometabolous insects - complete metamorphosis.

<u>Neuroptera</u> (Neurop'tera). Lacewings, ant-lions, dobsonflies

METAMORPHOSIS: Complete.

WINGS: Four membranous wings of nearly equal size, usually finely net-veined and held roof-like over the abdomen at rest.

MOUTHPARTS: Chewing, though secondarily adapted in many larvae for sucking the blood of other insects.

MISCELLANEOUS: Tarsi 5-segmented; larvae carnivorous, the aquatic forms usually possess abdominal gills; pupae exarate.

The family Sialidae includes the dobsonflies, the largest of the Neuroptera. They have a large, square prothorax, long antennae, and the male has very long mandibles. The larvae, called hellgramites, are aquatic, and well-known to fishermen, especially those fishing for bass. They live under water, usually under stones in swift currents, and are predatory on other insect life.

The Chrysopidae contains the lacewings. The green lacewing, <u>Chrysopa</u> <u>californica</u> Coq. (fig. 223), is a beautiful insect with gauzy green wings and golden eyes. The larvae are often important as predators of aphids. The larvae of the Myrmeleonidae (ant-lions or doodlebugs) · dig conical pitfalls in the sand to trap insects, mainly ants, which they devour as food. Only their mouthparts protrude at the bottom of these pits. After their victims are drained of their body juices they are cast far beyond the borders of the pit by these powerful little insects.

<u>Mecoptera</u> (Mecop'tera). Scorpionflies

METAMORPHOSIS: Complete.

WINGS: Four or none. When winged, has 2 pairs of similar net-veined and membranous wings carried longitudinally and horizontally in repose.

MOUTHPARTS: Chewing, and situated at the end of a deflexed beak or snout.

MISCELLANEOUS: Predaceous insects; small to medium sized and slender; antennae long, filiform, many-segmented; legs long, slender; compound eyes large, widely separated.

The common name of these insects, "scorpionfly", is derived from the shape of the upturned terminal segments of the males of the family Panorpidae, which resemble the corresponding part of a scorpion. The larvae and adults of the panorpids are scavengers, and are found in moist locations, usually near streams. The larvae resemble caterpillars. The members of the family Boreidae are wingless. The abdomen terminates in large, mandible-like claspers.

<u>Trichoptera</u> (Trichop'tera). Caddisflies

METAMORPHOSIS: Complete.

WINGS: Four similar membranous wings usually densely hairy and held roof-like over the back in repose.

MOUTHPARTS: Modified by reduction for chewing in the adult, the mandibles being absent, with the palpi well developed.

MISCELLANEOUS: Antennae long, setaceous; legs long, tarsi 5-segmented. Larvae aquatic, usually living in cases and breathing by means of abdominal gills.

The caddisflies are medium sized, moth-like insects, having hairy, brownish wings or dull-colored wings held roof-like at the sides of the body. They are found along streams and lakes, frequently in large numbers. The caddisflies attract popular interest primarily because of the cases which the larvae build about their bodies for protection. What may appear to be a mass of small sticks and pebbles at the bottom of a quiet pool might, upon further investigation, be found to be moving about, and the head, thorax and legs of the tiny larvae can be seen to be protruding from these protective cases. The cases of other species may be attached to a stone. There are a great variety of larval cases. Larvae of one family (Limnephilidae) construct the familiar "log cabin" type of structure out of tiny sticks.

Lepidoptera (Lep'idop'tera). Moths, skippers, and butterflies

METAMORPHOSIS: Complete.

WINGS: Wings usually well developed, rarely vestigial; fore and hind pairs dissimilar, covered with overlapping scales or hairs.

MOUTHPARTS: Siphoning in adults, chewing in larvae.

MISCELLANEOUS: Body of adult covered with scales and hairs, compound eyes large, antennae variable, often clavate or serrate, hooked or knobbed. Larvae or caterpillars cylindrical in shape, with 3 pairs of thoracic legs and from 2 to 5 pairs of prolegs on abdomen; pupae with appendages usually fastened to body, often in cocoons.

This order is among the largest in numbers of species and is also one of the most destructive of all insect orders. Among the destructive species may be mentioned the codling moth, oriental fruit moth, cutworms, corn earworm, European corn borer, the tomato hornworm, webworms, leaf rollers, tussock moths, the gypsy and brown-tail moths, flour and meal moths, and clothes moths. The silk moth is a species of considerable value to man. The general characteristics of the larvae and adults of the Lepidoptera are well known to everyone and further description for the sake of identification is unnecessary. However, certain features connected with the structure and habits of these insects are not so well understood. The very patterns and colors which constitute the chief attraction of the Lepidoptera to the majority of people are understood by only a few. The arrangement of the hairs or scales (see p. 15) that usually cover the wings and all other parts of the body accounts for their colors and color patterns. The scales vary in structure from simple hairs to broad, striated plates of various sizes and shapes. The Lepidoptera are also highly specialized in other ways. The coiled proboscis is probably the most highly specialized of all types of insect mouthparts.

The division of the Lepidoptera into two groups, the moths and butterflies (fig. 42) appears to the writer to be the most useful from the standpoint of the average person only casually interested in this group of insects, although not all authors recognize its validity. According to this method of classification the moths comprise two suborders, Jugatae and Frenatae, whose separation is based primarily on the nature of the structure which holds the fore wings and hind wings together, and also on the shape and venation of the wings. The antennae of the moths are varied in structure, but are never club-shaped at the tip. The pupae are often protected by cocoons which are spun of silk coming from the salivary glands. The moths are mostly night fliers. The wings are usually held either horizontally or roof-like at the sides of the abdomen when they are at rest.

The suborder Rhopalocera includes the butterflies and skippers. These are day fliers and have their antennae clubbed near the tip. The wings are usually held vertically above the body when at rest and the upper surfaces of the two pairs of wings are then in contact. The butterflies have no special structure to hold

their fore and hind wings together, this usually being accomplished by an expansion of the hind wing near its base. The skippers and butterflies have no ocelli, while the moths often have two.

The skippers comprise a single family and are so called because of their manner of flying. They dart suddenly from place to place. The abdomen of skippers is large in comparison with the remainder of the body. In the majority of skippers the enlarged tip of the antenna is pointed and curved like a hook. The caterpillars are usually naked and have a distinct constriction just behind their large heads.

Fig. 42. Lepidoptera. Above, achemon sphinx moth; below, a swallowtail butterfly. Photo by Roy J. Pence.

The caterpillars of the butterflies are usually not as hairy as those of the moths and they do not spin cocoons.

An advantage found by the amateur collector of butterflies is that the majority of species can be identified by means of illustrations in a number of books which are obtainable in the majority of libraries.

Coleoptera (Coleop'tera). Beetles and weevils

METAMORPHOSIS: Complete.

WINGS: Four or rarely none; when winged, the fore pair modified into horny or leathery "wing covers" or elytra that meet in a straight line along the middle of the back (fig. 43), beneath which membranous hind wings are folded when at rest.

MOUTHPARTS: Chewing.

MISCELLANEOUS: Adults usually heavily chitinized; larvae worm-like with 3 pairs of thoracic legs and not more than 1 pair of prolegs; pupae with appendages free, rarely in cocoons.

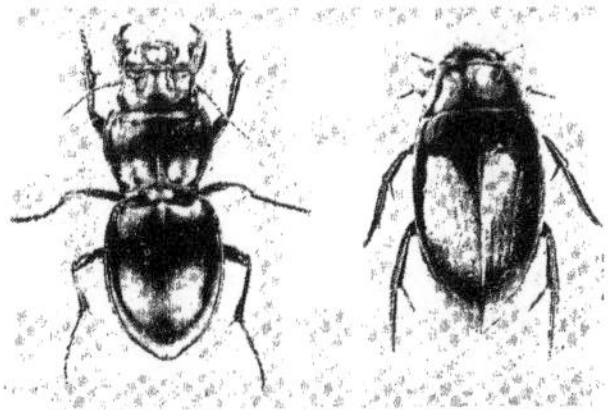

At least as far as the number of described species is concerned, the Coleoptera is the largest of the insect orders. Over a quarter of a million species have already been described. Thus about one out of every three described insects is a beetle or weevil. Like Hemiptera and Lepidoptera, Coleoptera contains large numbers of species injurious to agricultural crops, but among the beetles are

Fig. 43. Two predaceous beetles. Left, Pasymachus californicus (terrestrial); right, Hydrous triangularis (aquatic). Slightly enlarged. Photo by Roy J. Pence.

also found many predators, such as the ladybird beetles and ground beetles. Among the injurious species may be mentioned the June beetles and their larvae (white grubs), Colorado potato beetle, Japanese beetle, alfalfa weevil, cucumber beetles, flea beetles, Mexican bean beetle, bean and pea weevils, plum curculio, granary weevils, rice weevils, mealworms, and among the forest pests, the bark beetles, flat-headed borers and round-headed borers.

The Coleoptera may be divided into two suborders, the Adephaga and the Polyphaga (599). The Adephaga include two families of predaceous beetles (the tiger beetles and ground beetles, and two families of aquatic beetles (the diving beetles and the whirligig beetles). The latter are so called because of their strange gyrations when they circle round and round one another as they walk or skate on the surface of the water. The Adephaga are separated from the other Coleoptera by the separation of the pronotum from the propleura by a distinct suture, and also by

SUBTROPICAL ENTOMOLOGY

the first visible abdominal segment being deeply cleft by the hind coxal cavities.
Sclerites and sutures are often used in the taxonomy of Coleoptera, because the
sclerites are fitted together with remarkable exactness and precision. The larvae
of the Adephaga have 6 segments in each leg and 2 claws at the end of the leg.

The Polyphaga are divided into 7 superfamilies. All have the first ventral
segment of the abdomen in a single piece, not cleft by the hind coxal cavities.
The legs of the larvae have 5 or fewer segments and always end in a single claw.
The division of the suborder into its superfamilies is based mainly on the nature
of the antennae and on tarsal segmentation. The following is based on a system of
classification of the Polyphaga adopted by Metcalf and Flint (599).

1. Short winged beetles (Brachyelytra). These beetles have short elytra, exposing much of the abdomen, and they possess 5-segmented tarsi. They include the
rove beetles and the carrion beetles, which for the most part are scavengers.

2. Club-horned beetles (Clavicornia). The antennae are clavate and the tarsi
5-or 3-segmented. Includes the water scavenger beetles, the flat bark beetles
(Cucujidae), lady beetles (tarsi 3-segmented) and the dermestid beetles.

3. Saw-horned beetles (Serricornia). With serrate antennae and 5-segmented
tarsi. Includes the clerids, the fireflies or lightning beetles, the cantharids,
the flat-headed borers (Buprestidae) and the click beetles or wireworms.

4. Beetles with different-jointed tarsi (Heteromera). Five segments in tarsi
of first and second pair of legs; four segments in third pair. Includes the darkling beetles and blister beetles.

5. Leaf-horned beetles (Lamellicornia). These beetles have 5-segmented tarsi
throughout. The lamellate antennae consist of a cylindrical basal part and a
number of flattened leaf-like segments at the extremities. Includes the stag
beetles and the scarabs (June beetles, etc).

6. Plant-eating beetles (Phytophaga). The tarsi in this group are apparently
4-segmented. The long-horned beetles or rounded-headed borers (Cerambycidae), the
leaf beetles (Chrysomelidae) and the pea and bean weevils.

7. The snout beetles (Rhyncophora). The head often bears a long snout. In
this group the tarsi are again apparently 4-segmented. The antennae are clubbed
and elbowed. This superfamily includes the weevils and the bark beetles.

Strepsiptera (Strepsip'tera). Stylopids or twisted-winged insects

METAMORPHOSIS: Complete and both sexes undergo a hypermetamorphosis.

WINGS: Only males are winged; fore wings reduced to club-shaped appendages
(halteres); the hind wings relatively large, fan-shaped, with radiating wing
veins, and are folded longitudinally when at rest.

MOUTHPARTS: Vestigial or wanting.

MISCELLANEOUS: Adult female larviform and without legs, wings, eyes, antennae and
mouthparts.

Strepsiptera is a very small order and is primarily of interest because of the
peculiarities of the biology of its members. It was formerly considered as a
family (Stylopidae) of the order Coleoptera. The larvae are endoparasitic on bees,
wasps, and a few Hemiptera and Orthoptera. The female remains within the host,
and the males fly about, seeking a mate. Both males and females extrude the anterior portions of their bodies between the abdominal segments of their hosts, and
can be seen if an infested host is carefully examined. A few free-living Strepsiptera have been recently found, but nothing is known of their life histories.

44

CLASSIFICATION. THE ORDERS OF INSECTS

Hymenoptera (Hy'menop'tera). Ants, bees, wasps, parasites

METAMORPHOSIS: Complete.

WINGS: Four wings or none, dissimilar in size, membranous, usually with few veins.

MOUTHPARTS: Chewing or chewing-lapping.

MISCELLANEOUS: Ovipositor always present in females and modified for sawing, piercing, or stinging; larvae usually legless; pupae with appendages free; commonly in cocoons; many species are social.

Although at the present time the order Hymenoptera is third in size from the standpoint of the number of species which have been described and named, probably great numbers of inconspicuous species, especially among parasites, have escaped the attention of taxonomists, so it is likely that there are more species of this order in existence than of any other. Some authors place the Hymenoptera at the top of the list of insect orders because of the high degree of development of social organization found in this group, including the care of the young, shown by no other insects. The ability of some Hymenoptera to learn, or profit by experience, has been taken by some scientists to indicate that the rudiments of intelligence may be found among these insects. On the other hand, some authorities consider the flies and fleas as having reached a higher evolutionary status than the Hymenoptera, and they base their opinion on the greater degree of morphological specialization in these orders.

The social organization of the Hymenoptera has long excited the wonder and admiration of all people and has been the subject of much investigation. As with the termites, the reproductive function has been limited to a few specialized individuals, the kings and queens, but unlike the termites, the workers and soldiers of the Hymenoptera are exclusively females. The males are therefore called drones, their only function being the fertilization of the queen.

In the case of the bees, as the colony increases in size, the instinct for migration appears to be invoked. Special cells are then constructed by the workers for the production of males and queens. The males are reared in larger cells than the workers. They hatch from unfertilized eggs laid by the queen. Still larger cells are constructed for the queen, the partitions between adjacent cells being removed for this purpose. The eggs in two of the cells are destroyed and the "queen cell" is built over the remaining egg. While all larvae are fed a special food, the "royal jelly", for three days, all but the prospective queen are then fed with beebread or other food. The larvae in the "queen cells", however, are fed with "royal jelly" during their entire larval life.

The Hymenoptera are distinguished by having a highly specialized modification of the ovipositor into an organ known as the stinger. Only the females can sting, and the pain of the sting is due to venom ejected into the wound made by the stinger. Since the stingers of honeybees have a recurved hook, they can not be pulled out by the bees and, consequently, they tear loose a part of their viscera as they attempt to escape. This results in their death, so they can sting only once. Other species of bees, as well as wasps, locusts, and ants may sting repeatedly.

The order Hymenoptera may be divided into two suborders, the Chalastogastra and Clistogastra.

Suborder Chalastogastra. Sawflies. The anterior segments of the abdomen are as broad as the following segments and are joined to the thorax throughout their entire width (fig. 44, above). The trochanters are 2-segmented. The ovipositors are saw-like for cutting plant tissues in oviposition. The larvae are caterpillar-like but can be differentiated from lepidopterous larva in that they have 6 to 8 pairs of prolegs, while the lepidopterous larvae never have more than 5. Moreover the prolegs of the sawfly larvae never possess hooklets (crochets) like those of

45

Fig. 44. Hymenoptera.
Above, a siricid (sub-
order Chalastogastra),
below, an ichneumonid
(suborder Clistogastra).
Photo by Roy J. Pence.

lepidopterous larvae. The sawfly larvae have a single pair of ocelli, while caterpillars have several ocelli on each side of the head.

The suborder contains, among others, the families **Tenthridinidae** (sawflies) and **Siricidae** (horntails). It contains the majority of injurious Hymenoptera.

Suborder Clistogastra. The second abdominal segment is constricted into a slender petiole connecting the thorax to the abdomen (fig. 44, below; fig. 45). The ovipositor may be specialized for boring, placing eggs within other insects (in the case of the parasites), or it may be modified into a sting associated with poison glands. The larvae are always legless and grub-like, having reduced head and mouthparts, antennae and palps with at most one segment, and are usually without ocelli.

The Clistogastra may be divided into three groups; the wasps or wasp-like species have the hind tarsus usually slender and cylindrical and have no tubercles on the abdomen; the ants have the petiole of the abdomen with one or two swellings or tubercles (fig. 45), and the bees have branched or plumose hairs on the thorax, and they have broader and more hairy bodies than the wasps and ants. Nearly all individuals are winged.

Among the Hymenoptera in general there are few injurious species and the majority of these are in the Chalastogastra (sawflies). In the Clistogastra nearly all species are either beneficial to man, or harmless. The ants are an exception. They may be injurious because of their ability to keep predators and parasites away from injurious insects or in other ways caring for them. In return, they feed on the "honey dew" of the injurious species they protect, such as soft scales, aphids, mealybugs, and whiteflies. The "honey dew" is

Fig. 45. California harvester ant, Pogo-nomyrmex californicus. X 6. Photo by Roy J. Pence.

the excrement of the protected species and consists of practically pure sugar. Ants may also be phytophagous and occasionally do damage in this way. The Clistogastra, however, are predominantly beneficial. Hymenopterous parasites are the principal entomophagous species of the insect world. They are found among the aforementioned "wasp-like forms" of the suborder and include the families Ichneumonidae, Braconidae, Evaniidae, Chalcididae, and Proctotrupidae. The honey bee, of course, is of great value as a producer of honey and beeswax, as well as an important pollinater of fruit crops, many of which would be of no commercial value without pollination.

Diptera (Dip'tera). True flies, mosquitos, gnats

METAMORPHOSIS: Complete.

WINGS: Two membranous wings borne on mesothorax, hind pair of wings modified into club-shaped organs (halteres).

MOUTHPARTS: Sponging or piercing-sucking.

MISCELLANEOUS: Head, thorax and abdomen very distinct, larvae legless, usually maggot-like, with head greatly reduced and mouthparts replaced by a pair of mouthhooks; pupae with appendages commonly enclosed in a puparium or case.

The Diptera differ from all other orders in having throughout the order only one pair of wings (fig. 46). The vestigial second pair (halteres) are believed by some to be of use in orientation. A haltere may be seen on the asilid shown in

46

fig. 46. The few species in other orders that have only one pair of wings do not possess halteres. In the Diptera all species have halteres, even if they have lost the front pair of wings. Since only the fore wings are functional, and these are attached to the mesothorax, this segment makes up the larger part of the thorax. The thorax is highly developed, and the flies are among the most rapid fliers among the insects. Many species also show a marked specialization of the abdomen, with a great reduction in the number of segments.

The flies are not conspicuous among agricultural pests. Among the more important are the fruit flies, seed midges, and the hessian fly. The Diptera rank first, however, as vectors (carriers) of diseases affecting man and animals. Mosquitos are vectors of the casual organisms of malaria, yellow fever, dengue, filariasis, and encephalitis (experimentally). Horse flies can transmit tularaemia, filariasis and anthrax. The tsetse fly is the vector of sleeping sickness. Various diseases are transmitted by Diptera of the family Psychodidae (Phlebotomus spp.), others by the black flies (Simulium spp.) and the chironomids. The house fly is a mechanical carrier of

Fig. 46. Order Diptera. Above, a tabanid (parasite); below, an asilid (predator). Arrow points to haltere. Photo by Roy J. Pence.

the causal organisms of typhoid, dysentery, cholera, anthrax, yaws, conjunctivitis, and the eggs of several cestodes and nematodes. The larvae of house flies and flesh flies and others may cause myiasis.

The Diptera are divided into two suborders. In the suborder Orthorrapha the adults usually emerge from the pupal skin through a T-shaped or straight split down the back. The pupae are usually naked and the larvae often have a distinct head.

The suborder Orthorrapha contains the superfamily Nemocera, or long-horned flies, which have long and slender antennae, of 6 to 39 segments. The larvae have a distinct head, eyes, and true mandibles. In this group are included the crane flies, mosquitos, midges, gnats, black flies, and the moth flies (psychodids). Also included in this suborder is the superfamily Brachycera, the short-horned flies. In these flies the antennae are usually short and of three segments. The larvae often have the head invaginated and instead of mandibles they have mouthhooks which work in a vertical position. In this group are included the horse flies, soldier flies, snipe flies (or rhagionids), robber flies, bee flies, and the long-legged flies (dolichopodids).

In the second suborder, called the Cyclorrhapha, or circular-seamed flies, the larva pupates on the last larval skin, and the adult emerges by the anterior end, pushing off a circular lid. The antennae never have more than 3 segments. The larvae have typically greatly reduced heads which are invaginated into the pharynx. The first anal cell of the wings is always closed.

The suborder Cyclorrhapha may be divided into three superfamilies. The Aschiza includes flies in which the cap of the puparium is pushed off by the expansion of the face of the adult. In the superfamily are the humpbacked flies (phorids) and hover flies (syrphids). The Schizophora is a large group of flies in which a bladder-like structure, the ptilinum is forced out when the adult is ready to emerge. The cap of the puparium is forced off by the inflated ptilinum. The bladder is then withdrawn into the head. The Schyzophora include the ortalids, fruit flies, drosophila, anthomyiids, the house fly, flesh flies, the tachinids, and bot flies. In the Cyclorrhapha also, is found the superfamily Pupipara, louselike insects which are often wingless, with a very tough skin and indistinctly segmented abdomen. This group is of interest in connection with its metamorphosis. The larvae are born shortly before pupation, having been nourished from special uterine glands in the mother fly.

SUBTROPICAL ENTOMOLOGY

The pupipara are external parasites on mammals, birds, and insects. They include three families, the sheep tick and other louse flies (Hippoboscidae), the bat flies (Streblidae) and a small family of spider-like ectoparasites of bats (Nycteribiidae).

Siphonaptera (Siphonaptera). Fleas

METAMORPHOSIS: Complete.

WINGS: None.

MOUTHPARTS: Piercing-sucking, with 2 pairs of palps.

MISCELLANEOUS: Small, laterally compressed; hind legs enlarged for jumping, 5-segmented tarsi; adults are external parasites on warm-blooded animals; larvae slender, cylindrical, without legs or eyes, but with well-developed heads; pupae with appendages free, enclosed in a cocoon.

Fleas, upon microscopic examination, are easily distinguished from other insects (fig. 47). The laterally compressed body is in itself almost infallible as a taxonomic character. In addition the hard, polished body contains many hairs and short spines regularly arranged and directed backward. The fleas have no compound eyes and often not even simple eyes.

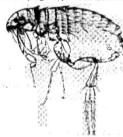

Fig. 47. The cat flea, Ctenocephalides felis. X 23. Photo by Roy J. Pence.

While fleas lay their eggs on their host like the lice, the eggs are not fastened, and will drop off, hatching on the ground or floor. The larvae feed on dead animal or vegetable matter and excrement, and pupate in a cocoon of silk to which particles of dirt usually adhere.

There are only five families in this small but important order, and two of these are of special importance. The Pulicidae, the common fleas, including the human flea, Pulex irritans, and the common pests of cats, dogs, and hogs. Fleas are most apt to be found in houses where pets are kept. They are vectors of bubonic plague. The chief vector is the oriental flea, Xenopsylla cheopis, which is found mainly on rats, but also attacks man. In the family Hectopsyllidae is found the sticktight flea, an important pest of poultry which also attacks many other animals. The chigoe flea, Tunga penetrans, is not found in the United States, but is found in many tropical and subtropical countries throughout the world.

CHAPTER IV
ORGANIZATION AND LEGISLATION IN ENTOMOLOGY

INTRODUCTION

It has been the writer's feeling that in the average college course there is generally insufficient emphasis on the scope and the professional applications of the subject being taught, be it in entomology or any other branch of agriculture. For the majority of people, nothing stimulates as much interest in the acquisition of knowledge as an occasional glimpse into the possible future applications of that knowledge. In this chapter some of the more important governmental and private organizations in which entomologists may be employed, their main activities and type of employment, and the entomological legislation, especially in the field of quarantine, survey, pest control and eradication, insecticide testing, and standardization, which is inevitably connected with the work of these organizations, will be discussed.

UNITED STATES DEPARTMENT OF AGRICULTURE, BUREAU OF ENTOMOLOGY AND PLANT QUARANTINE

Historical:- Prior to the year 1854, the Federal Government had become officially involved in entomology only to the extent of sponsoring a few exploring expeditions and natural history surveys, in the course of which insects were collected and classified. These insects were harbored in the collection of the Smithsonian Institution, which was later designated the United States National Museum, but which is nevertheless still under the direction of the Smithsonian Institution.

In 1854, Townsend Glover was employed by the U. S. Patent Office as Entomologist, one of his duties being to investigate the pests of citrus fruits. It was not until the creation of the Department of Agriculture under the act of May 15, 1862, that specific provision was made for an Entomologist. Mr. Glover became the first official Entomologist. In 1878 the service became a definite Division. C. V. Riley, following his work with the U. S. Entomological Commission, which had been organized under act of Congress in 1876 to study the Rocky Mountain locust, became Entomologist in 1878, but withdrew the following year. For two years the post was held by Professor J. H. Comstock, who then returned to his professorship in entomology at Cornell University where, as a teacher and in research, he was destined to have a profound influence on the development of American entomology. Riley returned to the Government in 1881 in charge of the Division, which had in the meantime acquired the services of L. O. Howard, one of Professor Comstock's students from Cornell. Howard was destined to become an outstanding figure among American economic entomologists.

It is interesting to note the early interest of the Federal entomological services in citrus insects. Glover had spent considerable time on citrus insects; Comstock, while in the Division, had studied citrus insects in Florida and California, and had published an extensive paper on scale insects in the annual report of 1880; H. G. Hubbard was sent to Florida to study citrus insects, and in 1885 his "Insects Affecting the Orange" was published by the Department of Agriculture as a special report of the Division of Entomology. This was the outstanding publication of its kind at the time, and long served as a pattern for similar types of publications. While in Florida, Hubbard worked out the so-called Riley-Hubbard formula for the preparation of kerosene emulsions, and kerosene emulsions prepared according to this formula were used in the control of scale insects and other types of sucking insects for many years. It was during this early period, also, that the famous introduction of the vedalia ladybird beetle into California to control the cottony cushion scale, the most serious of the citrus pests at that time, was made by Albert Koebele. These predators were received and distributed by D. W. Coquillett in 1888. Koebele and Coquillett were both employed by the U. S. Department of Agriculture.

49

Coquillett was also prominent in the development of hydrocyanic-acid gas fumigation in the control of scale insects attacking citrus trees in California. C. L. Marlatt, in 1903, wrote a "Farmers' Bulletin" (No. 172) on "Scale Insects and Mites on Citrus Trees".

In 1894 L. O. Howard became Head of the Division, which was finally organized as the Bureau of Entomology in 1904 and was divided into 12 Divisions. Howard retained his position as Chief of the Bureau until he retired in 1927, when he was succeeded by C. L. Marlatt.

The Plant Quarantine Act of 1912:- A milestone in entomological legislation in the United States was the belated Plant Quarantine Act of 1912, finally enacted after many previous efforts, dating back to 1897, had failed. The repeated finding of overwintering nests of brown-tail moths and egg masses of gypsy moths on imported nursery stock became the immediate incentive for the legislation. The bill had two main purposes: to prevent the entry of plant pests on imported plants, and to control and, if possible, eradicate any new pests having limited foothold within the United States. It is of interest to note that before the passage of the Plant Quarantine Act of 1912, at least 24 commercially important countries had prohibitions or restrictions on the entry of plants or plant products.

Federal Horticultural Board:- For the purpose of administration of the Plant Quarantine Act of 1912, the Federal Horticultural Board was established in the Department of Agriculture. This was an important administrative body of independent status, that is, it reported directly to the Secretary of Agriculture. The Board consisted of members of the Bureau of Entomology, the Bureau of Plant Industry, and the Forest Service. Its chairman, until the function of the Board was replaced by the Plant Quarantine and Control Administration in 1928, was C. L. Marlatt.

Under the Quarantine Act of 1912, the quarantine restrictions and regulations which were imposed did not represent the judgment of the Horticultural Board alone. The suggestion for a plant quarantine might come from any office of the Department of Agriculture or from any source outside the Government, as from states or individuals. Elaborate provisions for conference and discussion were made. This has continued to be the practice up to the present time.

During the 16 years the Federal Horticultural Board remained in existence, 64 quarantine laws were passed. The most controversial of these was the nursery stock, plant, and seed quarantine (No. 37), which forbade, except as provided in the rules and regulations, the importation into the United States of nursery stock and other plants and seeds from any foreign country, except for experimental or scientific purposes. "The quarantine allowed importation under permit and on compliance with the requirements of regulations, of certain bulbs, rose stocks, and fruit stocks, including cuttings, scions, and buds, and seeds of nut, fruit, forest and ornamental and shade trees and of hardy perennial ornamental shrubs. It provided for the importation, under special permits from the Secretary of Agriculture, of limited quantities of otherwise prohibited stock for the purpose of keeping the country supplied with new varieties of plants and stock for propagation purposes, not available in the United States". (931) The quarantine was bitterly opposed by practically all the associations of florists and nurserymen and workers concerned with ornamental plants. As a matter of fact, however, the United States lagged behind other great nations in the enactment of a quarantine of this type, and Marlatt in 1925 pointed out to critics that our country was paying the cost of this delay in about a billion dollars annual loss from pests imported from other countries.

The Plant Quarantine and Control Administration:- On July 1, 1928, on the recommendation of the Secretary of Agriculture, the Plant Quarantine and Control Administration of the U. S. Department of Agriculture was established. It took over the duties of quarantine and control work that had previously been the function of the Federal Horticultural Board, as well as the regulatory work relating to insect pests and plant diseases which was formerly performed by the Bureau of

ORGANIZATION AND LEGISLATION IN ENTOMOLOGY

ntomology and the Bureau of Plant Industry as agents of and in cooperation with
hé Federal Horticultural Board. In a reorganization effected in 1933, the P.Q.C.A.
as merged with the Bureau of Entomology under the combined title of Bureau of
ntomology and Plant Quarantine. In its brief existence, the P.Q.C.A. carried out
wo of the most remarkable and highly successful campaigns of pest eradication in
ntomological history, namely, date palm scale eradication in Arizona, California
nd Texas and Mediterranean fruit fly eradication in Florida.

Organization of the Present Bureau:- With reorganization in 1933 which re-
ulted in the present Bureau of Entomology and Plant Quarantine, Lee A. Strong be-
ame Chief of the Bureau, a position which he held until his death in 1941, when
he position was taken over by P. N. Annand, the present Chief. The Bureau of
ntomology and Plant Quarantine is without a doubt the largest single organization
n the world dealing with research, regulatory, and extension entomology. The ex-
ent of its activities can be ascertained from the "Directory of Organization and
ield Activities of the Department of Agriculture", published in 1939:

"The Bureau of Entomology and Plant Quarantine is concerned with investiga-
ions of insects and their economic relations; the development and application of
ethods for their eradication or control; the carrying out, in cooperation with the
tates, of necessary work to prevent the spread and to control or eradicate insect
ests and plant diseases that have gained more or less limited foothold in the
nited States; and the utilization of those species that are beneficial. These
ctivities include investigations on and direction of control campaigns against
he species injurious to agriculture and forestry; investigations on the species
ffecting the health of man and animals, or infesting human habitations or injuri-
us to industries; the culture and use of honeybees and beekeeping practices; in-
estigations on the natural enemies of insects and plant pests and the possibility
f using these as aids for control; the taxonomy, anatomy, physiology, and re-
ponses of insects; chemical and other problems relating to the composition, action,
nd application of insecticides; and the development of methods of manufacturing
nsecticides and materials used with them."

Although present efforts at reorganization of Federal bureaus may result in
changes in the existing organization of the Bureau of Entomology and Plant Quarantine, a listing of the present Divisions will give an idea of the scope of work encompassed by the present organization and they would probably be represented as
specific fields of endeavor in any reshuffled organization.

The Divisions of the Bureau as presently constituted, and which contain about
1500 scientifically trained workers, are as follows:

1. Finance and Business Administration
2. Personnel
3. Bee Culture
4. Cereal and Forage Insect Investigations
5. Control Investigations
6. Cotton Insect Investigations
7. Domestic Plant Quarantines
8. Foreign Parasite Introduction
9. Foreign Plant Quarantine
10. Forest Insect Investigations
11. Fruit fly Investigations
12. Fruit Insect Investigations
13. Grasshopper Control
14. Gypsy Moth Control
15. Insect Identification
16. Insect Pest Survey and Information
17. Insecticide Investigations
18. Insects Affecting Man and Animals
19. Japanese Beetle Control
20. Mexican Fruit Fly Control
21. Pink Bollworm Control
22. Plant Disease Control
23. Truck Crop and Garden Insects

FEDERAL REGULATORY AND CONTROL LEGISLATION

Quarantine against Mediterranean Fruit Fly and Melon Fly:- It had been specifically provided in the Plant Quarantine Act of 1912 that quarantine provisions
as applying to the Mediterranean fruit fly, Ceratitis capitata, should become effective upon its passage. Accordingly, on September 18, 1912, the Secretary of
Agriculture promulgated Domestic Quarantine No. 2, which, in its original form,
prohibited the shipment of all fruit and vegetables specified in the notice of

quarantine, from the Territory of Hawaii into the United States.. After a study by the Bureau of Entomology of the life history of the melon fly, Dacus cucurbitae, which attacks gherkins, tomatoes, and beans, Quarantine No. 2 was revised and amended as Quarantine No. 13, effective May 1, 1914, to protect the United States from both the Mediterranean fruit fly and the melon fly. Quarantine No. 56, effective November 1, 1923, extends the quarantine against the above fruit flies and some others to include all foreign countries except Canada. Certain fruits may be imported, however, under permit, when it has been determined that they are not carriers of fruit flies.

 Avocado Seed Quarantine:- Of special interest in relation to subtropical entomology are several other quarantines promulgated during the life of the Federal Horticultural Board. One was the foreign avocado seed quarantine of 1914, to prevent the introduction into the United States of the avocado seed weevil, Heilipus lauri Boh.[1] This quarantine forbids the importation into the United States of the seeds of avocado from Mexico and Central America. The avocado seed weevil is found only in the seeds, so by an order of the Horticultural Board, dated August 2, 1922, the entry of avocados from which the seed is removed, was permitted. These seeded avocados were sold at border ports for local consumption. Subsequent orders permitted the entry of avocados from Colombia at northern ports and from Cuba at all ports. Departments of Agriculture investigators had found that avocados in Cuba are free of pests which are likely to be injurious in the United States.

At the present time commercial shipments of avocados from Mexico are authorized only through the north Atlantic and north Pacific ports. They are not authorized, either with or without seed, through Mexican border ports adjacent to California. The thicker and rougher skinned Guatemalan races and hybrids are the only types authorized entry from Mexico or other foreign countries. The purple, very thick skinned Mexican race of avocados is not authorized entry except at certain ports of entry on the Mexican border east of California, and then only in small lots in the possession of individual passengers provided the seeds have been removed and inspection reveals them to be free of pest risk.

Mexican Fruit Fly:- In order to prevent the possible introduction of the Mexican fruit fly, Anastrepha ludena, Foreign Quarantine No. 5 was promulgated in 1913. This quarantine forbids the importation of oranges, sweet limes, grapefruit, mangos, ochras, sapotes, peaches, guavas, and plums from Mexico (fig. 48). Despite as strict as possible enforcement of this quarantine, in May, 1926, the Mexican fruit fly was reported to be widely distributed in the grapefruit orchards in Cameron and Hildago counties in Texas (931). The Bureau of Entomology and the Federal Horticultural Board, in cooperation with the Texas Department of Agriculture, immediately initiated a clean-up campaign which was supported the following year by a Congressional appropriation of $100,000. With the splendid voluntary cooperation of the local population,[2] a host-free period was maintained from March 1, 1927 and continuing for 7 months. All fruits in which the Mexican fruit fly larvae could survive were destroyed so as to starve out the insects. Great efforts were exerted to prevent further smuggling of Mexican fruit across the Texas border. The host-free period was maintained again the following year. The large grapefruit crop of 1927-28 in the previously infested district was harvested without the discovery of a single infested fruit. The success of the host-free period as a control measure is due to the fact that there are no flies emerging from the summer fruits in the valley to attack in large numbers the first ripening citrus (54).

---

[1] The quarantine at present is also directed against Heilipus pittieri Barber, recorded from Costa Rica, H. perseae Barber, from the Canal Zone, and Conotrachelus perseae Barber, from Guatemala.

[2] The cooperative spirit of the people on both sides of the Rio Grande was stated by Marlatt (587) to have been unparalleled in his experience. The Mexicans demonstrated a friendly and unselfish attitude which was again shown in the pink bollworm control campaign. A report on the success of this cooperative effort was written in the Journal of Economic Entomology in 1946 by Ing. Alfonso Delgado de Garay to mark the first appearance of a paper written in the Spanish language in the Journal (226). Again in 1947, in the citrus blackfly eradication campaign, an excellent cooperative spirit was shown by the Mexicans (994).

ORGANIZATION AND LEGISLATION IN ENTOMOLOGY

g. 48. Federal quarantine
spection at a border station.
urtesy Harold J. Ryan.

On August 10, 1927, a domestic Mexican fruit worm quarantine (No. 64) was promulgated which prohibits, except as provided in rules and regulations supplemental thereto,[1] the interstate movement of fruits of all varieties that are hosts of the Mexican fruit fly, in the raw or unprocessed state, from the regulated area of Texas. Mexican fruit flies are very rarely found in the regulated area, nevertheless a "vapor heat" treatment of the fruit described on page 99 is employed as an additional insurance against the shipment of Mexican fruit fly maggots from the formerly infested area.

Citrus Canker and the Citrus Quarantine:- A plant disease known as citrus canker, Bacterium citri, was introduced about 1908 with Japanese trifoliate orange stock and it was reported in the Gulf States in the summer of 1914. This serious disease was eventually eradicated. To prevent importation of this and other citrus diseases, Foreign Quarantine No. 19, effective January 1, 1915, forbade the importation into the United States from all foreign countries and localities of all citrus nursery stock, including buds, scions, and seeds. This was followed in 1917 by Foreign

arantine No. 28, forbidding the importation into the United States of all citrus uits from eastern and southeastern Asia, Malay Archipelago, the Philippine lands, Oceania (except Australia, Tasmania and New Zealand), Japan (including rmosa and other islands adjacent to Japan), and the Union of South Africa. Fruits the mandarin class may be imported under permit and upon compliance with certain nditions prescribed in the regulations. The quarantine is still in force as orinally promulgated.

Date Palm Scale Eradication:- In 1913 the Bureau of Entomology made an instigation of the two principal date scale insects, the date palm scale, Parlaoria blanchardi, and the red date scale, Phoenicococcus marlatti, which infested ite palms in certain sections of California, Arizona and Texas. The red date ale later proved to be of minor importance, but the date palm scale, which was troduced from Egypt and Algeria in 1890, was a serious pest and control was dificult. Domestic Quarantine No. 6 was promulgated on March 1, 1913, to prevent the read of these two date insects into other sections. Eradication work against the ite palm scale in Texas was begun in 1915 and eradication was completed in 1919. lthough an eradication campaign was started in the Territory of Arizona in 1907 d the original infestation at the experiment station was eradicated by 1914, her infestations in Arizona and California continued.

In 1922 a joint project by the Federal Horticultural Board and the Arizona periment Station was undertaken. The survey crew (fig. 49, left) first charted e infested palms. The leaves of the palms found to be infested were cut off and e trunk was seared with a gasoline torch (fig. 49, right). This treatment was fective in destroying the scale, and within a couple of years the trees were as ll foliaged as before the treatment. Serious new infestations were discovered 1927, and in the opinion of date growers and experts of the Department of Agriilture, the entire date industry of the United States was threatened. In 1928 e Federal Government made an appropriation of $40,000 for a cooperative eradicaion campaign with the states of California and Arizona. California appropriated 5,000.

A campaign of extensive and thorough inspection was well organized under the rection of B. L. Boyden of the Plant Quarantine and Control Administration. As

_____
[1]The fruits affected by the quarantine are admissable under certification by the U. S. Department of Agriculture, but as far as California is concerned, many of these fruits are excluded from is State by other quarantines.

Fig. 49. Date palm scale eradication. Left, inspecting palms; right, torching palms to destroy scales. After Boyden (120).

before, the infested palms were defoliated and "torched". It was remarkable how men could be trained to distinguish the small, inconspicuous, greyish colored scales from the myriads of blotches of desert dust of practically the same color, especially since the inspection of the leaves had to be, of necessity, rather rapid and cursory.

The last scale was found in Yuma, Arizona, in 1930; in Phoenix, Arizona, in 1932; in the Coachella Valley, California, in 1931, and in the Imperial Valley, California, in 1934. The campaign was terminated in 1936, and since no new infestations have been reported to date, the eradication of the scale can be considered to have been successful (120). Quayle (724) makes the following comment on the date palm scale eradication project: "Considering the painstaking work of inspecting large date palms and the extensiveness of the infestation at the outset, the result of the campaign of eradication of the date palm scale must be considered as one of the outstanding achievements in applied entomology."

Mediterranean Fruit Fly Eradication:- The fruit flies are small Diptera of the family Trypetidae (Tephritidae), the larvae of which infest many varieties of fruits, nuts, and vegetables. The Mediterranean fruit fly, Ceratitis capitata, has long been recognized as a potential menace to the fruit industry of the United States, as indicated by the fact that a quarantine against this insect was specified in the original Quarantine Act of 1912. It is a serious pest of deciduous fruits as well as citrus and many other subtropical fruits. It is distributed throughout the subtropical regions of the world, with the exception of southeastern Asia and North America. In 1929 the Mediterranean fruit fly was found in Florida, and Federal and State quarantine officials and entomologists were to receive a supreme test in the field of pest eradication which has never before or since been equalled in magnitude. Early in the eradication campaign a nationwide committee appointed by President Hoover issued the following statement: "In the event that the fruit fly should escape Florida, infesting the regions of the south and west, capital values invested in properties producing susceptible fruits aggregating $1,800,000,000, and producing annual incomes of $240,000,000 are threatened."

A brief history of the eradication campaign was recorded by Quayle (724). After the discovery of the Mediterranean fruit fly near Orlando, Florida, on April 6, 1929, a campaign of eradication was begun immediately by means of a State emergency fund available for this purpose, as well as a transfer of Federal funds from the pink bollworm appropriation. On May 2, however, the U. S. Government appropriated $4,250,000 and on June 7, the Florida Legislature appropriated $500,000. In charge of the campaign was Wilmon Newell, Florida State Plant Board Commissioner. The Bureau of Entomology and the Plant Quarantine and Control Administration of the U. S. Department of Agriculture, as well as State Agricultural officials, cooperated in the extensive campaign.

On May 1, 1929, a host-free period was begun and this was maintained until October 1. The planting or growing of vegetables that would mature or be subject to infestation by fruit fly during the host-free period had already been prohibited. The only fruits or vegetables permitted in the "protective zones" were citrus fruits on the trees which were not sufficiently mature to be subject to infestation, and host fruits and vegetables in storage or in retail sale for immediate consumption. Enormous quantities of host fruits were destroyed (fig. 50). The fruit flies were thus denied fruits or vegetables in which to oviposit.

Fig. 50. Destruction of citrus fruit in Mediterranean fruit fly campaign in Florida. Courtesy R. S. Woglum.

To prevent any carry-over of adults through the host-free period, poisoned bait sprays were used throughout the infested area. These were applied to the foliage of wild and cultivated plants at regular intervals. In the preparation of these poisoned bait sprays, 2,218,387 pounds of sugar, 299,309 pounds of lead arsenate, and 375,301 gallons of syrup were used. Sixty spray machines were used in these operations. Within the "eradication area" were located 72% of the bearing citrus trees of Florida (120,000 acres), as well as 160,000 acres of other fruits and vegetables.

A light infestation of Mediterranean fruit fly was found on November 16, 1929 and again on March 4, 1930, both of which were destroyed. Large sums of money continued to be appropriated for inspection work. On July 25, 1930, two pupae were found under a sour orange tree in St. Augustine and sources of infestation in that area were destroyed. On November 15, 1930, the quarantine was ended, and the Mediterranean fruit fly, after a known period of existence of 14 months, was not again found in Florida. In this campaign, the U. S. Government spent $6, and the State of Florida spent $381,475.95.

SOME OF THE NON-REGULATORY FUNCTIONS OF THE BUREAU OF ENTOMOLOGY
AND PLANT QUARANTINE OF SPECIAL INTEREST IN
RELATION TO SUBTROPICAL ENTOMOLOGY

Insecticide Investigations:- Possibly of most universal interest to entomologists is the work of the Division of Insecticide Investigation,

55

SUBTROPICAL ENTOMOLOGY

Roark, Haller, LaForge, Cupples, Gertler, Schlechter, Clark, C. M. Smith, Markwood, C. R. Smith, Goodhue, Jones and Siegler, all now connected or formerly connected with this Division, are well known to all those acquainted with the literature on economic entomology in the United States. Their fundamental work in insecticides and adjuvants is of interest to all economic entomologists, regardless of the diversity of their specific insect problems. An enormous amount of basic work was done in this Division on the chemistry of the insecticides of plant origin, such as rotenone, nicotine, and pyrethrum, and now the studies on the chemistry of the chlorinated hydrocarbons, organic phosphates and other new insecticides of high insecticidal effectiveness are keeping apace with the rapid tempo of developments in the field of synthetic organic insecticides (fig. 51).

Fig. 51. R.C.Roark (left) and H.L. Haller of the U.S.Bureau of Entomology and Plant Quarantine discuss the arrangement of the atoms in the molecule of a new preparation coming up for official tests. Courtesy U.S.D.A.

Among the many useful publications emanating from the Division of Insecticide Investigations is the "Review of United States Patents Relating to Pest Control" by R. C. Roark. This quarterly review, in mimeographed form, is sent to anyone requesting it. This helpful service of the Division of Insecticide Investigations was begun in 1927, at which time Roark (746) pointed out that patent literature, (1) often describes for the first time a valuable new insecticide, (2) gives the composition of some proprietary preparations sold under extravagant insecticidal claims, and (3) describes new apparatus and processes for the application of insecticides and for the destruction of insects by physical means.

Control Investigations:- Of interest to all economic entomologists is the work of the Division of Control Investigations, which deals with the testing and evaluation of insecticides, many of them from the laboratories of the Bureau of Insecticide Investigations. Hundreds of compounds are being annually "screened" in tests made against a wide variety of insects according to standardized laboratory and field techniques. Fundamental studies on the physiology of insects are in progress. The development and improvement of equipment for applying insecticides is also within the scope of the work of this Division. Much work has been done in recent years on aerosols. Such names as McGovran, Swingle, Yeager, McIndoo, Bulger, and Sullivan, all connected or formerly connected with this Division, are commonplace in the literature of economic entomology.

committee
event that the
and west, capital
gregating $1,800,0
ened."   tomological
                            , California.

A brief histor,

After the discovery d scale. In the
April 6, 1929, a ca

Fruit Insect Investigations:- More closely related to our subtropical fruit insect problems in California are two entomological laboratories of the Division of Fruit Insect Investigations. One, in Whittier, California (fig. 52), is in charge of A. W. Cressman, who is investigating insecticidal sprays for control of California red scale and miscellaneous citrus insects. In this same laboratory, H. R. Yust is investigating fumigation for red scale control and acaricides that may be used with HCN. B. M. Broadbent is investigating sprays for control of California red scale, including the residual effects of different oils. F. M. Munger is investigating the resistance of citrus thrips to insecticides, and the biology of the red scale. In addition he is studying the effect on the citrus red mite of sprays other laboratory, located at Fresno, California control of insects attacking fruits intended for drying is

56

eing investigated by Perez Simmons, D. F. Barnes and C. K. Fisher. Insect prob-
ems. of such subtropical fruits as the raisin grape, the fig, and the date are in-
estigated at this laboratory, sometimes in cooperative projects with the Univer-
ity of California.

Foreign Parasite Introduction:- In a later chapter it will be explained why
he importation and establishment of beneficial insects must be placed in the
ands of a relatively few highly trained and specialized entomologists. With ex-
eption of such privileges as are granted to a limited number of states and state
xperiment stations or in cooperative projects undertaken with the latter, this
ork has been taken over by the U. S. Department of Agriculture. Since the year
888, when an entomologist of the Bureau of Entomology, as then constituted, dis-
overed and successfully introduced the vedalia ladybird beetle, <u>Rodolia cardinalis</u>
Muls.), to control the cottony cushion scale on citrus trees in California, the
nvestigations of potential parasites abroad, and their importation and establish-
ent, continued to be an important activity of the Bureau and eventually led to
he establishment of a Division of Foreign Parasite Introduction.

Foreign countries have greatly aided the Bureau of Entomology and Plant
uarantine in its efforts to discover beneficial insects and to prevent the intro-
uction into this country of destructive pests. In return, the Bureau has rendered
onsiderable assistance to foreign countries in their efforts to control or exterm-
nate insect pests, and many important beneficial insects have been exported to
oreign countries.

Insect Identification:- The identification and classification of insects has
ormed an important part of the activities of the Bureau of Entomology and Plant
arantine from the date of the inception of this Bureau as a branch of the Patent
ffice. Taxonomic papers were prominent among the contributions of the earliest
ntomologists hired by the Bureau and show the importance that was attached to
his field of endeavor. Hundreds of papers of this type have since been published.
hrough many thousands of insect identifications annually, the Bureau has given
,reat assistance to other federal agencies, to state entomologists, quarantine
fficials, universities, and private workers interested in entomology. Such work
ls now concentrated in the Division of Insect Identification, and specimens sent
:o the Bureau for identification should be addressed to this Division. The insects or mites will be turned over for identification to a specialist in the par-
:icular taxonomic group to which they belong.

Publications:- A discussion of the Bureau of Entomology and Plant Quarantine
vould not be complete without mention of the numerous publications of the huge
rtaff of this organization that are available to the public. When available,
:hese are sold by the Superintendent of Documents, Government Printing Office,
Jashington, D. C., and with some exceptions, current issues are sent out, on appli-
:ation, by the Office of Information of the Department of Agriculture. In a
'Monthly List of Publications," sent free to persons requesting it, new publica-
:ions are announced. The following is quoted from No. 60 of the Service Monographs
)f the United States Government, written by G. A. Weber under the auspices of the
Brookings Institution and published in 1930:

"BUREAU PUBLICATIONS. During the period from the establishment of the Divi-
jion of Entomology in 1863, until July 1, 1913, the results of the work of this
3overnment unit were published in the annual reports of the Entomologist and in
)ther Bureau publications as well as in the Farmers' Bulletin and the Yearbook of
:he Department of Agriculture.

"The Bureau Publications issued in its own name prior to July 1, 1913, other
:han annual reports, were designated as Bulletins, Circulars, Insect Life, and
fechnical Series, and there were occasional miscellaneous unnumbered publications.
fwo series of Bulletins were issued by the Bureau, the old series being numbered
?rom 1 to 33, issued from 1883 to 1895, and the new series, from 1 to 127, issued
?rqm 1895 to 1913. Of the circulars there were also two series, the first, com-
)rising over forty numbers, were mostly circular letters; the second series,

numbered from 1 to 173 were issued from 1891 to 1913. Insect Life, a monthly publication, was first issued in July, 1888, and was discontinued in July, 1895. A general index to this series was issued in 1897. The volumes are numbered from 1 to 7. They consist of from five to twelve numbered parts each. The class of matter that appeared in this series was continued in the Bulletins and Technical Series. The Technical Series was begun on June 15, 1895, and was numbered from 1 to 27. All the series of Bureau publications, except the annual reports, were superseded in 1913 by new series issued from the Secretary's Office.

"Monthly Bulletin of Insect Pest Survey. The Bureau issues on the first of each month, from March to December, a multigraphed 'Monthly Bulletin of Insect Pest Survey,' which is distributed gratis to entomologists, agricultural colleges, and agricultural experiment stations. The Bulletin gives, each month, the outstanding entomological features in the United States and in Canada, and a detailed statement of insect conditions in the United States. At the end of the year a 'Summary of Insect Conditions' is issued.

"Posters. The Bureau has issued posters, for free distribution, for the use of schools, showing insect pests and their work, and methods of control. The posters that have thus far been issued deal with grasshoppers, the garden cutworm, potato beetles, the wheat jointworm, the chalcis-fly, the Hessian fly, the gypsy moth, the corn borer, the Japanese beetle, spraying of potato fields and orchard spray schedules.

"DEPARTMENT PUBLICATIONS: Since July 1, 1913, the Bureau of Entomology has published the results of its work in the annual reports of the Entomologist, and, like other bureaus of the Department of Agriculture, in the following series of Departmental publications: Farmers' Bulletins, Department Leaflets, Technical Bulletins (formerly Department Bulletins), Circulars, and Miscellaneous Publications (formerly Miscellaneous Circulars), and in articles and reprints from articles in the Yearbook of the Department of Agriculture and in the Journal of Agricultural Research, and articles in the Official Record.

"Annual Reports. Annual reports of the Entomologist have been included in the annual reports of the Department of Agriculture since 1863, except in 1878 when none was issued. Since 1879 the annual reports of the Entomologist have also been issued in separate form.

"Farmers' Bulletins. The Farmers' Bulletins cover a wide range of subject matter and are of interest to a great many classes of persons. They are issued for general distribution among farmers, and are, therefore, written in simple, nontechnical language, easily understood by the layman. Most of the recent publications of the Department of Agriculture dealing with entomological subjects are in this series, many of them, however, being reprints of previously issued Farmers' Bulletins.

"Department Leaflets. The series of Departmental Leaflets which was started in 1927, carries popular material of the same general character as the Farmers' Bulletin series, except that the leaflets are confined to specific practical directions and recommendations, remedies, and methods. They are brief and concise, are written in informal, popular style, and are limited to not more than eight pages.

"Technical Bulletins. The Technical Bulletins are a series of technical publications of the various Bureaus of the Department of Agriculture intended primarily for specialists and research workers. This series in 1927 superseded the Department Bulletin series, which was numbered from 1 to 1500. The Bulletins contain the results of the scientific and research work of these bureaus. In most cases only small editions are printed. A few issues each year are devoted to entomological subjects.

"Circulars. The Circular series carries the less technical and more informational material of the same general nature as that in the Technical Bulletin series. This series in 1927 replaced the Department Circular series numbered 1 to 425. Few of them are devoted to entomological subjects.

ORGANIZATION AND LEGISLATION IN ENTOMOLOGY

"Miscellaneous Publications. The series, Miscellaneous Publications, in-
des those publications of a miscellaneous nature that do not fall within any
the other series issued by the Department of Agriculture. V_{ery} few of the pub-
ations of the Bureau of Entomology appear in this form. This series in 1929
erseded the Miscellaneous Circulars, which were numbered from 1 to 110.

"Yearbook. The Yearbook of the Department of Agriculture contains a general
ort of the operations of the Department during the preceding calendar year.
rly all the articles in the Yearbook are reprinted in separate form. Occasional
icles by the Bureau of Entomology appear in the Yearbook.

"Journal of Agricultural Research. The Journal of Agricultural Research is a
dred-page semi-monthly periodical published by authority of the Secretary of
iculture, with the cooperation of the Association of American Agricultural
leges and Experiment Stations. It contains articles on technical agricultural
earch carried on by the Department of Agriculture or the state experiment sta-
ns and is, therefore, of primary interest only to agricultural scientists and
anced students. The editorial committee consists in part of employees of the
artment of Agriculture and in part of members of the Association. It was
ginally published monthly, and then weekly, and is now issued semi-monthly.
h article is reprinted as a separate. A considerable proportion of the publica-
ns of the Bureau of Entomology appears in this journal. It has been announced
at the publication of the journal will be discontinued after the issuance of the
st number of Volume 78 (1949).

"Official Record. The Official Record is a weekly publication containing
ficial orders and miscellaneous information concerning the activities of the
partment, and other matters of special interest to employees and persons having
do with the Department's work. Occasionally an entomological subject is dealt
th in this publication."

In addition to the journals listed above should be mentioned a useful series
photoprinted leaflets designated the "E Series," begun in 1915 and the "ET
ries", begun in 1934. The former deals with economically important insects and
eir control, and with insecticides, while the latter deals with entomological
paratus and techniques. Likewise, it should be pointed out that many scientific
pers written by personnel of the Bureau are published in the Journal of Economic
tomology and elsewhere.

The publication "Service and Regulatory Announcements" has been issued for
ch month, or for groups of months, since January, 1914. The publications con-
in "notices of quarantines and of amendments thereto; notices of the lifting of
arantines; plant quarantine decisions; Treasury decisions; orders and instruc-
ons concerning the exportation and interstate movement of specific plants and
ant products; notices of hearings on proposed plant quarantines; information
ncerning convictions for violations of the Plant Quarantine Act; warnings to
ssengers and crews concerning the transportation of quarantined plants and plant
oducts; and other information concerning questions and activities arising under
e Plant Quarantine Act and other acts enforced by the Plant Quarantine Control
ministration. Much of this information is also published in separate form,
ther multigraphed or in print. An index to the Service and Regulatory Announce-
nts is published annually."

A comprehensive idea of the contributions of the entire Bureau in a given
scal year may be obtained from the "Report of the Chief of the Bureau of Ento-
logy and Plant Quarantine" which is annually submitted on June 30 to the Secre-
ry of Agriculture.

The Experiment Station Record was begun in 1889, and was issued by the Office
Experiment Stations, but is included here for convenience. This was a monthly
blication of short abstracts of the papers written by U.S.D.A. and State Experi-
nt Station personnel. There were two volumes per year, and each volume was fol-
wed by an Index Number. The publication of the Experiment Station Record was

discontinued indefinitely with the issuance of the Index Number of volume 95, covering the period July-December, 1946.

Interested persons may obtain announcements of publications of the U. S. Department of Agriculture, as well as of the state agricultural experiment stations, by writing to the Department in Washington. The publications may also be obtained, although a small fee is charged for some of them. The price of the Yearbook of the Department of Agriculture is two dollars.

THE INSECTICIDE ACT OF 1910 AND THE PRESENT INSECTICIDE, FUNGICIDE, AND RODENTICIDE ACT

There is at present much less adulteration and mislabeling of insecticides and fungicides than in former times, thanks to effective, though belated, legislation. The Insecticide Act, was passed in 1910 and became effective on January 1, 1911. It pertained to imports and exports to and from the United States, interstate shipments, and sales within the District of Columbia and the Territories, such as Hawaii and Alaska. Actual standards were set only for Paris Green and lead arsenate, the most important insecticides of that period. The most important features of the act were summarized by Howard (448) as follows:

"(a) Definite standards for lead arsenates and Paris greens are stated, and it is required that all lead arsenates and Paris greens subject to the act shall conform to these rigid specifications.

"(b) All insecticides and fungicides (other than lead arsenates and Paris greens) which contain inert ingredients shall bear a statement upon the face of the principal label of each and every package giving the name and percentage amount of each and every inert ingredient contained therein and the fact that it is inert, or, in lieu of this, a statement of the name and percentage amount of each and every active ingredient which has insecticidal or fungicidal properties, together with the total percentage of inert ingredients.

"(c) For insecticides (other than lead arsenates and Paris greens) and for fungicides which contain arsenic or compounds of this metal, a statement must be made on the face of the principal label of the total arsenic, expressed as per cent of metallic arsenic, and total arsenic in water-soluble forms, similarly expressed.

"(d) No statement, design, or device appearing on the label of an insecticide, fungicide, Paris green or lead arsenate shall be false or misleading in any particular. It will at once be seen that all false or exaggerated claims relative to the efficacy of the article constitute misbranding, and the Government is empowered to institute criminal or seizure proceedings as outlined above.

"(e) All insecticides and fungicides (other than lead arsenates and Paris greens) must be up to the standard under which they are sold.

"(f) No substance or substances shall be contained in any insecticide or fungicide (other than lead arsenates and Paris greens) which shall be injurious to the vegetation on which such articles are intended to be used."

Because of the changes and advances which have been made in the insecticide field since the enactment of The Insecticide Act of 1910, it became increasingly apparent in recent years that a revision of the Act was necessary to meet present day conditions. Accordingly a new Federal Insecticide, Fungicide, and Rodenticide Act to regulate the marketing of economic poisons and devices was enacted by Congress and signed by the President on June 25, 1947, to replace the old Act. The principal changes and additions, as compared to the Insecticide Act of 1910, are summarized by McDonnell (547):

"1. An extension of its provisions to include rodenticides, herbicides, devices and substances intended for preventing, destroying, or repelling other forms of plant 'or animal life which the Secretary of Agriculture 'shall declare to be a pest.'

"2· A requirement that all products covered by the Act (except devices) must be registered with the Secretary of Agriculture prior to their sale or introduction into interstate commerce.

"3. A requirement that necessary directions for the use of the products be given on the labeling, and also a warning or caution statement, if deemed necessary by the Secretary, to prevent injury to man or other vertebrate and useful invertebrate animals, and vegetation (except weeds).

"4. A provision requiring poison labeling, including the skull and crossbones and an antidote statement, on products which contain any substance or substances in quantities 'highly toxic to man.'

"5. A provision that certain white powdered poisonous economic poisons must be 'distinctly colored or discolored' where 'feasible' and necessary for the protection of the public health.

"6. A requirement that there be a statement on the label of the net weight or measure of the contents of the package.

"7. Authority for the inspection of books and records relating to the delivery or holding of any economic poison or device by authorized Federal and State employees."

In compliance with one of the statutes of the Federal Act, the notices of judgment under the Act are published at irregular intervals and much can be learned of the practical legal aspects of the packaging and marketing of insecticides and fungicides by reading these notices. This publication is issued by the Production and Marketing Administration of the U. S. Department of Agriculture.

A few of the states had insecticide acts before the enactment of the Federal act in 1910, and now the majority have their independent insecticide acts, many patterned after the Insecticide Act of 1910. Some of the states, including California, have well organized inspection of the insecticides offered for sale, by which insecticide and fungicide samples are annually systematically collected and analyzed, and the results of the analyses are published.

PRODUCTION AND MARKETING ADMINISTRATION

The provisions of the Federal Insecticide, Fungicide and Rodenticide Act of 1947 are now administered by the Insecticide Division of the Livestock Branch of the Production and Marketing Administration of the U. S. Department of Agriculture. Enactment of this statute repealed the Insecticide Act of 1910. W. G. Reed, Chief of the Division, is in charge of the issuing of licences and registration of labels. Those wishing to have an insecticide licenced for sale must submit a statement of the composition of the material in terms of the per cent of each and every ingredient, and two copies of either a typewritten label, printer's proof or a printed label to the above Division. A typewritten label is ordinarily preferable, for the Insecticide Division may wish to make suggestions and return the proposed label to the licencee, after which the latter can make the changes and then send two printer's proofs or finished copies of the label to the Insecticide Division. The Division sends back to the licencee one of his printer's proofs or finished labels with a stamp of issuance of registration with date and serial number. No fees are charged for this service and the licence is valid for five years.

After a product has been registered, a field inspector of the Insecticide Division may at any time draw a sample for testing. If the product does not conform with the analysis submitted by the licencee, the licence can be revoked. Thus the issuance of registration is not a guarantee of the validity of the licence.

FEDERAL FOOD AND DRUG ADMINISTRATION

The name "Food and Drug Administration" first appears in the Agricultural Appropriation Act of 1931, although many of the functions we associate with the present Food and Drug Administration were carried on under different organizational titles since January 1, 1907, when the Food and Drugs act of 1906 became effective. The present Food and Drug Administration was transferred from the Department of Agriculture to the Federal Security Agency, effective June 30, 1940. It is charged with the enforcement of the Food, Drug, and Cosmetic Act, Tea Importation Act, Import Milk Act, Caustic Poison Act, and Filled Milk Act. The Food and Drug Administration endeavors to promote purity, standard potency, and truthful and informative labeling of the essential commodities covered by the provisions of the five acts mentioned above. It no longer is charged with the enforcement of the Insecticide, Fungicide and Rodenticide Act, this function being taken over by the Production and Marketing Administration.

The Federal Food and Drug Administration, however, is still in charge of the establishment and enforcement of legal tolerances for deleterious residues of economic poisons (insecticides, fungicides, etc.) on fruits and vegetables. Shipments of fruits or vegetables treated with economic poisons may be inspected by agents of the Food and Drug Administration at the point of shipment or destination, or inspectors may ascertain the residue situation at the point of origin by their own observations or in cooperation with the state bureaus of chemistry or the particular agency that does the enforcement work. In states in which economic poison residues are an especially important problem, there is close cooperation between the bureau of chemistry of the state and the Federal Food and Drug Inspectors in respect to the enforcement of established residue tolerances.

The Food and Drug Administration is concerned with the enforcement of the pure food laws. It is also concerned with the insecticidal residues remaining on treated food products. In recent years the tendency has been to pay more attention to the "accumulative effect" of minute amounts of poison, far below the minimum lethal quantity. This has been due to the recognition of the fact that minute amounts of poison, absorbed frequently, may lead to chronic ailments. Consequently, thorough pharmacological studies on laboratory animals must be carried out, usually for a period of years, before a minimum "tolerance" can be established by the Food and Drug Administration. Investigators have in the main closely cooperated with the Food and Drug Administration in withholding recommendations on the use of promising new insecticides until these pharmacological investigations and the accumulation of data gained from nation-wide experience and observation have shown under what conditions, if any, the insecticide may be used without hazard to public health.

Eventually tolerances are established, and treated fruits and vegetables offered for sale must meet the requirements of the official tolerance, which is enforced by the Food and Drug Administration. Thus the tolerance for lead arsenate is: lead (metallic), 0.05 grain per pound of fruit (7 parts per million); arsenic (arsenic trioxide), 0.025 grain per pound. The tolerance for fluorine is 0.05 grain per pound.

As a regulatory agency, the Food and Drug Administration is constantly on the alert to apprehend and institute action under the Federal Food, Drug, and Cosmetic Act against any interstate shipments of fruits which may be found to contain residues of the insecticidal poisons in excess of the tolerance. Such shipments as are encountered are removed from the market by seizure and either destroyed or the owners are required to cleanse the seized produce under the supervision of the Food and Drug Administration in order to remove the excess residue.

Much of the research relating to insecticides deals with the attempt to substitute safer preparations for those now being used, but for which a low tolerance has been set. For example, apples sprayed with lead arsenate must be washed with a dilute hydrochloric acid solution in order to meet the tolerance requirements. This not only entails considerable expense, but in addition, the lead arsenate

ust be applied in a manner which will be conducive to a satisfactory removal of
the residue, and this is not always the manner of application which is best suited
for maximum insecticidal efficiency.

CALIFORNIA STATE DEPARTMENT OF AGRICULTURE

The history of the State Department of Agriculture and of entomological
legislation in California is recorded in considerable detail by E. O. Essig in his
History of Entomology, and need not be repeated here, except for brief mention
of the State Quarantine Act of 1912[1] which provides the following:

1. Inspection of all plant products coming into the State.
2. Disinfection, removal, or destruction of all infested materials.
3. Marking of all shipments of plant materials passing through the State.
4. Proper sealing of containers of infested plant material passing through
   the State.
5. Prohibition of entry of hosts of the fruit fly family Trypetidae from all
   places where these flies are known to exist.
6. Prohibition of entry of English or Australian wild rabbit, flying-fox,
   mongoose, or other animals detrimental to agriculture.
7. Penalties for violation of provisions of the Act.

The former office of Commissioner of Horticulture was abolished by an act of
the State Legislature approved on July 22, 1919, and in its place was created the
present Department of Agriculture to be presided over by a Director of Agriculture
appointed by the Governor. The present Director of Agriculture for the State of
California is A. A. Brock. We are here concerned primarily with the Bureaus of
Entomology, Plant Quarantine, and Chemistry, all in the Division of Plant Industry.

The State Legislature, during the administration of the first Director, G. H.
Hecke, provided for a State Board of Agriculture, to consist of 9 members. The
members of this Board are selected from widely-separated areas of the State, and
their duties are purely advisory. They are of great value, however, in helping to
bring the agricultural problems of California to the attention of the Director of
Agriculture.

STATE BUREAUS[2] OF ENTOMOLOGY AND PLANT QUARANTINE

The work of the Bureau of Entomology and the Bureau of Plant Quarantine em-
braces the fields of insect survey, insect identification, control or eradication,
plant quarantine, nursery service, and apiary inspection. A good idea of the
practical benefits of this important public service may be obtained by reading the
annual reports of H. M. Armitage, the Chief of the combined Bureau of Entomology
and Plant Quarantine, as it was organized up to October, 1948. The following account refers, of course, to the work of past years while it was organized under a
single Chief, hence, the use of the singular "Bureau." These reports are published in the Bulletin. This brief account of some of the more recent work of the
Bureau in the various fields indicated above will give an idea of the scope of
activities.

Insect Survey:[3] The survey work undertaken by the Bureau's entomologists with
regard to the incidence, distribution, population density, migration, and potenti-

[1]This should not be confused with the Federal Quarantine Act of 1912, which represented the
beginning of Federal quarantine legislation, while the State Act was merely a revision and broaden-
ing of long-standing quarantine policy.

[2]Effective October 1, 1948, the former Bureau of Entomology and Plant Quarantine was divided
into two separate Bureaus: the Bureau of Entomology, whose chief is H. M. Armitage, and the Bureau
of Plant Quarantine, whose Chief is A. P. Messenger.

[3]It may be appropriate at this time to point out that the terms "survey" and "pest detection"
could be properly distinguished. Survey is a method of determining the actual or potential population density and distribution of a pest already known to occur in the area surveyed. In pest "detection" a search is made for a migrated or possibly newly established insect immigrant. However,
both fields of work are for convenience discussed under the heading of "survey" in accordance with
what has been the usual practice as indicated in the literature.

H.M. Armitage
Chief of the Bureau of Entomology of the California
State Department of Agri-
culture.

A. P. Messenger
Chief of the Bureau of
Plant Quarantine of the
California State Depart-
ment of Agriculture.

alities for epidemic outbreaks of the many important pest species in California is
an important part of the work of the Bureau.

In recent years survey crews sent out by the Bureau have inspected for Japanese beetle, cotton boll weevil, Colorado potato beetle, and the pink bollworm of
cotton, but none of these pests was found. Search for new potential pests led to
the discovery of new scale insects on deciduous trees and ornamentals and the discovery of the olive parlatoria scale in Los Angeles and San Joaquin Counties. Extensive surveys have delimited the distribution of the Mexican bean beetle. Survey
work has also been continued for citrus whitefly, pear psylla, cotton insects,
Comstock mealybug, white fringed beetle, and Howard and Putnam scales. The oriental fruit moth has not recently been found to be extending its range.

The Bureau has succeeded in its constant survey and control work in stamping
out recurring infestations of the citrus whitefly. It appears that an improved
method of HCN fumigation assures the eventual eradication of Hall scale, Nilo-
taspis halli (Green), which had been a threat to the peach and almond industries
of the State. Although the grape leaf skeletonizer, Harrisina americana (Guer.),
was found to be more widely spread than originally believed, progress is being
made in its eradication.

Pest survey is of practical value to agriculture in other ways than in connection with quarantine and eradication. Early season surveys on insect populations may guide growers in planting and pest control. As an example, warnings by
State officials regarding the great build-up of sugar beet leafhopper populations
in the breeding grounds of the pest in the winter of 1945-46 caused tomato growers
in northern counties of the State to delay planting until after the spring flight
of the insects in mid-April, and this resulted in the avoidance of great loss to
the canning tomato industry.

New Insect Pest Survey Plans for California:- On July 1, 1947, funds were
made available for new long range State insect pest survey plans in which the
county agricultural commissioners offices will be expected to play a major rôle.
The appropriation provides for the employment of three field entomologists and
three field pathologists on full travel status, who will give their entire attention to pest survey.

In the organization of the new insect pest survey plan, the State was divided
into three districts. Directly responsible for the survey work in northern Cali-

ornia is Henry T. Osborn, who is in administrative charge of the entire State
est survey activity. Directly responsible in the central California district is
obert P. Allen, located in Modesto. The corresponding official for the southern
alifornia district is Robert W. Harper, with headquarters in Los Angeles.

The three entomologists mentioned above are responsible for training the ex-
ensive staffs of the county agricultural commissioners as to the pests to be
earched for and the method of survey for each pest. Both classroom training and
ctual field survey practice under State supervision are provided in the new
rogram.

It is expected that the intensified survey program will make possible the detection of incipient infestations of newly introduced pests of known economic importance in sufficient time to permit the carrying out of eradication measures at
a minimum of expense and with some reasonable expectation of success. In addition,
on the basis of information obtained, protective plant quarantines which may now
be in effect might be modified to assure that the infested area is properly defined and to eliminate unnecessary restrictions. Incidental to the survey activity,
a complete check list of insect species in California will be obtained. This will
supplement the survey now being conducted by the University of California, and
plans have been made for an exchange of specimens and information between the Department of Agriculture and the University.

Insect Identification:- As was previously stated, much taxonomic work of the
highest merit and of wide academic interest has originated in the State Bureau of
Entomology and Plant Quarantine. The more practical aspects of this work, however,
and the basis upon which it receives its financial support as one of the functions
of the Bureau, is in relation to the identification of insects of economic importance. The greater part of the insect specimens received for identification
are received from quarantine officials from port and border stations. In 1946,
27,782 determinations were made for these agencies. Another large source of specimens sent in for identification is the oriental fruit moth survey, which, however,
is becoming more restricted as the threat of this insect to California agriculture
diminishes. In 1945, 31,277 determinations were made, but in 1946 only 9,304.
Specimens sent in from other survey projects annually account for a few thousand
more identifications. Also, general determinations made for the State Department
of Agriculture, the county agricultural commissioners, and the University of California, account for between 2,000 and 3,000 identifications per year.

Control and Eradication:- After the discovery of the Mexican bean beetle,
Epilachna varivestis (Muls.), by a representative of the Ventura County Commissioner's Office on July 22, 1946, eradication measures were immediately put into
effect. It was the first time this very injurious pest had been found west of the
Rocky Mountains. An intensive survey (fig. 53, D) revealed that 39 properties--
1,923 acres of lima beans--were infested in whole or in part, and 44,000 acres of
beans were threatened in the immediate district. Destruction of crops and the application of a cryolite-lethane dust by air or ground machines did not give as
good results as were hoped for because of the difficulty of covering the undersides of the foliage with the dust. However, reproduction of the beetles was kept
at a minimum. A more intensive program was drawn up for the following year. Applications of 1% rotenone dust are now made in all bean fields in the infested
areas as well as in barrier zones. After the destruction of plants within a 50-
foot circle surrounding each infested spot revealed by the surveys, the fields are
dusted--first by ground dusters (fig. 54, B), then by airplane. This has been a
cooperative project between the State Department of Agriculture and the Ventura
County Department of Agriculture. The Lima Bean Growers Association and some of
the individual bean growers have made financial contributions to partly defray
the expense of the bean beetle campaign. In surveys for bean beetles in other
districts of the state, the insect has not been found to date.

Work on the prevention of an outbreak of the aerial form of the grape phylloxera in the Government Experimental Gardens south of Fresno, by the destruction
of vines, was successfully completed in 1946. This marks the first time the

Fig. 53. Some state and county regulatory activities. A, nursery inspection; B, rodent control; C, spreading grasshopper poison; D, Mexican bean beetle survey crew. From various sources.

Fig. 54. State regulatory activities. A, killing wild grape with 2,4-D in grape leaf skeletonizer eradication; B, applying rotenone dust to beans in Mexican bean beetle eradication; C, fumigation of gardenia plants in a vacuum fumigator on compliance with quarantine regulations; D, trailer type fumatorium used in State regulatory work. A, photo by Charles Morse; B, C, D, Courtesy of State Dept. Agr.

66

aerial form of phylloxera had been found in California. The destruction of wild grapevines is a part of the grape leaf skeletonizer eradication program that is now being undertaken by the State (fig. 54, A).

Grasshopper control is another activity in which the State participates, although the Federal Government contributes a larger share of the cost of operations. Control operations in 1946, by bait spreading (fig. 53, C) and the use of chlordan and benzene hexachloride, is estimated to have saved $7,000,000 worth of crops as compared to a loss of approximately $1,000,000 from grasshoppers.

Much investigation has been aimed at finding the tolerance of various plants and fruits to insecticides, especially fumigants, which are used in the treatment of insects which infest them.

The State Bureaus of Entomology and Plant Quarantine are not empowered to engage in pest control research. Investigations of this nature that come to the attention of state officials are referred to the staff of the University of California. Nevertheless at times it is necessary to determine methods and materials that can be used in meeting quarantine requirements, or it may at times be necessary to quickly devise some means of solving an insect control problem peculiar to a special situation which would hardly fit into the University's research program. In this connection the State entomologists have for many years made noteworthy achievements, especially in the investigation of methyl bromide as a fumigant. Some of the pioneer work with methyl bromide, especially with reference to its use on living plants was done by D. B. Mackie in connection with the plant quarantine and regulatory work of the State. The latest investigations of methyl bromide by the Bureau have been in connection with the working out of treatments to meet quarantine restrictions regulating movement of quarantined material in connection with the Mexican bean beetle and the grape leaf skeletonizer. In each case the investigations resulted in effective methods of treatment.

Besides its investigations of new pest control methods and materials, the Bureau has found it necessary to demonstrate its findings to those with whom the State is cooperating in the various control and quarantine activities, such as the county agricultural commissioners and the nurserymen. Fig. 54, D is an example of the specialized type of equipment that must sometimes be used in this connection. With this mobile fumatorium the Bureau entomologists have (1) made fumigation demonstrations for agricultural commissioners and nurserymen, (2) cleaned up outbreaks of certain new pests in nurseries, (3) fumigated shipments of plants or fruits being held in quarantine where no other fumigation facilities were available and, (4) made experimental tests in the field when it was impractical to transport infested plants or fruits to Sacramento for treatment.[1]

Quarantine:- Quarantine is assuming an ever increasing importance with the continued increase in travel since the lifting of war-time restrictions. Interceptions of prohibited plant material are increasing both at the maritime ports and border stations. During the war, the oriental (mango, Formosan, Malayan) fruit fly, Dacus dorsalis, was introduced into Hawaii by air, and this species is proving to be a more injurious pest than the Mediterranean fruit fly, which it seems to be replacing in Hawaii. The oriental fruit fly attacks not only the fruits and vegetables attacked by the Mediterranean fruit fly, but also mature bananas and pineapples. It is known to be a serious pest of the avocado.

In 1946, fruit fly host material from Hawaii was intercepted 1,308 times. Three lots of material were infested with Mediterranean fruit fly larvae, 81 lots with melon fly larvae, and 4 with oriental fruit fly larvae. One string bean contained 31 melon fly larvae and one mango contained 75 Mediterranean fruit fly larvae.

Los Angeles port inspection of air arrivals from Mexico has become of greater importance than ever before because of the steady encroachment of the citrus black-

---

[1]From correspondence of October 26, 1948 from Mr. John B. Steinweden.

fly up the west coast of Mexico. Likewise the straw and plant litter used in packing is a potential hazard in relation to the possible introduction of hoof and mouth disease.

The extensiveness of California's quarantine service is indicated by the fact that the State Bureau of Plant Quarantine is at present maintaining 16 border

Fig. 55. Quarantine inspection stations at State border. Above, Truckee Station in winter; below, a desert station at Parker. After Fleury (336).

quarantine stations (fig. 55) permanently and 2 seasonally; 72 employees are employed permanently and 70 are seasonally employed from 2 to 9 months. Three maritime stations are operated by 44 permanent employees, executive and clerical. The Mexican border ports of California are manned by federal personnel. In 1946, 1,985,613 orchids were inspected at border stations, and 76,569 lots of material were intercepted in violation of provisions of the Agricultural Code. Interceptions found to be in violation of the quarantine against citrus pests numbered 12,030.

Nursery Inspection:- Much effort has been directed in recent years toward the establishment of uniform methods and procedures in the inspection of nurseries (fig. 53, A) in cooperation with the agricultural commissioners and their staffs. The counties have been encouraged to organize a program of inspection of nurseries for pests and recommendation of treatments, this program to be headed by an experienced man. The State Nursery Service has as one of its functions the training of county inspectors and the coordination of inspection methods throughout the State. Under a new policy which was established by the State Bureau of Entomology and Plant Quarantine and approved by the quarantine committee of the State Association of County Agricultural Commissioners, the cooperation of the commissioner's office and the nurserymen will be asked in an effort to eradicate new or uncommon pests, when they are found in nurseries, in addition to the pests known to be serious. Only the latter have been subject to this state-wide eradication policy in the past.

Apiary Inspection:- As of December 31, 1945, official records show 3,021 apiaries, representing 132,431 colonies of bees, registered in California. It is estimated, however, that there are approximately 461,000 colonies of bees in the State. In 1946 California led all states in the production of honey, with a production of 22,128,000 pounds. The total value of the honey and beeswax crop for the year was estimated at $5,000,000. In addition, 181,000 queens were shipped in the spring months. Also it should be borne in mind that bees have a great value as pollinators of many crops.

The State plays an important rôle in the assistance given to the counties in microscopic examinations for the presence of foulbrood in the hives and in cleaning up badly diseased apiaries. The percentage of diseased colonies has been increasing in recent years, indicating the necessity for a greater amount of inspection.

STATE BUREAU OF CHEMISTRY

The State Bureau of Chemistry, for many years under the able administration of Alvin J. Cox, contributed much toward the development of sound insecticide policies in California. Perhaps the most noteworthy accomplishment during the 13 years of Dr. Cox' leadership was the bringing together of all California legislation on agricultural chemicals into one group: Chapter 7, Division V of the Agricultural Code. The Chapter comprises three separate articles to provide jurisdiction over (1) spray residue on fruits and vegetables, (2) fertilizing materials, including commercial fertilizers, agricultural minerals, manures, auxiliary plant chemicals and soil amendments, and (3) economic poisons. The administration of these matters is a function of the Bureau of Chemistry.

Upon the retirement of Dr. Cox on July 1, 1945, the administration of the Bureau was taken over by Allen B. Lemmon. In his annual report for 1946 he discussed the nature of the services performed by the Bureau as well as some of its recent accomplishments (517). Those provisions of the State's Agricultural Code pertaining to labeling and sale of fertilizing material and economic poisons are administered by the Bureau of Chemistry. According to law, these products must be registered and properly labeled before being offered for sale. That portion of the Agricultural Code pertaining to spray residue on fruits and vegetables is also administered by this Bureau. The administration of the fertilizing materials and the economic poisons articles of the Agricultural Code is made self-supporting by means of registration fees and tonnage license taxes. The administration of the spray residue article, however, is supported by an appropriation from the Central Fund.

California is divided into 4 inspection districts, in each of which is a resident District Inspector and staff of field inspectors. District officers are located in San Francisco, Los Angeles, Visalia, and the main office and laboratories in Sacramento. At present 14 field inspectors draw official samples from agricultural chemicals which may be in the possession of the manufacturer, wholesaler, dealer, or still in unopened containers in the hands of a user. Usually duplicate samples are taken and one is left with the person present at the sampling so that he may have an independent analysis made if he so desires. A small entire package of an expensive material may be taken from a dealer's stock, in which case a receipt for the package is given the dealer so that he may arrange for replacement by the registrant.

In the main laboratories in Sacramento the physical characteristics and chemical properties of the samples are determined and the findings are compared with the guarantee on the label and in the application for registration. Bioassay of the samples supplements the chemical analysis and checks the efficacy of the material (fig. 56). A report of the State's findings is mailed to the registrant of the product and to any dealer or user concerned with the particular lot sampled. In a Special Publication of the State Department of Agriculture called "Economic Poisons", the tabulated results of analyses of official samples are issued. The Bureau of Chemistry's reports of analyses of official samples assure the user that insecticide, fungicide, and fertilizer materials conform to the guarantee. Alert manufacturers can utilize the Bureau's reports to improve their products.

In addition to the analyses of official samples, the Bureau of Chemistry cooperates with county, state, and federal officials by making analyses required in connection with the official duties of such officers. Analyses of fruit to determine whether it conforms to maturity standards, determination of the amount of

Fig. 56. Some typical activities of State Bureau of Chemistry. A, Inspectors samples are quartered down prior to weighing a portion for chemical analysis. B, Kjeldahl distillation apparatus used to measure the amount of nitrogen in organic chemicals containing nitrogen. C, Peet-Grady Chamber used in biological evaluation of complex insecticides when chemical methods fall short in indicating the toxicity of such products. Courtesy State Dept. Agr.

insecticide drifting from treated properties to others where the material may constitute a hazard, examination of bees, pollen, and combs from hives suspected of having been poisoned by insecticides used on or adjacent to nectar crops, or examination of the viscera of cattle or other farm animals suspected of accidental poisoning by insecticides, are examples of the work of the Bureau of Chemistry.

In California there are over 7,000 different economic poisons regulated for sale, an economic poison being practically any substance used for control of any pest. These must be registered by the Bureau of Chemistry before being offered for sale. The applicant submits the proposed composition with labels showing proposed claims (518). One such label, taken from a package of a proprietary insecticide, is shown below.

NAME OF BRAND

ACTIVE INGREDIENT:
     Hexachloro-cyclohexane, gamma isomer..................... 6.0%

INERT INGREDIENTS... 94.0%

A wettable powder for use as a spray to control the following insects:

    Aphids on walnuts, cotton, vegetable crops grown for seed such as cabbage, onion, garlic, and other seed crops; aphids on hops before formation of the fruit.

    Lygus bugs, stink bugs, flea beetles and grasshoppers on cotton.

    Colorado potato beetles and flea beetles on potatoes.

    Thrips on onion and cotton.

    Grasshoppers and crickets.

DIRECTIONS FOR MIXING AND USE:

Under ordinary conditions use 2 to 4 pounds of (trade name) per 100 gallons of water. Obtain thorough coverage. For specific recommendations as to dosage, time and number of applications, consult your local authorities.

This material may be absorbed and impart an odor or taste to vegetables and fruit. It should not be used on edible portions of vegetables or fruits where there is a possibility of residue remaining at harvest.

Buyer accepts free of warranty as to results.

4 LBS. NET WEIGHT

NAME OF MANUFACTURING COMPANY
ADDRESS

Skull and
Crossbones and
Antidote.   Caution!!!

Registration can be issued promptly if available data indicate a proposed economic poisons product is suitable for the intended use and that the label conforms to requirements of law. If not, the applicant can, if he chooses, submit data proving his claims. Proof is necessary, for the manufacturer's statement is not sufficient. The registration fee is $50. Applicants commonly submit a rough draft or printer's proof of their proposed label to the Bureau so that corrections can be made if necessary and violations avoided.

ORGANIZATION AND LEGISLATION IN ENTOMOLOGY

COUNTY AGRICULTURAL COMMISSIONER'S OFFICE
OR COUNTY DEPARTMENT OF AGRICULTURE

The term "county agricultural commissioner's office" is considered by the writer to be a preferred designation for the organization discussed in this section mainly because the possible alternative, "county department of agriculture", may easily be confused with the state or federal departments of agriculture. It appears that the only possible reason for using the latter designation is that Section 50 of the California Agricultural Code states, "there shall be the office of County Agricultural Commissioner in each county. Such commissioner shall be in charge of the County Department of Agriculture." In addition it may be stated that some of the counties officially use the designation "County Department of Agriculture". Nevertheless it appears that the farmers and the majority of those professionally connected with agriculture in California, speaking of the office in question, generally use the term "County Agricultural Commissioner" or a similar designation. In examining the California Agricultural Code, it will be found that on numerous occasions authority is given to "the Commissioner", but no authority is given to "the County Department of Agriculture". The latter term has the merit of implying the existence of a larger group (the inspectors), but so does the term "Commissioner's Office".

The county agricultural commissioner's office is unique to California counties and is comparable, insofar as its pest quarantine functions are concerned, to the office of State Entomologist in many states of the Union. In the words of Harold J. Ryan, Agricultural Commissioner for Los Angeles County, the commissioner "is the agricultural sheriff of the county, having the same duties to protect plants from injury by insects, diseases, noxious weeds or rodents that the actual sheriff has to protect human life and property against human criminals. The authority of the legislature to create the office of agricultural commissioner in California counties is derived from the police power inherent in all government."[1]

Of California's 58 counties, 50 have an agricultural commissioner and 15 have 10 or more full time employees in the department. The Los Angeles County Agricultural Commissioner's Office is the largest in the state, with 57 employees (in 1947), followed by Tulare, with 31; Orange 28; San Diego 23, and San Joaquin 20.

The California Agricultural Code, Statutes of 1933, now is the major source of statutory authority for the work of the commissioner's office, although authority may in some cases be also derived from the county charters or ordinances.

Ryan (759) prepared a report on the duties of the Los Angeles County Agricultural Commissioner's Office which, while pertaining specifically to Los Angeles County, nevertheless can for the most part be taken as a general description of the duties of the office anywhere in California, although in some of the smaller counties certain functions may be more limited in scope. The following is for the most part based on Mr. Ryan's report.

According to the Agricultural Code, the agricultural commissioner is an enforcing officer of all laws, rules and regulations relative to the prevention of the introduction into, or the dissemination within the state, of pests. He is also an enforcing officer of sections of the code regulating the sale and packing of fruits, nuts and vegetables, providing for the inspection of apiaries, standardization of honey, egg standardization, and the issuance of quarantine and standardization certificates.

In matters of statewide concern, the county officer, as stated before, is subject to supervision of the State Director of Agriculture. Since he is appointed, however, by the county board of supervisors and is working under a county appropriation, he is directly responsible to local interests. The present close relationship between the state and county in matters pertaining to quarantine had not been established in the early days, under the old Commission of Horticulture. The re-

---
[1] From a talk given at a conference on Los Angeles County Government on March 28, 1933.

sult was a lack of coordination with respect to state-wide problems, much confusion, and many injustices. (937)

State control has brought about a high degree of uniformity among the counties in matters of general interest, while at the same time local responsibilities have at times led to the development of autonomous protective and clean-up measures as needed in a county or community. Under the state code, wide discretionary authority is extended the county agricultural commissioner, and through appropriations and orders of the county board of supervisors, this authority can be applied in a manner which would hardly be feasible if it were confined to a purely state institution.

The primary functions of the county agricultural commissioner's office may be divided into 9 categories:

1. Plant pest quarantine
2. Plant pest survey
3. Plant pest control or eradication
4. Bee diseases quarantine and control or eradication
5. Prevention of deception in packing and marketing of fruits, nuts, vegetables, eggs and honey
6. Pest quarantine certification
7. Produce quality certification
8. Enforcement of seed labeling requirements
9. Agricultural statistics and other reports

A large part of the duties of the commissioner's office, however, involves the function of administration itself, with its clerical and stenographic phases, compilation of reports, assignment and instruction of field assistants, and the maintenance of a reference library and a reference collection of insect, plant disease, and weed specimens.

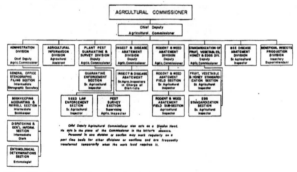

Fig. 57. Organization chart of the Los Angeles County Department of Agriculture. Redrawn from chart submitted by Harold J. Ryan.

The organization of the activities enumerated above can best be seen from fig. 57 which includes all the activities of the Los Angeles County Agricultural Commissioner's Office as of the year 1947. However, changes in the duties and functions of the commissioner's office may be made at any session of the legislature. In the following pages the activities of the agricultural commissioner's office which relate to entomology will be briefly discussed.

Plant Pest Quarantine:- This is a major functional activity, consisting of the enforcement of all laws and regulations designed to prevent the introduction

into the State or spread from one locality to another of insect, plant disease,
animal, or weed pests of economic importance (fig. 58, A & B).

All plant material intended for
propagation, as well as many other
classes of plant products, certain
fruits and vegetables, hay, grain, and
railroad cars from certain states where
cotton insects not established in this
state occur, must be inspected at the
point of their destination in California.
The importation from designated areas of
a variety of commodities likely to carry
such major pests as the fruit flies and
the cotton boll weevil is prohibited by
specific quarantines. Many insects of
lesser economic importance and not
specifically named in quarantines are
guarded against by inspection alone. In
1933-36, the number of incoming shipments subject to inspection was 105,147,
of which 1,788 were held because they
were infested with pests or did not
comply in some other manner with California quarantines. Of these, 496 were
infested with pests comprising 76 spe-

Fig. 58. County regulatory activities.
Quarantine: A, inspection; B, destruction of infested fruit. Survey: C, setting out lures; D, sweeping. Courtesy
L. A. Co. Agr. Comm. Office.

cies of insects, 6 of weed seeds, and 3 diseases.

To classify the relationships of the quarantine duties of County, State, and
Federal government, it may be stated that the county officers handle the inspection
of all interstate shipments in common carriers at points within the boundaries of
the State. County officers may assist the state and federal officers in maritime
port inspection. The State border inspection is done by State quarantine officers
except for the inspection done at the Mexican border, which is under federal jurisdiction and is handled by Federal officers. This subject is discussed in further
detail on pages 78 and 79.

Plant Pest Survey:- Since not every pest can be found by inspection, and an
embargo to insure complete exclusion of pests would stop commerce, surveys must be
made of plantings at the most suitable periods (fig. 58, C & D). Upon finding a
new pest, a survey crew is immediately concentrated in the affected area and the
limits of the infestation are determined. Then a council of representatives of
the farming interests and governmental agencies concerned is called and a plan of
action is worked out. Plant pest survey is considered to be a second line of defense.

Plant Pest Control or Eradication:- This is, from the standpoint of the
volume of work involved, the major function of the county agricultural commissioner's office. Eradication, as distinguished from control, may properly be considered as a third line of defense against pest attack. It has repeatedly been
demonstrated that many newly established pests whose area of infestation is small,
or whose rate of spread is slow, can be eradicated. In fact, eradication can also
be achieved against some established pests in accessible areas, as for example, in
ground squirrel eradication (fig. 53, B).

Let us suppose a new pest has become established, a survey is made and eradication is considered to be feasible. The immediate expenditure of a few thousand
dollars might result in wiping out a pest whose continuance might fix an annual and
much greater cost on the agricultural industry. Both the Agricultural Code and the
Governmental Code (formerly Political Code) provide legal authority for the eradication measures. Under the former, the agricultural commissioner may serve notice

73

on the owner of the property upon which the pest is found, giving him a reasonable period in which to exterminate the pest himself. If he fails to do so, eradication must be effected, at the expense of the general fund of the county, and if the cost of eradication is not paid for by the owner, it becomes a lien against the property and takes precedence over all other liens except that of taxes. The property may be sold to satisfy the lien.

Formerly under the Political Code and now under Governmental Code[1] the county board of supervisors has authority to provide for the destruction of pests, diseases, and noxious weeds. Prompt action is thus assured in case of a serious emergency or in case of a pest of such a nature that procedure under authority of the Agricultural Code would cause delay and thereby render eradication impossible or unlikely.

Consultation with the Grower:- The great majority of growers know that successful pest control is a prerequisite to successful fruit growing, and frequently they call upon the agricultural inspector acting for the agricultural commissioner's office, to give advice as to the most suitable pest control methods and the period they should be applied. The commissioner's office is thus seldom required to resort to authority of law to compel owners to use adequate control measures.

Fig. 59. County regulatory activities. Red scale inspection: A, in an orange orchard; B, in the packinghouse. Pest control inspection: C, inspection of a spray rig; D, inspection of spraying technique. Courtesy L. A. Co. Agr. Comm. Office.

[1]The State Law now giving this authority is in the Government Code, Part 2, Chapter 8, Article 3, Sec. 25892 - Control of Pests: "The Board of Supervisors may provide for the control or destruction of gophers, squirrels, other wild animals, noxious weeds, plant diseases, and insects injurious to fruit or fruit trees, vines, or vegetables or plant life." (Added by Stats. 1947, Ch. 424, Sec. 1).

There is also a Government Code, Part 2, Chapter 8, Article 1, Sec. 25801. Agreements with Director of Agriculture: "The Board may enter into agreements with the Director of Agriculture for the purpose of cooperating in the administration and enforcement of those provisions of law subject to the jurisdiction of the Department of Agriculture or County Agricultural Commissioners, County Sealers of Weights and Measures and County Livestock Inspectors. (Added by Stats. 1947, Ch. 424, Sec. 1).

ORGANIZATION AND LEGISLATION IN ENTOMOLOGY

It is in his work relating to pest inspection in the orchard (fig. 59, A), or sometimes in the packing house (fig. 59, B), or in relation to his consulting service, that the agricultural inspector is best known to majority of fruit growers. This particular service has become an important phase of his activities. Usually no one is better acquainted with the current pest situation in a given district than the district inspector, and his opinion on the proper time for treatment and the type of treatment best suited for a particular pest situation is usually highly respected and appreciated.

Granting of Pest Control Licenses:- An important function of the county agricultural commissioner's office has to do with the granting of licenses to commercial pest control operators. The agricultural inspectors inspect the equipment (fig. 59, C) and the operations of the crews of these concerns, and licenses of operators who do not properly carry out their work may be revoked. Likewise, operators who do pest control work without having secured a license or certificate are prosecuted. Examinations are given to test the knowledge of applicants for pest control licenses.

The Rearing and Distribution of Beneficial Insects:- The county maintains insectaries in which predators and parasites are reared and later distributed to combat insect pests. In the case of the citrophilus mealybug, it was also necessary for the county to institute a campaign against the Argentine ant because this insect was seriously interfering with the beneficial work of the predators and parasites. In Los Angeles County, both projects were authorized by the Board of Supervisors under authority of the Governmental Code and with an agreement with fruit grower's associations to reimburse the County for the expense. Growers cooperatives pay for the work on the basis of the number of boxes of citrus fruit sold by them.

In Los Angeles County, each spring the County-Wide Pest Control Committee, comprising representatives from the district fruit exchanges of the California Fruit Growers Exchange and representatives of the Mutual Orange Distributors, meet to discuss the general requirements of the insectary. A production program for the year is decided upon and returned to the cooperatives for their approval. The Insectary Superintendent arranges his plans for the year in accordance with the program adopted (482).

Bee Disease Quarantine and Control or Eradication:- This work is carried on under the authority of the Agricultural Code, and its main function is the prevention of the increase and spread of American foulbrood, a virulent bacterial disease of bees. In addition such minor diseases as European foulbrood, sacbrood and paralysis are attended to and vigilance is exercised to guard against the introduction of bee diseases not already prevalent in the area.

Standardization:- In addition to the county activities relating to entomology are the important functions which have to do with the maintaining of the standards set in the California Agricultural Code for size and quality and the labeling of the majority of the fruits, nuts, and vegetables, as well as eggs and honey. Inspections are made at packing houses and constant surveillance is maintained over the large wholesale markets. Retail markets are also visited, as well as highway stands, trucks and railway cars (fig. 60).

This work, which with respect to its importance and the time devoted to it ranks as a major activity of the county agricultural commissioner's office, has to do primarily with the prevention of deception. For example, fruit must be mature, but not over-ripe, and must be free from serious defects, with particular reference to such defects as are not readily discerned upon casual examination. Some years the inspection of oranges for frost injury becomes a major activity. The tolerance for frost injury is defined by law, and the lack of enforcement of such a tolerance would certainly be reflected in reduced market prices for the fruit. Another point which comes under the surveillance of the county inspectors is the matter of fruit or vegetable packs. Any kind of fruit or vegetable pack or display in which the exterior or exposed portion misrepresents the contents is deceptive and is pro-

Fig. 60. Fruit and vegetable standardization. Courtesy L. A. Co. Agr. Comm. Off.

hibited by law. Mislabeling is unlawful and placards having reference to the commodities must likewise not carry false or misleading statements. Similar legal requirements apply to eggs and honey.

PLANT QUARANTINE

Some Basic Principles:- It has been shown that quarantine constitutes a prominent part of entomological organization and legislation in federal, state and county agricultural departments. It behooves us to attempt to understand the basic principles which underlie such an important part of our agricultural program. Quarantine, for the majority of people, has meant an enforced halt at the border, the unpacking and inspection of luggage, repacking, and the resumption of the journey. Perhaps it has meant the loss of a bag of fruit or a bouquet of flowers. Thus many people have had questions about plant quarantine, but few have bothered to seek for the answers.

The term "quarantine" literally means a period of 40 days. In medieval times this was the period during which a ship suspected of being infected with a serious contagious disease was obliged to refrain from all intercourse with the shore. Quarantine had its inception in the attempts of early Europeans to exclude and prevent the spread of the dreaded plague or Black Death. The historic record shows that the present highly efficient and valuable public health measures maintained by all the progressive nations of the world were long in coming about, but it is likely that plant quarantine will attain a similar place in the affairs of civilized peoples much more rapidly. (284)

Denmark, in 1876, by prohibiting the importation of potatoes or parts thereof from North America, was the first country to enact a plant quarantine. In 1877, England passed a "Destructive Pest Act" and Victoria, Australia, enacted a quarantine, the purpose of which was to prevent the spread of grape phylloxera. In 1878, Germany, Spain, Austria, Hungary, France, Italy, Portugal and Switzerland took similar action with regard to the same pest. California passed its first quarantine act in 1881. As stated previously, our first federal plant quarantine act was not passed until August 20, 1912.

The "Principles of Plant Quarantine", a declaration of policy in plant quarantine as understood and accepted in the United States, was published by the National Plant Board (637). This declaration reads as follows:

ORGANIZATION AND LEGISLATION IN ENTOMOLOGY

"1. Definition. A quarantine is a restriction, imposed by duly constituted authorities, whereby the production, movement or existence of plants, plant products, animals, animal products, or any other article or material, or the normal activity of persons, is brought under regulation, in order that the introduction or spread of a pest may be prevented or limited, or in order that a pest already introduced may be controlled or eradicated, thereby reducing or avoiding losses that would otherwise occur through damage done by the pest or through a continuing cost of control measures.

"2. Basis in Logic. Since the ends to be attained by a quarantine and the measures required by it could not be undertaken by private individuals or groups, involving as they do restrictions on areas, persons, or activities for the benefit of wider interests or the public at large, resort to regulation imposed by public authority is logical.

"3. Necessity. Establishment of a quarantine should rest on fundamental prerequisites, as follows: (1) the pest concerned must be of such nature as to offer actual or expected threat to substantial interests; (2) the proposed quarantine must represent a necessary or desirable measure for which no other substitute, involving less interference with normal activities, is available; (3) the objective of the quarantine, either for preventing introduction or for limiting spread, must be reasonable of expectation; (4) the economic gains expected must outweigh the cost of administration and the interference with normal activities.

"4. Legal Sanction. A quarantine must derive from adequate law and authority and must operate within the provisions of such law.

"5. Validity. A quarantine established for the purpose of attaining an objective other than that which it indicates or defines is open to serious criticism, even though the actual objective is itself desirable.

"6. Public Notice. If the circumstances will permit, public notice of a proposed quarantine should be given and those interested should be invited to contribute facts in their possession. But if the objective would be defeated by the delay required for such notice and discussion, duly-constituted authorities should assume responsibility for the decision to impose or withhold quarantine action.

"7. Scope. The extent of restrictions imposed by a quarantine should be only such as are believed necessary to accomplish the desired end, but on the other hand the objective of a quarantine should not be jeopardized by omission of any necessary restriction.

"8. Relation to Eradication. If a quarantine is imposed in order that eradication of a pest from a given area may be undertaken, the restrictions involved may properly be relatively extensive, because of the importance of the objective sought, and because the time through which the quarantine will operate may be expected to be relatively limited.

"9. Relation to Retarding Spread. If a quarantine is imposed for the purpose of limiting or retarding spread of a pest, but without expectation of eradication, the restrictions imposed should be such as are in line with the objective of the quarantine and should recognize the fact that continuance of the pest in the area where it is established, or possibly its spread in time to new areas, is accepted.

"10. Cooperating Authorities. Since quarantines usually involve relations between public authorities, such as those of the government of one country with that of another, or of Federal and state governments, or of state government and local authorities, the cooperative relationship that is necessary to adequate enforcement should be clearly recognized and duly provided for.

"11. Cooperation of the Public. Because of the fact that the success of a quarantine requires that its restrictions be fully maintained, it is essential that all persons who are affected by it adhere to its requirements. In order that this

77

end may be attained the administration of a quarantine should seek the intelligent cooperation of the public affected, rather than exclusively depend on police powers, the imposition of penalties, or resort to court action.

"12. Clarity. In order that a quarantine may be administered readily and consistently, it should be designed with care, should be phrased clearly, and should be made as simple as is consistent with legal requirements and the objective to be attained.

"13. Information Service. Since the persons affected by a quarantine may not reasonably be expected to possess full or accurate knowledge of the circumstances that make it necessary, or the nature and importance of the aim sought, and since compliance with quarantine restrictions will be more complete if the objective and plans are understood, measures should be taken to set forth the conditions existing, the means to be employed, and the end to be attained, and these measures should be continued from time to time as the undertaking proceeds toward accomplishment.

"14. Research. If an emergency requires the establishment of a quarantine before satisfactory biological data are available, provision should be made as soon as possible for extending the fund of biological knowledge. The authority that exercises the right to establish a quarantine should command or secure the means for biological research, both in order that the quarantine may be made more efficient, and in order that the restrictions may be lessened where possible. The need for research, however, should not be permitted to delay the establishment of a quarantine believed by authorities to be desirable, thereby jeopardizing the objective that might otherwise have been attained.

"15. Modifications. As conditions change, or as further facts become available, a quarantine should promptly be modified, either by inclusion of restrictions necessary to its success or by removal of requirements found not to be necessary. The obligation to modify a quarantine as conditions develop is a continuing obligation and should have continuing attention.

"16. Repeal. If a quarantine has attained its objective, or if the progress of events has clearly proved that the desired end is not possible of attainment by the restrictions adopted, the measure should be promptly reconsidered, either with a view to repeal or with intent of substituting other measures.

"17. Notices to Parties at Interest. Upon establishment of a quarantine, and upon institution of modifications or repeal, notices should be sent to the principal parties at interest, especially to Federal and state authorities and to organizations representing the public involved in the restrictive measures."

In 1933 the results of an extensive study of the efficacy and economic effects of plant quarantine in California, made by a committee of University of California experts in the fields of Entomology, Plant Pathology, and Agricultural Economics, the Chairman of which was Professor H. S. Smith, were published in California Agricultural Experiment Station Bulletin 553, issued in July, 1933. This is a concise and effective treatise on the principles of and the operation of plant quarantine, with special reference to California.

The Interrelationship of Federal, State, and County Quarantine Services:- It will be noted that federal, state, and county agencies are operative in the protection of California agriculture by quarantine. There may be some confusion as to the part played by each and their interrelationship. This matter has been amply clarified in the report of the aforementioned University committee.

The federal quarantine agency is concerned with international quarantines or domestic quarantines dealing with pests not widely distributed in this country and, whose further dissemination would constitute a problem of national importance. The state agency is interested in preventing the entrance into the state of pests and diseases occurring in other states, but which the federal government does not con-

78

ider a nation-wide menace. The states may assist the federal government in en-
orcing a quarantine which is of special concern to the state in question, and
any interceptions of contraband material coming under the jurisdiction of federal
uarantines have been made by state inspectors.

For maritime port inspection, state inspectors work as collaborators under
ppointment by the federal government. In California this is a very important
unction of state inspectors because of the large number of plant pests and dis-
ases in countries bordering the Pacific which have not as yet become established
n this state.

Inspection at state borders is done by state inspectors and under state
uthority. Among the important pests and diseases occurring in other states which
ave not yet become established in California may be mentioned the cotton boll
eevil, the pink bollworm, the Mexican fruit fly, the cherry fruit flies, the
apanese beetle, the plum curculio, the Colorado potato beetle, the gypsy moth,
he white-pine blister rust, and the phony peach disease. These are often inter-
epted at the state border stations.

Inspections made at freight and express stations and port offices in Califor-
ia are made by county inspectors under authority of the Director of Agriculture.
Ithough few interceptions are made, this may be explained by the fact that the
hippers know of the quarantines and refrain from shipping contraband materials
hich would otherwise gain entry into the state in much larger quantities. The
mall amount of contraband intercepted at interior points thus becomes a measure
f the effectiveness of the quarantines.

The relative part played by the federal, state and county agencies in plant
uarantine enforcement in California is indicated by the costs of enforcement to
hese agencies in the fiscal year ending June 30, 1931. The cost to the federal
overnment was $16,342; to the state, $281,802; while the estimated cost to the
ounties was about $90,000.

Quarantined plant material is inspected and either passed or rejected in connection with four types of inspection activities: (1) maritime port inspection
(ships and water traffic), (2) border inspection (automobile, truck, and stage
traffic), (3) interior inspection (freight, express, and mail traffic), and (4)
airplane traffic inspection. Plants are inspected at 9 ports, 2 stations at the
Mexican border, and 11 permanent and 17 seasonal border stations.

A new problem has been posed by the increasing importance of air transportation since World War II. With respect to agricultural pests, the main problem is
in connection with (1) insects borne on plants and fruits and (2) those species
which are not likely to succumb to a DDT aerosol treatment in the plane. The
speed with which insects and diseases can now be transported makes it undesirable
to rely on port inspection alone as our only line of defense against their introduction. C. E. Cooley, who is in charge of the enforcement of foreign plant
quarantines on the Pacific Coast, believes that our "outer defense" should consist
of periodical pest surveys in those foreign countries from which we import plants
or unprocessed plant products, so as to familiarize ourselves with potential destructive species. The inspection of agricultural imports at the ports of entry,
as now instituted, would then become a second line of defense. Continual surveys
in the vicinity of ports of entry, to discover pests at the earliest possible
moment if and when they are established in the areas it is desired to protect;
should constitute a third line of defense. (191)

   The Value of a Quarantine Which May Fail to Permanently Keep Out a Pest or
Disease:- The argument has been used that since many, if not all, of the pests and
diseases that it is attempted to keep out will probably eventually gain entry into
the area supposed to be protected by quarantine, the quarantine might as well be
abolished so that the agricultural industry can adjust itself to its inevitable
fate. This argument has been answered, at least in part, in the above summary.
Following are statements of what are considered to be the benefits of quarantine
even if the quarantined pest or disease is not kept out of the protected area.

SUBTROPICAL ENTOMOLOGY

1. The quarantine is of economic benefit to the protected area as long as the pest is excluded, for the cost of the quarantine is only a small fraction of the probable cost of control measures once the pest is introduced and widely distributed.

2. The quarantine may limit eventual introductions to small incipient infestations which may be successfully eradicated. Thus quarantine and eradication might work together to prevent the permanent establishment and widespread distribution of a pest or disease, and the failure of quarantine per se need not necessarily mean that the quarantined pest will be permanently established in the protected area.

3. Even if a quarantined pest or disease eventually becomes established in a protected district, the delay in its entry may have afforded opportunity for experimentation on artificial control measures or on biological control with beneficial insects. A striking example is the oriental fruit moth which, although it has gained entry into California after years of quarantine, would now meet with highly effective artificial and biological control measures if it were ever to become a major pest anywhere in the state.

Insects Which It Is Attempted to Exclude from California by Plant Quarantine:-State and federal quarantine regulations attempt to keep the following insects from becoming established in California: Mediterranean fruit fly, Ceratitis capitata; melon fly, Dacus cucurbitae; oriental fruit fly, Dacus dorsalis; olive fly, Dacus oleae; West Indian fruit fly, Anastrepha fraterculus; Mexican fruit fly, Anastrepha ludens; cherry fruit flies, Rhagoletis cingulata and R. fausta; citrus white fly, Dialeurodes citri; the black fly, Aleurocanthus woglumi; rufous scale, Selenaspidus articulatus; gypsy moth, Porthetria dispar; Colorado potato beetle, Leptinotarsa decemlineata; sweet-potato weevil, Cylas formicarius; European pine shoot moth, Evetria buoliana; avocado seed weevils, Heilipus lauri, H. pittieri, H. perseae, and Conotrachelus perseae; European corn borer, Pyrausta nubilalis; Japanese beetle, Popillia japonica; alfalfa weevil, Hypera punctata; satin moth, Stilpnotia salicis; pink bollworm, Pectinophora gossypiella; cotton boll weevil, Anthonomus grandis; pecan leaf case bearer, Mineola indiginella nebulella; and the pecan nut case bearer, Acrobasis hebescella. In addition, quarantines are maintained against 18 plant diseases caused by fungi, bacteria, viruses, and nematodes.

A summary of state and territorial plant quarantines affecting interstate shipments has been made by Thompson (873).

INSECT ERADICATION

Two insect eradication programs have been described as being of special interest in connection with subtropical fruit insects, namely, the date palm scale eradication project in California, Arizona and Texas, and the Mediterranean fruit fly eradication project in Florida. Clay Lyle, in his presidential address before the 58th annual meeting of the American Association of Economic Entomologists in 1947, lists 26 different insect species as having been eradicated from areas ranging from small incipient infestations to others involving millions of acres. Among the eradicated insects were listed such serious species as the gypsy moth, cattle tick, San Jose scale, pink bollworm, sweet potato weevil, and the citrus whitefly. It is true that these pests are still with us in various places throughout the country, but the fact that they have been eradicated in some localities shows that they could be eradicated everywhere if sufficient interest and support were given to an eradication program. Among the principal successful eradication programs in foreign countries are the Colorado potato beetle (Germany, England), the brown-tail moth (Nova Scotia), the gypsy moth (Quebec), the codling moth (Western Australia), cattle grubs (Ireland), mosquitos Anopheles gambiae and Aedes aegypti (Brazil), and tsetse flies (Southern Rhodesia and Gold Coast). (543)

It has been suggested that in view of the many successful achievements in the past, with insecticides and equipment for the application of insecticides far less effective than those which are now in our possession, we should now consider eradication on a still greater scale, involving the complete elimination of many important species from the entire country. Among those which have been suggested for eradication are the house fly and the horn fly (Siphona irritans). Other insects which it is believed to be practical to eradicate are the cattle grubs,[1] cattle lice, screwworms, Argentine ant, and miscellaneous animal parasites. Eradication of mosquitos is believed to be possible, but the cost would be out of proportion to the danger involved. Eradication of human lice would be a simple matter, but a nation-wide compulsory physical examination would be necessary. There would probably be too small a per cent of our population concerned to justify a compulsory campaign against bed bugs. (543)

In cases in which an insecticide would have to be applied over a great continuous area, much study would be necessary with regard to the effect of the insecticides against beneficial species, such as parasites and pollinators. Some insecticides, such as DDT and benzene hexachloride, can create "faunal deserts", and it is not known whether the entomophagous insects can reestablish themselves as fast as the injurious species. Some injurious species, in fact, may be relatively unaffected by the treatment, while their natural enemies are eliminated (543). A case in point is the treatment of citrus thrips with DDT. This treatment has been found to increase the cottony cushion scale population by eliminating the vedalia lady beetles. The mite and aphid infestations are also often increased when DDT is applied.

The opportunity now undoubtedly exists, however, for much greater accomplishments in the way of insect eradication than could have been visualized even as late as 10 years ago.

THE REGIONAL AND NATIONAL PLANT QUARANTINE BOARDS

In May, 1919, G. H. Hecke, Director of Agriculture of California, organized the Western Plant Quarantine Board at Riverside, California. This Board consisted of members from the Western States as well as from British Columbia, Hawaii, and Lower California, Mexico. Its purpose was expressed in Article II of the Constitution and By-laws as follows: "It shall be the purpose of this organization to secure a greater mutual understanding, closer cooperation and a uniformity of action for the efficient protection of our plant industries against plant disease and insect pests." Through conferences, quarantine officials and entomologists of the various states became acquainted and discussed their mutual problems (219).

In 1925, regulatory officials of the Central States met and formed a similar board now called the Central Plant Board; in the same year the Middle Atlantic and Northeastern States Plant Conference Board was organized; and in 1926, a Southern Plant Board was organized.

On June 25, 1926, the National Plant Board was formed. It consisted of 2 representatives from each of the 4 regional boards. The National Plant Board has no legal status, but it represents the state inspection departments of all the states. It serves an important function as a correlating agency between the states and the federal quarantine organization. The federal government is not represented on the Board, but cooperates with the Board in the standardization of regulations. It has played an important rôle in the improvement of the general quarantine structure, both state and federal.

One of the important contributions of the National Plant Board was a declaration of the "Principles of Plant Quarantine". Copies of this set of principles, which are accepted as a statement of quarantine policy in the United States, were sent to all countries engaged in regulatory and control work and were also published in the June, 1932 number of Journal of Economic Entomology (see pp. 76 to 78).

[1] The annual loss in the United States from cattle grubs is estimated to be from 100 to 300 million dollars.

SUBTROPICAL ENTOMOLOGY

THE UNIVERSITY OF CALIFORNIA COLLEGE OF AGRICULTURE
AND AGRICULTURAL EXPERIMENT STATION

Little entomological work had been done by the states previous to the passage of the Hatch Act on March 2, 1887, which provided for the establishment in each state of an agricultural experiment station. With the organization of these stations in 1888, entomological work in this country had obtained a sound basis for its subsequent unparalleled development. According to Howard (448), in the period from the close of 1915 to September, 1929, of 7,311 entomological papers originating in the United States and reviewed in the Review of Applied Entomology, 2,115 were published by the federal government and 4,383 by officials of the state experiment stations. Probably the proportion of entomological papers written by experiment station entomologists is today even greater.

Interested persons may obtain from the U. S. Department of Agriculture announcements of current state agricultural experiment station publications, as well as the publications of the Department of Agriculture.

Many of the experiment stations operated in close connection with the state agricultural colleges organized under the Morrill Act of 1862 and this happened to be the case in California. In this state the College of Agriculture includes the academic departments of Agriculture, Forestry, Home Economics, and Veterinary Science, the Agricultural Experiment Station, and the Agricultural Extension Service. Its activities are conducted at Berkeley, Davis, Los Angeles, and Riverside, with extension offices and staffs in 44 counties of the State.

There are at present 61 entomologists on the staff of the University of California, not including laboratory technicians and laboratory assistants trained in entomological technique, and other personnel. The research of the University of California directed specifically toward the control of subtropical fruit insects is centered mainly at the Citrus Experiment Station (fig. 61) at Riverside, al-

though certain problems are being investigated by personnel located at Davis and Los Angeles. A well integrated program of laboratory and field research are required for the most efficient and thorough investigation of present day insect pest problems, both in insecticide and in biological control investigations (figs. 62 to 65).

In Berkeley the student may obtain not only his B.S. degree, but also his M.S. or Ph.D. degrees in the entomology major. The following courses are offered in the division of Entomology and Para-sitology: general entomology, insect morphology and histology, insect physiology, systematic entomology, forest entomology, helmin-

Fig. 61. Main entomology building, University of California Citrus Experiment Station, Riverside, California.

thology, economic entomology, insect vectors of plant diseases, medical entomology, insect ecology, insect toxicology, agricultural entomology, history of entomology, biology of aquatic and littoral insects, insect pathology (the relationship of microorganisms to insects), biological control, a summer practice and observation course, and seminars on the following topics: (1) systematic and economic entomology, and insect-borne plant diseases, (2) medical entomology and parasitology, and (3) insect toxicology, insect physiology, and insect pathology. In addition, of course, there are offered the special advanced study and research courses for graduate students.

At the Davis campus the student may obtain courses in general entomology, apiculture, veterinary parisitology, and economic entomology. At the Los Angeles

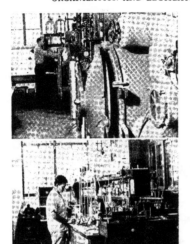

Fig. 63. Field research in entomology: walnut spraying. From U. C. Bul. 379.

Fig. 62. Laboratory research, University of California Citrus Experiment Station. Above, D.L.Lindgren in fumigation laboratory containing 2 vacuum fumigators; below, F.A.Gunther in one of the insecticide research laboratories.

Fig. 65. B.R.Bartlett, a specialist in the rearing of entomophagous insects, in a laboratory in the insectary at the Citrus Experiment Station at Riverside.

Fig. 64. Harold Compere, entomological explorer and taxonomist in the Division of Biological Control, at work on the description of a new hymenopterous parasite. By LIFE photographer J. R. Eyerman. Copyright Time Inc.

campus the following courses are offered: general entomology, insects affecting
subtropical fruit plants, insects affecting ornamental horticultural crops, and
medical entomology. At Davis or Los Angeles, students may take any or all the
courses in entomology that are offered, but if they wish to major in entomology
they must finish at Berkeley the requirements for graduation in the entomology
major. Likewise graduate students may work on research projects for the partial
fulfillment of the requirements for higher degrees at Davis or Riverside, if their
special interests make such a course desirable, provided they have first fulfilled
at Berkeley all other requirements for the higher degree.

STAFF MEMBERS OF THE DIVISION OF ENTOMOLOGY AND PARASITOLOGY OF THE UNIVERSITY
OF CALIFORNIA AND THEIR RESEARCH ACTIVITIES AND INTERESTS.

At Berkeley

E. O. Essig (Chairman of Division at Berkeley and Davis). Economic entomology;
all insects affecting agriculture; aphids, coccids.

Edward O. Essig
Professor of Entomology and Chairman of the Division
of Entomology and Parasitology at the University of
California, Berkeley and Davis.

M. W. Allen. Plant nematodes; soil fumigants.

A. D. Borden. Deciduous fruit tree insects; orchard sprays and dusts.

R. Craig. Insect morphology and physiology.

N. W. Frazier. Virus disease vectors. Leafhopper systematics, biology and control.

S. B. Freeborn. Assistant Dean, College of Agriculture. Medical Entomology.

J. H. Freitag. Virus diseases; insects transmitting virus diseases, especially
Pierce's disease of the grapevine; leafhoppers.

D. P. Furman. Parasitologist - medical entomology.

H. T. Gordon. Insect toxicologist.

W. M. Hoskins. Insect toxicologist; conducts experiments on residual effects of
sprays and dusts; etc.

84

ORGANIZATION AND LEGISLATION IN ENTOMOLOGY

D. D. Jensen. Virus disease insects; insects transmitting potato viruses; cucurbit viruses, etc; psyllids.

E. G. Lindsley. Forest and wood-boring insects; authority on wild bees and their habits; Cerambycidae.

A. E. Michelbacher. Economic entomology - alfalfa, tomato, asparagus insects; walnut insects; Symphyla.

W. W. Middlekauff. Economic entomology - insects affecting truck crops; mosquitoes.

A. E. Pritchard. Insects on ornamentals; insects infesting greenhouses; etc.; gall midges; mites.

D. J. Raski. Plant nematodes.

M. A. Stewart. Parasitologist - medical entomology.

H. H. P. Severin. Virus disease vectors; spittle bugs, leafhoppers.

G. L. Smith. Cotton Insects.

R. F. Smith. Alfalfa insects.

E. S. Sylvester. Virus disease vectors; insects transmitting big vein disease of lettuce, etc.

At Davis

S. F. Bailey. In administrative charge of the Division at Davis. Thrips of California, their biology and control on various crops, except citrus. Peach twig borer, miscellaneous insects on almonds.

R. M. Bohart. Mosquitoes of California; their distribution habits and taxonomy. Cherry insects.

J. R. Douglas. Animal parasites and cooperative projects with the Veterinary Science Division on various intestinal parasites. Barn and cattle spraying.

J. E. Eckert. Ants and their control. Apicultural problems.

H. H. Laidlaw, Jr. Breeding of bees.

W. H. Lange. Insects of truck crops and seed crops. Soil fumigants and wireworm control.

L. M. Smith. Red spiders on grapes and deciduous fruits. Berry insects, grape phylloxera, grape bud mite.

E. M. Stafford. Olive, fig, and grape insects.

At Riverside (Division of Entomology)

H. J. Quayle. Emeritus (Formerly Chairman of the Division). Subtropical fruit insects, cyanide fumigation, resistance.

A. M. Boyce. Chairman of the Division of Entomology at Riverside and Los Angeles. Insects attacking citrus and walnuts. Insecticides and insecticide equipment.

L. D. Anderson. Insects and mites attacking truck crops.

M. M. Barnes. Insects attacking deciduous fruits and grapes.

85

SUBTROPICAL ENTOMOLOGY

Alfred M. Boyce
Professor of Entomology and Chairman of the Division
of Entomology at the University of California, Riverside and Los Angeles.

G. E. Carman. Field investigations of insecticides used in red, yellow and purple
scale control (exclusive of tent fumigation).

R. C. Dickson. Quick decline of oranges and mosaic of cantaloupes. Aphids.

W. H. Ewart. Field investigations on citrus thrips, citricola, and black scales.

P. D. Gerhardt. Fumigation in enclosures. Control of ants and snails.

F. A. Gunther. Chemical investigations on insecticides in relation to all projects

L. R. Jeppson. Biology and field control of mites on citrus.

D. L. Lindgren. Laboratory and field investigation of fumigation in enclosures.
Insects affecting dates. Control of snails.

R. B. March) Physiological, biochemical and toxicological investigations of in-
R. L. Metcalf) secticides other than petroleum oils and fumigants (in enclosures).

H. T. Reynolds. Insects and mites attacking truck crops.

L. A. Riehl. Laboratory and field investigations of petroleum oil.

At Riverside, Albany, and Berkeley (Division of Biological Control)

At Riverside

H. S. Smith. Chairman of the Division. Biological control of insects and weeds,
insect populations. (see p. 323)

B. R. Bartlett. The culture of insects (hosts and parasites)

A. J. Basinger. Assistant to the Chairman of the Division. Biologies of parasites
tachinid parasites, snails, orangeworms.

Harold Compere. Entomological exploration, taxonomy of the chalcids.

P. H. DeBach. In charge of parasite liberation and the field study of parasites.

ORGANIZATION AND LEGISLATION IN ENTOMOLOGY

. E. Flanders. Biological research in insect culture, the biologies of parasites, sexual and reproductive phenomena among the Hymenoptera.

. A. Fleschner. Field population studies of parasites.

. H. Timberlake. General insect taxonomy, identification of Hymenoptera, curator of collection.

At Albany

. L. Doutt. In charge of parasite liberation and the field study of parasites.

. L. Finney. The culture of insects (hosts and parasites).

. K. Holloway. Assistant to the Chairman of the Division. Biological control of weeds.

. B. Huffacker. Plant ecology, liberation of insects attacking weeds.

At Berkeley

. A. Steinhaus. Insect microbiology. In charge of the culture of diseases.

At Canton, China

. L. Gressitt. Parasite collector.

At Los Angeles (Division of Entomology)

. Ebeling. In administrative charge of the Division at Los Angeles. Insects attacking avocados and minor subtropical fruits.

. N. Belkin. Medical Entomology.

., R. Brown. Insects affecting woody ornamental crops (shrubs, shade trees, etc.).

:. N. Jefferson. Insects affecting ornamental horticultural crops.

In addition to their research, the majority of the above staff members at Berkeley, Davis, and Los Angeles are teaching one or two courses in entomology.

The entomological research on insect problems north of the Tehachapi Mountains, with the exception of the work on citrus insects, is done largely by the Berkeley and Davis Divisions. This includes the greater part of the work on deciduous fruit insects and truck crop insects and all the work on insects attacking man and animals. At Riverside the research on citrus and other subtropical fruit insects is centered, as well as some research on deciduous fruit insects and vegetable insects as it applies to southern California problems. The work on avocado insects and certain other minor subtropical fruit insects is done from the Los Angeles campus. The research on insects attacking ornamental shrubs and flowers is centered at the Los Angeles and Berkeley campuses. Biological control investigations are centered at Riverside, Richmond, and Berkeley.

UNIVERSITY PUBLICATIONS CONTAINING ARTICLES ON ENTOMOLOGY

University of California Publications in Entomology. 1906-
University of California College of Agriculture Bulletins. 1884-
University of California College of Agriculture Circulars. 1903-
University of California College of Agriculture Extension Service Circulars. 1926-
Report of the College of Agriculture and Agricultural Experiment Station. 1887-
Hilgardia. 1925-
California Agriculture. Dec., 1946-
Entomology News Letters (Riverside). 1933-
Entomology News Letters (Berkeley and Davis). Issued by Agricultural Extension at Berkeley. 1946-

SUBTROPICAL ENTOMOLOGY

The Agricultural Extension Service:- By means of the Agricultural Extension Service, the results of the research of the Experiment Station, as well as that of the United States Department of Agriculture, are extended to the farmers of the State. The 252 members of the Extension Service include the Farm Advisor and Assistants in 48 counties of the State and the Home Demonstration Agents in 39 counties. Some counties in California have only one or two of these extension specialists or none, while 6 counties have from 10 to 15.

The Extension Service is a cooperative undertaking by the U. S. Department of Agriculture, the state land-grant institutions and the farmers. It dates from the enactment of the Smith-Lever Act of May 8, 1914. The cost is borne cooperatively by the national government, the states and the counties. A part of the money obtained from the national government is available to states and counties only when it is matched by an equal appropriation by the state. The amount of federal money allotted to any particular state is determined by the ratio of that state's rural population to the rural population of the United States.

The advising of farmers on entomological problems is, of course, only a small part of the Farm Advisor's work. In California much of the burden of entomological advice falls to the county agricultural commissioner's office because of the close relation of advisory and regulatory work. Nevertheless, the Farm Advisor keeps in touch with new developments in entomology, and is particularly well informed on the experimental projects of the research workers of the Experiment Station and the manner in which the more promising of the new developments can be utilized to best advantage by the growers under conditions peculiar to their particular district or orchard. The Farm Advisor may visit the orchards when requested by the growers to do so, but for the general distribution of information the Extension Service depends on published and mimeographed matter and numerous meetings in the local districts arranged by the Farm Advisor. At these meetings either he or some authority on a specific problem of interest to the growers in the district will speak.

ENTOMOLOGICAL SOCIETIES AND CONFERENCES OF NATIONAL AND LOCAL INTEREST

The American Association of Economic Entomologists:- The founding of the American Association of Economic Entomologists in 1889 was an event the importance of which could hardly have been visualized by its founders. It is significant that the Association was formed a year after the establishment of the state experiment stations; in fact, the first regular meeting was held in connection with the meeting of the Association of Agricultural Colleges and Experiment Stations at Urbana, Ill. in November, 1889, and joint meetings with this group were held for some years. The organization grew rapidly in size and in influence. At present there are well over 2,000 members, including many Canadian entomologists. There are also 33 foreign members from 20 countries. The Association is one of the numerous "Associated Societies" of the American Association for the Advancement of Science, being affiliated with Section F (Zoological Sciences) of this organization. It holds its annual meeting either in conjunction with the annual meetings of the A. A. A. S. or separately.

Application for membership in the Association may be made by anyone having an interest and a background in economic entomology. The candidate must be endorsed by two members, each of whom must furnish a letter giving the background of the candidate. Election of new members takes place at the annual meeting of the Association. The annual dues of $7.50 must accompany the application. These dues also pay for the bimonthly official organ of the American Association of Economic Entomologists, called the Journal of Economic Entomology. This Journal, begun in 1908, contains many of the papers read at the annual meetings of the National Association and its various branches. In addition many other papers of an economic character are published in the Journal, and it constitutes the most important single source of information on the progress of economic entomology in the United States.

In California, a Pacific Slope Association of Economic Entomologists was organized in 1909, and this society was affiliated as the Pacific Slope Branch of

ORGANIZATION AND LEGISLATION IN ENTOMOLOGY

American Association of Economic Entomologists in 1915. It holds its annual
tings in June at various locations representing different geographical régions
the Pacific Slope States. A Cotton States Branch affiliated in 1926, and an
tern Branch in 1928. A midwestern branch resulted from the informal annual con-
ence of the entomologists of the northern Mississippi Valley states, who began
ir meetings in 1921. This group was not formally organized until 1945 when it
cted its first officers, and in 1946 the members voted to become affiliated with
American Association of Economic Entomologists as the North Central States
nch.

The American Association of Economic Entomologists has had a great influence
the development of economic entomology in the United States. In the Association
ere are at present official committees on the following subjects: membership,
on names of insects, publications, program, codling moth, insecticide termin-
ogy, popular entomological education, cooperation with pest control operators,
lation of entomology to conservation, and grasshopper research. These indicate
me of the phases of public service to which the Association is dedicated. His-
rically, the Association can be connected with many important events in the de-
lopment of economic entomology in the United States. Among other things, it has
ways been active in the promotion of policies connected with state and federal
gislation for inspection and quarantine, and was an important factor in the
tablishment of the Federal Bureau of Regulation, through the Federal Horticul-
ral Board. The National Plant Quarantine Board, which has so effectively served
e interests of plant quarantine in this country, was formed as a result of
opositions adopted at a meeting of the American Association of Economic Entomolo-
sts in 1925.

The sponsoring of the national and regional meetings of entomologists is, of
urse, a noteworthy contribution of the Association, as is also the publication
the Journal. In addition the Association publishes the Index of American
onomic Entomology. Seven such indexes have been published. The seventh index,
cently published, covers the years 1940 to 1944, inclusive. The eighth index,
w in preparation, will cover the period from 1944 through 1948.

The Index to the Literature of American Economic Entomology is a joint proj-
t of the Library of the U. S. Department of Agriculture, the Bureau of Entomol-
ry and Plant Quarantine, and the American Association of Economic Entomologists.
; is compiled in the Division of Bibliography of the Library.

An important contribution of the Association is its periodic publication of
ie common names of insects which have been approved by the association. This
.st is generally accepted as an official authority for those who have occasion to
ie common names in their scientific papers, and as such, has greatly reduced the
onfusion which formerly existed in entomological literature because of lack of
:andardization in this respect. The last list was published in August, 1946.(628)

The Eastern Branch of the American Association of Economic Entomologists
irned the gratitude of all economic entomologists and pest control operators when
iey began in 1937 to publish Entoma, the now well known directory of pest control.
i this directory, which has been published at irregular intervals of several years
7 the Eastern Branch, are listed insecticides, fungicides and seed disinfectants
id chemicals used in their manufacture, insecticide and fungicide manufacturers,
isecticide machinery, entomological supplies and equipment, biological testing
iboratories, consulting entomologists, insect pest control companies, entomologi-
il societies of the United States and Canada, other associations, societies,
ialers, directories, yearbooks, hand books, magazines and leaflets of interest to
ionomic entomologists, the composition of the U. S. Department of Agriculture,
ith the administrators, directors or chiefs of its various branches, and the
gricultural experiment stations and their directors. Entoma also contains basic
iformation on insecticides and fungicides and valuable dilution and conversion
ibles. In the 1947 issue of Entoma, the names and addresses of 416 insecticide
id fungicide manufacturers and accessory companies are listed. Also listed are
ie names of about 2200 trademarked insecticides, fungicides and their adjuncts

together with the names and addresses of the companies handling their materials, and the names and addresses of about 900 pest control operators, listed according to states.

Other Societies:- It will not be attempted here to mention more than a small percentage of the large number of entomological societies existing in the country today. The Entomological Society of America, organized in 1906, "was a response to an evident need for an organization of an international character which would provide for a large number of amateur and professional entomologists who were not included in the American Association of Economic Entomologists on account of the distinctive economic character of that society" (668). The society holds an annual meeting, usually in conjunction with the American Association of Economic Entomologists, and the Proceedings of the meeting are published in the Annals of the Entomological Society of America. Other similar societies in this country and Canada and their publications are: The American Entomological Society (Transactions), The Entomological Society of Ontario (Canadian Entomologist), The Brooklyn Entomological Society (Bulletin), The Cambridge Entomological Club· (Psyche), The Washington Entomological Society (Proceedings), New York Entomological Society (Journal), The Entomological Society of Hawaii (Proceedings), The· Florida Entomological Society (The Florida Buggist), The Society of Natural History of San Diego (Transactions), The Pacific Coast Entomological Society and The California Academy of Sciences at San Francisco (The Pan Pacific Entomologist), The Lorquin Entomological Society (connected with the Los Angeles County Museum).

California Entomological Clubs:- The Entomological Club of Southern California was organized in 1926, and at present the membership is 380. There are usually from 100 to 150 members in attendance at the meetings, which are held on the first Friday in March, June, September, and December at the Y.M.C.A. in Alhambra, California. Papers and informal talks of a widely varied nature, but usually of · economic significance, are presented. A club of similar purpose and organization was organized in 1930 for the northern part of the State, called the California Entomological Club.

National Pest Control Association and Regional Pest Control Conferences:- The National Pest Control Association has done much to increase the cooperation of its members with federal and state experiment station entomologists in the interest of better pest control. Upon request, it provides lists of its members in the various states to the state and county officials. In 1936 the National Association started the movement to provide regional educational conferences for the benefit of pest control operators through the nation. In 1937 the first Pest Control Operators (P.C.O.) conference was held at Purdue University under the direction of Professor J. J. Davis.

In 1938 the first Regional West Coast Conference of the P.C.O. was held at Palo Alto, and in succeeding years at the University of California, at Berkeley. In 1943 the most important papers presented during the first 6 years of the West Coast P.C.O. conferences were incorporated into an excellently prepared and illus-trated lithoprinted book. This book contains 36 papers which are divided into the following general headings of subject material: General Entomology, Wood Destroying Organisms, Household Insects, Dooryard Pests, Household Rodents, Insect Control, Pest Control Literature, and Glossary. It can be seen that the subject matter dealt with in these Regional Conferences is largely confined to matters pertaining to the biology and control of structural and household pests.

A Southern Conference was inaugurated at Louisiana State University in 1938, · · an Eastern Conference at Massachusetts State College in 1940, and a Canadian Con-ference at the University of Montreal in 1943.

In addition to the P.C.O. Conferences mentioned above, there are more local-ized meetings of pest control operators. Pest Control Operators of California, Inc. meets in annual conventions in which entomological subjects are discussed, also mainly in relation to structural and household pests. · The Southern California Fumigators' Society meets annually not only to discuss mutual problems relating to commercial fumigating and spraying, but also to hear invitational talks given by local entomologists and others on subjects of a wide variety of interest.

Each year a number of Spray Conferences are held throughout the country, and the interest in these conferences tends toward orchard and truck crop pests and insecticides used in their control. In the West, the Western Cooperative Spray Project is an official conference of entomologists, chemists, plant pathologists, and horticulturists engaged in state or federal research work on orchard spraying problems in the region involved. Recommendations on the use of sprays are made at the annual conference of this group that are used by some states as a basis for their own official recommendations. The reports of the meetings, which are sent to members, contain useful, up-to-date information on the composition of many of the newest proprietary insecticides and the results of the latest experiments and observations on the use of these materials on a wide variety of pests.

COMMERCIAL ENTOMOLOGY

The chemical industry in the United States is undoubtedly the largest in the world. Of the thousands of new compounds which are annually being developed, and among the countless by-products of this vast industry, a certain percentage will annually be found to be of some value in pest control, either as insecticides or as adjuvants. Accordingly the majority of the large chemical companies and many of the smaller ones, have organized entomological research departments where the compounds of special interest to the company may be "screened" to segregate those

which show some promise as insecticides. Many companies, also, are engaged entirely in the manufacture of insecticides. An increasingly larger percentage of men and women trained in entomology have found employment with the commercial companies in recent years. The entomological laboratories of these companies are often large and well equipped, and in equipment, personnel, and quality of work, they are second to none in the country (fig. 66).

As is to be expected, the number of new insecticides being developed by the research staffs of the commercial companies is now far greater than those being developed by federal or state agencies. The latter are gradually assuming a secondary rôle in the actual synthesis of new compounds of insecticidal value, but are doing considerable chemical research with compounds which the various exploratory techniques of the commercial companies have indicated to be promising as insecticides. This research may be along the line of molecular rearrangement, new combinations, or related compounds which may be selected or synthesized in an effort to obtain even greater insecticidal efficacy. Likewise laboratory and field experiments on the effect on man and warm blooded animals, insecticidal value, phytocidal effect, compatibility with commonly used insecticides, the effect on beneficial insects, special equipment for application, and various practical problems which must be worked out separately for the various crops and climatic conditions, are to a large extent handled by federal and state entomologists whose training and experience has made them especially competent in biological appraisal.

Fig. 66. Above: aerial view of an agricultural laboratory owned and operated by private industry. Below: L.D.Glover applies an oil spray to a potted orange tree in the above laboratory. Courtesy Shell Oil Co.

The obvious relationship of the work of the commercial companies and that of the Universities conducting entomological work, has led to the establishment of grants and fellowships in which the work of entomologists, often graduate students, is financially supported by the companies. Much important work has been accomplished in this manner in fields of research of special interest to the companies concerned in these cooperative projects.

The Crop Protection Institute:- The Crop Protection Institute was organized under the auspices of the National Research Council and in affiliation with the American Association of Economic Entomologists. A report of the Crop Protection Institute is each year published in the Journal of Economic Entomology.

This organization has been instrumental in securing cooperation between commercial companies and experiment stations and other biological research institutions for the solution of problems connected with the development and use of insecticides and the equipment for their application. Dr. W. C. O'Kane, of the New Hampshire Agricultural Experiment Station, had an active part in the organization of the Institute and has been the chairman of the Board of Governors since its beginning. This Board consists of 3 entomologists, 3 plant pathologists and 3 chemists, these being selected from various state universities and experiment stations and other research institutions.

In 1946, 21 commercial companies financially supported organized research through the Crop Protection Institute and 28 research associates, assistants and technicians were employed. Hundreds of new organic compounds from various industrial laboratories were studied in a preliminary way and a considerable number were more intensively studied. (660)

The National Association of Insecticide and Disinfectant Manufacturers:- This organization, commonly known as the NAIDM, was organized in 1914 (403). Together with federal and state officials, who are welcome at its meetings, the NAIDM discusses and helps to solve mutual problems. The interests of the group tend toward the field of household insecticides. The NAIDM established the Peet-Grady Method for evaluating liquid household insecticides and established standard specifications for these products which were also adopted by the National Bureau of Standards and the U. S. Department of Commerce. It has always striven to promote the development of new products and the means for standardizing and testing these products. The organization has recently completed the compilation of all economic poisons laws and regulations in a single book which it intends to keep up-to-date by annual supplements. Copies of this compilation can be obtained at a nominal cost.

The National Agricultural Chemicals Association (Formerly the Agricultural Insecticide and Fungicide Association):- A large number of American chemical companies either directly or indirectly interested in insecticides and fungicides formed an association in 1943 to further the common interests of the groups in keeping abreast of developments pertaining to entomology, plant pathology, and chemicals used on pest and disease control and, in addition, to keep the interested public informed in regard to these matters. Through meetings, and more especially through its bimonthly publication, AIF News, the association has successfully attained this purpose, and has in addition provided a welcome source of up-to-the-minute news of nation-wide interest to economic entomologists. Often entomological news items regarding new insecticides and fungicides, new types of pest control equipment, the insecticide supply outlook, research development, pest problems of various parts of the country, meetings, insecticide legislation, and matters of general concern to the chemical industry can be found in AIF News before they appear in any other publications. In the AIF News for May-June, 1947 (Vol. 5, No. 7, p. 7) 80 companies were listed as belonging to this association.

The AIFA has recently changed its name to the "National Agricultural Chemicals Association" (NACA) and its headquarters were transferred to the Barr Building, Washington, D.C. (35). Along with the change in name and headquarters will come an expanded membership to include the following groups: (1) basic producers, (2) reprocessors and re-mixers, (3) custom applicators, (4) equipment manufacturers, (5) suppliers, (6) regional associations, (7) individuals (scientists, dealers, government workers, etc.), and (8) allied industries.

ORGANIZATION AND LEGISLATION IN ENTOMOLOGY

Among its many functions, the NACA will offer a means for the exchange and
rdination of information on technical matters such as: (1) toxicological data,
compatibility of insecticides and fungicides, (3) identification of trade name
ducts, (4) packaging and labeling, (5) nomenclature of new and complex materi-
, (6) analysis methods, (7) residue determination and removal, (8) ground and
application methods, and (9) adaptability of application equipment to new
micals.

California Fruit Growers Exchange, Bureau of Pest Control:- The great majority
citrus growers in California market their fruit through a large marketing or-
ization called the California Fruit Growers Exchange, and depend upon the

ter for many services, including advice on
t control. This service has been extended
the citrus growers through the Bureau of
t Control. This Bureau has been staffed
many years by four entomologists:[1]
S. Woglum (Chief), J. R. LaFollette, W. E.
don, and H. C. Lewis. The latter three
e each been responsible for the gathering
information relating to the current pest
trol situation, each in some definite
tion of the citrus area of southern Cali-
nia. Data on the abundance and the stage
development of the various citrus insects
mites are thus brought together, notes
compared, and a monthly report is pre-
ed under the direction of the Chief of
Bureau.

Russell S. Woglum
Chief Entomologist for the Pest Control Bureau of the California Fruit
Growers' Exchange, from Aug. 1, 1920
to Jan. 1, 1949.

For eleven years, beginning in 1924, the
t Control Bureau issued an annual Handbook
Citrus Pest Control but the need for cur-
t advice and information as the season

)gressed led to the issuing of monthly reports in 1935, and these became an im-
rtant source of up-to-the-minute information and recommendations in citrus pest
itrol. These reports were published in a photoprinted circular called the
:hange Pest Control Circular, which was sent free of charge to all members of
: California Fruit Growers Exchange and to those officially connected with ento-
logy in southern California. In this circular the practical pest control prob-
as of the month were discussed and recommendations were made as to the pests to
treated, the type of treatment to be used, dosages, precautions, and other in-
rmation relating to the immediate practical problems of the month. Occasionally
. entomologist from outside the organization was invited to write an article for
: Circular on a timely subject relating to pest control in which he happened to
an authority.

Besides the regular numbers of the Circular, there occasionally appeared a
ibject Series" number on some new development in pest control of wide general in-
rest to citrus growers. The first of these, "Subject Series No. 1", appeared in
ly, 1944, on the subject "Boom Sprayers in Citrus Orchards". The other "Subject
ries" numbers which have been issued to date are: "Tent Pulling Apparatus in
lifornia Citrus Groves", November, 1944; "Compatibility of Materials Used as
rays and Dusts on Citrus Trees in California", October, 1945; "The Chemical
2atment of Psorosis (Scaly Bark)", June, 1946; and "Service Rigs in Citrus Spray-
3", March, 1947.

Effective January 1, 1949, the publication of the Pest Control Circular as
>h was indefinitely discontinued, as an economy measure, but the information
rmerly contained in the Circular now appears monthly in the California Citro-
iph.

[1] Wm. E. Shilling joined the Staff of the Bureau of Pest Control in 1948, which resulted in the
tention of four entomologists on the staff upon the retirement of Mr. Woglum on January 1, 1949.
. Woglum was succeeded as Chief of the Bureau by J. R. LaFollette.

Citrus Protective Districts:- An important item of entomolgical legislation
in California are the "protective districts", "leagues", or "associations" which
are organized in citrus areas in which the red scale, which is the most serious of
the citrus pests in California, has either not become established or, if it has
become established, is present in only a small percentage of the orchards. The
principle involved in the protective districts is that those growers whose orchards
have not as yet become infested with red scale may insure the continuance of a
scale-free condition in their orchards by contributing to the cost of the unusually
severe measures employed to get rid of existing infestations. The measures employed may be in excess of what the growers of the infested orchards would be
likely to employ merely to keep their orchards "commercially clean." Likewise a
systematic inspection service may be employed by the district to locate incipient
infestations. This inspection is an important adjunct to pest control designed to
eradicate or confine the limits of the spread of a pest, and it could not be had
without a cooperative organization.

The first protective district to be organized for the purpose of voluntary
cooperative red scale abatement was the Ventura County Citrus Protective League,
organized in 1921. The purpose, the organization, and the accomplishments of this
organization were discussed by Hardison (407). While the red scale has not been
eradicated from the area embraced by the Protective League, the League has been
eminently successful in preventing its spread, as is attested by the fact that the
fumigations for red and purple scale have cost, considering the district as a
whole, an average of only $1.98 per acre per year.

Red scale was known to occur in the Redlands-Highlands citrus district as
early as 1932, and probably existed in the area for a considerable period before
this date, but was unrecognized because of being masked by the more predominant
yellow scale. In 1936, plans were made for a voluntary protective district, but
by 1938 it was recognized that the problem in the Redlands-Highlands district was
too severe to be satisfactorily handled by a voluntary group. As a result of the
difficulties experienced by the Redlands-Highlands Voluntary District, an effort
was made by the citrus growers of the region to form a legal red scale control
district, to include all citrus growers, in which the funds for an eradication
attempt could be equitably assessed. This led to the Citrus Pest District Control
Act being placed in the California Statutes in 1939. In the Redlands-Highlands
District the assessment was $2.50 an acre for trees 5 years old or less and $5.15
per acre for older trees (176).

The Citrus Pest District Control Act provides that on petition signed by the
owners of 51% of the citrus acreage in a proposed district and the filing of such
a petition with the Board of Supervisors, a hearing shall be held within 20 to 40
days. If the Board of Supervisors gives approval, the district is declared organized and a Board of five members, who must be citrus orchard owners, is appointed.
The Board, serving for terms of four years without compensation, is guaranteed
broad authority to take all the steps necessary to organize and carry on the work
of eradicating red scale or any other pest for which the district may be organized.
The funds required for this work are raised by levying a special tax on the assessed value of the citrus trees within the district, after an estimate of required funds has been made by the Board of Directors and after a hearing has been
held at which a tentative budget is presented and a final budget is adopted. The
assessing and collection of this special tax occurs at the same time as the general county taxes. The district may be dissolved by a petition of the owners of
60% of the citrus acreage within the district.

No assessments have been made in the Redlands-Highlands Pest Control District
since the fiscal year 1942-1943, and the activities of this District were discontinued in 1945 although it has never been formally dissolved. The cessation of
the activities of the District was caused by the fact that the red scale had spread
over so much of the area involved that the original reason for the formation of the
district, that is, the protection of large numbers of uninfested properties, no
longer obtained. The District had, in fact, been started too late to reap the full
measure of benefit ordinarily resulting from such community pest control districts.

Several other pest control districts have been completely dissolved for the same reasons, the latest being the Moreno Valley District, which was discontinued in 1947. At the present time there are five pest control districts organized under the Citrus Pest District Control Act of the State of California: the Southern Tulare County Citrus Pest Control District, the Tulare County Red Scale Protective District, the Coachella Valley Red Scale Eradication District, the Hemet-San Jacinto Pest Control District, and the Fontana Red Scale District.

The red scale protective districts of Ventura County are compulsory, but independently organized. The names of these districts are: Ventura County Citrus Protective League, the Fillmore Citrus Protective League, and the Simi Valley Protective District.

In addition to the above protective districts should be mentioned several recently established in northern California, which, while they are not formally organized, are working in cooperation with the county agricultural commissioner toward the eradication of red scale, the growers being assessed to pay the costs. Such districts have been functioning in the Oroville and Orland citrus regions. (995)

An example of the severe measures that are sometimes considered necessary to stamp out a red scale infestation is the eradication work being conducted in the legal "Citrus Pest Control Districts" of Coachella Valley (formed on September 18, 1946) and in the Hemet Valley (formed on November 12, 1946). The first money under the legal districts was available in July, 1947. The methods employed in surveys and in the treating of the infestations are described in a report made by F. R. Platt (700), and may be briefly summarized as follows:

After a complete survey of the areas involved, infested trees are pruned to remove as much as possible the infested branches in contact with the ground, thus reducing the infestation to the lowest possible limits before starting treatments. The infested and adjoining trees are given 2 sprays of 2% light medium oil and one fumigation with the 24 cc schedule, then the entire cultivated unit in which the infested trees occurred is fumigated.

Red scale found on the fruit or trees is reported by growers and packinghouses and rewards are paid for the discovery of new infestations. In addition, commercial orchards are given a tree-to-tree inspection. Once an infestation is found, the infested property is surveyed annually and is considered to be free of scale only if no scale is found throughout a 3-year period. Even fruit displays in stores and markets are constantly being checked for pests. All citrus nursery stock being shipped into the protective districts must either have a certificate of freedom from red scale or a vacuum fumigation with HCN or an atmospheric fumigation with methyl bromide.

Considering the difficulties involved, much progress has been made and there is every reason to believe that under the presently organized protective districts in Coachella and Hemet Valleys, eradication will be effected. Constant vigilance must be exercised, however, to bring about this goal.

Besides using chemical control measures, protective districts may also rear and liberate beneficial insects in insectaries. The Fillmore Citrus Protective District rears and liberates the predator, Cryptolaemus, for mealybug control and the hymenopterous parasite, Metaphycus helvolus, for the control of black scale (fig. 67). The acreage requiring fumigation or spraying for black

Fig. 67. Part of the insectary of the Fillmore Citrus Protective District.

scale is thus reduced by thousands of acres each year. The Cryptolaemus beetles are reared on citrus mealybugs, which are in turn reared on potato sprouts.

M. helvolus is reared on black scale, which in turn is also reared on potato sprouts. The rearing of M. helvolus began in 1937 and now about a million parasites are reared per year.

PEST CONTROL ORGANIZATIONS IN CALIFORNIA

By far the greater portion of the 330,000 acres of citrus trees in California are treated for pests by private or cooperative pest control organizations. The majority of the larger citrus fruit companies have their own pest control equipment and do the spraying, dusting, or fumigating themselves, but it would not pay the owners of small orchards to buy the expensive motorized equipment which is necessary for effective and efficient work, and their skill in the application of the insecticides would generally probably not be as great as that of the experienced commercial pest control operators. Consequently, only a very few of the small growers do their own pest control work. To the owners of small orchards belongs by far the greater portion of the citrus acreage. The same situation pertains in the case of other subtropical fruits, especially the Persian walnut, avocado, and the date.

There appear to be about 30 cooperative pest control associations (fig. 68, above) doing pest control work on citrus in California. Some are groups of growers who have formed an organization strictly for pest control purposes. They have boards of directors which elect their own managers. The growers own shares of stock in the cooperative, on the basis of either one share, a part of a share, or more than a share per acre of trees owned by the grower. A more complete idea of the nature and organization of their cooperatives may be obtained by reading the Articles of Incorporation and By-Laws which are published by these cooperatives as well as the annual Statement of Assets, Liabilities and Capital which is distributed at the annual meeting of the members.

The cooperatives own the pest control equipment and handle the fumigation, spraying, dusting, and other pest control activities at actual cost, and if at the end of the year there has accumulated any profit, this is refunded to the growers.

Fig. 68. Equipment yard of a cooperative pest control association (above) and a private pest control company (below).

There are 8 or 10 cooperatives of this type among the citrus districts of California. In addition there about 20 cooperative pest control organizations in which the growers have been organized primarily for the purpose of picking, packing, and handling the fruit, and pest control is an additional and incidental activity. Some of the associations do any kind of pest control work for their members, some only fumigation, spraying, or dusting. In some cases a pest control foreman is hired, but in other cases someone from the existing packing house organization assumes this responsibility.

There are at least 15 large orchard companies doing their own pest control. They have the equipment and trained personnel for this specialized type of orchard work. The remainder of the citrus pest control work is done by private operators who may have from one to as many as a dozen spray rigs as well as dusters and tent pullers, and one to several strings of tents for fumigation (fig. 68, below).

ORGANIZATION AND LEGISLATION IN ENTOMOLOGY

REGULATIONS PERTAINING TO PEST CONTROL OPERATIONS

All commercial pest control operators must comply with regulations as set forth by the State Director of Agriculture, and it is the duty of the county agricultural commissioner's office to see that the regulations are complied with. The regulations as they pertain to (1) definitions, (2) grounds for revocation of certificate, (3) agricultural pest control, (4) lethal and/or seriously injurious materials, (5) tree, vine and plant fumigation, (6) tree and crop spraying and dusting, (7) chemical weed control, and (8) other agricultural pest control operations, are found in a set of "Rules and Regulations" of Title 3 (Agriculture) of the California Administrative Code. These were issued on September 5, 1946. The Counties of Los Angeles, Orange, Riverside, San Bernardino, San Diego, Santa Barbara and Ventura prepared a set of (1) general recommendations, (2) citrus fumigation recommendations and (3) citrus spraying recommendations. Under Sections 3075, 3081, and 3082 of the aforementioned Code, violations of any of these recommendations constitutes grounds for revocation of Pest Control Operators Certificates. The recommendations are as follows:

General Recommendations

1. Each operator shall notify the Agricultural Commissioner prior to start of operations, giving the names and locations of properties to be treated in such manner as shall be designated by the County Agricultural Commissioner.

2. Each operator shall submit by the 10th of each month a report to the County Agricultural Commissioner showing for each piece of work performed during the previous month: date, grower's name and address, location of orchard or field, number and kind of trees or acreage and kind of crop, weed, pest, kind and dosage of material used, and amount of undiluted material used.

Citrus Fumigation Recommendations

1. All fumigating tents shall be clearly and legibly marked with figures not less than three inches in height in accordance with the Morrill system.

2. Each tent shall be placed so that one line of numerals runs over center of top of trees. Tents shall be kicked in to hang perpendicularly from outer limbs of tree to ground.

3. Work shall not be carried on when wind is strong enough to cause any appreciable movement of tent walls.

4. Each tree requiring more than a four-unit charge shall be taped and correct dosage given as called for upon standard dosage chart established by the University of California, Citrus Experiment Station.

5. Each tree shall be given not less than 45 minutes exposure regardless of dosage, and work of tent pullers shall be regulated accordingly.

6. Trees shall not be fumigated with cold liquid gas when the temperature is below 50° F. Trees shall not be fumigated with any apparatus delivering a heated gas when the temperature is below 40° F. or above 70° F. in coastal and intermediate districts, or below 37° F. or above 80° F. in interior districts.

7. Thermometers shall be so placed as to record the temperature in that portion of the grove being treated.

8. Summer fumigation shall be discontinued when moisture begins to form a film on the fruit.

Citrus Spraying Recommendations

1. The foreman of each spraying crew shall be equipped with a thermometer which has been properly tested at 90° F. and certified as to its accuracy by a competent agency.

SUBTROPICAL ENTOMOLOGY

2. Temperature readings must be made within the orchards being treated, and at least 20 feet from a paved roadway, waterway, open water filling place, or recently sprayed trees.

3. Spraying operations shall not be carried on when there is recognized danger of tree or crop injury because of climatic or other conditions. Experience indicates severe damage has occurred at high temperatures. Danger limits may be reached in coastal areas at 80° F., intermediate areas at 86° F. and interior areas at 95° F. Spraying should not be carried on when relative humidity is 35% or lower in coastal areas, 30% or lower in intermediate areas, or 20% or lower in interior areas.

4. Spray gun discs processed for hardening shall be used. The disc size must not be so large as to cause improper coverage or over spraying, and in no event shall a disc of a number 10 size or larger be used.

·CHAPTER V
ARTIFICIAL CONTROL

PHYSICAL AND CULTURAL CONTROL

When considering the artificial control of insects, the tendency in modern times is to think in terms of chemical control, although in actual practice purely physical or cultural measures still play an important rôle in the control of some species. These will be briefly discussed before taking up the chemical control easures.

Among the physical (or mechanical) conrols are such measures as jarring insects rom trees, handpicking, destruction of egg asses, and removal by a strong stream of ater.

Insects may be excluded from a crop which it is desired to protect by barriers, such as are provided by sticky banding material, cotton, glass wool, etc., or cultivated units may be surrounded by deep furrows or trenches in which crawling insects may be trapped and destroyed. Likewise metal barriers have been extensively used in this connection (fig. 69).

Fig. 69. Metal barriers against locusts in Argentina. From Argentine Dept. Agr. Misc. Pub. No. 205.

Fig. 70. A bait trap for Japanese beetles which attracts, kills, and allows beetles to fall free. Courtesy G. S. Langford.

Insects may be lured to traps (fig. 70) by light or olfactory lures, usually the latter. They may then be destroyed. Such traps have been of practical value mainly as a means of ascertaining the abundance of insects or determining their periods of emergence or oviposition preparatory to the proper timing of insecticide treatments.

Crops particularly favored by insects may be planted near the crop which it is desired to protect, so that the insects may be lured away from the latter. The insects may then be destroyed on the trap crop by insecticides or some other means.

Insects may be destroyed by temperature regulation. Both low and high temperatures have been used to control insects. The artificial cooling or heating of stored products, or storage bins, warehouses, and mills in which they are stored, is commonly practiced. Practically no development of insects takes place below 40° F, and control measures are usually aimed at preventing damage and increase in numbers rather than killing the insects.

For the sterilization of citrus and other fruits infested with fruit fly larvae, a "vapor-heat process" has been worked out which is dependent on the circulation of air, saturated water vapor, and water in the form of a fine mist, in large volume throughout a load of fruit. Heat is supplied by electric heating elements or live, low-pressure steam is injected into a stream of air to which a water spray is added if necessary. The process is dependent for its success on the latent heat of evaporation released by condensation. On the basis of extensive experiments on Mexican fruit fly infesting oranges, it was concluded "that the vapor-heat process applied to oranges at a temperature of 110° F, with an approach period of 8 hours and an exposure period of 6 hours, will guarantee the death of any larvae or eggs that may be present in the fruit, even if the population should be relatively large." (51) Later experiments made with mangos infested with Mexican fruit fly larvae showed the high rate of security of the vapor-heat process (56).

The hot water or steam in which dried figs are processed before they are packed is sufficient to destroy all insect life without changing the fruit beyond condition which the packers consider prime for packing. In some experiments temperatures of 180° to 200° F within the figs in the center of the lots that were being processed were reached (777).

Courtney et al (199) have written a paper on hot-water tanks for heating bulbs and other plant materials for the control of bulb flies, mites, nematodes, etc. One of the types of tanks in common use nowadays is shown in fig. 71.

Flooding has been employed on a limited scale against certain soil inhabiting insects, and conversely the draining of swamps and marshes is practiced for the control of insects such as the mosquitoes.

The utilization of cultural practices inimicable to insect development has figured prominently in the control of many serious pests. Certain modifications in cultural practices may in some instances offer suffici-

Fig. 71. Hot-water tank for treating bulbs and other plant materials.

ent control, in other instances they may be used in conjunction with mechanical or chemical control to increase the effectiveness of the general pest control program. Cultural control may be divided into the following categories: (1) crop rotation, (2) host free periods, (3) proper timing of the periods of planting and harvesting, (4) destruction of crop wastes and overwintering host plants, (5) soil tillage, (6) maintenance of plant vigor by proper irrigation, fertilization, and cultivation, (7) resistant varieties.

Of greatest interest to us in relation to subtropical entomology is the utilization of host-free periods as it has been practiced in eradication campaigns. A host-free period has resulted in the reduction of the Mexican fruit fly population in the Rio Grande Valley of Texas to negligible proportions. There had never been many deciduous fruits in this region and a destruction of these fruits was made possible on a purely voluntary basis. The fruit fly was thus deprived of hosts to tide them over from one citrus fruit season to another. The inauguration of a host-free period was also an important factor in the success of the Mediterranean fruit fly eradication campaign in Florida in 1929.

ATTRACTANTS AND REPELLENTS

The fact that insects are attracted or repelled by various stimuli such as odors, heat, light, moisture, air currents, and the physical nature of their substratum, must have been known since man first began to observe and to reflect on nature. Insect repellents have been utilized since very early in human history, but means of attracting insects aroused little interest until comparatively recent times. Up to the present the research on attractants and repellents has been pursued largely on a "hit or miss" basis. If the recently proposed hypothesis (608) that infrared absorption is involved in the olfaction of insects is substantiated by further investigation, however, a new experimental approach to the subject of attractants and repellents may have been found which could greatly enhance the possibilities of more systematic and more fruitful investigation, especially as regards the rôle of chemical structure.

Dethier (234) has grouped the compounds which are being used as attractants into the following categories:

ARTIFICIAL CONTROL

1. Essential oils, resins, and related substances.
2. Fermentation products (alcohols, acids, aldehydes, esters, carbinols).
3. Protein and fat decomposition products (fatty acids, amines, ammonia, carbon dioxide).

Dethier considers the following to be the more important uses of attractants: "(1) to lure insects into traps or to poisons to decimate the population; (2) to sample local populations; (3) to offset the repellent properties of certain sprays; (4) to lure insects away from crops; (5) to act as counteragents against which to test the efficiency of repellents."

It appears that the use of chemicals to lure insects to traps, so as to sample populations, constitutes the greater part of the work with attractants at present.

Physical and chemical stimuli in the same categories mentioned in connection with attractants will elicit avoiding reactions from insects. However, while attractants are outstandingly specific, being largely such substances as lead an insect to its food, mate,- or oviposition site, repellents are notably non-specific (234). Repellents warn an insect of enemies, improper food and inimical surroundings, and these dangers, generically considered, are nonselective.

Dethier (234) divided repellents into the following categories:

1. Repellents extracted from plants. Certain insecticides derived from plants possess insecticidal as well as repellent properties. The most important of these are pyrethrum, derris, nicotine, and oil of citronella.

2. Synthetic repellents. The following are among those which have been used against plant feeding insects: hydrated lime and aluminum sulfate against Japanese beetle; derris against Mexican bean beetle; phenothiazine on grape foliage; tetramethylthiuram disulfide; whitewash against potato leafhopper; Bordeaux mixture plus nicotine sulfate against melon fly, Dacus cucurbitae; an emulsion of pyridine, naphthalene, or creosote used against cauliflower pests; molasses against Queensland fruit fly, Strumeta tryoni; furfural, Dippel's oil, ammonium sulfide or amyl acetate against oriental fruit moth, Grapholitha molesta; paraffin or turpentine against flea beetles; and copper stearate and copper resinate against tent caterpillars.

Many repellents have been used to repell mosquitoes and blood-sucking flies. The early types were chiefly the essential oils (citronella, eucalyptus, pennyroyal, rose geranium, cedarwood, thyme, wintergreen, clove, lavender, cassia, anise, bergemot, pine tar, bay laurel, and sassafras), certain acetates (linalyl, santalyl, terpinyl, geranyl), glycols, indalone, and pyrethrum. Until World War II, the most successful repellent was dimethyl phthalate, and this was used as a control in the testing of new materials. During World War II the best repellent proved to be a mixture of dimethyl phthalate, 2-ethyl-hexanediol-1,3 (Rutgers #612), and indalone in a proportion of 6:2:2.

ADJUVANTS

An adjuvant or accessory substance is a material which, while not itself toxic, is added to an insecticide to improve its physical or chemical characteristics. Adjuvants are most commonly used to improve the insecticide-depositing and weathering properties of sprays and dusts. They will be classified in the following discussion according to the purpose for which they are primarily intended, although it should be borne in mind that they may overlap considerably in this respect. Thus blood albumin is a good spreader, a fair wetting agent, a fair emulsifier, a fair sticker, a good stabilizer for solids in suspension, and a good deflocculating agent. In the present discussion it is classed as a spreader. Calcium caseinate is a fair spreader, a fair wetting agent, a good emulsifier for quick-breaking oil emulsions, and in some cases a good sticker and deposit builder.

It is considered here mainly as an emulsifier. Petroleum oil is primarily a contact insecticide, but it is also a good carrier for other insecticides, and may be used as a sticker and deposit builder. A helpful discussion of the usual terminology encountered in the literature dealing with insecticides and fungicides, and including the above terms, is given by Cox (201).

SURFACE FREE ENERGY, SURFACE TENSION, AND INTERFACIAL TENSION

Among the adjuvants are the spreaders, wetting agents and emulsifiers. An attempt will be made to explain some of the physico-chemical phenomena involved in the use of this particular class of adjuvants.

Spreading, wetting, and emulsification cannot be readily visualized without some concept of what is meant by surface free energy and surface tension. A good discussion of the subject is given by Adams (3). In liquids, the cohesional forces between molecules keep them close to one another. Beneath the surface, each molecule is surrounded by others on every side and is subject to attractions in all directions. At the surface, however, the molecules are attracted inwards and to each side by adjoining molecules, but encounter very little attraction from above. Above the liquid is air, in which the molecules are far apart and few in number at any particular place. There is consequently little outward attraction to balance the inward pull. Every surface molecule is subject to a strong inward attraction perpendicular to the surface. This causes a contraction of the surface until it has become the smallest possible for a given volume. For this reason water forms in spheres when falling through space.

The extent to which the forces of attraction of the molecules at the surface of a liquid are not satisfied by molecules existing above the surface indicates the amount of free energy possessed by the liquid. Surface free energy is a fundamental property of surfaces, and surface tension is simply its mathematical equivalent. A hypothetical tension is presumed, acting in all directions parallel to the surface and equal to the surface free energy. This tension, in the case of liquids, can be measured by various means. The du Nouy Tensiometer (244) is commonly employed. With this instrument a platinum ring is lowered into the surface of a liquid and withdrawn. When it is withdrawn the liquid tends to adhere to the ring, and before it is broken away, an appreciable pull is exerted. The torsion of a wire is used to counteract the tension of the liquid film and to break it. A value is read from a vernier and multiplied by a correction factor of 0.943 to obtain the surface tension in dynes/cm.

Water possesses more surface free energy (72 dynes/cm.) than petroleum spray oil (32 dynes/cm.). If a soap is dissolved in water, a molecular layer of the soap will so orient itself that the CH_3 groups will form the new surface of the liquid, the polar metal ends of the molecules having been drawn into the water. The new surface of CH_3 groups will have less free energy than water, as might be expected from the difference in the free energy of water and oil, for the latter also presents a surface of CH_3 groups. The addition of the soap has consequently reduced the surface tension of the water by forming a surface of radicals of characteristically low energy.

Surface tension might more accurately be called interfacial tension, for the "surface" of a liquid is actually an interface liquid/air. When dealing with liquid/liquid interfaces the term interfacial tension is used. If a surface active solute is added to either the oil or the water so that its molecules will orient themselves at the interface, the polar (hydrophilic) ends in the water and the non polar (hydrophobic) ends in the oil, the surface tension may be greatly reduced. The dissimilarity of the two liquids is reduced by the layer of molecules, which form a "bridge" between them.

Employing a du Nouy Tensiometer, the writer found the interfacial tension at 72° F of water and a 92 U.R. light medium spray oil to be 31.4 dynes/cm 30 seconds after the interface was formed. By adding 1% glycol oleate (one OH group) to the oil, the interfacial tension water/oil was reduced to 16.6 dynes/cm. By adding

1% glyceryl monoöleate (two OH groups) to the oil the interfacial tension was reduced to 1.5 dynes/cm. When 1% glyceryl monoöleate was added to the oil and calcium caseinate at 1.1 g to 100 ml was added to the water, the interfacial tension
water/oil became so small that it could not be measured.

SPREADING AND WETTING, SPREADERS AND WETTING AGENTS

The terms "spreading" and "wetting" are often used interchangeably, but actually there is a distinction between the two phenomena. Hensill and Hoskins (426)
have offered the following definitions:

"A spreader is a material which increases the area that a given volume of
liquid will cover on a solid or another liquid."

"A wetting agent is any substance which increases the readiness with which a
liquid makes real contact with a solid - i.e.; wets it, if necessary by replacing
a previous contaminant on the solid."

It should be remembered that spreading pertains to the extension of a liquid
over a surface and wetting pertains to the retention of the liquid by the surface.

A good spreader is not necessarily a good wetting agent, and vice versa.
Blood albumin is a good spreader. Complete spreading of water over the surface of
a citrus leaf may be effected by the addition of 8 ounces of blood albumin spreader
(1/4 blood albumin and 3/4 inert diluent) to 100 gallons of water. However, soon
the water gathers into droplets scattered about in various locations over the leaf.
Although the leaf may have been uniformly covered with a film of water, usually
less than half the surface is covered with water a few moments after the leaf has
been sprayed. The water can be said to have spread over the entire leaf, but to
have wet only that portion of the leaf upon which it has been retained.

When a drop of liquid is placed on a horizontal solid surface and allowed to spread until it
comes to rest, a number of forces have reached a
state of equilibrium. These forces (fig. 72) are
the surface tension of the liquid, γl, the interfacial tension of the solid liquid interface, γsl,
and the surface tension of the solid, γs. It
should be mentioned in this connection that although no means have been discovered to measure the
surface tension of a solid, it nevertheless is known to have surface free energy.

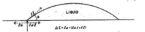

Fig. 72. Forces involved when
a drop of liquid comes to rest
on a solid. Explanation in
text.

At equilibrium the force γs is opposed by the force γsl working in the opposite direction, as well as the horizontal component of the force γ l, namely, the
cosine of the angle made by a tangent to the surface of the liquid at the point of
contact, p, of the liquid and the solid. This angle, designated by θ, is called
the angle of contact or contact angle. When equilibrium has been reached, γ
s = γl cos θ + γsl. Any change in energy which may occur in the system is represented mathematically by the relation, $\Delta E = \gamma s - (\gamma sl + \gamma l)$.

The angle made by a liquid on a horizontal surface is called the advancing
contact angle. If some of the liquid is removed, causing the drop to recede, a
smaller contact angle will be formed, and this is called the receding contact angle.
While the advancing contact angle is a measure of the spreading ability of a liquid,
the receding contact angle is a measure of its wetting ability.

The penetration of a liquid into capillary tubes or porous solids may also be
expressed in terms of surface tension and contact angle. It has been shown that a
reduction in surface tension may increase the penetration of an aqueous solution
into the spiracles of insects or through masses of waxy threads such as cover wooly
apple aphids and mealybugs, while at the same time reducing the penetration of the
liquid into the porous substrate upon which these insects rest (254).

103

The insecticidal efficiency of a spray can be increased only to a certain point by improving its spreading and wetting properties. The spray should spread and wet sufficiently well to promote ease of application and an effective coverage of the tree or plant. Beyond this point increase in spreading and wetting will usually decrease the insecticidal efficiency of a spray. In general, maximum retentivity of a spray occurs just at the point that run-off begins, and care should be exercised to not further increase the concentration of a spreader (290).

SOAPS

Soaps have been used as spreaders or wetting agents for centuries, this being, in part, their function in such domestic uses as washing and shaving. Soaps are long chain compounds with a polar metal or organic base end and a non-polar ($-CH_3$) end. They can be used successfully at very low concentrations. Salts of long chain fatty acids with from 12 to 18 carbon atoms are generally used in the production of soap. Sodium oleate may be taken as an example. In oleic acid there are 17 carbon atoms.

$$\begin{array}{ccc} & \text{H H} & & & & \text{H H} \\ & | \ | & & & & | \ | \\ \text{H-C-C} & - - - & \text{CO}\boxed{\text{OH} + \text{Na O}}\text{H} \longrightarrow & \text{H-C-C} & - - - & \text{COONa} \\ & | \ | & & & & | \ | \\ & \text{H H} & & & & \text{H H} \end{array}$$

    Oleic Acid   Sodium Hydroxide   Sodium Oleate

Thus it can be seen that some soaps are salts of fatty acids. Fatty acids occur in nature as glycerides. When boiled with such bases as NaOH or KOH, then cooled, a soap is formed. Salt is added, which helps in the recovery of the soap. This process is called "salting out". Soaps may also be esters produced by neutralizing organic acids with organic bases such as triethanolamine.

The majority of soaps are solids; certain potassium and ammonium soaps are liquid. The majority of the soaps are insoluble in water, but the ammonium and alkali soaps are "soluble" [1] As between the hard (sodium) soaps and the soft (potassium) soaps, the latter are more satisfactory for spraying purposes, for they are more readily miscible with cold water. Soap may be precipitated in "hard" water, losing its ability to act as a spreader, wetting agent or emulsifier. Hardness in water is usually caused by calcium, magnesium or other alkali earth salts. The sodium or potassium in the soaps is replaced by metals which produce insoluble soaps. These are, of course, of no value in sprays. The water may be "softened" by the addition of trisodium phosphate and certain other compounds, and this may aid to some extent in making hard water more suitable for spraying purposes.

SULFATED AND SULFONATED COMPOUNDS

This group of compounds are similar to soap in their action. Like the soaps, they have a long hydrocarbon chain solubilized, usually through a sulfate, sulfonate, hydroxyl or ether group, so as to orient to the surface. They have the advantage, however, of being more stable than soaps in acid solution and of having soluble calcium and magnesium salts. They can thus be used more effectively in "hard" water than the soaps.

In soaps and soap substitutes of fatty materials, the polar group is generally a primary one and it is located at the end of a long, straight-chain hydrocarbon molecule. It is this group which reacts so readily with the salts in the water, and its chemical alteration was found to be desirable. It was recognized that if the polar end of the fatty acids could be modified so that the calcium and magne-

----

[1] Physical chemists have shown that, at concentrations at which they are useful as wetters and detergents, soaps exist as colloidal electrolytes and are agglomerated into micelles of groups of probably 5 to 20 molecules. Decreased chain length or increased temperature may result in a true solution with loss of surface activity, and conversely with increased chain length and decreased temperature the soap concentration is so reduced as to be below that which is necessary for good results. (546)

sium salts were sufficiently soluble to prevent their precipitation in hard water, a product could be made which would have many of the characteristics of soaps without some of their bad features. One way of accomplishing this result has been by the hydrogenation of fat and fatty acids, followed by sulfation of the resulting alcohols. This important development was first accomplished in Germany about 20 years ago by the copper-chromium reduction of the fatty acids to the alcohol, followed by a sulfation and neutralization in order to form the polar end of the molecule. In this manner sodium lauryl sulfate was made from lauric acid. The product was introduced into the United States in the early thirties under the trade name of "Gardinol" and was successful (546). The primary alcohol sulfates closely resemble the normal soaps, prepared from the same raw products, in surface activity. They are, however, less reactive to "hard water" (951).

It is of interest to note that it is not necessary for the polar group to have an end position on the molecule. A comparison has been made of the wetting properties of a group of compounds comprising soaps, fatty alcohol sulfates, and sulfated fatty acid esters and amides, in which the polar or water soluble grouping is a primary one, with another group of compounds comprising secondary alcohol sulfates, sulfated esters of higher alcohols and dibasic acids, and the numerous alkyl derivatives of aryl or aromatic sulfonates, each of which contains its polar group in a secondary position with the non-polar portion extending from it in two directions. The first group of compounds were found to have somewhat superior emulsifying properties, but the second group were much superior as wetting agents and penetrants. (951)

The sulfated alcohols are esters of sulfuric acid, formed when straight-chain fatty alcohols ranging from 8 to 18 carbon atoms are treated with sulfuric acid, chlorosulfonic acid, or sulfur trioxide, at a temperature of not over 50º C. At higher temperatures some sulfonates may be formed. (748) Some entomologists have been careless in the use of the terms sulfate and sulfonate. In the sulfated compounds the sulfur atom is linked to carbon through the oxygen atom ($R-O-SO_3Na$) while in the sulfonated compounds the sulfur is linked directly to carbon ($R-SO_3Na$). (748)

A common household detergent known as Dreft, as well as the same and other closely related products under different trade names, is composed principally of sodium lauryl sulfate. A synthetic higher secondary alcohol is sulfated in the production of Tergitols 4 and 7. Sodium salts of sulfated fatty acid amides and esters of sulfated fatty acids are also used as spreaders and wetting agents. Since fatty acids occur in nature and are much cheaper than the corresponding alcohols, this would appear to be a logical development, but so far these products have not had extensive use in agriculture. (444)

Sulfonates may be prepared by the action of sulfuric acid on cheap hydrocarbons, as in the case of Ultrawet and a large number of other sulfonated petroleum products. They have been used not only as spreaders and wetters, but quite extensively in the preparation of emulsive spray oils. As compared to the soaps, they are relatively unaffected by hard water.

Sulfonation takes place especially well with aromatic hydrocarbons, and products such as Santomerse, Areskap, and Aresket are examples of the products formed in this manner.

Still more surface-active are the sulfonates of dibasic acids and secondary alcohols. Among this group are the Aerosols. Aerosol OT (Vatsol OT), which is an ester of sulfonated dicarboxylic acid, gained considerable popular recognition when pictures in Life Magazine (November 1941) showed some very surprised ducks all but submerged in a solution of this highly effective wetting agent (fig. 73). The surface tension of the water had been reduced to such an extent that it penetrated the envelope of air normally kept adjacent to the body by the feathers.

$$NaO_3S-C\begin{matrix}H & O\\|&\|\\ -C-OC_8H_{17}\\H-C-C-OC_8H_{17}\\|&\|\\H&O\end{matrix}$$

Aerosol OT
A dioctyl ester of sodium sulfo succinate.

Fig. 73. The near submersion of ducks in water of greatly re-duced surface tension. Courtesy American Cyanamid Co.

This resulted in such loss of their accustomed buoyancy that the ducks could hardly paddle fast enough to keep from sinking. In one investigation Aerosol OT was found to be the most effective of all wetting agents tested in reducing the contact angle of water on citrus foliage, although on beeswax, sodium oleate was more effective. This is an example of the fact that the relative effectiveness of two or more wetting agents on one surface may not necessarily obtain on other surfaces of other composition. (255)

In the Triton group of wetting agents, the surface active constituents are sulfonated ethers. Other sulfonated products are the mixed alkyl aryl derivatives (Alkanol HG, Nacconol N.R., Aresklene) and esters of fatty acids (the Atlas group).

BLOOD ALBUMIN

Blood albumin is a good spreader and a fair wetting and emulsifying agent. Smith (818) found blood albumin spreader to be the most effective of all spreaders tested in the development of the tank mixture method of using oil spray in citrus spraying. Blood albumin has a high dynamic spreading ability. Water with 8 ounces of blood albumin spreader will spread over the entire surface of young citrus leaves, but the water will soon recede and form into droplets. Blood albumin has the advantage of being relatively unaffected by hard waters.

Proprietary blood albumin spreaders consist of 1 part of powdered blood albumin mixed with 3 parts of fuller's earth or similar forms of natural earth. Pearce and Chapman (681) suggested specifications for an acceptable grade of blood albumin as follows: Total nitrogen content, 14.0 ± 0.5%, solubility in water of not less than 90%, and not less than 90% passing through a 100 mesh screen.

CALCIUM CASEINATE

Calcium caseinate has been widely used, not only as a spreader and wetting agent, but also as an emulsifier, especially in the production of the quick breaking emulsions. As compared to the soaps, calcium caseinate is relatively unaffected by hard water, but more so than blood albumin.

Casein is made from skim milk. The milk is run into large vats and there it is mixed with dilute acid in order to produce a curd. The curd is then neutralized, pressed into cakes, dried at a moderate temperature, and ground to various degrees of fineness.

Calcium caseinate is a mixture of hydrated lime and casein in the proportion of 4 parts of hydrated lime (calcium hydroxide) to 1 part of 60-80 mesh casein. Casein dissolves in a weak alkaline solution. The purpose of the lime is to produce this type of solvent. If the water is fairly soft, not much lime is required. With increased hardness, however, a portion of the lime is neutralized and it is necessary to add an excess to supply sufficient alkalinity to dissolve the casein. It may be noted in this connection that spray materials made up with the excess of lime, such as Bordeaux mixture, many fluorine preparations, and lime sulfur, may be used in hard water without adverse effects.

Ammonium caseinate has replaced calcium caseinate to some extent. About the year 1930, growers in the Yakima, Washington apple districts began to substitute ammonia for lime in the preparation of oil emulsions. Several advantages were found in the use of ammonia. The ingredients were cheaper, they did not deteriorate before use, they mixed more readily with water, and the resulting emulsion (flowable emulsion) was thinner and hence more easily handled. (645) The ammonia acts as a preservative for the casein in addition to making it miscible with water.

EMULSIFIERS AND EMULSIFICATION

If two liquids which differ considerably in density and are almost entirely insoluble in each other are mixed, they will separate into distinct layers or "phases". If the liquids are agitated, one will momentarily become uniformly dispersed through the other in the form of small globules. The liquid which forms the globules is called the disperse phase and the liquid through which the globules are distributed is called the continuous phase.

The stability of the mixture is supposed to be due to the high interfacial tension between the two phases. It is the function of emulsifiers to lower the tension by coating each globule of the dispersed liquid with a colloidal layer. If oil and water are shaken together, as in a flask, the oil and water globules will at first be uniformly mixed, the water being the continuous and the oil the disperse phase. If a surface active solute, such as soap, is added to the mixture, the interfacial tension water/oil will be reduced. Thus the two liquids will act more as if they were the same liquid, and the rate of "layering out" will be greatly reduced; an emulsion has been formed.

It is also supposed that among the soluble emulsifiers one end of the molecule will dissolve in one phase of an emulsion and the other end in the other phase, thus keeping the two phases together (444). Most emulsifiers are of a colloidal nature and soluble, but insoluble non-colloidal substances in a fine state of division, such as talc and clay, may also act as emulsifiers.

It appears likely that emulsions may be formed as a result of a variety of interfacial phenomena. Bancroft (57) believed that all that was necessary for emulsification was that the globules of the disperse phase be surrounded by a coherent film of some kind to keep them from coalescing and "layering out". That this may often be the case is suggested by the fact that emulsification can be induced by finely divided particles and by Seifriz' (768) observation that membranes which surround the dispersed globules of many emulsions are obviously not monomolecular layers, but are optically visible and may under certain conditions separate off as persistent structures.

Finkle et al (309) came to the conclusion that (1) lowering of the interfacial tension, (2) the electric charge of the dispersed globules, and (3) the mechanical action of the film at the interface in keeping the globules apart, all played a part in stabilizing an emulsion.

For methods of preparing oil emulsions see page 185.

SUSPENSIONS

Much that has been written above concerning emulsifiers and emulsification pertains also to suspensions of solids in liquids. A suspension as used for spraying is an intimate mixture of finely divided solids with water. As in the case of oil emulsions, the suspended particle may be kept uniformly mixed with water simply by agitation, but ordinarily an emulsifying agent or stabilizer is added to the mixture. This substance collects or is adsorbed on the surface of the solid particles and prevents them from coalescing.

DEFLOCCULATING AGENTS

When finely divided suspended solids gather together in clumps, they are said to "flocculate", and "defloculators" must be added if flocculation is to be prevented. Proteinaceous materials acting as protective colloids, such as glue gelatin, milk products, and gums are most commonly used, but many of the substances mentioned in the discussion on spreaders, wetting agents, and emulsifiers could be successfully used. Nowadays the powders to be suspended are finely ground, or as in the case of some brands of lead arsenate, they are already "deflocculated" by the addition of a suitable material in the prepared product, so the spray operator seldom needs to add substances for the purpose of stabilizing his insecticide suspensions.

SUBTROPICAL ENTOMOLOGY

ADHESIVES OR STICKERS

Often, as in the case of lead arsenate, there is no special tendency of the insecticide particles to remain attached to a surface, and <u>adhesives</u> or <u>stickers</u> may be added to bring about this result. The adhesive often also increases the deposit of the insecticide on the tree or plant surface. In one investigation it was found that the most effective adhesives for lead arsenate on apple leaves were summer spray oil, bentonite-sulfur, and milk. These caused not only greater initial deposit but also greater weathering ability. For the deposition and retention of sulfur, dry skim milk proved to be the most effective substance (930).

Adhesives are usually sticky, gummy, or varnish-like at high concentrations or when they dry. Proteinaceous materials such as glue, gelatin, casein and flour are often used as stickers. Bordeaux mixture has been successfully used in this capacity also. Oils may be used as stickers, the drying oils, such as linseed oil, being particularly effective. Some wetting agents and emulsifiers may also serve as adhesives.

As much as 10% of mineral oil, vegetable oil or fish oil can be incorporated into a dust by means of suitable mixing equipment, and as much as 5% of an oil is often used in dusts to increase their adhesion to the dusted plants.

DEPOSIT BUILDERS

The deposit of solids from a spray suspension can sometimes be greatly increased by the addition of "deposit builders". These generally consist of a combination of free and fatty acids and in some cases petroleum sulfonates, together with certain metallic salts. It is believed that insoluble metallic soap forms on the surfaces of the insoluble insecticide or fungicide particles and that the molecules are oriented with the nonpolar ends on the outside. Thus the solid particles are given a sticky, oily character, which may be further enhanced by the addition of a small amount of mineral oil.

A characteristic of deposit builders is that they cause a number of solid particles to adhere together, forming a "floc". Flocculation is conducive to heavy deposit. When spraying, a large mass of material is less apt to be carried off the tree surface with the water than small particles which remain in suspension as discrete particles. The water that runs off the tree is practically clear.

The deposit of practically any finely divided solid in a spray can be increased by using deposit builders, and the insecticidal effectiveness of the spray may be greatly increased. The deposit is not only greater in quantity, but the solid material is also more uniformly deposited. The appearance of the deposit suggests that an improvement of spreading has occurred, but this may not necessarily be the case. What happens is that the material sticks to the leaf or fruit surface at the moment of its original distribution and does not run off with the water to be deposited at random as the water droplets evaporate. An increase in the adhesiveness of the solids also is caused by deposit builders. It appears that all deposit builders increase adhesiveness, but not all adhesives increase deposit.

It is generally acknowledged that the most effective as well as the least expensive deposit builder is an emulsive oil. An emulsive oil has the qualities already mentioned as being desirable in deposit builders, namely, giving the solid particle a sticky exterior and causing an agglomeration of particles (flocculation). In addition the oil is itself a hydrophobic phase of the emulsion, readily adhering to the plant and insect surfaces and carrying the insecticide solids with it, while the water runs off. This subject, however, had perhaps better be discussed under the heading of "preferential wetting."

PREFERENTIAL WETTING

In the Pacific Northwest, the steadily increasing seriousness of the codling moth problem on apples led to considerable investigation of the problem of increas-

108

ing the deposit of lead arsenate and making it more resistant to weathering. The success of these efforts is attested by the fact that in later years a similar amount of effort was expended in finding an efficient way of removing the heavy and tenacious lead arsenate residues. The inverted or "dynamite" mixtures were prepared to enable the lead arsenate to be carried to the tree surface by oil instead of by water (589), thus increasing not only the deposit of lead arsenate but also its sticking ability. In the "dynamite" sprays the principle of "preferential wetting" was employed.

The principle of preferential wetting was utilized in working out a formulation combining oil and cryolite or oil and barium fluosilicate in the treatment of a combination of insects often found together in a citrus orchard, namely, black scale, Saissetia oleae, and the orangeworms, Argyrotaenia citrana and Holcocera iceryaella. (687)

It was discovered that the emulsive oils were much more efficient in spreading and depositing cryolite and barium fluosilicate than were the other oil types. Some emulsive oils were more effective than others. When emulsive oils were used in the spray mixture, the solid particles were preferentially wet by the oil and they were thus drawn into the oil phase. Since oil wets and deposits on a tree surface more efficiently than water, solids in the oil phase would obviously be deposited more evenly and in greater quantities than if they were in the water phase. Without certain solutes in the oil to cause the preferential wetting, the solids were for the most part in the water phase. This was the case in all sprays in which emulsions or tank mixtures were used. (687)

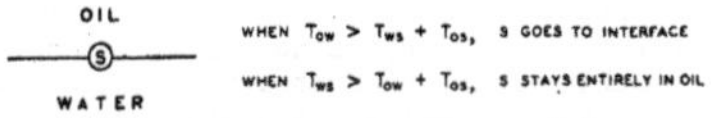

Fig. 74. The interfacial tension relationships of a system containing a solid particle at the interface oil/water. Explanation in text.

The interfacial tension relationships in preferential wetting are shown in fig. 74. When the interfacial tension oil/water is greater than the sum of the interfacial tension water/solid and oil/solid, the solid goes to the interface oil/water. When the interfacial tension water/solid is greater than the sum of the interfacial tension oil/water and oil/solid, the solid particle goes into the oil, which is the usually desired goal in preferential wetting.

CORRECTIVES OR SAFENERS

In many cases in the use of insecticides the margin of safety between adequate control of the insects and reasonable safety to the plant is precariously small. The possibility of injury to the plant is commonly accepted as a risk inevitably involved in the use of the insecticide. Needless to say, every possible effort is made to reduce this risk as much as possible by precautions used in the application of the insecticide, by providing that the ingredients are chemically adjusted so as to insure a minimum of phytocidal effect, and whenever possible, by the addition of substances which might counteract the chemical changes which result in injury. The latter are called "correctives" or "safeners". Hydrated lime or zinc sulfate as a corrective for arsenate sprays and ferrous sulfate as a corrective for lime sulfur, and also to reduce arsenical injury, are in quite general use as correctives.

DILUENTS

Diluents (carriers, extenders, fillers, dispersents) as used in the preparation of insecticide dusts, facilitate the passage of finely divided insecticides through the dusting machinery and to the dusted trees or plants and make possible an even distribution of the minimum quantity per unit area of insecticidally effective dosages of the insecticide. All other things being equal, the least expensive diluents would be chosen in the preparation of a dust. In actual practice, however, the choice of material depends on a number of factors aside from cost, namely, (1) whether the diluent is compatible with the toxicant from a chemical standpoint, (2) whether or not it is too abrasive, (3) whether the diluent functions satisfactorily in dusting machinery, (4) whether, when mixed with an insecticide, the two will remain as a homogenous mixture both before and after discharge from the duster, (5) the flowability of the diluent and the type of dust cloud it forms, that is, whether the dust settles quickly or slowly, and also how evenly the dust is distributed, (6) the tenacity and weathering ability of the diluent-toxicant mixture on the foliage and fruit on which it is to be applied, and (7) the effect of the diluent on the electrostatic charge of the dust. Sometimes a sorptive dust is desired and under other conditions sorption is an undesirable characteristic; sometimes an alkali dust is required, and in other cases a neutral or acid composition is required. Some insecticides require the addition of certain kinds of diluents to facilitate grinding. Sometimes there may be only one or two diluents that are capable of aiding in the grinding operation. DDT dusts are prepared by grinding technical DDT with talc, clay, pyrophyllite, sulfur, or certain other diluents.

Having decided on the class of diluent, the problem remains of selecting the proper particle size. Ordinarily it is required that 90% of the diluent must pass through a 325 mesh screen and that not more than 2 or 3% should be of 200-mesh size or larger.

The user of insecticide dusts is often impressed with large, billowy clouds of dust, but too large a cloud may be wasteful and a material of greater density might be more efficient. Dusts of too low a density increase the possibility of unwanted drift of dust onto neighboring properties. Some states have a law requiring that poisonous dusts must contain 1% oil to increase the settling rate and reduce the amount of drift from one property to another. A number of diluents for "high bulk density" and "low bulk density" DDT dust mixtures have been suggested. (158)

Sometimes several diluents may be satisfactory for a certain insecticide, as far as compatibility, cost, etc., are concerned, but actual field trials may reveal that one particular diluent or combination of diluents gives much better control than any of the others because of better deposit, distribution, or other contributing factors.

Watkins and Norton (918) worked out a system of classification of diluents, which is shown in table 1. It will be noted in this table that the diluents fall into two major groups, the botanical flours and minerals. Among the minerals may

Table 1
Classification of insecticide dust diluents and carriers*

Botanical Flours	Silicates
Soybean flour	Mica
Tobacco flour	Talc
Walnut shell flour	Pyrophyllite
Wheat flour	Clays
Wood flour	Montmorillonite group
	Montmorillonite
Minerals	Saponite
Elements	Nontronite
Sulfur	Beidellite
Oxides	Kaolinite group
Silicon	Kaolinite
Tripolite	Nacrite
Diatomite	Dickite
Calcium	Anauxite
Calcium lime	Attapulgite group
Magnesium lime	Attapulgite
Carbonates	Sepiolite
Calcite	Phosphates
Dolomite	Apatite
Sulfates	Indeterminate
Gypsum	Pumice

*From Watkins and Norton (918).

110

be found a number of manufactured materials placed in accordance with their physical and chemical characteristics.

Watkins and Norton (919) also prepared a very useful looseleaf booklet of data sheets with 21 entries on each of 242 commercial diluents or carriers classified according to the aforementioned system of classification. Some of the entries are filled by the authors on the basis of information they had available; others are left blank, to be filled in by the investigator or user on the basis of his own tests or information he may be able to obtain. The 21 entries are: (1) mineralogical content, (2) average chemical analysis, (3) where mined, (4) how processed, (5) how packed, (6) price, (7) color, (8) hardness, (9) specific gravity, (10) apparent density, (11) average particle size, (12) range of particle size, (13) dry screen analysis, (14) wet screen analysis, (15) particle shape, (16) angle of slope, (17) oil absorption, (18) pH, (19) wettability, (20) other physical or chemical properties, and (21) special features. This booklet was reproduced and is distributed by the National Agricultural Chemical Association (formerly the Agricultural Insecticides and Fungicide Association), Barr Building, Washington, D. C.

DILUENTS FOR SPRAYS

Diluents are commonly used to effect a satisfactory dispersion of a powder to be mixed with water. As in the case of dusts, the diluents must be compatible with the toxicant. Often a wetting agent is added to dust mixtures in the preparation of proprietary wettable powders. The diluent may likewise be selected on the basis of its own wettability. Diluents are also added to wetting agents and spreaders to effect their rapid dispersion in water. Blood albumin spreader contains 75% of diluent and Vatsol OTC contains 90%.

INSECTICIDE DUSTS AND COMMONLY USED DILUENTS

1. CALCIUM ARSENATE. Diluents: Talc, diatomaceous earth, sulfur, bentonite, clay. High magnesic hydrated spray lime found desirable and high calcic lime undesirable (414).

2. LEAD ARSENATE. Diluents: Hydrated lime the common diluent. Also talc, clay, etc., and fungicides such as Bordeaux or sulfur.

3. PARIS GREEN. Diluents: Hydrated lime, sulfur.

4. FLUORINE COMPOUNDS (sodium fluoaluminate (cryolite), sodium fluosilicate, calcium fluosilicate, barium fluosilicate). Diluents: Talc, pyrophyllite, diatomaceous earth, sulfur. Lime is not compatible.

5. SULFUR. Diluents: Used as a dust without diluents except for "conditioned" sulfur, which contains small amounts of gypsum, bentonite, talc, pyrophyllite, etc., to make the sulfur flow more freely. Sulfur is often used as a diluent for other insecticides, such as calcium arsenate, pyrethrins, rotenone, and DDT.

6. NICOTINE SULFATE. Diluents:
   1. Sorptive materials. Kaolin, bentonite, pyrophyllite, fuller's earth, diatomaceous earth, talc. Such dusts keep well in storage but are not as insecticidally active as other forms of dust. The evolution of nicotine fumes is restricted.
   2. Inert crystalline materials. Gypsum, sulfur, slate dust. Little effect on evolution of nicotine fumes other than by increasing evaporation surface.
   3. Alkaline materials. Hydrated lime, calcium carbonate, dolomitic limes and limestones. These react with nicotine sulfate to form free nicotine. Greatly increase rate of evolution of nicotine fumes, making the dust more insecticidally effective, but difficult to store without loss. Desirable when rapid contact action of the insecticides is required.

111

7. PYRETHRINS. Diluents: The more highly sorptive diluents such as kaolin, bentonite and diatomaceous earth release pyrethrins more slowly than less sorptive materials, such as gypsum, talc, pyrophyllite, and walnut shell flour. Anti-oxidants are sometimes added. Hydrated lime is probably incompatible, for pyrethrins are not stable in alkali, but reaction is not rapid when the dust is dry and lime and other alkaline materials have been used satisfactorily, but are not recommended for use in the preparation of dusts that must be stored. The ground flowers of pyrethrum can also be used with diluents in the preparation of dusts.

8. ROTENONE. Diluents: Same as for pyrethrins. The ground roots of derris and cube can also be used with diluents in the preparation of dusts. Dusts prepared with alkaline diluents such as hydrated lime, diatomaceous earth, certain talcs and clays, bentonite, and limestone deteriorate under humid conditions or if they must be stored for a long period.

9. LETHANE. Diluents: Diluents which are satisfactory for derris or pyrethrum will be found to be satisfactory for Lethane.

10. RYANIA. Diluents: Compatible with all of the commonly available inert diluents including talc, sulfur, pyrophyllite, clay, and diatomaceous earth.

11. DINITRO-O-CYCLOHEXYL PHENOL. Diluents: Powdered pumice, walnut shell flour, redwood bark flour. Acid diluent desirable. Hydrated lime, bentonite, talc, etc., react, forming salts, and therefore not suitable as diluents.

12. DDT AND OTHER CHLORINATED HYDROCARBONS. Diluents: Talc, clay, sulfur, pyrophyllite. Alkaline materials such as lime, fuller's earth, and Bordeaux should be avoided, for they will cause a chemical deterioration.

13. PARATHION. Diluents: In the preparation of dusts, parathion is sprayed into the dust mixer as a liquid. Consequently at least a small amount of the diluent is ordinarily a highly adsorptive material. One company uses Attaclay and ground bentonite. After the parathion has been adsorbed by this material, further dilution can be made with a number of neutral or nearly neutral materials such as frianite, pyrophyllite, neutral talcs and clays. Basic materials should not be used.

14. TEPP. Diluents: TEPP hydrolyzes rapidly on contact with moisture, so the diluent must be a material that will be preferentially wet, that is, have a greater affinity for water than TEPP. An anhydrous form of calcium sulfate has been used. This should have little or none of the "dead burnt", the alpha form of anhydrous calcium sulfate, which is not hydroscopic. Also a very dry volcanic rock diluent, prepared at Friant, California, has been satisfactorily used.

COMPATIBILITY

As Roark (750) has pointed out, the terms "compatibility" and "incompatibilit as applied to insecticides, are subject to criticism on the basis that combination of substances may be chemically incompatible and yet be suitable or even highly de sirable as insecticide combinations. An example is the combination of nicotine sulfate and lime. These materials are chemically incompatible, for a chemical reaction takes place and free nicotine and calcium sulfate are formed. The toxicity of the mixture is increased, however, by the liberation of the free nicotine, and the combination of lime with nicotine sulfate is desirable when rapid action is required.

Probably the most complete exposition of the subject of compatibility of insecticides to appear in the literature is the one written by A. J. Cox, which appeared in a special publication of the California State Department of Agriculture 200). The following discussion on the meaning of compatibility is taken from his report:

"The term 'compatibility' is loosely used to mean several different things. t is used here to refer to the suitability of pest control materials for simultaeous or successive applications to the same plant. Although some substances may eact with each other and still be compatible, the question of compatibility is argely one of chemistry, and involves a consideration of the probability of retion between materials and the possible effect of reaction-products.

"A mixture of several chemicals used for pest control or their separate application to the same plant within a short interval of time may be (1) desirable, 2) usable, (3) undesirable, or (4) injurious as follows:

"1. The components of such a mixture may have a beneficial synergistic effect ncreasing the efficacy of some of the ingredients or decreasing some hazard inolved in their use.

"2. The components or their efficacy may be unaffected by such usage and imply give the additive effect of the separate application of each component. he sole reason for using such combinations is the economy of application.

"3. The efficacy of one or more of the individual components may be somewhat essened by such usage, but not sufficiently to prohibit altogether the use of uch a combination.

"4. Some mixtures may be definitely injurious to plants, or through chemical eaction may result in the complete destruction of the efficacy of some of the omponents.

"Further complications arise from the fact that a certain mixture may be used n fruits or plants under optimum weather conditions in certain regions, but not inder adverse conditions. It may be used on certain plants but not on others. It

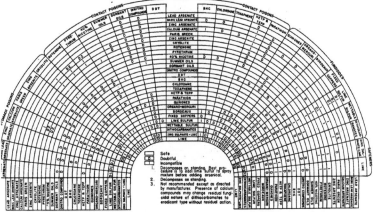

Fig. 75. Compatibility chart for insecticides and fungicides. Adapted from "American Fruit Grower" 68(2):40, (1948).

may be satisfactory when mixed just prior to use on plants in leaf, but not be
suitable for long storage. It may be hazardous to plants in leaf, but suitable
for application to dormant deciduous plants. It may afford satisfactory control
of the pests concerned, but cause unsightly spotting of fruit or foliage, or
render spray residue removal too difficult. The components may not be usable in
a combination spray but may be applied separately within a short time. Some
materials might be compatible from a chemical standpoint, but are similar in
effect and are therefore not used together."

A comprehensive compatibility list, from the standpoint of insecticidal
effectiveness and tree hazard, for the various insecticides, fungicides, and
nutritional deficiency elements now commonly used in the citrus orchard was prepared by J. R. LaFollette, Entomologist for the Bureau of Pest Control of the
California Fruit Growers Exchange in cooperation with A. M. Boyce, J. R. Allison,
D. D. Penny, and Paul R. Jones (500). Whether a particular combination of materials is compatible or not can be most readily ascertained by examining a
"compatibility chart", such as shown in fig. 75.

CHAPTER VI
INSECTICIDES USED PRIMARILY AS STOMACH POISONS

PRESENT DAY SIGNIFICANCE OF THE TERM "STOMACH POISON"

The separation of insecticides not used as fumigants into stomach poisons and contact poisons does not now have its former significance. There has never been a sharp line of demarcation between the two, but the proportion of contact insecticides which have considerable value as stomach poisons has sharply increased among the new synthetic insecticides produced during the last decade. Also the residual effect of the recently developed contact insecticides, a property not ordinarily possessed by the plant derivatives, may result in their being equally effective against sucking and chewing insects. Nevertheless, the separation of certain insecticides into those predominantly used as stomach poisons and those predominantly used for their contact action may still serve a useful purpose, at least for anyone who is attempting to organize and systematize in a relatively short period a large body of independent facts.

Stomach poisons are applied to the surface of the food consumed by the injurious insects, providing they have mandibulate or "chewing" mouthparts (fig. 76).

Fig. 76 (at left). Chewing mouthparts. The plant is eaten from the outside. Photo by Roy J. Pence.

Fig. 77 (at right). Sucking mouthparts. Aphis rumicis feeding on avocado leaf stem. Above, winged form; below, wingless form. Arrows point to points of insertion of mouthparts. X 26. Photo by Roy J. Pence.

They are used to poison "those types of insects that (a) bite off, chew, and swallow, portions of the food plant, (b) rasp the surface of the food plant and suck up the resulting pulp, (c) lap up food or moisture from surfaces that can be poisoned, (d) can be induced to feed upon artificial foods such as poisoned baits, or (e) possess the habit of using the mouthparts to clean the body and appendages, and so may swallow poisons placed where the insects must encounter them. In all such cases, the poison is taken into the alimentary tract." (508)

If the insects to be controlled have hausteliate or "sucking" mouthparts (fig. 77) they will pierce the layer of insecticide and the plant epidermis upon which it is deposited and suck out the unpoisoned plant sap. Unless the poison also has contact value it will be useless against such insects. Obviously, insecticides used predominantly as stomach poisons are not ordinarily applied against sucking insects, although a few may have some contact action against certain species. In this respect, at least, the old distinction between stomach poisons and contact insecticides has retained its original significance. For the majority of insecticides, with the exception of fumigants, the type of mouthparts possessed by an insect may still be an important consideration in the selection of the most suitable insecticide.

115

The question may properly be asked as to why stomach poisons should be used at all when there are so many insecticides that will kill by contact action. It must be borne in mind, however, that until the advent of DDT, no contact insecticide other than oil had an appreciable "residual effect" against the insects not directly hit by the spray, or their progeny. The stomach insecticides are sometimes referred to as "protective insecticides" because they can be applied to a plant, often with no effort at thorough coverage, and the insect eventually consumes some of the poisoned plant surface in foraging about for food. The plant may thus be protected for prolonged periods. This same "protective effect," however, is now possessed by many of the new synthetic organic compounds which are used primarily for their contact action. The insects not directly hit by the spray, or their progeny, may succumb weeks and even months after the spray is applied, merely by crawling over the sprayed plant surface. The term protective, formerly applied to stomach poisons, could now be equally well applied to many insecticides noted primarily for their contact action. In some cases these also perform as stomach poisons. A disadvantage in their use, however, is that they also destroy the beneficial entomophagous insects.

The insecticides limited to use as stomach poisons are for the most part inorganic compounds of low lipoid solubility and not having the power of penetrating through the outer cuticle of the insect, nor even the walls of the fore-or-hind intestine, for the latter, being of ectodermal origin, is lined with chitinous cuticle. These inorganic compounds may, however, readily penetrate the walls of the mid-intestine or "stomach" of the insect.

A stomach poison may be chosen in preference to a contact insecticide because of its "protective" value or its lack of phytotoxicity to the particular plant being sprayed, in some cases its greater economy or the ease of application compared to certain contact insecticides which might be successfully used, or its usual lesser effect on beneficial insects compared to contact insecticides with residual effect. However, certain factors limit the applicability of stomach poisons. A compound may be distasteful to the insect, and less than the lethal dose may be consumed by some species of insects. Some stomach poisons may repel the insects completely (659), in which case the latter are not destroyed.

The newly-hatched codling moth larva rejects for the most part the portions of skin cut out in making its entrance into the apple, ingesting only a few particles. If it were not for this unfortunate habit of codling moth larvae, the control of the pest would be much less difficult. (816)

The rapidity of toxic action may be a factor of importance. If an insecticide is not sufficiently poisonous to kill an insect in a reasonable period, too much destruction of foliage or fruit may take place despite the eventual death of the insect.

Presuming the insecticide to be ingested, it may then be much more soluble in the intestinal fluids than in water. Low solubility on the plant surface, resulting in low phytotoxic effect, coupled with high solubility in the insect gut, resulting in high insecticidal efficacy, is a fortunate combination of circumstances. Obviously, the pH of the spray liquid, the pH of the stomach fluids of the insect, and the buffering power of the contents of the stomach, all play a part in determining the effectiveness of a stomach poison (446). Some insects appear to have the ability to regulate the buffer value of the intestinal contents, which may aid them in reducing the toxic effects of certain stomach poisons (835).

THE ARSENICALS

The arsenicals have long occupied a leading position among insecticides. Their importance was particularly enhanced in this country by the enormous amounts of lead arsenate used for the control of codling moth and other insects and of calcium arsenate used in control of the boll weevil on cotton. It was unofficially estimated that 60,000,000 pounds of lead arsenate was used in 1938 (344). It is now evident, however, that the newer synthetic organic insecticides will in many instances replace arsenicals.

INSECTICIDES USED PRIMARILY AS STOMACH POISONS

In this chapter the arsenicals will not occupy a space proportionate to their general importance, for it happens that the only appreciable volume of arsenicals used against the subtropical pests is in connection with the control of the codling moth on walnuts, certain pests of the pecan, and the peach twig borer on almonds. The arsenicals cannot be used on citrus because they result in a reduction in concentration of both acid and solids in the fruit (610).

White arsenic:- This is the common name for arsenic trioxide (As_2O_3), which in aqueous solution forms arsenious acid. The reactions involved are as follows:

$$As_2O_3 + H_2O \longrightarrow 2 \ HAsO_2 \text{ (slightly ionized)}$$

$$As_2O_5 + 3 \ H_2O \longrightarrow 2 \ H_3AsO_4$$

White arsenic is the base for the manufacture of all arsenicals used as insecticides. It is a white solid which may occur in a vitreous form. On standing, however, the vitreous material changes to an octahedral crystalline form. A monoclinic form may be prepared by heating for some time at 200° C. In the United States white arsenic is a by-product from the smelting of various ores.

White arsenic may be used as such in the preparation of poison baits, ant sirups and wood preservatives, but its high phytotoxic action prevents its being used on plants. About 25% of the white arsenic produced in the United States is used in the manufacture of weed killers.

The Arsenites:- With the exception of Paris green, arsenites are made by dissolving As_2O_3 in alkalies. The soluble salts of the types Na_3AsO_3 (sodium orthoarsenite) and $Na \ AsO_2$ (sodium metarsenite) are highly hydrolyzed, due to the weakness of the acid. The arsenites are less stable than the arsenates, and will set free more acid by hydrolysis. In addition, arsenious acid is more phytotoxic than arsenic acid. Consequently, the arsenites are commonly used in preparations such as baits and ant sirups, which are not used on plants, while the arsenates are used in sprays or dusts applied to foliage.

Paris Green

Paris green, usually written $(CH_3COO)_2Cu.3 \ Cu(AsO_2)_2$, originally used as paint pigment, first came into use as an insecticide when applied as a dust against the Colorado potato beetle in the western United States in 1867. It is not desirable as a spray because its particles are coarse and heavy and do not remain suspended well, and, of course, its instability and high arsenic content make it unsafe to foliage. It was used mainly in baits for cutworms and grasshoppers. It has also been used rather widely against mosquitoes, in the control of which it may be dusted on the surface of the water. It is less expensive than oil and easier to apply. In this connection, however, it is being largely replaced by DDT.

Sodium Arsenite

Sodium arsenite, which is considered to be either sodium orthoarsenite (Na_3AsO_3) or sodium metarsenite ($NaAsO_2$) or a combination of the two, is used in making baits, poison fly papers, cattle dips, ant sirups, in termite control, and as a weed killer. In the control of subterranean termites, a 10% solution may be sprayed on the ground at the rate of 5 gallons per 100 sq. ft. of ground area, or the solution may be poured into a trench 6 inches deep, dug along the foundation and the pier blocks, using 1 gallon per 15 linear feet. Some of the solution should also be applied to the top of the refilled trench. (581) Care should be exercised in not getting this poison on soil planted to trees, shrubs, flowers, etc., or where it is expected that planting may later take place.

Zinc arsenite, calcium arsenite, Scheele's green and London purple have been used from time to time on a very limited scale.

The Arsenates:- As stated before, the arsenites are made by treating As_2O_3 with an alkali. In the preparation of arsenates, however, the As_2O_3 must first

117

be oxidized to As_2O_5. Usually the arsenates are prepared by the direct action of a solution of arsenic acid on a metallic oxide. The oxides of lead, zinc, calcium magnesium, and copper are the bases most used in the manufacture of arsenates. (190) Being less active chemically than the arsenites, the arsenates are safer for use on plants. Some injury may occur with these compounds also under certain conditions.

Acid Lead Arsenate

Acid or standard lead arsenate ($PbHAsO_4$) is the most widely used of the stomach poisons. It was first prepared in 1892 by F. C. Moulton, a chemist for the Gypsy Moth Commission in Massachusetts. At first it was largely homemade, then a paste was manufactured commercially, containing from 45 to 50% water, while today the powder form is used almost exclusively. Although the above formula is the one generally used to represent acid lead arsenate, the various combinations between lead and arsenic result in the production of a number of compounds. These have been discussed in a series of papers by McDonnell and Smith (549, 550, 551).

Acid lead arsenate is light and fluffy and stays in suspension well. Once dried on foliage, it adheres very well.

Beginning in 1938, lead arsenate and other white arsenicals were colored pink to differentiate them from white flours which are used as food.

The two most important chemical determinations on lead arsenate, from the practical standpoint, are total arsenic oxide and water-soluble arsenic oxide. In the dry lead arsenate the total arsenic oxide varies from 31 to 33%, with as much as 0.25% water soluble.

Although acid lead arsenate is relatively insoluble, it sometimes forms enough soluble arsenic on plant surfaces to result in injury. Hydrolysis of the arsenate may take place, resulting in the formation of arsenic acid (551):

$$5 \ PbHAsO_4 + HOH \longrightarrow Pb_4(PbOH)_3(AsO_4)_3 + 2H_3AsO_4$$

Acid lead arsenate is sufficiently stable so that it can be used with many other compounds without breaking down. With lime sulfur, arsenic acid is formed, this reaction being retarded, however, by the addition of lime.

The acid lead arsenate powder is generally used at 2 to 3 lbs. per 100 gallon of water. Various adjuvants are often added to increase spreading, sticking, depositing ability, or as safeners. When used in walnut spraying, lime, or zinc sulfate may be added as safeners, that is, to retard the formation of water-soluble arsenic. Apples are washed with a dilute hydrochloric acid solution to remove the lead arsenate residue. The tolerance for As_2O_3 is 0.025 gram per pound. The greatest single use of this material is in spraying for codling moth on apples, or which as many as 6 to 10 applications may be made each year.

By adding a properly prepared oil (see p. 109) to a lead arsenate spray mixture, the arsenate-depositing ability of the spray is greatly increased. The resulting mixture is called an "inverted" spray or "dynamite" spray described by Marshall (589) as "one in which a suspended solid initially wetted by water becomes wetted by oil prior to or at the moment of impact upon a sprayed surface." Marshall considered this phenomenon, as applied to the insecticide-fungicide field as a "new principle of considerable promise." Inverted mixtures were widely used in lead arsenate spraying for codling moth on apples despite the difficulty encountered in removing the residue.

Acid lead arsenate may also be used as a dust, usually diluted with from 1 to 3 parts of sulfur, hydrated lime, or some other diluent.

118

INSECTICIDES USED PRIMARILY AS STOMACH POISONS

Basic Lead Arsenate

Basic lead arsenate [$Pb_4(PbOH)(AsO_4)_3 \cdot H_2O$], as can be seen from the formula,[1] has a higher percentage of lead than acid lead arsenate, and it is accordingly heavier. Its heaviness, coupled with its relative coarseness, results in its being physically less desirable than acid lead arsenate as a suspension. It contains only about 23% of the arsenic oxide, and kills more slowly than acid lead arsenate. It is less chemically active and is therefore less apt to "burn" tender foliage, but by the same token it is less readily acted upon by the digestive juices of the insect's stomach, and a part of it is apt to pass through the intestinal tract undissolved. In regions of high humidity, where acid lead arsenate sometimes breaks down and causes "burning," as in the spraying of apples in the coastal districts of California, basic lead arsenate is used. From 4 to 5 pounds of the powder is used to 100 gallons of water.

Calcium Arsenate

Calcium arsenate ($CaHAsO_4$) came into extensive use during World War I when the high price of lead salts made as great as possible substitution of calcium arsenate for lead arsenate especially desirable. It has been used in enormous quantities for the dusting of cotton in boll weevil control since 1919. Because of its greater injury to foliage, calcium arsenate has never extensively replaced lead arsenate in codling moth control.

Basic Copper Arsenate

Basic copper arsenate, $Cu(CuOHAsO_4)$, was found to have high toxicity and to possess excellent adherent qualities when tested against the fifth instar larvae of the southern armyworm, Prodenia eridania and the velvetbean caterpillar, Anticarsia gemmatilis. It compared favorably with acid lead arsenate, calcium arsenate, and synthetic cryolite. (276)

Manganese arsenate has been used as an insecticide, although it is less toxic than lead arsenate.

Organic Arsenicals:- Some organic arsenicals, such as m-chlorophenylarsonic acid, phenylarsonic acid, 2-amino-1-naphthalene-arsonic acid and a few other compounds have been found to have very promising insecticidal properties in preliminary investigations (452).

THE FLUORINE COMPOUNDS

Soluble compounds of the highly active and poisonous element fluorine have long been prominent as household insecticides. Within the last 25 years the less soluble fluorine compounds have also gained wide favor as agricultural insecticides. However, they have been subject to residue tolerances nearly as stringent as those for lead arsenate, the tolerance for fluorine being, like the tolerance for lead, about 7 parts per million. The physiological effects of fluorine against warm-blooded animals have been reviewed by McClure (545). Chronic dental fluorosis ("mottled teeth") occurs in areas having an excess of fluorine in the drinking water, notably parts of the southwestern United States and Argentina. As little as 1 p.p.m. in water can sometimes cause this condition in humans. However, teeth ith mottled enamel seem to be no more susceptible to caries than normally calci-ied teeth, and an optimum concentration of fluorine in teeth aids in resisting decay (770).

A review of the literature on fluorine compounds as insecticides by Carter and usby in 1939 contained 700 titles (150). Fluorine compounds are often used in lace of arsenicals when the latter are found to be unsafe to foliage or injurious o the soil, but sometimes the fluorine compounds are found to be more toxic

---

[1]This formula, however, is only an approximation. Basic lead arsenate is a mixture of several compounds.

against the insect in question, as in the case of the Mexican bean beetle, walnut husk fly, cutworms, and grasshoppers. They generally produce very little injury to foliage with the exception of a few plants or trees, such as the peach.

Sodium Fluoride, NaF:- The fluorides are effective as insecticides, but they are too soluble for use on foliage. The only inorganic fluoride of importance as an insecticide is sodium fluoride, which is used almost entirely as a dust, without any diluent, against household insects and poultry lice. Roaches need only to run through the powder, which sticks to their legs and other parts of the body. Some penetrates through the integument, acting as a contact insecticide in this respect, and causes some irritation. The roach attempts to remove the powder by licking it and in so doing it swallows some and eventually succumbs to the poison.

Sodium fluosilicate, Na_2SiF_6:- This compound was found to be effective against the Mexican bean beetle and was later tried against a number of pests, but its commercial use on plants has been rather limited, primarily because of its phytotoxicity. A 0.5% solution of sodium aluminum fluosilicate is used for mothproofing clothing and furniture and sodium fluosilicate is also widely used as a mothproofing agent, particularly because of the great affinity that wool possesses for this compound (344).

Certain investigators at one time advocated the addition of lime to the fluosilicates to neutralize their acidity in solution, but it was shown that this would result in the formation of insoluble calcium fluoride (149):

$$Na_2SiF_6 + 3\ Ca(OH)_2 \longrightarrow 3\ CaF_2 + 2\ NaOH + Si(OH)_4$$

Barium Fluosilicate, $BaSiF_6$:- The solubility of this salt is 0.025 g per 100 ml at 25° C as compared to 0.762 g per 100 ml for sodium fluosilicate, so it can be used on foliage with a greater margin of safety than the latter compound. Barium fluosilicate was found to be nearly as toxic to adult insects as sodium fluosilicate and considerably more toxic than cryolite, although the latter was about as effective against Mexican bean beetle as barium fluosilicate in field trials (582). Both gave good control either as sprays or as dusts. A similar rate of toxic action was found against the walnut husk fly from barium fluosilicate and cryolite (100).

Cryolite (Sodium Fluoaluminate), $3NaF.AlF_3$ or Na_3AlF_6:- The word cryolite comes from the Greek and means "icestone". It is believed that the compound is so named because a lump of the mineral in water resembles a lump of ice. The only known commercial deposit of cryolite is in the southwest coast of Greenland, but large quantities are made synthetically. It is estimated that natural and synthetic cryolite is being used for insecticidal purposes at the rate of 16 million pounds per year. (130)

As seen from the formula, cryolite is a double salt of sodium fluoride and aluminum fluoride in the proportion of 60 to 40 by weight. Its specific gravity is 2.96. The natural cryolite is only very slightly soluble in water (1 part to 2800 parts of water), while the synthetic product is somewhat more soluble (1 to 1639). (130)

Cryolite is not acutely toxic to warm-blooded animals and no human fatality has ever been recorded. Except in California, no poison labels are required on cryolite insecticides. A sheep was given a pound of cryolite in 2 quarts of water at one time, but showed no signs of ill health or loss of appetite. (130)

Brunton (130) lists uses for cryolite against insects attacking 19 important crops in the United States, but states that "DDT will probably affect the use of cryolite for the control of certain insects such as the gypsy moth, white-fringed beetle, potato beetles, bollworm, and codling moth." Among subtropical fruit insects, cryolite is used in the control of orangeworms and the walnut husk fly. Cryolite can be used for codling moth control on apples under arid conditions. It has been stated that in 1944, in the State of Washington alone, nearly 5,500,000 pounds of cryolite was used as compared to about 11,300,000 pounds of lead arsenate

INSECTICIDES USED PRIMARILY AS STOMACH POISONS

Cryolite is compatible with a greater number of insecticides than the other lorine insecticides, but in combination with lime it will form the highly inluble calcium fluoride (583):

$$Na_3AlF_6 + 3Ca(OH)_2 \longrightarrow NaAlO_2 + 3CaF_2 + 2NaOH + 2H_2O$$

In one experiment it was found that a number of commercial fixed copper comnds used in combination with natural cryolite did not reduce the toxicity of latter, but Bordeaux delayed and decreased toxic action against Mexican bean etle larvae. In a single field experiment with Colorado potato beetle, Bordeaux layed toxic action of the cryolite without, however, reducing final mortality en compared to cryolite-fixed copper combinations (938).

Silicofluorides. These compounds in aqueous solutions now make up the greater t of the moth proofing products. Two commercial products contain combinations magnesium silicofluoride and ethanolamine silicofluoride, and two others conln sodium aluminum fluosilicate and lithium fluosilicate respectively. (580)

TARTAR EMETIC

Tartar emetic, potassium antimonyl tartrate $[K(SbO)C_4H_4O_6 \cdot 1/2\ H_2O]$, is a lsonous white crystalline salt which has been used in dyeing as a mordant and in licine as a sudorific and emetic. However, 325 to 650 mg is the usual fatal se for adults, and as low as 48.6 mg (0.75 grain) for a child and 129.6 mg (2 lins) for an adult have been said to be fatal (770). It is soluble in water at rate of 1 part to 12 - 14 parts of water at ordinary temperatures.

Good results with tartar emetic and white sugar were reported in extensive lals made in the control of the citrus thrips, Scirtothrips citri, on lemon trees ring 1937-38. With a spray-duster (p. 280) only 1.5 lb. of tartar emetic and lb. of sugar in 25 to 50 gallons of water were applied to an acre of orange es in the spring after the majority of the petals had fallen. On lemons about ice as much material per acre was required (116). This treatment was very effecve until in 1941 it was noticed that it was beginning to fail in some areas ling to the development of resistance among the thrips. At present tartar emetic be successfully used in only a few districts.

OTHER INORGANIC STOMACH POISONS

A number of other stomach poisons have been used against rodents [thallium lfate, Antu (alpha naphthylthiourea), diisopropyl fluorophosphate, sodium fluoretate], and in limited quantities against insects (cuprous cyanide, cuprous thioanate, phosphorous, lead salts, mercury salts, and zinc phosphide).

ORGANIC STOMACH POISONS

Fixed nicotines:- Although nicotine is very toxic when ingested or absorbed rough the skin, it decomposes rapidly on exposure to light and air. To be effecve as stomach poisons, nicotine formulations must be so prepared that the nicole will retain its insecticidal value for a sufficient period after being desited on the plant surface, for a stomach insecticide must remain toxic until ch time as the insect feeds on the foliage or fruit upon which it is deposited. nicotine must also remain undissolved in the spray tank so that it may be ficiently deposited on the plant surface. As might be expected, the search for ostitutes for lead arsenate in codling moth control has been an important factor the development of new types of stomach poisons, including the nicotine formuions.

Success has been reported in the use of nicotine tannate in codling moth conol. This combination was made in the spray tank by adding 3 pounds of tannic ld and 1 pint of "Black Leaf 50" to 100 gallons of water (419). A "nicotine at" combination has been prepared, apparently involving the formation of a watersoluble nicotine salt (585).

121

nicotine bentonite mixtures are prepared by adding the nicotine to the bentonite suspensions in the spray tank immediately before application. A wetting agent may be added. The nicotine is held very tenaciously by the bentonite particles and is therefore available over a considerable period. Certain tests have been suggested to determine the desirability of different bentonites for fixing nicotine. It was found that the bentonites must have high swelling properties and must be highly sorptive (399).

Codling moth control programs have been developed with nicotine bentonite formulations which were found to be comparable in effectiveness to the usual lead arsenate program in the regions in which the experiments were made, and they possessed certain advantages (838, 457). In addition to its effect as a stomach poison the nicotine bentonite combination also has contact value, as indicated by the experience in the Northwest since the use of DDT in codling moth control. The use of nicotine bentonite periodically throughout the season reduces the wooly apple aphid infestations (647). The addition of oleic acid and aluminum sulfate to nicotine bentonite has been found to greatly increase deposit (5).

Other Organic Stomach Poisons:- Hundreds of organic compounds have been investigated in relation to their value as stomach poisons, and some have shown considerable promise, but have not as yet gained prominence as insecticides. A considerable number of materials known primarily as contact insecticides, such as sulfur, phenothiazine, many nicotine preparations, derris and cubé, hellebore, and many of the new chlorinated hydrocarbons and phosphates, notably DDT, benzene hexachloride and parathion, may also be used as stomach poisons and have been so used in the control of certain insects.

CHAPTER VII
SULFUR COMPOUNDS

INTRODUCTION

The source of sulfur was at one time primarily in the vicinity of extinct or
ctive volcanoes. Much of the sulfur obtained in this way came from Sicily. Be-
inning in 1903, the Frasch superheated steam process[1] (fig. 78) of raising sulfur

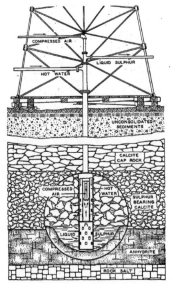

Fig. 78. Sulfur well piping. Courtesy Texas Gulf Sulphur Co.

o the surface from subterranean deposits near the Gulf Coast in western Louisiana
nd Texas resulted in the opening up of an important additional source of this ele-
ent. Sulfur may also be obtained from pyrites, of which Spain is an important
ource. It was estimated that 4,450,000 tons of sulfur was moved in the United
tates in 1947 (33). About 50 million pounds of sulfur are annually used for crop
usting in this country and about half that amount in lime sulfur preparations.

Sulfur occurs in 2 allotropic forms, the monoclinic, a needle-like form, which
s the stable form between 96° and 110° C, and the rhombic or crystalline form,
hich is the only stable form below 96° C. Ground sulfur will again take on a
hombic form if a stabilizer is not added.

[1]Three concentric pipes are driven into subterranean sulfur deposits. Live steam is forced
etween the outer set of pipes to melt the sulfur, which is then forced up between the inner set
f pipes by compressed air delivered down through the center pipe. The molten sulfur is collected
n large wooden bins to cool and solidify. The resulting enormous blocks of sulfur are broken up
ith dynamite and the "crude" product, which is over 99.5% pure, need only be ground to the proper
article size for use as an insecticide.

Sulfur may be used in the elemental form as a finely divided dust; made wettable and used as a spray; combined with nicotine, rotenone, calcium arsenate, etc.; or combined chemically with calcium or other elements and used as a spray. Sulfur is a constituent of a large number of insecticidal compounds, especially among the organic compounds.

INSECTICIDAL FORMS OF ELEMENTAL SULFUR

The greater part of the sulfur used for insecticidal and fungicidal purposes is used in the elemental form. After it is obtained from its source, it must be reduced to a powder by grinding, and is then called ground sulfur.

Sulfur may be ground by conventional grinding methods, either a fixed hammer mill or a centrifugal roller impact mill. Wetting agents (when wettable sulfur is desired) and conditioning agents may be added to the sulfur introduced into the mill. The conditioning agents may be such materials as gypsum, bentonite, talc, lime, calcium stearate, magnesium carbonate, neutral clay and tricalcium phosphate. They may be added in quantities ranging from a mere trace to 15% or more. They are added to make the sulfur flow more freely if it is used as a dust. A stream of air containing 10 to 15% carbon dioxide is forced through the machine by a fan and carries the sulfur upward through sets of rapidly revolving blades that throw the coarser particles out of the air stream by centrifugal force. These fall back to be reground while the finer particles move upward into a collector. Such sulfur is said to be "ventilated". In some processes the sulfur is not "ventilated" but is ground continuously until the entire lot is of acceptable fineness (387).

Sulfur may also be ground by the "micronizer method", a special air-grinding method producing very small particles. The grinding is accomplished by the impact of particle against particle. The larger particles are ovoid in shape and serve to identify the micronized sulfur products.

Sulfur may be extremely finely divided into spherical particles by the Grinrod process, a process of emulsification and atomization of molten sulfur. The emulsifying agent is adsorbed on the surface of the sulfur particles and serves to make the finished product wettable. The finished product may be a paste or a dry wettable powder. The sulfurs produced by the Grinrod process have been found to have the smallest and most uniform particles of all artificially fractionated sulfurs. (387)

A powder of small crystals is produced by the condensation of the vapor of sulfur in the process of refining by distillation, and is known as flowers of sulfur or refined flowers of sulfur. The average particle size of this type of sulfur is larger than that of the other forms and it is consequently not as widely used as formerly. Flowers of sulfur have been recommended as being especially suitable for burning in SO_2 fumigation (217). The term flowers of sulfur should not be confused with sulfur flour, which is merely another name for ground sulfur.

Flotation sulfur is obtained as a by-product in the confined heating or destructive distillation of coal to obtain coal gas and coke. The coal gas must be purified by removing the small amount of hydrogen sulfide present before it can be pumped into the mains.

Colloidal sulfur is a term applied generally to the hydrophylic (wettable) sol made by passing hydrogen sulfide through a saturated solution of sulfur dioxide in water until the odor of the SO_2 can no longer be distinguished (227):

$$2\ H_2S + SO_2 \longrightarrow 3\ S + 2\ H_2O.$$

The reaction of sodium thiosulfate and sulfuric acid or other mineral acid will also result in this type of hydrophylic colloidal sulfur:

$$Na_2S_2O_3 + H_2SO_4 \longrightarrow Na_2SO_4 + SO_2 + S + H_2O.$$

SULFUR COMPOUNDS

Colloidal sulfur sols made according to the above reaction are sometimes called Oden sols after the Swedish investigator Sven Oden.

A hydrophobic colloidal sulfur may be prepared by precipitating lime sulfur solution by adding concentrated hydrochloric or sulfuric acid. There is a tendency for the sulfur particles to aggregate to the point of precipitation in a few days. The addition of other colloids such as glue or gelatin delays this aggregation of particles especially if added before the acid. This type of sulfur has been known as precipitated sulfur (227).

The hydrophylic colloidal sulfurs have been found to be the more effective, and as stated before, the term colloidal sulfur ordinarily refers to the hydrophylic or wettable type. The term is sometimes used loosely to apply to any extremely finely ground sulfur and to sulfurs which are not actually colloidal.

The colloidal and flotation sulfurs and those atomized by the Grinrod process have the smallest particle size of the elemental sulfurs and are generally near or below 5 microns in diameter. The sulfurs from the micronizer reduction mill are not quite as fine, but are generally below 5 microns, while in the conventionally milled products the majority of the particles are below 12 microns in diameter and are high grade products, but are not fine enough for some purposes (387).

Sulfur to which oxidizing agents such as potassium permanganate have been added is called oxidized sulfur. (515)

Sublimed sulfur is made by heating crude sulfur in the presence of air until a red liquid is formed. The sulfur vapor from this liquid is forced against a cold surface where it condenses and returns to the solid state. The resulting particles range from 2 to 40 microns in size.

Bentonite sulfur is produced by the fusion of elemental sulfur and bentonite. Enough heat is applied to keep the sulfur in the molten state. The elemental sulfur is adsorbed on or within the lamellae of the bentonite, thus causing the mixture to act like bentonite when dispersed in water. The processed product is granular, free flowing, approximately one-third sulfur and two-thirds bentonite. It is added directly to the spray tank by sifting it on the water while it is being agitated. This product does not require a conditioner.

Making Sulfurs Wettable:- In order that the individual sulfur particles may be surrounded by water, a wetting agent must be added to reduce the high interfacial tension sulfur/water. At first glue and soap were the most widely used materials for this purpose, and later calcium caseinate, plant gums, resins, saponins, licorice root extract, and sulfite liquor were used. Nowadays a large number of complex organic surface active agents, including higher alcohol sulfates and sulfonic and succinic acid derivatives, are used (387). These may be added either to the sulfur or to the water, but are usually included in the finished "wettable sulfur" product. The sulfur should be wetted rapidly and completely when added to the water. The wetting agent should not cause the sulfur to lump or cake before it is used and should not induce excessive foaming in the spray tank or cause excessive run-off of the spray. The wetting agent should be stable in hard water, a property not possessed by the majority of soaps. It should not react with or interfere with the action of the other insecticides or fungicides which may at times be used along with the sulfur.

The grower may have occasion to prepare his own wettable sulfurs. A small batch of wettable sulfur may be easily prepared as follows:

Let us suppose that the amount of wettable sulfur required for a particular purpose is 5 lbs. per 100 gallons and we wish to prepare enough of the material for a 400-gallon tank. Pour 2 gallons of water into a large bucket. Stir in 2 lbs. of calcium caseinate. Add 20 lbs. of dusting grade sulfur and stir to a smooth paste. An effective stirring device can be made by bending bailing wire back and forth into a "paddle" of wire about a foot long. The resulting paste can be poured into a 400 gallon tank of water.

Toxicology:- The theories on the nature of the toxic action of sulfur may be designated as (1) the volatilized sulfur, (2) the sulfur dioxide, (3) the penta-thionic acid, and (4) the hydrogen sulfide theories (770, 344). Toxic action may, of course, involve more than one of these phenomena at a time. Regardless of the mode of action of sulfur, however, investigators are agreed that insecticidal effectiveness of this element increases with decrease in particle size to a certain point, both through better adherence of the sulfur to the foliage and through increased toxicity. Because of their more rapid disappearance from the foliage, the more finely divided sulfurs may not retain their insecticidal effect as long as might be desired, in which case the addition of a certain amount of sulfur of larger particle size may prolong their usefulness. Thus ordinary dusting sulfur, or even flowers of sulfur, may be added to colloidal sulfur, flotation sulfur, or lime sulfur to prolong their effectiveness.

The higher the temperature the greater is the insecticidal efficacy and phyto-toxicity of sulfur. It is believed to be ineffective below a temperature of 70° F. On the other hand, the higher temperatures encountered in California are

apt to result in severe injury (fig. 79). In a rather extensive investigation of the problem of phytotoxicity (949) it was concluded that the fungicidal action of sulfur cannot be ascribed to pentothionic acid as had previously been reported, and that hydrogen sulfide is highly toxic to the spores of fungi. It was postulated that the vapor given off by elemental sulfur is reduced to H_2S in fungus spores or leaves and that H_2S is the phytotoxic compound. The leaves and spores of all species of plants tested produced H_2S when in contact with sulfur.

Fig. 79. Lemons showing "sulfur burn" from a sulfur dust. Photo by Roy J. Pence.

An exhaustive investigation of the phyto-toxic action of sulfur has recently been made by Turrell (897). It was shown that citrus fruit produces hydrogen sulfide in relatively large amounts, sulfur dioxide in relatively small amounts, and sulfuric acid. Turrell and Chervenak (898) using radioactive sulfur (S^{35}), demonstrated that when lemons are dusted with sulfur and incubated at warm temperatures (106°F), a large proportion of the S in the H_2S formed is derived from the S applied to the fruit. If relatively large amounts of S are applied, the SO_4 formed in the peel may be derived largely from the S applied, but the SO_2 formed is derived largely from a source within the fruit. Elemental S vapor penetrates lemons and produces compounds similar to those produced by fruit in contact with elemental S. Radioautographs were made of the peel of lemons dusted with radioactive S and incubated at 106° F. These suggest the incorporation of the S into the tissue proteins (898). Phytotoxic action was believed to be due to both the action of sulfuric acid formed in the tissues from the sulfur applied and the effect of high temperature. Apparently the combination of high temperature and sulfuric acid damages the protein components important in metabolism.

Uses:- Elemental sulfurs used as dusts or in sprays have been particularly useful as acaricides, but they are also used against a number of other species of insects. Sulfur is often used as a diluent for other insecticides or as a combination diluent and insecticide. As a diluent-insecticide some workers believe sulfur has not yet been exploited as extensively as its usefulness would justify. For example, sulfur will kill several species of aphids (460). This has apparently been overlooked by previous investigators. As an aphicide the sulfur must make contact with the insects and the temperature must be near 70° F or higher. Its action against aphids is very slow.

In the control of citrus thrips, a dust is being used containing 2% DDT and 85% sulfur. This combination is much more effective than either insecticide alone used in the same amount per acre. Sulfur dust has been used alone for the control of citrus thrips, citrus red mite, and citricola scale, and as a supplementary treatment in the control of black scale, but in recent years has largely given way to more effective treatments. Great quantities of sulfur are used in the control of grape mildew.

LIME SULFUR

In preparing lime sulfur solution in the laboratory, 50 grams of quicklime (CaO) may be added to a liter of water in a glass beaker. This is called "slacking" the lime. Heat is produced and calcium hydroxide, $Ca(OH)_2$, is formed. Then add 100 grams of sulfur and boil while agitating until the ingredients have gone into solution. A reddish brown liquid will be formed, known as lime sulfur, and a yellowish sediment will settle to the bottom of the beaker. According to Tucker (894), under laboratory conditions the following reaction takes place:

$$3\ Ca(OH)_2 + 12\ S \longrightarrow 2\ CaS_5 + CaS_2O_3 + 3\ H_2O.$$

Commercial lime sulfur, however, contains only a fifth of the amount of calcium thiosulfate indicated in the above formula, owing to a secondary reduction of the thiosulfate to elemental sulfur by H_2S. Also the ratio Ca:S is closer to 1:4.7 than to the 1:5 that would be inferred from the above formula. The calcium di- tri-and tetrasulfides are apparently found under special conditions.

In boiling, the solution goes through a gradation of colors. The boiling should be continued until the deeper shades are attained. The intensity of color is largely a measure of the amount of polysulfides and these are considered to be the most valuable constituents. The structures of the polysulfides of lime sulfur are discussed by Tucker (894) and by Frear (344).

The specific gravity of sulfur is most commonly expressed by the Baumé scale. On the Baumé scale, zero indicates a density of 1.0. A 33° Baumé solution has a specific gravity of 1.295. Homemade lime sulfur solutions usually are 27° to 28° Baumé, and consequently should not be diluted as much as the commercial solutions in the preparation of the dilute spray. Since specific gravity is correlated with polysulfide sulfur content in pure lime sulfur solutions (886) it would appear that the Baumé reading is a valuable criterion of the effectiveness of a solution.

Toxicology:- Oxidation of lime sulfur after it is deposited on the plant surface yields elemental sulfur, which would presumably result in toxic action equivalent to that of colloidal sulfur. However, the lime sulfur solution has good adhesive properties, and the possibility of initial caustic action against the insect, owing to the high alkalinity of the lime sulfur, must be considered. The possibility has been considered that when lime sulfur is applied to scale insects, oxygen may be used up so rapidly by the lime sulfur as to deprive the insect beneath the scale covering of this gas (769). In one experiment it was discovered that adult female San Jose scales surviving a lime sulfur spray did not produce young (648). Likewise it has been noted that while the eggs of the forest tent caterpillar may hatch after a lime sulfur spray, the larvae fail to develop (470).

Lime sulfur is known to be more toxic to foliage than is elemental sulfur.

POTASSIUM-AMMONIUM-SELENO-SULFIDE

This compound, having the formula $(KNH_4S)_5Se$, probably owes its toxic action more to the selenium than to the sulfur, but it acts much like a sulfur compound. It has been used primarily as an acaricide. The proprietary product Selocide, a 30% solution of a mixture of potassium hydroxide, ammonium hydroxide, sulfur, and selenium in proportions corresponding to the above empirical formula was used in dilutions of 1:600 and 1:800 against the citrus red mite, Paratetranychus citri, on citrus and the Pacific red spider, Tetranychus pacificus, on grapes. The effectiveness of the material was increased by the addition of lime sulfur, wettable sulfur, or oil. Despite the fact that it was shown that no health hazard was involved when Selocide was used in the recommended manner (445), the material never gained official acceptance on citrus and grapes and was never widely used on these crops.

PHENOTHIAZINE

This compound, also called thiodiphenylamine, is an organic sulfur compound. It is crystalline, colorless when pure, but ordinarily appears light yellow, or it may be greenish in commercial preparations. It has a melting point of 180° C. It is neutral, insoluble in water, but is slightly soluble in petroleum and the common organic solvents.

Phenothiazine was found to be more toxic to mosquito larvae than rotenone, being very effective at a concentration of 1:1,000,000 (144). It was found

Phenothiazine

to have considerable promise against codling moth by some investigators, but has generally been found to be erratic in performance, especially as between materials obtained from different sources. Consequently it never attained a prominent place in codling moth control. It has been used successfully for the control of the horn fly and the screw-worm, insects attacking domestic animals, and as an anthelmintic, especially against the nodular worm of sheep. Phenothiazine has the advantage of not being very toxic to human beings.

THIOCYANATES

It was pointed out by Murphy and Peet (633) that ordinarily minor alterations in the organic insecticide molecule result in compounds of practically no insecticidal value and that it is usually difficult to establish a relationship between chemical structure and insecticidal action. They suggested as an exception to this general situation the insecticidal activity of the thiocyanate group -S-C≡N, which persists despite important changes in the molecule to which this group is attached. This, they believed, permits of modifications making possible the development of insecticides possessing special properties for special purposes. They showed that an aliphatic thiocyanate preparation they prepared could successfully compete in insecticidal effectiveness with nicotine and pyrethrum against a number of representative insect types. In addition the concentrated insecticide did not deteriorate with heat, age, and other conditions which result in the deterioration of such insecticides as pyrethrum and rotenone.

In accord with the wide latitude of insecticidally active molecular arrangements in the thiocyanate molecules mentioned by Murphy and Peet, a large number of thiocyanates of the aliphatic, and to a lesser extent of the aromatic, series have been developed as floricultural, horticultural and household insecticides. Their effectiveness against mealybugs (631) is especially noteworthy because of the difficulty ordinarily experienced in killing these troublesome pests in greenhouses. This fact, together with their effectiveness against mites (632), makes them a particularly useful group of insecticides for greenhouse use. Their phytocidal action, however, makes it necessary that they be used with care, and in some cases this has been the limiting factor in their commercial use. They were a fortunate addition to our supply of effective contact insecticides during World War II when the stocks of rotenone and pyrethrum were seriously depleted in this country. The latter insecticides could often be used with the thiocyanates to reduce the quantity required for effective pest control. In some cases they could be replaced completely. Only a few of the more important of the thiocyanates will be discussed here.

Toxicology:- The thiocyanates are known to be paralytic poisons affecting the cerebral axis of both warm-blooded and cold-blooded animals (869). Hartzell and Wilcoxon (411) dissected the nerve cord of a mealworm larva (Tenebrio molitor) killed by ϒ-thiocyanopropyl-phenyl-ether applied externally and found a cellular degeneration in cross sections of the abdominal ganglia by using the toluidine blue histological technique. Although, as stated above, the thiocyanates are known to be toxic to warm-blooded animals, they are relatively non-toxic to human beings under the conditions of their ordinary commercial usage.

128

SULFUR COMPOUNDS

Formulations:- Lauryl thiocyanate, $CH_3(CH_2)SCN$, which may be produced by the reaction of lauryl chloride and sodium thiocyanate, is marketed under the trade name Loro. It was considered the most effective of the aliphatic thiocyanates against aphids (97).

Beta-butoxy-beta-thiocyanodiethylether, $C_4H_9OCH_2CH_2OCH_2CH_2SCN$, marketed under the trade name Lethane, has been used effectively against aphids, mealybugs, thrips, leafhoppers, mites, house flies, and many other pests. For house flies it has often been combined with pyrethrum and rotenone.

Among the aromatic thiocyanates, a terpene thiocyano ester, isobornyl thiocyanate, is perhaps the best known, being marketed under the name of Thanite. Thanite is a blend of thiocyanoacetates of secondary and tertiary terpene alcohols. It is prepared by reacting monochloroacetic acid with camphene. It is chemically stable under all normal storage conditions. Thanite is soluble in all the usual organic solvents.

Thanite

In one experiment no benefit was derived from the addition of various pyrethrin, rotenone, Lethane, or pure oil preparations to Thanite. The Thanite-pyrethrum combination was not as effective as Thanite alone besides being subject to loss of toxicity from decomposition. (696, 598) Thanite is used most commonly in household and livestock sprays, being effective against a wide variety of pests. Residual effect lasts as long as 7 hours after spraying.

Fenchyl thiocyanoacetate is another aromatic thiocyanate, similar to Thanite in its properties.

Extensive investigations of both acute and subacute toxicity of Thanite showed that this insecticide will produce primary irritation to laboratory animals, but no progressive local damage. Upon removal of the animals from exposure, recovery from the ill effects is rapid. In an independent toxicological laboratory, human subjects were exposed for 2 weeks in a dense fog of Thanite spray for 1/2 hour each day without recognizable injurious effects. It has been concluded that Thanite in concentrations recommended in household or industrial sprays produces no detectable injurious effects on human beings.

129

CHAPTER VIII
INSECTICIDES OF PLANT ORIGIN

A considerable number of organic contact insecticides derived from plants have had a prominent place in the control of insect pests and are still widely used, although since World War II they have gradually declined in importance in comparison with the synthetic organic compounds such as the chlorinated hydrocarbons and organic phosphates. Along with the sulfur compounds and the oils, they may be said to be almost exclusively of value because of their contact action and are rarely used as stomach poisons. With the exception, of course, of the fumigants, they are unique among insecticides in having practically no "residual effect" in insect control, since they decompose quickly in exposure to air and light. Obviously, direct contact with the insect is nearly always essential to mortality.

NICOTINE

Introduction:- Water extracts of tobacco and the dried powdered leaves were noted for their insecticidal effect, particularly against aphids, even before it was known that nicotine ($C_{10}H_{14}N_2$) was the toxic principle. In 1941, it was estimated that 2,406,000 lbs. of nicotine insecticides calculated in terms of nicotine sulfate solution of 40% nicotine content, was used in the United States (567). Since World War II, the amount of nicotine available has never been sufficient to meet the demand, and although various preparations appear to be quite generally available in small-package quantities for garden use, it is often not possible to obtain sizeable quantities for large scale pest control work. In the United States the stems, leaf midribs, and low grade leaves of Nicotiana tobacum were once considered to be unsuitable for making smoking or chewing tobacco and were used as a source of nicotine, but now they are sent to Europe for use as smoking tobacco, thus seriously limiting the domestic supply of nicotine. In Russia, Nicotiana rustica, which contains a high per cent of nicotine (10-12%), is cultivated as a source of this alkaloid.[1]

Chemistry:- The alkaloid, 3-(1-methyl-2-pyrrolidyl) pyridine, with the empirical formula $C_{10}H_{14}N_2$, was first synthesized by Pictet and Rotschy in 1904 (344). It is a colorless, odorless liquid boiling at 247.3°C. However, one almost invariably sees a darkened solution, for the compound darkens upon exposure to air, and may become dark brown or nearly black.

Nicotine (β-form)

Nicotine salts may be obtained from high-nicotine tobacco or from tobacco waste by extraction with water, after which the nicotine is liberated with alkali and steam-distilled. There are also other methods in commercial use for obtaining nicotine from tobacco. Various methods of purification are employed to remove small amounts of related alkaloids from the extract.

In the manufacture of nicotine sulfate, two processes may be used: (1) Ground tobacco stems and other by-product tobacco may be placed between layers of lime and charged with steam. The lime activates the tobacco and nicotine passes off in volatile form. It is then collected, condensed to a liquid, and charged with sulfuric acid. (2) Sulfuric acid is added to a mixture of by-product tobacco and kerosene. After vigorous agitation the kerosene is drawn off from the top of the mixture. (578)

---

[1] "Accurately defined, an alkaloid is a naturally occurring, usually heterocyclic, optically active, nitrogenous base of relatively high molecular weight and showing marked physiological activity ..... Alkaloids usually occur in the plant as salts of the organic acids characteristic of that plant" (770). Some examples of alkaloids, other than nicotine, are coniine, the alkaloid obtained from hemlock which Socrates was forced to drink; quinine, from the Chinchona tree; atropine, from nightshade; cocaine, from cocoa leaves; morphine, from the poppy; and strychnine, from the seeds of Strychnos voxomica.

INSECTICIDES OF PLANT ORIGIN

Formulations:- The commercial extracts of nicotine may be obtained in solutions containing 40%, 50% or 95% free nicotine or in nicotine sulfate $[(C_{10}H_{14}N_2)_2 \cdot H_2SO_4]$ solutions containing 40% nicotine. In the former the nicotine is more volatile and kills more quickly. It is also more dangerous to human beings. The nicotine sulfate has a longer residual effect, although this is short in comparison with some of the new synthetic organic insecticides which are to be discussed later. A common concentration for nicotine sulfate is 1-800, that is, one gallon to 800 gallons of water, such as is used in citrus aphid control. Against some pests, however, concentrations as low as 1-1600 are used. In general, basic compounds release the free alkaloid, which is generally much more effective than nicotine sulfate (738). A basic spreader, such as calcium caseinate or soap, affords not only the advantage of higher pH, but also an improved wetting and spreading.

To prolong the effectiveness of nicotine, "fixed nicotine" preparations (p. 122) have been used.

In the preparation of dusts, the dry roots, stems, and wastes from the tobacco plant are finely-powdered to form "tobacco dusts"; which are of unknown and variable nicotine content. More commonly used are the "nicotine dusts" which consist of a suitable diluent mixed with nicotine sulfate or nicotine extracts. The diluents may be (a) adsorptive carriers, such as bentonite, talc and fuller's earth, which decrease the volatilization of the nicotine, (b) inert carriers, such as gypsum and sulfur, which are neutral or inactive toward nicotine or (c) active carriers such as hydrated or air-slaked lime, which free the nicotine in nicotine sulfate in the presence of moisture. The dusts are most effective on a hot dry day and with a minimum of air movement.

Nicotine may also be used as a fumigant. To generate the toxic fumes, tobacco may be burned or boiled in water. Concentrated extracts of nicotine may be vaporized by heat, as on a hot plate, in buildings or in portable machines (usually with the aid of a canvas hood dragged over the fumigated plants). Six ml of "Black leaf 50" (50% nicotine) vaporized under a tent covering a large apple tree was sufficient to kill all adult codling moths, while 30 ml of HCN was required for equal effectiveness (820).

Nicotine can be dissolved in spray oils at a concentration of close to 1% at room temperatures. Greater concentrations can be obtained by the use of intermediary solvents. Such solutions have been atomized on plants for the control of insects and mites (742).

At lower concentrations the solubility of nicotine in oil is about the same as its solubility in water (657). The difficulty in the use of nicotine-oil in a dilute spray is that there is usually about 50 times as much water as oil in the spray tank when the oil is used at concentrations that the tree will tolerate. Since the nicotine is about equally soluble in oil and water, nearly all the nicotine will be taken up by the water phase of the emulsion. It is of interest to note, however, that in one experiment, a 95% nicotine solution was added to spray mixtures containing 1, 1.5 and 2% of a solution of 90% spray oil and 10% of a dibutyl phthalate-trichlorethylene solution in the control of California red scale. The solvents were added to make the test comparable to one in which they were used to dissolve derris extractives in the oil solution. The nicotine was added to the spray at the rate of 1 part of nicotine (100%) to from 1,500 to 2,000 parts of spray. Substantial increases in mortality of the scale were obtained (209).

When a tree is sprayed, the oil is for the most part deposited on the foliage and other tree surfaces while the water nearly all runs off and is lost. If nicotine could be satisfactorily held in the oil phase of an emulsion it could be efficiently deposited and would probably be effective in the control of many insect pests.

Nicotine can kill insects by penetration of the integument, even by way of parts as remote from the vital centers as the tips of the antennae (663). Nicotine

SUBTROPICAL ENTOMOLOGY

may act as a contact poison, a stomach poison, and as a fumigant. A number of investigators have shown that nicotine is more effective than its salts, thus the free alkaloid is more effective than nicotine sulfate. The addition of an alkaline material to nicotine sulfate will release free nicotine molecules and thus increase the insecticidal effectiveness of the spray. Thus one may add soap to a nicotine sulfate solution to increase its potency. If nicotine and nicotine sulfate are injected into the body cavity of the insect, they are found to be equal in toxicity at a given molarity, which indicates that the superiority of the nicotine to the nicotine sulfate is caused by the greater speed of penetration of the nicotine molecules (275).

Toxicology:- Nicotine is very poisonous to man and other warm-blooded animals. About 40 mg is a fatal dose for man. Nicotine preparations used as contact poisons kill insects by paralysis of the nervous system, acting (1) through the body wall, (2) by particles of spray or dust entering the spiracles, or (3) by vapor entering the spiracles.

NORNICOTINES

The d and l nornicotines correspond to the d and l nicotines except that a hydrogen atom replaces the methyl group attached to the pyrrolidine ring. Laboratory tests conducted by the U.S.D.A., Bureau of Entomology and Plant Quarantine, have shown that nicotine and nornicotine are similar in toxicity (89). Nornicotine may be present in considerable amounts in nicotine sulfate solutions and might be expected to increase the insecticidal effectiveness of the latter (98).

ANABASINE

Neonicotine was found in 1926 to be the one of 6 possible isomeric dipyridyl derivatives which had the greatest toxicity (739). Later this compound was found in the weed Anabasis aphylla Linn., a perennial of the family Chenopodiaceae growing in southern Russia and adjacent regions and was named anabasine (344). Anabasine was found to be present in the leaves and roots of the tree tobacco, Nicotiana glauca, which was introduced from South America and now may be found throughout southwestern United States along stream beds and in newly cleared areas where it is not in competition with native vegetation (785). It has caused the death of several persons who consumed the leaves as "greens". Anabasine is about equal to nicotine for the majority of insecticidal purposes, and commercial anabasine sulfate appears to be equal or superior to nicotine sulfate as an aphicide (345).

OTHER ALKALOIDS

Hellebore:- This is a product of the false hellebore (Veratrum), which belongs to the lily family. It is primarily located in the rhizomes and rootlets. The veratrin alkaloids, of which there are 5, are the toxic principle, but only one, protoveratrin ($C_{32}H_{51}NO_{11}$) is very poisonous to insects.

Sabadilla:- Sabadilla seeds are obtained from another lilaceous plant, Schoenocaulon spp. Interest in sabadilla was greatly stimulated during the war years because of the critical shortage of insecticides at that time. An extensive study of the subject was made by Allen et al (8).

The seeds of Schoenocaulon grow on a spike which superficially resembles those of the barley plant. In fact, the Mexicans sometimes use the term cebadilla or cebadilla mexicana in referring to sabadilla, the term cebadilla being the diminutive form of the word cebada, meaning barley. Our word sabadilla apparently derives from the Spanish word cebadilla. Three species are known in the United States, 19 from Mexico and one, S. officinale, from Costa Rica, El Salvador, Guatemala, Honduras, Peru and Venezuela. Most of our supply of seeds comes from Venezuela.

According to Allen et al, sabadilla has been found to be highly effective against many species of insects in many parts of the world, and has been used as a

132

contact insecticide, stomach poison and fumigant. Its action in the latter capacity was made possible by the vapor from heated seed, which was found to be almost as effective as SO_2 against bedbugs, cockroaches, and mosquitos. The toxic ingredient is said to be veratrin, a mixture of several alkaloids. The empirical formula for one of these alkaloids, cevadin, is $C_{32}H_{49}NO_9$. (770) When extracting in petroleum, heat must be used to make available the toxic constituents. Also the addition of alkaline materials increases toxicity when the seeds are ground to form insecticidal dusts. In this country, sabadilla has been particularly useful as a control for squash bugs, Anasa tristis, Lygus bugs and pentatomid bugs, being about as effective as DDT.

Sabadilla applied with sugar as a bait spray is showing considerable promise as a control for citrus thrips, and extensive experimentation with the insecticide in this connection is now underway (293).

Ryania:- This insecticide obtains its name from a genus of tropical American shrubs and small trees of the family Flacourtiaceae. A cooperative research project between the Research Laboratory of Merck & Co. and the Department of Entomology of Rutgers University led to the discovery that these plants contain insecticidally effective active principles. These active principles are alkaloids, and a compound known as ryanodine was found to be of primary importance. The insecticidal principles of Ryania spp. were found to be located in the roots and stems, with no significant concentration in the leaves. The principal source of the insecticidal material has been the stems of Ryania speciosa.

Ryania is toxic both as a contact and as a stomach poison. The toxic action appears to be as a general depressant, causing cessation of feeding, locomotion, and reproductive activity, but permitting the insect to remain in a semi-moribund state for a relatively long period before it dies. This delayed mortality, such as that which is frequently found when using DDT, should be recognized as a typical reaction when evaluating the effectiveness of the material.

An advantage possessed by Ryania is that it is much more stable on exposure to light and air than pyrethrum or rotenone, and has a longer residual effect than is ordinarily obtained from insecticides of plant origin. It is stable under dry storage conditions. Ryania has been successfully used as a spray with sulfur and copper fungicides and in dusts with all of the commonly available inert diluents. Its activity appears to be partially masked when used in sprays containing mineral oils. There have been no reports of plant damage from Ryania and experiments have shown it to have, in general, a lower acute oral toxicity than derris or cubé to a number of warm-blooded animals with which tests have been made. Ryania shows no outward evidence of cumulative toxicity.

To date, no proved method for chemical assay of Ryania has been worked out, but biological assay is employed, using clothes moth larvae as the test insects and the damage to woolen fabrics as the criterion of effectiveness.

Ryania is used commercially under the trade name of Ryanex. One formulation is a dust containing 40% ground stem of Ryania speciosa and the other is a wettable powder containing 100% ground stem of R. speciosa. Pepper and Carruth (685) found Ryanex dust gave an "excellent degree of practical control" of the European corn borer under commercial conditions. Ryania insecticides have also been found to be toxic to many other insects, including the cotton bollworm, Japanese beetle, corn earworm, codling moth, oriental fruit moth, chinch bug, Mexican bean beetle and elm leaf beetle.

PYRETHRUM

Introduction:- The toxic principle of pyrethrum is obtained from the flower head of certain plants of the family Compositae, genus Chrysanthemum. We find in the Arabian Nights that the Arabs used pyrethrum flowers for killing insect pests in ancient times, but the modern history of pyrethrum seems to begin about 1800 in the Caucasus region of Asia Minor. Since the toxic principle of the flea-and louse powders prepared in that region was kept a secret for about 50 years, exorbitant prices were asked for them. After it became widely known that these powders consisted of ground pyrethrum flowers, the price became reasonable and huge quantities were used. Pyrethrum flowers were first introduced into the United States in 1861. Before the last war, as much as 20,000,000 pounds was imported in a single year. An exhaustive treatment of the origin, culture, sources, chemistry, commercial preparation and uses of pyrethrum is contained in a book entitled Pyrethrum Flowers by C. B. Gnadinger.

By far the most important of three insecticidal species is Chrysanthemum cinerariaefolium, which resembles the field daisy. It is 18 to 24 inches high. The flower heads consist of a rounded receptacle bearing a disk of numerous yellow flowers and a circle of white or cream-colored ray flowers. Japan, Yugoslavia, Kenya and Italy have been the most important pyrethrum-producing countries, but since the last war nearly all of the flowers imported into this country have come from Africa, especially Kenya.

The flowers are harvested by hand or mechanically, after which they are dried and baled for shipment. The maximum pyrethrum content is found in the flower head at about the time the flower becomes fully open. From dried and powdered pyrethrum, the insecticidal principles, pyrethrum I ($C_{21}H_{28}O_3$)[1] and pyrethrin II ($C_{20}H_{28}O_5$), may be lost to the extent of 9 to 20% (depending on the original content) in 6 months, from 12 to 25% in 12 months, and up to 40% in 24 months (934). When pyrethrum is exposed to air and light in thin layers, the loss of toxicity is especially rapid, but it can be retarded by the addition of antioxidants (866).

Chemistry:- After many attempts to discover the toxic principle of pyrethrum had failed, Staudinger and Ruzicka succeeded in isolating 2 active principles. These were chrysanthemum monocarboxylic and chrysanthemum dicarboxylic acid mono methyl esters and were referred to by the above investigators as pyrethrins I and II. They published a series of papers describing their work in the Swiss chemical journal Helvetica Chimica Acta in 1924. These studies have been discussed by Gnadinger (362). Pyrethrin I and Pyrethrin II are viscous liquids, the former boiling at 150° C under high vacuum and the latter decomposing when distilled in vacuum. Pyrethrin I has usually been found to be a little more toxic to insects than Pyrethrin II. Neither is soluble in water, but they are both soluble in the usual organic solvents. Being esters, the pyrethrins in contact with water are partially decomposed into their component parts, alcohol and acid. The presence of an alkali causes this reaction to be irreversible, for the liberated acid is bound by the alkali as fast as it is formed.

Pyrethrin I

It has been advised that soaps should not be used with pyrethrum sprays (747). In actual practice, however, soaps are being used. Obviously, in such cases the diluted spray mixture should be used immediately and should not be allowed to stand. Also, non-aqueous extracts have been safely combined with soaps and the dry powdered pyrethrum flowers have been prepared into dust formulations with hydrated lime as a carrier.

Pyrethrin II

[1]The empirical and structural formulas of the pyrethrins were obtained by correspondence of January 26, 1949, from Dr. F. B. La Forge.

INSECTICIDES OF PLANT ORIGIN

Formulations:- The powdered pyrethrum flowers may be used as dusts alone or diluted with a suitable diluent. Such dusts are effective against some insects, but the efficiency of the product is reduced because the pyrethrins are held within the cell walls of the flower material. These pyrethrins may be extracted with kerosene or alcohol and then mixed with the diluent, resulting in a much more effective dust.

Extracts of pyrethrum in kerosene, alcohol or acetone may be diluted with water and used as sprays.

More pyrethrum is used in fly sprays and mosquito sprays than in any other way. Such solutions may be made by extracting the toxic principles from the coarsely ground flowers by means of kerosene or other hydrocarbons and the solutions are then usually diluted to a uniform pyrethrin content of 0.1%. According to another method of preparation, an extraction of the pyrethrins and oleoresins is first made by means of a more highly efficient solvent, such as ethylene dichloride. This solvent is then evaporated and the remaining residue is diluted with kerosene or other hydrocarbon. In the control of leafhoppers on grapes, such solutions have been widely used, usually with the addition of a heavier hydrocarbon besides the kerosene. Other compounds may be added, such as antioxidants, stabilizers, perfumes (in household solutions) and compounds which have been found to increase the insecticidal effectiveness of the pyrethrins. Pyrethrum cannot be packed in copper, zinc, or galvanized lined containers because contact with these metals causes decomposition. Bags must exclude air and moisture to be suitable containers.

Pyrethrum remains unsurpassed to this day from the standpoint of harmlessness to warm-blooded animals and its ability to cause rapid paralysis of insects at extremely low concentrations. However, to obtain a high kill from pyrethrum, rather than merely a paralysis, the concentration required may in some instances make the treatment too expensive. Much has been accomplished along the line of extenders which become highly effective when used with only a trace of pyrethrins. The most effective pyrethrin extenders may be found among certain organic compounds possessing a methylenedioxyphenyl group, which are readily obtained from safrole, a constituent of the root bark of certain Lauraceae, and from piperine obtained from the black pepper. In addition, the sulfoxides and sulfones produced by oxidation of the mercaptansafrole and isosafrole addition products have proved to be highly effective as extenders (865).

The demand for pyrethrum was greatly stimulated by the development of the liquified gas aerosol bomb. In the original formula there was 4% pyrethrum extract (20% pyrethrins), 6% sesame oil and 90% Freon-12 (dichlorodifluoromethane). About 2/3 pound of pyrethrum flowers was used per bomb. Later, when DDT was added to the solution, the pyrethrum extract was reduced 1/2%. In 1947 it was anticipated that 15,000,000 high pressure 1-pound bombs and 30,000,000 low pressure 12-ounce cans, all containing some pyrethrum, would be necessary for civilian use.[1]

Toxicology:- Because of its harmlessness to warm-blooded animals, pyrethrum has been widely used under circumstances in which poisonous insecticides, such as the arsenicals and fluorine compounds, could not be used. Pyrethrum exerts its toxic action against insects as a neuromuscular poison and paralytic agent and effects morphological changes in the hypodermis, muscles and nerves (770). The toxic principle of pyrethrum may result in the death of an insect by penetrating through the body wall, even in areas as remote from the vital centers as the tarsi. There is an axial gradient in toxicity to pyrethrum from the head to the caudal region of the body of the insects. A greater period is required to evoke toxic symptoms as one passes toward the posterior end of the insect (410).

It has been shown that pyrethrum may act as a stomach poison, but not as a fumigant (770).

---

[1]Data from U. S. Department of Agriculture.

One of the desirable characteristics of pyrethrum is its rapid toxic action. Even when insecticides such as DDT are used, which result in a higher per cent mortality, pyrethrum may be added to obtain a more rapid action by the spray. Thus in the preparation of DDT fly sprays, pyrethrum is added to effect a more rapid "knock down" of the flies than is afforded by DDT. The DDT is more apt to result in the death of the flies, but its toxic action is slow in comparison with that of pyrethrum.

ROTENONE

Introduction:- Rotenone ($C_{23}H_{22}O_6$) is a toxic compound known to occur only in certain tropical and subtropical leguminous plants. The natives in the various regions where these plants grow used them as fish poisons, despite the fact that rotenone possesses an extremely low water solubility. The poisoned fish are edible.

One of the principal genera of rotenone-bearing plants is Derris, and the principal species of this genus is elliptica. Derris elliptica is cultivated in Malaya and the East Indies, besides growing in the wild state, and is known there as tuba or toeba. Selection of highly productive clones has resulted in increasing the per cent of rotenone to as high as 13% and the total ether extract (containing besides the rotenone other less toxic principles such as deguelin, tephrosin, toxicarol and other rotenoids) to as high as 30%. The genus Lonchocarpus is the principal source of rotenone in South America. The most important species are L. nicou, L. utilis, and L. chrysophyllus. These plants are referred to as barbasco in Spanish-speaking America, except that in Peru the name cube is generally used. In Brazil and Ecuador they are referred to as timbo and in British Guiana as haiari. World War II resulted in our temporary dependence on the South American sources of rotenone, and as a result interest in the cultivation of rotenone-bearing plants in that part of the world was greatly stimulated. In the United States, rotenone has been found in "devil's shoe string", Tephrosia virginiana, and strains of this plant have been developed which may prove to be good sources of rotenone. Considering the fact that commercial importations of rotenone-bearing plants did not begin until about 1931, the development of the use of this material for insecticidal purposes has been remarkably rapid in this country and it has likewise been attended with great success.

Gunther and Turrell (395) investigated the location and state of rotenone in the plant cell. They found that it exists in the form of a resin in discrete particles in certain cells in the xylem rays, xylem parenchyma, phloem parenchyma, and pericycle, but not in the vessels or in the fibers. They found that the particles of whole derris resin are in partial solution in an etherial oil, and when the cell wall is broken, the globules of resin and oil are suspended, "probably by means of slowly homolyzing saponins which were shown to be present, in the sap of the plant." The milky sap exuding from freshly cut roots of Derris elliptica, and the milky aqueous extracts of either living or air-dried roots, were found to be acidic suspensions containing the particles of whole resin, ranging from 0.8 micron to 3.9 micron in diameter, and particles of starch, ranging from 7 micron to 38 micron in diameter.

Chemistry:- The principal insecticidally active compounds which have been isolated from the aforementioned fish-poison plants, have been named rotenone, deguelin, toxicarol and tephrosin. Rotenone is not only usually present in greatest amount, but it is also the most potent as an insecticide. Rotenone was isolated from Lonchocarpus nicou by Geoffroy in 1895, and given the name nicouline (359) and later by Nagai in 1902, who gave it the name rotenone (636), which is the term which is now accepted for this compound. Extensive investigations were made on the chemical structure of rotenone (501, 502) from which the structural formula presented herewith was derived (see p. 137). Rotenone and other insecticidally active extractives may be extracted from finely ground plant material by means of chloroform, ethylene dichloride, trichloroethylene, benzene, and other solvents. The color of the total extractives varies, depending on the solvent used for the extraction. The melting point of the rotenone is 163° C for the ordinary form, but the melting point of a form of rotenone, differing in its

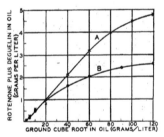

Rotenone

crystallographic axial ratios from the ordinary form, was found to be about 180° C. (367). When crystallized from alcohol, pure rotenone exists in the form of white six-sided plates belonging to the orthorhombic system (344). Although rotenone possesses a water solubility of only about 1 part in 6,000,000 (344), the water extracts of fish-poison plants have long been used as fish poisons. Rotenone is soluble in petroleum spray oil to the extent of only 1.6 grams per liter at 25° C. (269).

Formulations:- The American manufacturers of rotenone usually obtain the roots of Derris or Lonchocarpus in bales, and find that the rotenone content varies greatly between different species and even within a single species. However, they must grind the root material, and can then blend the ground root from different sources so as to make a product of uniform rotenone content. The ground root usually contains from 4 to 5% rotenone, the exact percentage being stamped on the container. The grower or manufacturer may then dilute the ground root to form a 0.75% or 1.00% rotenone dust by diluting with a suitable diluent (see p. 112).

Concentrated rotenone solutions may be obtained from the manufacturer and diluted to the required concentration by the grower. Acetone solutions of rotenone lend themselves very well for the preparation of aqueous suspensions and protective colloids may be added to stabilize the suspensions.

The extractives of rotenone-bearing roots may be incorporated into petroleum by means of mutual solvents which may be divided into two classes: solubilizers result in a colloidal solution of the extractives in the oil and oleotropic solvents result in a true solution. Finely ground root material (200 mesh) may be soaked in petroleum oil to obtain a concentration of total extractives sufficiently high to increase substantially the insecticidal efficiency of spray oil. Emulsive oils are especially efficient in extracting these toxic principles (fig. 80). By soaking 1 pound of ground cube root in 2 gallons of emulsive oil for 20 minutes, with occasional stirring, a concentration of 0.17% by weight of extractives was obtained. (269) Ground cube root is being used in this manner to increase the effectiveness of spray oils used to control the citrus blackfly in Mexico (see p. 507).

Fig. 80. Amount of Derris extractives dissolved in A, emulsive oil and B, straight oil at various concentrations of ground cube root in the oil. After Ebeling et al (269).

Toxicology:- One of the advantages in the use of rotenone preparations is their relative safety to human beings as they are applied in commercial usage. Also from the standpoint of the effect of residues, the rotenone products are considered to be safe firstly because of their low toxicity to man[1] and secondly because decomposition of the toxic principle may take place in a few days to a week, although it was reported that rotenone sprays or dusts retained a high insecticidal effectiveness at the end of 2 weeks when applied to bean plants which were kept in shade (892). Nevertheless, rotenone is considerably more toxic to human beings than is pyrethrum. Up to 0.2 g per kg of body weight was found to be non-

---

[1]It is estimated that 12.6 lbs. of 1% rotenone dust would have to be taken orally to result in acute poisoning, and vomiting would take place long before this amount could be consumed.

toxic to dogs (202). However, the cortex from a piece of fresh derris root less than 1/4 inch in diameter by 3 inches long is sufficient to kill an adult human being if consumed (436). Likewise, the contact of rotenone-containing dusts and mists with the throat and lung tissue may cause serious complications, at least with some individuals. When inhaled over long periods derris dusts can be fatal to experimental animals (396).

The extractives of rotenone-bearing roots act both as contact and stomach insecticides, but not as fumigants. They are remarkably toxic to some species of insects, but other species are quite resistant, as is shown by the fact that about 45 species of insects and other anthropods feed on the dried roots of rotenone-bearing plants.

Rotenone usually exerts its toxic effects against insects less rapidly than pyrethrum, but the affected insects are less likely to recover. The less rapid decomposition of rotenone enhances its insecticidal value as compared to pyrethrum, but as compared to the new synthetic chlorinated hydrocarbons, the rotenone loses its insecticidal effectiveness comparatively rapidly.

CHAPTER IX
CHLORINATED HYDROCARBONS

INTRODUCTION

The discovery of the outstanding insecticidal properties of DDT ushered in a new era in insecticide chemistry. Previously it had been considered almost axiomatic that halogenated compounds were highly phytocidal. Few had been found to be of value as insecticides, and halogenated compounds were always regarded with greatest suspicion as far as their potentialities in this direction were concerned. The advent of DDT, with a chlorine content of 51.8% by weight, as well as its bromine and fluorine analogs, all remarkably free of phytocidal properties, compared to many other widely used insecticides, abruptly changed the trend of thought along these lines. DDT was followed by benzene hexachloride, with 73.2% chlorine; chlordan, with 69.2%; and Toxaphene, with 68.6% (percentages refer to pure compounds).

The chlorinated hydrocarbons dominated the field of new developments in insecticides until quite recently, when the organic phosphates began to vie for the attention of insecticide chemists and entomologists. All these new developments have had the effect of increasing the importance of the field of organic chemistry as the probable source of the greater part of the future development along the line of insecticides. Figures from the U. S. Tariff Commission reveal that from the production in the United States of 287,510 lbs. of synthetic organic insecticides in 1940, the production rapidly increased to 40,870,000 lbs. in 1945 - over a 142-fold increase in 5 years (345). It will be noted that not only are the organic compounds increasing in importance as insecticides, at the expense of inorganic compounds, but also the synthetic organic compounds are becoming more important than the natural products. This development parallels the experience in the fields of the dyes and the drugs, and for the same reason, namely, that the synthetic compound can be produced when and where needed, without regard to season and locality, and can be manufactured with a uniform composition. Eventually, if not immediately, economic considerations also favor the synthetic product. Synthetic organic insecticides have also surpassed the natural products in insecticidal effectiveness, especially with regard to long term residual effect.

The prolonged residual effect of the chlorinated hydrocarbons, although valuable from an insecticidal standpoint, has resulted in new public health problems. Although rotenone, one of the natural insect poisons, has been found to be even more toxic to warm-blooded animals than DDT, its use on edible plants does not constitute a public health problem, for the toxicity of the compound is lost soon after it is applied. The chlorinated hydrocarbon insecticides remain toxic on the plant surface for some weeks to several months, and may also penetrate into the plant tissue. They may also be stored temporarily in the animal body, especially in fatty tissue, and in the milk of animals feeding on treated herbage.

The chlorinated hydrocarbons are similar to other halogenated compounds, such as chloroform, in that large doses affect the heart or the central nervous system and small doses, if administered over a prolonged period, injure the liver and kidneys.

DDT

Introduction:- In 1874, Othmar Zeidler, a German graduate student, synthesized a number of organic compounds incidental to his doctorate research (1030). Among these was a compound now popularly known as DDT. This obscure student, working in his laboratory, had no way of knowing that he had in his possession a compound which was destined to become one of mankind's greatest boons and, in fact, a compound which was in due time to play a major rôle in the shaping of human history, in a war of world-wide dimensions.

139

About a half century later, in the laboratories of J. R. Geigy, A. G. in Switzerland, a program of research was initiated with a view of finding better mothproofing compounds. Among the promising compounds were the substituted diphenyl sulfones, and by substitution of various chemical groups in the molecule of one of these compounds, Läuger, Martin, and Müller (512) arrived at the fantastically effective compound chemically designated, according to the American system of nomenclature, as 1,1,1-dichloro-2,2-bis(p-chlorophenyl)-ethane and according to the European system β,β,β-dichloro-α,α-bis(p-chlorophenyl)-ethane. In the aforementioned paper, through a maze of 127 structural formulas depicting the evolution of the author's central idea, the reader is led at last to the statement (translated): "From the highly efficacious stomach poison dichlorodiphenylsulfone, through the substitution of the SO_2 group by the group

$$Cl - \overset{\overset{\displaystyle CH}{|}}{\underset{\underset{\displaystyle Cl}{|}}{C}} - Cl$$

a contact poison came into being." And so came into being also a new epoch in man's struggle against his insect foes. This re-synthesis of DDT took place in 1939. It was not the result of accident, as great discoveries are popularly, but erroneously, believed to be, but of years of painstaking and methodical research by teams of well-trained scientists utilizing the laboratories and facilities of a large chemical company. In 1948 Paul Müller received the Nobel prize in physiology and in medicine for having revealed the insecticidal properties of DDT.

Swiss entomologists found the compound, as a 1% dust, to have striking effectiveness against the Colorado potato beetle, consequently it was to be of inestimable value in Switzerland's war-enforced struggle for self-sufficiency, especially since the formerly used insecticide, lead arsenate, had to be imported and was not available in sufficient quantity. A high degree of effectiveness against the stable fly, granary weevil, apple-blossom weevil, and many other insects, emphasized the wide applicability of this new insecticide. It was unique among insecticides in having a combination of these important properties: (1) it was both a contact and a stomach poison; (2) the effectiveness of the residue left on the plant ("residual effect") remained for as long as several months, as contrasted to the very short period of residual effect of other contact insecticides, mainly of plant origin, which were available at that time; and (3) it was effective at extremely small concentrations. Despite the high degree of insecticidal effectiveness possessed by DDT, however, entomologists have constantly warned the potential users of the product that it is not a "cure all"; in fact, it is singularly ineffective against some insects and mites. A number of the more recent insecticides have been even more effective than DDT against many pests. In addition, the public is warned that the use of DDT may in some instances have to be avoided or sharply limited because of the harmful effects of the insecticide on the natural balance between injurious insects and mites and their natural enemies.

In August, 1942, the Geigy Company in the United States received 100 lbs. each of a dust and a spray material containing DDT from the parent company in Basle, Switzerland. Then 100 kilos of the active ingredient was imported so that research might be continued and manufacture of the compound be studied in the United States. The preliminary entomological research work, which was carried on mainly in the Orlando, Florida, station of the U.S.D.A., confirmed the spectacular effectiveness of the insecticide and it became desirable to manufacture it in the United States. Its effectiveness against insects affecting man and animals appeared to be particularly noteworthy at that time. In the Pacific campaigns in World War II, entire islands were dusted or sprayed with DDT by planes to eliminate completely mosquitoes and other noxious insects. One part of DDT per 10 million parts of water was found to be sufficient to kill larvae and pupae of many species of mosquitoes, with somewhat higher concentrations needed for others. In 1945, 33,000,000 lbs. of DDT was produced in the United States, the production having increased from nothing to nearly 3 million pounds a month in less than 2 years. Only 5% of this was allotted by the War Production Board for civilian use, but after the war civilian use was enormous, not only for pests attacking man and animals and the household pests, but also for a large number of the pests of the field, garden and orchard.

Chemistry:- The popularized symbol "DDT" was derived from the descriptive word dichlorodiphenyltrichloroethane, which is in actual fact, however, a generic term that may apply to any one of 27 different compounds. Specifically DDT refers to 1,1,1-dichloro-2,2-bis(p-chlorophenyl)-ethane [(p-ClC$_6$H$_4$)$_2$CHCCl$_3$]. In the manufacture of DDT, the compound is ordinarily prepared from 1 molecule of chloral hydrate ("knockout drops") and 2 molecules of monochlorobenzene in the presence of sulfuric acid or other condensing agent. The DDT crystallizes from the reaction mass and the sulfuric acid is washed out. The chemical derivation of DDT from its original commercial sources, two compounds which we have in great abundance, can be visualized by inspection of the following structural formulas:

| Ethyl alcohol | Chloral hydrate | Benzene | Monochlorobenzene |

By condensing 2 molecules of monochlorobenzene with 1 of chloral hydrate there is obtained

DDT

The product crystallized from the reaction mass is called technical DDT. It is a powder, or lumpy material, and may be sticky or oily or waxy. It is light brown to tan in color and has a faint, not unpleasant odor. It contains from 60 to 95% of the isomer[1] of principal insecticidal value, namely, the p, p-isomer or p, p-DDT. In technical DDT, the para-para, ortho-para, and ortho-ortho isomers, as well as a number of other compounds have been found. The principal impurity is o, p-DDT, which is present to the extent of about 15 to 25%. The p, p-isomer may be obtained by the recrystallization of the technical product from ethyl alcohol. To obtain a pure product, several recrystallizations are required.

The p,p-DDT is a white crystalline substance with a melting point of 108°– 109° C and a density of 1.556. It ordinarily occurs in the form of long biaxial needles. It is nearly insoluble in water. Its solubilities in some common organic solvents, petroleum oils and vegetable oils, in grams per 100 ml at 27° to 30° C are as follows: cyclohexanone, 116; benzene, 78; trichloroethylene, 64; orthodichlorobenzene, 59; acetone, 58; xylene, 52; carbon tetrachloride, 45; dimethyl phthalate, 34; cottonseed oil, 11; crude kerosene, 8-10; refined, odorless kerosene, 4; Velsicol AR-50 (chiefly mono- and di-methylnaphthalenes), 55 (24). DDT is about 50% more soluble in asphaltic base than in paraffin base petroleums. Anthracenes or heavy aromatic naphtha are used to maintain complete solution.

Although one of the advantageous characteristics of DDT is its relative stability, certain types of contamination, such as by ferric oxide, iron and aluminum chlorides, alkaline substances, and exposure to high outdoor temperatures, possibly accompanied by a certain amount of photodecomposition from far-ultraviolet

---

[1]Isomers are compounds composed of the same elements united in the same proportion by weight, but because of differences in molecular structure they may differ in one or more properties.

radiation, will result in degradation of the p, p-DDT into insecticidally less effective compounds by splitting off hydrochloric acid. The tendency of DDT to split off HCl was utilized by Gunther (389) in working out a quantitative estimation of DDT and of DDT spray and dust deposits. Gunther has recently discussed the various techniques used at the Citrus Experiment Station for the sampling and manipulation of the different varieties of fruit for the analysis of DDT residues by his dehydrohalogenation method, modified somewhat when sulfur residues are present. A mass production efficiency is provided by the measurement of leaf areas by means of a photoelectric aerealimeter and the adoption of an electrometric titration for the chloride ion liberated from the DDT. (391)

Toxicology:- DDT being a relatively stable compound, it is likely that its effect in the insect body can be studied much more effectively than that of the insecticides of plant origin, such as pyrethrum, rotenone, and nicotine. Despite the short period since DDT was first known as an insecticide, considerable toxicological work has been done, largely because of the interest of the armed forces in the compound.

It appears from the literature that there are two opposing theories on the chemistry and physiology of the toxic action of DDT. According to Laüger et al (512) the ClC6H4 groupings constitute the toxic portion of the DDT molecule and the CCl3 grouping the lipoid-soluble portion, while according to Martin and Wain (592a) the opposite is true, that is, the ClC6H4 is the lipoid-soluble and the CCl3 is the toxic portion of the molecule. Both theories agree with the concept generally held by insect physiologists that both lipoid-soluble and toxic groupings must be present in an effective contact insecticide. Subsequent workers have been unable to determine which of the two theories is correct. Moreover, there is increasing evidence that they may both be an oversimplification, and that the molecule as a whole, with special reference to the spatial relationship of the constituent atoms or, in other words, the shape of the molecule, is the pertinent factor in toxic action.

A characteristic of the toxic effect of DDT on insects is that it is slow. A treated insect may not give any indications of discomfort for a while, but finally loses its power of coordination, trembles, goes into convulsions, then dies. This usual sequence, provided the insect has obtained a lethal dose of DDT, may last only from one to several hours or it may last for days before the insect succumbs. These reactions indicate that DDT is a "nerve poison" to insects.

It has been shown that DDT increases the metabolism of insects and depletes reserves. This is probably caused by an overstimulation from the central nervous system. Other evidence indicates that DDT probably interferes with the calcium balance in the nerve sheaths (see 736).

DDT can be fatal to warm-blooded animals, but generally the dose required to kill them with external application is several hundred times greater, on the amount per body weight basis, than the dose required to kill insects or cold-blooded animals. When the DDT is subcutaneously administered, the difference in susceptibility of warm-blooded animals and insects is not so great. The skin of warm-blooded animals is a much more effective barrier to the entrance of DDT than the cuticula of the insects.

DDT and chlordan have been found to be about equal in their toxicity to white rats (458). On the other hand, Lehman (516) found the acute toxicity to rats of a number of the new synthetic organic insecticides to be in the ratio indicated at top of following page.

In Lehman's investigation, chlordan was shown to be only half as toxic to rats as DDT. The order of toxicity of the compounds will vary with different test animals, but in general it is more or less similar to that indicated in the following table.

Lehman (516) has listed the more important pathologic lesions produced by some of the compounds listed as follows: "DDT produces liver damage and focal

CHLORINATED HYDROCARBONS

Insecticide	LD 50* Mg/kg	Ratio
DDT	250	1
DDD	2500	1/10
DMDT	6000	1/24
Benzene hexachloride		
alpha isomer	500	1/2
beta isomer	6000	1/24
gamma isomer	125	2
delta isomer	1000	1/4
Toxaphene	60	4
Chlordan	500	1/2
HETP	7	35
TEP	2	125
Parathion	3.5	70

*Dosage required to effect a 50% mortality

muscle necrosis. DDD in addition to being a liver toxicant appears to have special predilection for the adrenal glands. Limited data on DMDT indicate that this substance is a kidney poison. Benzene hexachloride and its isomers are principally liver poisons. This also holds for toxaphene and chlordan. The organic phosphate compounds (limited data) appear to produce their greatest damage in the colon; necrosis of the gall bladder has been noted also."

In making comparisons of insecticides on a pharmacological basis it should be borne in mind that the minimal lethal dose is not the only factor that should be considered. The least toxic compound may not necessarily be the safest, because it may have to be used at a high dosage to be insecticidally effective. Also relative differences in immediate toxicity do not necessarily indicate the relative differences in the chronic toxicity of different compounds. It should also be remembered that the solvent in which an insecticide is dissolved may itself have considerable toxicity.

It has been estimated that it would require 5 to 10 grams of DDT to provide a fatal dose for an adult human being. Since DDT is relatively light, this would be about a handful of the powder. As is characteristic of the chlorinated hydrocarbon insecticides as a group, the higher doses would affect the heart or the central nervous system while small doses, if administered over a prolonged period, would produce liver and kidney damage. DDT is unusual in one respect, however, in that it stimulates the central nervous system, causing convulsions, whereas the characteristic reaction of other halogenated hydrocarbons is to depress the central nervous system (427).

DDT poisoning from insect-control operations is not likely if the worker will follow reasonable precautions. Stammers and Whitefield (834) made an extensive study of the toxicity of DDT to man and warm-blooded animals in a comprehensive review of the literature, including reports of suicides and accidental deaths from DDT formulations. It is significant that they arrive at the following conclusions: "In the light of our own experience and the data presented in this paper we consider that DDT when used as an insecticide, with reasonable intelligence and the precautions normal to the use of modern insecticides, is harmless to man and animals. It is, nevertheless, possible by unskilled formulation, by the use of unsuitable solvents and by misuse in its application, to incur risk to man and animals. The possibilities of cumulative effects from the storage of DDT in the milk tissues of sheep and cattle require further investigation." A survey of the literature shows that the above report is representative of a consensus on the subject to date.

Tests with animals have shown that DDT is present in the urine long in advance of any detectable signs of poisoning, and it is therefore recommended that

workers exposed for long periods to inhalation or absorption of DDT
frequent and regular checks made of their urine. Examination of th
reveal the existence of danger long before any trouble could develo
be remembered in this connection that DDT can be taken into the ski
tions, but this is not likely to happen to any appreciable extent w
pound is in contact with the skin as a solid, that is, as a dust or
left by an aqueous suspension.

Phytotoxicity:- Non-oil formulations of DDT have been on the w
non-injurious to plants, although certain cucurbits may be injured.
to certain varieties of camellias has resulted from a spray of 1 lb
table DDT powder to 100 gallons (683). Injury to citrus foliage fr
emulsions has been reported (268, 147). If DDT is to be used in a
tion, it is recommended that preliminary tests be made to explore t
ties of injury before applications are made on a large scale.

Residual Effectiveness:- The insecticidal action of the residu
spray is known as its "residual effect". The intensity or period o
effect depends on the susceptibility of the insect species in quest
ture as it affects volatility, aeration as it affects volatility, w
plant growth, and the nature of the formulation. If an insect is h
ceptible to an insecticide, the residual effect will be prolonged.
temperature and the greater the degree of aeration the less will th
residual effect. Regional and seasonal variations in the degree of
from rain and sandstorms, cause variability in the performance of a
as far as residual effect is concerned. As a plant grows, the surf
by the insecticide expand, resulting in the appearance of unprotect
Formulation also may be an important factor affecting the residual
of an insecticide. For example, in the experimental work with DDT
given amount of DDT per unit of surface area will be far more effec
applied in the form of a dilute kerosene emulsion than as a wettabl

One of the properties that make the chlorinated hydrocarbons o
their effectiveness against insects is their relatively high chemic
The residue left on the treated surface can kill many species of in
siderable periods after treatment. The comparative residual effect
number of chlorinated hydrocarbons and a few other organic insectic
house flies and mosquitoes was observed over a 26-week period. It
the insecticides tested ranged in order of their long-term effectiv
DDT>benzene hexachloride>chlordan>toxaphene>DDD. Pyrethrins with p
hexenone and piperonyl butoxide had comparatively little residual e
and di-2(ethylhexyl) phthalate showed no residual effectiveness (30
the above compounds might not necessarily fall in the same order of
tiveness against other insects, DDT is consistently outstanding in

Many plant products such as the pyrethrins, rotenone, and nico
ing highly insecticidal, quickly lose their toxicity in the presenc
through oxidation. Laüger et al (512) reflect on the catastrophy w
resulted if Nature had endowed such powerful natural insecticides w
stability. They conclude, however, that "Nature is devoted to life
death". Man, with his newly found tool of organic synthesis, may w
the havoc which Nature has been able to avoid, in bringing about a
the equilibrium of animal populations. It may be safely assumed, h
man will have the intelligence to harness eventually his newly foun
nature to his best advantage.

Formulations:- DDT may be used in a large number of formulatio
in the form of solutions, emulsions, wettable powders, dusts, aeros
impregnated fabrics and papers (940, 1031). A review of DDT formul
discussion of methods of preparation and suitable adjuvants (with t
representative products) for water dispersable powders, dust mixtur
and emulsions is now available (158).

CHLORINATED HYDROCARBONS

Solutions

As previously stated, DDT is soluble in a large number of organic solvents. Highly refined kerosene, being odorless, colorless and relatively inexpensive, is the most widely used of the solvents for the preparation of proprietary solutions and also for experimental tests. Refined kerosene will generally dissolve 5% by weight of technical DDT, but it should be borne in mind that low night temperature will cause the crystallization of much of the DDT. This can be avoided by adding to the kerosene a small amount of some more effective solvent. Certain petroleum products with a high content of alkylated naphthalenes are often used for this purpose. Likewise if more than 5% DDT is desired, the addition of other solvents to the kerosene will be necessary. If the solution is to be used as a household spray, the solvent added to the kerosene either must be free of objectionable odor or the odor must be masked by a perfume.

Emulsions

In the preparation of emulsions, the dissolving of the DDT in the solvent involves more than the preparation of a stable solution. The solvent, if used alone or if added to kerosene or other petroleum fractions, must be one which will not result in too tight an emulsion and too much "run-off" of the insecticide solution when the emulsion is diluted with water and applied. Xylene, methyl naphthalenes, and tetrahydronaphthalenes are some of the solvents that have been successfully used to add to kerosene or spray oil to form quick-breaking sprays for use on trees. For the spraying of barns, outbuildings, etc., the emulsions have usually been prepared with a suitable wetting agent or detergent to emulsify the DDT-solvent solution and water. The emulsions contain from 25 to 50% DDT.

If a volatile solvent is used, such as xylene, or even kerosene, a deposit of DDT crystals remains after spraying. If a relatively nonvolatile solvent is used, such as petroleum spray oil, the sprayed surface is left coated with a solution of DDT in oil after the water evaporates. This difference in the nature of the residual film may be of great practical importance. The relative effectiveness of various insecticides may be different with dry films than with oily films (143). DDT was found to be 4 to 6 times as toxic to the aphid <u>Macrosiphum solanifolii</u> when dissolved in a water-insoluble solvent (benzene) and used in an emulsion, than when dissolved in a water-soluble solvent (acetone-carbitol) and used in a suspensoid (868).

Emulsions are advantageous for use on a large scale because the high concentration of DDT results in a reduction of shipping costs. They are not as convenient to use on a small scale, as in the home, because the emulsion must be diluted with water before use.

Wettable Powders

Although DDT is a waxy material which cannot be ground by itself, it can be mixed with inert diluents and ground into small particles. The resulting powder can be made wettable by adding a suitable wetting agent. The wettable powders usually contain 50% DDT. They have a wide usage for the spraying of buildings on which a white residue is not objectionable. The wettable powders are the most popular form of DDT formulation for use in horticultural sprays. From 1 to 2 lbs. of the 50% wettable powder is used to 100 gallons of water. An adjuvant to increase the deposit and sticking qualities of the DDT can often be used to advantage.

It should not be assumed that the effectiveness of a residue will be in direct proportion to the amount of actual DDT deposited per unit of surface regardless of the formulation used. For example, it has been shown that a DDT-kerosene spray has a much greater residual effect than a spray in which a wettable DDT powder is used when the two are applied, with the same amount of DDT per 100 gallons, in the control of red scale. The former leaves a residue of DDT crystals

145

while the latter leaves a residue in which the DDT particles are partly mas
the inert diluent. (268) Against some other insects, for example the grape
beetle, it apparently makes no difference whether the DDT is applied as a s
when dissolved in kerosene or as a wettable powder.

Colloidal Dispersions

It has been shown that colloidal suspensions may be effectively used a
quito larvicides (536a). Some "colloidal dispersions" of DDT in water have
prepared with the majority of the particles about 1 micron in size but with
as long as 30 microns and 2 to 3 microns wide (484). The DDT was transform
the colloidal state in a colloid mill. Flocculation was prevented by means
dispersing agents. The dispersions formed were of two types: one containi
an emulsifying agent and water, and the other identical except for a suitab
solvent. The latter proved to have the more satisfactory physical properti
The dispersions were of a stable, creamy, semi-solid consistency, but poure
readily and did not cake. When dilute sprays were applied to surfaces, dep
of DDT could be seen as fine, evenly distributed, silvery particles. Littl
noticeable residue remains on plant surfaces, and this fact was considered
favorable to the use of the dispersions in the treatment of ornamental plar
Little agitation is required to keep the DDT in suspension (484).

In laboratory toxicity tests, a colloidal DDT dispersion was found to
favorably with emulsifiable DDT and micronized DDT wettable powder and was
toxic than pulverized wettable powder. It was safer to foliage than DDT em
and compared favorably in safety with wettable powders. Field trials showe
colloidal dispersions to give good initial insect kill, but poor residual e
tiveness as compared to emulsions and wettable powders. However, this may
be an advantage in cases where there may be a residue problem on fruits and
vegetables.

Dusts

As stated before, DDT can be ground with a diluent such as pyrophyllit
talc. When dusts and wettable powders were first used, it was believed tha
least 75 to 80 parts of diluent were required for satisfactory grinding, ma
dust of 20 - 25% DDT, but now the usual concentration of DDT is 50%. Condi
are added to the dusts to make them free-flowing and reduce the tendency to
For agricultural dusts, the concentration of DDT ordinarily varies from 1 t
depending on the susceptibility of the insect to be controlled.

Dusts may also be prepared by impregnating the diluent with a solution
in a volatile solvent such as acetone or benzene. The solutions and diluen
be mixed and the solvent allowed to evaporate, after which the mixture is g
According to another method the solution is sprayed into the diluent during
grinding process. Nonvolatile solvents may also be used if the amount of s
required is not enough to impair the dusting qualities of the finished prod

Aerosols

The nature of aerosols and equipment used for their generation are dis
in Chapter 15. DDT has been a popular constituent of aerosols, either alon
combination with other insecticides. In aerosol bombs, for example, a comm
formulation had been 3% DDT and 2% pyrethrum extract (20% pyrethrins), freq
with a suitable synergist. Nonvolatile liquids may be added to regulate pa
size. The low solubility of DDT in Freon necessitates the addition of an a
iary solvent such as cyclohexanone or certain aromatic hydrocarbons.

The prolonged residual effect of DDT often makes it desirable to use a
ration which will deposit an appreciable amount of the insecticide. This c
be accomplished with the aerosol.

During the war, DDT aerosols were applied by plane, utilizing the prop
wash, venturi tubes, breaker fans, and other methods for breaking up DDT so

into extremely fine particles (24). At present they are being applied by planes principally to large insect-infested forest areas (242).

Paints and Polishes

Certain types of paint into which DDT has been incorporated will leave enough of a deposit of DDT on the surface to be lethal to many kinds of insects. The ordinary oil paints and varnishes have not been found to be satisfactory for this kind of formulation. The dried paint film is not very insecticidally active. However, some of the oil-bound water paints, synthetic resin finishes, cold water calcimines and casein paints, have given good results, and for some the claim has been made that they retain their insecticidal effectiveness for several months. With these, the DDT gradually migrates to the surface of the film to take the place of that which has volatilized. One such product, a colorless, synthetic-resin liquid with 6% DDT, can be brushed on screens, walls, stairs, etc., and quickly dries to an almost invisible film which will withstand considerable rubbing. It is claimed that the product will remain insecticidally effective for at least 3 months inside and at least a month outside (1031). Whitewashes not containing lime or other basic materials also show promise as DDT carriers. A "DDT wallpaper" is now on the market.

Impregnated Fabrics and Papers

Cotton garments treated with 0.05% DDT in volatile solvents and emulsions have remained effective against body lice for more than a week of continuous wearing, and higher concentrations were effective for much longer periods. Garments treated with 1% DDT were completely effective after 4 washings and still highly effective after 5 washings. Higher dosages increased the period of effectiveness. Cotton garments treated in the same way were effective against body lice after 1 dry cleaning. Wool garments treated with DDT in volatile solvents or emulsions withstand washing, without loss of toxicity to lice, about as well as the cotton. When the DDT-treated garments were stored in the dark for one year, they showed no decrease in toxicity to lice (466).

It is likely that DDT can be utilized for impregnating paper used for packaging food materials. Experiments have shown DDT to be the most effective of a large number of chemicals tested for this purpose. Complete protection of otherwise insect-infested flour was obtained for a period of 2 months. (197)

COMPOUNDS RELATED TO DDT

DDD(TDE, Rothane):- This compound, which can be chemically designated as 1, 1-dichloro-2, 2-bis (p-chlorophenyl)-ethane, was shown to be even slower in "knock-down" than DDT, and to have much less residual effect, in the original work on DDT and related compounds at the U.S.D.A. laboratory at Orlando, Florida. However, later work showed DDD to have greater toxicity to Anopheles mosquito larvae than DDT (232). These preliminary investigations were followed by many others which established the fact that DDD compares favorably with DDT against many pests.

Methoxychlor:- The methoxy (or anisyl) analog of DDT, called di(p-methoxy-phenyl) trichloroethane or 1,1,1-trichloro-2,2-bis (p-methoxyphenyl)-ethane, as a 10% dust, has been found to be as effective as 0.5% rotenone dust against the Mexican bean beetle and has also given excellent control of grasshoppers as a 20% dust at 20 lbs. per acre. Against adult white-fringed beetles, 2 lbs. in 100 gallons of water per acre resulted in rapid "knock-down" but slow mortality. Although the beetles remained active they could not coordinate their action or feed. Applied to the soil, the methoxy analog of DDT was highly toxic to the larvae. The compound was found to be almost nontoxic to the European corn borer and far less effective against Anopheles mosquito larvae than DDT. (89)

Bromine analog:- The bromine analog of DDT, 1,1,1-trichloro-2,2-bis (p-bromo-phenyl)-ethane [(BrC$_6$H$_4$)$_2$ CHCCl$_3$], was found to be somewhat more toxic than DDT to Anopheles mosquito larvae, about one-tenth as toxic as DDT to house fly maggots,

147

but equally toxic against the adults (89). Among other insects, the bromine analog is reported to control flies, potato and tomato psyllids, tuber flea beetles, and leafhoppers. It does not interfere with nitrogen-fixing by bacteria and thus has an advantage over DDT when used where legumes are grown to increase the nitrogen in the soil. DDT is said to have a depressing effect on the formation of the bacterial nodules (28).

Fluorine Analog:- The fluorine analog, 1,1,1-trichloro-2,2-bis-(p-fluorophenyl)-ethane [(FC$_6$H$_4$)$_2$CHCCl$_3$], which may for convenience be called DFDT, was used commercially as an insecticide in Germany during World War II and was known by the trade name Gix.

DFDT can be produced by the condensation of 1 mole of chloral hydrate with 2 moles of fluorobenzene in the presence of sulfuric acid. Technical grades, containing as high as 90% para-para isomer are semisolids. The purified p,p-DFDT, however, crystallizes from alcohol as white needles with a melting point of 44° - 45° C and has a very pleasing odor of ripe apple. It is extremely soluble in organic solvents, and no mutual solvent would be necessary to incorporate the material into petroleum oils in any practical concentration. It is also soluble in certain oil-soluble emulsifiers. By blending DFDT with suitable diluents, conventional dusts and wettable powders can readily be made. The limited data available indicate that DFDT is even less toxic than DDT to mammals (603).

From a review of the literature, Metcalf (603) concluded that, on the average, the toxicity of DFDT to insects was comparable to that of DDT. He made further tests of his own, using 12 species of test insects, with the following results: The residual effect of DFDT against California red scale crawlers was much less than that of DDT, probably because of its greater volatility. The effectiveness of DFDT was much greater than that of DDT against the German cockroach, Blatella germanica, and the large milkweed bug, Oncopeltus fasciatus. In initial toxicity, DFDT was on the average about equal to DDT, but was more rapid in its toxic action. The residual effect of DDT lasts much longer than that of DFDT, probably because of the greater volatility of the latter compound. This factor should be taken into consideration when pest control problems are encountered in which it is desirable that the residue should rapidly lose its poisonous properties.

Metcalf (603) found DDT and DFDT to be considerably more toxic to insects than DDD or DFDD, the corresponding products remaining after dehydro-chlorination. Likewise, these 4 compounds were the only DDT analogs of 14 that were tested against fruit flies and greenhouse thrips which had adequate insecticidal properties.

BENZENE HEXACHLORIDE

Introduction:- Benzene hexachloride, known to chemists as 1,2,3,4,5,6-hexachlorocyclohexane, with the formula C$_6$H$_6$Cl$_6$, was first synthesized in 1825 by the famous Michael Faraday. In 1912 the existence of 4 isomers, alpha, beta, gamma, and delta was established and their melting points determined (906). The existence of 2 of these had already been demonstrated. A fifth isomer, called the epsilon isomer, was described in 1946 (477), and a sixth (zeta) isomer in 1947 (77).

The period elapsing between the original synthesis of benzene hexachloride and the discovery of its insecticidal properties was even greater than that of DDT. Its outstanding insecticidal effectiveness was unknown to pest-ridden humanity for 116 years after its discovery before Dupire, in France, found the compound to be toxic to clothes moths in 1941 and submitted it for evaluation against agricultural insect pests (402). The insecticidal properties of benzene hexachloride were discovered in England in 1942 and were made public in a lecture given by R. E. Slade in 1945 (780). It was later established that the insecticidal effectiveness of the compound was also independently discovered in Spain (365a) and at the New Hampshire Agricultural Experiment Station in the United States (661). Thus the insecticidal effectiveness of the compound was independently discovered 4 times in 4 countries within a period of 4 years.

148

CHLORINATED HYDROCARBONS

Early in 1943, research workers with the Imperial Chemical Industries in England discovered that the toxicity of benzene hexachloride was due almost entirely to the gamma isomer, which was found to be more effective than any other insecticide they had ever used against the test insects (weevils) which they happened to be working with at the time. The gamma isomer was called "Gammexane". It was present to the extent of 10 - 12% in the crude reaction product of benzene hexachloride. Later Gunther (390) found that a product containing 42% of the gamma isomer can be obtained by treating benzene with chlorine in the presence of actinic light and a dilute aqueous sodium hydroxide solution. This observation, however, has not been confirmed, although a number of laboratories have attempted to duplicate the synthesis. That Gunther actually had the gamma isomer was confirmed by infrared analysis.

Benzene hexachloride is easily prepared from benzene and chlorine, two compounds that are readily available. Consequently several American companies began manufacturing the material, and by 1947 about a dozen companies were making it in large quantities. It has found a wide usage and is more effective than DDT against many insects. Benzene hexachloride has been most extensively used on cotton, for it is effective against all cotton pests but the bollworm. In addition, no residue problem is involved in the treatment of cotton.

<u>Chemistry</u>:- Benzene hexachloride is prepared by the chlorination of benzene in the presence of light. This results in an addition product with the formula $C_6H_6Cl_6$. The isomers of benzene hexachloride can be most easily visualized if it is assumed, in accord with the older concept presented in some textbooks on organic chemistry, that all the carbon atoms of a saturated benzene ring lie on one plane. Two valences of each carbon atom are directed so that one is above and one below the plane. Then in benzene hexachloride it

Benzene
hexachloride

would be theoretically possible to have all the chlorine atoms above the plane and all the hydrogen atoms below, accounting for 1 isomer. There may be an arrangement in which there would be 5 chlorine atoms above and one below, 2 arrangements with 4 above and 3 below, and 3 arrangements with 3 above and 3 below. Thus it can be seen that theoretically there would be 8 isomers. Slade (780) pointed out that a number of workers have offered X-ray evidence that the benzene (or cyclohexane) ring has a centro-symmetrical form, the carbon atoms being disposed in tetrahedral fashion. If, in accord with this concept, cyclohexane itself may

exist in 2 forms, there are 16 theoretically possible configurations for benzene hexachloride. Slade (780) believed, however, that because of the close proximity of the chlorine atoms in many of the 16 configurations, there would probably be only 5 strainless forms. The probable configuration of 3 of these isomers is indicated by the models shown in fig. 81. The possible structures of 5 isomers of 1,2,3,4,5,6 - hexachlorocyclohexane have been determined by the methods of infrared spectroscopy (213).

Technical benzene hexachloride melts over a considerable range of temperatures owing to the fact that the 5 isomers have widely different melting points.

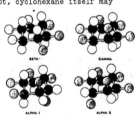

Fig. 81. The beta, gamma, and alpha isomers of benzene hexachloride. After Slade (780).

Isomer	Melting Point (°C)
Alpha- - - - - - - -	157.5-158
Beta - - - - - - - -	309
Gamma- - - - - - - -	112.5
Delta- - - - - - - -	138-139
Epsilon- - - - - - -	218.5-219.3

The crude benzene hexachloride approximates the following percentages of isomers: alpha 65-70%, beta 5%, gamma 10-12%, delta 6%, and epsilon 4%. In addition there is 4% heptachlorocyclohexane and 0.6% octochlorocyclohexane. The latter materials are the result of chlorination of monochloro-and dichlorobenzenes (602). The density of the gamma isomer is 1.85.

Benzene hexachloride does not appear to undergo dehydrochlorination when heated with traces of certain metals or their salts, in this respect differing from DDT. Like DDT, however, dehydrochlorination takes place in the presence of alkalies to afford essentially 1,2,4-tri-chlorobenzene (392). All of the isomers of benzene hexachloride are practically insoluble in water, but they are quite soluble in a number of organic solvents such as acetone, benzene, chloroform, carbon tetrachloride, cyclohexanone, dioxane and ethylene dichloride. The gamma and delta isomers are more soluble than the others in the majority of solvents.

Toxicology:- Against the majority of insects the gamma isomer of benzene hexachloride is more toxic than DDT. It has value as a stomach poison and fumigant as well as being a contact insecticide of extremely high effectiveness. Against greenhouse thrips, the gamma isomer of benzene hexachloride has been found to be about 10 times as toxic as p-p' DDT, 25 times as toxic as toxaphene, 50 times as toxic as chlordan, and 100 times as toxic as hexaethyltetraphosphate (main toxic ingredient tetraethylpyrophosphate). Other isomers of benzene hexachloride had no practical degree of toxicity, nor did they show any synergistic or antagonistic effect when applied with the gamma isomer. (602)

Compared with DDT for long-term insecticidal effectiveness, benzene hexachloride has the disadvantage of possessing a shorter period of residual effectiveness. This may be due to its much higher vapor pressure (9.4×10^{-6}mm Hg as compared to 1.3×10^{-7}mm Hg for p-p' DDT, both at 20°C). The high vapor pressure of benzene hexachloride also probably accounts for the fumigant action of the material which has been reported by a number of investigators.

It has been estimated that the gamma isomer of benzene hexachloride is about 4 times as toxic to warm-blooded animals as DDT while the other isomers are about equal to DDT in toxicity. A respirator is recommended when using benzene hexachloride not only because of its disagreeable odor, but more particularly because it may cause headaches and smarting of the eyes.

The phytotoxic properties of benzene hexachloride are also greater than those of DDT. Care should be taken that the minimum effective dosage should be tested in a small way on the plants to be treated before larger scale application is undertaken.

Formulations:- Benzene hexachloride is obtainable in the form of a wettable powder containing 12% gamma isomer, as an emulsive concentrate containing 1 or 2 lbs. of gamma isomer per gallon, or a 12% gamma isomer dust concentrate. The dust concentrate may be blended with suitable diluents to make a dust of any lower concentration of gamma isomer. For example, 1.5 to 2% of gamma isomer is often used in dusts. The 12% wettable powder is normally used at 2 or 3 lbs. per 100 gallons of water. The emulsion concentrates are used under special conditions by airplane applicators in a sort of concentrated vapor spray.

One of the drawbacks in the use of the technical benzene hexachloride is that it leaves a persistent unpleasant odor which may be objectionable on products to be used within a reasonable period for human consumption. The gamma isomer, however, does not have this odor. Products are now being manufactured containing around 95% gamma isomer, and these may be used in cases in which the odor left by the spray residue might be objectionable. The gamma isomer may be called "lindane".

CHLORDAN

Introduction:- This is a chlorinated hydrocarbon having the empirical formula $C_{10}H_6Cl_8$, and was discovered by Dr. Julius Hyman early in 1945. It was for some years called <u>Velsicol 1068</u>, the numbers deriving from the formula. This insecticide was found to be more toxic than DDT or benzene hexachloride to some insects (478). It has for some time been recognized as especially well suited as a general household insecticide, but is rapidly gaining favor for use against many agricultural pests and is a superior insecticide for ant control. It is very efficient in the control of a number of major soil pests including wireworms, mole crickets, and the larvae of the Japanese beetle, green June beetle, and the white fringed beetle.

Chemistry:- Chlordan is known to chemists as 1,2,4,5,6,7,8,8-octachloro-4,7-methane-3a, 4,7,7a-tetrahydroindane. This compound makes up between 60 and 75% of the liquid currently produced as the technical chlordan. The 99% pure liquid is viscous, colorless, nearly odorless, and boils at 175° C at 2 mm pressure. It is insoluble in water and soluble in all proportions in the majority of organic solvents such as aromatic, aliphatic, and chlorinated hydrocarbons, ketones, ethers, esters, and alcohols. Technical chlordan has a density of 1.61. The complete miscibility of the compound with deodorized kerosene and other petroleums commonly used in insecticide formulations is of special interest from a practical standpoint. Like the other chlorinated hydrocarbons, chlordan dehydrochlorinates readily in the presence of weak alkali to form products of much lower insecticidal effectiveness. It has a higher volatility than DDT, but lower than benzene hexachloride.

Chlordan

Toxicology:- It is believed that chlordan and DDT act on the nervous system of insects in a somewhat similar manner. When coming in contact with DDT, however, the insects exhibit greater nervousness or excitability than they do with chlordan. Besides acting as a contact insecticide and a stomach poison, chlordan has been found to have considerable fumigating action against some insects.

An extensive study was made of the comparative toxicity of chlordan and DDT, using the white rat as a test animal (458). Comparing chlordan and DDT weight for weight, they were found to be of approximately the same order of toxicity to white rats. Both compounds produced anorexia, loss of weight, hyperexcitability, and tremors. The period required to kill the rats was longer for chlordan than for DDT. The latter caused less pulmonary injury but greater injury to the liver.

Formulations:- Chlordan preparations are available as wettable powders, oil concentrates, water miscible concentrates, and dusts. Each is especially suitable for certain specific purposes as has been previously discussed in connection with the corresponding DDT formulations. However, the water miscible concentrates have proved to be the most versatile of the chlordan formulations. They are not only useful for general agricultural application, but also in household, industrial, and veterinary applications. The concentrations of chlordan in the formulations, as well as the amount applied, vary greatly depending on the pest to be controlled and a number of other considerations. Inquiries regarding the use of the material for a specific problem in a given locality should be directed to the agricultural officials of that locality.

TOXAPHENE

Introduction:- In a program of cooperative investigation by the University of Delaware and Hercules Powder Company, the fourth of the series of chlorinated hydrocarbons described in this chapter was developed. This compound is a chlorinated camphene known as toxaphene. In the first report on the potentialities of this new compound it was stated that the material competed successfully with DDT as a mosquito larvacide (836). It was then found to be effective against household insects in space-spray formulations. Like DDT, toxaphene is slow in its action, and quick-acting paralytic agents have been added to the space sprays. It has been found to be as toxic as DDT to potato leafhopper and as effective as

rotenone against the Mexican bean beetle (837), and has given excellent control of
several of the more important cotton insects (461, 678). Against some of the
other insects this compound is not as toxic as DDT. Like all other insecticides,
toxaphene was found to have special merit against certain insects, and like the
others, also, it falls far short of being a "cure-all", a goal not yet reached
even by these so-called "superinsecticides". Its residual effect is not as prolonged as that of DDT, probably because of its volatility, which is greater than
that of DDT, but not as great as that of the gamma isomer of benzene hexachloride.

Toxaphene is readily formulated into oil-soluble, water-dispersable, wettable-
powder and dust concentrate forms. The promising results of such investigations
as have been made to date, together with the fact that terpenes are available in
large quantities, appears to assure a prominent position for toxaphene in the
household and agricultural insecticide field.

Chemistry:- Toxaphene with an approximate empirical formula of $C_{10}H_{10}Cl_8$, has
been designated as a chlorinated bicyclic terpene (837). The technical product,
which is the insecticide base, is "a cream-colored, waxy solid with a mild piney
odor; it contains 67 to 69 per cent chlorine; it melts in the range of 65° to $90^{\circ}C$;
it has a density of 1.6; and is highly soluble in common organic solvents" (678).
Of special practical interest is the fact that toxaphene is highly miscible with
deodorized kerosene or other mineral oils. It is not soluble in water.

In common with DDT, the gamma isomer of benzene hexachloride, and chlordan,
toxaphene evolves HCl upon heating. The rate depends on temperature and the
amount of catalytic impurities. However, the technical grade of material can be
held for 2 weeks at 50° C with no increase in HCl content. Toxaphene will liberate HCl at a slightly greater rate than DDT upon exposure to ultraviolet light.
The evolution of HCl takes place more slowly under diffused light. The solid form
of this material can be stored in cardboard cartons or paper bags for at least a
year without deterioration, and the same can be said for solutions stored in suitable cans. (678) Like DDT, toxaphene has one labile chlorine atom and is attacked
by bases. As with all the other chlorinated hydrocarbons, strongly alkaline
diluents should be avoided.

As a 25% wettable powder, toxaphene was found to be compatible with lead
arsenate, lead or calcium arsenate, nicotine sulfate, summer oils, sulfur, Thanite,
pyrethrum, rotenone, DDT, neutral emulsifying and wetting agents, Fermate, Bordeaux
mixtures, Yellow Cuprocide and copper Compound A (837, 678).

Toxicology:- On the basis of patch test studies on 200 unselected human subjects, it was established that toxaphene is neither a skin irritant nor a skin
sensitizer. Likewise it appears that toxaphene does not cause allergenic action.
When administered orally in deodorized kerosene, it was found to be considerably
less toxic to white rats than when dissolved in corn oil. In the latter it was
compared with DDT and found to be more toxic to the rats. This is to be expected
in view of the digestibility of the corn oil. The dosage of toxaphene for a
L.D. 100 (that dose which killed all of a series of 10 or more animals) was found
to be 50 mg/kg when dissolved in corn oil and administered orally to dogs. This
compound appeared to kill the dogs by the stimulation of the central nervous
system. Sodium pentobarbitol, a central nervous system depressant, saved 5 of 6
dogs given a dosage of 60 mg/kg of toxaphene, normally enough to kill all the test
animals.

Toxaphene solutions are readily absorbed through the skin and adequate protection should be provided for persons whose skin is likely to be exposed. In
solid form, however, it is not readily absorbed. This compound can cause toxic
effect on the nervous system when ingested, inhaled, or absorbed through the skin.

Toxaphene is phytotoxic to cucurbits, but so far no other plants tested have
suffered appreciable injury from the wettable powders and dusts used at concentrations of insecticidal value. It appears to have practically no fungicidal
action. (837)

152

CHLORINATED HYDROCARBONS

Formulations:- Toxaphene is used in the form of dusts, wettable powders, oil solutions and emulsion concentrates. Dusts may be prepared from a 40% concentrate aving a minimum of 95% passing through a 325-mesh wet sieve. The wettable powder ontains a 25% concentration of toxaphene. In the oil soluble concentrates, deodorized kerosene, crude kerosene, or aromatic petroleum hydrocarbons are satisactory solvents, the concentration of toxaphene being usually 25%. Since the ater miscible concentrates are usually used because the oil is for some reason bjectionable, the object should be to get as high a concentration as possible of oxicant. As high as 75% of toxaphene in deodorized kerosene is possible in the reparation of the emulsion.

COMPOUND 118[1]

A chlorinated hydrocarbon recently developed by Julius Hyman & Company, known s Compound 118, has been found by the manufacturers to equal or exceed the gamma somer of benzene hexachloride in insecticidal effectiveness and to be comparable o chlordan in residual activity. No common ame has as yet been given to this compound.

Chemistry:- Compound 118, chemically esignated as 1, 2, 3, 4, 10, 10-hexachloro-:4, 5,:8-diendomethano-1, 4, 4a, 5, 8, 8a-exahydronaphthalene, has the empirical ormula $C_{12}H_8Cl_6$. This is a white crystaline solid melting at 100°- 103°C. It is oluble at room temperature in all the ommon organic solvents, including the ighly refined paraffinic hydrocarbons, and is insoluble in water. At room temperture Compound 118 is nearly odorless, but when warmed it has a mild, pine-like dor.

Compound 118

Compound 118 appears to be compatible with the other insecticides and fungicides. In contrast to the previously discussed chlorinated hydrocarbons, Compound 118 is completely stable in aqueous and nonaqueous solutions and suspensions of alkaline materials. The volatility of Compound 118 is slightly greater than that of chlordan.

Toxicology:- At equally effective doses, Compound 118 appears to be slightly slower in its insecticidal action than chlordan. However it has been found that only 0.15 lb. of Compound 118 per acre is sufficient to control grasshoppers.

Preliminary investigations have indicated that the LD/50 to white rats, in single doses, may be 15 mg per kg of body weight. Pending further toxicological investigations, the same precautions should be observed in using this compound as with other chlorinated hydrocarbons discussed in this section.

Plants appear to have a high tolerance for Compound 118.

Formulations:- For preliminary testing by Federal and state research agencies, the following formulations have been prepared: (1) an emulsifiable concentrate containing 1 pound of actual Compound 118 per quart, (2) a 25% wettable powder, and (3) a 1% or a 2.5% dust. This new material is at present intended only for strictly experimental use.

COMPOUND 497[1]

Another chlorinated hydrocarbon compound, known merely as Compound 497, also exhibits a high degree of insecticidal activity. It is stated to be comparable to the gamma isomer of benzene hexachloride in its insecticidal activity. It is slower in its action, but has a much longer residual effect, being comparable to DDT in this respect.

[1] Information courtesy of Julius Hyman & Company, Denver, Colorado.
Compound 118 is now called "aldrin" and Compound 497 is called "dieldrin."

Chemistry:- The chemical name of Compound 497 is 1, 2, 3, 4, 10, 10-hexa-chloro-6, 7-epoxy-1, 4, 4a, 5, 6, 7, 8, 8a-octahydro-1, 4, 5, 8-dimethanaphthalene. It has the empirical formula $C_{12}H_8OCl_6$. Compound 497 is a white, crystalline solid. When absolutely pure, it melts at 175 - 176 C. As ordinarily supplied it melts above 150 C. It is essentially odorless. At 26 C the following solubilities, in grams per 100 grams of solvent, have been determined: in methanol, 4.9; in acetone, 54.0; in benzene, 75.0; in hexane, 7.7; and in petroleum oil, 4.3.

In common with Compound 118, Compound 497 is completely stable to the action of alkalies and acids normally encountered.

Compound 497

Toxicology:- In a comparison of the effectiveness of the various chlorinated hydrocarbons against 10 species of insects, Compound 497 was found to have the greatest insecticidal efficacy and the most prolonged residual effect. The relative toxicity of the compounds was as follows: Compound 497 > Compound 118 = heptachlor[1] = gamma benzene hexachloride > chlordan > chlorinated camphene > DDT. The relative residual effectiveness of the above compounds appeared to be as follows: Compound 497 > DDT > Compound 118 > heptachlor = chlordan > gamma benzene hexachloride. (478a)

Formulations:- Julius Hyman & Co., the discoverers of Compound 497, suggest that this insecticide may be formulated for insecticidal application as follows:

1. SPRAYS: "(a) Solutions in oils or other organic solvents.

"(b) Emulsions prepared by dissolving the compound in a solvent and dispersing the solution in water with the aid of a suitable emulsifying agent, such as Atlox 1045A, Triton X-100 and Duponol O.S.

"(c) Suspensions made by precipitating the compound from acetone solution by the addition of water.

"(d) Suspensions may be prepared by the addition of wettable powders to water with agitation.

2. DUSTS: "(a) Finely ground compound thoroughly mixed with a suitable diluent such as pyrophyllite, talc or diatomaceous earth.

"(b) Impregnation of diluent with solution of compound in a volatile solvent which is allowed to evaporate and the mixture finely ground."

Because of the prolonged residual effectiveness of the compound, its use is discouraged on crops which are to be used as human or animal food. It is believed that the insecticide appears most promising in the control of "(1) flies and mosquitoes, (2) moths and carpet beetles, (3) cotton insects, (4) forestry pests, (5) termites, (6) pests in soil, (7) pests of lumber products and (8) industrial pests not actually infesting food products."

---

[1] (or 3a), 4, 5, 6, 7, 8, 8-heptachloro-4:7-methano-3a, 4, 7, 7a-tetrahydroindene. This compound was isolated by R. B. March and C. W. Kearns from the mixture known as technical chlordan. Heptachlor is now being tested by various state and federal experiment stations and is said to show a high degree of promise as an insecticide.

CHAPTER X

ORGANIC PHOSPHATES

INTRODUCTION

The insecticidal developments in the field of the chlorinated hydrocarbons, one quickly following the other, would in themselves have kept insecticide chemists and entomologists well occupied for years to come. Never before had there been such feverish interest, and thousands of technical, semi-technical, and popular articles testify to the enormous amount of work that has been done along this line of activity. However, the termination of the war in Europe, with the subsequent seizure of German chemical patents, revealed the fact that another group of synthetic organic compounds, the phosphates, gave evidence of being equally spectacular as insecticides. It has developed, indeed, that parathion, not reported until late in 1947, bids fair to surpass any of the chlorinated hydrocarbons in insecticidal effectiveness. In addition, it is a good acaricide, and the chlorinated hydrocarbons are generally conspicuously ineffective against the majority of the acarina (mites). The organic phosphates are contact and stomach poisons and their vapors act as fumigants.

Unfortunately parathion, as well as the other insecticidal phosphates that have been developed to date, are far more toxic to man and warm-blooded animals than the chlorinated hydrocarbons. They all have the characteristics of attacking the intestinal tract and of destroying cholinesterase. In a recent test, men who had been applying parathion preparations in orchard experiments for prolonged periods were found to have a 30% reduction in blood cholinesterase. It is not yet known to what extent this disadvantageous characteristic of the phosphates will limit their use as insecticides. However, some commercial use of the phosphates is already being made in connection with treatments in which the poisonous residue is of no concern and where proper safeguards can be exercised to protect those applying the insecticide.

HEXAETHYL TETRAPHOSPHATE

Introduction:- In a German patent applied for by G. Schrader in 1938, a process for the production of hexaethyl tetraphosphate (HETP) was described (764, 765). A proprietary product known as Bladan, containing 60% of the crude compound, was used as an insecticide in Germany in World War II.

HETP was of special interest because it was reasonably toxic to mites and aphids, the majority of which were not killed by DDT. For example, HETP in dusts and in aqueous solutions was found to be more effective than nicotine sulfate and nicotine alkaloid against both the cabbage aphid and the pea aphid (127). In California, HETP was widely known commercially in the proprietary product Vapotone, used for control of mites and aphids. Some spraying of citrus trees for aphids was done with this material. HETP was also used to a considerable extent as an aerosol in greenhouses (425). However, HETP was also found to be very toxic to warm-blooded animals and was absorbed through the skin readily. Toxic symptoms also followed the inhalation of the vapors. It was necessary to use extreme care in handling this compound.

Chemistry:- The original process for producing HETP as described by Schrader, involved the reaction of triethyl phosphate and phosphorus oxychloride. Later a similar material was prepared by the reaction of triethyl orthophosphate and phosphorus pentoxide (1013). It is now known that "hexaethyl tetraphosphate" does not exist in this form, but is instead a "fortuitous mixture of organic phosphates which has an average molecular weight and combustion analysis percentages agreeing with the assigned formula" (406). Hall and Jacobson (399) concluded that the so-called HETP, which was given the empirical composition $(C_2H_5)_6P_4O_{13}$, comprises tetraethyl pyrophosphate, ethyl metaphosphate, and possibly also pentaethyl triphosphate.

155

HETP is an oily, somewhat viscous liquid of a light amber color. It has a specific gravity of 1.28. It is miscible with water and many organic solvents, but not petroleum. It is stable when undiluted, but hydrolyzes readily in contact with water. It is important when using HETP that the contents of the spray rig be sprayed out within about 30 minutes to avoid appreciable loss in toxicity. HETP is corrosive to metals, particularly galvanized iron, black iron and tin plate.

Toxicology:- The minimum lethal dose of HETP by oral ingestion has been found to be as low as 5 mg per kilogram of body weight for some species of warm-blooded animals. Absorbed through the skin, 5 - 10 mg per kilo has been fatal, although symptoms of systemic poisoning are noted at dose levels as low as 1 mg per kilo. The systemic effects noted "progress from gastro-intestinal tract upset, anorexia and severe diarrhea to a characteristic "head-drop," a great weakness and apathy, depending on the severity of the poisoning" (31).

The principal mode of action of HETP is by its destruction of cholinesterase, and the symptoms experienced by humans exposed in spraying operations, such as shortness of breath with a sense of constriction in the chest, are to be expected from such an action. Workmen have also experienced a temporary blindness when exposed to too great a quantity of HETP. Human beings must avoid ingestion, skin contact or inhalation of HETP. The following precautions have been recommended by a subcommittee of the Interdepartmental Committee on Pest Control. They are of interest not only in connection with the use of HETP, but more especially for TEPP, which has superseded it in pest control, and presumably other organic phosphate materials, such as parathion. The following rules of precaution should be followed with these materials at least until much more information is obtained than is available at present.

1. "Avoid contact with the skin, especially when handling the concentrated material. Gloves impervious to hexaethyl tetraphosphate should be worn. If skin is accidentally contaminated, wash carefully with soap and water immediately. Individuals should be required to keep shirts buttoned at the neck, sleeves down and buttoned at the wrist.

2. "Avoid the inhalation of hexaethyl tetraphosphate mist, dust or aerosol by wearing a respirator or mask approved by the United States Bureau of Mines.

3. "Adequate personal hygiene and cleanliness of the operation is necessary. At the end of the operation the clothing should be removed, followed by a thorough bath with warm water and soap.

4. "Avoid contamination of food; smoking, eating and chewing tobacco should be prohibited in the operating areas.

5. "Any persons developing symptoms of headache or tightness of the chest when using hexaethyl tetraphosphate should be removed from the exposure. In the case of ingestion of hexaethyl tetraphosphate an emetic, such as mustard or warm soapy water, should be used immediately and the patient referred to a physician.

6. "A dye should be added to hexaethyl tetraphosphate in such a concentration that it will be readily detectable in the final insecticide solution."

Present Status:- As stated before, HETP is believed to be a mixture of two or more compounds, one of which is tetraethyl pyrophosphate. The latter is a far more effective insecticide and does not hydrolyze as rapidly. In addition, tetraethyl pyrophosphate has a higher vapor pressure, which makes possible its distillation at reasonable temperatures, thus enabling the manufacturer to make a pure product. In view of these facts, it is likely that HETP will be replaced completely as an insecticide by tetraethyl pyrophosphate.

TETRAETHYL PYROPHOSPHATE

Introduction:- This compound, commonly known as TEPP, appears to have been first prepared and described by Clermont, in 1854. It was prepared by the re-ion of ethyl iodide and silver pyrophosphate. The hydrolysis of the compound water was noted (169).

Increasing difficulties with mites and aphids following the use of DDT stimu-ed keen interest in the development of insecticides showing promise in the con-l of these pests. The experience with HETP indicated that certain organic de-atives of phosphorus showed considerable promise in this respect. In an evalu-on of a number of phosphorus compounds, tetraethyl pyrophosphate was mentioned being one of the most promising (542). It was one of 8 phosphorus compounds, luding HETP, which were superior to nicotine alkaloid in toxicity to aphids. utions of TEPP as low as 18 ml/400 ml of water have been used for greenhouse igation with good results against red spider young and adults, but not against eggs (1032). Likewise, mealybugs could be controlled or greatly reduced in bers with repeated treatments, but the eggs could not be killed. Laboratory ts with sprays, using a dilution of 1 volume to 1600 of water resulted in good l of Aphis rumicis and red spider young and adults. A related compound, hexa-hyl tetraphosphate (HMTP), at 1 to 2000 resulted in "perfect kill" of red ders. In one experiment, TEPP was isolated from the so-called HETP and was nd to be 3 to 5 times as toxic as the latter (399). It would appear from this t practically all the insecticidal activity of HETP is due to TEPP.

Chemistry (409):- The formula for tetraethyl pyrophosphate is $(C_2H_5O)_4 P_2O_3$. determination of the structure of pyrophosphoric acid, considered to be the ic molecule of which tetraethyl pyrophosphate is an ester, is extremely diffi-t. The structural formula is presumed to be either one of the two presented

ewith. Despite a series of investiga-ns by European workers, there has been conclusive evidence for either struc-e. Another arrangement, proposed by Reinicke, considers the phosphorus m as the center of a tetrahedron, but uires a three-dimensional model for per visualization.

A purified form of TEPP can be dis-led under vacuum. This has been nd to have a specific gravity °/25°C) of 1.19; approximate boiling nt 100°C at 0.01 mm Hg and 150°C at mm Hg (decomposition at higher peratures); molecular weight 290.20;
refractive index (N_D^{25}) of 1.47. It is a colorless, mobile liquid. It is cible with water, acetone, alcohol, benzene, carbon tetrachloride, chloroform, cetone alcohol, ethyl acetate, glycerine, orthodichlorobenzene, pine oil, uene, xylene, alkyl naphthalenes, etc., but is not miscible with petroleum er, kerosene, or other petroleum oils at 36.5°C. The period required to de-pose one-half of the tetraethyl pyrophosphate is about 7 hours.

Based on chemical considerations, Harris (409) states that "It would be ex-ted that tetraethyl pyrophosphate would be compatible with most organic ma-ial such as DDT, chlordane, benzene hexachloride, oils, etc. Organometallic gicides and 'fixed copper' compounds might be considered doubtful; and basic erials such as zinc, calcium and basic lead arsenate and any material contain-lime would be expected to be incompatible."

Like HETP, TEPP attacks black iron, galvanized iron, and tin plate. There is tle or no evidence in preliminary tests that the compound attacks nickel, stain-s steel and aluminum. For packaging the insecticide, glass and certain types lacquer-lined iron containers have been found to be satisfactory. It is be-yed that the very dilute solutions used in spray equipment would have a negli-le corroding effect on most materials of construction. The possibility of this roding effect by the acid products of hydrolysis could be minimized by adding

Tetraethyl pyrophosphate

SUBTROPICAL ENTOMOLOGY

the TEPP to the spray tank after first filling with water. An additional precaution would be to flush out the tank, pump, and attachments with water after
using the equipment. However, before assuming the resistance of the materials of
construction in spray equipment, observations should first be made on the possible
corroding effects of the TEPP (409).

A method of chemical assay of TEPP has recently been worked out by Hall and
Jacobson (399).

Toxicology:- TEPP has been found to be effective against mites, aphids and
other insects at concentrations of from 1:5,000 to less than 1:20,000. Unfortunately the compound is also one of the most toxic of the insecticides to warm-
blooded animals. Experiments with rats, mice, and rabbits have shown that TEPP is
extremely toxic to warm-blooded animals, either administered orally or absorbed
through the skin. Two mg/kg of body weight is the approximate lethal dose for
rats when administered orally. With intraperitioneal injection the LD 50 for mice
is 0.7 mg/kg. Rabbits are killed in less than an hour when TEPP is applied to the
skin at the rate of 48 mg/kg.[1]

TEPP undergoes hydrolysis rather rapidly, with a proportional reduction in
toxicity. In 24 hours a 1% solution of TEPP at room temperature loses 9/10 of its
toxicity to rats. Obviously, the toxicity of residues does not present a public
health problem. Proper precautions in the application of the material are important, however, because of its high toxicity to human beings. The operator
should be shielded as much as possible from the chemical in any kind of formulations, including the vapor. The use of protective clothing, masks and respirators
is advisable. Any of the insecticide accidentally coming in contact with the skin
should be removed by washing with soap and water. See the set of rules for proper
precautions in the use of HETP (p. 156), which apply equally when using TEPP.

Phytotoxicity:- HETP and TEPP have been applied to plants as vapors, sprays,
and as solutions in the soil. The vapors from the heated chemicals in closed containers injured foliage and later caused epinastic responses (curving downward).
HMTP (hexamethyl tetraphosphate) did not cause this response when used in the
same way. HETP and TEPP in water solutions diffused by means of a Funeral Dif-
fuser did not injure the leaves. Water solutions of TEPP with more than 25 mg/50
ml applied to the soil severely injured or killed tomato plants growing in the
soil. Epinasty of leaves was caused by lower dosages (1032).

As ordinarily used and at recommended dosages properly formulated, TEPP preparations have not been injurious to the great majority of plants.

Formulations:- Proprietary liquid formulations may contain as much as approximately 40% tetraethyl pyrophosphates. These may be used in aqueous solutions,
emulsions, dusts, and aerosols. The TEPP solution may be added directly to water
in the proper concentration and wetting or spreading agents may be added if desired.. It may also be dissolved in a suitable solvent, such as xylene and, with
the addition of a suitable emulsifying agent, can be added to water in the desired
concentration to form an emulsion. Dusts may be prepared by adding the TEPP solution to an inert diluent. The latter should be as dry as possible and the finished dust should be used the day it is prepared. Aerosols may be prepared, but
their preparation requires special techniques and equipment for compounding the
formulations and charging the aerosol dispensers.

PARATHION

Introduction:- One of the most recent acquisitions among the synthetic organic compounds of high insecticidal activity is another organic phosphate for
which the term "parathion" has been registered as the common name. This compound
is not only extremely toxic to insects but in addition is generally very effective
against mites. No species of insects or mites have been found to be resistant to

[1]Data from Monsanto Technical Bulletin No. O-46, August 1, 1948.

parathion at concentrations comparable to those now used with any insecticide (361). Extremely small concentrations of this compound are insecticidally effective. At dilutions of 1 to 100,000, parathion will give 80 to 100% kills of Aphis rumicis in 48 hours, and even greater dilutions are effective if a good wetting agent is added.

It is noteworthy that 3 pests that are relatively resistant to DDT, namely the Mexican bean beetle, the German cockroach, and the red spider, are highly susceptible to parathion. A 100% kill of all stages of the Mexican bean beetle can be obtained with a 0.2% dust in the laboratory. Parathion in dilutions as great as 1 to 40,000 gives 100% kill of red spider, and aerosols with 0.5 g per 100 cu. ft. have been effective against this pest, as well as aphids, thrips and mealybugs. A 100% kill of the German cockroach has been obtained in 24 hours with 50 mg of a 0.2% dust. (361)

Parathion has the serious disadvantage of being highly toxic to warm-blooded animals. Nevertheless, it is possible that the extremely dilute concentrations ordinarily required, when used with suitable precautions, may make possible the application of this material with a sufficiently wide margin of safety. Parathion has considerable residual effect, and while this further enhances its insecticidal effectiveness, it makes necessary a very cautious attitude toward the commercial use of the material on edible fruits, vegetables and forage crops.

Chemistry:- The chemical name for parathion is O, O-diethyl o-p nitrophenyl thiophosphate. The pure compound is a yellow liquid, the boiling point of which is estimated at 375°C at 760 mm. The specific gravity is 1.26 and the refractive index at 25°C is 1.5360. The solubility of parathion in water is only 20-25 p.p.m., but it is completely miscible with a wide range of esters, alcohols, ketones, ethers and aromatic and alkylated aromatic hydrocarbons.

$$O_2N-\text{(benzene ring)}-O-P{\underset{O-C_2H_5}{\overset{O-C_2H_5}{\longrightarrow}}}S$$

Parathion

It is only slightly soluble in kerosene and refined spray oils (about 2% in the latter), but the concentration can be increased with the aid of suitable mutual solvents.

Parathion differs from HETP and TEPP in being not readily hydrolized in ordinary tap waters and in acid solutions, but it decomposes in a short period in an alkali solution of pH 11 or such fungicidal mixtures as Bordeaux 8-8-100 or lime sulfur. However, it is compatible with "neutral coppers" as well as wettable sulfurs and the majority of insecticides. It has been found that the insecticidal activity of parathion is not destroyed by atmospheric oxygen nor by ultra-violet light.

Toxicology:- Parathion is believed to be a nerve poison to insects, inhibiting one of the important functional enzymes of nerve-impulse transmission. Insects succumb to parathion by contact, by ingestion of the compound into the alimentary tract, or by fumigation. Preliminary toxicological studies indicated a "muscarine-like effect" on warm-blooded animals, that is, a stimulation of the parasympathetic nervous system.

Studies on animals indicated that symptoms of acute poisoning in man might consist in "abdominal pain and discomfort with possible vomiting and diarrhea, a sense of constriction of the chest, constriction of the pupils, and possible evidence of peripheral spasms or even convulsions. Atropine in therapeutic doses appears to be a physiological antidote, along with the usual measures employed in cases of acute poisoning, but this should be administered by a physician."

In addition to sharing the toxicological properties of HETP and TEPP, it should be borne in mind that parathion also has a prolonged residual effect and

[1]Information obtained in mimeograph form from the American Cyanamid Company.

may be transported into the plant or fruit. This causes the public health problem connected with the use of parathion to be even more serious than with the more readily decomposed phosphates.

In general, insecticides toxic to man are much safer when applied in dry form than when in solution. Parathion and nicotine are exceptions. They are also highly toxic in dry form.

Some idea of the approximate minimal lethal doses of various formulations can be obtained from the following data:

1. Oral administration: albino mice and rats 3-5 mg/kg of body weight; guinea pigs 5-10 mg/kg.

2. Skin absorption: when applied to the shaved skin of rabbits, covered by rubber sheeting, with an 18-hour exposure; technical compound, 63 mg/kg; technical 20% in certain organic solvents, 92-229 mg/kg.

3. Inhalation of dusts. Animals in a closed chamber of 2 cu. ft. were exposed to dust under continuous forced circulation. Fifty grams of dust was used with an exposure of 3 hours. With a 1% dust, 1 mouse died out of 14 (expressed as 1/14); rats, 0/4. With 15% dust: mice 20/20; rats 2/2; guinea pigs 0/2.

4. Exposure to mists. Rats, mice and guinea pigs were exposed to concentrations of parathion as a mist for 2 or 3 hours. In a series of tests, concentrations were increased up to 32.5 mg in an 8 cu. ft. chamber without fatalities. Examination of mice and guinea pigs after 2 exposures which did not cause death, disclosed hemorrhagic areas on the lungs.

5. Effect on pupillary response of rabbit eyes. The introduction of a drop of technical parathion into the eye sac caused lacrimation, mild hyperemia, and constriction of the pupil. The pupil returned to normal overnight. This effect was not produced by the introduction into the eye sacs of 7 rabbits of a 15% wettable powder.

6. Toxicity of vapors. Preliminary experiments with albino mice indicated that the vapors of parathion are toxic if inhaled over a considerable period.

7. Chronic toxicity. In a long-range test which has not yet been consummated, albino rats which had been receiving food containing 50 p.p.m. of parathion in their food experienced normal weight gains.

The same precautions should be used in handling parathion as any other insecticide known to have considerable toxicity to warm-blooded animals. After applying or working with the material, the hands and other exposed parts of the body should be washed thoroughly with soap and water. It should be borne in mind that the skin, as well as the mouth, is an avenue of entry of toxic materials into the human system. An effort should be made to avoid exposure to the material in a confined area. A respirator which bears the approval of the U. S. Bureau of Mines for use with toxic dusts and mists should be used. Avoid contamination of foodstuffs. See further precautions on page 156.

Residual Effect:- Parathion is more volatile than DDT or the gamma isomer of benzene hexachloride. However, as stated before, the period of residual effect depends in part on the susceptibility of the insect species in question. Parathion, being often many times more toxic to insects than the chlorinated hydrocarbons, may in some cases have a more prolonged effect despite its greater volatility. The residual effect of this compound has varied from 3 days to over a month.[1]

---

[1] Information from Technical Bulletin No. 2 of the Insecticide Department of the Agricultural Chemicals Division of the American Cyanamid Co.

ORGANIC PHOSPHATES

Residue Tolerance:- The present tentative residue tolerance on parathion set by the Food and Drug Administration is 2 parts per million, with the provision that the tolerance level might be raised if such action is justified by future information. Many residue investigations on edible crops have shown that, with normal dosages and under a wide variety of conditions of application, the parathion residue will be well below the above tolerance even if the insecticide is applied as late as 10 days before harvest. Nevertheless, the present tendency is to recommend that parathion should be applied to edible crops at least a month before the time of harvest.

Formulations:- Parathion has been used experimentally or commercially as a 15% wettable powder at from 1 to 4 lbs. to 100 gallons and as a dust with concentrations of from 0.25 to 2.0%, and in these formulations it has proved to be effective against a remarkably wide range of insect and mite pests. Some experimental work has also been done with a 20% liquid concentrate.

Experiments have been made with parathion in aerosols used in greenhouses (788). Ten per cent of the technical product was used with either acetone and methyl chloride 1 to 8 or methyl chloride alone as the propellant. Aerosol bombs were used to generate the aerosol. The dosage recommended as the result of these experiments was 10 grams to 1,000 cu. ft. or 1 lb. to 50,000 cu. ft. of space. When applying parathion aerosols, a tight-fitting, full-face gas mask equipped with a suitable canister is worn. After each application the worker should thoroughly wash his hands and face with soap and water. Aerosol solution that may be accidentally spilled on the skin should be immediately removed. It is stated that several companies have been licensed and authorized to produce parathion aerosols for sale.

Much work needs to be done with parathion to determine how and when it is to be applied in cases in which a public health hazard is involved or in which the destruction of beneficial insects might offset whatever advantage might be gained by the use of this new material.

Although information on the compatibility of parathion with other materials is not complete, it has been stated that the compound appears compatible and safe with wettable powders and dust blends of DDT, benzene hexachloride and toxaphene, with rotenone, pyrethrins, insoluble coppers, wettable and dusting sulfurs, insoluble metal salts of dithiocarbamic acids, wheat flour, skim milk, pyrophyllite, fuller's earth, and diatomaceous earth and bentonites, provided the pH of the latter two is below 8.5. (412)

CHAPTER XI
MISCELLANEOUS ORGANIC INSECTICIDES

DINITRO-O-CRESOL

One of the earliest of the synthetic organic insecticides was dinitro-o-cresol. Its extensive use in the German dye industry made this compound commercially available as an insecticide in large quantities in Germany as early as 1892. The chemical name for the compound is 2,4-dinitro-o-cresol or 3,5-dinitro-2-methyl phenol. It is also known as DNOC.

The pure compound is a yellow solid melting at 85.8° C and with a specific gravity of 1.4856. It is only slightly soluble in water, but the sodium, potassium and ammonium salts are soluble and are used not only in pest control but also for killing weeds. DNOC is very toxic to insects and mites, but it is so highly phytocidal that its use in pest control is practically limited to dormant spraying on fruit trees, especially as an ovicide. DNOC products contain 40% active ingredients, the remainder being usually talc and a wetting agent. They are ordinarily used with dormant oils in the control of mites or mite eggs, aphid, leafroller or bud moth eggs, and San Jose and oyster shell scales.

4,6-Dinitro-o-cresol

DINITRO-O-CYCLOHEXYLPHENOL

Another dinitro compound, dinitro-o-cyclohexylphenol (DNOCHP) or, more specifically, 2,4-dinitro-6-cyclohexylphenol, was investigated by Kagy and Richardson (473), whose experiments indicated the value of this compound dissolved in dormant oil in the control of many species of insects or their eggs. Kagy (472) found the parent compound as well as its calcium, magnesium, lead, and copper salts, to be several times more toxic as a stomach poison to the corn earworm than acid lead arsenate. The calcium salt was 4.4 times more toxic than acid lead arsenate to the corn earworm, 17 times more toxic to the armyworm and significantly more toxic to the cabbage worm. Structurally related compounds had little toxicity as stomach poisons.

Chemistry:- DNOCHP is a nearly odorless, yellowish-white, crystalline compound having a molecular weight of 266. Crystallized from alcohol and certain other solvents, including petroleum, it is in the form of needles; from acetone it crystallizes in the

2,4-Dinitro-6-cyclohexylphenol

form of hexagonal plates. The pure compound has a melting point of 106° C and specific gravity of 1.2717. The solubility at 300° C of DNOCHP in kerosene and spray oils was found to vary from 2.29 to 3.71 wt.% and in dormant oil (U.R. 71%) to be 16.00 wt.%. The phase distribution of DNOCHP in dilute emulsions of spray oil in water varied with the pH. Over 95% of the DNOCHP remains in the oil phase of a dilute emulsion when the pH of the aqueous phase is less than 5. (111)

Toxicology:- A number of papers have been reviewed which deal with the toxicology of DNOCHP with reference to warm-blooded animals. In one of the investigations the minimal lethal dose range of the compound for single subcutaneous administration in olive oil to mice and guinea pigs was between 20 and 45 mg per kilogram of body weight. For oral administration the minimal lethal dose was between 50 to 125 mg per kilogram. (451)

MISCELLANEOUS ORGANIC INSECTICIDES

It was evident for the toxicological work done with mammals that DNOCHP is toxic only in comparatively large doses. Unlike 2-4-dinitrophenol, the 2-4-dinitro-6-cyclohexylphenol does not stimulate the metabolism of man.

DNOCHP preparations as dusts or as sprays have not presented a public health problem, but even the dusts have occasionally caused damage to foliage.

Formulations:- Following several years of investigation of DNOCHP for the control of the citrus red mite (117), a proprietory product (DN-Dust) appeared on the market in 1938. This product contained 1% DNOCHP adsorbed on walnut shell flour and was used at the rate of about 1 lb. per tree. It gave fairly satisfactory control of red mite, if properly used, and was very valuable to growers who wished to use HCN gas for scale insects and did not wish to use even the low dosage of oil required for red mite control. The material was also satisfactory for the control of mites on walnuts, almonds and peaches. (107, 118) An important factor in the success of the DN-Dust was the development by University of California investigators of improved multi-vane fish tail type of orchard power dusters with mechanically or manually operated deflecting devices at the orifices (118).

In 1940-41 the product DN-Dust D-4 was recommended for citrus red mite control. This formulation contained 1.7% dinitrocyclohexylamine in Frianite, a volcanic ash. It was found that 1.68% of this salt was equivalent to 1% of the parent compound. This material was less apt to cause injury to foliage, an improvement especially needed in the dusting of grapes and walnuts. The new formulation had the advantage of having a longer residual effect against the mites than its predecessor. It also was used at the rate of 1 lb. per tree.

The next advance in the development of DN dusts for mite control was DN-Dust D-8, which was the same as DN-Dust D-4, except that it contained 2% light medium oil in the dust to increase the adhesion of the dust to the trees. DN-Dust D-8 was not as easy to apply as DN-Dust D-4, and the latter is still used today to some extent with the smaller dust machines. Either dust gives good control of the active mites on the tree at the time of treatment, but does not kill the eggs. The larvae as they hatch from the eggs, however, succumb to the residue left by the dust treatments, providing it has not been largely removed by a rain. The individual grower must judge from his previous experience whether a single dust treatment will control the citrus red mite in his orchard. Sometimes, in some localities, a single treatment may suffice, but it is more likely that 2 dustings per year may be necessary, the second preferably following 10 to 20 days after the first. Sometimes even an oil-sprayed orchard will require a supplemental treatment in the form of a dust if the red mite is to be adequately controlled. See page 373 for precautions to be observed in the use of DN products on citrus trees.

In 1944 a DN product appeared on the market which could be used in a spray. This material is known as DN-111. DN-111 contains 20% of the dicyclohexylamine salt of 2-4-dinitro-6-cyclohexylphenol, which is water-soluble. Some growers believe they obtain better results from a spray. Others may happen to have their own spray equipment and may wish to do the mite control work themselves, in which case they would use the DN-111, applying a "skeleton spray" with a concentration of 3/4 to 1 lb. of DN-111 to 100 gallons of water.

DN-111 has been successfully used with DDT in combination codling moth-European red mite control in the Pacific Northwest. The DN-Dusts may also be combined with DDT to counteract the usual tendency of the latter insecticide to increase the mite population.

K-1875 (NEOTRAN)

K-1875, chemically known as bis(p-chlorophenoxy) methane, is the toxic constituent of Neotran, a proprietary product used principally as an acaricide. Neotran is a water-dispersible powder containing 40% by weight of K-1875. K-1875 is a product of Dow Chemical Company which was found to have special merit in the control of the citrus red mite (464). In aqueous suspensions it has been found effective against the eggs as well as the active stages, and the residue left by the spray is toxic to the mites not contacted by the spray, or hatching from surviving eggs, for one to several weeks after treatment. As the result of extensive field experiments in 150 orchards comprising 1400 acres, Jeppson (464) recommended the use of Neotran in spray-duster applications, using 7.5 to 10 lbs. per acre applied in 100 gallons of water or as a dilute spray with conventional equipment., using 1.5 to 2 lbs. per 100 gallons of water. The material was found to be compatible with cryolite, nicotine sulfate, derris or cube, zinc oxide, zinc sulfate, manganese sulfate, soda ash, lime, Fermate, Zerlate; and oil.

Neotran has also been found to be effective against the European red mite, Paratetranychus pilosus (C. & F.), the two-spotted mite, Tetranychus bimaculatus Harvey, the Pacific mite, T. pacificus McG., and the six-spotted mite, T. sexmaculatus Riley.

Chemistry:[1] Bis(p-chlorophenoxy) methane has the empirical formula $C_{13}H_{10}O_2Cl_2$. It has a molecular weight of 269.13 and specific gravity of 1.3589. It is insoluble in water. Its solubility in wt./% at 25° C in a number of organic solvents is as follows: acetone, 189; ethyl ether, 87; benzene, 40; carbon tetrachloride, 28; methanol, 0.5; VMP. Naphtha, 0. Bis(p-chlorophenoxy) methane is stable in neutral and alkaline media, but may be hydrolyzed by boiling in dilute aqueous acids.

Bis(p-chlorophenoxy)methane

Toxicology:- Neotran has been used safely on apples, cherries, peaches, prunes, and plums as well as on citrus. On the latter no indications of injury to foliage or fruit have been found. Some stunting of foliage has been noted on certain succulent plants.

Research conducted by the Biological Research Laboratory of Dow Chemical Company indicates that K-1875 is of a low order of toxicity to warm-blooded animals, being only about 1/10 as toxic as DDT. It is not readily absorbed by or significantly irritating to the skin. In fact, K-1875 is considered to be one of the least toxic insecticides ever developed.

DI-2-ETHYL HEXYL PHTHALATE

This compound, also known by the code number 899, is an amber-colored, viscous liquid. Interest in the compound is enhanced by the fact that it will dissolve 25% by weight of DDT and is also a good solvent for rotenone, pyrethrum, and other insecticides and some fungicides. At 1 quart to 100 gallons of water, 899 has generally resulted in control of citrus bud mite on lemons for a longer period than oil spray, and against citrus red mite the same concentration has usually been as effective as oil spray. It has been found to be effective against the six-spotted mite if applied before the characteristic "cupping" of the leaves. Concentrations effective against bud mite and citrus red mite have not controlled citrus rust mite, scale insects, thrips, and certain other citrus pests (464).

HYDROXYPENTAMETHYLFLAVAN

This compound, chemically known as 2-hydroxy-2,4,4,4, 7-pentamethylflavan ($C_{20}H_{24}O_2$), has shown some promise against the European red mite and the spider mites and consequently has been considered as a possible adjunct to DDT. It cannot, however, be said to have attained commercial importance to date.

---

[1]Data from Dow Chemical Company.

CHAPTER XII
SPRAY OILS

CLASSIFICATION AND COMPOSITION
OF CRUDE PETROLEUM OILS

The word "petroleum" means "rock oil", being derived from the Greek petros (rock) and oleum (oil). Petroleum is in the main an oily liquid mixture of numerous hydrocarbons, the existence of which is believed to be due to the decomposition of the remains of animal and vegetable marine organisms. Gravitation, capillarity, pressures, or other causes, are believed to have caused the liquid to be moved through porous rocks and stone until it reached some impervious formation (fig. 82). Upon encountering the impervious barrier, the oil, if in sufficient quantity, completely saturated the formation below; usually sand. The oil thus collected in "pools" varying greatly in extent and in depth. The oil-producing areas are called "fields" and regionally these fields are often referred to as Eastern, Mid-Continent, and Western or Californian.

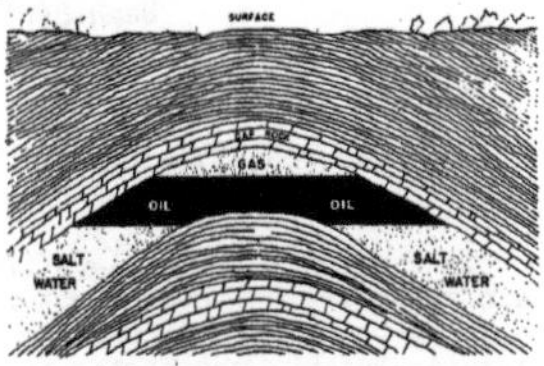

Fig. 82. Rock strata, showing location of oil-bearing sand.

As might be expected, a given oil-bearing sand ordinarily contains a crude oil of reasonably uniform chemical and fractional composition, in the same or even in widely separated pools. There are cases, however, of oil fields close together producing quite different crudes from the same stratum. Thus two Grozny (Russian) oil fields only 10 miles apart produce crude oils from the same tertiary sands, yet one is highly paraffinic and the other highly asphaltic (760).

In the same pool, different oil-bearing sands may produce similar or different types of crudes. In Pennsylvania, strata ranging from Devonian to Pennsylvanian formations commonly produce crude oils of similar chemical and fractional composition. On the other hand in California, in a given pool, crude oils are usually heavier and more asphaltic from the upper sands, while in Baku, Russia, the opposite is usually the case. Geological identification of crudes from fields in which only a single stratum is being utilized is comparatively easy, while in fields in which several oil-bearing strata are exploited, identification is more difficult.

The classification of crude oils is usually based on their chemical composition. In the United States, crude oils are designated as paraffin-base and asphaltic-base,[1] depending on the content of paraffin wax or asphalt. The majority of crudes do not belong distinctly to either of these two types, but have intermediate properties and are called "mixed-base" crudes.

In the East the crudes are predominantly paraffin-base in character, in California they are predominantly asphaltic-base, while the Midcontinent crudes tend to be of the mixed-base type, but exceptions may be found among individual wells in each of these areas.

The hydrocarbons of petroleum are broadly classified as paraffins, naphthenes, aromatics, and unsaturates. This provides a sharp distinction for low and medium molecular weight hydrocarbons, for among these an aromatic or naphthenic ring or a

---

[1]The term "naphthenic-base" is frequently seen in the literature, but this is an outgrowth of the earlier belief that asphaltic-base crudes consisted predominantly of cyclic or "naphthenic" hydrocarbons. This is rarely the case, and in the United States true naphthenic crude oils are represented only by some California crudes. All other asphaltic-base crudes contain a higher percentage of combined paraffins, aromatics, resins and asphaltenes than of naphthenes (760, p. 416).

double bond imparts characteristic properties. The high molecular weight hydrocarbons, however, might contain several structures (paraffinic, aromatic and naphthenic) in a single molecule without manifesting the characteristic properties of the dominating structure. For these compounds Waterman's "ring analysis" may be used to reveal significant data. Let us take, for example, a paraffinic crude oil of the following composition (760):

Paraffin hydrocarbons	40%
Naphthenes	48%
Aromatics (sulfonatable)	10%
Resins and Asphaltenes	2%

Ring analysis gives the following figures for the same crude:

Paraffinic side chains	78%
Naphthenic rings	16%
Aromatic rings	6%

The four broadly classified types of petroleum hydrocarbons mentioned above may be briefly described as follows:

1. _Paraffins_. A well-defined class characterized by the aliphatic (non-cyclic) chain structure. Other structures are absent. An enormous number of isomers is possible, the number increasing with the size of the molecule. The higher the molecular weight, the higher the boiling point, specific gravity, and viscosity. The first 4 members of the paraffin series are gases, the next 10 are liquids at ordinary temperatures, and beginning with pentadecane ($C_{15}H_{32}$), the remaining paraffin hydrocarbons are solids. The paraffin series has the type formula C_nH_{2n+2}.

Examples:

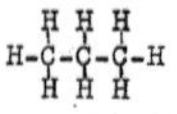

| Propane | Butane | Iso-butane |

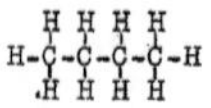

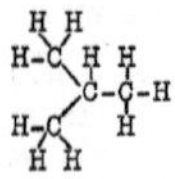

2. _Naphthenes_. Saturated ring hydrocarbons. In low-boiling fractions the naphthenes are monocyclic and in high-boiling fractions they are polycyclic. The usual type formula for the naphthenes is C_nH_{2n}, but saturated compounds of the formula C_nH_{2n-2} or C_nH_{2n-4} may be found. These may represent 2 or more polymethylene rings bound by simple bonds.

Examples:

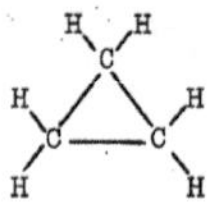

| Cyclopropane | Cyclohexane |

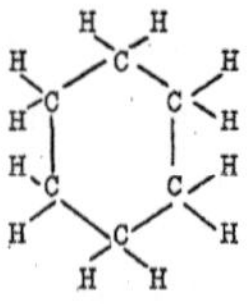

3. Aromatics. Ring hydrocarbons with conjugated double bonds. They may be, like the naphthenes, monocyclic or polycyclic. The aromatics occur principally as benzene and its derivatives and also as methyl compounds of benzene (toluene, xylene, etc.). Naphthalene has been found in various oils.

166

Examples:

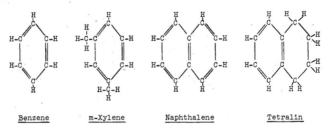

Benzene   m-Xylene   Naphthalene   Tetralin

It will be noted that cyclohexane, a naphthene, differs from benzene, an aromatic, only in the number of hydrogen atoms attached to the carbon atoms.

Combined naphthenic and aromatic rings may exist in a single compound, but even one aromatic ring in a polycyclic hydrocarbon imparts the aromatic properties, such as greater solubility in solvents, and the capacity for sulfonation. It should also be borne in mind that cyclic hydrocarbons may have paraffinic (aliphatic) side chains, sometimes rather long.

4. Unsaturates. All hydrocarbons, aliphatic or cyclic, which have one or more active double or triple bonds. They include, therefore, not only olefines, acetylenes and similar series, but also partially hydrogenated cyclic hydrocarbons; for example, terpenes are included. They are characterized as a group in regard to their capacity for forming addition compounds with halogens, their behavior with concentrated acid, their being easily oxidized, and their tendency to polymerize.

Examples:

Propylene   Isobutylene   Styrene

The importance of the unsaturates, from our point of view in the study of spray oils, is that a double bond is a potentially reactive center or point of attack. It will later be shown that the tendency of the unsaturates to oxidize to asphaltogenic acids is the principal non-physical factor involved in phytotoxicity.

Other Compounds. In addition to the above groups, petroleum contains resinous and asphaltic compounds. These are amorphous constituents which may contain oxygen, sulfur and nitrogen. They occur mostly in high-boiling fractions.

The chemical classification of the crude used in the manufacture of spray oils is of practical importance, for it has been shown that paraffinic base oils have a much greater insecticidal efficiency, at least against some insects and insect eggs, than asphaltic-base oils (684, 682, 155, 154). A joint spray oil research project between certain members of the divisions of entomology at Cornell University and at the University of California Citrus Experiment Station is now in progress which is designed, among other things, for the investigation of the relation of the chemical composition of petroleum and synthetic oils to insecticidal efficiency.

DISTILLATION OF CRUDE OIL[1]

The principles of oil refining are simple, but actual practice in the refining industry is quite complicated, owing to the many different modifications which have been made in the elementary processes of distillation and purification. Centuries ago the fact was discovered that by heating a quantity of petroleum to successively higher temperatures, vapors are given off which, when condensed, form a series of products. These products differ from one another in regard to volatility, specific gravity, viscosity, and molecular composition.

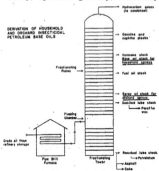

For the preliminary separation of crude oil at the refinery into fractions of different distillation ranges, a process is employed which is known as fractionating. The equipment used in this process is schematically shown in fig. 83. The petroleum crude oil is piped from the refinery storage tanks and is pumped under pressure through pipe coils located in a furnace and heated by flame or by hot combustion gases. The oil must be forced very rapidly through the heated coils in order to avoid decomposition. From the furnace the hot crude is forced into a portion of the fractionating tower known as the flashing chamber, which is an empty space between the upper and lower series of fractionating plates. The reduced pressure in this chamber results in a vaporization or "flashing" of the oil. The vapors then ascend the fractionating tower, and on their way upward they encounter a series of fractionating "plates" or "trays", each perforated with many holes so that vapors may pass through. Usually 20 to 30 plates are located

Fig. 83. Schematic sketch of a pipe still furnace and fractionating tower. Adapted from Hockenyos and Patton (430).

above the flashing chamber, but more may be added when fractions having boiling points very close together must be prepared. Below the flashing chamber there are seldom more than 6 plates. Fractionating towers are often 25 feet in diameter and 120 feet high and may hold 50 plates. The pressure exerted by the vapors keeps the liquid that may condense on the tray from falling back through the holes. Extending up from each hole is a short pipe called a "chimney". Fitted over the chimney is a "bubble cap" (fig. 84). Each plate is equipped with a "down-pipe" extending from above the plate down through the plate and reaching nearly to the tray below. Since this pipe projects above the plate, no liquid can return to the tray below until a considerable amount has collected.

Fig. 84. Diagram of a "bubble cap". Explanation in text.

The liquids are allowed to flow back through the "down-pipe" a number of times, and with each recondensation a more homogeneous fraction is being segregated on the plate. The higher the vapors ascend in the fractionating tower the lower their temperatures, and thus progressively lighter fractions will condense to form cuts of progressively lower boiling points. These cuts may be removed through the pipe at the side of the plate (fig. 84) called a "side-draw". Asphalt, lube stock, gas oil, and kerosene are removed in this way. The lightest vapor does not condense in the tower, but passes from the top of the tower through a pipe called the "vapor line". It is conducted to a condenser and condensed into straight-run gasoline. Further processing is required for the production of high-grade motor fuel.

[1] Good popular and semi-technical books on petroleum refining and technology have been written by Kalichevsky (475), Perry (686) and a number of others. These may be referred to by anyone desiring more detailed information.

A cut of reasonably narrow boiling range may be taken at the appropriate place on the tower, or a number of cuts may be drawn off together to form blends of various boiling or distillation ranges.

It can be seen from fig. 83 that the lightest fractions that can be drawn off from the fractionating tower as liquids are the gasoline and naphtha stocks. Still lighter fractions pass off as oil vapors and are later condensed in another unit. The next to the lightest liquid fraction is kerosene, which along with many other common uses, forms the base for household fly spray, livestock sprays, and even some agricultural crop pest sprays. The next heavier fraction is the fuel oil or gas oil fraction, followed by mineral seal oil, which has limited applications in agricultural spraying. The oil bases for the usual summer and dormant orchard spray oils are taken from the lighter lubricating oil fractions. The residuum, the heavy material withdrawn from the bottom of the fractionating tower, may be used as fuel, or if the crude is suitable, it may be reworked into asphalt. The distillation relationships of the various petroleum fractions are graphically indicated in fig. 85.

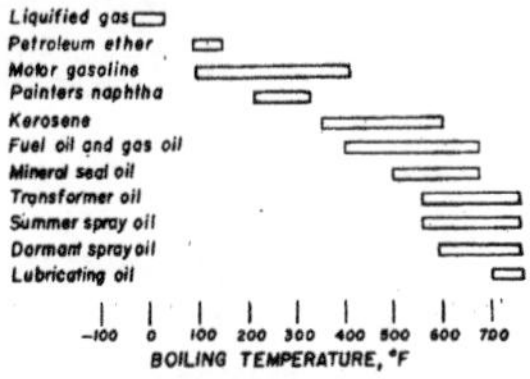

Usually distillation comprises two steps: the first fractionating tower operates under atmospheric pressure; then the heavier fractions are transferred to a second still. Here distillation is carried on not only under partial vacuum, but in the presence of large quantities of steam. This reduces to a minimum the distillation temperatures and eliminates oil decomposition.

Fig. 85. The distillation ranges of various petroleum fractions. All fractions "crack" at about 750°F unless fractionating is done under vacuum.

CHEMICAL REFINING OF OIL

After the required cuts are obtained from the fractionating tower they must be subjected to chemical treatments to remove undesirable substances. Although gasoline is the most important of the petroleum fractions, its refinement is of no interest to us in connection with spray oils. It is interesting to note, however, that while in refining gasoline an attempt is made to retain as far as possible the aromatic and other unsaturated hydrocarbons, because of their anti-knock value, in refining kerosene the object is to completely eliminate these constituents because they cause kerosene to smoke when it is burned, and to have a damaging effect on the lamp wick.

The refining of kerosene, spray oil, and lubricating oil is somewhat similar. Although these fractions are frequently refined with sulfuric acid, there is an increasing tendency to employ for this purpose a process involving the use of sulfur dioxide, which is known as the Edeleanu Process. The oil and sulfur dioxide are mixed in a pressure system to avoid the loss of reagent, which is a gas at ordinary temperatures. The mixture is then cooled to obtain separation of two layers. The lower layer (extract) contains the aromatic and other unsaturated hydrocarbons dissolved in the liquid sulfur dioxide. In the upper layer (raffinate), sulfur dioxide is dissolved in the remaining hydrocarbons. The refined oil is taken from the raffinate, while the extract, which contains the aromatic hydrocarbons, becomes a source for industrial solvents. The lighter portions of the extract, when kerosene is being refined, may be added to gasoline to improve its antiknock qualities. This is an obvious advantage when compared with the sulfuric acid refining process, in which the hydrocarbons can not be recovered from the extract layer.

The sulfur dioxide or Edeleanu process cannot remove as high a percentage of the undesirable constituents as is required in the preparation of the more highly refined spray oils. These require further sulfonation by means of strong sulfuric acid. The remaining unsaturated and aromatic compounds dissolve in the sulfuric acid, forming a dark sludge. This sludge is drawn off at the lower part of the treating tank. Then the oil is neutralized by washing with an alkaline solution and finally it is washed with water.

169

DISTILLATION RANGE

The distillation range indicates the minimum and maximum temperatures between which a given oil distils, and also the per cent of oil distilling at various temperatures within the range.

Fig. 86. Apparatus for determination of the distillation range of a spray oil.

Procedure. A standard apparatus (fig. 86) is used for testing spray oils for their distillation range. At the left of the distillation apparatus is a gas burner or electric heater within a shield. Over the burner or heater is placed 100 ml of the oil sample in a 250 ml Engler flask. A few pieces of broken glass or similar substance are placed in the flask to prevent "bumping" of the oil, and a high temperature thermometer also is placed in the flask. The top of the flask, bearing the thermometer, can be seen in fig. 86 at the top of the compartment at the left. As soon as the oil vapor begins to condense on the neck of the flask, the heat is decreased so that about 5 ml of oil per minute is distilled. The temperatures when the first drop and each subsequent 5 ml of oil distils over are recorded. This is continued until the temperature drops and the residue is gummy. The reason the temperature at 90 per cent distillation is given instead of the end point is that the temperature required to obtain the end point would be too high. Above 400° C (752° F) oils have a strong tendency to crack, that is, to break up into various compounds not originally in the oil as such. This could be prevented by vacuum distillation.

EVAPORATION

An officially recognized method of determining the rate of evaporation of an oil is explained in the following description of an experiment made with kerosene and two summer spray oils - one of the "Light" grade and the other of the "Heavy" grade:

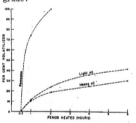

Fig. 87. Evaporation curves for 3 petroleum fractions according to a standardized test.

Procedure:- One gram of oil was placed in a 10 cm aluminum dish containing 9 g of 20-40 mesh sand and spread evenly over the bottom of the dish. This gives an oil surface approximating the amount of surface exposed when oil is sprayed on a leaf. The dish with contents was weighed and placed over an 8 cm hole on top of the boiling water bath. The dish and contents were cooled, dried and weighed at 0.5,1,2,4 and 6 hour intervals. A heavy and a light spray oil (see pp.174,175) and a kerosene were evaporated according to this procedure.

The per cent of oil evaporated, plotted against time, is shown in fig. 87.

VISCOSITY

Viscosity is the resistance of a liquid to change in shape or to fluid motion. It pertains to the flowing quality of an oil. Usually the viscosity of a spray oil, or other oil of the lubrication type, is given as "viscosity Saybolt Universal at 100° F" or simply as "Saybolt viscosity". According to this method, viscosity is expressed in terms of the number of seconds required for 60 ml of oil at a temperature of 100° F to flow through a small orifice (about 2 mm in diameter) in an instrument known as the Saybolt Universal viscosimeter (fig. 88).

SPRAY OILS

Procedure:- The bath vessel of the viscosimeter is filled nearly full of water or oil and an
electric heater is placed in the proper receptacle to heat the water or oil to the temperature
required for the test. About 100 ml of oil is
strained into a beaker and heated to a little
above the required temperature, then poured into
the oil tube, which has been stoppered below with
a cork. With one thermometer in the water bath
and the other in the oil, the bath is stirred by
means of turntable handles and the oil is stirred
by means of the thermometer immersed in the oil
tube. The required temperature is allowed to remain constant for about a minute, the thermometer
is removed from the oil, and the cork is removed
from below the oil tube. The time required for
60 ml of oil to run into the tube is determined
with the aid of a stop watch. The procedure is
repeated until results agree within an error of 2%

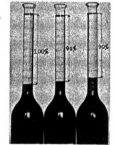

Fig. 88. Saybolt Universal Viscosimeter.

UNSULFONATED RESIDUE (THE
SULFONATION TEST)

As stated before, in commercial practice petroleum oils are refined by the
use of sulfuric acid or liquid sulfur dioxide. When still greater refinement is
desired, as in the production of summer spray oils, hot sulfuric acid is then used
in the final stages of refinement. The saturated hydrocarbons react very slightly
with the acid and when agitation is completed they rise to the top of the tank,
thus separating from the acid sludge. Treatment with various amounts of acid
produces oils of varying degrees of refinement. Out of this practice has grown a
laboratory test which is one of the measures of the safety of an oil to be used on
either dormant or foliaged trees. This is known as the sulfonation test. By this
test one may determine the percentage of oil which will not react with sulfuric
acid or, in other words, the unsulfonated residue (U.R.).

The procedure used in the determination of U.R. can perhaps be best presented
by describing an actual sulfonation test made to determine the U.R. of four oils:
a highly refined light medium summer spray oil, a medium summer spray oil, a heavy
dormant spray oil, and kerosene.

Procedure:- The specific gravities of the oils at 25°
were determined by means of a specific gravity hydrometer
for liquids lighter than water. From the data obtained,
the weights of the oils equivalent to 5 ml were determined.[1] Five ml of each oil was placed in each of 4
Babcock cream test bottles (fig. 89). Twenty ml of
37N H_2SO_4, in 4 equal portions, was slowly added to each
of the oil samples. The bottles were gently shaken or
rotated and care was taken to prevent a rise in temperature to above 60°C by cooling the contents of the bottles
in cold water when necessary. When all the H_2SO_4 was in
and thoroughly agitated, the bottles, except for the one
containing kerosene were immersed in a water bath at
100°C for 1 hour, and were shaken every 10 minutes for 10
seconds. The kerosene was not immersed in the boiling
water because it would volatilize at 100°C. The bottles
were then removed from the water bath, filled with concentrated H_2SO_4 until the oil was all in the graduated
neck of the bottles, centrifuged at 1500 r.p.m. until the

Fig. 89. Babcock cream
test bottles showing
varying amounts of oil
recovered after sulfon-
ation. After Smith (818)

[1] Unless great accuracy is desired, the oil may be pipetted into the Babcock cream bottle.
The pipette, after the preliminary draining, may be drawn several times through the flame of a
Bunsen burner and again drained thoroughly.

171

oil volume was constant (about 5 minutes). The oil was then cooled to 25°C and the volumes of the unsulfonated residues were then read in the graduated necks of the bottles (fig. 89). The values obtained were the number of milliliters of unsulfonated residue remaining of the 5 ml oil which was poured into the bottles. These were multiplied by 20 to obtain the per cent U.R., as shown in the following table:

Oil	Specific gravity	Grams per 5 ml	Unsulfonated residue in ml	Per cent U.R.
Kerosene	0.797	3.707	4.90	98
Light Medium	0.860	4.000	4.85	97
Medium	0.874	4.370	4.60	92
Heavy Dormant	0.921	4.605	3.50	70

It is unfortunate that citrus trees, at least in California, will not tolerate a higher per cent of unsaturates and aromatics in the spray oil. The added treatment, in addition to the usual lube oil sulfonation, to which spray oils must be subjected to purify them to the extent that is found to be necessary in this State, greatly increases the cost of the final products.

DETERMINATION OF THE PERCENTAGES OF THE MAJOR HYDROCARBON STRUCTURES OF PETROLEUM

A method for determining the percentages of the major hydrocarbon structures of lubricating oils, often referred to as the "Waterman analysis" was developed by Vlugter, Waterman and van Westen (910). According to this method, values for molecular weight, refractive index, density, and analine point are used to calculate the percentage of aromatic rings, naphthenic rings, and paraffins (free paraffins and paraffinic side chains). No chemical or physical separation between the hydrocarbon groups is necessary. The Waterman analysis, though suitable for the analysis of the dormant oils (682) is not so applicable for the analysis of the lighter summer spray oils (155). However, such properties as density, analine point, refractive index and specific dispersion are indicative of the proportions of the major hydrocarbon groups present.

Another method of ring analysis, called the Lipkin-Kurtz method, makes possible the determination of hydrocarbon structures by a comparison of the densities of the various oils to be tested at different temperatures. Refractive index may be substituted for density because of the simple relation $\Delta n = 0.60 \Delta d$ existing between the increment of refractive index and the increment of density; or more precise correlations can be computed with the aid of other more complete equations. (538)

RELATION OF PHYSICAL AND CHEMICAL PROPERTIES OF THE OIL TO INSECTICIDAL EFFICIENCY AND PHYTOTOXICITY

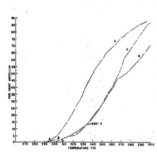

Distillation Range: Relation to Standardization:- If the per cent of oil distilled is plotted against temperature, the resulting curve gives us much information regarding the physical characteristics of the oil. In fig. 90 are shown three such curves, based on the State Division of Chemistry spray oil distillation chart for 1930, which indicate the distillation ranges[1] of three spray oils used in citrus spraying that year. Curve A represents a light medium spray oil. This oil began to distil at

Fig. 90 (at left). Distillation curves for 3 cuts of spray oil. B and C have the same per cent distilled at 636° F (335.6° C).

[1] The complete distillation ranges were not obtained because the temperatures required would "crack" the oil.

293°C (559.4°F) and 88% was distilled at 400°C (752°F). This was a relatively "narrow-cut" oil in comparison with other oils of the same grade in 1930, there being a range of only 107°C (193°F) between the initial distillation temperature and the temperature at which 88% distilled. Other proprietary oils of the same grade were more "wide-cut", usually beginning to distil at from 10° to 20° C lower temperatures.

The uniformity of curve A shows that the oil was "straight-cut" that is, it was not prepared by blending a number of oils of widely differing distillation ranges. Curve B represents a "wide-cut" oil which, from the broken character of the curve, appears to be a blend of a very light, a medium, and a heavy fraction. It was sold as a "Heavy" grade of summer spray oil and was widely used in citrus spraying. It does not appear to be, like oil A, a product of continuous distillation. Although the initial boiling point is lower than that of oil A, only 70% was distilled at 400°C (752°F). According to present day standards, the "heavy ends" in the oil were far too heavy and present in too great a quantity for safety in citrus spraying. In fact in 1946-47, the average temperature for 90% distillation of the "Heavy" grades of proprietary spray oils was only 726.6°F. The range was from 706 to 738°F. Needless to say, oil B caused great injury to citrus trees. It was, of course, very effective against the red scale because of the long-continued residual action of the heavy fractions contained in the oil.

It will be noted that oil C, also a "Heavy" grade of spray oil, had a higher initial boiling point than oil B, yet 87% distilled at 400°C as compared to 70% for oil B. Oil C was a much narrower cut than oil B, and did not have as high a percentage of "heavy ends", yet it was too heavy, even for the spraying of lemons, according to present day standards.

Curves B and C show why several points along the distillation curve must be given before an adequate picture of the characteristics of an oil is presented. Often the per cent of oil distilled at 636°F is given to indicate the heaviness of an oil. It so happens that with oils B and C (fig. 90) the same per cent of oil distilled at 636°F, yet oil B had a greater amount of insecticidally ineffective light ends and a greater amount of heavy ends decidedly dangerous to the tree. The superior quality of oil C would certainly not be indicated by the "per cent distilled at 636°F". In the early days of the use of oil sprays for citrus pest control, kerosene was sometimes blended with dangerously heavy fractions, yet such undesirable blends were not detected because complete distillation data were not required.

Figure 90 indicates the lack of uniformity and standardization among the proprietary spray oils in the early days of the quick-breaking emulsions, a situation which has since been remedied.

It is generally recognized that the heavy spray oil fractions have a greater long-term effectiveness against scale insects and mites, primarily because of their more prolonged "residual effect". Likewise they are more harmful to trees, presumably because of their lower volatility and consequent greater persistence in the foliage and other tree tissue. It behooves the grower and spray operator, therefore, to be able to specify the oil fraction he wishes to use for a given purpose. Ordinarily, at least in citrus pest control, the heaviest fraction which can be reasonably safely used on a given variety of citrus in a given region and season of the year, will be requested.

Five grades of oil, based on the range of temperature at which they distil (distillation range) are prepared for use in citrus spraying in California. Annually the State Bureau of Chemistry publishes in a Special Publication called "Economic Poisons" the temperatures at which 5%, 50%, and 90% of the oil is distilled in the various proprietary brands sold as summer spray oils. The per cent of oil distilled at 636°F is also given. The figures given below were obtained by averaging the data from "Economic Poisons" for 1946-47 for four leading brands of oil emulsions and emulsive oils offered for sale in California during the fiscal year ended June 30, 1947. These samples were drawn from original unopened con-

tainers in possession of dealers, registrants, or users, by inspectors from the
State Bureau of Chemistry.

Grade	Temp. (°F) for distillation of			Per cent of oil distilled at 636°F
	5%	50%	90%	
Light	555	617	675	66.2
Light Medium	571	628	703	55.4
Medium	582	643	715	43.2
Heavy Medium	585	656	728	39.1
Heavy	612	671	727	18.0

Light Grade:- This grade was at one time used to some extent on citrus in the
interior sections, but is now considered too light for satisfactory results
against red and yellow scales and mites. It is no longer used for citrus pest control and appears to be used in California only in the oil-lead arsenate-nicotine
combination applied on Newtown Pippin apples in coastal areas. In this combination it is used at 1 to 1.5% concentration.

Light Medium Grade:- Years of experience has shown that the light medium
grade of spray oil is the lightest that can be used anywhere in California with
reasonable assurance of success when the red scale is involved as a pest. In the
interior districts[1] it is recommended for use on oranges for the control of red,
yellow, black and citricola scales and for the control of mites. Fumigation may
be necessary, in addition to the oil spray, especially if red scale is abundant.
Light medium oil is also often used on oranges in coastal and intermediate areas,
especially the latter. It affords good control of citrus red mite and bud mite.

There is a continually increasing use of light medium oil in the "two-spray
program" for the control of red scale on lemons. Usually one spray is applied in
March or April, and the other in the regular fall spray season. Newcomb (643) reported on experiments begun in the Corona district in 1942 in which two sprays per
year on lemons with 1.7% light medium oil, were compared with two fumigations. In
the fumigated plots, 3 or 4 treatments per year with DN-Dust were necessary to control citrus red mites. Production records for a five-year period showed just as
good yield in the sprayed plots as in the fumigated plots except in one pick in
which a spray applied over ripe lemons caused a heavy drop of fruit. The two-spray
program resulted in no marked loss in quality of fruit during the five years of
the trials. Likewise, control of red scale and citrus red mite was satisfactory.

In 1942, the writer began an experiment in a lemon orchard in San Fernando
Valley in which the two-spray program was compared with 1-3/4% heavy medium emulsive oil applied in October in a series of three replicated plots involving a total
of seven acres. The once-sprayed plots required an auxiliary fumigation in the
second year of the experiment in order to avoid excessive damage from red scale.
These plots, despite the extra fumigation, produced a large percentage of scaly
fruits. Tree ripe fruit and fruit colored in the packinghouse had a coppery discoloration usually found on lemons sprayed with any oil heavier than a light medium.
Although over the four-year period, the twice-sprayed plots produced seven per cent
less fruit[2] than the once-sprayed plots, they yielded a greater net profit because
of their superior fruit and the absence of cullage caused by red scale. The trees
appeared in better condition at the end of four years because of the absence of
deadwood, caused by red scale, such as was to be found in the once-sprayed plots.

Medium Grade:- Medium oil is used in intermediate and coastal sections in the
spraying of oranges for red, purple and black scales and for mites where light
medium oil has proved to be ineffective. It may also be used in the spraying of
lemons and grapefruit in any climatic zone. The latter are more tolerant to oil
spray than the orange. Medium oil, besides resulting in better scale control;

[1]See map on page 352 for location of climatic zones.

[2]Records on file at the Division of Entomology, University of California, Los Angeles, Calif.

also is more likely to control the citrus red mite than is the light medium oil. Despite the better scale control, fumigation must sometimes be used in addition to the medium oil when red scale is very abundant. Recommended dosage: emulsives, 1-3/4% and emulsions, 2%.

Heavy Medium Grade:- Used only on lemons for control of red scale in orchards in which this insect is so abundant that it is unlikely that a medium oil would result in effective control. Has good "carry-over" effect in the control of red mite.

Heavy Grade:- Used only on lemons for the control of red scale. This grade of oil was once generally used in spraying for red scale on lemon trees, but has fallen into disuse in the majority of districts because of adverse tree and fruit reaction. It is now rarely used except in the Chula Vista district, where any lighter oil fraction would not result in satisfactory control of red mite.

With even the heaviest oils which may be safely used on lemons, an auxiliary treatment with HCN gas may occasionally be necessary to keep the red scale under control.

Distillation Range: Relation to Insecticidal Efficiency:- In comparative tests for insecticidal efficiency of spray oils used against the California red scale, cognizance must be taken of the profound effect of the substrate upon which the insects are located. Red scales situated on the branches are less apt to have their ventral surfaces wet by oil from the usual oil spray application than when situated on other morphological parts of the tree. The mortality from oil spray is consequently the lowest on the branches. Table 1 shows the results of counts made on lemon trees sprayed with three concentrations of oil according to the tank mixture method and using blood albumin spreader at 4 ounces to one hundred gallons of spray. The data shown in Table 1 are the summation of experiments made in five lemon orchards in Orange County in which 215,641 adult female scales were counted.

Table 2

Per Cent of Red Scale Surviving Different Concentrations of Heavy Oil Spray on Branches, Twigs, Leaves, and Fruit of Lemon Trees.

Unit Examined	Per Cent of Scales Surviving Sprays with oil concentrations of		
	1.5%	2.0%	2.5%
Branches	44.1	17.3	9.6
Green Twigs	14.2	1.4	0.4
Leaves	2.5	0.7	0.1
Fruit	15.7	7.0	6.6

It can be seen from Table 2 that there is a decreasing per cent kill from oil spray on the different morphological parts of the tree in the order leaves > twigs > fruits > branches. The per cent survival is proportionately decreased as the oil concentration is increased from 1.5 to 2.0 to 2.5% except in the case of the fruit, on which the increase in dosage from 2 to 2.5% results in very little increase in the effectiveness of the spray. The reason for this is apparent when the counts are made. It is difficult to wet the back sides of the fruit or areas between two or more fruits located close together--often actually touching one another. When the survival is caused by incomplete coverage of the spray, a group of scales are found alive in the unwet area without dead scales amongst them (see fig. 91). By far the greater part of the survival on the fruit in plots sprayed with 2% and 2.5% oil was due to incomplete coverage of spray. Otherwise, there is no reason to believe the per cent survival on the fruit would be any higher than on the leaves and green twigs. The leaves and green twigs are more apt to be completely covered by the spray than the fruits. The branches are also easily wet, but it is difficult to kill the scales on the branches, probably because (1) the branches are readily wet by both oil and water, resulting in less "preferential wetting" by oil

175

Fig. 91. Red-scale-infested lemon sprayed with 1-3/4% heavy medium oil. Scales not killed by the spray were removed, revealing the white wax from the scale body. If this wax were wet by oil it would not be white.

than occurs on the green twigs, leaves and fruits, and (2) the rapid absorption of the oil by the porous bark, resulting in less of the oil creeping under the bodies of the scales.

Among the five grades of spray oil used for citrus spraying in California, there does not appear to be any significant correlation between the heaviness or viscosity of the oil and the oil-depositing ability of the spray, provided, of course, that other factors such as U.R. and emulsifier are uniform. One would expect this to be the case, for at a given U.R., and with the same emulsifier or spreader, the controlling factor in oil deposit would be the interfacial tension oil/water as affecting the size of the oil globules. Measurements of surface and interfacial tension of the various grades of spray oil, by means of a du Noüy tensiometer, showed no difference between the grades of spray oil ordinarily used in citrus spraying.

A comparison was made of oil deposit from sprays with medium, light medium and light oil, all used at 1-2/3% concentration. Four orchard tests were made, each with a different emulsifier. In three of the four tests, there was no difference in oil deposits, but in the fourth test the oil deposit increased with increasing heaviness (and viscosity) of the oil. (9)

In another experiment no difference was found, either in laboratory or field experiments, in the amount of oil deposited by a light medium oil spray and a heavy oil spray when the two were applied at the same per cent concentration, providing corrections were made for relative loss from evaporation between the two grades of oil during the period between application of the spray and gathering of the leaves. (209)

L. A. Riehl made a study of oil deposit of eight of ten fractions of an asphaltic base spray oil stock used for citrus spraying in California.[1] The heaviest fractions of this oil (99.2 to 283.2 seconds viscosity Saybolt) deposited more oil and the lightest fraction (56.3 seconds viscosity Saybolt or less) deposited less oil than the intermediate fractions, but the latter, representing the range of spray oils now used in citrus spraying, deposited uniformly.

Since among the usual citrus spray oils, there does not appear to be any relation between heaviness and oil deposit, it might be expected that the per cent kill from sprays with different grades of oil would be similar. This, however, has not proved to be the case. It was shown, on the basis of many field experiments involving the counting of about a half a million insects, that in the range of spray oil fractions used in citrus spraying, the heavier the petroleum fraction, the greater the per cent kill of California red scale (818, 247). A summary of all field work showed, for example, a significantly higher kill of red scale on the branches and fruit of lemon trees from heavy spray oil than from medium oil when both were used at the same percentage concentration (table 3). Cressman (209), also in orchard experiments, found heavy oil to be significantly more effective against red scale than light medium oil. Swingle and Snapp (864), experimenting with paraffin base oils, obtained no kill of San Jose scale from a 47 seconds Saybolt viscosity oil, but obtained a 96% kill with a 76 seconds viscosity oil. Both were used at 3% concentration. Again the heavier oil proved to be the more effective. Marshall (590) found dormant oils of 200 to 220 S.S.U. viscosity at 100°F to be more effective against San Jose scale and European red mite than oils of 100 to 110 S.S.U. viscosity, but they were less effective against apple mealybugs.

[1]Unpublished data on file at the Citrus Experiment Station, Riverside, California.

Table 3

Summary of the Results of Spraying Lemon Trees with Different
Concentrations of Heavy and Medium Spray Oil During the Years
1930 and 1931 (after Ebeling, 247).

	Percentages of insects surviving treatments on		
	Branches (1930)	Branches (1931)	Fruit (1931)
Tank Mixtures			
Heavy oil, 1-1/2%	29.2	18.9	6.7
Heavy oil, 1-2/3%	----	13.3	5.0
Heavy oil, 2%	14.5	7.3	3.7
Heavy oil, 2-1/2%	5.8	---	---
Medium oil, 2%	19.8	8.6	4.6
Medium oil, 2-1/2%	9.7	5.5	2.2
Proprietary emulsions, 2%	26.6	16.6	5.3

The writer[1] applied a light oil at 2-1/2%, a light medium oil at 2% and a
heavy oil at 1-1/2% to lemon trees infested with California red scale. Counting
approximately 10,000 scales per treatment on the branches, the following per cent
survival was found 2 months after the application of the sprays.

Treatment	Per cent survival
Light oil, 2-1/2%	8.3
Light medium oil, 2%	8.1
Heavy oil, 1-1/2%	7.9
Untreated	64.6
L.S.D. at odds of 19:1	1.4

In the above experiment, the three grades of spray oil were approximately
equal in insecticidal effectiveness despite the wide variation in per cent of oil
in the spray.

In order to demonstrate the relation of distillation range to per cent kill
with the entire range of citrus spray oils, the writer in October 1936 used as
test insects red scale infesting lemons. Each of the 5 grades of oils in use in
1936 was used at 0.66%, with blood albumin spreader added to the spray at 1.1
gram per gallon. Forty lemons were sprayed with each oil and the experiment was
repeated. Altogether, 53,010 scales were counted in this experiment. The low
percentage of oil was used so that the kill would not be too great for most advantageous comparison of treatments. The results of the experiment are shown in
Table 4.

Table 4

Per cent kill from sprays containing the 5 grades of
spray oils commonly used for citrus spraying in 1936

Grade of oil	% distilled at 636°F	Saybolt Viscosity	% kill Series I	% kill Series II
Light	70	61	39.6 ± 0.52	43.6 ± 0.29
Light medium	62	65	42.4 ± 0.46	46.1 ± 0.35
Light medium	53	71	48.5 ± 0.86	50.4 ± 0.71
Medium	40	78	54.6 ± 0.65	54.8 ± 0.53
Heavy	13	95	62.4 ± 0.61	60.5 ± 0.70

In the previously mentioned tests made by Dr. Riehl, red scale-infested grapefruit were sprayed with 8 of 10 narrow fractions distilled from an asphaltic base
spray oil. A concentration of 0.6% oil was used through the experiment. Three
tests were made with each of the 8 fractions used in the experiment. The per cent
kill from fraction No. 2 (Saybolt viscosity 56.3) averaged 70.8; from fraction
No. 3 (Saybolt viscosity 62.9), 79.5; and from fraction No. 5 (Saybolt viscosity
77.3), 95.4. Fractions 6 to 10 inclusive gave no higher kill than fraction 5.

[1]Data on file at the Division of Entomology University of California, Los Angeles.

The kill from fraction 10 (Saybolt viscosity 283.2) was 95.5%. It will be noted that the kill was considerably lower with those fractions representing the light ends of the spray oil, the part of a spray oil normally lost in a few days by evaporation.

Oil deposit being the same, the question arises as to why the heavier grades of oil should result in better kills of red scale. Kerosene can be expelled from the tracheae of the red scale, but the fractions in the range of the spray oils

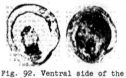 cannot be expelled from the tracheae. It might be expected that per cent kill would be related to deposit and would not vary unless oil deposit varied. This would probably be the case if the spiracles of all the scales hit by the spray were occluded by the oil, resulting in suffocation. In actual practice, however, on the branches only about half of the adult scales may receive enough oil to cause an occlusion

Fig. 92. Ventral side of the red scale. Left, unsprayed; right, sprayed with oil.

of the spiracles. The remainder show by their appearance that the oil has not seeped under the armor and wet the wax on the ventral scale covering (see fig. 92). A large percentage of these individuals will slowly be killed by the oil on and in their bodies. The per cent kill gradually increases for about two months, and that is why counts of per cent kill from oil spray are not made until 6 to 8 weeks after spraying. The lighter the oil the greater will be the amount leaving the bodies of the scales by evaporation. Cressman (209) found on lemon leaves 30% loss of light medium oil and 9% loss of heavy oil within a day or two after spraying. It is possible that a similar differential might exist in the evaporation of the two grades of oil from the bodies of scale insects. Among those scales which are not initially killed by the spray, it might logically be expected that those with a given amount of heavy oil on and in their bodies would be more apt to eventually die than those having lighter and more volatile oil, a certain percentage of which is lost by evaporation.

While it has been shown that the heavier grades of oil result in a better kill of red scale, this is not the most important factor accounting for their superiority when considered from the standpoint of long term control of red scale. An even greater difference can be shown in the residual effects of different grades of oil against the young (crawlers) which are born of the insects not killed by the spray. In an experiment in which medium and light oils were applied to orange seedlings at 1-2/3% concentration, it was found that after 28 days only 37.0% as many red scale "crawlers"[1] transferred from heavily infested lemons, were able to settle and form "whitecaps" on the leaves sprayed with medium oil as on those sprayed with light oil (and 25.6% as many as on untreated leaves). The effect of the oil residue on the tree surface does not end, however, with the inhibition of the settling of crawlers. Among those insects born after a tree is sprayed, a higher mortality occurs among scales on an oil-sprayed tree than on an unsprayed tree even among those scales surviving beyond the whitecap stage. In the above experiment 48.0% as many insects which had reached the whitecap stage survived on leaves sprayed with medium oil as on those sprayed with light oil (and 34.3% as many as on untreated leaves). It should be pointed out that not all investigators agree that increased residual effect will be obtained from each successive increase in the weight of the spray oil. In one experiment no consistent difference was found in the residual effectiveness of light medium and heavy medium spray oil on the branches of lemon trees from 6 to 12 months after the application of the sprays, but it was found that an "extra light"[2] oil was much inferior to either oil with regard to its residual effect against red scale and citrus red mite. Ten months after treatment, 7 times as many red scales were found in the "extra light oil"-sprayed plot as in the plot sprayed with light medium oil, and 8 months after

[1]"Crawlers" are the newly born young of the red scale. They are motile for several hours before settling on their host. After settling they exude waxy threads which cover their bodies, and are then known as "whitecaps".

[2]I.B.P. 514°F, 50% distilled at 570°F, and 90% distilled at 643°F. This oil is a little lighter than "light" spray oil as listed in the State Department of Agriculture Special Publication "Economic Poisons".

treatment, 5 times as many red mites were found in the "extra light oil"-sprayed plot as in the plot treated with light medium oil. (210)

Years of field experience in California has led to the generally accepted conclusion that the residual effect of spray oil against scales and mites is progressively greater with increasing heaviness of oil, and even the difference between a light medium and medium oil is considered to be an important factor in this connection. All recommendations are based on the above conclusion.

Another residual effect of spray oil is its effect on the reproductive capacity of adult females not killed by the spray. A greater percentage of dead crawlers may be found grouped about the pygidium of the adult females. A rather high percentage of these remain enveloped in a membrane known as the amnion. This phenomenon, as well as dead crawlers under the scale covering, may be occasionally seen among untreated insects, but occurs to a much greater extent among sprayed insects which have not been killed by the oil (251). In producing this effect, it is reasonable to suppose that the heavier grades of oil would have the greater influence, for they would leave the body of the scale by evaporation more slowly than the lighter oils, but no data have been procured on this subject.

Distillation Range: Relation to Phytotoxicity:- The factor determining the grade of oil to be used on a given variety of citrus is the effect the oil will have in the tree and on the appearance and quality of the fruit. Oil spray injury will later be discussed in more detail (p.198). At this point it is wished merely to emphasize the increasing phytotoxicity with oils of increasing heaviness (higher distillation range). In areas, for example, where a light medium oil causes some defoliation, fruit drop, deadwood, etc., yet can be annually used with a reasonable margin of safety, from a commercial standpoint, the use of a medium oil, even for a single year, might prove to be disastrous. In such areas, if an annual application of a light medium oil fails to control red scale, the wise grower will not change to a heavier grade[1] of oil or use a higher percentage of oil in the spray, but will switch to fumigation. If fumigation also fails to offer adequate control, or results in a red mite problem that can not be satisfactorily coped with, the grower will spray and fumigate during the same season. This should reduce the red scale population to such an extent that an annual application of light medium oil at the usual dosage will again offer satisfactory control of the red scale for at least a number of years.

To obtain some measure of the effect of heaviness on phytotoxicity, counts were made of the number of leaves that had dropped under 10 trees in each of 42 plots in a red-scale-infested lemon orchard sprayed with medium and heavy oils of different unsulfonatable residues, and using different percentages of oil and different amounts of blood albumin spreader. The leaves were counted about 3 weeks after treatment. The results of the leaf counts are graphically shown in fig. 93.

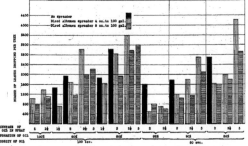

Fig. 93 (at right)

Relation of dosage, U.R., viscosity (or distillation range), and quantity of spreader to defoliation of oil-sprayed lemon trees   After Ebeling (247).

[1]The possibility is not excluded, however, that the use of heavier grades of oil may be made safe by the addition of 2,4-D as is already indicated by considerable commercial experience.

It can be seen from the figure that defoliation was found to be proportional to the heaviness of the oil and the percentage of oil used in the spray, and inversely proportional to the amount of spreader used and the degree of refinement of the oil. (247)

With respect to their effects on the appearance (see pp. 204 and 205) of the fruit, the spray oils also become distinctly more harmful as they increase in distillation range.

Viscosity:- The average Saybolt viscosities at 100°F of the five grades of spray oils of 4 leading brands refined from California crudes and sold as foliage sprays in California in 1946-47, were as follows: light, 59.3; light medium, 68.5; medium, 77.8; heavy medium, 86.7; and heavy, 103.2. It can be seen that viscosity increases with distillation range and specific gravity.

Viscosity figured prominently in the early days of the quick-breaking refined oils as a physical criterion of insecticidal effectiveness and phytotoxicity. If one deals entirely with oils from one region, as for example from California wells, viscosity is both a convenient and a reliable criterion of relative effectiveness and relative phytotoxicity, provided of course, that the oils in question are reasonably narrow-cut. However, viscosity figures could be misleading if one were to compare, for example, spray oil obtained from a Pennsylvania crude with one from a California crude, for at a given distillation range, the Pennsylvania oil would be considerably lower in viscosity. To cite a specific example: a light medium spray oil refined from a Pennsylvania crude, and having a distillation range similar to that of the average California light medium spray oil, was found to have a Saybolt viscosity of 44.2 compared to a range of viscosities of from 63 to 74 (average, 68.5) for the California oils. At a given distillation range, the oils predominantly paraffinic in character have a lower viscosity than oils predominantly naphthenic and aromatic in character.

Another drawback to viscosity data is that, like per cent distilled at 636°F, they do not indicate the distillation range of the oil, and consequently do not indicate what proportion of the oil is composed of the ineffective light "ends" and the excessively phytotoxic heavy "ends". The viscosity of an oil far too heavy for safe use on trees in foliage, and with a Saybolt viscosity, for example, of 130 seconds, could be reduced in viscosity to 70 seconds by the addition of a small amount of kerosene. A small amount of a light fraction greatly influences the viscosity of a heavy fraction, but the converse is not true. The resulting oil, despite a viscosity similar to that of the usual asphaltic base light medium spray oils, would be composed primarily of injurious heavy fractions and an insecticidally ineffective kerosene.

Since the grade of a spray oil is based entirely on its distillation range, it would appear that the presentation of viscosity data, as in spray charts and labels, is superfluous.

Volatility:- It may reasonably be assumed that it is not "heaviness' per se' which makes the heavier grades of oil more injurious to trees and fruit than the lighter grades, but rather the fact that the less volatile fractions of oil remain in the tree for greater periods and the heavier grades of oil have a larger per cent of these less volatile fractions. It has been found that some of the less volatile portions of all but the lightest spray oils remain in citrus leaves throughout their lifetime, which may be more than two years (752). Eighty per cent of a light medium oil applied to citrus trees in the fall at 1-3/4% concentration can be expected to leave the foliage in 17 to 25 days. A medium oil will require about 40 to 45 days for this degree of loss from the foliage, and a heavy oil will require 4 to 5 months.[1] Presumably the loss is mainly by evaporation, although absorption by dust probably accounts for much loss of oil under usual orchard conditions. It will be remembered that Cressman (209) found a loss of 9% of a heavy oil and 30% of a light medium oil, applied in 1.5% concentrations to lemon trees, in 24 hours. The greater defoliation occurring in plots sprayed with

---

[1]Correspondence from Mr. J. R. Allison, March 3, 1948.

heavy oil as compared to those sprayed with medium oil (247) was probably due entirely to the greater amount of the heavy oil that was in the foliage during the three weeks from the time of spraying to the time the counts of the fallen leaves ere made.

It would appear from the above that a standard test for volatility at somewhere near the temperatures in the orchards would be a most useful criterion of insecticidal effectiveness and phytotoxicity, but it does not appear to be practicable because of the great period which would be required before the more volatile fractions would be volatilized. The State of California, in connection with its analyses of the spray oils for 1931, gave figures for per cent evaporation in 2, 6, and 24 hours at 100°C for the four fractions recognized as distinct grades that year. The evaporation tests were made in the manner previously described (see p. 170). In 24 hours the following percentages of the four grades of oil had evaporated: light oil, 85.8; light medium oil, 72.4; medium oil, 64.2; and heavy oil, 52.9. Since the 100°C temperature at which these tests were made is far above the normal orchard temperature, and since the figures for per cent evaporation merely express in inverse relation the relative distillation ranges of the oils, they may be considered, like viscosity, as being superfluous as long as the distillation ranges of the various oils are given. The figures for evaporation at 100°C were not given in the annual spray analysis charts after 1931.

Unsulfonated Residue:- As stated before, sulfonation of spray oils requires an additional treatment with sulfuric acid after the usual treatment given the lubricating oils with liquid sulfur dioxide. Ordinarily sulfuric acid is known as a drying agent and a strong acid. When hot, however, the acid has another property, that of an oxidizing agent. These three properties of hot sulfuric acid make it impossible for any but the most inert substances to go through the sulfonation process unaltered. Any compounds made acidic by the sulfuric acid are washed away and made into soaps in the commercial processes. The oil is then neutralized by washing it with an alkaline solution. It is then again thoroughly washed with water and dried by blowing air through the oil.

The compounds acted upon by sulfuric acid may be summarized as follows:

1. Unsaturated hydrocarbons - straight chain or aromatic hydrocarbons with one or more unsaturated bonds. It is commonly supposed that these compounds react with sulfuric acid, but there is evidence to indicate that they are merely dissolved in the acid without undergoing any changes in composition (936). Gray and de Ong, in 1915-16, were the first to demonstrate that the unsaturated hydrocarbons were the most important cause of the phytotoxicity of petroleum and that their removal makes spray oil relatively harmless to plants if it is properly used (379). This important investigation marked the beginning of a new era in the development of oil sprays.

It was at first believed that unsaturated hydrocarbons are injurious to foliage because of their tendency to become saturated by means of addition reactions at the expense of essential constituents of the foliage. However, the investigations of Tucker (895) indicated that the unsaturated hydrocarbons are not toxic to foliage until they are oxidized to oil-soluble asphaltogenic acids. He found that this takes place at ordinary temperatures only in the presence of air and light. Apricot leaves smeared with petroleum oils containing 25% or more of unsaturated hydrocarbons were protected from injury by being covered with manila bags. No indications of "burning" were noticed at the end of 15 to 18 days, although uncovered leaves showed typical effects of injury in 2 or 3 days. Injury was noticeable with light intensities of 25 or more foot candles or lumens per square foot. The toxic threshold was reached when approximately 0.5% of asphaltogenic acids was formed. Not all oils of the same U.R. showed the same resistance to oxidation, indicating that oils from crudes of different origins may be expected to cause different degrees of injury to foliage.

Frear (344) points out the possibility that dissolved or suspended insecticides, such as are often incorporated into an oil spray, may be adversely affected by the acids which may develop in the oil. He suggests that the addition of anti-oxidants might be of assistance in this connection.

2. Oxygen compounds. These compounds are removed by combination with alkali or by washing out. They may be present to the extent of 0.1 to 3%. They are generally either phenols or naphthenic acids.

3. Sulfur compounds. These are usually present to the extent of less than 1%. They may be of the paraffin type, such as mercaptans and pentamethylene sulfides, with the type formula CH_3CH_2---S---CH_2CH_3, or unsaturated ring compounds such as thiophene. These compounds add to the phytotoxic effect of petroleum. They are removed in the sulfonation process.

4. Nitrogen compounds. Example: quinoline, a nitrogen derivative of petroleum with the structure   Nitrogen derivatives are basic and are therefore removed by sulfuric acid

The saturated hydrocarbons are acted upon by sulfuric acid to only a very slight extent and they compose what is known as the unsulfonated residue. They are aliphatic (straight chain) or, to a limited extent in western oils, aromatic compounds of the paraffin or naphthene series which go into chemical combination with other substances only by means of substitution reactions.

Ideally, spray oil should have a U.R. of 100% to be as safe as possible for use as summer or foliage sprays. It would then be similar in refinement to medicinal oil. The cost of such a degree of refinement would be excessive, however, and there is reason to compromise in the degree of refinement between the cost and the possibility of injurious effects. The percentages of unsulfonated residue (U.R.) for the various grades of foliage spray oil, as finally standardized for California conditions, are as follows: light oil, 90; light medium, 92; medium, 92; heavy medium, 92; and heavy, 94. A U.R. of 92, for example, means that 92% of the oil was not sulfonated in the standard sulfonation test for spray oils (p. 171).

The above standards were agreed upon for California, but other regions need not necessarily adopt the same standards. It is known, for example, that oils as low as 80 U.R., which would be excessively injurious to citrus trees in California, can be safely used in Texas, Florida and Mexico. It appears that in more humid regions a citrus tree is not as adversely affected by oil as in California, and that oils of a lower degree of refinement can be safely used.

Among insects of various orders which were tested, those immersed in oils of low U.R died more rapidly than those immersed in oils of higher U.R. (265). The most complete investigation was made with larvae of the potato tuber moth, Gnorimoschema operculella, which, when immersed in either paraffinic or asphaltic base oils, died progressively more rapidly as the U.R. of the oil was reduced. In other experiments, lady beetles (Hippodamia convergens) were immersed in oil. While it required an average of 268 seconds to kill beetles in an oil of 99.4% U.R., the period decreased with each successive reduction in unsulfonated residue until in an oil of 90% U.R., an average of only 46.2 seconds was required to kill the beetles (228). On the other hand, in the spraying of armored scale insects, no difference was found in the effectiveness of oils of different U.R. (864, 247, 211). Finally, in experiments with the eggs of the oriental fruit moth and the codling moth, it was discovered that in the case of both paraffinic and asphaltic base spray oils, those of higher percentage unsulfonated residue resulted in better kill (155). It can be seen that (1) greater toxicity, (2) no effect, and (3) reduced toxicity has been attributed, under different conditions, to the presence of unsaturated hydrocarbons in spray oil.

Paraffinicity:- As previously noted, Penny (684) found paraffinic base spray oils to be greatly superior to asphaltic base spray oils when used against the eggs of the fruit tree leaf roller, Archips argyrospila Wlk., in California and that he stated that ·G. M. List had a similar experience with the two types of oil in Colorado. The most exhaustive studies on the influence of paraffinicity on insecticidal efficiency seem to have been made by Pearce et al (682) with respect to dormant oils and by Chapman et al (155) with respect to summer spray oils. These investigators also found that the greater the paraffinic character of the oil the greater its efficiency against the eggs of the fruit tree leaf roller, as well as the eggs of the·oriental fruit moth, Grapholitha molesta (Busk), the codling moth, Carpocapsa pomonella (Linn.), and the eye-spotted bud moth, Spilonota ocellana (D & S). Paraffinic base spray oils were also found to be more efficient than naphthenic base oils for the control of the San Jose scale in delayed dormant spraying (154). Laboratory experiments[1] conducted by L. A. Riehl have also definitely established the greater insecticidal efficiency of paraffinic base oils used against California red scale. It should be pointed out, however, that this relationship does not necessarily hold true with all insects and mites; for example, aphid eggs are very susceptible·to low concentrations of tar oils which are predominantly aromatic in character (155).

No difference was noted in the period required for 50% mortality when larvae of ·the potato tuber moth were immersed in paraffinic base and asphaltic base spray oils of similar physical properties (except for viscosity) and with similar percentages of unsulfonated residue. The suggestion was advanced that one might not necessarily find differences in mortality when·the test insects are completely immersed in the two types of oil even though differences might be found when only a film of oil is involved, as when the insects have been sprayed. It was discovered that if oil-covered insects are kept in an oxygen free environment, as when completely immersed in oil, they live much longer than· when merely covered by a film of oil, which is not impervious to air. (265)

In experiments with sprays, differences have not always been found in the effectiveness of paraffinic and asphaltic base oils, but the spectacular differences that have been found in some cases have created much interest in recent years in the relation of the chemical constitution of petroleum oil to insecticidal efficiency, and this phase of the oil spray problem is being investigated with special thoroughness.

SPRAY OIL FORMULATIONS

Historical:- As an insecticide, petroleum seems to have been first used in the form of kerosene. It may have been used before written accounts as to its efficacy were kept, but records from Gardener's Monthly, an old publication, show that it was used in the control of scale insects on orange and lemon trees as early as 1865. It was applied·directly to the tree without dilution, as, for example, by brushing it on with feathers. By 1870 kerosene was used diluted with water, and with whale oil soap, possibly as an emulsion (540). In 1878 A. J. Cook prepared an emulsion which, however,.contained only 20% kerosene, and in 1883 C. V. Riley and H. G. Hubbard announced the formula for an emulsion containing 67% kerosene. This was possibly the first spray oil emulsion somewhat comparable in the percentage of oil, and in its stability, to present day emulsions. It was known as the R. & H. formula, and was used against citrus scale insects·in Florida by Hubbard. The formula for the emulsion was as follows: kerosene, 2 gallons; common or whale oil soap, 1/2 pound; and water, 1 gallon. The kerosene was stirred in the hot, soapy water and the mixture was pumped through a spray nozzle one or more times to produce a thick emulsion.

Kerosene was used as high as 25% on dormant trees and 10% on foliaged trees including citrus, and was comparatively safe except for that portion which soaked beneath the soil line. Not·being allowed to evaporate, the kerosene penetrated into the bark of the trunk and crown roots below the surface of the soil, and often killed the bark and·the entire tree. There has been occasional resurgence of interest in kerosene as a spray up to the present day. In citrus pest control it is being used at 1-1/2 or 2% concentrations as a carrier for DDT in the combined citricola scale-citrus thrips control program in Tulare County, California (296), and

---

[1] Work in progress at the Citrus Experiment Station, Riverside, California.

for the control of black scale on olives (828). DDT-kerosene formulations at 3% concentrations were found to be highly effective in the control of purple scale (533). In large scale experimental tests, 2 sprays of 3% DDT-kerosene, one following the other by about 6 weeks, have proved to be highly effective in the control of the California red scale (147). The concentration of the DDT in the kerosene is 9.4 wt./%, made possible by the use of mutual solvents. Three per cent DDT-kerosene has been found to be the most effective treatment for the blackfly on citrus in the control program worked out in the eradication campaign in northwestern Mexico (997).

A good historical review of the development of spray oils in California is given by Essig (285) and Smith (818). In citrus pest control, the rather widespread use of kerosene, usually as a tank mixture (unemulsified), by 1905 gave way almost entirely to petroleum distillates of about 28° Baumé, used at 3% concentration. However, severe injury to trees often resulted, and by 1915 kerosene was again in use, this time in the form of emulsions. In the meantime crude oil, 16° to 22° Baumé and direct from the wells, was being used, in the form of emulsions emulsified with fish oil, as a dormant spray for deciduous fruit pests. The homemade crude oil emulsions were nearly entirely replaced by more standardized proprietary emulsions by 1920.

In the aforementioned tank mixtures and emulsions, the separation of the oil and the water in the spray tank was a continual source of trouble, for mechanical agitation had not reached a sufficient degree of perfection to insure, as it now does, a reasonably good mixture of broken emulsions. This particular difficulty was overcome by the miscible oils. These contained oil soluble emulsifiers, usually carbolic or cresylic acid. The resulting dilute emulsion in the spray tank was very "tight" and consequently high percentages of oil had to be used in comparison with the emulsive oils of the present day. The latter contain oil, soluble emulsifiers, but they are of a type that allow for "quick-breaking" and efficient oil deposit.

HIGHLY REFINED LUBRICATING OIL EMULSIONS

The increasing importance of the San José scale as a deciduous pest, and its apparent resistance to lime sulfur, and the ever increasing resistance of the California red scale to hydrocyanic acid gas, resulted in a renewal of interest in oil emulsions. This interest was heightened by wide publicity given in 1922-23 to a U.S.D.A. formula for boiled lubricating-oil emulsion and a formula for cold engine-oil emulsions from the Missouri Experiment Station. The essential contribution of these formulas was the use of lubricating oil instead of the various oils used almost exclusively in the past. Lubricating oil had previously been used for years in citrus spraying in Florida, but had not gained wide recognition. With the knowledge gained in 1915-16 by G. P. Gray and E. R. de Ong, at the University of California, regarding the relation of purity to safety, it was soon possible to successfully prepare highly refined lubricating oil emulsions comparatively safe for use as summer or foliage sprays. The tendency at first was to use too heavy fractions, but gradually oils of lower distillation range were used, until at present only the lighter lubricating oil fractions, comparatively narrow cut, are being used as summer spray oils.

At first the use of highly refined white oils was restricted by the excessive cost of the spray at the high percentages that it was necessary to use with the tight emulsions of that period (circa 1922-23). To make practicable the use of highly refined white oils for citrus spraying, it was necessary to develop an oil emulsion which when diluted would more efficiently deposit the oil on the tree surface. A spray consisting only of oil and water, without previous emulsification would be expected to have the most efficient oil depositing ability, for the separation of oil and water upon striking the tree would be unhindered by a colloidal film surrounding the oil globules. However, agitation in spray tanks was not sufficient at the time to maintain a uniform mixture of oil and water. The solution arrived at is best expressed by the words of de Ong, Knight and Chamberlin (232) who began their investigations on the subject of quick-breaking emulsions in

1924.[1] They believed that "by very slightly emulsifying the oil, ordinary spray tank agitation is capable of overcoming the natural buoyancy of the separated oil droplets and of maintaining a fairly uniform suspension of oil throughout the body of the liquid. The emulsifying agent is used in quantities just sufficient to separate the oil into relatively large droplets. The interfacial membrane is consequently weak and easily broken, thus liberating the enclosed oil. There is no danger that the stability of the system will be sufficient to withstand rupture upon impact with the leaf or fruit surface, and the maximum amount of oil is consequently freed and made available. On the other hand, the oil in the tank is maintained in the form of isolated droplets, there being no continuous sheet of oil to be broken up as would be the case with a mechanical mixture."

In the preparation of quick-breaking oil emulsions there are many formulas used by the different manufacturers. Casein dissolved with ammonia is the emulsifier that has been used most extensively. A formula that has been used with success is as follows:

$$
\begin{array}{ll}
\text{Ammonia (28\% NH}_4\text{)} \ldots\ldots\ldots & \text{3 pints} \\
\text{Casein (finely ground)} \ldots\ldots & \text{5 pounds} \\
\text{Water} \ldots\ldots\ldots\ldots\ldots\ldots & \text{17 gallons} \\
\text{Spray oil} \ldots\ldots\ldots\ldots\ldots & \text{83 gallons}
\end{array}
$$

The proportion of ammonia and casein may be varied according to the stability desired in the emulsion. The emulsion is prepared commercially by placing the ammonia-casein solution in the water and running the oil and water simultaneously into a large emulsifying chamber. The emulsifying chamber is equipped with motor-driven blades which revolve at speeds ranging up to 800 or 1,000 r.p.m. The emulsion is being produced as the oil, the emulsifier and the water are introduced into the chamber. The mixture may be withdrawn from the bottom of the chamber through a pipe and forced through a gear pump. The latter forces the mixture up to the top and back into the chamber through a narrow aperture and at a high pressure. The pressure depends on the r.p.m. of the gear pump and the size of the aperture. This aids in emulsification. When the chamber is full the blades are allowed to run for 1 or 2 minutes to insure homogeniety of the emulsion.

The above formula will produce a quick-breaking emulsion. A more stable emulsion, depositing less oil on the tree, can be made by using a larger amount of emulsifier, or it may be made by using potassium fish-oil soap along with the ammonia-casein.

The quick-breaking emulsions are usually used at the rate of 2 gallons to 100 gallons of water in spraying citrus trees. Such a spray will result in a uniform distribution of oil, covering the entire tree.

Newcomer and Carter (645) describe the preparation of oil emulsions in the Yakima Valley, Washington, by growers who used their own spray rigs for this purpose. The proportion of ingredients in these emulsions was as follows:

$$
\begin{array}{ll}
\text{Ammonia (28 per cent)} \ldots\ldots & \text{1 quart} \\
\text{Casein (finely powdered)} \ldots & \text{3 pounds} \\
\text{Water} \ldots\ldots\ldots\ldots\ldots\ldots & \text{33 gallons} \\
\text{Oil} \ldots\ldots\ldots\ldots\ldots\ldots & \text{100 gallons}
\end{array}
$$

"Grower A has a 600 gallon portable spray outfit. He puts 132 gallons of water into it, adds 1 gallon of ammonia, and then slowly sifts in from 10 to 12 pounds of casein while the agitator is running. The oil comes in 50-gallon drums, and the grower has provided a 1-1/2-inch pipe which reaches to the bottom of a drum through the large bung in one end. He connects this pipe to his spray pump by means of a piece of hose, and pumps the oil into the tank with the agitator running. The drum does not have to be lifted, and its contents can be emptied in

---

[1] By 1924 a quick-breaking commercial emulsion of highly refined petroleum oil was already being made by W. H. Volck.

less than two minutes. When 400 gallons of oil have been added in this manner, a suitable hose has been attached to the overflow, and the mixture, which is already partially emulsified by the agitation it has received, is pumped through the overflow into empty drums, under a pressure of 250 pounds, and is ready for use. The whole operation takes about 1-1/2 hours, and requires the services of two men."

Flowable emulsions:- The emulsions manufactured before 1937 were of the "mayonnaise" or "paste" type, these terms referring to their consistency. These emulsions did not flow; it was necessary to remove them from a drum or barrel by means of a scoop or ladle. In the year 1937, the "flowable emulsions" began to appear on the market. These may have the same ingredients in the same proportion as the mayonnaise or paste type of emulsion, yet are of such a consistency that they will freely flow out of the bung of the spray drum or the attached spigot. The oil is thus more accurately measured in the measuring can and the fluid nature of the emulsion obviates the necessity for ladling out the emulsion, which is always objected to by those who have enjoyed the convenience of handling emulsive or straight oils (tank mixture). Practically all the emulsions now used in spraying citrus trees in California are of the flowable type.

Flowable emulsions may be prepared in a variety of ways, using the same equipment described on page 185. The same ingredients used for the paste type of emulsion may be retained, but the emulsion is formed under lower speed agitation during emulsification, or lower pressure, if pressure is employed to supplement agitation. There are various emulsifiers that can be used to make both flowable and paste emulsions. Some of these materials are superior for flowable emulsions, but may be also suitable for the preparation of paste emulsions. Casein products can be used equally well for both flowable and paste emulsions.

In the preparation of flowable emulsions, at least one insecticide company in California uses both an internal (in the oil phase) and external (in the water phase) emulsifier in the same emulsion. For example, diglycol oleate, petroleum sulfonates, or sulfonated long-chain alcohols at 1 or 2% may be used in the oil phase and blood albumin, casein, sulfonated petroleum or other types of water-soluble sulfonated products may be used in the water phase. The emulsification is done in such a way that the water is not beaten into a froth such as usually forms in the preparation of paste emulsions. Homogenizers or centrifugal pumps of high speed may be used in bringing about this type of emulsification.

As will be explained in the section on "emulsive" oils, certain advantages are gained by the use of surface-active substances in both the oil and the water phases of an emulsion. A better wetting of the tree may be accomplished without a reduction in the oil-depositing ability of the spray. Besides, the oil can be more readily mixed in the spray tank than is possible with present types of emulsive oils; thus the risk of injury to the trees in the occasional instances of inferior agitation or careless mixture of the oil in the spray tank, are reduced.

DETERMINATION OF THE OIL CONTENT OF AN EMULSION

Since it is the oil, in an oil preparation, which is the insecticidally active ingredient, recommendations as to dosage should take into account the actual amount of oil which is present in the preparation in question. In 1946-47, 51 brands of summer oil emulsions of all five grades which were used in California ranged in per cent of petroleum oil from 66.9 to 90.0, the remainder being water and emulsifier. The average per cent of petroleum oil was 82.4. On the other hand, the emulsive oils ranged in per cent of petroleum oil from 97 to 99, the remainder being oil-soluble emulsifier. Although it has been argued that an emulsion, despite its lower oil content, may be so prepared that it will deposit as heavily as the average emulsive oil, it does not appear that this should invalidate the general rule that the recommended dosage should take into consideration the per cent of oil present in the oil preparation. The emulsive oils can also be made to deposit more heavily, but with either type of spray oil, an increase in the oil-depositing ability of the resulting spray mixture decreases the uniformity of the oil deposit and therefore increases the chances of adverse tree and fruit reaction.

186

SPRAY OILS

In practice, therefore, where 1-3/4% of an emulsive oil or tank mixture (straight oil) is recommended, 2% of the average emulsion is recommended to give on the average, a similar oil deposit.

The grower or pest control operator, upon purchasing an emulsion, may wish to determine whether or not it conforms in oil content, within reasonable limits, to the average emulsion. A number of substances when placed in an emulsion will cause it to separate. These substances may act as electrolytes and change the charge of the emulsion and consequently break it, or they may be dehydrating agents, removing the water from the emulsifier and rendering it ineffective. Sulfuric acid is a good dehydrating agent, and so is sodium hydroxide plus alcohol. These compounds might also act directly on the emulsifier, breaking it down and removing it from the interface. As examples of effective electrolytes, NaCl and Ba(OH)$_2$ may be cited. These compounds coalesce the oil globules by neutralizing their charge. The two types of compounds may also be used together, as in the following example.

Procedure:- Ten grams of the specified emulsion is weighed in a Babcock cream bottle and is then diluted with about 10 ml of hot water. Then 5-10 ml of sulfuric acid is added. To hasten the separation of the oil, the bottle is set in hot water for about five minutes. Sufficient NaCl solution is then added to bring the oil layer into the graduated neck of the bottle. The bottle and contents are then centrifuged at 1200 r.p.m. for five minutes and allowed to cool. The volume of oil may be read in the graduated neck of the bottle. Then from the density of the oil, the percentage composition of oil in the emulsion may be obtained. If fatty acids, phenols, etc., are present in the oil, this will be reflected in its density, and the per cent of non-petroleum liquid in the oil may be calculated (174).

The per cent of oil in a dilute spray mixture in the spray tank may also be determined in a similar manner. Equipment which may be used for such tests, using warm sulfuric acid, is shown in fig. 94 (818).

Fig. 94. Equipment required for determination of per cent of oil in oil spray mixtures. A, sulfuric acid; B, washing bottle with hot water; C, 8-ounce bottle containing sample of spray mixture (note layer of dyed oil); D, cage holding 25 ml test tubes; E, Babcock milk-test bottles; F, centrifuge. (After Smith, 818).

TANK MIXTURES

Certain advantages in the use of oil and water, without previous preparation to produce an emulsion, were recognized by early investigators, and, in the case of kerosene, "tank mixtures" were commonly used. Until relatively recent years, however, uniformity of the mixture of oil and water throughout the spray tank was poor because the engines used with the spray rigs did not have sufficient power to provide the necessary r.p.m. for adequate agitation. In addition, the need for the proper number, shape and arrangement of agitator blades (fig. 95) for optimum efficiency was not appreciated.

Fig. 95. Various types of agitator blades. B, E, and H were
found to be unsatisfactory for use with oil sprays. Type F, a
flat blade, was the most efficient (After Smith, 818).

Ralph H. Smith, who began his oil spray investigations at the Citrus Experiment Station in the summer of 1926, soon discovered that a great variation existed
in the performance of the commercial oil emulsions (818). He found, for example,

Ralph H. Smith (1888-1945)
Professor of Entomology, University of California. Well known
for his contributions toward the standardization of summer spray
oils and the development of "tank mix" for citrus pest control.

that among 40 proprietary summer oil emulsions used at 2% concentration in sprays
applied to glass sections, the most efficiently depositing emulsion deposited 7.5
times as much oil as the least efficiently depositing emulsion. Later, Dr. Smith
found that among 11 proprietary brands of oil emulsion tested, the deposit of oil
from a 2% spray ranged from 15.6 to 50.7 mg per 25 sq. in. of surface when applied
on orange leaves and from 6.9 to 21.0 mg per 25 sq. in. of surface on glass. The
ratio of oil deposit of the various emulsions on orange leaves and on glass was
quite similar.

It was obvious at the time of Smith's early investigations that with the
large number of brands of oil emulsion on the market, each subject to change in
composition at any time, any grower's experience with oil spray was left to the
whims of the emulsion manufacturer. By choosing a particular brand merely on the
basis of the manufacturer's claims, the grower might obtain a spray far too low in
its oil-depositing properties, or one depositing too much oil and causing excessive
tree reaction or, by chance, the spray might have the qualities conducive to optimum performance. The extreme variability encountered in the performance of the
various brands of oil emulsion was an undesirable factor in experimental studies
with proprietary emulsions. The investigator might standardize on a single brand
of emulsion for the entire season's work, but could not depend on uniformity in
performance even for such a limited period. Besides, it is unsatisfactory to carry
on an investigation with a material of unknown composition.

Emulsification achieved by agitation in the tank in the presence of specified
amounts of an emulsifier of known composition was first conceived as a means of

providing a standard oil spray mixture for experimental purposes. It was then presumed that these very qualities of the mechanical mixture might best serve the purposes of the fruit grower. The mechanical problem of adequate agitation had been solved by increased power, increased r.p.m. of the agitator shaft, and a greater number of more efficient blades on the agitator shaft.

At one time it was considered that the size of the oil globules in the spray tank was an important factor influencing the performance of the spray. Smith discovered, however, that regardless of the size of the oil globules in the spray tank, they were broken up and reduced to a uniform size by passing through the spray nozzle under normal spraying pressures.

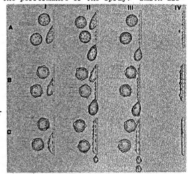

The mechanical mixture eliminated to a large extent the variability in results obtained from the proprietary emulsions as they were at that time developed. Unfortunately, however, mechanical mixtures were also subject to considerable variation in the amount of oil deposited. These variations resulted largely from differences in the period of exposure of the sprayed surface to the spray stream. This was due to the extremely quick-breaking nature of the spray. In a mechanical mixture of only oil and water, the oil adheres to the foliage and the water runs off quite completely (fig. 96)--more completely than when the spray contains a prepared emulsion. Wherever a surface of undiluted oil occurs on the leaf, another layer of oil is readily deposited, and if a leaf is struck by the spray stream several times, as must inevitably occur

Fig. 96. The mechanism of oil "build-up". A, very stable spray mixture leaving too little oil; B, extremely quick-breaking mixture, causing "build-up" of oil; C, a medium condition in which sufficient water remains on the surface to prevent excessive "build-up" of oil. Explanation in text. After Smith (818).

in thorough spraying, there is a considerable "build-up" of oil. Since some parts of the tree are struck by the spray more times than other parts, an unevenness of oil deposit results. This unevenness is most striking when water and oil alone (mechanical mixture) are used. In the case of a conventional emulsified oil there is greater tendency for the "initial deposit" of spray to be a mixture of oil and water (fig. 96) rather than oil alone. The tendency of successive films of oil to adhere to a mixture of oil and water is less than the tendency to adhere to undiluted oil. For this reason successive streams of spray result in less "build-up" if an emulsion is used than if a mechanical mixture of oil and water is used to form the spray. This disadvantage of the mechanical mixture was believed by Smith to be overcome by the addition of a suitable spreader. The spreader also has the advantage of increasing the spreading of the water over the foliage, a function which may or may not be performed by an emulsifier, for a good emulsifier may not necessarily be a good spreader.

The spreader was believed to perform the same function as the emulsifiers of the oil emulsions in reducing "build-up". Either the "tightening" of an emulsion or the addition of a greater quantity of spreader will have the effect of decreasing build-up, but will also have the effect of increasing the "run off" of the spray, that is, the tendency of the spray to fall to the ground without having deposited a sufficiently large percentage of the oil (fig. 96, A). However, the oil-depositing efficiency of a quick-breaking spray is desirable, so a "happy medium" must be found between oil-depositing efficiency and safety to the tree. In the tank mixture spray, sufficient spreader is added to prevent excessive "build-up" without unduly interfering with the oil-depositing ability of the spray. The aim, in making the decision as to the quantity of spreader which should be added for a

189

tank mixture, was to arrive at a spray which would perform, as far as could be observed and quantitatively determined, in a manner which could be considered about average for the emulsions used in citrus spraying. It was desired to avoid the inadequate oil deposit of the tightest emulsions (fig. 96, A) and the excessive oil deposit of those of the highest oil-depositing ability (fig. 96, B), but to leave a uniform deposit adequate for a reasonably high kill of the pest (fig. 96, C).

Blood albumin was decided upon as the most suitable spreader. Grade A Powdered Blood Albumin is approximately 96 to 98% water soluble, has a moisture content of about 5%, and from 90 to 95% will pass through a 100-mesh screen. In the preparation of the spreader, 1 part of blood albumin is mixed with three parts of some diluent, such as fuller's earth. The diluent facilitates the rapid dispersal of the blood albumin in the water. Years of field trials established the commonly recommended amount of blood albumin spreader, 4 ounces to 100 gallons, as the optimum quantity, considering both insecticidal efficiency and safety to the tree.

Blood albumin proved to be a very effective spreader. It has excellent dynamic wetting ability without corresponding emulsifying quality. It is relatively unaffected by variations in the hardness of waters in different districts. It has stood the test of time, and is now more or less a standard spreader wherever tank mixture oil sprays are used in the United States.

One of the advantages of the tank mixture was that for every gallon of spray material paid for, the grower obtained a gallon of oil, as compared to an average of about 0.82 gallon of oil in the case of the emulsions. Pest control advisors take this difference into account in recommending 2% oil emulsion for sprays calling for 1.75% straight oil. In addition it should be stated in this connection that emulsions may "break down", with a separation of oil and water, before arriving at their destination, especially when long in storage or when left standing in an unshaded location or in freezing weather. Straight oil will not be adversely affected under these conditions.

A disadvantage of the tank mixture method is that it is less "fool-proof" than emulsions, that is, equipment must be kept in better working condition than would be necessary for the effective use of emulsions. Likewise it is necessary to add an extra ingredient to the spray (the blood albumin spreader) besides the oil, and neglect to do so results in a spray of inferior performance. It should be said on behalf of the tank mixture method, however, that during the years of its greatest usage in citrus spraying in California, these factors proved to be of negligible importance.

By 1933 over half of the spray oil applied for citrus pest control in California was of the tank mixture or "tank mix" type. For dormant spraying of deciduous trees a "Tank Mix Grade A" oil of a U.R. of 70 or above and a viscosity of 100 to 120 seconds Saybolt at 100°F was recommended by Borden (94). The saving in cost was stated by Borden to be about one-half, and the "coverage and oil-depositing properties" of the tank mixture sprays were considered to be better than those of any of the proprietary products then in use. Tank mixture continued to be widely used in dormant spraying, but in the spraying of citrus its use gradually diminished in California after the peak year of 1933, and at present very little tank mixture is being used.

When tank mixture was first used commercially, a considerable price differential existed between the straight oil and oil emulsions, despite the fact that the latter contain about 15 to 20 per cent water. This price differential soon narrowed down to a few cents per gallon. Thus one of the incentives for the use of tank mixture was practically eliminated. In the meantime emulsions were improved so as to be more uniform and dependable in their performance. Since there remained no incentive to the grower for the use of tank mixture, and since for various reasons there was no incentive for the spray manufacturing companies to exploit the tank mixture idea, it is understandable that this manner of using spray oil should gradually be discontinued, at least in some regions.

SPRAY OILS

Among persons qualified to pass an opinion, some believe tank mixtures to be inherently inferior to emulsions, while others still agree with Ralph H. Smith's contention that the fundamental considerations in the use of oil sprays are the grade and purity of the oil and the "oil-depositing quality" of the spray, and that the emulsification of oil before it is placed in the spray tank results in no advantages in regard to these considerations.

In evaluating the benefits that have accrued to the citrus grower from the tank mixture investigation, the following appear to the writer to be worthy of note:

1. The controversial nature of the "tank mix" problem initiated much investigation, observation, survey, and reappraisal of the situation, as well as numerous conferences among entomologists and manufacturers. The result was a more adequate standardization of spray oils as to grade, purity, uniformity of oil deposit, and more careful recommendations for usage to conform with the varied requirements of the different districts, seasons, varieties, and other conditions; more adequate agitation in spray rigs; greater attention to the size of the spray nozzle aperture and the technique of application, and in general, a greater interest in bringing to light all the factors pertaining to the use of oil sprays.

2. Many facts concerning spray oils formerly unknown or only vaguely understood by citrus growers and others interested in pest control became reasonably well understood. Many growers and cooperatives purchased straight oil of known composition directly or indirectly from the refiners and prepared their own sprays until the price differential no longer made such action worth their effort. This could be done again if ever the price situation should warrant such action.

Granting the wholesome effect of the tank mixture investigation, there was evident, in the writer's opinion, a tendency toward over-simplification of the oil-spray problem by the tank mixture advocates. An emulsion is a very complex physico-chemical system, and its very complexity gives a certain latitude for improvements which may or may not be within the scope of the tank mixture method. For example, one may make an emulsive oil by dissolving 1% glyceryl monooleate in the oil. This emulsive oil can be used at 1.75% concentration on citrus trees. However, by adding 4 ounces of blood albumin spreader to 100 gallons of this spray, better spreading and wetting will be attained than by either an emulsive oil or a tank mix spray used separately. By adding surface-active substances to both the oil phase and the water phase of an emulsion, the interfacial tension oil/water is reduced far more than by the addition of such a substance to only one phase of the emulsion, yet strangely enough, one does not necessarily obtain a reduction in oil deposit. In the above instance, better spreading and wetting of the foliage and fruit is obtained, but without sacrifice of oil-depositing quality; in fact, oil deposit is somewhat increased and, presumably, the oil is more evenly distributed because of the improved wetting.

In the determination of interfacial tension, the ring of the du Noüy tensiometer is placed at the interface oil/water. The addition of an interfacial tension depressant to the oil phase may reduce the energy required for the upward pull of the ring, that is, from water to oil, without much effect on the energy required for a downward push, that is, from oil to water. The addition of a suitable solute to both the oil and the water phase results in reducing the tension in both directions.

It appears that the advantageous effect of surface-active substances in both the oil and the water phase of a spray mixture is being utilized by some of the manufacturers of the "flowable emulsions" (see p. 186), and the possibility should not be discounted that certain improvements in oil sprays may be attained by such proprietary preparations which could not be attained by the tank mixture method.

EMULSIVE OILS

The "emulsive" or "emulsible" oils may in a sense be considered as tank mixtures. As in the case of the tank mixture, the emulsive oil is not emulsified before it is poured into the spray tank. However, at least in their original form, emulsive oils contained oil soluble emulsifiers which aided their dispersion in the water, with moderate agitation, because of the reduction in interfacial tension oil/water affected by the solutes used for this purpose.

Emulsive oils differ from the "miscible" oils in that they do not form a spontaneous emulsion upon coming in contact with water, but must be agitated. Their advantage, as compared to the miscible oils, is that they form a quick-breaking rather than a tight emulsion, and consequently form a spray with much greater oil-depositing capacity. As compared to emulsions, the emulsives have the advantage of not containing water, and as compared to tank mixtures they have the advantage of being "one package" instead of "two package" preparations, that is, only one material need be added to the spray tank, while with tank mixture a spreader must be added in addition to the oil. The directions for mixing, however, are the same as for tank mixtures. The oil is added while filling the spray tank with water. When the water (15-20 gallons) has reached to about the height of the agitator shaft, and with the agitator in motion, the oil is added. Ideally, 2 or 3 minutes should elapse before the filling of the tank with water is resumed. This gives the oil a chance to be forced through the pumps and back through the "overflow" into the spray tank. Since only a small amount of water is present at the time, complete emulsification is rapidly attained and the oil globules are uniformly distributed when the tank is full of water.

DEPOSIT AND DISTRIBUTION OF THE OIL

When a quick-breaking oil spray strikes a leaf, the oil is deposited on the leaf surface as an "initial deposit" and a "secondary deposit" (818), the two terms referring, respectively, to a dynamic and a static phenomenon. As the spray rushes over the leaf surface, much of it adheres to the leaf cuticle, despite the preponderance of water in the spray, because oil wets leaf cuticle more readily than does water. This oil deposition while the spray is in motion over the leaf is called the "initial" or "primary" deposit. Water does not wet lemon leaves as well as orange leaves (255) and consequently the "preferential wetting" of the lemon leaf by oil is greater than that which takes place on the orange leaf. In one experiment, oil deposit[1] from orchard spraying was found to be less on orange leaves than on lemon leaves in all of six separate tests with six different emulsions. The average oil deposit on orange leaves was only 74.1% as great as on lemon leaves (9). With successive contacts of the spray stream on any given leaf in the course of an ordinary spray application, a perceptible "film"[2] of oil is built up

on the leaf surface. After the leaf is no longer struck by the spray stream the spray water will collect into lenses (contact angle 60° to 75° on clean young leaves or lower surfaces of any leaves; much lower on upper surfaces of old or dirty leaves) each bearing many oil globules. When the water evaporates, the oil in the drop is deposited to form the "secondary deposit". If after spraying a tree, one circumscribes each drop of spray on the

Fig. 97 (at left). Photomicrograph of patches of spray oil on the surface of a clean oil-sprayed orange leaf. Greatly enlarged.

[1] The amount of oil deposited on citrus foliage can be determined by any one of a number of techniques. All methods involve extraction of the oil from the surface of the leaf, or from both the surface and inside of the leaf, with a suitable solvent or by a steam distillation process. Then, except when colorometric methods are utilized, the solvent is evaporated and the oil is volumetrically or gravimetrically measured. Probably the most satisfactory method of oil residue analysis is a steam-distillation process (393).

[2] While the oil may appear to be in a film, microscopic examination will show that the oil exists on a clean leaf in irregular patches with intervening spaces free of oil (fig. 97). This is because the leaf cuticle has a higher degree of polarity than is commonly recognized, and spray oil may have a contact angle of around 20° on a clean citrus leaf (255). The term "film" is retained here for convenience and because it is so well established in the literature.

underside of a leaf, where the stomata occur
and penetration is rapid, he will find a dark,
oil soaked area in each spot where drops or
lenses of spray mixture remained after spraying. This causes the mottled condition on the
underside of the leaf after spraying. Eventually the oil migrates internally to an area
along the midrib and along the margins of the
leaf (fig. 98) where its presence is indicated
by a distinct dark discoloration months after
the spray is applied, especially if the leaf
is allowed to wilt.

OIL SOLUBLE SOLUTES

The solutes in the emulsive oils as
originally conceived by Hugh Knight were to
be "glycerides of higher fatty acids or other
hydroxy esters of high molecular weight organic acids" (497). Glyceryl oleate, glyceryl
monoöleate, glycol oleate, diglycol oleate
and triethanolamine oleate, usually at 1%
concentration, were examples of the earlier
solutes, and some of these are still used.

Fig. 98. Orange leaves showing
areas of spray oil accumulation
within the leaf tissue according to
the characteristic pattern. Above,
excessive oil deposit; below, normal deposit.

Petroleum sulfonates and oil soluble soaps were eventually used to some extent.
Sometimes solutes were added in addition to the emulsifier to increase the persistence of the oil film on the tree surface. Aluminum dinaphthenate was used in
the original commercial emulsive oils for this purpose.

The addition of solutes to spray oils opened new possibilities in oil spray
research beyond those envisioned by Knight. Certain phenomena which have been made
possible by the physical and physico-chemical changes in the oil affected by the
solutes will be briefly summarized.

When an insect is placed in an emulsive oil for varying periods, depending on
the species, droplets of water gather on the surface of the cuticle. With straight
oil this appearance of water either does not occur, or is only feebly effected.
The appearance of the water on the surface indicates the oil has penetrated through
the external hydrophobic lipoid layer of the cuticle and made contact with the internal hydrophilic chitin complex of the insect cuticle.

If the more rapid and more complete appearance of the water on the surface of
the insect indicates a more rapid and more complete penetration of the emulsive oil
through the cuticle, this should be translated into terms of practical results in
the killing of the insects. This was shown to be the case in repeated experiments
with various solutes, and using potato tuber moth larvae Gnorimoschema operculella
(Zell.) as test insects (265). Tuber moth larvae wet with emulsive oils survived
less than half as long as larvae wet with straight oil. For example in one experiment tuber moth larvae were wet with straight light medium spray oil and the same
oil with 14 different solutes. Fifty larvae were used for each oil solution. The
median lethal periods (M.L.P.) in minutes for a few of the solutes which have been
used in the preparation of emulsive oils, were as follows: light medium spray oil,
275; with 1% glyceryl monoöleate, 93; with 1% glyceryl dioleate, 111; and 1%
glyceryl trioleate (glyceryl oleate), 120. In an analysis of variance of the data
from all 15 trials, the least significant difference at odds of 19:1 was found to
be 45.5 minutes. In field experiments on the black scale, a higher per cent kill
was obtained with emulsive oils than with straight oils (emulsions and tank mixtures) even though the sprays containing the emulsive oils did not deposit as much
oil. The writer has had no indications from experimental work, however, that the
emulsive oils are more effective than emulsions and tank mixtures against the red
scale.

In three Navel orange orchards, commercial brands of emulsive oil, emulsion,
and tank mixture were compared for a two-year period. In each orchard there were

three randomized sub-plots sprayed with each of the three types of sprays, the emulsive and "tank mix" oils at 1.67% and the emulsions at 2%. The same spray crew applied all three types of oil in a given orchard. Each type of oil spray was applied on from 75 to 102 trees divided into three randomized sub-plots in each orchard. During the summer following the second year of spraying with the same materials, a count was made of the number of adult black scales on each of 10 units (twigs 10-15 inches in length) on the north side of every tree in each plot in each of the three orchards. The results of the experiment are shown in table 5.

Table 5

Average number of adult Black Scale* per tree after
Two Years of Treatment with Three Types of Spray Oil

Type of Oil Spray	Concentration	Average Oil Deposit ml/cm^{2}x10^6	Average number of live adults per tree			
			Orch.A.	Orch.B.	Orch.C.	Average
Emulsive	1.67	67	0.41	0.48	0.31	0.40
Emulsion	2.00	89	0.76	0.82	0.45	0.68
Tank Mixture	1.67	106	0.80	0.58	0.40	0.59

*While the greater effectiveness of the emulsive oil spray against the black scale was quite striking, there was no evidence that this type of oil spray was superior to the others against the California red scale.

Fig. 99. Lower surface of lemon leaf showing no penetration (left) of an emulsive oil and complete penetration (right) of a straight oil, 5 minutes after application.

In the spraying of citrus trees in California the treatment is usually directed against the red scale and its effect against this insect is the factor upon which its efficiency is based. Other scale insects are incidentally well controlled if the treatment is properly timed. Thus the emulsive oils have not benefited from a practical standpoint from a certain inherent superiority against some insects which they possess.

In connection with this discussion it should be pointed out that C. R. Cleveland found emulsive oils to be more effective than emulsions in the control of codling moth and pear psylla (497).

Fig. 99 shows the difference in the rate of penetration of a light medium emulsive oil and a light medium straight oil such as would be used in the preparation of an emulsion or tank mixture. The oils were applied by means of a camel's-hair brush. The emulsive oil, which contained 1% glyceryl monooleate, was painted on the left half of the lower leaf surface (fig. 99, left) and the straight oil on the right half (fig. 99, right). A photograph made 5 minutes later showed the light, normal color of the leaf, indicating no penetration of oil, where the emulsive oil was applied, compared with the dark discoloration of the leaf, indicating the presence of oil within the leaf tissue, where the straight oil was applied. The same difference can be shown on the upper leaf surface, but penetration is much slower, requiring several days, unless a very light petroleum fraction is used.

All emulsive oils which have been tested by the writer have been found to penetrate leaf cuticle less rapidly[1] than straight oil. However on the lower

---

[1]Rate of penetration was increased, however, by the addition of 1% Aerosol OT.

surface of the citrus leaf practically all of the oil penetrates anyway, for penetration is comparatively rapid during the daylight hours (fig. 100). This is the case even if emulsive oils are used. Consequently the advantage, if any, of the emulsive oils must be confined to the upper leaf surfaces. This retardation of penetration of emulsive oils into leaf cuticle may be one of the factors accounting for their greater insecticidal efficiency against certain leaf-and-fruit-inhabiting soft scales. The red scale is most difficult to kill on the rough bark, and emulsive oils penetrate porous solids as rapidly as straight oils, so no advantage could be expected from this standpoint when emulsive oils are used in the control of the red scale.

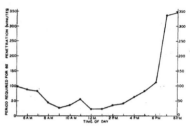

Fig. 100. Diurnal change in rate of penetration of spray oil into the undersides of lemon leaves. After Ebeling (254).

ADDITION OF INSOLUBLE SOLIDS TO OIL SPRAY MIXTURES

Sometimes it is advantageous to add finely-divided solids to an oil spray. Thus in the control of orangeworms, cryolite may be added to an oil spray even though the latter is directed against an entirely different group of insects. By combining the oil and the cryolite, a separate treatment for orangeworms is obviated, for orangeworm control is combined with, let us say, red scale control. Furthermore, a properly prepared oil makes a good carrier and sticker for cryolite. The amount of cryolite necessary per tree is reduced by the addition of the oil. The spread and adhesiveness of the cryolite is improved. Two pounds of cryolite is added to 100 gallons of spray along with the usual concentration of oil. The spray is applied with a thoroughness necessary for the control of scale insects or the citrus red mite.

The principle of preferential wetting has been discussed in a previous chapter (pp. 108 and 109). The solutes that are ordinarily added to oil in the preparation' of emulsive oils result in a preferential wetting of cryolite particles. Only emulsive oils or flowable emulsions containing suitable solutes in the oil as well as the water phase, can be successfully used with cryolite.

Zinc oxide may be added to an emulsive oil in the correction of zinc deficiency of citrus trees, but may also be used with flowable emulsions containing certain solutes in the oil phase. As in the case of the oil-cryolite combination, the recommendations of the manufacturers of the emulsive oil or flowable emulsion should be followed when an oil-zinc oxide combination is used, for the finely divided solids may cause an alteration in the oil depositing quality of the spray (usually an increase in deposit) and the concentration of oil in the spray may have to be modified accordingly.

Because of the preferential wetting of organic solids caused by the solutes added to emulsive oils, there is a resulting improved penetration into the solid particles and improved solubility of their oil-soluble constituents in the oil. Thus if ground cubé root is added to spray oil, using 120 grams to a liter of oil, nearly twice as much rotenone and deguelin is extracted from the ground root if an emulsive oil (1% glyceryl monooleate) is used than if a straight oil is used as the solvent (269). Thus if ground cubé root is to be used with oil, there is a decided advantage if the oil is of the emulsive type, for a greater concentration of the cubé extractives will then be dissolved in the oil (see p. 137 and fig. 80).

PRESENT TRENDS IN SPRAY OIL FORMULATIONS

As stated before, practically all oil spraying of citrus trees is being done with emulsives and emulsions. After the severe injury from oil spray in inland

Navel districts in 1943, a distinct but temporary trend back to emulsions was noted, the "flowable" type being especially favored. It was felt by some investigators that emulsives were not as "fool-proof" as emulsions. Manufacturers of spray oil preparations generally believe that although emulsive oils will perform very well when the rules for mixing are strictly followed, too often these rules are ignored. Then the oil may not be evenly mixed, allowing for a certain amount of stratification in the spray tank. The last trees to be sprayed with a tank of spray may receive a very high concentration of oil and injury may result. On the other hand, sometimes an emulsive oil, when not properly mixed, becomes increasingly "tighter" as the volume of spray in the tank is reduced, and this results in lack of uniformity in the performance of the spray.

The tendency during the last 8 or 9 years has been to increase the oil-depositing ability of the emulsive oil spray. This has been brought about by the use of oil-soluble emulsifiers that have the ability to invert an emulsion, that is, to cause the oil to be the continuous phase instead of the water. Not enough emulsifier has been added to bring about an inversion of the emulsion, but enough has been added to cause an "unstable equilibrium" between the oil and water phases (495). The principle of "unstable equilibrium" has been utilized to increase the oil-depositing ability of the emulsive oils, but this has been done with the sacrifice of uniformity of performance of the spray.

An oil emulsion is a complex system. When the oil spray reaches the tree surface, other surface and interfacial forces are involved besides those inherent in the emulsion. The emulsifier or possibly added solid ingredients, upon being deposited on the tree surface, might conceivably affect the uniformity of deposition of the oil. A drop of spray oil will ordinarily cover about 80 square inches of leaf surface as it is deposited by the spray. It is conceivable that the uniformity of deposition of this extremely tenuous "film" of oil might be affected by the manner of emulsification or the nature of the emulsifier. Some workers believe this to be the case, but as far as the writer is aware, they have so far based their conclusions on casual observations, and have not produced photographic or quantitative evidence that the uniformity of the "film" of oil is affected by the nature of the emulsion. It would certainly indicate a defeatist attitude, however, to assume a priori that the uniformity of the oil deposit could not be improved, and it is possible that such a degree of deposit control could only be attained by emulsification prior to use.

It would appear to the writer that any means of increasing the spreading and wetting ability of a spray without reducing its oil-depositing capacity would result in a more uniform distribution of the oil over the tree surface per unit of oil deposited. Efforts to attain such a desirable goal should certainly be encouraged.

At present (1948) probably about 85% of the citrus acreage is being sprayed with emulsive oil, and the remainder almost exclusively with flowable emulsions. The paste type of emulsion and the tank mixture type of spray are used on a relatively small acreage.

HOW OIL KILLS INSECTS

Investigations made with potato tuber moth larvae immersed in petroleum oil showed that death takes place much more rapidly in highly refined oil than in nitrogen (i.e. by suffocation), which shows that even a highly refined oil may kill by virtue of its toxicity rather than by suffocation. As the per cent of unsaturated hydrocarbons increases, the toxicity of the oil usually increases, if the insects are immersed in the oil (265). If insects or insect eggs are sprayed with an oil spray, the more refined oil may be the more effective (155). In the latter case, however, a very tenuous film is involved, and physical rather than chemical factors may play an important rôle. In experiments with potato tuber moth larvae, it was found that kerosene kills more rapidly than heavier fractions of similar U.R. This may be due principally to its more rapid penetration and perhaps abetted by a certain amount of volatilization of toxic vapors. Kerosene kills the mealworm (<u>Tenebrio</u>) over 7 times as rapidly as a white oil of 80 to 85 seconds viscosity Saybolt (662). Lowered viscosity, however, may also result in reduced effectiveness. With California red scale kerosene, as a dilute spray, is less effective than spray oil because the red scale can expel the less viscous kerosene from their tracheal system (231).

The red scale being the most important of the California citrus pests, the action of oil on this insect has been studied especially thoroughly. If oil reaches the spiracles, it is drawn by capillarity through the tracheal system. This can be studied especially well when the red scale is in the "gray adult" stage, for then the armor can be easily detached from the body of the adult female (fig. 101). When a red scale is sprayed with a heavy dosage of oil, the oil causes

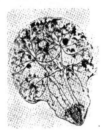 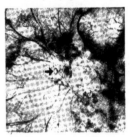

Fig. 101 (at left). Tracheal system of the red scale. After de Ong, Knight and Chamberlin (231).

Fig. 102 (at right). Photomicrograph of the left anterior spiracle (shown by arrow) of a red scale. Spray oil has entered the spiracle and a short distance into the tracheae, changing them from dark to translucent.

the ventral covering of the insect to be translucent and under these conditions the spiracles and adjoining tracheal trunks can be seen clearly under the microscope. Before the oil enters, the low refractive index of the air-filled tracheae causes them to show as black lines, when viewed with the aid of a microscope. When the tracheae become filled with oil they change to a translucent appearance, only the walls of the tubes showing dark. This makes it possible to see exactly to what point the oil has penetrated (fig. 102).

When California red scale are covered with oil they live longer than they do in nitrogen, so the oil can hardly be said to be toxic. It appears that if the spiracles are occluded by the oil, the red scale dies from suffocation. Even the addition of unsaturates or toxic substances to the oil does not cause death to occur more rapidly. However, when red scales are treated with an oil spray, a considerable percentage of those individuals which are hit by the spray do not receive enough spray on and around their bodies to allow for a penetration of oil to the spiracles (fig. 91). A large percentage of these insects will be killed rather slowly by the oil on or in their bodies, apparently by a prolonged impairment of their physiological processes.

Some insects are paralyzed or killed very rapidly by oil. The adults of some species of aphids, beetles and ants, immersed in spray oil, will become completely motionless in from 1 to 5 minutes, with no resumption of motility and no recovery, even if the oil is quickly removed from their bodies. The insect <u>Carausius morosus</u> can be paralyzed immediately by applying oil to the dorsal surface of the mesothorax (1029).

It can be seen from the above that spray oil may kill insects by suffocation or by toxic action. The latter is not well understood, although a number of conjectures regarding the toxic action of oil have been made (see 770).

The effect of petroleum oil on the eggs of the oriental fruit moth, Grapho-litha molesta, has been investigated. It was demonstrated that although penetration of the oil through the egg shell evidently occurs if the eggs are immersed sufficiently long, the oil exerts its lethal effect without penetration. The normal respiratory activity of developing eggs is markedly reduced. There was considerable evidence that oil exerted its ovicidal effect through a mechanical interference with the normal gaseous exchange mechanism. It was suggested that the prevention of ready elimination of toxic metabolites, resulting in their accumulation in lethal amounts within the egg, may be the means by which oil kills the eggs. (786)

As previously stated, one of the functions of spray oil in relation to the control of citrus insects and mites is the inhibition of settling and further development of the young of those insects not directly killed by the spray. Red scale crawlers placed on lemon foliage a day after a spray of 2% heavy oil was applied were immediately mired in the oil on the surface of the leaf and could not move about. Since the crawlers of scale insects are very sensitive to the effects of the oil, these were killed by the oil. For the next 2 days all the 40 crawlers transferred to the sprayed foliage could move about, but adsorbed a film of oil within 30 seconds. On the third day after spraying, 8 out of 20 adsorbed a film of oil in 30 seconds and all the crawlers, out of 20 transferred, adsorbed a film of oil in less than a half hour. Not until 14 days after spraying could red scale crawlers be transferred to the sprayed foliage without adsorbing a film of oil to their bodies in less than a half hour. Considerable inhibition of settling of crawlers may be noted a month after treatment with oil spray, and even those scales which settle and continue development through the whitecap stage are less apt to reach maturity than scales settling on unsprayed foliage. (251)

The inhibition of settling of red scale crawlers caused by oil spray takes place on all the morphological parts of the tree (branches, twigs, leaves and fruit). The residual effect is least on the lower sides of the leaves, where the oil penetration is the most rapid. On the upper sides of the leaves the residual effect is more prolonged than anywhere else on the tree. On the fruit there is a great variation in residual effect depending on the rate of penetration of the oil into the rind. The oily residue is surprisingly effective on the branches. Although the oil film appears to penetrate quite readily, it appears that a minute quantity is drawn back to the exterior and keeps the bark surface moist with sufficient oil to inhibit the development of red scale crawlers.

In an experiment made on two lemon trees, one sprayed with 1.75% light medium and the other sprayed with 1.75% heavy medium emulsive spray oil, the inhibition of settling of red scale crawlers was 90.8% as great on branches 0.5 to 1.0 inch in diameter as on the upper leaf surfaces and on the average it was considerably more effective than on the fruit. In this experiment, effectiveness of inhibition of settling was based on the per cent surviving on treated tree parts as compared to parts of a similar nature and consistency on the same tree which were not sprayed. In the entire experiment, 8171 red scale crawlers reared on banana squash were used.[1]

HOW OIL INJURES PLANTS

Although citrus trees are less sensitive to spray oil than foliaged deciduous trees, the "margin of safety" is precariously small, and even "safe" spraying with a dosage sufficient to control such citrus insects as the California red scale, Florida red scale and purple scale results at least in impaired coloration and reduced quality of the fruit.

Spray oil enters rather rapidly through the stomata of the lower part of the citrus leaf, but penetrates very slowly through the upper cuticle, which not only is thicker than the lower cuticle, but also has no stomata. The oil also enters the stomata of green twigs and fruit. It penetrates into young fruit more rapidly

[1]Ebeling, W. 1941. Effect of oil spray residue in inhibiting the settling and development of red scale crawlers. Manuscript on file at the Division of Entomology, University of California, Los Angeles.

than into old fruit, probably because of partial occlusion of the stomata of the latter. Penetration of oil through the stomata takes place increasingly more rapidly from morning until noon, then gradually the rate of penetration decreases. from about 1 o'clock until nightfall. (254) Apparently the rate of penetration of the oil is proportional to the size of the openings of the stomata and is at a minimum when these are closed.

When an oil spray is used in accordance with recognized commercial practice, it does not penetrate deeply into the plant tissue, and does not enter the plant cells that contain protoplasm, but remains between the cells immediately beneath the epidermis of the leaf, twig, or fruit. It penetrates to a depth of over a half a dozen plant cells only when too great a quantity of oil is applied. The heavier "ends" of the foliage spray oils may remain in the leaf for a year or more, or throughout the life of the leaf. (752)

During the period of oil penetration and initial translocation, there occurs in citrus leaves a decrease of transpiration and increase in respiration. There also occurs a temporary cessation of photosynthesis. This might be caused by the effect of the oil on the chloroplasts, for intercellular oil is turned green by the extracted chlorophyll. Recovery from the effects of the oil is most rapid with the lighter oils, and is indicated by a return to normal of transpiration and respiration rates, and by the accumulation of carbohydrates in the leaves. Abnormally large amounts of carbohydrates accumulate, owing to the occlusion by oil of the conducting vessels. (496)

Oil-sprayed foliage exposed to the direct rays of the sun regains its normal transpiration rate much sooner than that of shaded foliage. It has been noted that in the orchard, after a period ranging from 1 to 6 or 8 weeks after an oil spray, the transpiration rate may equal or even noticeably exceed the rate before spraying. (65)

SYMPTOMS OF OIL SPRAY INJURY

1. Leaf Drop:- At all times there are a large percentage of old, senile, and probably comparatively nonfunctional leaves on a citrus tree. The force of the spray stream will remove many of these leaves and others will fall over a period of several weeks after treatment. The loss of these leaves is not considered to be of practical importance. Sometimes, however, a large number of mature leaves showing no signs of senility are caused to drop by oil spray. The ground may in some cases be completely covered by such leaves and the sparseness of the foliage remaining in such trees is striking when compared with normally foliaged trees. In several months the tree will have forced out much "sucker" growth and new foliage, but in the meantime there is obviously a reduction in photosynthesis and the elaboration of starch, and the new flush of growth can hardly be considered as a sign of "stimulation" from the oil spray.

Leaf drop may begin a week or 10 days after treatment and continue for several weeks. Certain meteorological or soil conditions, often of an obscure nature, may predispose a tree to heavy leaf drop following an oil spray. Infestation by insects or mites is also a predisposing factor. Light or even moderate leaf drop is considered to be a normal and unavoidable reaction to oil spray and to have little practical importance. Severe leaf drop, however, leads to crop reduction and the dying back of twigs and branches.

Fig. 103. Average leaf drop per tree caused by 2% heavy oil spray of 86 U. R. (left), 95 U. R. (center), and 100 U. R. (right). Explanation in text.

Leaf drop constitutes a rather accurate index of the phytotoxicity of the oil. In fig. 103 are shown three piles of leaves showing the average leaf drop in each of 3 plots of lemon trees sprayed with 2% oil according to the tank mixture method. The small pile (378 leaves) represents the average drop per tree in the plot

sprayed with 100 U.R. oil, the medium sized pile (980 leaves) is the average per
tree in the plot sprayed with 95 U.R. oil, and the largest pile (2530 leaves) is
the average per tree for an 86 U.R. oil. The 100 U.R. oil had a Saybolt viscosity
of 100 and the other two had a Saybolt viscosity of 95. The heavier oil would
cause the greatest leaf drop if the U.R. were the same for all 3 oils.

Often when oil causes leaves to fall it turns them slightly yellowish or
brownish in color, and oils low in U.R. may cause a "burn" of both leaves and
fruit, but this effect is seldom seen nowadays with oils such as are recommended
for citrus spraying. It may occasionally occur, however, on leaves or fruit exposed to the sun, (fig. 104).

Fig. 104. Characteristic "burn" from spray oil
sometimes occurring on oil-sprayed leaves and fruit
exposed to the sun. Above, "burn" on leaves; below, "burn" on oranges.

Orange leaves are sometimes affected by a malady known as "mesophyll collapse".
Certain regions of the soft mesophyll tissue of the leaf collapse and dry out.
Several separate places on the lower side of a leaf, often bounded by two main
lateral veins, show at first a slight change in color, then lose their chlorophyll
and become dry, turning light gray and finally brown. When mesophyll collapse is
in progress, leaf drop from oil spray is usually greatly accentuated. Later the
collapsed layers become corked over and the edges of the areas become healed. The
leaves appear to recover and are no longer specially susceptible to leaf drop following oil spray. Fig. 105 shows two leaves in which the collapsing of tissue had
long terminated. These leaves did not fall following an oil spray.

SPRAY OILS

The resistance of the principal varieties of citrus trees in California to defoliation by oil spray is, in descending order, as follows: lemon > grapefruit > Valencia orange > Navel orange > tangerine > lime. Lemon trees are usually sprayed with a heavier grade of oil than other citrus varieties, for the grower wishes to obtain the best possible control of his pests compatible with reasonable safety to his trees. Grapefruit trees sprayed with medium oil usually suffer less defoliation than orange trees sprayed with light medium oil. Lime trees are sometimes killed by a light medium oil spray. Leaf drop, as well as other symptoms of tree reaction, are usually less severe near the coast than farther inland. Likewise tree reaction to oil spray is less in Florida, Texas and other Gulf States than in California.

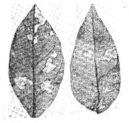

Fig. 105. Mesophyll collapse in orange leaves; a physiological malady that predisposes the leaves to drop after an oil spray.

2. <u>Dead Twigs and Branches</u>:- The citrus tree is more or less a hollow shell of foliage. The "shell" is very heavily foliaged while relatively few leaves occur in the interior of the tree. This condition is less marked on lemon trees, especially when they are "short-pruned". If a branch of a mature orange tree is examined from its point of origin to the periphery of the tree, it will be found that few, if any, leaves occur throughout the length of the branch until the outermost twigs are reached.

Sometimes a long branch will have at its end only a small tuft of foliage to elaborate its food (fig.106) If this tuft of foliage is covered with a film of oil, considerably reducing the normal physiological activity of the leaves, and particularly if these leaves should drop, the small twigs at the extremities of the branch are apt to die. New twigs and new foliage may be forced out farther back on the branch, but this is not apt to occur on a branch as obviously debilitated as the one shown in figure 106.

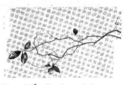

Fig. 106. Dead wood in an orange tree caused by oil spray.

If two orchards are growing side by side and one is consistently fumigated while the other is consistently oil sprayed, the latter is more likely to have an excessive amount of "deadwood", usually consisting of only small twigs but with occasional limbs measuring 1/4 to 1/2 inch in diameter. There is apt to be less fruiting wood in the interior of an oil-sprayed tree.

Some years deadwood resulting from oil spray is especially severe. The dying back of twigs and branches was particularly severe in southern California during the fall of 1943. Navel orange trees in the interior Navel belt of southern California suffered the most damage, but Valencias in the same areas also were injured more than usual. Orange trees appeared to be weakened by adverse conditions peculiar to the 1943 season, and this was indicated by the ease with which the leaves were dislodged by the spray stream. September was an abnormally hot month, with prolonged periods of high maximum temperatures and low humidity. There had been a heavy growth of foliage during the summer, a heavy crop of Navel oranges, and the rate of growth of fruit during the months of September and October was below the normal rate for this period. The various symptoms of tree weakness were most apparent in orchards with light soil, where soil moisture is most rapidly lost. Three other years, since general oil spraying of citrus trees started in 1926, were characterized by weather conditions similar to those of 1943. During these years there also was greater than average damage to orange trees from oil spray (978).

The rather general injury from oil spray in 1943 consisted of leaf drop, fruit drop and dying back of twigs and branches. The death of branches an inch in diameter only 2 or 3 months after treatment was quite common in the most severely injured orchards.

SUBTROPICAL ENTOMOLOGY

In company with Wm. E. Shilling, County Inspector for the San Fernando Valley Red Scale Control District, the writer made an extensive survey of the oil spray damage in 26 oil-sprayed orchards in·San Fernando Valley from November 29 to December 1, 1943. A striking correlation was found between oil spray damage and citrus mite damage of the previous year. No severe damage was found in any orchard oil-sprayed in 1942, with resulting good control of citrus red mite. All orchards sustaining severe injury from oil spray in 1943 had been fumigated or un-treated in 1942. Some had a subsequent treatment with DN-Dust for mites, with only partial control, and others no treatment for mites. Usually mite injury was still evident on the damaged trees. It appears that at least in the San Fernando Valley, Navel orange trees weakened by the attacks of the red mites in 1942-43, because of inadequate control of this pest, were predisposed to injury from oil spray in the fall of 1943.[1]

Fig. 107. Green lemons which dropped after the application of 2% heavy medium oil emulsion.

3. Fruit Drop:- A certain percentage of small fruit drops each year in the normal "June drop" period, but the "June drop" has usually terminated before the first oil spraying in late July and August. Beginning 1 or 2 weeks after spraying and continuing for a month or more, an-other drop of green fruit may occur (fig. 107), this time caused or accentuated by the oil spray. The loss from drop of green fruit may amount to as much as 50% or more of the crop. Small lemons may drop, as well as oranges, following an oil spray. As in the case of leaf drop and deadwood formation, a physiological condition of the tree induced by the meteorological peculiarities of the season or by cultural conditions cause the fruit to be in the weakened condition that pre-disposes it to drop after being sprayed with oil. Fruit drop, like other adverse effects of oil spray, is much more common some years than others. In the fall of 1943 when leaf drop and the dying back of twigs and branches were so common, much drop of green fruit was also noted on both Navels and Valencias. The most severe fruit drop in Los Angeles County appears to have been in the Rivera-Downey-Pico districts, where there was reported a drop of fruit on about 1,500 acres of oil-sprayed orange trees, mostly Valencias. About one box of fruit per tree dropped from approximately 1,000 acres and between 2 and 3 boxes per tree on the remaining 500 acres. Eight orchards lost between 3 and 4 boxes per tree and in three orchards about 85% of the crop dropped to the ground.

It is difficult to connect any meteorological or cultural factors with the predisposition which trees occasionally have to drop fruit after an oil spray. Some years the tendency to drop is greater than others, but in a given year adjoining orchards with apparently similar vigor will differ radically in their response to oil spray as far as fruit drop is concerned. The following statement was made by R. H. Smith, who gathered many field data on the subject of fruit drop after oil spray:

"Drop has occurred when the soil moisture was high, when it was moderate and when it was low. In many cases groves appearing to be in excellent condition have suffered as heavy drop as those which appeared to be in a poor state of vigor" (817).

Despite our ignorance of the predisposing factors relating to the physiology of the tree, we may still reduce fruit drop by attention to factors pertaining to the oil spray program which are within our control. It is known that spraying oranges when they are too small increases the risk of fruit drop. This risk of premature spraying is taken by those who spray in July, or even early August.

[1]Ebeling, W. and W. E. Shilling. Oil spray damage in the San Fernando Valley in 1943. Mimeographed copy on file in the Division of Entomology, University of California, Los Angeles.

202

Also in general the factors such as high temperatures, low humidity, "electric" conditions, dry soil, insect and mite damage, etc., which make for defoliation and deadwood following oil spray, also accentuate fruit drop, although as stated before, fruit drop may also occur under apparently favorable conditions for oil spraying.

When green fruit drops following an oil spray, one may sometimes notice a breakdown of the rind, which is apparently caused by the penetration of oil into the rind, with a resulting interference with the normal functioning of the cells. The rind becomes depressed in limited areas, then, especially in the case of oranges, the collapsed area usually spreads in extent and the affected part slowly dries and becomes brown. Affected oranges nearly always drop, and sometimes the collapsing begins after the fruit has dropped. In the case of lemons, rind breakdown sometimes becomes serious only after the fruit has reached the packinghouse. In the case of both oranges and lemons, rind breakdown is observed much more frequently some years than others, but even during those years when the fruit is in the physiological state which appears to predispose it to this particular type of reaction to oil spray, it is found in only a few scattered orchards.

A drop of ripe oranges may also be caused or accentuated by oil spray. Again in this respect, certain little understood physiological conditions appear to predispose the tree to adverse reaction to oil spray, and loss is greater some years than others. As far as oranges are concerned, the occasion for spraying over ripe fruit is most apt to occur in the case of the summer and early fall spraying of Valencias. The obvious solution to the problem is the picking of the ripe fruit before spraying, if this can be arranged.

Mature lemons also are often caused to drop by an oil spray, and again picking is advised prior to treatment. High temperatures greatly increase the chances of drop of both mature and very small fruit, and since there are no objections to late spraying of lemons, they may be sprayed after the oranges have been treated. October is a good month to start spraying lemons, and spraying may continue through November and December.

Addition of 2,4-D to Oil Sprays to Reduce Leaf Drop and Fruit Drop:- The drop of mature citrus leaves and fruit can be considerably reduced by adding to the oil minute quantities of the plant growth regulator 2, 4-dichlorophenoxyacetic acid, commonly called "2,4-D". This is the same compound which, at higher concentrations, has been widely used as a weed killer. It has also been used in an aqueous solution sprayed on the trees to reduce the preharvest drop of citrus fruit. When added to oil, the reduction in defoliation is especially marked when low U.R. oils are applied (846). Figure 108 shows the difference in leaf drop between 2 branches of a Navel orange tree to which a 60 U.R. light medium petroleum oil had been applied by means of a brush. Oils of such low U.R. are, of course, not used in citrus spraying. Where the straight oil was applied, a complete defoliation occurred, while where 0.1% 2,4-D was added to the oil, there was practically no leaf drop.

Fig. 108. Orange branches from a tree sprayed with a 60 U.R. oil. One-tenth per cent 2, 4-D was added to the oil applied to the branch on the left. After Stewart and Ebeling (846).

After 3 successive annual applications of 2, 4-D, in extensive field trials, Stewart and Riehl (848, 849) were able to report as follows:

"Adding 2,4-D to oil sprays reduced fruit-drop, mature leaf-drop, fruit-stem die-back, and "black-button" formation during citrus fruit storage.

"The drop of immature, green navel oranges that sometimes occurs in the fall was reduced as well as the subsequent preharvest drop of mature fruit that occurs in the spring. The 2,4-D in oil applied in September was still effective in reducing fruit-drop the following March."

An oil soluble ester of 2,4-D may be dissolved in oil at 250 parts per million of the free acid equivalent of 2,4-D (4 p.p.m. in terms of the total volume of spray). This is as effective as 8 p.p.m. of a water-soluble salt of 2,4-D. The oil-soluble ester may be incorporated into the oil before it is poured into the spray tank. In the case of emulsions, it is preferable that this be done before emulsification.

Reduction of Bloom and Set of Fruit:- A reduction in the number of blossoms of oil-sprayed trees, as compared to untreated or fumigated trees, has frequently been observed. The reduction in bloom is especially noticeable on the north sides of the trees, where the oil residue from the spray remains longest. A reduction in bloom does not necessarily result in reduced amount of fruit, and trees sprayed with oil, and having as a result fewer blossoms, usually have about as much fruit actually setting as adjoining fumigated or untreated trees.

A severe reduction in bloom is most often associated with the use of too heavy an oil, or in the case of oranges, with spraying too late in the fall. October spraying appears to be especially deleterious in this respect, more so than spraying at an even later date. Reduction of bloom and reduction of the following years crop by oil spray is most apt to occur on oranges and in the interior sections.

Impaired Appearance:- The most striking effect of spray oil on the appearance of the fruit is in connection with its color. Oil-sprayed fruit will attain normal color later than fumigated or untreated fruit and may need to be subjected to the ethylene gas treatment before sufficient color is attained. Even after treatment with ethylene gas, lemons sprayed with a medium or heavy medium oil nearly always have a darker, less desirable color than untreated or fumigated fruit.

Aside from interfering with normal coloration, the oil spray has the effect of causing the fruit to appear as if its "bloom" were removed and this is considered to be another of the undesirable effects of oil spray. "Bloom" is the term used to designate the wax coating with which fruit is covered. It consists of a layer of flattish scales (fig. 109) roughly equal in area to the varying surfaces of the epidermal cells of all types. These scales are formed by the fusion of rodlets of wax secreted by the epidermal cells (767). The coating of wax causes the slight but definite dullness of the surface associated with unsprayed fruit. When the fruit is sprayed the wax becomes translucent and the fruit appears as if the layer of wax scales had been removed. One often hears it stated that oil spray "removes the bloom", but this is not the case. The oil neither dissolves nor removes the coating of wax--it merely penetrates the wax and makes it translucent. Neither does the oil inhibit the subsequent secretion of wax, for it continues to be secreted on the fruit, which is usually still growing and expanding at the time the trees are sprayed (767).

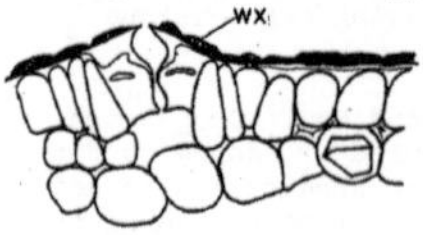

Fig. 109. Wax scales on the cuticle of a Navel orange. After Scott and Baker (767).

The writer, in company with Harold Lewis, an entomologist for the California Fruit Growers Exchange, made a survey of 86 Valencia orange orchards in Orange County, California, from July 25 to July 30, 1932, to determine the effect of the pest control treatment of the previous year on fruit appearance and quality. The treatments represented among these 86 orchards were oil emulsion sprays, tank mixture sprays, fumigations, spray-fumigation treatments, and two orchards were untreated. Among the factors affecting appearance, the following were studied: softness, puffiness, roughness, and stem end growth. A normal fumigated or untreated orange is rather firm to the touch unless, of course, it has been partially dessicated. Moderate pressure from the fingers will not cause the peel to be indented. With a soft fruit, one may easily press a hole into the fruit from the pressure of the fingers. When fruit is "puffy", the peel is loose and is puffed out in various places. "Roughness" refers to the nature of the surface of the rind. The most desirable characteristic is the smooth rind. To grade a fruit as to varying degrees of roughness requires, of course, a knowledge of the range of variation to be expected with regard to this physical characteristic of the rind. "Stem end growth" refers to a puffing out of the rind in the vicinity of and surrounding the stem. It is a tendency occasionally slightly developed on fumigated and untreated fruit, but often greatly accentuated if the fruit is oil-sprayed. The fruit in each orchard was graded with respect to the degree to which the various undesirable characteristics mentioned above had developed. The results of the survey, as it pertained to fruit characteristics affecting appearance, are shown in table 6.

Table 6

Relative Effect of Oil Spray and Fumigation on Valencia Oranges as Indicated by Numbers of Orchards* Suffering Various Degrees of Deleterious Effect.

Degree of Seriousness
Number of Orchards in which malady was

Treatment	None	Slight	Moderate	Pronounced	Extreme
Softness					
Oil Spray	15	30	10	3	1
Fumigation	16	10	0	0	0
Puffiness					
Oil Spray	21	25	9	4	1
Fumigation	25	1	0	0	0
Roughness					
Oil Spray	32	18	1	6	0
Fumigation	23	4	0	1	0
Stem end growth					
Oil Spray	24	15	3	6	1
Fumigation	27	1	0	0	0

*The survey included the examination of 86 orchards. The data from two untreated orchards is included under "Fumigation" and data from four orchards treated with the spray-fumigation treatment is included under "Oil Spray".

It can be seen from table 6 that with regard to the effect of pest control treatments on (1) softness of fruit, (2) puffiness of rind, (3) roughness of the surface of the peel, and (4) the incidence of stem end growth, the oil sprays were more injurious than fumigation or no treatment. The ungrouped data showed relatively more orchards in the "pronounced" and "extreme" categories in the orchards sprayed with emulsions than among those sprayed with tank mixtures. It was assumed, however, that this could be due to chance, for there is no reason for assuming that the emulsions, which deposited about the same amount of oil as the tank mixtures, would result in any greater effect on the appearance of the fruit.

Parker et al (676) made an analysis of data provided by about 150 orange growers from Orange County with reference to the production and quality of their fruit for the years 1935 to 1942. Their analysis also showed a greater amount of puffy fruit from oil-sprayed orchards, as well as a greater amount of fruit of inferior color.

Impaired quality:- The inferior flavor of oil-sprayed, as compared to fumigated or untreated, fruit usually is so striking that a person not knowing the origin of a group of oil-sprayed and unsprayed oranges can usually separate the sprayed from the unsprayed fruit on the basis of the difference in flavor. Thus W. E. Baier of the California Fruit Grower's Exchange conducted a test with fruit obtained from 18 fumigated and 23 oil-sprayed Navel orange orchards. The strained juice was scored as to flavor by several observers without knowledge of the identity of the samples. Of the fumigated fruit, 88.9% were judged to be above average in flavor and 11.1% below average. Of the oil-sprayed fruit, 43.5% were judged to be above average in flavor and 56.5% below average. The "Brix test", a hydrometer reading indicating per cent of sugar, was 13.1 for the fumigated-fruit juice and 12.2 for the oil-sprayed-fruit juice. The average per cent of citric acid was 0.95 for the fumigated and 0.85 for the oil-sprayed fruit (48).

Lewis and Ebeling, in the previously mentioned survey, found a marked difference in flavor between oil-sprayed and fumigated fruit in the 86 Valencia orange orchards examined. The percentage of fruit considered to be "flat" or insipid in oil-sprayed as compared to fumigated orchards is shown in table 7.

Table 7

Relative Effect of Oil Spray and Fumigation on the Flavor and Texture of Valencia Oranges as Indicated by Numbers of Orchards Suffering Various Degrees of Deleterious Effect.

Number of Orchards in which malady was "Flatness" of Taste.

Treatment	None	Slight	Moderate	Pronounced	Extreme
Oil Spray	24	29	4	2	1
Fumigation	22	4	0	0	0

Granulation

Treatment	None	1 out of 12	2 out of 12	Extreme
Oil Spray	45	8	5	1
Fumigation	26	0	0	0

In table 7 the proportion of fruit, if any, showing granulation also is shown. "Granulation" refers to an abnormality in the pulp of Valencia oranges, usually in the stem end. It refers to an enlarging, hardening, and turning gray of the juice sacs (69). It is not evident from the exterior, and is also usually not evident when the fruit is cut open, until the fruit is mature or nearly mature. Granulation is more common in certain strains of Valencias than others, and more common in rapidly growing fruits than in slow-growing fruits. A larger percentage of fruit is affected some years than others. It can be found in untreated or fumigated orchards, but it will be noted from table 7 that it was not found in any of the 26 fumigated orchards included in the survey made by Lewis and Ebeling. Tests made over a 7-year period by the Division of Plant Physiology of the University of California Citrus Experiment Station[1] showed that oil sprays usually doubled the severity and amount of granulation in Valencia oranges.

Sprays ranging from 0.25% to 1.75% light medium oil caused a reduction in the total soluble solids and in reducing and total sugars of the juice of oranges. Oil spray also showed a tendency to reduce the titratable acidity. Only slight differences were noted in chemical composition of fruits sprayed with oil concentrations ranging from 0.75% to 1.75%. Climate, soil type, and period of the year when the oil spray was applied did not appreciably affect the influence of the oil on the chemical composition of the fruit. (778)

The Plant Physiology Division at the University of California Citrus Experiment Station at Riverside has accumulated a vast amount of information with regard to the influence of pest control treatments on citrus fruit quality. Of 100 oil-sprayed orange fruit samples gathered over a period of many years, 81% showed a reduction (2.5 to 26.0%) in total soluble solids, and all but 2 or 3 showed a reduction (2 to 32%) in acids. In 12 to 18 determinations, a reduction in Vitamin C was noted. Eighty-one per cent of 74 determinations showed that oil sprays caused a reduction in dry matter in orange and grapefruit leaves. Kerosene sprays had no effect on soluble solids and acids. Some effect on soluble solids (sugars) and acids and on the amount of granulation was noted the same year after oil spraying.[1]

[1]Correspondence from Dr. E. T. Bartholomew, dated March 1, 1948.

Water Spot:- Sometimes in California rain will fall intermittently for a week or two without ceasing sufficiently long to permit a drying of the trees and fruit. Such continuous periods of wet weather usually occur in January and February, when Navel oranges are mature and susceptible to a malady known as "water spot" (fig. 110). The rind imbibes enough water to cause it to swell, and often minute cracks appear in the cuticle in the affected portions. If wet weather continues sufficiently long, the affected parts of the rind become water-soaked and turgid, and may later become covered with dark green to black saprophytic fungi, usually a mixture of Cladosporium sp. and Alternaria sp. Colletotrichum, Alternaria, the blue and green penicilliums (Penicillium italicum and P. digitatum) and other fungi and bacteria may later invade the affected tissue and cause a "water rot". If, however, the early stages of water spot are followed by dry weather, "no soft decay results but the tissues collapse and dry, making depressed, hardened areas which darken in color through various shades of brown until they approach the chestnut and auburn brown of Ridgway" (303).

 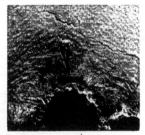

Fig. 110 (at left). Water spot of navel oranges. Above, navel end water spot; below, "shoulder spot."

Fig. 111 (at right). Cracks near the navel end of an orange caused by immersion in water for 12 hours. Greatly enlarged. Photo by Roy J. Pence.

Incipient water spot can be induced in a few days by immersing Navel oranges in water (fig. 111) and severe water spot occurs in less than a week. Typical water spot can be most rapidly produced by injuring the rind, allowing ready access to water, then immersing the fruit in water. Navels are much more susceptible than Valencias, and besides, Valencia oranges are not mature during the periods of greatest rainfall in California. Thompson Navels are much more susceptible than the more coarse-skinned Washington Navels. As one procedes eastward from the Navel areas in western Los Angeles County, where the greatest losses are suffered from water spot, the fruit becomes increasingly more resistant. Untreated Washington Navel oranges from Covina, Pomona, and Riverside were placed in a "rain chamber" and kept wet for 6 days. The following percentages of water spot were found among the fruits from the 3 districts: Covina, 72.3%; Pomona, 41.6%; and Riverside, 8.5%. Untreated Navel oranges picked in Riverside were in a better condition after being in the rain chamber for 2 weeks than fruit picked in Covina was 1 week. (270)

Aside from areas adjacent to wounds, such as made by pricking and abrasion during wind storms, the areas surrounding the navel end of the Navel oranges are most apt to be affected by water spot. The stem end, or the "shoulder" portion of the orange, is also apt to suffer the same type of injury, and this is called "shoulder spot". The latter type of injury is greatly accentuated by the formation of ice crystals from drops of water collecting on the orange. It can be caused merely by the collection of dew on the stem end of the fruit, and consequently a certain amount of loss from this source is suffered every year.

Usually there is a much greater percentage of water spot on the north than on the south side of a tree and more on the upper than in the lower half. The first recorded case of water spot was reported to have followed a 10-day period of wet weather in February, 1927. Oil spray was suggested at the time as one of the accentuating factors (302), and has since been definitely established as the chief accentuating factor. A survey was made of 81 Navel orange orchards in the most severely affected areas in 1936. The orchards included in the survey had been sprayed with emulsions, emulsives or tank mixtures of the light medium grade applied in August, September and October, 1936. The date of spraying appeared to have no effect in the severity of the water spot. In the orchards sprayed with the regular concentration of oil an average of 25.1% of the fruit was found to be affected by water spot. It appeared that the concentration of oil was not an important factor, for an orchard sprayed with 1/2% oil (plus toxicant) suffered a 25.8% loss of fruit from water spot. Oil spray was found to accentuate "shoulder spot" to some extent, but it was not as great a contributory cause as in the case of water spot. Among the fumigated orchards examined the loss from water spot averaged 4.7% (270).

In the season of 1937-38 the Divisions of Entomology, Plant Pathology and Orchard Management of the University of California Citrus Experiment Station, in cooperation with the Farm Advisor's office, selected two mature orange orchards in the Claremont area in which a number of pest control and other supplementary treatments could be applied (271). Fifteen treated plots and one untreated check were arranged in each orchard according to a modified system of random distribution, using 4 sub-plots of 16 trees each for each treatment. The fumigation was done on August 31 and September 6, 1937 and the sprays were applied on September 20 and 24.

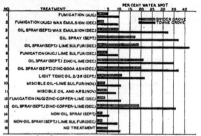

Fig. 112. Per cent of water spot in plots of Navel oranges with various pest control treatments, some containing oil spray and others containing no oil. After Ebeling et al (271).

The results of the experiments are shown in fig. 112. In one of the orchards (the Snyder Orchard) the fruit was picked on February 23 and in the other (the Towne Orchard) the fruit was picked on March 9 and 10. Between the two dates of picking 10.8 inches of rain had fallen in the district in which the orchards were located and, consequently, as shown in fig. 112, the incidence of water spot was much greater in the fruit picked at the later date. It can be seen from fig. 112 that in the Towne Orchard, which suffered the greatest loss from water spot, none of the plots sprayed with the regular dosage of oil had less than 32.1% loss from water spot. Where 0.66% light oil plus derris extractives was used, the loss was 21.8%. This shows that, within the limits of the regular spray oils, either reducing the heaviness of the oil or reducing the concentration of oil in the spray, even to an extreme degree, does not overcome the tendency of the oil to accentuate water spot. A low deposit of very light oil, as in the miscible oil-lime sulfur combination does not accentuate water spot. It has also been found that kerosene, even at a 10% concentration, does not accentuate water spot unless the rain should occur within a day or two after the kerosene is applied. Within a couple of days the kerosene which has penetrated the rind has apparently evaporated. This demonstrates an interesting possibility in connection with the influence of petroleum on the incidence of water spot. It indicates that the petroleum must be present in the rind to accentuate this malady. Apparently it is not physical or chemical changes caused by the oil which make the fruit more susceptible to water spot, but the presence of the oil per se. Scott and Baker (767) found no structural changes in the cuticle of Navel oranges attributable to oil spray.

In the untreated plot or the treated plots in which no oil was used, the per cent of water spot was never above 10.8%, and the greater part of the loss in these plots was due to wounds on the fruit caused by a windstorm prior to the rainy period. A wound--even the smallest rupture of the cuticle--was always surrounded by a water-soaked area. It will be noted that a spray of wax emulsion contributed very little in the way of protection to the fruit. Also other auxiliary materials which have at times been stated to have some preventative value, such as lime sulfur, zinc lime, zinc-soda ash, and zinc-copper-lime, proved to have no value.

It should be pointed out in connection with this discussion of water spot, that among late-picked Navel oranges the loss from dropped fruit in the fumigated orchards may in some cases be as great as the earlier loss from water spot in the oil-sprayed orchards. The loss from fruit drop is greatly accentuated if citrus red mite occurs in the orchard in appreciable numbers. Orchards which have not received an oil spray are almost certain to have enough red mite to weaken the fruit if no non-oil red mite treatment, such as DN-Dust, is used. Even if a non-oil mite treatment is applied, the control of mites is not always sufficient to prevent an accentuation of fruit drop.

Ebeling and Klotz (270), in connection with their survey of the water spot situation in 1936, made the following statement in connection with the loss from fruit drop in late-picked orchards:

"Although the water rot situation this year happened to be an unusually severe indictment against oil sprays, it is considered advisable to mention the late drop of fruit in many fumigated groves so that there will be no tendency to evaluate the relative merits of the two types of pest control treatment in the basis of the water rot situation alone. On March 21 and 23, many groves were visited in the Pomona, Claremont, La Verne, and San Dimas districts. Counts of the number of fallen fruits were made in the majority of these groves. The average loss from the late drop was possibly several times as great in fumigated and untreated groves as in sprayed groves and ranged as high as 30 percent of the total crop in some instances.

"In areas where citrus red mite is a problem, it is almost impossible to dissociate the excessive late drop of fruit from mite injury. In many cases where the grower has fumigated he has delayed treatment for citrus mite until a certain amount of injury has been done. In some cases when treatment was sufficiently early a reinfestation of mites has occurred. It is the belief of the writers that the differences in the amount of fallen fruit in sprayed and fumigated groves from about the time of the beginning of the blossoming to the last pickings is due largely to the presence of red mites in the fumigated groves. It appears that even a light infestation of mites is sufficient to weaken the fruit appreciably, especially in a year like this, when the fruit is over mature and weak at the time of blossoming."

No appreciable loss from water spot had occurred before the season of 1926-27, the first season of heavy rains after the general use of quick-breaking, white oil sprays for the treatment of citrus pests in California. The average loss of fruit from water spot that season was estimated by R. S. Woglum to exceed 40% of the crop in eastern Los Angeles County (961). Water spot was greatly accentuated by wounds in the fruit resulting from a severe windstorm late in the autumn. In non-oil-sprayed orchards, loss from "shoulder spot" and from the drop of fruit in late-picked orchards tended in many orchards to approach the loss of fruit from water spot in the oil-sprayed orchards.

In the 1928-29 season heavy losses from water spot again occurred in oil-sprayed orchards following a protracted wet period in February.

Particularly severe losses from water spot have been experienced in the years 1927, 1932, 1936, 1937, 1938, 1940, 1941, and 1944. During these years, losses as high as 60 to 80% of the crop were recorded in oil-sprayed orchards. Likewise losses were not entirely absent during some of the years not listed above, especially in closely-planted orchards of large Thompson Navel orange trees. Until 1940, the water spot was for the most part caused by rains occurring in February, but in 1940 the damage was due to a rainy period lasting from January 3 to 12, inclusive. In the 1940-41 season a small amount of cracking of the cuticle was observed after rains occurring from December 16 to 24, 1940, inclusive, but very few fruits were lost at that time. This was the earliest that the initial stages of water spot had ever been observed. Rainy periods again occurred January 22 to 26, 1941, causing a further extension of water spot, amounting in some cases to as much as 10% of the fruit in the north sides of the trees. The period from February 11 to March 4 was a period of many rains, with little opportunity of the fruit to become dry. The result was a serious loss from water spot--as great as the loss during the two worst previous years, 1927 and 1932. Also the malady extended in a severe form over a greater territory than ever before.

The last year of severe losses was 1944. Losses were not as great as in some of the previous years, but some orchards suffered great damage. Shoulder spot, apparently accentuated by ice formation on the stem end of the fruit, was common in 1944. In 1947-1948 it was again common despite a very dry winter. The winter was very cold, however, and ice formation on the fruit was noticed a number of times following moist conditions. The fruit was advanced in maturity and Navel oranges were especially tender and of a fine texture. Oil sprays accentuated the susceptibility of the fruit to shoulder spot following ice formation.

EFFECT OF OIL SPRAY ON YIELD AND SIZE OF FRUIT

The question as to whether oil spray applied to citrus trees will decrease production and reduce fruit sizes has been subject to much controversy, but to surprisingly little experimentation, considering the importance of oil sprays in the citrus pest control program. Some well qualified observers serving in various capacities in the citrus industry are convinced that the continued use of oil sprays over extended periods has the inevitable result of decreasing production and may be one of the factors involved in the serious reduction in fruit sizes of Valencia oranges in southern coastal districts of California. Other equally well qualified observers do not share this opinion. Analyses have been made of yield records of large numbers of oil-sprayed as compared to fumigated orchards, and these have generally shown a higher production for the latter (914, 676). These analyses could be variously interpreted. They show that among the orchards surveyed, those sprayed with oil for 1 or more years previous to the years of the harvest records had, on the average, less fruit than those not receiving oil spray (usually fumigated). This need not necessarily be interpreted as indicating that the difference in production was caused by the adverse effect of the oil sprays. The writer is inclined to deduce from the data presented in the analyses merely that orchards likely to be fumigated are on the average more productive than those which are likely to be sprayed. This is to be expected, for growers who have the best orchards have earned greater profits and are probably the growers who, on the average, would be most apt to invest in fumigation. This treatment, although more expensive than oil spray, and requiring one or more extra treatments for citrus red mite, is known to generally result in fruit of better appearance and quality. During periods of low prices, the tendency to use the less expensive pest control measures on orchards of lower yield is especially marked. Also growers who are keeping orchards on a minimum subsistence basis, with the idea of selling or subdividing at the earliest opportunity, are apt to use oil spray as being the most economical treatment.

In the writer's opinion, the most reliable indication of the relative effect of different pest control treatments is obtained if the treatments occur in randomized plots within a single orchard or cultivated unit, so that it may be assumed that the average potential performance of the trees receiving the different treatments is about equal. Pre-treatment records of the yields of the various plots further increase the significance of the data and the conclusions which may be drawn from the experiment. The writer, from 1935 to 1940, conducted a series of 6 such experiments[1] in Valencia and Navel orange orchards in Orange, Los Angeles, and San Bernardino Counties, the experiments extending from 2 to 4 years in duration.

No differences in production between plots sprayed with 1.66% or 1.75% light medium oil and the fumigated plots were found except in one experiment in which 2 of the oil sprays, applied on August 18, 1936, caused excessive defoliation and a certain amount of fruit drop, apparently because of the high temperature and low humidity at the time they were applied. An unexpected development occurred in an experiment made at East Highlands in an orchard apparently free of pests except for a very light infestation of citrus thrips which had never resulted in appreciable damage. Here for a 4-year period in a series of plots of 2 rows each replicated 4 times for each treatment, both the sprayed and fumigated trees yielded significantly more fruit, in relation to the yield of the base year of 1936, than the untreated trees. There was no significant difference in production between the plots sprayed with the oil at 1-2/3% concentration and the fumigated plots. The greatest yield of all was from the plots sprayed with a reduced dosage of light medium oil--1% plus derris powder. No attempt is made to explain these interesting data. It is believed that the experiment is worthy of replication elsewhere in some pest-free citrus area.

In recent years there has been a marked tendency toward abnormally small size of oranges in California, particularly in the case of the Valencia variety and in the southern coastal districts. Hodgson (431) has discussed the climatic, cultural and other factors which may have contributed to this undesirable condition. In all the above experiments, whenever records of fruit size were obtained, it was found that oil-sprayed fruit was larger than fumigated fruit. The apparent explanation is that oil spray reduced the number of blossoms and even the number of fruit set, but the fruit became somewhat larger than the fumigated fruit in adjoining blocks, resulting in approximately equal tonnage, on the average, from sprayed and fumigated plots.

Newcomb (643) made orchard experiments on oranges and lemons for 5 years in which he also found no difference in yield between sprayed and fumigated plots when spraying was avoided on days of high temperature. A grower in the Azusa district reached a similar conclusion in experiments in a Navel orange orchard extending over a period of 12 years.

In the interpretation of the above experiments it should be noted that the results are at variance with the conclusions which have been reached on this problem by many growers, packinghouse managers and agricultural officials, and in view of this fact, the writer believes that the experimental evidence available is far too limited to be conclusive. In addition, it should be borne in mind that in nearly all the experiments referred to above, the citrus red mite was a factor, and although attempts were made to control the mites with existing measures in the unsprayed plots, there were still more of them in these plots than in the oil-sprayed plots. Even the relatively low numbers of mites remaining in the unsprayed (fumigated) plots may have had an adverse effect on production which might not occur if a more effective mite control program were carried out. It is probably safe to conclude, however, that when judiciously used, under favorable weather conditions, oil sprays need not necessarily adversely affect either yield or size of citrus fruit.

Precautions In the Use of Oil Spray on Citrus:- Assuming the grade of oil, the oil-depositing properties of the emulsion, the percentage of oil used, and the method of application to be satisfactory, there still remain a number of precautions to be heeded when using oil spray on citrus. Spraying is usually practiced as soon as possible after irrigation in order that the foliage may be as turgid as possible. Weather conditions should be carefully watched and spraying should be

---

[1]Ebeling, W. 1948. Relation of pest control to yield and size of Valencia and Navel oranges. Unpublished manuscript on file at the Division of Entomology, University of California, Los Angeles.

discontinued as soon as it is evident that temperatures will rise above 80-95° F, depending on the locality. In southern California it has been the practice to discontinue spraying in the coastal area when the temperature reaches 80° F, in the intermediate areas at 86° F, and in the interior area at 95° F. In southern California during the autumn spraying season, high temperatures are often accompanied by low humidity and winds from the desert known as "Santa Anas". Oil-sprayed trees are most apt to be adversely affected if the spraying is followed by hot "Santa Ana" winds, and spraying should be discontinued when it is known that such weather is approaching. In no case should oil spray be applied when the relative humidity is 35% or lower in coastal areas, 30% or lower in intermediate areas, or 20% or lower in interior areas.

Oil spraying during cold weather is disadvantageous mainly from the standpoint of the effect on the fruit. In the case of lemons there occurs an intensified coloration (bronzing) of the fruit, retarded growth of fruit, and sometimes a certain amount of fruit drop. Generally, oil spraying should be completed by December, although close to the coast, as at Chula Vista and Carpinteria, California, where lemons are immature during the month of December, oil spray affects the fruit less adversely than it does farther inland. The adverse effect of oil on the coloring of oranges also will be increased by cold weather. In addition it has been noted that sub-freezing temperatures following oil spray may cause severe defoliation. (972)

With dormant oils applied to deciduous trees, low temperatures have no adverse effect. Beran (83) has recently shown that the use of dormant oils on apple, pear, quince, and cherry during sub-zero weather is as safe as at higher temperatures and, in fact, is to be recommended, because the heavy oil deposit at such low temperatures makes possible a 50% reduction in the amount of oil in the spray.

THE TECHNIQUE OF SPRAYING CITRUS TREES

The scale insects constitute the major citrus pest problem in the majority of the citrus-growing areas of the world. Since the scales are sessile insects, a spray, to be effective, must reach each individual insect. The situation is not the same as when the insects move about and eventually come in contact with the insecticide and succumb either by contact or by feeding on the sprayed fruit or foliage. In addition to this difficulty it must be remembered that citrus trees support a dense foliage which is difficult to properly penetrate and thoroughly wet with the spray. It is especially difficult to wet the lower sides of the leaves and the back sides of the fruit. Yet by using a sound technique of application, and with some experience, a remarkably thorough spray coverage can be attained. It has been the experience of commercial spray operators that a man without experience, but willing to learn a systematic and logical spraying technique, will soon be a better sprayman than one who has had years of experience with haphazard methods.

A good job of spraying cannot be done without good spraying equipment, a subject to be discussed in a later chapter. Presuming the possession of adequate equipment, there still remains the necessity for proper technique in the application of the spray.

Tower Spraying:- In California, nearly all spray rigs used for spraying citrus trees are equipped with a tower (fig. 3), making it possible to strike the tree with a direct stream of spray from above. Except for small trees, it is not possible to adequately wet the upper half of the citrus tree, especially the rather flat top above the "shoulder" of the tree, without the aid of a tower. One or two "tower men" are used for each rig. The spraying of the tops of the trees is relatively simple compared to the task remaining to the "ground crew." Consequently, the "tower men" are usually the least experienced men on the spray crew, or men who are not physically able to effectively continue spraying from the ground throughout the long working day of the average spray crew. Only a limited degree of manipulation is possible from the stationary position of the man spraying from the tower. It is nevertheless essential that the "tower man" be alert and that he apply the spray systematically and efficiently.

Position of the Hoses:- The spray rig is driven between two rows of trees and these are completely sprayed, so it is only necessary to drive down every second "middle." The ground men begin their spraying of a tree in such a position that when they are through with a tree, the spray hose will be unwound from the trunk of the tree and only a few more steps will suffice to bring them into position for beginning to spray the next tree (fig. 113). The tower man has finished spraying the top of the tree (fig. 114, A) before the ground man commences.

Inside Spraying:- The citrus tree tends to produce a dense foliage at its periphery. Behind this wall of dense foliage is found a network of supporting branches, but very little foliage. Certain scale insects, including the California red scale and the purple scale, may be found anywhere on the branches and even on the trunk of a tree as well as on the fruit, foliage and twigs.

By finding places on opposite sides of a tree where the spray gun can be thrust through the foliage, the inside branches and the inside of the wall of foliage can be rather easily hit by the spray (fig. 114, B). The sprayman should either find a place where he can see inside the tree, or part the foliage with one hand while he is spraying, so that he can observe the application of the spray and see that the inside spraying is systematically and thoroughly done. Special attention should be given to the inside of the top of the tree.

Outside Spraying:- The proper spraying of the outside of the tree from the ground is the most difficult of all the spraying procedures. The spray gun is usually held in such a position that the handle will rest on the arm in the manner shown in fig. 115. The other hand is left free to aid in pulling the hose around the tree. This has been found to be the best way of holding the gun with maximum comfort consistent with effective use of the gun.

A high pressure is necessary to penetrate the dense mat of twigs and foliage and turn the leaves, so that the spray can hit both leaf surfaces. Pressures of from 500 to 600 pounds per square inch at the pump are ordinarily used in citrus spraying in California. It should be borne in mind, however, that there is a loss of about one pound of pressure per foot of hose. Some spray crews utilize a pressure of 600 pounds at the gun, but they use special guns that give only a broad spray stream. A narrow stream with such pressure would knock off some fruit and

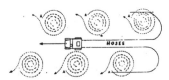

Fig. 113. Positions of the hoses in the spraying operation. Adapted from Camacho (139).

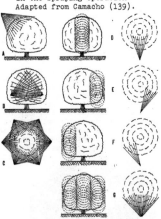

Fig. 114. Spray patterns for citrus trees. A, spray stream directed from tower; B, interior of tree sprayed from opposite directions; C, complete pattern of outside spraying from six points. In outside spraying, the spray stream, for any particular "panel" of tree surface, is directed straight ahead (D), to the right (E), and to the left (F), resulting in the completed pattern for the panel (G).

Fig. 115. Position of spray gun in the arm. After Camacho (139).

213

foliage. The spray stream must be kept narrow enough to insure sufficient velocity for the penetration of the outer wall of foliage, but, of course, the widest stream consistent with proper penetration of the tree will result in the most efficient application of the spray.

The opening in the disk in the spray nozzle is usually 8/64 or 9/64 of an inch in diameter (No. 8 or No. 9 disk), and it requires experience and alert spraying to utilize this size of disk opening without waste of material. A beginner could use a 7/64 inch disk opening more efficiently.

Several methods of citrus spray application have been advocated by experienced spraymen. Some prefer a revolving motion of the spray stream (fig. 114, D-G) some prefer the "wig-wag" system of spraying horizontally from side to side, and some prefer the vertical motion. Some sprayers will use a combination of two or more systems or change their system on trees of different shapes, sizes, or density of foliage.

An essential factor in spraying the outside of the tree from the ground is to spray every portion of the tree from as many angles as possible. The following technique is more or less in accord with spraying practice in California as it has evolved through decades of practical citrus orchard spraying experience. A man should station himself at 5 or 6 points about the tree (fig. 114, C) and from each of these points he should spray all portions of the tree that are visible. It is better to spray a wide swath of foliage from one position than to continuously spray directly ahead while slowly moving around the tree. From one point a man should spray first a "panel" on one side of the section of the tree seen from his point of vantage, let us say the middle panel (fig. 114, D) until it is system-

atically and completely sprayed from top to bottom. Then he might spray the right panel (fig. 114, E) and finally the remaining panel at his left (fig. 114, F). Not until this entire section of the tree is completely sprayed (fig. 114, G) should he move on to the next position. This same sequence is shown photographically in (fig. 116).

When following the above method of spraying, every portion of the tree is sprayed first from the inside toward the outside, then from 3 directions from the outside. In addition, the top of the tree is sprayed from a tower. The effectiveness of such a system is most apparent when it is taught to spraymen working in regions where a systematic spray technique has never been developed. The same spraymen who previously had left large sections of a tree unsprayed can, with the same number of gallons per tree, do such an effective job of spraying that it is difficult to find a dry spot on the tree even amongst the most dense mats of foliage. Recognition of the benefits of systematic spraying in improving the results of oil spray against scale insects attacking citrus trees in Mexico led to the description and illustration of the above method of spraying in the Mexican agricultural journal "Fitófilo" (139, 761a).

COAL TAR DERIVATIVES

Fig. 116. The 3 directions of the spray stream from each standing position of the sprayman.

Coal tars, also known as carbolineums, tar creosotes, tar distillates, and tar oils, were first used as insecticides in Germany or Holland about 30 years ago. They were then introduced

Coal tar and its derivatives are used in dormant spraying of some pests of deciduous trees, but have not been used in the spraying of any of the subtropical fruit trees in this country.

CHAPTER XIII
FUMIGANTS

INTRODUCTION

Fumigants may be applied in enclosed spaces such as tight fumigation chambers, with or without vacuum; in buildings and greenhouses; under tents covering trees or other plants; in burrows of animal pests; or for soil fumigation, especially with soil injection equipment.

Fumigants were among the first insecticides used by man, the fumes of burning sulfur, for example, having been used since the period of earliest historical record. Many substances may be used as fumigants, and they may be in the form of a solid, liquid or gas, but regardless of their physical state when applied, they must have the ability to volatilize readily and to form a toxic concentration of vapor in an enclosed space, or, if applied out-of-doors, within a short distance of the point of application. The less the toxicity of a compound, the more volatile it must be to have practical value as a fumigant. Among organic compounds only those of simple structure are sufficiently volatile to be fumigants. In a homologous series of organic compounds, toxicity usually increases as the series is ascended, but volatility soon becomes so low as to render the compound ineffective as a fumigant.

PRECAUTIONS

All fumigants that have been used successfully against insects have been found to be also poisonous to warm blooded animals, including man. A concentration of 0.03 vol./% (300 parts per million) of hydrocyanic acid gas is said to be sufficient to quickly kill human beings. In the fumigation of buildings, gas masks, correctly charged for the particular gas being used, are often employed in applying the fumigant and when entering the building afterward to open doors and windows, if these cannot be opened from the outside. In outdoor fumigation, the work is arranged so that the drift of air is away from those applying the fumigant. Precautions must be taken to avoid damage to persons or animals apt to be in or near the area to be fumigated. In the fumigation of citrus trees with HCN, for example, it has been found that fowl of all kinds are especially sensitive to the gas, and preparation for the fumigation of an orchard sometimes involves the laborious and time-consuming process of removing domestic fowl until the fumigation is completed.

Compounds which are gaseous at or immediately above ordinary fumigation temperatures, such as hydrocyanic acid, methyl bromide and ethylene oxide, must be confined in strong metal cylinders. The dangers involved in storing, transporting, and releasing such fumigants make fumigation operations involving their use especially hazardous to all but specially trained and experienced persons.

In working with fumigants one should bear in mind that gases may be absorbed through the human skin. For example, it is not safe to work in an atmosphere containing 2 vol./% hydrocyanic acid gas even though an effective gas mask is used, for sufficient gas can be absorbed through the skin at this concentration to cause poisoning. Carbon disulfide, hydrogen sulfide, carbon tetrachloride, ethylene dichloride, formaldehyde, and nicotine vapor may also be absorbed through the skin.

The fire hazard is another factor making necessary the utmost care in the use of fumigants. Hydrocyanic acid gas, carbon disulfide and ethylene oxide are especially hazardous from the standpoint of their flammability. Carbon dioxide may be mixed with fumigants to eliminate the danger of explosion. Different ratios by volume of carbon dioxide are required for the various fumigants. (see 469)

Fumigants are also injurious to plants, and great care must be exercised in order to apply a fumigant so that it might be used in concentrations and under

216

FUMIGANTS

conditions of temperature, humidity and period of exposure so as to be effective against insects without at the same time being injurious to plants. As in the case of some of the sprays and dusts, the "margin of safety" between insecticidal and phytocidal effect may be rather small, and requires. on the part of the fumigator a thorough knowledge of his pest control operation and of the risks and the responsibilities that are envolved. Fortunately, dry food substances are not damaged by the common fumigants, but there is danger of poisoning foods that contain considerable quantities of water. Likewise in the fumigation of living plants with fumigants which dissolve in water, like HCN, the danger of injury is increased by the accumulation of water on the plants. For this reason it is important to avoid a falling temperature during the exposure, for moisture may condense on the foliage, absorb some of the gas, and result in injury to the plant.

TOXICOLOGY

Fumigants, whether or not they are respiratory poisons, gain entrance mainly through the respiratory system. As might be expected, factors affecting the opening or closing of the spiracles or the rate of respiration generally influence toxicity, although there are important exceptions to this general rule. Carbon dioxide may increase the rate of respiration and has been added to certain fumigants to increase their effectiveness. Increase in temperature increases the rate of respiration and generally increases the effectiveness of fumigants. In the fumigation of stored products, the effectiveness of the fumigation may be decreased by sorption of the gas by the fumigated materials, and sorption decreases with rise in temperature. Fumigation in enclosures is usually done at temperatures ranging from 70° to 90° F. Within this range of temperatures the concentration of the toxicant usually can be reduced to a minimum. If temperatures are lower, artificial heat might be economically desirable because of the reduced concentrations of gas required.

Methyl acetate was added to liquid hydrocyanic acid by Brinley and Baker (126) to increase the toxicity of the gas. They believed the added toxicity to be due to the fact that a small amount of methyl acetate kept the spiracles open, while in pure HCN the spiracles were quickly closed. Hardman and Craig (408) discovered that the California red scale closes its spiracles within 3 to 5 minutes after HCN gas reaches them. In the race that is resistant to HCN the spiracles remain closed for at least 30 minutes, providing the gas is present, but in the non-resistant race the spiracles remain closed for only about 1 minute. This observation indicated that a difference in spiracular activity under the influence of HCN gas might account for the difference in the tolerance for this gas of the resistant and non-resistant races of the California red scale. However, Quayle (729) pointed out that a difference in tolerance of the two races is noted even if they have been exposed to the gas for only 2.5 minutes, before the spiracles are closed in either race, and that also for a number of other reasons it is not likely that the difference in spiracular activity noted by Hardman and Craig is entirely responsible for difference in resistance.

The amount of gas adsorbed or absorbed by the fumigated material affects the concentration remaining to act upon the insects, and this fact, of course, has long been recognized by those familiar with fumigation practice. Lindgren and Sinclair (535) have shown that the sorption of HCN by house fly pupae decreased as the age of the pupae increased from 1 to 3 days, then increased again, becoming progressively greater until the pupae were about 8 days old and the adults were about ready to emerge. These investigators were impressed by the similarity of the HCN sorption curve for house fly pupae and the mortality curve of pupae of Tribolium confusum worked out by Lindgren (524).

Since a method for accurately recovering small amounts of HCN from fumigated insects had been developed (779) an experiment was made to determine how much cyanide could be recovered from fumigated resistant and non-resistant red scale (534). It was found that more HCN is recovered from fumigated non-resistant than from the fumigated resistant red scale, and the same relation was found to be true with respect to non-resistant and resistant black scale. Likewise, more HCN was recovered from red scale than from the walnut husk fly pupae, on a weight basis, and the pupae of the walnut husk fly are also more resistant to HCN gas.

PENETRATION - VACUUM FUMIGATION

Often the rate of penetration of the gas into the material to be fumigated is an important factor in fumigation. This is of course related to the adsorptive capacity of the fumigated material, which varies with different substances. The rate of penetration of the fumigant is greatly increased if fumigation is done in an air-tight chamber from which the air has been previously exhausted. This type of treatment is called vacuum fumigation, and appears to have been first used by D. B. Mackie in 1914, using CS_2 as the fumigant in the treatment of cigars for cigarette beetles, Lasioderma serricorne (F.), and by E. R. Sasscer and L. A. Hawkins, who used vacuum fumigation independently of Mackie and at about the same time. They used hydrocyanic acid gas as the fumigant (285, pp. 495-97). The effect of vacuum is not only to increase the rate of penetration of the fumigant, for the toxicity of the fumigant may also be increased under vacuum, as was demonstrated with reference to methyl bromide used against the confused flour beetle, Tribolium confusum (315).

The fumigant may be removed quite completely from the fumigated material by drawing a second vacuum after the proper period of exposure of the gas and then breaking this vacuum with air. This process is known as "air washing."

Another advantage of vacuum fumigation is that the workmen are not exposed to the gas.

Vacuum fumigation can be used to advantage when speed is an important factor, for usually the period required for effective kill of insects protected by plant materials, as for example in baled cotton, is reduced to a small fraction of the period required under atmospheric conditions. The amount of fumigant required is greatly increased, however, when much plant material is present, because of greater adsorption of the gas by the fumigated materials. The larger quantity of gas required, as well as the original cost and upkeep of the equipment, constitute the disadvantages of vacuum fumigation.

It appeared at one time that vacuum fumigation would be far more important than we find it to be today, and the de-emphasis in vacuum fumigation has been brought about largely by methyl bromide. The great penetrating ability of this fumigant has in the majority of cases obviated the necessity for the use of vacuum. Vacuum fumigation with HCN is sometimes employed, however, for the fumigation of nursery stock in the observance of quarantine regulations.

HCN FUMIGATION

HCN is generally the most toxic to insects of all fumigants and is considered to rank with methyl bromide as the most important commercially of all fumigants. It is especially valuable for the fumigation of buildings that are not particularly tight and in which it is impossible to maintain a high concentration of gas for a sufficient period. Under such conditions it is necessary to have a quick-acting gas, and HCN is the only gas available that will kill insects quickly. The same thing pertains to fumigation under tents, which are not gas-tight.

Characteristics:- HCN is a colorless liquid boiling at 80° F. Its liquid density is 0.682 at 77° F. At 70° F, 650 ml of HCN gas weighs 1 pound. Its vapor density is 0.9348 at ordinary room temperatures, being nearly that of air. The gas is inflammable. It has a characteristic pungent odor, said to be similar to that of bitter almonds. The odor may serve as a warning of the presence of the gas, but some people cannot detect it. The commercial product is 96.5% to 98% HCN, the remainder being water. Decomposition and polymerization of HCN are accentuated by alkaline substances and retarded by acids. Yellow discoloration of the liquid is evidence of decomposition. The color changes to brown and black with increasing decomposition, the black substance finally precipitating. Ammonia, CO_2 and heat are generated, pressure in a container may rise to 1000 pounds per square inch, and an explosion can occur under these conditions. Great caution is used to prevent contamination by alkalis. Sulfuric acid at 0.005% concentration acts as a stabilizer.

HCN is extremely toxic to warm blooded animals. Fifty milligrams is considered to be a lethal dose for human beings. No more than 50 to 60 p.p.m. can be inhaled for one hour without serious consequences. A concentration of 3000 p.p.m. rapidly results in death.

With respect to plants, Moore and Willaman (613) found that the reduction in the activity of the oxidases and catalases, and therefore in respiratory activity, which was caused by absorbed HCN, resulted in an inhibition of photosynthesis and translocation of carbohydrate and in the closing of stomata. Another result of HCN observed by these investigators was "an increase in the permeability of the leaf septa, which causes less rapid intake of water from the stems and more rapid cuticular transpiration." The physiological conditions under which the leaves and fruits of Valencia oranges absorb different amounts of HCN were investigated by Bartholomew et al (67).

According to present day knowledge, the cyanogen compounds combine with the catalysts in the living cells of the body tissues containing iron and sulfur, thereby exerting an inhibiting action on tissue oxidation. After cyanide poisoning, the venous blood is red, for the oxygen is not taken from the blood in the tissues and it must, therefore, return in the veins in the same condition as in the arteries (424).

Sources:- Although potassium cyanide was first used in the manufacture of HCN, the sodium salt is at present more widely used because it is less expensive and has a greater proportion of CN. The action of sulfuric acid on sodium cyanide results in the following reaction: -

$$2\ NaCN + H_2SO_4 \longrightarrow Na_2SO_4 + 2\ HCN$$

One ounce of NaCN yields about 20 ml of HCN at 60° F.

Manufacture:- Liquid HCN may be manufactured in the following manner: A brine tank is charged with 1250 gallons of water and 5000 pounds of NaCN. The resulting solution is allowed to flow through a meter to a generator, where it combines with the proper quantity of sulfuric acid, also flowing through a meter. The mixture is agitated to prevent local heating, and steam is introduced to accelerate the driving off of the HCN. The resulting vapors are 40% HCN and 60% steam. These vapors are led into a cooler and later the HCN vapors are led to a condenser held below freezing temperature. About 650 pounds of HCN are generated and condensed per hour (146). Several new methods of manufacture have been developed, however, and may be in use at this time.

The HCN is stored in 18-gauge steel, heavily tin-lined drums of 80 pounds capacity (fig. 117). When transported, the drums are packed in ice. HCN may also be absorbed on discs of porous cardboard or on other materials, such as diatomaceous earth.

Fig. 117. Drum in which liquid HCN is transported to orchard. Also shown is vaporizer and portion of a mechanical tent puller. Courtesy American Cyanamid Co.

219

Uses:- The principal use of liquid HCN, in regard to subtropical fruit pests, is in the fumigation of citrus insects. The trees are covered with tents and the work is done at night. Details of the practice of orchard fumigation with HCN are given in the following chapter.

Hydrocyanic acid gas may be generated "on the job" in commercial fumigation of dwellings, and greenhouses, using the "pot method". Dilute acid is used in order to generate the maximum amount of HCN, for with a concentrated acid, much carbon monoxide would be formed. The following proportions of ingredients are used.

Commercial (60° Bé) H_2SO_4............... 1.5 pints
Water .... 3 pints
Sodium cyanide .... 1 pound

The required amount of water is poured into an earthenware crock and the acid is carefully added. When the crock has been placed in the enclosure to be fumigated, the proper number of sodium cyanide "eggs", each egg weighing 1 ounce, are dropped into the acid. The operator should be provided with a gas mask. For efficient fumigation all possible effort should be made to prevent leaks from the fumigated room or enclosure. Ordinarily 1 pound of NaCN (16 one-ounce "eggs") is sufficient to fumigate 1000 cubic feet of space, and exposures are usually from 8 to 12 hours. Such a concentration will successfully control cockroaches, bedbugs, clothes moths, carpet beetles, and the various stored products pests (Indian meal moth, Mediterranean flour moth, confused flour beetles, saw-toothed grain beetles, etc.). Fleas, body lice, and silverfish can be controlled with 8 ounces of sodium cyanide to 1000 cu. ft. and rats and mice are killed by a dosage of 4 ounces of sodium cyanide to 1000 cu. ft. for an exposure of 4 hours. In the early days citrus trees also were fumigated, under tents, with the pot method (see p. 232).

For large spaces, such as mills and warehouses, liquid HCN under air pressure is often used. This may be piped to various parts of the building.

Portable cyanide generators which one man can carry are available in which the HCN gas can be generated as in the pot method (fig. 118). The hot gas can

then be pumped into a building or enclosure and the operator does not have to enter the building until ready to ventilate. Three generators are made in sizes making possible the fumigation of 10,000 cu. ft. or 25,000 cu. ft. of space with one charge of the generator. Such equipment can also be conveniently used in the fumigation of rodents in their burrows.

Fig. 118. Placing acid and calcium cyanide eggs in a machine used for building or warehouse fumigation. Courtesy du Pont de Nemours & Co.

When calcium cyanide dust is used, as in greenhouse fumigation, the HCN is generated by the action of atmospheric moisture on the calcium cyanide according to the following reaction:

$$Ca(CN)_2 + 2H_2O \longrightarrow 2\ HCN + Ca(OH)_2$$

When the dust is spread in thin layers the gas is evolved even if the relative humidity is 25% or less, and the rate of evolution increases with relative humidity. Commercial preparations of calcium cyanide yield about 55 wt./% of HCN.

A convenient method of fumigating is with the wafer-like discoids that are now available which consist of liquid HCN absorbed in some inert material. These discoids are sealed in cans. Each discoid contains about 0.5 oz. liquid HCN, which evaporates on exposure to air. The cans may be precooled with dry ice before opening to retard the generation of gas, thus providing greater safety in their use. The discoids should be placed on several thicknesses of paper to avoid injury to the floor.

FUMIGANTS

Precautions:- HCN is highly toxic to man and warm-blooded animals as well as insects, and this is one of its drawbacks. It should be used on a large scale only by experienced persons. Some cities have ordinances governing its use, and in a few cities its use is prohibited. Precautions to be used in orchard fumigation with HCN are given on page 249.

Table 8

A Comparative Rating of Hydrocyanic Acid and Methyl Bromide

		Rating*	
Desirable Characteristics		HCN	Methyl Bromide
1.	Violently toxic to insects	E	E
2.	Nontoxic to plants	P-F	E
3.	Nontoxic to vertebrates	P	F
4.	Easily and cheaply generated	E	G
5.	Not readily condensed to a liquid	E	E
6.	Not soluble in water	P	E
7.	Great power of diffusion	E	E
8.	Easily detected by senses	G	P
9.	Harmless to foods	P-F	P-F
10.	Efficient penetrating qualities	F	E
11.	Non-persistent	E	E
12.	No fire or explosive hazard	F	E
13.	Non-corrosive to metal and harmless to fabric	E	P-F

*E - Excellent; G - Good; F - Fair; P - Poor.

Per cent of approximation of certain fumigants to the hypothetical ideal fumigant according to the system of rating shown above:

Methyl bromide 60-75
HCN 60
Chloropicrin 50
Ethylene oxide 30
Carbon disulfide 30
Sulfur dioxide 20

METHYL BROMIDE

Methyl bromide (bromomethane, CH_3Br) is one of the most recent of fumigants, yet it ranks with the most effective and useful, such as hydrocyanic acid, chloropicrin and ethylene oxide (table 8). It is considered by some to be the most universal fumigant yet developed. Apparently the first use of methyl bromide as a fumigant was in France, in 1932, when it was used to eliminate the fire hazard of ethylene oxide, but was found to be even more effective than the latter. It was later found to be more toxic to the granary weevil than HCN. Where it is found to be sufficiently toxic to the insects, it has certain advantages over HCN, such as having greater penetrating ability, being generally less harmful to plants and noninflammable. The rapid acceptance of methyl bromide as a fumigant is due largely to the energy and enthusiasm of D.B. Mackie and his associates in the California State Department of Agriculture, who did much early experimental work with this material.

David B. Mackie (1882-1944)
Formerly Chief of the California State Bureau of Entomology and Plant Quarantine. Well known for his achievements in vacuum fumigation and the development of methyl bromide as a fumigant.

Characteristics:- Methyl bromide is a gas at ordinary temperatures, the boiling point of the liquid being 4.6°C, lower than that of any other important fumigant with the exception of sulfur dioxide, and this makes the gas particularly useful for low temperature fumigation. The molecular weight of methyl bromide is 94.94, the specific gravity of the liquid is 1.732 at 0°.C (32° F) and the specific gravity of the gas (air = 1.00) is 3.27 at 0° C and 760 mm pressure. The weight per gallon is 14.4 lbs. at 0° C. At ordinary temperatures methyl bromide is a colorless gas. It is only slightly soluble in water, but highly soluble in oils. It is noninflammable at concentrations ordinarily used, except that ignition can be produced with a very intense electric spark in the restricted range of concentration of 13.5 to 14.5%. Methyl bromide, being not very active chemically, is less absorbed and is more penetrating than any other known fumigant.

The above characteristics are very desirable, although it should be remembered that the great penetrativity of methyl bromide results in its escaping readily from imperfect enclosures. This is a disadvantage, from the safety standpoint, in unlike HCN and some other gases, methyl bromide is not readily detected by its odor.

Methyl bromide is the least injurious to plant tissues of all the important fumigants and does not affect the germination of seeds. Green vegetables and other foliaged plants are regularly being fumigated with methyl bromide as a part of standardized quarantine practice.

One of the characteristics of methyl bromide is that it often kills very slowly. In one recorded instance, potato tuber moth larvae entered a fatal coma 24 hours after fumigation, but did not die in some cases until 5 or 6 days had elapsed (574). Methyl bromide is toxic to insects in all stages of development, and in this respect it has an obvious advantage over such fumigants as are characteristically ineffective against the eggs and pupae.

Manufacture:- While the exact process of manufacture of methyl bromide is a trade secret, in general it consists of reacting bromide with methyl alcohol in the presence of sulfuric acid, then refining.

Uses:- The manifold uses of this most versatile fumigant are discussed by Steinweden (844). Methyl bromide is used for disinfestation of propagative plant material imported through maritime ports. Thousands of carloads of fresh fruits and vegetables, nursery stock and green house plants as well as fruits and field boxes, shipped out of quarantined areas are fumigated with this gas. Also in the enforcement of quarantines, Irish potatoes are fumigated for tuber moth; sweet potato plants and cuttings for sweet potato weevils; strawberry plants for crown borer; tarsonemus mite, and Paria canella; various host plants for European corn borer; grapevines for grape phylloxera; tomatoes for pin worm; deciduous and ornamental nursery stock for olive scale (Parlatoria oleae); and camellia plants for several scale insects. In commercial insect control, pears in refrigerator cars and at canneries are fumigated for so-called invisible worms. Grains, milled flour and feeds, dried fruits, nut meats, dried beans, peas and seeds are fumigated for storage insects, and many other food products such as dehydrated vegetables, chocolate, dairy products, dried milk, cheese, butter, and margarine are likewise fumigated with methyl bromide. Nursery stock, greenhouse plants, and bulbs are fumigated for a wide variety of pests, some of them resistant to other gases.

In the fumigation of buildings or chambers it is important to make them as nearly as possible air tight, for as stated before, methyl bromide escapes more readily than other gases. It is practically odorless at ordinary fumigation concentrations and for this reason leaks are hazardous to operators, who may be unaware of the presence of the gas. A concentration of thirty parts per million is considered harmful to man. Likewise, the escaping of gas may reduce the insecticidal effectiveness of the treatment, for dosages are calculated on a time x concentration basis and no allowances are made in these calculations for leakage. Leaks in a fumigating chamber may be detected by a halide leak detector, an adaptation of the Beilstein test for organic halide vapors. Satisfactory prefabricated

FUMIGANTS

plywood fumigation chambers are being made which are proving to be suitable for nursery and warehouse use, for pest control operators, and for the county agricultural commissioner's office. These may be mounted on trailers, thus increasing their usefulness (fig. 54, D).

Methyl bromide may be used in vacuum fumigation, but the great penetrativity of the gas in most cases obviates the necessity for this type of fumigation, the chief virtue of which is to insure better penetration. Vacuum fumigation may be used to advantage, however, when the material to be fumigated must be handled quickly and not kept tied up for extended periods. Ordinarily 1.5 to 2 hours exposure is sufficient.

Gas-tight tarpaulins are sometimes placed over piles of grain, boxes, fruit, or nursery stock, and the gas is released under the tarpaulins. Much greater care must be exercised to prevent leakage from under the tarpaulin than in the case of fumigation with HCN gas, which is not only much lighter, but less penetrating than methyl bromide. "Sand snakes", made of canvas tubing 5 or 6 feet long and 3 to 4 inches in diameter, are filled with sand and are carefully arranged to keep the canvas tightly pressed to the floor around the pile of fumigated material. Since there is no way of controlling temperature, fumigation under a tarpaulin should not be done in direct sunlight, for the temperature will rise too high for safety in the fumigation of fruit and plants.

Steinweden (844) recommends temperatures of from 80° to 85° F for fumigation with methyl bromide, although by increasing the dosage, fumigation may often be successfully done at lower temperatures (844). Humidity has little effect on insecticidal effectiveness, but sufficient humidity should be maintained to prevent dessication of plants. Exposures are seldom for less than 90 minutes, for methyl bromide acts slowly and its toxicity is increased with increase of exposure period. The usual period for the fumigation of foliaged nursery stock, greenhouse plants, fruits and vegetables is 2 hours, while dormant deciduous nursery stock and dormant bulbs are treated for 3 or 4 hours. For grains, beans, peas, cereals, raisins, and dried fruits, low dosages are generally used, usually 1 lb. per 1000 cu. ft., for 24 hours. Air circulation is important to prevent stratification of the gas. An initial period of 10 to 15 minutes of circulation in a chamber is generally sufficient. For exposures of 24 hours or longer, as in warehouses, fans are not generally used, for the long period of exposure offsets the lack of circulation.

The State Department of Agriculture has prepared mimeographed booklets containing useful practical information on the treatment of agricultural commodities, mostly with regard to methyl bromide. These treatments have been checked by the Department or are based on the published work of recognized authorities. The first booklet (575) deals with nursery stock, the second (576) deals with stored products such as grains, seeds, dried fruit, nuts, cereal products in package and bulk, etc., and the third (577) deals with fruit and vegetables and includes treatments for potatoes, tomatoes, pears, peaches, plums, nectarines, persimmons, and citrus.

Precautions:- Precautions to be followed in the use of methyl bromide are to be found in a mimeographed announcement from the State Department of Agriculture (E-27, November 28, 1947) containing information to be used in the fumigation of walnuts in packinghouses, but applying generally to methyl bromide fumigation.

CHLOROPICRIN

Chloropicrin, CCl_3NO_2, is a colorless liquid with a boiling point of 112.4°C, specific gravity of 1.692 0°/4°C, and a vapor pressure of 23.9 mm at 25°C (77°F). The solubility in water is 0.162 grams at 25°C. This liquid and its vapors are noninflammable.

In greenhouses, lath houses, hotbeds, cold frames, and truck crop areas, chloropicrin is used for sterilizing soil for nematodes, insects, diseases, and in weed control. It has fungicidal as well as insecticidal properties. However, because of its high toxicity to foliage, it cannot be used in greenhouses when plants are present.

223

A concentration of chloropicrin of 2 to 3 parts per million in air will cause weeping, but this is a desirable characteristic of the vapor, for it results in fumigated areas being shunned by unprotected persons.

CARBON DISULFIDE

Carbon disulfide, CS_2, is a colorless liquid with a boiling point of $46.3^\circ C$ ($114.8^\circ F$), with a specific gravity of 1.263 $20^\circ/4^\circ C$, and the vapor is 2.63 times as heavy as air. It evaporates on exposure to air. The vapor has a disagreeable odor, is extremely inflammable, and mixed with air it is explosive over a wide range of concentrations, being an even greater fire and explosion hazard than gasoline or ether. The vapor may be exploded by sparks from metal striking metal, and sparks from an electric light switch. According to Back and Cotton (44), carbon disulfide is applied "by pouring the liquid into the evaporating pan of a vault at the rate of about 5 pounds per 1,000 cu. ft. of space. If the vault is not tight, more fumigant should be used. It should not be used at temperatures below $70^\circ F$, and the exposure should be at least 12, and preferably 24 hours."

If breathed for a sufficient period, carbon disulfide vapors are poisonous to man. For continuous working conditions, concentrations must be kept well below 0.1 mg per liter and preferably below 0.03 mg per liter (10 p.p.m.). (770)

SULFUR DIOXIDE

Sulfur dioxide, SO_2, is one of the oldest fumigants, for it can very easily be generated by the burning of sulfur. This is the way SO_2 is usually generated in pest control work, although it can be obtained as a liquid in cylinders. The "flowers" of sulfur have been recommended as being the form most suitable for burning in mushroom house fumigation (217). In the presence of moisture, it forms sulfurous acid and has a corrosive effect upon metals. It is also under such circumstances injurious to dyes. Consequently sulfur dioxide has lost favor in competition with many newer fumigants without such disadvantages. However, because it can be so easily and so inexpensively generated by the average person, it is believed that SO_2 could often be recommended for home use under suitable conditions. (770)

This gas is highly toxic to warm-blooded animals, a short exposure to 400 to 500 p.p.m. resulting in inflammation and edema of the human respiratory tract, but on the other hand it is so irritating that very small amounts (20 p.p.m.) cause coughing, and, as a result, cases of acute poisoning have been rare. Sulfur dioxide is as toxic to granary weevil as hydrocyanic acid gas or chloropicrin. (770)

CARBON TETRACHLORIDE

Carbon tetrachloride, CCl_4, is a colorless liquid boiling at $76.7^\circ C$. The specific gravity is 1.65 at $25^\circ C$ and the vapor pressure is 114.5 mm at $25^\circ C$. Carbon tetrachloride is noninflammable and is often used in fire extinguishers. As far as this property of the gas is concerned, it is a very desirable fumigant, but unfortunately it is not very toxic to insects. Its ready availability and noninflammability sometimes cause it to be used in homes, for on a small scale high concentrations may be used without regard to expense. CCl_4 should be used at the rate of at least 30 pounds per 1000 cu. ft. and at a temperature of $75^\circ F$ or higher.

It should be borne in mind by household users of carbon tetrachloride that the vapors of this liquid are poisonous, causing dizziness, nausea and vomiting, followed by other systemic disorders (953). Injury to the liver and kidneys results from continued or repeated exposures to low concentrations (over 100 p.p.m.).

Carbon tetrachloride is the least expensive compound which may be mixed with other fumigants to make them noninflammable, and this is the most important use for the compound in the field of fumigation.

FUMIGANTS

ETHYLENE DICHLORIDE (DICHLOROETHANE)

Ethylene dichloride, CH_2Cl-CH_2Cl, is a colorless liquid boiling at 83.5^oC, with a density of 1.26 $20^o/4^oC$, and vapor pressure of 79.6 mm at 25^oC. It is soluble in water only to the extent of 0.9% at 30^oC. On exposure to air the liquid evaporates, and the gas that is formed is over 3 times as heavy as air. Its odor is similar to that of chloroform.

Ethylene dichloride is non-corrosive to metals and is not considered to be dangerously inflammable, but is sometimes mixed with CCl_4 or trichloroethylene in a ratio of 3 to 1 by volume to reduce the fire hazard. Inexperienced persons can use these mixtures with few precautions and with little equipment.

Propylene dichloride, $C_3H_6Cl_2$, is chemically closely related to ethylene di-chloride and is nearly as toxic to insects as the latter compound. Its boiling point is 96.4^oC, compared to 83.5^o for ethylene dichloride, hence its volatility is somewhat lower. This compound is sometimes used as a fumigant in mixtures with CCl_4.

TRICHLOROETHYLENE

Acetylene tetrachloride may be treated with various bases to form trichloro-ethylene, $ClHC=CCl_2$, which is a colorless, noninflammable liquid with a boiling point of 87^oC, a density of 1.46 at $25^o/4^oC$, and a vapor pressure at 73 mm at 25^oC. Back and Cotton (45) suggested a mixture of 1 part of trichloroethylene and 3 parts of ethylene dichloride as a fumigant. This is a noninflammable mixture and is an effective fumigant at the rate of 14 pounds to 1,000 cu. ft. of space.

TETRACHLOROETHANE

Tetrachloroethane, $Cl_2HC-CHCl_2$, is formed by passing acetylene and chlorine into antimony pentachloride. The symmetrical form of this compound has a boiling point of 146^oC, and a density of 1.60 at $20^o/4^oC$. This compound was first used as a fumigant against the whitefly in 1915, and has been used with some success against a variety of pests, although it is toxic to certain plants (344).

b-METHYLALLYL CHLORIDE

b-Methylallyl chloride, C_4H_7Cl, has a boiling point of 72^oC. It is one of the recently developed fumigants for grain fumigation. One pound of b-methylallyl chloride is diluted with sufficient CCl_4 to make one gallon of fumigant mixture. Two gallons of the mixture, under tight bin conditions, is sufficient to fumigate 1000 bushels of grain.

ETHYLENE OXIDE

Ethylene oxide, C_2H_4O, is a colorless liquid boiling at 10.7^oC, (57.2^oF), with a density of 0.89 at $7^o/4^oC$, and is consequently a gas at ordinary room tempera-tures. It must be handled commercially in metal cylinders. It is very reactive at high temperatures and is inflammable except in concentrations below 3.5 pounds per 1,000 cu. ft. of space. Two pounds per 1,000 cu. ft. with an exposure of from 10 to 20 hours, will give good results in a tight vault (45).

The value of ethylene oxide as a fumigant was discovered by Cotton and Roark (198). It is now used to a considerable extent in the fumigation of warehouses, mills, etc., and in vacuum fumigation, but is mixed with carbon dioxide at the rate of 1 to 9 to eliminate the fire hazard.

Shepard (770) quotes Waite et al as stating that ethylene oxide is less harm-ful to man and higher animals than the fumes of hydrochloric acid and sulfur di-oxide, more harmful than chloroform and carbon tetrachloride, and similar to am-monia. This gas causes moderate though distinct irritation to the eyes and nose in comparatively safe concentrations, but to avoid serious injury, this irritation must be taken as a warning of a dangerous exposure.

SUBTROPICAL ENTOMOLOGY

DICHLOROETHYL ETHER

The b, b-form of dichloroethyl ether, $C_4H_8Cl_2O$, is a liquid which is very slightly soluble in water, has a boiling point of 178°C, and because of its high boiling point it has been particularly useful where long retention of the gas is an important factor, as in soil fumigation.

Dichloroethyl ether is recommended for application to the soil to kill plum curculio in the larval and pupal stages. It is applied as an emulsion prepared stirring 9 parts by volume into 1 part of potash fish-oil soap, then slowly stir ring in 9 parts of water. This is done in the open or in a well-ventilated room This concentration is used at 1.5% against larvae to 45% against pupae.

NAPHTHALENE

Naphthalene, $C_{10}H_8$, a coal tar constituent, is a white, glistening crystall solid with a melting point of 80°C, boiling point of 218°C, and a vapor pressure of 0.10 mm, at 25°C. It sublimes slowly on exposure to air, forming a toxic vap It is obtained from coal tar distillates by cooling and centrifuging, the purer grades being subsequently sublimed. Naphthalene is slightly soluble in hot wate but in alcohol, benzene, and other organic solvents, it is readily soluble. In light motor oil it is soluble to the extent of about 10%. Large quantities are used in the United States, often in the familiar form of "moth balls", in protecting woolen goods from insect injury. It is also used against pests attacking insect and other animal collections in museums.

In greenhouse fumigation, naphthalene has been most effective against thrips and red spiders (Tetranychidae). Whitcomb (946) has discussed the practical problems connected with greenhouse fumigation with naphthalene. Naphthalene has also been used as a soil fumigant, broadcast in carrot fields to control the rust fly, and stored with gladiolus corms to kill thrips.

PARADICHLOROBENZENE

Paradichlorobenzene, $C_6H_4Cl_2$, often designated as P.D.B., is a by-product in the chlorination of benzene. It is, like naphthalene, a white crystalline compound, and is used in large quantities for similar purposes. An advantage in the use of P.D.B. in the fumigation of fabrics, as compared to naphthalene, is that th fumigated materials lose their odor more quickly. Its boiling point is 173°C, melting point, 56°C, and its vapor pressure is 1.0 mm at 25°C. It sublimes slowly on exposure to air. P.D.B. is more volatile than naphthalene and a much greater quantity is necessary to saturate the atmosphere (0.5 lb. per 1000 cu. ft. of space as compared to 0.04 lb. for naphthalene). The vapor is noninflammable as ordinarily used.

The vapors of paradichlorobenzene are not considered to be very harmful to man, but it is not advisable to remain in a concentration of the gas for long periods (196).

Certain other substituted benzene ring compounds such as orthodichlorobenzene and monochloronaphthalene have also shown promise as fumigants, particularly in the forms of soil insecticides and greenhouse fumigants.

NITRILES (229)

Acrylonitrile, CH_2:CHCN, has a boiling point of 77.3°C, a freezing point of 82°C, and a specific gravity of 0.801 at 25°C. Its solubility in water is 7.0 wt./%. It is inflammable in air when present in from 3 to 17% by volume. It is noninflammable , however, when mixed with equal parts of carbon tetrachloride. It compares favorably with some of the better known fumigants in toxicity. When equal volumes of acrylonitrile and carbon tetrachloride are mixed, the mixture is noninflammable, and besides, has a greater efficiency than either of the gases used alone.

FUMIGANTS

Monochloroacetonitrile, ClCH$_2$CN, and trichloroacetonitrile, CCl$_3$CN, may be used as fumigants, but for practical reasons are not as good as acrylonitrile. In preliminary tests, certain aromatic nitriles (halogenated benzonitriles) have shown promise.

NICOTINE

The properties of nicotine have been discussed in a previous chapter (p. 130). While nicotine and its compounds are better known in connection with sprays and dusts, considerable quantities are used every year in the fumigation of greenhouses.

Liquids containing 40, 50 or 95% free nicotine alkaloid may be painted on steam pipes or vaporized on hot plates. They should be painted on the pipes while they are cool; then the steam is turned into the pipes to drive off the nicotine vapor. Usually 1 fluid ounce of nicotine to 2,000 cu. ft. of space is used in older greenhouses, whereas only half as much is needed in newer or tighter houses. The vapors are usually left in the greenhouse overnight.

Porous papers or coarse granular material impregnated with nicotine and mixed with various chemicals that support slow combustion may be burned in the greenhouse to generate nicotine fumes.

SOIL FUMIGANTS

Many crop pests attack the portion of a plant situated beneath the surface of the soil, or, if they attack other parts of the plant, they may use the soil as an abode or place of refuge. Many entomologists the world over have attempted to find satisfactory controls for a host of subterranean insects and other plant pests, but what might at first thought appear to be the simplest pest control operation has proved to be one of the most difficult (374). Much progress has been made in relatively recent years, but the best of the soil insecticides are still far from perfect. Nearly all soil insecticides have been used as fumigants. The majority of these have already been mentioned in this chapter in connection with their other uses; in addition to these, two soil fumigants of relatively recent origin, DD and EDB, are primarily used for soil fumigation, although EDB is also used as a grain fumigant. De Ong (230) has recently summarized the practical aspects of soil fumigation with the various compounds now generally used for this purpose.

Application:- Most soil fumigants are applied as liquids. Their application is facilitated by various kinds of "injectors", either hand operated (fig. 119) or mechanically operated (fig. 120), by which the required amounts of liquid can be applied at regular intervals and at the desired depth in the soil. In the injectors shown in fig. 120, the fumigant is led down each "tooth" or "chisel" of the cultivator-like equipment into the furrow made by that tooth. Often a roller or drag (fig. 120, bottom) is pulled after the machine to cover the fumigant to the desired depth.

Carbon Disulfide :- Employed as an insecticide in connection with soil fumigation by the French chemist Thenard in 1872. For the control of the phylloxera of the grapevine he injected the required amount of CS$_2$ into the soil at various places around the vine. The treatment

Fig. 119. Hand application of soil fumigant. Courtesy Dow Chem. Co.

Fig. 120. Mechanically operated injectors for application of soil fumigants. Courtesy Dow Chem. Co.

proved to be very successful and was used on a large scale in the control of phylloxera in France. In the United States, carbon disulfide has been used as a soil fumigant mainly in the control of the Japanese beetle. It may be poured into holes 12 inches apart that are 1 to 2 inches deep. Twenty-one cubic centimeters of liquid is poured into each hole; 6 pounds to each 100 square feet. Carbon disulfide can also be used in the form of a dilute emulsion. It is first emulsified with a suitable emulsifier, then diluted at the rate of 90 parts to 10,000 parts by volume of water. Two and a half gallons of this dilute emulsion is applied per square foot of soil surface. These procedures have been used mainly for the treatment of soil in plots, cold frames, and hotbeds (334).

Chloropicrin:- This fumigant, sold as Larvacide, has been widely used as a soil fumigant. It is injected 5 to 6 inches deep in holes 12 inches apart. The holes are filled and the soil is tamped to prevent too rapid escape of the fumigant. Three fourths to 1 pound of the liquid are applied per cubic yard of earth for compost soil. If applied by "continuous drip" apparatus, from 300 to 600 pounds are applied per acre (650). After applying the liquid to a given area, the soil is leveled and sealed by sprinkling sufficiently to wet it to a depth of about 1 inch. For best results, this sprinkling may be repeated as needed for 3 days. The efficiency of the treatment may be further increased by adding wetted peat moss as a mulch, or old burlap sacks, canvas, etc. Various gas-tight covers are available which may be used to seal the ground. The effectiveness of the fumigation is more or less proportional to the degree of gas retention in the soil. It is important to refrain from planting until all traces of the gas have left the soil; this may require from 5 to 24 days.

Chloropicrin kills most insects, as well as nematodes and fungi (especially Fusarium and Verticillium, the damping off organisms).

Methyl bromide:- Mixtures of methyl bromide for soil fumigation may be purchased on the market. They contain 10 or 15% by volume of active fumigant dissolved in cheap, harmless, nearly odorless solvents (650). These can be used within a foot or two or most plants. They are particularly useful under conditions of low soil temperatures (below 65°). Methyl bromide being very volatile, a seal by means of water or otherwise must be applied promptly. It is best applied with injectors, the holes being immediately stopped up. Against root-knot nematode, the commercial preparations of methyl bromide are used at 15 to 30 lbs. per 1000 sq. ft. (500 to 1200 lbs. per acre).

D D Mixture:- D D is a dark liquid consisting chiefly of 1, 2-dichloropropane and 1, 3-dichloropropene at the ratio of 1 to 2. Following are some of its physical properties:

Boiling range (°C)............................. 95-100
Specific gravity............................... 1.2
Pounds per U.S. gallon......................... 10
Vapor pressure, mm of mercury (25°C)........... 40
Flash point (°F).............................. 75-85
Solubility in water (approx. wt./%)........... 0.2

FUMIGANTS

In 1940, Walter Carter, an entomologist employed by the Pineapple Research Institute in Hawaii, discovered the value of D D. He found D D to be particularly advantageous in an area where the beetle _Anomola orientalis_, nematodes and pythiaceous fungi had previously resulted in serious crop failures. He stated that the benefits were at least equal to those derived from chloropicrin.

Carter (151) states in relation to his experiments with D D: "It is probably true that the broad function of treatments such as these is to amend the biological complex of the soil so that the end result expressed in terms of plant health and plant yield is favorable. Biological complexes in the soil may be radically changed through the elimination of some specific organism and the suppression or stimulation of others. Such changes may be as significant for the end result as the initial effect on the specific organism, particularly if the crop in question is slow in maturing."

D D has proved to be a much less expensive control for root knot nematodes and wireworms than chloropicrin or methyl bromide, and it performs as well even without the sealing of the soil following its application. It is not as good at killing weed seeds as chloropicrin, but will kill some if the soil has been moistened a few days before application. It has rather limited usefulness as a soil fungicide. (650)

Large quantities of D D are now being used, especially where nematodes and wireworms are important problems. The common dosage rates are 20 to 40 gallons (200 to 400 pounds) per acre. According to Lange (507), "Tests conducted in the Salinas and Sacramento valleys since 1943 indicate that the 400-pound (40-gallon) treatment is the most satisfactory from the standpoint of crop yields, wireworm control, and cost. To achieve a perfect kill one often has to use 600 pounds per acre; but the per cent of wireworm kill varies according to soil texture, soil moisture, soil temperature, and many other factors. The soil should be in good friable condition, ready for planting, but not ridged up (listed), and not too wet or too dry. The soil temperature at an 8-inch level should not be below 50°F and preferably should be higher. The interval of spacing should be 12 to 15 inches, and the fumigant should be drilled to a depth of 6 to 8 inches. If the fumigant is applied at a depth of 6 to 8 inches it will diffuse downward some 16 to 18 inches, upward to the surface of the soil, and horizontally 8 to 10 inches. Usually it is not advisable to treat the soil immediately after plowing under a considerable amount of plant debris, or a covercrop. Often it is advantageous to fill in the grooves made by the chisels drilling in the fumigant by using a drag behind the applicator, or by using a round roller. With shallow-seeded crops, such as lettuce, seeding can be done 10 to 15 days after treatment. If a strong odor of the fumigant is detected in the soil, planting should be deferred. With beans it is better to wait 20 to 25 days after treatment, especially in heavier types of soils. The planting date will depend upon the soil type and upon the moisture content at the time of treatment and thereafter."

Ethylene Dibromide:- This colorless liquid, more familiarly known as "EDB", is odorless, non-inflammable, and non-corrosive. It has the following physical properties:

Specific gravity......................... 2.17
Pounds per U.S. gallon................. 18.0
Boiling point (°C)...................... 131.6
Solubility in water (approx. wt./%)..... 0.4

This fumigant is supplied as a 10% (by volume) solution dissolved in a suitable solvent (Isobrome D) or a 20% (by volume) solution (Dowfume W-40) or a 5% solution (Garden Dowfume). Like D D, it can be applied to garden plots, seedbeds, or in a greenhouse by means of hand applicators or to fields with special field application equipment which can treat about 20 acres per day. When the hand applicator is employed, injecting the liquid at 12-inch intervals and using 20 gallons of 10% EDB (by volume) per acre, about 2 ml is applied per injection. Perfect control of root knot nematode is reported with 12 pounds of the 10% product per 1000 square feet at a cost of about $2.00 (650). On the West Coast, large quantities of EDB are being used for wireworms and nematodes. The dosages are 10 to 20 gallons of a 20% solution per acre.

According to tests made by Lange (507), "2 to 4 gallons of technically pure ethylene dibromide per acre are needed to control wireworms, the amount depending upon the soil type, soil texture, soil moisture, and other factors. In general, the soil should be in good friable condition, and not too wet or too dry. EDB should be used before a crop is planted. It is best to wait 7 to 15 days before planting, although with beans and certain other crops immediate planting is often possible with no damage. Certain plants, such as tomatoes, may be injured if set too soon after the soil treatment.

"As with D D, it is better not to treat with EDB immediately after the plowing under of weeds, debris, or a covercrop. The treatment is most effective at temperatures above 45°F. The spacings of the points of injection should not be greater than 12 inches, and the depth of injection should be 6 to 8 inches. EDB, when applied at 6 to 8 inches in the soil, diffuses downward 6 inches, upward to the surface, and laterally about 6 inches."

Recently EDB has been used as a soil fumigant in the form of bead-like capsules and was found to be about as effective in reducing root-knot nematode as it was according to the conventional methods of application (781).

Other Soil Fumigants:- In addition to the above soil fumigants, dichloroethyl ether, ethylene dichloride, paradichlorobenzene, naphthalene, calcium cyanide and sodium cyanide have been used in soil fumigation, but they are not primarily, or at least not extensively, used for this purpose. Some of the principal uses, experimentally or commercially, to which these compounds have been put in soil fumigation are given in connection with the preceding discussions of the compounds as general fumigants. Gough (374) has made a review of world literature dealing with soil insecticides, and the various means in which the fumigants mentioned in this chapter, as well as a host of less important compounds, have been used.

CHAPTER XIV
ORCHARD FUMIGATION

INTRODUCTION

The type of fumigation with which we are concerned in the treatment of citrus pests, is called "tent fumigation". According to this method, an octagonal sheet of tent material from 30 to 70 feet in extent, is pulled over the tree to confine the gas (HCN), which is injected under the tent. The great advantage of this method is that all insects, except for a few which happen to be situated at a point where the tent lies against the foliage or fruit, are reached by the gas. Before the red scale developed a resistance to HCN gas, the results from orchard fumigation of citrus trees were so highly satisfactory that fumigated trees usually did not need to be retreated for several years. Today there are very few orchards remaining in which such a high degree of control can be obtained by fumigation. Nevertheless, fumigation remains one of the important measures used in the control of scale insects on citrus trees. Combined with oil spray in the "combination treatment" or "spray-fumigation treatment," it constitutes a highly effective means of combating any of the citrus scale insects, whether they be of the resistant or the non-resistant races.

HISTORICAL

Essig (285) divided the history of the development of orchard fumigation with hydrocyanic acid gas in California into five periods.

First or Experimental Period, 1886-1888:- The initial impetus to the first experimental work on orchard fumigation appears to have been afforded by the introduction of the cottony cushion scale, Icerya purchasi Mask, into the citrus orchards of California in 1876, its subsequent rapid spread, the seriousness of the injury it caused to citrus trees, and the ineffectiveness of insecticides available to combat the insect. D. W. Coquillett was appointed in 1885 by the Division of Entomology of the U. S. Department of Agriculture to investigate the cottony cushion scale and to devise a means of combating it. For several years previous to Coquillett's appointment, experiments had been made with gases released under tents which had been pulled over the trees and either fastened around the trunk or allowed to fall upon the ground. Carbon disulfide was the gas most commonly used. The most energetic investigator and supporter of investigations was J. W. Wolfskill, a Kentucky explorer and trapper who became the first commercial grower of citrus in California. His 70-acre orchard (fig. 121) where the earliest fumigation experiments were made, was located on the site occupied today by the Union Depot in the heart of Los Angeles (734). Wolfskill was the first to cover an orange tree with a tent, using heat to kill the black scale. Coquillett was im-

Fig. 121. J. W. Wolfskill in his famous orchard, in the year 1886. Courtesy First National Bank, Los Angeles.

pressed by this method of treatment, so in September, 1886, he began experimenting in the Wolfskill orchard and found that HCN gas, liberated under a tent, was the most promising method of killing cottony cushion scale as well as all other scale insects. The University of California sent a chemist, F. W. Morse, to investigate the problem in 1887, about six months after Coquillett's discovery, and Morse independently reached about the same conclusion as Coquillett and his co-worker. Morse (618) was the first to publish, and his bulletin is the first published account of the use of hydrocyanic acid for the fumigation of citrus trees.

In the early work, bicarbonate of soda was mixed with potassium cyanide dissolved in water, and this mixture was added to sulfuric acid. The carbonate of soda, which evolved carbonic acid gas, was found to reduce injury to the tree, but made a longer exposure necessary. Since the relationship of sunlight to HCN injury had not yet been discovered, the early attempts at fumigation resulted in much injury to the tree. Unwieldly derricks were used to cover the trees with the tents and cumbersome methods were used to generate the gas.

Second Period, 1889-1906. Pot Generation, Night Work:- The red scale became as injurious as the cottony cushion scale, and experimental work was begun in Orange County with HCN gas used against this pest. This work was done by Coquillett and A. D. Bishop, in the latter's orchard. Within two years there occurred the following important developments, as summarized by Essig (285):

(1) The use of a black tent during the day.
(2) The substitution of night work for day work.
(3) Pot generation of gas beneath the tent.
(4) A change of formula and reduction of dosage to the proportions of one ounce of dry potassium cyanide, one fluid ounce of sulfuric acid, and two fluid ounces of water.
(5) The use of duck tents.

These discoveries were described by Coquillett (192). In addition it may be stated that the measuring of the volumes of the trees by means of numerals marked on the tents and accurate calculation of HCN dosage with the aid of "dosage tables" was being investigated by C. W. Woodworth, of the University of California (1016).

Fumigation by night was an important step in the evolution of present day fumigation procedure, and fumigation of citrus trees would probably never have become a practical pest control measure if the relation of sunlight to HCN injury had not been discovered. A. D. Bishop, in fact, was granted a patent on fumigation "in the absence, substantially, of the actinic rays of light," but the patent was not held to be valid when contested in court.

Fig. 122. "Pot" for "pot fumigation". X 0.14

According to the "pot method" of HCN generation, one ounce of dry potassium cyanide, one fluid ounce of sulfuric acid and two fluid ounces of water were used to generate the gas in a container made of lead and having the shape of a water pail. The tent was moved to the next tree in fifteen minutes. Later a glazed earthenware vessel of about two gallons capacity (fig. 122) was used to generate the gas. This vessel has a restricted neck to prevent to some extent the splattering of acid onto the tent material. It was placed near the trunk of the tree. The required amount of water was poured in, followed by the acid. The cyanide, which was in lumps of about one ounce, was then placed in the vessel. The residue remaining after the interaction of the ingredients was dumped in the center of four trees that had already been fumigated.

Third Period, 1907-1912. Standardization:- Requests from state and county agricultural officials, as well as growers, were directed to the Bureau of Entomology for a systematic and thorough investigation of the whole matter of fumiga-

tion for citrus insects. The following points were to be investigated (265):
(1) Elimination of guesswork in determining a more exact dosage.
(2) Study of the chemistry of fumigation.
(3) Determination of the physiological effect of HCN on the foliage and fruit.
(4) Improvement of mechanical equipment employed in fumigation.

R. S. Woglum was prominent among those who made contributions during this important period in the development of citrus fumigation (954). Basing his calculations on the premise that the space occupied by a citrus tree approximated that of a hemisphere set upon a cylinder, Woglum worked out more reliable dosage tables. He found that there was a higher per cent kill when large trees were fumigated than when small trees were fumigated and that this was caused by a greater leakage of HCN gas away from the small trees, through the tent, than from the large trees. Consequently, in the dosage schedule finally developed, the differential in leakage of gas from trees of different size was taken into consideration, as well as the volume of space under the tent. Woglum's calculations as to the proper dosage for trees of different sizes, coupled with the adoption of a system of tent marking to obtain the distances over the trees, worked out by A. W. Morrill in Florida, provided the basis for dosage determination for the individual trees as they are to this day. While in California fumigation was enthusiastically encouraged and was soon developed into the most important citrus pest control measure, in Florida it had not developed to any important extent for another twenty years, then later it was nearly completely discontinued.

In 1909, sodium cyanide, first used by C. P. Lounsbury in South Africa in 1905, was substituted for potassium cyanide. Woglum had demonstrated that when free from the impurity sodium chloride, this compound was satisfactory for citrus fumigation. The amount of sulfuric acid used for generating the gas had to be increased and the formula for pot generation of the gas became the following:

Sodium cyanide (51-52% cyanogen) -- 1 ounce
Sulfuric acid (66° Baumé) --------- 1-1/2 fluid ounces
Water ---------------------------- 2 fluid ounces

The advantage in the use of NaCN is that the pure compound contains 1/3 more cyanogen than pure KCN. Accordingly the dosage tables had to be revised.

Fourth Period, 1913-1916. Generation of Gas by Machines:- The "pot method" of HCN generation had resulted in frequent injury to the tents because of the splattering or accidental spilling of acid on the tents. The development of portable machines for the generation of gas (378, 1021) was therefore hailed as an important contribution to fumigation procedure, although in actual fact the use of machines marked a return to the methods employed by Coquillett in the earliest days of fumigation. The machines, however, were considerably improved. Although the machines were a boon to the fumigators, they did not, of course, increase the effectiveness of the fumigation against the insects, and the cost of fumigation to the grower remained the same. The principal advantage of machine generation was the lengthening of the life of the tents, for acid burns in the tents were almost entirely eliminated.

Fifth Period, 1916-1927. Liquid HCN:- The first use of liquid hydrocyanic acid for the control of plants in the field appears to have been made by C. W. Malley in South Africa (717). Malley used the liquid HCN in the fumigation of the mealybug Pseudococcus capensis on grapes. In California, Wm. Dingle independently experimented with liquid HCN at about the same time, and it was used experimentally in 1916 and on a commercial basis in 1917 for the fumigation of citrus trees. The characteristics of liquid HCN have been described in the preceding chapter (p. 218). With the development of safe containers and efficient machines for the application of the fumigant, the use of liquid HCN gained rapid approval and widespread use in California. Twenty ml of the liquid at 60°F yielded about as much HCN gas as one ounce of sodium cyanide (51 to 52% cyanogen). Since the liquid HCN must be transported from the factory to the orchard under conditions in which it can be kept cool, its use is limited, from a practical standpoint, to areas in

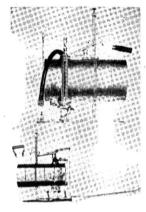

Fig. 123. Two types of "cold gas" applicator. Courtesy du Pont de Nemours & Co.

which there is a large acreage of citrus within a relatively small area. The drums of liquid are transported from the factory to the orchard in a truck, and usually under blocks of ice. Professor H. J. Quayle introduced the liquid HCN method into Spain in 1923, and Spain appears to this day to be the only region, aside from California, where the liquid HCN is used.

The machines required for the application of liquid HCN were light and readily portable, compared to the bulky and heavy equipment used for the pot and machine methods of gas generation. They have been loaned to the fumigator by the company manufacturing the gas he uses. The design and construction of these machines fulfilled the requirement for (1) correct measurement of the liquid, (2) the conversion of the liquid into a uniformly distributed gas or vapor under the tent. One type of appliance called the "cold gas" applicator (fig. 123), atomized the liquid into the tented enclosure as a fine mist. Three to five gallons of HCN was contained in the tanks of the cold gas applicators, and the machines could be carried by hand. This type of applicator was used exclusively for years, but in the meantime the study of gas concentration and distribution under tents disclosed the limitation of the atomizing principle of liquid HCN application. However, some of these machines are still being used in Tulare County. Subsequent developments in gas-dispensing machines were initiated by the findings of several investigators in relation to the subject of "protective stupefaction".

Protective Stupefaction

Gray and Kirkpatrick (380, 381) showed that red and black scales are "stupe-fied" by short exposures to sublethal concentrations of HCN and are then killed with greater difficulty. This phenomenon became known as "protective stupefaction". Protective stupefaction could be brought about by (1) a slow distribution of HCN gas under the tent or, (2) the drifting of gas which leaks from the tents to the trees of adjoining rows. Gray and Kirkpatrick believed that a more rapid distribution of gas throughout the tented enclosure could be obtained by an appliance which would distribute the lethal concentration of the gas quickly and uniformly. The stupefaction resulting from "drift gas", they believed, could be avoided by pulling the tents in the direction opposite to that from which the air is drifting. A type of appliance was finally designed and perfected which converted the liquid into a warm gas prior to its introduction into the tent. This machine was called a "vaporizer" (figs. 124 and 125). Gas concentration studies (fig. 126) showed that whenever the temperature of the ground under the tent is lower than the temperature of the air above and consequently no convection currents are present, a rapid and uniform distribution of HCN gas can be obtained only by using the vaporizer or "hot gas" machine. The vaporizer proved to be particularly valuable under winter daylight[1] conditions when the atomizers did not produce sufficiently rapid distribution of HCN gas to pre-

Fig. 124. Vaporizer used in orchard fumigation. Courtesy du Pont de Nemours & Co.

[1]When the temperature is sufficiently low, and particularly when the sky is overcast, fumigation may be safely undertaken several hours before sunset.

Fig. 126. Determination of HCN concentration at different points under a fumigation tent. The gas is withdrawn from the tent by a water displacement method. U.S.D.A. Yearbook, 1943-47.

Fig. 125. Hose of vaporizer in place under tent as HCN gas is introduced. Courtesy American Cyanamid Co.

vent protective stupefaction. However, the vaporizer generally gives higher initial concentrations and more uniform distribution of the gas at all seasons of the year. (705)

The vaporizers consist of a gas tank, metering pump, water boiler containing vaporizing coils and heated by a gasoline heater, all mounted wheelbarrow-like and capable of being pushed and operated by one man, although horse-drawn machines were also made and used. Although these machines began to be used commercially in 1921, they did not completely supplant the atomizers for about fifteen years. They are now used exclusively for applying liquid HCN in California.

Spray-Fumigation Treatment

In 1925, W. M. Mertz, a grower and manager of citrus properties, made the observation that in an orchard which happened to be fumigated a week or two after the application of oil spray, an exceptionally good control of red scale was obtained. Soon it became recognized that the spray-fumigation treatment, more commonly known as the "combination treatment", results in a control which is superior to that which would be expected merely from the summation of the results of the two independent treatments. Thus the "combination treatment" became widely used in orchards where red scale was especially abundant and difficulty had been experienced in its control in previous years. The reason why a combination of oil spray and fumigation should be superior to two sprays or two fumigations, or the summation of the calculated independent results of oil spray and fumigation, is obvious when one considers the fact that oil spray is especially effective on the periphery of the tree - on the green twigs, leaves and fruits - while fumigation is especially effective on the rough bark of the larger twigs and the branches. The two treatments therefore are complementary, one being the more effective where the other is the weaker.

The two treatments are also complementary in their action in another sense. With oil spray the adult females are the most difficult to kill of all stages and instars; the remaining instars are relatively easily killed. With HCN fumigation the second molt insect is the most difficult to kill. Thus the oil spray is apt to kill the younger instars but to leave a relatively large percentage of live adult females which might succumb to the fumigation (731, 527). The extent to which this complementary effect of the combination or spray-fumigation treatment might be utilized in the pest control program for red scale has been demonstrated in extensive field experimentation (731).

Whether the spray precedes the fumigation or _vice versa_ appears to make no significant difference in the end results as far as red scale control is concerned. However, since August and September are the preferred months for spraying oranges, while fumigation can also be done during the winter months, the usual practice is to spray first and fumigate later. Also the presence of citrus red mite on the trees in the fall would favor spraying at that time.

Calcium Cyanide Dust

The flakes of calcium cyanide, $Ca(CN)_2$, ground into a fine powder, may be used as a source of HCN gas for the fumigation, under tents, of citrus pests (719).

This compound liberates HCN gas upon exposure to the moisture in the atmosphere. In southern California, conditions are often too humid for the safe use of this material; leaf drop and fruit burn may result. Tests made by Professor Quayle in Australia, however, showed that cyanide-dust fumigation could be successfully used until the period of winter rains. This method of fumigation soon took the place of older methods and is still used in Australia as well as in South Africa and Egypt. The trees are covered with tents and these are measured as in liquid HCN fumigation. Hand operated machines (fig. 127) are used for applying the cyanide dust. They consist of a hopper for holding the charge of dust and a rotary fan for blowing it under the tent.

Fig. 127. Machine for blowing calcium cyanide dust under fumigation tent. Courtesy American Cyanamid Co.

Under conditions in which the fumigation operations are not sufficiently extensive to justify the purchase of a blower, a very simple method of applying the cyanide dust may be used. The weighed dosage of cyanide dust for a given tree is placed in a small can which has been perforated and to which a long wooden handle has been attached. The dust is scattered on the ground under the tent in a manner similar to the use of a salt shaker. The gas is readily released by contact of the dust with the moisture of the air and the soil, and diffuses throughout the space enclosed by the tent. This method of fumigation has been used in the treatment of avocado trees for latania scale, _Hemiberlesia lataniae_ (569, 724).

Sixth Period, 1928. Basic Research:- Following the five periods in the history of fumigation as conceived by Essig (285), we may describe the developments which followed as belonging to what might be considered as the Sixth Period - the period from 1928 to the present day. Up to the beginning of this period many important discoveries had been made which were as yet imperfectly understood, even though practical applications had already been made. There followed a period of basic research on physical, chemical, physiological, and even genetic problems, which had accumulated without a solution over the 40-year period since the beginning of commercial fumigation.

Quayle (713a) had called attention to the fact that it was becoming more difficult to kill red scale in the Corona area. Red scale from this area were fumigated in the same chamber with red scale from other areas where this increasing tolerance to HCN gas was not noticed, and it was proved that the Corona scale were more resistant to the gas. This resistance to HCN was next noted near Orange, in Orange County, and gradually throughout the following years the same phenomenon was noted in various areas throughout southern California. By 1915 resistance to HCN gas among black scale in the Charter Oak area had become quite evident; it had been noted by Woglum as early as 1912 (724).

The development of resistance to HCN, especially with red scale, was recognized as one of the most important problems with which the citrus industry was faced and, as might be expected, the problem of resistance alone provided subject matter for much basic research. After it had been well established that two strains or races of red scale exist in the orchards, one resistant to HCN and the other non-resistant, it became obvious that basic research could be carried out only if the investigator knew with which race he was dealing in an experiment. The research at the Citrus Experiment Station was thus placed in a sound and scientific basis when Lindgren (525) began rearing resistant and non-resistant races of red scale in insect-proof rooms in the insectary at the Citrus Experiment Station. The original resistant stock was obtained in Corona and the non-resistant stock from the foothills east of Glendora, in August 1926.

The rearing of scales in large numbers made possible more careful studies on the effectiveness of HCN against the red scale, as well as making possible accurate quantitative studies on the nature of resistance. More detailed studies on "protective stupefaction" were facilitated by the rearing of the test insects, and these will be described in a later section. Working with these insects of known ancestry, Dickson (235) was able to demonstrate that resistance to HCN in the California red scale is a sex-linked character depending on a single gene (or groups of closely linked genes) in the X-chromosome. Many basic contributions on what appear to be physical and physiological differences in the resistant and non-resistant races of red scale, insofar as these affect results from HCN fumigation, were made in the course of the investigations at the University of California Citrus Experiment Station at Riverside, the U.S.D.A. Entomological Laboratory at Whittier, and by investigators employed by the cyanide manufacturers. These contributions will later be discussed in the section dealing with the biology and control of the red scale (pp. 455 to 458).

While it is generally true that fumigants are more effective against insects at the higher temperatures, and this was also proved to be true in the case of the black scale (493), an interesting exception to this rule is to be noted in the fumigation of the California red scale. A number of extensive investigations showed that a low temperature increases the effectiveness of HCN against the red scale, especially under gas-tight fumigation conditions (612, 732). It was found that when red scale-infested lemons were preconditioned for four hours or longer at the same relative humidity, but at three temperatures, 50°, 75°, and 95°F, the best results from HCN were obtained with those insects preconditioned at the lowest temperature (732). It was also demonstrated that as between relative humidities of 22 to 52% compared to relative humidities of 80 to 89%, there was a better kill from red scales preconditioned and fumigated at the lower humidities. The difference in kill amounted to 10%. However, no difference in kill was found when red scales were preconditioned at 50 to 50% relative humidities as compared to 80 to 89%.(732). In orchard fumigation, however, better kills may be obtained at the higher humidities, for the moisture causes the tent to become more nearly gas-tight.

Greater injury to rooted lemon cuttings was obtained when these were preconditioned and fumigated at 70% relative humidity than when they were preconditioned and fumigated at 90% relative humidity. Also at a given concentration of gas, a greater number of the lemon cuttings were injured when growing in dry soil than when growing in wet soil. (732)

Methods have been worked out for the accurate determination of HCN and investigations have been made of the factors affecting the recovery of HCN from fumigated citrus tissues (66, 67). Utilizing the information gained in these studies, Bartholomew et al (68) measured the amounts of HCN absorbed by citrus tissues during fumigation under the controlled conditions of the fumatorium and in the field. The following is in part their summary of this investigation:

"Considerably more HCN was absorbed by fruits preconditioned overnight at 43°F before fumigation than by those preconditioned at 80°, and green fruits absorbed an average of 5.4 times as much HCN as mature fruits.

"Under laboratory conditions the absorption of HCN by fruits was retarded by the application of oil spray, but both fruits and leaves sprayed under field conditions absorbed as much HCN as unsprayed fruits and leaves. In the laboratory none of the fruits were injured by the HCN; in the field none of the unsprayed, but about 6 per cent of the oil-sprayed fruits were injured.

"Less HCN was absorbed by leaves and fruit on trees that had not been recently irrigated than by those on trees that had been recently irrigated, but there was no appreciable difference in the amounts of HCN absorbed by turgid and nonturgid leaves and fruits fumigated in the laboratory. The turgid fruits were more severely injured than the nonturgid fruits. Fruits sprayed with water and a spreader and fumigated at once, absorbed less HCN than similar fruits whose surfaces were dry.

"In 1939-40, green fruits from inland areas absorbed less HCN than green fruits from coastal areas; but in the similar experiment in 1940-41, the inland fruits, absorbed more HCN than the coastal fruits. The coastal fruits were much more severely injured than the inland fruits both years.

"Green fruits fixed or chemically changed absorbed HCN so that it could not be recovered and determined by the usual methods.

"Leaves and fruits of trees fumigated during the day absorbed approximately the same amounts of HCN as those fumigated at night, but were much more severely injured. In these experiments recoverable HCN was retained by mature leaves for at least 60 hours, by green fruits 35 to 40 hours, and by mature fruits 20 to 25 hours. An average of 6.3 times as much HCN was recovered from the green fruits as from the mature fruits.

"The stomata are apparently not important in governing the rate of entrance of HCN into citrus leaves and fruits .

"The physiological condition of the tissues rather than environmental influences or the amount of HCN absorbed seems to determine whether they will or will not be injured by HCN after fumigation at night; injury after day fumigation appears to result from the effects of sunlight, which raises the temperature and influences the physiological condition of the tissues.

"The results of laboratory fumigations may, but do not always, indicate the results that will be obtained when the fumigations are made under field conditions."

The scarcity of labor during World War II led various commercial operators to devise mechanical contrivances (see Chapter 15) to pull the tents over the trees. Within three years after the first "tent puller" was built by Ben Aldrich of Santa Paula, California, 80 of these contrivances had been built. It is likely that mechanical tent pullers have come to stay, representing still another advance in the art of orchard fumigation, although, on the other hand, if the new nylon tents should prove to be acceptable, their lighter weight might be an inducement toward a return to manual tent pulling.

FUMIGATION PROCEDURE

Tents:- Octagonal-shaped tents are now used to cover citrus trees in orchard fumigation. The tents may be made entirely of 7 or 8-ounce U.S. Army duck, or the central strip may be made of 7 or 8-ounce U.S. Army duck, while the sides of "wings" are composed of a lighter material called 6-1/2-ounce drill. It is the center strip of tent which must withstand the greater strain and greater wear, for the pull is exerted against the center strip in hoisting the tent over the tree. The central portion of a fumigation tent is the part that extends the full diameter of the tent in the direction in which the strips of fabric run, and the tent is always pulled in this direction. The sizes of tents generally used are 36, 40, 45, 48, 52, 55, 64, 72 and 80 feet in diameter.

By following a definite procedure, two men working together can fold fumigation tents (fig. 128), and then roll them into a compact roll. The tents are hauled to an orchard and unloaded along the first row of trees to be fumigated - one to each tree. If the fumigation is not finished in one night, the tents are

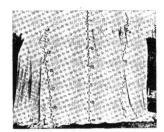

Fig. 129. Numbers marked on tent, from which distance over the tent can be computed. Courtesy American Cyanamid Co.

Fig. 128. The proper folding of a tent prior to rolling it up. Courtesy American Cyanamid Co.

pulled off the last row of trees to be fumigated and are left lying on the ground, ready to be pulled onto the next row of trees the following night, if weather permits.

In order to determine the amount of gas required for a given tree, the tent, after it is pulled over the tree, must be measured both as to the distance over the tent and as to its circumference. To facilitate the calculation of the distance over, the tent is marked in feet, beginning at the center and proceeding in opposite directions parallel with the seams. Another line of figures is stamped on the tents parallel with the center line, 3 to 5 feet to the left, and 3 to 5 feet to the right (fig. 129). Thus the tent does not have to be accurately centered over the tree, for one of three lines of figures may be used for the measurement. By adding the number read nearest the ground on one side of the tent to the number, on the same strip of numbers, nearest the ground on the opposite side of the tent, the distance over can be calculated.

Tents made of nylon are now being tested for orchard use, and they appear to have many advantages over the fabrics now used (530). A nylon tent is only about one-third as heavy as the 8-ounce duck and one-half as heavy as the 6-1/2 ounce drill. Nevertheless, the nylon material is twice as resistant to tear as 8-ounce duck. Also the nylon has a hard finish and slides over the tree easily. It has a further advantage of not being subject to mildew. The present type of canvas can increase in weight 120% when wet, while nylon increases less than 50%.

For a given dosage, under dry conditions, the concentration of HCN is not as great under a nylon tent as under the usual cotton tents (530, 1028). The explanation has been offered that shrinkage of cotton tents vitiates the accuracy of computations of dosages based on the numbers stenciled on the tents, so that a higher dosage of gas is injected under cotton tents than would be justified by their actual cubical contents if they were measured at the time of fumigation. On the other hand, during damp weather the concentration of HCN under nylon tents was 38.5 per cent higher than under cotton tents. Tests with a porosity meter showed that practically no air penetrated wet nylon fabric with a pressure of 3/4 inch of water. It was found that wet duck and drill cotton fabrics remain more porous than wet nylon (1028).

Yust et al (1028) have stated the disadvantages of the nylon tent as follows:

"Although the light weight of nylon has many advantages, it has some disadvantages. The tents float when pulled from tree to tree by machine, and time must be allowed for them to settle on the tree. This delay slightly retards the rate of pulling. A light breeze increases the work of pulling nylon tents on the trees. The light nylon is supported by air when the tents are spread out to make them into compact rolls for hauling. This floating tendency delays rolling. The rolled tents fall apart too readily because the cloth is slick. The light cloth is also objectionable in a grove with tall cover crop, because the tent skirts do not weigh down the plants. Several fumigators have made tents with nylon centers and cotton drill skirts, and this type of tent appears to be a promising compromise.

Fig. 130. Repairing fumigation tents.

"The normal life of nylon tents may be the most important consideration in making a change from cotton to nylon cloth. To be practical at present costs, the nylon must withstand wear and retain its strength at least as long as cotton drill and duck tents. Although new nylon cloth is tear-resistant, after being exposed to the weather for six months on the roof of a building it could be torn apart more readily than duck or drill cotton cloth. The relative deterioration of nylon and cotton tents in normal fumigation usage must be determined before the financial risks involved in changing over to nylon tents can be estimated."

Tents are subject to tear when being pulled over the trees, and since every hole is a source of loss of gas, great care is taken to keep the tents in good repair (fig. 130). Mildew-proofing is also an essential operation, especially when the tents are used during the months of greatest rainfall. Various solutions have been used in mildew-proofing fumigation tents, including hot tannin solution and a zinc sulfate solution. Various copper compounds would be effective in combating mildew, but even a very small amount of copper on the tent or on the tree would result in severe injury to the tree when combined with HCN. The following requirements have been suggested as being essential for a good fumigation tent antiseptic:

Fig. 131. Pulling a fumigation tent over a tree by means of poles. Courtesy American Cyanamid Co.

good antiseptic qualities; water insolubility (to prevent leaking); it must be relatively nonvolatile; it must not react with hydrogen cyanide; it must not close up the pores of the tent so as to effect its pliability or greatly increases its weight; it must have no detrimental effect on the fibers of the canvas; it must be free from industrial health hazards; it must be simple to apply; and it must be inexpensive in the concentration required for efficient and lasting protection. A phenyl mercury oleate treatment has been recommended as being a highly satisfactory compound for the mildew-proofing of tents. It is stated that the treatment should be effective for two or more seasons. (862)

Poles and Mechanical Tent Pullers:- The poles used for pulling the tents over the tree (fig. 131) are 14, 16, or 18 feet long and 2 to 2-1/2 inches in diameter. They are made from straight grained and well seasoned pine. About 6 inches from one end of the pole a cotton rope is fastened. The "puller" must find a ring, of which there are two on each tent. The end of the pole is inserted in the ring and the rope is looped over the ring to hold it in place. The "puller" then inserts the other end of the pole, which is bluntly sharpened, into the ground at the side of the tree and begins to pull the rope, which is only a little longer than the pole. His partner must simultaneously pull on his rope at the other side of the tree and tent is thus pulled over the tree (fig. 131). The rings of the tent are then slipped off the pole and the "puller" is ready to proceed to the next tree. As many as 60 trees can be covered by strong and experienced men during the 45 minutes they are allowed for removing a "string" of tents from one row of trees to the next row.

As stated before, the mechanical tent pullers are now employed for much of the fumigation work (fig. 132). These will be described in Chapter 15, p. 296.

Taping:- The "pullers" having pulled the tents over the tree, the "taper" must then determine the volume of the space enclosed by the tent. He fastens one end of a measuring tape onto a fold of the tent and pulls the remainder of the tape around the tree, kicking in the "skirts" of the tent as he goes, in order to insure the smallest possible volume under the tent and to see to it that the tent lies firmly against the ground and is not affording an outlet for the gas which will soon be injected (fig. 133). The skirts can be kicked in because a "dog hole" (fig. 134) has been constructed by the taper which lets out the air when the tent is compressed. This "dog hole" also serves as an opening for the delivery pipe of the vaporizer. After encircling the tent and returning to the point at which the measuring tape was fastened, the "taper" reads from his

Fig. 132

Fig. 133

Fig. 134

Fig. 132. Pulling a fumigation tent over a tree by means of a mechanical tent puller. Courtesy R. S. Woglum.

Fig. 133. Fumigation operations. Left, "Gunning"; right, "Taping". Courtesy American Cyanamid Co.

Fig. 134. A "dog hole" into which the hose leading from the vaporizer is placed. Courtesy American Cyanamid Co.

tape the distance around the tent, and from the figures stamped on the tent he determines the distance over the tent. With these two figures he can determine from a chart the volume of the tent. In present day practice it is more likely that a measuring tape will be used on which the volume of the tent can be read on the tape at or near the figure corresponding to the distance around the tree (fig. 135).

Fig. 135. Orchard fumigation measuring tape with distance around (48) and number of units (11-25) corresponding to various distances over the tent (28-52). Adapted from photograph by American Cyanamid Co.

Thus, if the distance around the tree were 48 feet and the distance over were 36 feet, one would find the figure 16 to the right of the figure 36 at the 48 foot marker on the tape. The volume of such a tree would be 2033 cu. ft., but due to the fact that the larger trees retain a proportionately higher concentration of gas, because of the reduced ratio of tent surface per unit volume of gas as the trees increase in size, the number indicated by the tape is 16, as if the cubic content were only 1600 cu. ft. This insures a uniform concentration of gas for trees of all sizes. Such a tree would be designated as a "16-unit tree". The "taper" calls the figure 16 to the "gunner" who applies the gas.

The "Gunner":- The "gunner" is the man who operates the vaporizer. He places the hose leading from this machine under the tent through the "dog hole" then closes the aperture by allowing the tent to drop to the ground (fig. 125).

The "gunner" has been informed by the taper as to the number of "units" of space under the tent. The amount of liquid HCN to be used, however, depends not only on the number of "units", but also on the "dosage schedule". For every "unit" of space under the tent, there will be required a certain amount of liquid HCN, the amount on the "schedule" being employed in the particular orchard in question. If an 18 cc schedule is being employed, this means that 18 cc of liquid HCN is introduced for every "unit" of space. In the early days of fumigation practice, what is now called an "18 cc schedule" was known as the "100 per cent schedule". If 16 cc per "unit" were used, this would be called a "90 per cent schedule," 20 cc would constitute a "110 per cent schedule", etc. Nowadays the schedule is designated as the number of cubic centimeters of liquid HCN introduced under the tent per unit of space. The schedule employed may be as low as 14 or 16 cc, and as high as 26 or 28 cc. The schedule employed in a particular orchard depends on various factors which will be discussed later.

The regulation of the discharge of liquid HCN from the vaporizer to meet the requirements of the "dosage schedule" is effected by using measuring racks graduated in various denominations, as 16, 18, 20, 22, etc., which can be set to deliver the dosage required. The "gunner" is thus guided only by the number of "units" determined for the tree in question by the taper. In the example used above, the number of "units" was 16. If the handle of the pump which forces out the liquid is turned to the right, the full stroke of the pump will be long and will discharge

ORCHARD FUMIGATION

enough gas for 5 units at whatever dosage schedule for which the measuring rack was set. If the handle of the pump is turned to the left, the full stroke of the pump will be short and will deliver only enough liquid HCN for one unit. Therefore, to deliver enough HCN for a 16-unit tree, the "gunner" will employ three long strokes and 1 short stroke of the pump.

CHOICE OF DOSAGE SCHEDULE

The choice of "dosage schedule" depends on the following factors:

1. <u>Variety</u>. Higher schedules are employed in the fumigation of lemons than in the fumigation of oranges and other citrus varieties which are more susceptible to injury.

2. <u>Season</u>. Higher schedules are employed in winter than at other periods of the year, for the trees can then tolerate a higher concentration of gas.

3. <u>Locality</u>. Citrus trees are most susceptible to injury from HCN under humid conditions, so in California higher dosage schedules can be employed in interior sections than in coastal sections.

4. <u>Species of Insect to be Controlled</u>. Of the citrus pests, only scale insects are controlled by fumigation. Fumigation for soft bodied scales, such as black scale and citricola scale, may be timed so as to be directed against the instars most susceptible to HCN gas. Purple scale are best fumigated while in the susceptible "fuzz stage". Under such conditions the "dosage schedule" of HCN need not be as high as in the fumigation of red and yellow scales, among which the instars and stages overlap to such an extent that the fumigation must be directed against the most resistant stages, which happen to the second molt and the adult.

5. <u>Resistance</u>. Black, citricola and red scale have developed resistant races over certain parts of their range; in the case of the red scale by far the greater part of the citrus area in California contains resistant insects. In areas in which a species is resistant to HCN a higher "dosage schedule" must be used than in areas in which the same species is nonresistant.

HOW TO CALCULATE THE AMOUNT OF HCN REQUIRED TO FUMIGATE AN ORCHARD

Given a ten acre orchard of 900 trees of a size requiring an average of 17 units per tree, and using an 18 cc dosage schedule, one may calculate the number of pounds of liquid HCN that will be needed to fumigate the orchard.

> Amount of HCN per tree.....17 x 18 cc = 306 cc
> Amount for 10 acres.......900 x 306 cc = 275,400 cc
> 275,400 ÷ 650 cc (amount of HCN per pound) = 423.7 lbs.
>
> Add the usual 2% for unavoidable loss during the fumigation operation and you have a total of 432.2 lbs. or 0.48 lb. per tree.

An experienced fumigator can, by walking through an orchard, get a good idea of the average number of "units" per tree on which to calculate the amount of liquid HCN which will have to be transported to the orchard prior to the fumigation. Also, if the orchard has been fumigated before within a reasonable number of years, the previous fumigation record will form a good basis for the calculation of material. If the orchard is mature, there will be very little difference in tree sizes over a period of several years; in fact, average tree size might be reduced by severe "topping back" in pruning or by removal of trees that are unthrifty or have low yield.

To cite a specific example, one fumigator decided that 55 foot tents were needed in a certain orange orchard of large trees, that is, he decided that 55-foot tents would cover the largest trees in the orchard. There were 1524 trees in

243

this orchard and 1172 pounds of liquid HCN was used. A 20 cc dosage schedule was employed. It may be calculated that the average tree in this orchard was a 25-unit tree and required 0.77 lb of liquid HCN. In another orchard, this time of lemon trees, 48-foot tents were used. With a 22 cc dosage schedule, 400 pounds of HCN was used in fumigating 724 trees, an average of 0.55 pound per tree.

COST OF FUMIGATION

In the case of the orange orchard mentioned above, $0.45 per tree (April, 1948) was charged for applying the HCN. The HCN cost $0.44 per pound or $0.339 per tree. The cost of fumigating the orchard can thus be seen to be about $0.79 per tree or about $71.10 per acre. Some fumigators add an additional charge of $15.00 per job for hauling the tents into the orchard and removing them.

In addition to the fumigation, 1 or 2 dustings are required for control of citrus red mite in areas where this pest occurs, which is by far the greater part of the citrus area of California. In the above orange orchard two treatments of DN-Dust-D8 were needed. The cost of this material (1948) is $8.75 per 100 pounds. In the first dusting 1525 pounds was used (exactly 1 pound per tree) and in the second dusting 1500 pounds was used. Figuring 90 trees per acre, the cost of red mite treatment for the year was $6.27 per acre per treatment for the dust and $2.00 per acre is charged for application. The total cost for red mite treatment was $16.54 per acre for the year.

Occasionally a grower must either fumigate twice or fumigate and spray the same year for red scale. A citrus grower's pest control bill may for some years amount to well over $100 per acre.

METEOROLOGICAL AND ORCHARD CONDITIONS REQUIRED
FOR SAFE AND EFFECTIVE FUMIGATION

Like with so many pest control measures, the "margin of safety" in the fumigation of citrus trees with HCN gas is rather small. One attempts to use as high a concentration of gas as possible without undue risk of injury to the trees. However, some risk of injury is always taken, even under "favorable" conditions. Nevertheless an attempt is made to avoid certain meteorological and orchard conditions known to be adverse.

1. Light. As previously stated, the relation of light to fumigation injury was known as early as 1890. In 1920, Woglum (957) showed that light before and after fumigation must be taken into consideration as well as light during the exposure to the gas. He stated that the postfumigation influence of light appears to be somewhat greater than the prefumigation influence and may approximate the effect of light during exposure to HCN. Light following fumigation may result in destruction of the foliage and discoloration of the fruit. Some effect from light, according to Woglum, may be noticed as long as two hours after fumigation. The injurious effect of light is accentuated by rise in temperature.

It was found that diffused light before, during, or after fumigation exerts no more deleterious influence than darkness (957), and this observation, coupled with the knowledge that low temperatures result in minimizing the effect of light, resulted in daylight fumigation during the winter months. Such fumigation may often safely begin in midafternoon. This is an important factor in view of the fact that nowadays there is more fumigation done in winter than at any other time of the year.

2. Temperature. In California, fumigation is done at temperatures ranging from $40°$ to $80°F$, the latter temperature being the maximum for safety in the drier interior sections, while $70°F$ is considered the maximum in the coastal sections. The minimum temperatures, $40°$ to $45°F$, may be too low if they occur too early in the evening, for a further drop of 15 to $20°F$ the same night may result in injury even if the fumigation is discontinued after the temperature falls below $40°F$.

ORCHARD FUMIGATION

In California, the U. S. Fruit Frost Service, with headquarters at Pomona, gives the regular nightly forecasts of minimum temperatures and dew points for the various citrus districts of the State. This government service to the growers begins on November 15 of each year. The forecasts are released over Radio Station KFI every night at 8:00 PM, but the forecasts can be obtained earlier by telephoning the Fruit Frost Service, Pomona 1074, or from the district agents located in 9 towns in various parts of California (987).

3. <u>Humidity</u>. Because of the lesser diffusion of gas through a moist tent, fumigation injury is more apt to occur under humid conditions. Fumigation is discontinued when the dew point is reached. When the tents become wet, particles of soil adhere to them and as the tents are pulled over the tree an abrasion of the fruit may occur, further increasing injury. The mere presence of moisture on the tree in itself, however, causes no ill effect; in fact, it was found that in a fumitorium there were many more lemon cuttings injured when they were fumigated at a relative humidity of 27 to 42% than at a relative humidity of 80 or 90% (732).

Experienced fumigators can tell when to discontinue fumigation merely by feeling the tents for moisture.

4. <u>Soil Moisture</u>. It will be remembered that one attempts to apply an oil spray as soon as possible after an irrigation. Not so with fumigation. The association of fumigation injury with wet soil has often been made, and it is recommended that fumigation should be done before irrigation. On the other hand, if the soil has been too dry for a prolonged period, so as to result in reduced vitality of the tree, fumigation injury may be increased.

5. <u>Previous treatment with Bordeaux</u>. Bordeaux spray, which contains copper sulfate, is often applied to the lower portion of the tree to prevent brown rot infection. Bordeaux paste is sometimes applied to the trunk and lower branches of citrus trees in brown rot gummosis control. Copper and HCN are incompatible, for severe injury results when HCN fumigation follows even months after Bordeaux spraying. Even Bordeaux spraying immediately after fumigation may result in injury, and it is recommended that 2 or 3 days should elapse after fumigation before the fumigated trees are sprayed with Bordeaux. Five or six months must elapse, even if only the "skirts" of the tree (2 to 4 feet from the ground) are sprayed, before the tree can be fumigated with absolute safety, although some growers fumigate as soon as 3 months after a "skirt spray" of Bordeaux. This practice, however, is attended with considerable risk, especially if the fumigation is done under humid conditions. If the entire tree is sprayed with Bordeaux, an entire year should pass before fumigation is attempted. (959)

Haas and Quayle (398) found that citrus trees that have year after year received more injury from fumigation than trees in other orchards, contained greater amounts of copper. They recommended that where copper is applied for the cure of exanthema, the treated trees should not be fumigated until growth, soil precipitation, and leaching have had time to reduce the copper concentration in the tree.

Spraying with Bordeaux is usually done in November and December, before the period of greatest rainfall. If fumigation is not contemplated, a spray consisting of 3 pounds of copper sulfate and 3 pounds of hydrated lime per 100 gallons of water may be used in San Diego, Ventura, Santa Barbara, and eastern San Bernardino Counties, but this formula has caused injury to foliage in recent years in Los Angeles, western San Bernardino, and Orange Counties (993). In these areas it is recommended that the following mixture be used: 5 lbs. of zinc sulfate, 1 lb. of copper sulfate and 4 lbs. of hydrated lime to 100 gallons of water. Not less than 0.25 lb. of metallic copper (the amount of copper in 1 lb. of copper sulfate) should be used in 100 gallons of spray. Proprietary zinc-copper-lime mixtures should be used on this basis. Spreader, 0.5 lb. calcium caseinate to 100 gallons, should be added.

If the trees are to be fumigated within a few months after spraying, only the zinc-copper-lime formula should be used for brown rot control, for this is an effective brown rot spray despite the low concentration of copper, and it greatly reduces the possibility of damage from fumigation as compared to the regular Bordeaux spray.

In the control of brown rot, the "skirts" of the tree should be thoroughly sprayed, inside the tree and outside, up to a distance of 3 or 4 feet from the ground. The tree trunk should also be sprayed thoroughly to reduce the possibility of brown rot gummosis. Usually 4 to 6 gallons of spray are sufficient to spray an average sized citrus tree for brown rot. In Tulare County, septoria fungus, zinc deficiency (mottle-leaf) and leafhopper control are combined in one spray. For leafhoppers enough additional lime is added to whiten the trees, and this amounts to 15 to 20 lbs. to 100 gallons of spray. (991)

SYMPTOMS OF INJURY

An excellent pictorial record of 13 types of fumigation injury to citrus trees and fruit was made by Klotz and Lindgren (491). These are shown in fig. 136 and will be described in the order they are numbered in this figure.

1. Pitting:- "Fumigation pits with green to brown margins and tan to gray bottoms, the surface of which may be scabby and cracked. The rind of the stem of the orange is more susceptible than that of the distal (blossom) half."

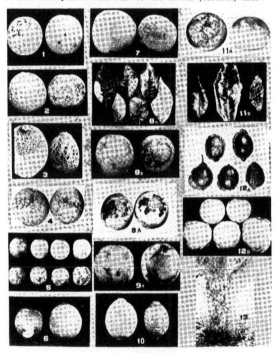

Fig. 136. Various types of injury from HCN fumigation. Explanation in text. After Klotz and Lindgren (491).

ORCHARD FUMIGATION

2. Rind Collapse:- "Fumigation may cause the white portion of the rind to collapse in irregular areas without always injuring the rind surface and oil-bearing portion (left). Frequently large surface areas are affected becoming sunken to various depths and dry, cracked and scabby."

3. Scabbing:- "Surface crusts or scabs caused by fumigating young lemon fruits The injury is first manifested by the exudation of gum (see 2). A new rind surface is formed below the scabs."

4. Canker:- "Fumigation of small green Valencia oranges sometimes causes injuries which later develop into slightly elevated canker-like spots resembling citrus canker caused by the bacterium Phytomonas citri."

5. Spotting of green fruit:- "Fumigation injury on green, half-grown Valencias, showing difference in susceptibility of Riverside fruit (top) and Orange County fruit to exposure of same concentration of gas. Specimens from E. T. Bartholomew, W. B. Sinclair and D. L. Lindgren. This difference disappeared as the fruit from the two localities matured."

6. Gumming of fruit:- "Gumming of the rind of young green Valencias caused by cyanide fumigation. Scabs may later appear in the affected areas."

7. Rubbing of tent:- "Fumigation injury thought to follow rubbing by canvas tents. The darker areas of the ripe oranges are greenish."

8. Chlorosis and necrosis of leaves and fruit:- "Injury following fumigation in presence of sunshine or when latter follows too closely the exposure to the gas. The sunshine apparently activates the chemical so that it produces mild chlorosis and necrosis in fruit and leaves. The margins of the affected areas of the orange rind are green to brown. Very little of the rind surface of the fruit on the left has been killed while a large necrotic area shows on the one on the right. The light colored portions of the leaves are yellow, the darker areas are a normal green."

9. Injury caused by fumigation at low temperatures:- "A. The darker areas are a reddish brown and are surrounded by injury due to liberation of rind oil. B. The darker areas of the rind surface are green; the effect may be indistinguishable from injury due to HCN-sunshine (see 8)."

10. HCN-copper injury:- "Type of injury that may result when fumigation follows a copper spray. The spray in this case was the dilute 1-1-100 bordeaux mixture. A month or at least a moderate rainfall, or better both, should have separated the spraying and fumigation. For greatest safety fumigate before the application of copper."

11. HCN-copper-zinc-manganese injury:- "Injury to grapefruit from fumigation (January 29, 1942) following a spray on December 1, 1941 of zinc sulfate (2-1/2 pounds), copper sulfate (1 pound), manganese sulfate (2-1/2 pounds) and soda ash (4 pounds) in 100 gallons of water. A. The light-colored affected areas are grayish brown and the darker margins are brown. B. The light, dried portions of the leaves are a grayish brown. Valencia oranges were similarly affected. A manganese sulfate-soda ash spray has also been observed to increase fumigation injury. Zinc sulfate is thought to decrease the tendency of copper to increase fumigation injury."

12. Ridges and wart-like elevations:- "Fumigation just preceding or during blossoming may cause ridges or wart-like elevations on the rind. The rind of the affected fruit is thicker and the size may be larger than unaffected fruit. A cockscomb effect is shown on the green lemons of the upper photo. In the lower photo the two upper unaffected grapefruit are normal in size. Oranges are similarly affected. Recently A. F. Kirkpatrick discovered that olives may show similar deformities following HCN fumigation in the spring."

13. Gumming of bark:- "Gum stimulation by cyanide fumigation in presence of high temperature and soil moisture. Tendrils of gum are forced from the bark. Scales form from death of small areas of the surface bark where gum extrudes; new tissue quickly grows and fills these pockets, however, and no apparent permanent damage results. Liquid cyanide coming in contact with the trunk will kill bark and outer wood; pink pustules of a fungus (Fusarium sp.) appear on the dead surface".

In addition to the symptoms listed above, there are several other types of fumigation injury.

Leaf drop:- After any fumigation a few of the old senile leaves may drop, but this is not considered to be of any importance. Sometimes, however, a severe defoliation may follow fumigation. It may begin as early as a couple of days after treatment or may be deferred for a week or more. When not associated with the presence of copper on the trees or in the soil, leaf drop may be accentuated by too wet soil, too dry soil, adverse meteorological conditions, or it may occasionally occur, like leaf drop caused by oil spray, under conditions that appear, on the basis of our existing knowledge, to be favorable for treatment. Leaf drop from fumigation may occur in conjunction with fruit pitting and a burn of the tips of tender young sprouts, but often these latter symptoms do not occur. Sometimes when there is considerable defoliation after fumigation, small necrotic areas may be noticed on a small percentage of the leaves.

Tip burn:- New flushes of growth of all citrus varieties are very susceptible to fumigation injury. In practically every treatment, if young, tender growth occurs, the tips are burned back to some extent, often for 2 or 3 inches. Unless this type of injury is very severe, it is not considered to be of any economic importance, but is looked upon as a normal consequence of fumigation with adequate concentrations of gas.

HAZARDS AND PRECAUTIONS

When hydrocyanic acid vapor is mixed with air at a concentration of at least 10%, it is inflammable and explosive. Even in the highest concentrations in commercial fumigation this figure would never be deliberately reached, but presumably it could be reached accidentally. In the fumigation of citrus trees with an 18 cc dosage schedule, the maximum concentration of HCN is 0.4% (510). In case of fire, it should be remembered that the products of HCN combustion are harmless, being nitrogen, carbon dioxide, and water vapor. If HCN is spilled, or if one should wish to dispose of a left-over quantity for which no satisfactory means of storage is available, the simplest means of disposal is by burning.

Ordinarily in orchard fumigation the tents are pulled against the direction of air drift, so that the gas is being blown away from the workmen. Also, after the usual 45-minute exposure, when the tents are ready to be pulled onto the adjacent row of trees, the gas has nearly completely left the tent by diffusion through the fabric, aided by the drift of air. However, a reversal of air drift during the night may sometimes result in sufficient concentration of gas to cause some workmen to become ill. Likewise, it should be borne in mind that liquid HCN can be absorbed through the skin. If liquid HCN should be accidentally spilled on any portion of the body, any impediment to rapid evaporation, such as clothing, should be removed. If the liquid should run inside the shoes or boots of a workman, for example, these should be immediately removed. Running water helps to remove cyanide. Never allow cyanide to come in contact with open wounds or skin abrasions. The safety and first aid rules should be known by all those working with HCN.

The following instructions on the recognition of symptoms of HCN poisoning and the administration of first aid, are copied from an instruction card supplied to commercial orchard fumigators.

ORCHARD FUMIGATION

INSTRUCTIONS

Hydrocyanic Acid Poisoning

Symptoms:
    The first symptoms of hydrocyanic acid poisoning are dizziness, faintness, weakness of the legs, rapid heart, breathing difficulty and nausea. In severe cases, the patient collapses and becomes unconscious.

Procedure:

Inhaled Gas

    In mild cases, the patient should be taken at once into the open air away from any hydrocyanic acid fumes, and made to sit or lie down. He should breathe ammonia fumes, but also some fresh air. He will generally recover quickly under this procedure.

    In a case in which the patient has lost consciousness, get him into the open air away from any hydrocyanic acid fumes as quickly as possible. Break one amyl nitrite pearl in a handkerchief or other piece of cloth, and hold loosely to the nose of the patient. Have him lie on his side or stomach, not on his back. Wrap him up to keep him warm. If he does not revive promptly, send for a physician.

    If he stops breathing, or if his breathing is at long intervals, start artificial respiration at once and have an assistant administer amyl nitrite. Continue artificial respiration until the patient is able to breathe without assistance. Watch for relapse, and if relapse does occur, lose no time in continuing artificial respiration. Patient should remain quiet for one to several hours after he has regained consciousness.

Liquid Hydrocyanic Acid Spilled on Clothing or Skin

    Thoroughly soak with water all of the area of clothing and skin in contact with the liquid just as quickly as possible. Then remove the clothing from the affected area, and wash the skin thoroughly with water. Next, treat the patient in the same manner as described above for inhaled gas.

Things Not To Do:

DO NOT walk patient around or try to move him except to get him out of the gas.

DO NOT put ice or cold water on his head or back. (Water should be used only when liquid hydrocyanic acid has been spilled on the patient.)

DO NOT take patient to doctor's office or hospital. (Have doctor come to patient).

DO NOT give patient whiskey, or any alcoholic liquor.

DO NOT attempt to give patient anything by mouth while he is unconscious.

DO NOT administer more than one amyl nitrite pearl to a patient.

WORK CAREFULLY AND AVOID ACCIDENTS

    In case of an emergency, use calm judgment and act quickly.

CHAPTER XV
EQUIPMENT FOR APPLICATION OF INSECTICIDES

INTRODUCTION

It has already been noted that in the earliest use of kerosene against scale insects, the liquid was applied directly to the scale-infested branches, without dilution, by means of feathers. This was as late as 1865, less than a century ago. Brushes such as shown in figure 1, were used for applying Bordeaux mixture up to the beginning of the 20th century. Equally primitive devices, such as bellows and blowing tubes, were used for dusting. The history and present status of insecticide application equipment have been reviewed by Irons (459) and Campbell (145), the former pointing out the desirability of fundamental engineering studies to coincide with the investigations of entomologists and plant pathologists. As pointed out by Metcalf and Flint (600), the development of insecticide application equipment has been largely an American achievement.

Insecticides may be applied as sprays, dusts, gases, and as aerosols or mists. A simple device of low capacity is not necessarily obsolete, for pest control equipment will be used which best suits the needs and the pocketbook of the user, whether he be an amateur engaged in controlling his own dooryard or garden pests, or a large commercial grower or spray operator. Therefore all types of equipment obtainable at present, from the simplest and least expensive to the largest, most expensive, and most specialized power-driven equipment, will be discussed.

Among the sources of information on insecticide application equipment are Lodeman (540) (historical), Mason (594), Knapp and Auchter (492), Metcalf and Flint (599, 600), French (349), Howard et al (450), McClintock and Fisher (578), and especially the catalogs of insecticide application equipment manufacturers. For aerosol applicators, scattered short articles of recent origin are available. The present chapter is based largely on the above sources of information.

SPRAY EQUIPMENT

Fig. 137. Atomizer or "flit gun".

Atomizer:- The atomizer or "flit gun" (fig. 137) is the smallest and simplest type of hand-sprayer, and may be effectively employed for household spraying and even for garden spraying, for often on garden plants the insects are confined to limited areas, as for example, aphids grouped about the stems of rosebuds. The atomizer consists of an air-compressing pump fastened to a small tank of from 1 pint to 2 or 3 quarts capacity. When the air is compressed by the pump, it passes over the end of a tube extending down into the tank, and this causes the liquid to be sucked up out of the tank and to be blown out of the nozzle of the atomizer with the air stream, by which it is broken into a mist. Agitation of the emulsion or suspension is accomplished by the motion incidental to the spraying or, in the case of weakly emulsified or suspended materials, by continual deliberate shaking.

Chisholm et al (159) adapted an improved hand duster, which they had constructed, so that it could be used for atomizing liquids (see pp. 271, 272). It would appear that this would be a highly effective device for spraying purposes, especially in enclosures.

Hose Attachment: The venturi effect from the compressed air in the atomizer described above can be obtained as well from a stream of water. In the sprayer shown in fig. 138, the water from a garden hose, to which the device is attached, passes over a tube extending into a jar. The concentrated spray solution in the jar is drawn into the stream of water in concentrations depending upon adjustments made with a screw. The spray stream is atomized and spread by striking the area of the venturi tube outlet with high velocity.

Bucket Pump:- Designed especially for garden and dooryard spraying, the bucket pump (fig. 139) consists of a single- or double-acting pump which may be clamped into the bucket containing the spray. In the case of the single-acting pump there is only 1 cylinder, which exerts pressure on the down stroke. The double-acting pump has 2 cylinders, thus the pressure is exerted on the spray stream by the up stroke as well as the down stroke. This results in a more even pressure. Bucket pumps may be equipped with an air chamber to further increase the uniformity of pressure and flow of the spray stream. Agitation may be furnished by a rod connected with the handle. To this rod is attached a brass plate. Agitation may also be furnished by a jet of liquid forced through an

Fig. 138. An atomizer attached to a garden hose.

orifice at the bottom of the pump into the pail, causing a continuous motion of the liquid and suspended or emulsified materials.

Fig. 139. The bucket pump. Courtesy F. E. Myers & Bros. Company.

Knapsack Sprayer:- The same in principle, but more easily operated and more expensive than the bucket pumps are the knapsack sprayers. Several acres of low-growing crops, and even small fruit trees, may be sprayed with this type of equipment. These sprayers are equipped with straps and are carried on the back like a knapsack. The shoulder straps are adjusted to fit the operator's body.

With the knapsack compressed-air sprayer (fig. 140) air within the tank is compressed until a pressure of 50 to 75 pounds per square inch is obtained. When the discharge nozzle is opened, the spray liquid is forced out of the nozzle by the air pressure. Agitation is maintained by the motion incidental to walking, often supplemented by occasional deliberate violent shaking of the sprayer. As the liquid level decreases, the air expands and its pressure decreases, so spraying pressure is constantly decreasing and the pressure must be built up again by more pumping. When uniform pressure is desirable, this type of sprayer is not suitable.

Knapsack sprayers may also be obtained with constant-pressure pumps of the plunger and the diaphragm types, both of which are positive-displacement mechanisms. With these types, the tank does not have to withstand any pressure; it merely holds the spray liquid. Constant pressure is maintained by means of an air chamber. The pump is mounted inside the tank, which may hold from 3 to 5 gallons of liquid. Agitation of the spray mixture is maintained by an attachment to the handle. The handle of the pump extends over the shoulder or under the arm of the operator, who

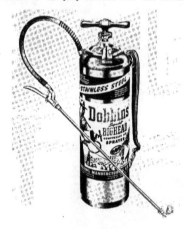

Fig. 140

Fig. 141

Fig. 142

Fig. 143

Fig. 140. Knapsack compressed-air sprayer. Courtesy Dobbins Mfg. Co.

Fig. 141. A wheelbarrow-type hand-operated sprayer.

Fig. 142. Motor-driven wheelbarrow sprayer.

Fig. 143. Two types of hand-operated barrel sprayers.

EQUIPMENT FOR APPLICATION OF INSECTICIDES

is able to pump with one hand and spray with the other. An advantage over the compressed air sprayer and bucket pump is that a fairly high pressure may be uniformly maintained by continuous pumping. It should be borne in mind, however, that this is a very tiring operation if continued over a period of hours, for those who are not used to this kind of work.

Wheelbarrow Sprayer:- The wheelbarrow sprayer is designed to make possible the use of equipment too heavy and bulky to be carried by the operator. One may find a variety of designs, but the basic principle of the pump mechanism is similar to that of the barrel sprayer. The wheelbarrow sprayer may be hand-operated (fig. 141) or motor-driven (fig. 142), and may or may not have a compression chamber, although the latter feature is desirable. The capacity of the tank varies from 5 to 18 gallons in outfits now obtainable.

Barrel Sprayer:- This type of sprayer may be hand-operated (fig. 143) or motor-driven, and is suitable for the treatment of a small orchard for family use or a large garden of several acres. The tanks vary from 5 to 50 gallons and the pressure from 100 to 400 pounds, depending on the size of the pump and the power available for driving it. A good barrel sprayer should allow for easy access to both the upper and lower valves and to the plunger packing. The sprayer should also have a good agitator, preferably with steel or brass agitator blades.

Barrel sprayers are mounted in wheelbarrow types of carriages, on 2 wheels, and some are mounted on skids, to be dragged over the ground.

Traction-type Sprayer:- The first traction sprayer was built in Ohio in 1887 and this type of sprayer has been used to a limited extent to the present day. In this type of outfit 1-, 2-, or 3-cylinder pumps secure their power for operation from the movement of the wheels of the chassis and the connecting gears. Spray is discharged only while the outfit is in motion, so this type of sprayer is unsuitable for spraying of trees, for a stop must be made at each tree. It is most suitable for such row crops as grapes and berries, for cotton and tobacco and for vegetable crops such as celery, potatoes, tomatoes and melons.

In the traction sprayer the pump and tank are usually balanced on a bed bolted to the axis of a 2-wheeled chassis. Customarily the pump is mounted in front, resulting in the preponderance of weight being borne on the tongue or shafts when the tank is empty. Some traction sprayers have tanks of 100 to 200 gallons capacity, but many carry 50 gallon barrels. They may produce 250 to 300 pounds of air pressure. Some have 6 to 12 nozzles mounted on a wide boom, each discharging about a gallon per minute. In vegetable spraying usually at least 100 to 125 gallons of spray liquid is applied per acre.

For the best type of equipment, it should not be necessary to increase the rate of driving to increase discharge. In an ideal outfit, a set of gears which the operator could shift would make possible the regulation of the rate of delivery of spray material at all times. It is also desirable that a traction sprayer be able to develop maximum pressure almost immediately after the wheels start moving. This greater latitude with respect to pressure has usually, in practice, been provided by changing from traction-driven sprayers to outfits with small auxiliary engines.

Power Sprayers:- Most satisfactory for the commercial orchardist or vegetable grower are the power sprayers, so called because the spray pumps are driven by engines. The first power sprayers appeared simultaneously in California and Connecticut in 1894. These were powered by means of a steam engine, but about 1900 the first complete power sprayers operated by gasoline engines appeared on the market. The pressure regulator, a very important contribution, appeared in 1911, and the spray gun in 1914 (145). Power sprayers as used today may range in size from small, inexpensive, 1-cylinder outfits with a capacity of 3 or 4 gallons per minute to the large outfits with 400 to 600 gallon tanks and a 60-gallon-per-minute capacity such as are generally used in citrus spraying.

253

Some very effective small power sprayers which may be drawn by one man have been developed in recent years. Fig. 144 (above) shows such an outfit which was

designed for use around homes, estates, small farms, dairies and resorts. It has a 15-gallon steel tank, a jet agitator to agitate the spray mixture, a pressure control valve with a range of from 10 to 200 lbs. of pressure per square inch, and a twin-cylinder pump which can discharge 3 gallons per minute. The pump is powered by a 1-cylinder, air cooled gasoline engine. Equivalent performance can be obtained from a 1/2 H.P., single phase, 110 volt electric motor. The weight of the outfit, with gasoline engine, hose and gun, is 118 lbs. Fig. 144 (below) shows an "Estate" sprayer of greater capacity than the above outfit. The tank holds 50 gallons. The pump has a capacity of 4 gallons per minute and the maximum pressure is 400 lbs. per sq. in. The weight of the outfit is 510 lbs.

Among the larger power sprayers, a pressure of 1000 pounds may be developed, and the usual pressure in citrus spraying is from 500 to 600 pounds. In the spraying of pineapples in Hawaii, a pump with a capacity of 80 gallons per minute, utilizing 1000 lbs. pressure, is being used (fig. 145). The spray is distributed over a wide area by means of a 60 ft. boom with 160 nozzles.

Fig. 144. Above, a small power sprayer weighing 118 lbs. that can be drawn about by hand; below, an "Estate" sprayer with a capacity of 4 gallons per minute and which can develop a pressure of 400 lbs. Courtesy Bean Cutler Div. of Food Machinery Corp.

A constant pressure is maintained by means of the pressure regulator. Any required degree of agitation of the spray mixture can be maintained. Three or 4 leads of hose are used in citrus spraying, 2 for the ground crew and 1 or 2 on the tower, for the spraying of the upper part of the tree.

Fig. 145. A power sprayer, with a 60 ft. boom, spraying pineapples in Hawaii. Courtesy Hardie Mfg. Co.

EQUIPMENT FOR APPLICATION OF INSECTICIDES

Remarkable advances have been made in power sprayers in recent years, principally in the improvement of lubrication by enclosing working parts. This keeps out dust and makes possible a more nearly self-oiling system. New designing for easy accessibility of valves and plungers has greatly simplified the maintenance of the pumps and the use of new high grade metals has prolonged the life of working parts. The trend has gradually been toward greater capacity and higher pressure.

The Power Unit

On the small types of power sprayers the spray pump is operated by either a small gasoline motor or a small 1-cylinder stationary engine. The power for large spray pumps may be supplied by an engine, usually of 4 cylinders, installed on the spray outfit in the factory, or an automobile engine is sometimes installed by the dealer selling the spray equipment or even by the grower or spray operator. The most popular type of orchard sprayer at present is the complete outfit powered with an auxiliary engine and mounted on a truck chassis or a chassis that can be drawn by a tractor or by horses. In citrus areas, at least in California, the types of sprayers mounted on motor trucks are the most common, because spraying is done almost exclusively by commercial operators, cooperatives, packing houses, or by large orchard companies, which can afford to have expensive equipment. Often considerable distances need to be traveled, so truck-mounted spray outfits are the most practical. The 500-gallon spray tank has recently come into quite general use in citrus spraying, particularly when a "service rig" is used to haul the water and oil. These are usually mounted on one and one-half ton trucks equipped with reduction gears and large or dual tires.

The engine should be able to operate the pump at a maximum output without being loaded to more than 75% of the rated horsepower of the engine. In some sprayers the cooling system consists of pipe coils installed in the spray tank. This obviates the necessity for radiators or fans, and permits a more complete enclosure of the engine from dust and spray. The pipe coils in the spray tank may interfere with agitation of the spray mixture, but not necessarily so, for effective agitation can readily be provided by utilizing the proper r.p.m. of agitator shaft and the proper number and shape of agitator blades.

Fig. 146. Spray rig operated by a power take-off from a tractor.

The initial cost of spray equipment can be reduced by allowing the spray pump to be operated by a tractor with a direct drive assembly shaft (fig. 146). This is especially advantageous if a tractor is to be used anyway for pulling the spray outfit. Each time the tractor stops and the clutch is disengaged to shift to neutral, the power for the operation of the spray pump is momentarily cut off. In addition the r.p.m. of the power take-off shaft is reduced when the tractor is moving from tree to tree. To avoid this disadvantage, some manufacturers have provided a type of sprayer with a transmission having 2 gear ratios so that with any given power-take-off speed, a choice of 2 speeds is available for the pump. Possibly the main objection to the operation of a sprayer from a power take-off is that the short turns normally required in orchard spraying cause severe strains and even breakage of universal joints. Also the length of the entire assembly may make turning difficult at the end of each row, especially if irrigation ditches or standpipes are present, or if the trees are large and the rows are close together. However, maneuverability of the power-take-off units is considerably enhanced by the fact that they are generally mounted on a 2-wheel trailer-type chassis.

The Spray Pump

As shown in figs. 147, 148, and 149, spray pumps vary considerably in design and construction. However, certain features commonly found in spray pumps should be understood.

The pumps are of the displacement, single-acting, reciprocating type.

The Cylinders

The cylinders (fig. 150) develop the pressure. Nowadays these are usually coated with acid-resistant porcelain to retard corrosion and abrasion. If abrasion occurs, new plunger packings cannot prevent leakage.

The spray liquid is forced through the cylinders by plungers or pistons (fig. 150). These fit tightly inside the cylinder by means of packing. For the plunger to be effective, provision must be made for easy tightening or renewal of the packing. A plunger may be fitted with an expanding type of packing, known as the plunger cup, which moves with the plunger, as shown in fig. 150. In other types of pumps the plunger operates through stationary packing which serves as the cylinder wall of the displacement chamber. This type of packing may be so tight as to cause scoring. A slight leakage of spray liquid indicates that the packing is not too tight.

The most common mechanisms for providing the reciprocating motion of displacement plungers are the following: (1) crankshaft and connecting rods (fig. 147), (2) eccentric and connecting rods (fig. 148), and (3) the Scotch-yoke assembly (fig. 149). Quoting from French (349), "The only difference between the crankshaft and eccentric is in the diameter of crank pins: with a crankshaft the diameter of the pin is less than the throw; with an eccentric the pin diameter is greater than the throw. Since the Scotch-yoke assembly permits the operation of two opposed plungers from one reciprocating mechanism, sprayers using this system are built with either two or four cylinders. Whether the cylinder should be vertical or horizontal is largely controversial."

Capacity of a Pump

Manufacturers rate their pumps as to maximum capacity in gallons per minute at a given pressure. The capacity, in the case of reciprocating pumps, depends upon (1) the number of cylinders, (2) the diameter of the cylinders, (3) the length of stroke of the plunger, (4) the number of strokes of the plunger per unit of time, and (5) volumetric efficiency of the pump. Volumetric efficiency is determined by dividing the volume discharged by the plunger displacement. It should be 90 per cent or higher. This degree of efficiency requires that there be very little leakage past the valves or plunger packings.

Capacity should not be confused with pressure. Low capacity pumps can be built to deliver high pressure and conversely high capacity pumps can be built for low pressures. In citrus spraying both high pressure and high capacity are necessary for good penetration of the tree and thorough wetting. The most powerful pumps on modern spray rigs such as are generally used in citrus spraying have a maximum capacity of 50 to 60 gallons per minute and can generate a maximum pressure of 800 to 1000 lbs. per square inch. Under the usual conditions of commercial operation, the pumps are required to discharge about 30 gallons of spray liquid per minute at a pressure of from 500 to 600 lbs. A 600-gallon tank is sprayed out in about 20 minutes.

The Valves

In modern power sprayers the valves are usually of the ball type. The valves and valve seats direct the flow of spray. Grains of sand may become lodged under the valve ball or an accretion of certain types of solids may build up on the surface of the ball, or the ball may become worn. All these factors may result in a reduction of pressure.

Fig. 147 (at left). Vertically built sprayer pumps with crankshaft and connecting rods. Courtesy Hardie Manufacturing Company.

Fig. 148 (at right). A horizontally built sprayer pump with eccentric and connecting rods. Courtesy Bean Cutler Div. of Food Machinery Corp.

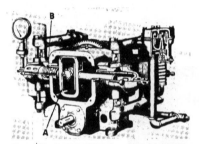

Fig. 149. Scotch-yoke assembly. A, drive; B, plunger and outside-type packing. Courtesy Friend Manufacturing Company.

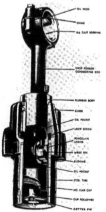 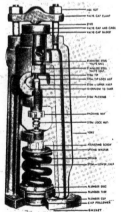

Fig. 150. Cylinder and piston of a sprayer pump.

Fig. 151. Pressure regulator of a modern spray rig.

The Air Chamber

The <u>air chamber</u> equalizes the pressure and removes excessive strain on the pump. Its volume depends on the capacity of the spray pump. At the bottom of the air chamber is located the discharge opening which leads to the nozzles. The spray liquid enters the bottom of the chamber and compresses the air which acts as a cushion in maintaining an even flood of spray liquid to the nozzle. The <u>pressure gauge</u> is mounted on the air chamber (fig. 147).

The Pressure Regulator

This device (fig. 151) performs the following functions: (1) it is a safety device; (2) it makes possible a uniform pressure at the spray nozzle, being aided in this respect by the air chamber; and (3) it allows the pump to operate at a greatly reduced load when the spray guns are shut off. Near the bottom of the pressure regulator (see fig. 151) is located either a spring-loaded diaphragm or a plunger which will lift a valve ball and permit excess liquid to by-pass to the supply tank if the pressure of the spray liquid exceeds the resistance of the compression ring. By <u>increasing</u> the resistance offered by the spring, the pressure of the liquid exerted against the valve ball which permits excess liquid to flow back to the tank will be reduced. Consequently less liquid will be discharged through the overflow pipe and more through the spray guns. Thus to increase pressure you increase the tension on the spring located on the pressure regulator. This is accomplished by means of an adjusting screw immediately above the spring.

If the needle on the pressure gauge is unsteady, this indicates the pressure regulator is not operating effectively and that possibly one of the valve balls is stuck.

Suction Hoses for Filling Tank

If an overhead pipe or hydrant of sufficient capacity for supplying water is not available, the water is drawn into the spray tank by suction provided by the spray rig pump. Permanently mounted tank refillers are favored in the citrus districts of California. These are the injector or siphon type (fig. 152) and are built for capacities of 40 to 120 gallons per minute, depending on the size and speed of the pump with which they are to be used. The suction hose, leading to a weir box or canal, is pushed onto the intake pipe and the turning of a valve (fig. 152) will start the flow of water to the supply tank. Fillers working on the same principle are used on "service rigs" or "nurse rigs" when these are used to haul the water to the spray rig. Quickly detachable tank fillers are also available which can be left at the source of water. These are hooked onto the metal flange of the filler hole (fig. 153).

Fig. 152. Permanently mounted tank refiller of injector or siphon type.

Fig. 153. Detachable tank filler.

The Filter

The spray mixture coming from the tank must pass through the filters (fig. 154) before reaching the pumps so that particles of gravel or insoluble spray materials likely to score the pump cylinder walls will not gain access to the pumps. The filter consists of a brass wire screen strainer with a large straining area. It

should not clog quickly and should be easy to clean by washing or brushing. The covers of the screens are held in
place by a clamp and are easily removed by loosening a set
screw. The screens occasionally become clogged, especially
if a large quantity of solid material of large particle
size is used in the spray. They must then be removed and
cleaned.

The Hose:- Spraying at the high pressures used in the
application of oil spray to citrus trees places tremendous
demands on the spray hose. Likewise the oil is injurious
to rubber. Fig. 155 shows the concentric sections of material making up one of the common brands of spray hose.
In addition to 3 plies of rubber there are 3 plies of heavy
wrapped duck and 2 braided plies of strong hawser yarn in
this hose. Hoses of this type will withstand pressures of
800 pounds or more. The usual length of hose in citrus
spraying is 65 feet. Such a hose will have a 1/2 inch inside diameter. If a greater length of hose is required a
hose of 3/4 inch inside diameter should be used to avoid
excessive loss in pressure. Loss in pressure is caused by
the friction of the liquid against the inside surface of
the hose. It varies with the square of the velocity of
flow, the length of the hose, and the roughness of the inside surface of the hose. Fig. 156 shows average pressure
losses for various sizes of hose with different volumes of
liquid flowing, as compiled by French (349). In a 10-foot

Fig. 154. Three types
of filters for spray
rigs.

bamboo rod there is a great loss of pressure caused by friction, but there is
comparatively little loss in the short "guns" which are now almost universally
used. Pressure may also be reduced by using too small fittings.

Fig. 155. Showing the concentric sections
of material in a spray hose.

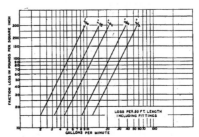

Fig. 156. Friction losses in different sizes of spray
hose at various rates of flow. After French (348).

The Gun and Nozzle

A type of "short spray gun" commonly used in citrus spraying in California is shown in fig. 157. In this gun the nozzle is of the eddy-chamber type (fig. 158).

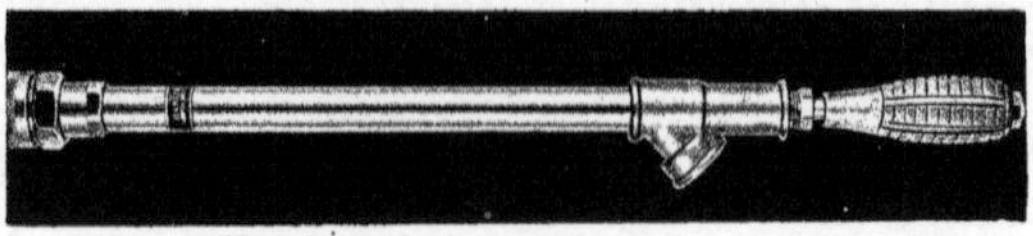

Fig. 157. One of the types of short spray gun.

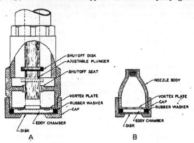

Fig. 158. Spray nozzles of the eddy-chamber type. A, variable-depth type; B, fixed-depth type. After French (349).

In these nozzles the liquid flows through a "vortex plate" or "whirl plate" with spiral or tangentially arranged channels which whirl the liquid in the eddy chamber. This tends to break up the liquid before it leaves the nozzle orifice and also directs the liquid at an angle which will cause it to spread out into a cone after leaving the orifice. A center hole in the vortex plate results in a solid cone. An adjustable plunger (fig. 158) varies the depth of the chamber and this in turn changes the angle of the spray cone. A shallow chamber will result in a wide angle of spray, for a large part of the liquid will be forced through the vortex plate and this liquid leaves the orifice at an angle. If the eddy-chamber depth is increased, more of a jet type of stream will be emitted from the orifice, that is, the liquid will not "fan out" into as wide a cone.

The orifice is located on a circular plate of metal known as the disk. The orifice is of various sizes, but the sizes most commonly used in citrus spraying are 8/64 or 9/64 inch in diameter (No. 8 or No. 9). The disk orifice is subject to enlargement even though the disk is in some cases made of stainless steel. The disk may be easily replaced by unscrewing the cap of the nozzle (fig. 159).

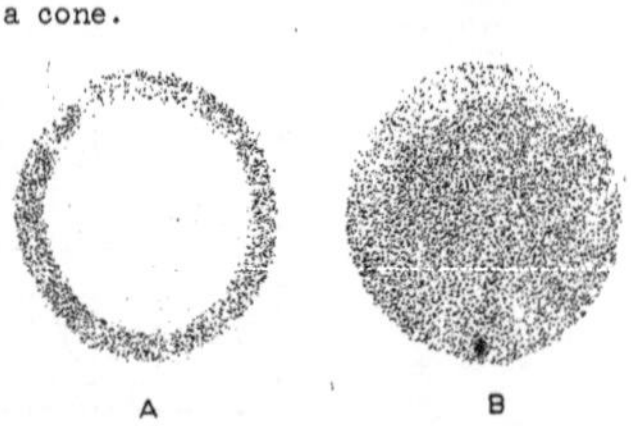

Cone sprays produce 2 types of spray pattern, a ring or a solid pattern (fig. 159). As stated above, the solid pattern is obtained when the whirl plate, besides its

Fig. 159. Two types of spray pattern: A, ring; B, solid. After French (349).

vortex openings, has a central orifice directly in line with the spray-disk orifice and of about the same diameter. Through this central orifice sufficient liquid is discharged straight ahead to fill the center of what would otherwise be a cone.

EQUIPMENT FOR APPLICATION OF INSECTICIDES

The rate of discharge of spray liquid with a short gun, with different sizes of disk orifice and different pressures, is given in table 9.

Table 9

Rate of discharge of a "short gun."*

Pressure in lbs. per square inch	Gallons per minute discharge using disks with the given sizes of orifice						
	4/64 inch	5/64 inch	6/64 inch	7/64 inch	8/64 inch	9/64 inch	10/64 inch
300	1.50	2.25	3.00	4.50	5.50	6.00	6.75
400	1.60	2.57	3.60	5.15	6.17	8.02	9.00
500	1.75	2.85	4.01	5.63	6.67	8.38	10.50
600	----	3.17	4.43	6.28	7.55	9.76	----

*Data from catalog of Hardie Manufacturing Company.

Broom Guns

Broom guns, with multiple nozzles, are capable of delivering a great volume of liquid in a broad spray stream and with remarkable carrying power. In fig. 160, 4 such guns are shown with 3, 4, 6, and 8 nozzles per gun. The nozzles come equipped with No. 4 (4/64") disks. The discharge in gallons per minute of these guns with No. 4 disks and with various pressures is as follows:

Pounds pressure	3 nozzles	4 nozzles	6 nozzles	8 nozzles
300	4.1	5.5	8.2	11.0
400	4.7	6.3	9.5	12.6
500	5.4	7.2	10.8	14.3
600	5.9	7.9	11.9	15.9

With No. 5 (5/64") disks the capacity of the guns is increased by 1/3 and with No. 3 (3/64") disks the capacity is decreased by 1/3.

Fig. 160. Broom gun heads with 3, 4, 6 and 8 nozzles.

When moderately good coverage of the tree suffices, broom guns can be advantageously used. Usually the operator stands on the spray rig (fig. 161) and sprays with an up and down movement of the guns as the rig is driven slowly down the row without stopping. The speed of the spraying operation is thus greatly increased.

If thorough spray coverage is necessary, however, the broom gun is not so well adapted. For the most thorough coverage, the spray stream, regardless of its volume, must be directed from inside and out, and must be directed from all possible angles. The broom guns are so long and heavy that they could not be handled with the dexterity necessary for this type of spraying, and even if they could be manufactured to be suitable for such purposes, too much spray material would be required per tree. The broom guns have thus adapted themselves best to the spraying of deciduous trees and to certain types of citrus tree spraying in which only an "outside coverage" is necessary, as

Fig. 161. Broom guns in action.

261

in thrips, aphid and mite control or in the application of sprays for minor element deficiency correction.

Pressure

French (349) made an experiment with a short gun adjusted for both close and long-range spraying with pressures varying from 200 to 1000 pounds per square inch. He found that with the gun adjusted for close-range "open-cone" spraying the distance the spray droplets carried increased rapidly with increased pressures up to 600 pounds, but very little with further increase in pressure. With the gun adjusted for long-range "solid-cone" spraying, the maximum quantity of spray carries much farther than with the wide-angle cone sprays. The carrying distance increased with pressure in much the same manner as with the wide-angle cone spray, but at pressures above 800 pounds the carrying distance actually decreased. French explained this decrease in carrying distance at the highest pressures by the fact that the pressure finally caused the entire stream to be broken into small droplets which lacked sufficient momentum to carry. He showed that a fourfold increase in pressure reduces droplet diameters one half. To secure additional carrying distance after an increase in pressure to 800 pounds with a solid-cone spray one would have to change to a larger disk orifice to increase the rate of discharge.

The use of high pressure is advantageous, for with higher pressures the liquid is broken into many more droplets with the result that a more uniform coverage is obtained at a more rapid rate. The tendency in recent years has been toward increasingly higher pressures. In this same connection it should be mentioned that, contrary to popular belief, French (349) found that with a given nozzle and a constant pressure, decreasing the size of disk orifice does not decrease the size of the spray droplets. This is accomplished only by increasing pressure. Pressure as high as 600 pounds is being used in citrus spraying but will cause injury to some varieties of deciduous fruits.

The Spray Tank

Spray rigs have tanks of various shapes and ranging from 100 to 600 gallons capacity. They may be of wood or steel construction, although the latter now predominate. Wooden tanks, when not in use, must be kept full of water to prevent them from drying out and leaking. Steel tanks should be washed out after use and spray liquid should not be left in them when they are not in use. At the end of the spray season they should be cleaned and bare spots should be wire-brushed and repainted with a suitable metal paint. The proper type of paint for this purpose can be supplied by the sprayer manufacturer.

Agitation

The problem of agitation of the spray mixture in the spray tank has been rather thoroughly investigated (818, 94, 349). Agitation insures a uniform concentration of emulsified or suspended material in the water from the time the spraying of a tank of spray mixture is begun until the tank is empty. In the case of oil spray, if insufficient agitation is provided, some of the oil will rise to the top of the tank and float on the water. This results in a lowered concentration of oil when the spraying of the tank of spray material is begun, for the spray liquid is withdrawn from the tank from the bottom. When a tank of such an imperfectly mixed dilute oil emulsion is nearly empty, the concentrated oil emulsion or free oil floating on the water will be in a position to be withdrawn and the spray will then have an abnormally high concentration of oil, resulting in injury to the last few trees to be sprayed. This is especially likely to result when broken or improperly prepared proprietary emulsions are used or when emulsive oils are not properly mixed.

Suspended solids may settle on the bottom of the tank if the spray mixture is not properly mixed.

The common means of agitation for power sprayers are 2 or more propellers or paddles mounted on the "agitator shaft" which runs lengthwise of the tank and is

situated sufficiently close to the
bottom so that the paddles reach to
within 1/2 inch of the bottom of
the tank (fig. 162). The agitator
shaft is driven by chain and sprock-
ets or by gears from the spray pump.
Oil sprays, particularly those of
the tank mix type, require greater
agitation than the majority of the
solid suspensions. Flat blades are
more efficient for agitation than
propellers (see fig. 95). The
greater the capacity of the tank
the greater the number of paddles
required. Ordinarily an extra

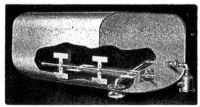

Fig. 162. Spray tank showing location of 4
flat-type blades on agitator shaft.

paddle is added for every 100 gallons increase in tank capacity. Power consumption for agitation in tanks of different size and shape has been discussed by
French (349). In citrus spraying the agitator shaft usually turns at the rate of
125 r.p.m. when oil emulsions and emulsive spray oils are used.

Towers or Platforms

In Chapter 12 it was stated that for the effective use of oil spray against
scale insects, every portion of the tree must be hit by the spray stream from
every possible angle. The ground crew can spray the tree from all angles except
from above. It devolves upon the man in the tower to spray the upper half of the
tree surface from above. This is especially important in the case of citrus trees,
for they often have a more or less flattened or at least a broadly rounded top,
and it is not possible for the ground crew to effectively spray over the "shoulder"
of the tree and thoroughly wet the top.

In general, in deciduous spraying a tower or platform is built onto the top
of the supply tank, or attached in various other ways to the spray rig, so that
the operator's head will be about as high as the top of the average mature tree.
Various types of solid shields covering a 3-legged tower will prevent damage to
the branches and fruit when the trees are closely planted. Towers or platforms
should have a railing about 40 inches high, a nonskid platform surface and toe-
boards around the edge of the platform in case the sprayman should slip. Towers
may be hinged so that they can be lowered when not in use.

In citrus spraying, many rigs are equipped with hydraulic towers operated by
the hydraulic pressure generated by the sprayer pump. The masts consist of several
sections of telescoping steel tubing. The sprayman in the "crow's-nest" or "catwalk" can control the height of the tower by opening or closing a valve attached
to the protective rail. Fully extended hydraulic towers generally lift the sprayman about 30 feet above the ground. The "crow's-nest" is sufficiently large for
2 men, and "catwalks" (fig. 3) make it possible for a sprayman to walk a distance
of 7 feet on either side of the mast so as to more thoroughly spray the half of
the tree away from the center down which the spray rig is driven. When not in use,
the masts can be lowered and the catwalks can be turned in a position parallel
with the spray rig and the "catwalk" can be hinged down to further reduce over-all
height. A cable-operated extension tower has been constructed (977).

How a Power Sprayer Operates

Using fig. 163 as a guide, the course of the spray liquid from the supply
tank to the spray gun will now be traced. In the present instance we will be dealing with a sprayer in which the reciprocating motion of the displacement plunger
(p) is obtained by means of an eccentric (e) and connecting rod, and with each
plunger moving in a horizontal plane. As the plunger is withdrawn, the liquid
flows past the suction valve ball (sv) and fills the portion of the cylinder
vacated by the withdrawing piston. It will be noted that all the liquid leaves
the supply tank at the bottom of the tank and must pass through a filter which

263

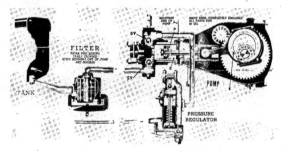

Fig. 163. Diagram of the mechanics of a spray pump and the flow of the
spray liquid from the spray tank to the spray hose. C, cylinder of pump;
DV, discharge valve; E, connecting rod drive; P, movable plunger-cap
packings; SV, suction valve. Adapted from illustrations provided by Bean
Cutler Div. of Food Machinery Corp.

takes out all material not in a very finely divided state. The liquid fills the
cylinder to the point indicated by the line in fig. 163. As the plunger is pushed
forward, the suction valve (SV) closes, for the liquid under pressure forces the
valve ball back into its seat. On the other hand the discharge valve ball (DV) is
pushed up and liquid continues to flow through the valve until the suction caused
by the receding plunger draws this valve ball back into its seat. The liquid
which is forced forward by the plunger flows to the lower valve of the pressure
regulator. If the spray gun nozzles are open, the pressure of the liquid lifts
the valve ball and the liquid flows through to the spray gun. The liquid builds
up a pressure against the diaphragm which is held down by the compression spring
in the pressure regulator, the operation of which was explained on page 258.

What to Do if Pressure Drops

A loss or variation of pressure may come from several causes and the pressure
regulator should not be adjusted until the following points are checked:[1]

1. Clogged suction screen in both suction strainer and suction well (filter).
These must be clean, or liquid cannot enter the pump.

2. Loose connection or missing gaskets in suction line, or partially opened
suction valve, broken or leaky suction hose, or suction pipe. Drain cocks not
closed.

3. Leaking outlet valves. Leaking tank filler valve.

4. Worn gun or nozzle disc.

5. Too slow pump speed. Often increasing pump speed is all that is necessary.

6. Suction valves temporarily stuck. Remove cap and loosen with screw driver.

7. Worn out valve seats or valve balls. If ball when placed on seat and held
to light does not shut off all light, replace with new seat or ball as indicated by
this inspection.[2]

[1]Courtesy Hardie Mfg. Co.

[2]Tools for the removal of parts, such as valve balls and valve seats may be obtained from
sprayer manufacturers. Also instructions for the care of equipment and replacements of worn parts
may be obtained from these companies. (Author's note).

8. Worn out regulator seat permitting too great an overflow. To replace see No. 7. Improperly adjusted regulator.

9. Worn plunger cups. Worn out cups will leak and give immediate evidence of their condition.

Boom Sprayer

The "boom sprayers", like other labor saving devices such as the service rigs and tent pullers, were born of necessity during the period of extreme labor shortage immediately prior to and during World War II. The first boom was built and attached to a spray rig by one of the sprayer manufacturers in 1939, and boom sprayers received their first commercial testing in 1940. Their subsequent development took place largely at the hands of commercial spray operators and was greatly stimulated by the rapidly increasing labor shortage following the entry of the United States into the war.

The first spray booms consisted of a single upright pipe or "boom" to which a varying number of nozzles were fitted, depending upon the capacity of the spray pump, and spraying was done in one direction only (fig. 164, above). Later booms were built to spray in both directions. These could be moved from side to side by hand, and a tower was available from which a man could supplement the mechanical application. Fig. 164 (below) shows a boom sprayer with the nozzles arranged in a semicircle.

The first mechanically oscillating spray boom to operate in California citrus orchards was built in the spring of 1940 by D. A. Newcomb, the pest control manager for the Corona Foothill Lemon Company. This marked the beginning of a wide variety of devices deriving their motion from eccentrics powered by independently mounted motors or from the pump shaft. After a series of modifications; Newcomb's sprayer consisted of a trailer-mounted 500-gallon tank and a pump delivery of 59 to 61 gallons of spray liquid per minute at 500 pounds pressure. It is powered with a 35 horse-

Fig. 164. Above; boom sprayer with stationary boom spraying in one direction; below, boom sprayer with nozzles arranged in a semicircle. Courtesy Hardie Manufacturing Company.

power motor. The boom itself is stationary but has 6 movable short orchard spray guns which oscillate vertically 60 times per minute. Mr. Newcomb believes the vertical movement is the most efficient, both in manual spraying with a single gun and for the boom sprayers. The motive power for the oscillation of the gun is derived from a 1 horsepower

motor housed behind the supply tank. Number 8 disks are used in the nozzles and these are set 2.5 feet apart and at angles to one another of about 60 degrees. There are 8 nozzles, one reaching over the tree and spraying down at an angle and the lowest nozzle delivering a backward stream for branches reaching toward the center of the row in close-set lemon orchards. This rig is driven at 2/3 miles per hour and fifty 500-gallon tanks are sprayed out per day with, of course, the aid of a nurse rig. (980)

Another boom rig used at the Corona Foothill Lemon Company has 2 booms, each supporting 7 guns. With this outfit the spray can be directed to 2 rows at one time, or all the spray streams can be directed toward 1 row of trees to deliver a particularly drenching spray, as in fig. 165. Fig. 166 shows a highly effective boom sprayer in which the vertical motion of the guns is hydraulically operated, but the side to side motion is manually operated. The spray stream may thus be directed back at a tree just passed and forward at a tree which is being approached. Thus one of the weaknesses of purely mechanical operation is partially eliminated.

265

Fig. 165 (at left). A boom sprayer used at the Corona Foothill Lemon Company. Courtesy R. S. Woglum.

Fig. 166 (at right). A boom sprayer with guns obtaining their motion from both hydraulic and manual operation. Courtesy R. S. Woglum.

Fig. 167. A boom sprayer equipped to apply the spray in a rotary motion. Courtesy R. S. Woglum.

The boom rig does not give as good results against red scale as obtained from the conventional type of red scale spraying with a ground crew, but a saving in cost of 30 to 50% is effected and the work is much more rapidly completed. The boom spraying is advantageous even against red scale, as a supplemental treatment. It is especially adaptable as a treatment supplemental to fumigation, for the latter treatment is especially effective in the interior of the tree while the boom sprayer is most effective on the outer surface, just where most of the scales surviving a fumigation are located. Likewise the boom sprayer is effective against citrus red mite, which is not controlled by fumigation. In general it may be said that in citrus pest control boom sprayers give satisfactory results against pests and diseases demanding only an outside coverage, such as thrips, citrus red mite, bud mites, aphids, leafhoppers, orangeworms, brown rot and mottle leaf.

The majority of spray booms work the nozzles with a vertical movement, but some are equipped to apply the spray in a rotary motion. On one patented boom, known as the Rotary Orchard Spray Boom, ten guns are set in an upright frame in such a way that all 10 guns can apply the spray in one direction (fig. 167, above) when spraying for scales and red mites, or 5 guns can be made to spray in one direction while the others spray in the opposite direction (fig. 167, below)

266

when spraying for thrips or aphids, or when applying minor elements. For large trees the spraying of the tops of the trees can be supplemented by a man spraying from a tower (fig. 167, above). Tower spraying can, in fact, be made supplemental to boom spraying with any type of boom sprayer.

In the above outfit the motive power for the operation of the boom is derived from the pump shaft through a gear box. The pressure at the gauge registers 750 to 900 pounds. The nozzles make approximately 100 revolutions per minute. (988)

It is likely that the boom sprayers have reached the limit of their capacity in the direction of thoroughness of application in citrus spraying, yet they have not attained the degree of coverage considered necessary for red scale control. Nevertheless they will no doubt continue to play an important rôle in the citrus pest control program in the application of supplemental or rapid emergency treatments for red scale, for the control of a number of other pests and diseases, as listed above, and in the application of minor element sprays for the cure of deficiency diseases. Automatic spraying from 25-foot towers is being successfully employed in the control of walnut pests (fig. 168).

Fig. 168. Automatic spraying of walnut trees from a 25-foot tower. Photograph by Mr. C.C. Anderson.

SERVICE RIGS (NURSE RIGS)

Advantages:- Service rigs, also called "nurse rigs" have greatly increased spraying efficiency. It is because of this increased efficiency that the cost of application of oil spray to citrus orchards has increased only about a third (from the pre-war price of 3/4 cent to the present price of 1 cent per gallon), while costs of labor and equipment have at least doubled.

The service rigs haul the water and spray material from the "commissary", located at the source of the water, to the spray rig. Formerly the spray rigs were driven to the commissary. The use of service rigs makes for greater specialization of equipment: lighter, less expensive equipment is engaged full time in a task formerly engaging much of the time of a much more expensive piece of equipment. Likewise the driver of the service rig is now employed at a task which formerly engaged much of the time of a 4- or 5-man spray crew. The acreage per day sprayed by the normal 4- or 5-man crew is usually increased 50 to 75% by the employment of one additional man with the service rig.

The service rigs, being lighter and smaller than the spray rigs, pack the soil less and have the additional advantage of tearing off fewer twigs and less fruit than the spray rigs as they pass in and out of the orchard. The use of "catwalk" towers is on the increase, and with these it is frequently not practical to be continually driving in and out of the orchard for refilling the spray tank.

Design and Operation:- The design and operation of service rigs have been discussed in detail (1002). A service rig consists of a tank and a pump mounted on a light truck. Self priming centrifugal, gear rotary and van rotary pumps are used, and these are usually mounted in front of the supply tank (fig. 169) and are operated by means of a power take-off from the truck transmission or by a small air-cooled auxiliary motor. If the pump is mounted at the rear of the supply tank, an auxiliary motor is used. As a source of power, the power take-off from the truck is usually preferred because of its simplicity and because it eliminates an additional motor as a possible source of trouble. Some spray operators desire agitation of the oil and water in the service tank while hauling the spray material, so

267

as to obviate the necessity for stopping spraying operations while the spray tank is being filled. Such operators may mount an auxiliary motor on the service rig (fig. 170), for it is not practical to provide agitation in the service rig from a power take-off while the rig is moving.

Fig. 169. Service rig drawing water from a canal. Courtesy R. S. Woglum.

Fig. 170. Auxiliary motor mounted on some service rigs to provide agitation for spray mixtures while driving to the spray rig. Courtesy R. S. Woglum.

One service rig is generally required for each spray rig, and when boom sprayers are used, two service rigs are required to supply one spray rig. Many spray rigs are now of 500-gallon capacity so tanks of similar capacity are commonly used on service rigs, although some operators prefer keeping the size of the service rig tank down to a 400-gallon capacity if the spray material is to be mixed in the service rig. This makes for better mixing of the spray mixture, lighter equipment, and less packing of soil, while on the other hand it entails no loss in time, for it makes no difference whether the spray rig tank is ever filled to capacity or not. However, with a 400-gallon service rig, 25% more trips per day need to be made between the commissary and the spray tank than with a service rig of 500-gallon capacity.

Some service rigs carry the oil and water mixed in the tank while others carry only water in the main tank and the oil in a small auxiliary tank which seldom needs to be of more than 10-gallon capacity. The service rigs that carry both oil and water in the same tank may or may not provide continuous agitation for the spray mixture while moving from the commissary to the spray rig.

268

EQUIPMENT FOR APPLICATION OF INSECTICIDES

The outfits which carry oil and water in separate tanks may be provided with a spray injector. When supplying the spray tank, the injector may pull oil and water through a single line (fig. 171, above). Two lines are sometimes used, however, the injector drawing oil and some water in one line while the service rig pumps water directly into the spray tank (fig. 171, below). Other types of service rigs carrying oil and water separately are not provided with injectors. The oil and water are pumped by the service pump into the spray rig tank or, occasionally, the oil is poured into the tank by hand.

Tests were made with spray rigs serviced by all four types of service rigs described above. No difference was found in the per cent of oil in the spray at the beginning of the spraying of a tank and at the end, regardless of how the oil and water were boosted into the spray rig. In all cases, however, when the oil and water were hauled in separate tanks, all the oil was boosted into the spray tank before more than a small part of the water had been transferred, thus insuring an adequate mixture of oil and water by the time the spray tanks were filled and spraying was resumed. (1002)

STATIONARY SPRAY PLANTS

Sometimes the mixing and pumping of the spray mixture is done at some central point in the orchard and the spray liquid is forced through pipe lines, usually underground, to hydrants on these pipe lines located at convenient places. From each hydrant a block of trees can be sprayed. This system of spraying is especially advantageous when the ground is wet and motorized equipment cannot be drawn through the orchard; also in the spraying of steep hillsides. On large acreages the original cost of the equipment and installation is about the same as for portable equipment, but after the initial outlay, spraying costs are greatly reduced.

It is advantageous to install a pump and supply tank specially designed for the stationary spray plant. A typical example is shown in fig. 172. In the example shown, the electric motor furnishes 3 H.P. for driving an agitator in a 1200 gallon tank. The same outfit may be equipped with a gasoline engine drive.

Fig. 171. Service rigs boosting water and spray oil to a spray rig through a single hose (above) and through separate hoses (below). Courtesy C.F.G.E. Pest Cont. Bur.

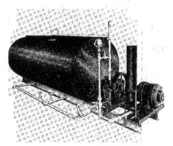

Fig. 172. A stationary spray pump.

The question is often asked as to the possibility of separation of the suspended or emulsified material in the pipelines. According to O. C. French, formerly Agricultural Engineer in the College of Agriculture of the University of California, if there is a two-foot flow per second in the pipelines, there is enough surging and turbulence to keep both solids and oil in a uniform suspension. Oil emulsions, including tank mix, as well as lead arsenate, wettable DDT powder, and sulfur preparations have been successfully used in stationary spray systems in many orchards in central California. A. D. Borden[1] determined in his early investigations of the tank-mix method of using oil spray, that the per cent of oil in the liquid taken from the spray gun did not change even after the liquid had stood in the pipelines for periods up to an hour.

Stationary spray plants have not been used in citrus spraying, but there is apparently no reason why they should not be, and in some circumstances, particularly when hilly land is involved, they could probably be used to good advantage.

DUSTING EQUIPMENT

As a general rule, if dusting equipment is available and it is known that the pest in question can be controlled by dusting, this method of treatment is preferred, for it has many advantages. These advantages were discussed by Mason (594), and may be summarized as follows:

1. Perhaps the greatest advantage in dusting is the great speed with which a treatment can be applied. One hundred acres of citrus can be dusted in one night with a modern high-powered multivane fish-tail type of duster. This is a great advantage when a large acreage must be treated within a relatively limited period in order to obtain an optimum timing of the treatment.

2. Usually fewer men are required to operate a duster than are required to operate an orchard sprayer and this, coupled with the much greater acreage treated per day, result in a great saving of labor. This is often an important consideration, especially in periods of labor shortage or in agricultural areas near large industrial districts.

3. Usually as much tree area can be covered with 2 lbs. of dust as with 10 gallons (80 lbs.) of spray. The hauling of water involves a large proportion of the labor and equipment in spraying, and the fact that this is not necessary in dusting constitutes one of its important advantages. Materials for a half a day (or night) of dusting can usually be carried along on the dusting equipment or be placed throughout the orchard, vineyard, or field in convenient locations. When water is not readily available or must be hauled over long distances, this particular advantage of dusting becomes of special importance.

4. Since less motive power is required for the hauling of a duster, and since treatment by dusting requires less time than spraying, tractors or horses which would be required in a spraying operation are released for other work.

5. Dusting equipment is less expensive than spraying equipment.

6. Dusting usually subjects the treated crop to less hazard than spraying. For example, sulfur dust is less hazardous than lime sulfur spray yet often accomplishes the same results.

7. Dusting equipment is lighter than spraying equipment, and since no water needs to be carried, the difference in weight of the two types of equipment when loaded with the spray material is especially great. Light equipment requires less motive power for hauling and does not pack down the soil as much as heavy equipment.

8. Dusting can be done effectively when trees are wet with dew or rain, while spraying is usually discontinued under such circumstances, although often data to show why spraying should be discontinued are lacking.

9. In fields, vineyards and orchards of small trees, considerable acreages can be treated with knapsack dusters or other hand equipment, while comparable equipment in the line of sprayers is much less adapted to large scale operations.

---

[1] The information pertaining to stability of emulsions and suspensions in stationary spray systems was supplied by Mr. A. D. Borden in correspondence of May 6, 1948.

EQUIPMENT FOR APPLICATION OF INSECTICIDES

10. In the treatment of certain low-lying crops, a dust can be applied to the under sides of foliage which cannot be reached by spray.

The disadvantages of dusting may be summarized as follows:

1. With a given insecticide, the effectiveness of a spray is nearly always greater than that of a dust, presuming that both are properly applied.

2. In dusting, the sides of the leaves receiving the direct blast of the dust are heavily covered while the opposite sides of the same leaves have little dust.

3. Often the greater cost of material in dusting is not offset by the saving in labor.

4. If a grower must invest in spraying equipment for his most difficult problems and for dormant spraying it may pay him to do all his pest control with this equipment.

5. Dusts are readily removed by rain or wind.

6. In many cases dusting can be done only at night or early in the morning when the air drift is negligible.

TYPES OF DUSTERS

HAND DUSTERS

Shaker Type:- As in the case of sprayers, dusters should suit the needs of the users, and very simple and inexpensive equipment may be satisfactory for occasional limited home garden use. In fact, in the absence of any manufactured equipment, dust can be applied to low growing plants by shaking a dust-laden cloth over the plants and allowing the dust to settle on them. Better yet, one may place dust inside of burlap or cheese-cloth sacks and shake it out over the plants. An obvious modification of such a device would be a can the bottom of which is perforated with holes. Dust may be placed in such a can and shaken over the plants.

Telescope Type:- The telescope type hand duster comprises the package in which the dust is sold. It consists of one cardboard cylinder within another. By pulling the 2 tubes apart, air is sucked through a valve, and on compression the air passes through a half-inch cardboard pipe, sucks the dust through a hole in the pipe, and carries the dust from the carton to a distance of about 3 feet.

Bellows Type:- The dust is held by a small container and is carried from the container into the air passage below by gravity. A bellows forces a current of air through the passage below the container and carries the dust to the plants.

Plunger Type:- In the plunger type duster (fig. 173) the dust container may be in the plunger tube or it may be an enlargement at the end of this tube. When the plunger is drawn back it reaches a point behind the intake hole in the piston tube. This allows air to enter the piston tube. The forward action of the plunger forces a current of air down the tube, carrying with it the dust. A continuous supply of air can be maintained by a double plunger

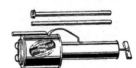

Fig. 173. Plunger-type hand dusters.

tube. Dust caps can be attached to the outlet tube so as to direct the dust downwards or upwards for overleaf and underleaf dusting.

Improved Hand Duster:- In the course of a study of methods for the distribution of DDT in refrigerator cars, Chisholm et al (159) eliminated certain de

271

ficiencies of conventional hand dusters. In the conventional hand dusters, because of the small volume of air compressed during each stroke of the plunger and the low velocity of the entrained insecticide, the area treated with a single stroke is rather small and the treatment of large areas is laborious and time-consuming. Another disadvantage is that agglomerates of partially dispersed dust may be deposited, and if the plunger is operated more rapidly to increase the area of effectiveness, the tendency toward the deposition of agglomerates is increased.

Fig. 174. An improved hand duster. Left, complete unit; right, one of the insecticide containers. Explanation in text. After Chisholm et al (159).

In one of the dusters constructed by Chisholm et al (159), air compressed by means of a plunger is forced into a delivery pipe attached horizontally to the top of an insecticide container. A scoop diverts about a half of the air from the pipe and leads it into the container. Into this diverted air stream the dust is entrained and conducted through an outlet into the pipe. Here the remainder of the compressed air, which has passed by the scoop, further disperses the dust and forces it through the outlet (fig. 174).

By the proper relation of the rate of introduction of the air to the diameter of the pipe and to the proportion of the air that is diverted by the scoop, the duster can be adapted to specific requirements. Experiments showed that it was possible to increase the area treated with a single stroke of the plunger and at the same time effect a more uniform distribution of dust and reduce the proportion of agglomerates deposited, as compared with those deposited by a conventional hand duster. By substituting a cylinder of liquified gas, such as carbon dioxide or Freon-12, for the plunger, and attaching the pipe at the bottom of the insecticide container to facilitate its charging, the results were still further improved. Carbon dioxide was found to be especially well adapted because it is inexpensive and readily available. A thoroughly disintegrated insecticide mixture was delivered with a high velocity throughout a refrigerator car in a few seconds. In the experimental model used (fig. 174, A), two insecticide containers and two outlets were installed to provide for delivery of the insecticide into both halves of the car simultaneously.

KNAPSACK DUSTER

Knapsack dusters can be effectively utilized in dusting a flower garden, nursery, a few acres of vegetable crops and even young fruit trees. It is possible to treat a larger area than that which is practicable with knapsack sprayers. Different devices for generating the air currents are used, including bellows (fig. 175) and blower fans (fig. 176). The bellows duster is operated by a lever, extending to the front and side of the operator. Air is forced through the discharge chamber by the extention and compression of a bellows attached to the top of the duster. The blower duster is operated by a crank operating a blower fan and agitator. The blower forces a draft of air through a chamber to which the dust hopper is connected. The agitator in the hopper feeds the dust into the air chamber, after which it passes down through the delivery tube. A constant flow of dust is maintained, but by the same token the required continuous cranking is rather tiring.

The bellows type of duster discharges the dust in puffs and is well adapted to hill crops. The blower type of duster is not as well adapted to hill crops, because a steady discharge of dust continues as one walks from one plant or group of plants to the next, but on the other hand it is better adapted to the dusting of row crops.

Fig. 175. Knapsack duster with bellows for generating air current. Courtesy Naco Manufacturing Company.

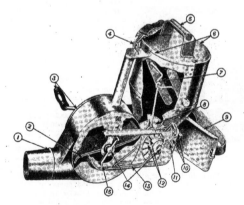

Fig. 176. Sectional view of rotary type knapsack duster. 1, outlet tube; 2, fan blades; 3, crank; 4, carrying strap; 5, cover handle; 6, cover; 7, upper agitator (swinging type); 8, lower agitator (rotary screw type); 9, body brace; 10, micrometer feed control valve; 11, feed regulator guide; 12, adjustable feed slots; 13, pusher agitator (prevents clogging in air flue); 14, bearings; 15, beater agitator on fan shaft (forces powder into fan). Courtesy Hudson Manufacturing Company.

TRACTION DUSTERS

Machines developed for field dusting in which the fan is geared to the wheel or wheels on which the duster is mounted are called "traction dusters". These may

be mounted on a wheelbarrow-type carriage (fig. 177). The operation of these dusters is on the same principle as that of the blower-type knapsack duster except that the power comes from the movement of the wheel. The operator is relieved of turning a crank and carrying the dust.

Traction dusters mounted on a 2-wheeled carriage drawn by horses or a tractor have been successfully used in extensive dusting operations, especially for ground crops. Rotation of the axle causes a small gear and chain to start the "bull gear" moving. From this one large central gear all the gears used to speed up the blower are driven. From the hopper the dust is fed, by means of an agitator, into the air chamber or directly into the fan and blower unit and hence to the delivery tubes. The nozzles and tubes are attached to steel rods which are connected to a supporting frame. These can be adjusted for various widths of rows. From 2 to 6 rows may be dusted at once. The deposit of dust can

Fig. 177. "Traction duster" on wheelbarrow-type carriage. Courtesy Naco Mfg. Co.

be increased by the use of a canopy covering the plants as they are being dusted, and this makes possible dusting during moderate winds.

POWER DUSTER

With power dusters a larger acreage of vegetable or field crops can be treated per day than with traction dusters, and in addition they are suitable for the more exacting requirements of orchard dusting. They vary in size and design from small

wheelbarrow-type dusters to the large high-powered fish-tail dusters powered by 6 cylinder motors that are used in citrus dusting.

A wheelbarrow-type duster that can be used for row crop dusting is shown in fig. 178. The hopper in this outfit holds about 20 pounds of dust. It is powered by a rope-started, air-cooled, 3/4 horse-power engine. The diameter of the fan is 10 inches and the fan speed is 3300 r.p.m. It develops 1.75 horse-power at duster speed. The feed adjustment is for 5 pounds to 30 pounds per acre. The net weight of the duster is 150 pounds.

Fig. 178. A wheelbarrow type row crop duster.

For more extensive operation a larger outfit of somewhat similar design may be obtained with a hopper capacity of about

50 pounds of dust. This duster has a fan of 12" diameter and a fan speed of 3250 r.p.m. The engine has a rope starter and a governed speed of 2900 r.p.m. The feed adjustment allows for the application of from as little as 5 pounds to as much as 50 pounds of dust per acre. The boom clearance is adjustable to 18 inches above or below the duster bed plate level. This outfit, mounted on a tractor, is shown in fig. 179.

For orchard dusting, equipment of greater capacity should be used when possible. Figs. 180 and 181 show types of dusters which may be used for the dusting of trees. The duster shown in fig. 180 has a hopper capacity of 3.25 cu. ft.,

179. Duster mounted on a tractor for row crop dusting.
Courtesy Naco Manufacturing Company.

Fig. 180 (at left). A small orchard duster. Courtesy Naco Manufacturing Company.

Fig. 181 (at right). An orchard duster with a hopper capacity of 4.35 cu. ft., a fan diameter of 16 inches, r.p.m. of 3600 and air velocity of 185 m.p.h. Courtesy Niagara Chem. Div. of Food Machinery Corp.

275

holding about 100 pounds of dust. The fan is 19-5/8 inches in diameter and has a speed of 3215 r.p.m. The air velocity from the fan is 175 miles per hour. This duster has an air-cooled, rope-started engine with a governed speed of 2800 r.p.m. The feed adjustment is from 5 pounds to 80 pounds per acre.

Fig. 181 shows a duster in which the hopper capacity is 4.35 cu. ft., holding 150 to 200 lbs. of dust. The fan diameter is 16 inches, with a speed of 3600 r.p.m. It is powered by a VF-4 Wisconsin motor which can be obtained with or without starter and generator. The weight of this outfit is about 950 lbs. Equipment of similar capacity, with the power take-off from a tractor, is also available.

With all engine driven equipment, as well as the traction duster, the rate of discharge of the dust is controlled by the size of the opening from the hopper into the discharge chamber, and can be maintained very uniformly. In the larger engine-driven dusters the hopper discharges directly on the fan instead of into a separate chamber. The advantage of this method of discharge is that the dust is given a somewhat higher velocity and the force of discharge can be more easily regulated by increasing or diminishing the r.p.m. of the fan.

The self-mixing type of dusting machine was invented by Smith and Martin (815) to fulfill the special requirements for dusting with nicotine dusts. In this outfit (fig. 182) an agitator is placed in the hopper so that dust mixtures can be prepared in the desired combinations and concentrations in the hopper and immediately prior to use. This is of special advantage when volatile substances, such as nicotine, are to be used in the dust, for it insures the use of an absolutely fresh dust mixture.

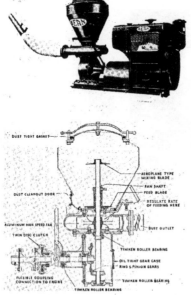

Fig. 182. A self-mixing duster with diagram of parts.

276

EQUIPMENT FOR APPLICATION OF INSECTICIDES

High-Powered Multivane Fan Type Citrus Duster:- About 20 years ago, great strides were made toward important improvements in engine-driven dusters to meet the exacting needs of citrus dusting. It was estimated that in those days about 26,000 acres of citrus in California and Arizona was dusted with sulfur every year, the greater part of it for citrus thrips control, but some for the citrus red mite, soft scales and rust mite. The usual type of duster contained one or two flexible delivery-tubes such as are commonly used in deciduous orchard dusting to this day (figs. 183 and 184). Unless the operator kept the hose moving continuously, the application of dust was not uniform or efficient. Some parts of the tree received too little dust while other parts had such a heavy load of dust as to greatly accentuate the possibility of burn during periods of hot weather. Likewise, since the citrus foliage is so thick, the air blast was frequently too small to drive the dust through the typically thick foliage of the citrus tree. It was found that unsatisfactory results were obtained when the dust was allowed to merely drift through the tree; it had to be blown through (520).

To fill the needs for citrus dusting, powerful blowers mounted on auto trucks and capable of treating 100 acres of citrus in one night were developed. First the efficiency of the fans was increased by increasing motive power. Four-horse-power engines were replaced by 8 or 10 horsepower engines and then by 4-cylinder automobile engines. Next the hose-like discharge tubes were abandoned and replaced by "fish-tail" stationary spreaders, so as to envelop a large part of the tree at one time in a cloud of dust. First one-way, and later two-way fish-tail spreaders were built. This required more powerful blowers. Also multivane fans were beginning to be used to insure greater uniformity in the spread of the dust. By 1932, 6-cylindered automobile motors were used for motive power; the fans were operating at from 3000 to 3700 r.p.m. and the motor from 1200 to 1500 r.p.m. Soon all the fish-tails were equipped with vanes to spread the dust evenly from the bottom to the top of the tree. Great ingenuity was shown by commercial operators in the construction of their equipment, and gradually all factory-built parts were discarded. A high-powered blower assembly on skids was built for around $400. Fig. 183 shows one of these fish-tail type dusters.

Fig. 183.. One of the older fish-tail type dusters.
Courtesy R. S. Woglum.

Present Types of Fish-Tail Duster:- In the early investigations of DN-Dust for the control of citrus red mite, it became evident that still greater improvements would have to be made in dust application to obtain satisfactory mite control (110). This led to a cooperative project between the Division of Entomology of the Citrus Experiment Station, Riverside, and the Division of Agricultural Engineering at the University of California, Davis, with the aim of developing an improved duster. The present high-powered, movable multivane fish-tail citrus duster resulted from this investigation. The principal improvement over the

277

former types of dusters was in the so-called "flippers" or deflectors on the fish-tail discharge of the duster, which slowly move the dust-laden air stream backward and forward. This backward and forward motion moves the foliage and fruits and enables the air stream to strike the foliage from various angles. As in spraying, this results in a more thorough coverage.

As stated before, dust does not deposit well on the side of a leaf or fruit opposite to that from which the air stream is coming. However, with the present type of citrus dusters, the dust is blown through the tree and is deposited on the lower surfaces of the leaves on the opposite side of the tree. Since the duster moves down both sides of a row of trees, the lower (and inside) surfaces of the leaves are quite satisfactorily covered with dust. In applying dusts for mites, the duster should be moved through the orchard at a speed of not more than 3 to 4 miles per hour to insure the required coverage and deposit of dust. In the application of nicotine dusts for aphids, much greater speeds may be employed. The speed employed in applying contact dusts is very important. The requirements differ, depending on the pest to be controlled.

The fish-tail type dusters reported by Lewis in 1932 had fans operating at 3000 to 3700 r.p.m. The "paddle wheel" type high pressure-small volume fans were used. The present types of citrus dusters are equipped with fans possessing the forward curved blade designed for slower speeds. On one of the citrus duster types (No. 4) being made at present, the fan r.p.m. is 850 to 900 and the volume of air from the fan is approximately 20,000 cu. ft. per minute with an outlet velocity of approximately 60 miles per hour. The other type (No. 5) possesses a fan operating at 1000 r.p.m. It has about the same velocity as the No. 4 unit, but the volume is approximately 30,000 cu. ft. per minute. Some dusters have been made with a volume of 50,000 cu. ft, per minute.

Motors used on citrus dusters at present may be of 4 or 6 cylinders, usually the latter. These have an r.p.m. of 1800 to 2000 as compared to an r.p.m. of 1200 to 1500 when fish-tail dusters were first used in the early thirties. The Ford V-8 motor used in citrus dusters is usually operated at an r.p.m. of about 2200.

Fig. 184. Streamlined multivane fish-tail citrus duster. Courtesy C.F.G.E., Pest Cont. Bur.

The high volume of dust discharged by modern citrus dusters and the effectiveness of its penetration through the citrus tree are indicated in fig. 242. A recently developed streamlined version of the citrus duster is shown in fig. 184. The front wheels are shielded, the windshield is narrow and the lights are close-set. With this equipment it is possible to crowd against trees on turns without breaking branches or tearing off fruit.

AIR ATOMIZING SPRAYERS

Although lacking the flexibility of hydraulic sprayers, various types of air atomizing sprayers have been used for many years and relatively recently they have undergone rapid change and development and have been found to be adaptable to an increasingly greater number of uses. With air atomizing sprayers, particles of concentrated spray are diluted with air instead of water. The air stream is also utilized to carry the atomized particles of concentrated spray to the plant surface.

At first a dilute liquid spray was used in this type of equipment, with concentrations of insecticide similar to those used with regular spray equipment, and this was atomized and carried to the tree by means of a high-velocity air stream (fig. 185).

The more recent tendency has been toward air atomizing equipment for the application of concentrated liquid sprays, with small quantities of spray well distributed throughout the tree. The tendency is toward greater volume of air at lower velocity. The first large-scale test of the efficacy of such equipment in California was in connection with the treatment of a serious infestation of grape leafhoppers in the San Joaquin Valley during the period 1930-1932. Small quantities of pyrethrum-oil concentrate were successfully atomized and carried to the grapevines by means of air, with good results against the leafhoppers. (677)

Quantity of Liquid Required:- In the "vapo-dusting" of grapes it was found that only 2 to 4 gallons of concentrate per acre was necessary, while for tree crops from 12 to 25 gallons per acre was used. Thus air atomizing equipment can be seen to have one of the advantages of dusting equipment, namely, that only a small amount of liquid per acre is necessary.

Fig. 185. A "Liqui-Duster". One of the early types of air-atomizing equipment. Courtesy Food Machinery Corporation.

This permits of smaller sprayer units and saves much time because refills are less frequently required. Also work can be carried on in areas where the transportation of large quantities of water would be impractical.

COMPRESSED-AIR SPRAYERS

Among the various types of air atomizing sprayers are the compressed air sprayers. Air compressors of various sizes, depending on the number of nozzles devised, are driven by gasoline engines. A displacement of 5 cu. ft. per minute is ordinarily satisfactory for a single nozzle of the type shown in fig. 186. Two such nozzles could be operated, with 80 pounds air pressure per square inch, by means of a compressor capable of displacing 10 cu. ft. of air per minute. A 4-horsepower engine would suffice for such a compressor. The spray liquid flows to the nozzles by gravity from a container mounted quite high on the sprayer. Usually the individual nozzles are operated by hand (fig. 187), although they may be fitted to a boom for certain types of truck crop spraying.

Fig. 186. Nozzle for atomizing liquid with a compressed-air sprayer. After French (349).

BLOWER OR FAN TYPE SPRAYER

With this type of air atomizing equipment, fixed nozzles are utilized, usually of the fish-tail type, in the center of which liquid nozzles are located. A large volume of air at relatively low pressure flows through the fixed nozzles and atomizes and disperses the liquid from the liquid nozzles,

Fig. 187. Compressed-air atomizing sprayer supplying 4 hand-operated nozzles. After French (349).

and carries it into the vine or tree. With a "fish-tail" type of nozzle a band of spray can be produced sufficiently wide to cover a tree. If the blower is sufficiently large, a fish-tail nozzle, together with a series of liquid nozzles, can be supplied on both sides of the sprayer, thus spraying 2 sides at once (fig. 188). The liquid is supplied to the liquid nozzles located in the fish-tail air discharge openings by means of a small pump. If the pressure of the liquid is too small to cause any atomization, the air velocity should be 125 to 150 miles per hour, but if the spray liquid is forced through atomizing nozzles at 60 to 100 pounds pressure, the air velocity may be reduced to 65 to 85 miles per hour at the point of discharge, provided a sufficiently large volume of air is maintained.

In spraying grapes, the fans should produce 3,000 to 4,000 cu. ft. of air per minute to operate 2 nozzles; for spraying deciduous trees, using single "fish-tail" sprayers, at least 7,000 to 8,000 cu. ft. of air per minute, and preferably more, should be produced; and the "spray-dusters" used on citrus require 16,000 to 18,000 cu. ft. per minute.

The "Mist-Duster" shown in fig. 189 has two fish-tail nozzles with openings 40 inches long and 3 inches wide. They can be adjusted up or down for various sized trees. The spray liquid is injected into the air stream through a longitudinal brass atomizing tube. The outfit carries 115 gallons of spray material, which is enough to spray 10 to 12 acres of deciduous trees. It sprays at the rate of 6 to 15 acres per hour. A dry dust hopper may be added to the outfit for dry dust application.

The versatility of air-atomizing applicators has been increased by equipping them with dry-dust hoppers. Such outfits can apply either an air-atomized liquid spray or a dust or even both spray and dust at the same time.

The blower or fan type sprayers, equipped with power units of from 20 to 40 horsepower and driven at ground speeds of 3 to 6 miles per hour, are capable of covering from 30 to 100 acres per night.

THE CITRUS SPRAY-DUSTER

The construction of the modern fish-tail multi-vane citrus duster and its effectiveness in forcing a large volume of dust-laden air through a citrus tree have already been discussed. The same outfit can be equipped for air-atomizing purposes by installing vertical booms with a series of nozzles in the throat of the air discharge outlets of the duster. It is then known as a "spray-duster" (fig. 190). The small hydraulic pressure required can be generated by means of a small pump, and a tank for holding the liquid is mounted on the duster.

Persing and Boyce (690) prepared a "News Letter" with detailed information on the equipment needed to adapt a modern citrus duster for use as a spray duster. Their investigation was directed primarily toward developing an equipment with which a small amount of tartar emetic-sugar solution could be rapidly applied to citrus trees in amounts as low as 20 gallons per acre in citrus thrips control. They found that for best results the liquid should be atomized to a finely divided mist with special nozzles and with a hydraulic pressure of 60 pounds per square inch. Even greater pressures, up to 200 pounds, are now used for certain types of application. Nevertheless, the main reliance, both for the atomization of the liquid and for distributing it throughout the tree, is placed in the stream of air generated by the standard citrus duster. Nowadays, in the control of citrus thrips with wettable DDT powder, about 100 gallons of spray is applied per acre.

Pumps:- From the standpoint of simplicity and ease of maintenance, the centrifugal or turbine pumps were considered by Persing and Boyce (690) to be the most practical for use on the spray-duster. Either type of pump can be used for suction-filling of the spray tank with very little additional equipment. Bronze, brass, stainless steel, or other noncorrosive metals can be advantageously utilized in the construction of these pumps in order to decrease the possibility of corrosion. For the dual purpose of filling and spraying, the pump should have a minimum capacity of 30 gallons per minute at a pressure of 60 pounds per square inch. The turbine pumps are slightly more expensive than the centrifugal pumps, but are more compact and have the advantage of a greater range of pressure.

Fig. 188.

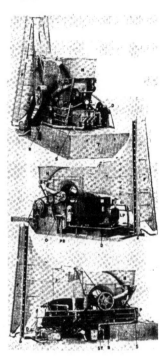

Fig. 189.   Fig. 190.

Fig. 188. Fan type of air atomizing equipment often used in vineyards. Courtesy·
Shell Oil Company.

Fig. 189. A "mist duster" applying atomized oil to dormant trees. Courtesy Bean
Cutler Div. of Food Machinery Corporation.

Fig. 190. Citrus spray-duster with cover removed to show parts. A, air chamber;
B, breather; D, discharge pipe; F, fin of "fish-tail"; G, pressure gauge; H, dust
hopper; LP, low pressure pump for refilling tank; P, pump; PR, pressure regulator;
S, suction side of pump; ST, strainer; T, tank for spray liquid. Courtesy Master
Fan Corporation.

If the pump is located below the supply tank, the liquid in this tank can be used for priming when the pump is used for suction-filling of the tank, provided 3 or 4 gallons of liquid remain in the tank. In case the tank should be emptied completely, extra water can be carried along for priming or an auxiliary prime tank or an automatic primer can be installed. The centrifugal or turbine pump would then not have to be located in such a position as to be dependent on gravity flow from the supply tank for priming.

Gear pumps are smaller and less expensive than centrifugal or turbine pumps, but are subject to greater wear, especially when silt or other abrasive material is present in the water. However, very small-mesh brass or copper screens installed on the suction side of the pump will aid in preventing such material from entering the pump.

From the standpoint of performance, reciprocating pumps possess certain advantages. The greater pressures that are made possible by these pumps increase the range of usefulness of the spray-duster. However, they are the most expensive type of pump. If a reciprocating pump is used, it is desirable to obtain a completely encased model so that the moving parts will be protected from abrasive materials when the spray-duster is used for applying dusts. Persing and Boyce suggested a capacity of 10 gallons per minute and a pressure limit of 300 pounds as being a desirable performance requirement for the reciprocating type of pump.

There is at present a great difference of opinion as to what type of pump and what pressure is required for best results. The trend seems to be toward the reciprocating type of pump with pressure ranging up to 500 pounds. The pump shown in fig. 190 has a capacity up to 35 gallons per minute.

If a small gear pump or reciprocating pump operating at low pressure is used, it will be necessary to make some provision for suction-filling of the liquid supply tank. This can be done by installing a rotary pump with a capacity of 40-50 gallons per minute, equipped with a clutch assembly. This type of pump can make a 10-foot lift.

Filters and Strainers:- A strainer equipped with a 20-mesh brass or copper screen can be used advantageously to remove the larger solid particles from the spray water and thereby reduce pump wear and the possibility of clogging the disc orifices of the nozzles. Such a strainer should be placed in the suction line to the pump, as shown in fig. 190, ST. When it is necessary to fill the tank by suction, the suction hose should be equipped with a 20-mesh screen strainer.

Clutch on Spray Pumps:- Occasionally a spray duster is used for applying dust exclusively, in which case it is not necessary for the liquid pump to be operating. The spray pump can be disengaged by means of a small, inexpensive clutch. The machine shown in fig. 190 is equipped with such a clutch.

The Tank:- The size and location of the liquid supply tank in the spray duster depends on available space on the duster truck and also the fact that it may be desirable to remove the tank when liquids are not being used. Since a 200-gallon tank will usually furnish enough liquid to treat a 10-acre block of oranges, it is the size that is commonly used (fig. 190). Steel tanks are usually used and should be painted with a nonmetallic paint or similar protective substance. For solutions, like tartar emetic-sugar solution, agitation is not necessary, but for the application of wettable DDT powder, for example, agitators are required, so they should be installed in the supply tank.

Nozzles:- Spray nozzles designed to produce a fine, mist-like spray are mounted in a vertical series in the throat of the duster discharge (fig. 190). The liquid is led to the nozzles by a 1/2-inch pipe placed on the back of the fish-tail so as not to interfere with the discharge of the air. One half to 1/4-inch "T" connections are fitted in the pipe at intervals and holes are drilled through the side of the fish-tail to admit a 1/4-inch nipple which is screwed into the "T" connection. A 1/4-inch strut elbow connection is inserted into the nipple and the

spray nozzle is screwed into the
strut elbow. The angle of the nozzle
is thus adjustable. The disk apertures in the nozzles vary from 0.038
inch to 0.052 inch in diameter.

The nozzles can also be mounted
on the "flippers", if the duster is
equipped with these. This provides
the added advantage of manual manipulation of the flippers in wide set
orchards, in which some degree of
manipulation may be desirable. The
nozzles should be spaced at a distance from one another which will
permit the spray cones to intersect
about one foot from the nozzle (fig.
191) at the operating pressure. This
will insure good coverage of the out-

Fig. 191. Spray-duster in operation.
Courtesy R. S. Woglum.

side foliage in closely set orchards. The number of nozzles depends on the vertical distance in the throat of the discharge and the angle of the spray cone of the
type of nozzle that is being used. On the spray-duster shown in fig. 190, 18
nozzles were installed on each side.

Spray Capacity Regulation:- Variation in the performance of a spray duster
can be minimized by maintaining a constant liquid pressure and a constant orifice
size of the spray disk. For pressure control, a globe valve in the by-pass pipe
can be manipulated or an automatic pressure regulator can be installed (fig. 190,
PR). It is essential to have an accurate pressure gauge equipped with a gauge
cook or a shock absorber diaphragm and a dial calibrated in increments of 5 to 10
pounds per square inch. Change in sizes of the orifices in the spray disks can be
decreased by the use of stainless steel disks.

SPEED SPRAYERS

The speed sprayer (fig. 192),
which has undergone continued development during the past 15 years,
utilizes the principle of blowing
the spray liquid through the trees
by an air blast generated by a propeller. This particular type of
equipment has demonstrated that a
large volume of air at a relatively
low velocity can be both efficient
and economical for certain types of
spraying. Although the coverage
problem does not seem to have been
satisfactorily solved in certain
types of citrus spraying, as for red
scale, very good coverage is being
obtained on deciduous trees. For
citrus pest problems other than the
red scale in California, and apparently for all citrus pest problems

Fig. 192. A speed sprayer. Courtesy
Bean Cutler Div. of Food Machinery Corp.

in Florida, the speed sprayer has given good results. Even in red scale control
in California the speed sprayer has been employed as a partial treatment, supplementary to some other control measure. Promising results have been obtained with
speed sprayers in walnut aphid control experiments (666, 667).

According to R. V. Newcomb, DiGorgio Fruit Corporation, the speed sprayers
have incorporated 3 essential requirements for the successful use of an air blast
as a carrier for tree sprays: "(1) sufficient air volume to provide rapid and

complete replacement of the air in and around each tree with spray laden air; (2) high enough air velocity to disturb or set in motion the foliage and smaller twigs which mask a large part of the area to be covered; and (3) uniformity of air pattern to provide thorough coverage for any type or size of tree" (644).

The speed sprayer is powered by a six-cylinder gasoline engine driving a metal propeller which delivers a great volume of low velocity air. The spray liquid is introduced into this air stream by means of a centrifugal pump and is carried into the tree in the form of a driving mist. The round steel storage tank has a capacity of 500 gallons. One man can operate both the tractor used to haul the equipment and the controls of the speed sprayer. He can turn off the entire discharge head or one side or the other. Since the 500-gallon tank can be sprayed out in 6 to 8 minutes, it is advisable to have available at least two service rigs for refilling. Speed sprayers are capable of delivering over 50,000 cu. ft. of air per minute. Borden (95) found that equipment with a 30-40,000 cu. ft./min. capacity has been quite satisfactory for trees of average size, but for large trees, 50,000 cu. ft./ min. or more is desirable. While the standard speed sprayer equipment has 80 nozzles discharging into the air stream (fig. 195), Borden (95) was able to reduce the number of spray nozzles to 42 and also reduced the disk apertures from 96/1000ths to 62/1000ths in dormant spraying. The rate of travel may be 0.8 to 2.5 miles per hour in the application of prebloom sprays to large pear trees (95). The rate of travel is regulated so as to allow for the application of enough liquid to cause a slight drip from the treated trees. This method of application has resulted in a saving of 70% in labor and 50% in material compared with the conventional "bulk spraying".

The chief disadvantages of the speed sprayer are its high initial cost and great weight. The latter is of special importance in wet soils.

EQUIPMENT FOR GENERATING INSECTICIDAL AEROSOLS (SMOKES, FOGS) OR MISTS

Definition of "Aerosol":- The word aerosol was suggested by the physicist F. G. Donnan as "a convenient term to denote a system of particles of ultra-microscopic size dispersed in a gas" (749). A great variety of insecticides may be used as aerosols. If solids are dispersed in air in particles of colloidal dimension, a smoke is formed, whereas if liquid insecticides are dispersed in colloidal form, they are called fogs. Aerosol is a generic term for either of these types of suspensions in air or other type of gas. The trade name of a well-known and widely used wetting agent happens to be Aerosol, and this has led to some confusion, but the two words, one a generic physical-chemical term and the other a trade name pertaining to a specific product, relate to entirely different phenomena.

Particle size is usually measured in units known as microns. A micron is 0.001 millimeter. A thousand microns (1 mm) is 0.040 inch. A hundred microns (0.1 mm) is 0.004 inch, or about the thickness of an ordinary piece of paper. Aerosols generated for pest control have varied from about 1 to about 100 microns in diameter. Larger particles constitute a mist, but the same equipment may generate either a fog or a mist, so the generation of mists is included in the present discussion. In either case, small quantities of insecticide may be rapidly applied over large areas.

In the case of liquid aerosols, the particle size decreases with a decrease in the surface tension of the liquid and an increase in the velocity of the propellent gas. If the amount of liquid to be broken up is large in comparison with the amount of propellent gas available, particle size increases rapidly with the viscosity of the liquid.

Types of Aerosol-Generating Equipment:- Aerosol-generating equipment may be divided into various types, depending on the method of generating and propelling the aerosol. As might be expected, the high degree of development of the "smoke screen" for military purposes gave rise to hopes that a similar type of aerosol could be used as an insecticide, but until recently this approach to the aerosol problem had not been especially productive. The aerosol "bomb" was highly effective, but expensive. The effectiveness of an aerosol containing a given concentration of insecticide is principally a function of particle size and is not influenced by the method by which it is generated. It is therefore to be expected that much attention should have been directed toward less expensive methods of generating aerosols than those which involve the use of liquified gases. The machines that have been utilized include several which generate cold air as the propellent force for breaking up and dispersing the insecticides (types 3, 4 and 5, below) and several types in which heat is used so that the viscosity of oil solutions can be reduced (types 6 and 7). The various types of aerosol-generating equipment will first be listed, then discussed more fully.

Type 1. Generators for Smokes, Fumes or Vapors:- Insecticide solutions are sprayed against hot surfaces or into a stream of hot gases.

Type 2. The Aerosol Bomb:- A gas is used which is liquid under pressure of its own vapor This liquid, together with the dissolved insecticides, is confined in a "bomb". As the mixture is released, the liquid insecticide is atomized by the resulting gas, which also acts as a propellent force.

Type 3. Cold-Air Atomizer:- This type of machine is represented by the paint sprayer or DeVilbiss-type air gun. It can break up small quantities of oil solutions to which low-viscosity and volatile solvents are added.

Type 4. Fan-or Blower Fog-or Mist Generators:- A large quantity of air is forced out of a fan-or blower duct at high velocity, atomizing and propelling the insecticide.

Type 5. High Pressure Fog-or Mist Generator:- High pressure generated by compressed gas is utilized to force a liquid insecticide through a "turbulance or aerosol chamber", thus producing the atomization. The energizing force is separated from the liquid being dispersed, so that nothing but the insecticide solution, without propellent, is released.

Type 6. "Dry Type" Fog-or Mist Generator:- The exhaust gas from an internal combustion engine is used as the atomizer and propellent of the insecticidal liquid, or the exhaust mechanism may heat air and the expanding air then acts as the atomizer and propellent.

Type 7. "Wet Type" Fog-or Mist Generator:- Steam is generated to both atomize the insecticide and propel the resulting aerosol.

The types of aerosol-generating equipment outlined above will now be discussed in greater detail.

Type 1. Generators for Smokes, Fumes, or Vapors:- Smokes are probably among the oldest means of combating or repelling insects. Smoke was utilized by primitive man to repel noxious insects from his domicile. Some smokes, such as those produced by burning pyrethrum flowers or tabacco, are not only repellent, but are actually insecticidal. Fumes or vapors can be produced by spraying solutions of insecticides on a hot surface. Vapors condense, forming aerosols of extremely small particle size. Nicotine was perhaps the first insecticide to be treated in this manner, utilizing the heat of internal combustion engines. Even such unstable substances as pyrethrum and rotenone can be suspended in air by this method with satisfactory results (858). Suitable machines have been patented (368). Smith et al (820) used a "nicofumer" in which finely atomized nicotine passed through two copper pipes heated to 350°C by means of a gas burner. The atomized nicotine changed to dense fumes which were carried through a blower to the vegetation. A large sheet was dragged behind the machine to confine the fumes to the treated plants (fig. 193).

An improved aerosol nozzle for use on engine exhausts was described and illustrated by Yeomans (1019). This type of nozzle constructed from pipe fittings

 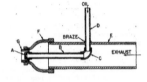

Fig. 193 (at left). A nicotine vaporizer with long sheet for confining fumes to the treated plants. Photo by R. H. Smith.

Fig. 194 (at right). Aerosol nozzle for use on engine exhausts. After Yeomans (1019).

(fig. 194), is claimed to be more efficient and more easily constructed than previous types. It can be used on small trucks and tractors. It utilizes the heat of the exhaust to lower the viscosity of the oil in the aerosol solution, and the velocity of the exhaust gases to break up the oil. Particle size is determined by the speed of the engine, the rate of flow of the solution, and the amount of constriction at the nozzle.

Interest in generators for smokes, fumes, or vapors gradually diminished, mainly because the small particles generated were not conducive to adequate deposit of the insecticide on the plant surfaces. Recently, however, interest in this type of generator has been greatly stimulated by the finding that if the particles are sufficiently small, they apparently act in the manner of a gas, and are effective with no appreciable deposit of insecticide on the plant or insect surface.

The insecticide which has been used to special advantage in the new type of "smoke generator" is tetraethylpyrophosphate (TEPP). The insecticide is dissolved in Diesel oil and the oil solution is introduced into the hot exhaust fumes from the manifold of a motor. The gases must be very hot. A pipe 3/8 to 1/2 inch in diameter and 3 or 4 inches long is attached to the manifold. The insecticidal oil is introduced into this "vaporization chamber", and since considerable pressure exists within, the oil must be forced in under pressure, which is supplied by a small gear pump. The oil enters through a needle valve adjusted so that a dry white smoke is generated. The vaporization chamber leads to an expansion and cooling chamber 1-1/2 to 2 inches in diameter and 6 inches long and then into a smaller discharge pipe. Where this smoke generator has been used to date on a commercial scale, it has usually been attached to a tractor motor. However, the smoke has also been generated by an independent motor, and this has the advantage of uniformity of operation with a consequent greater uniformity of dosage. It has been found necessary to envelop the trees in a smoke for 2 or 3 minutes to obtain the "fumigation effect". In the dry inland areas the device has been received with great enthusiasm wherever it has been tried commercially with TEPP as the toxic agent, and is replacing other types of aerosols, as well as sprays and dusts, especially for the treatment of mites (916). However, in more humid areas the smoke generator, using TEPP, has been less successful, possibly because of the effect of the moisture in the atmosphere on the TEPP, although no information is available on the point.

Only about 1/2 to 1 pint of the TEPP is used per acre in the control of mites and 1 pint per acre in the treatment of aphids. Since the treatment leaves practically no residue, it must be repeated to kill the insects or mites which happen to be in the egg stage at the time of treatment.

Type 2. Aerosol Bomb:- Goodhue (368) pointed out that the use of heat for the production of aerosols is not always practical because of the lack of facilities or the danger from fire. To overcome this objection, he developed a method for the dispersion of insecticides as aerosols, without the use of heat, by dissolving the

[1]Goodhue and Sullivan were joint patentees of the aerosol bomb.

286

EQUIPMENT FOR APPLICATION OF INSECTICIDES

insecticide in some low-boiling solvent, such as Freon (dichlorodifluoromethane, the gas that is used in household refrigerators) or methyl chloride, then placing the liquid in a suitable container and allowing it to escape under its own vapor pressure. The solvent is called the propellant. Pyrethrum and sesame oil were used as the insecticidal ingredients of the first dispensers. Freon is now almost universally used as the solvent because it is non-toxic to man. The container, modified in various ways, but working according to the same principle as the one originally used by Goodhue, is now known as the <u>aerosol bomb</u>. During World War II the entire production of aerosol bombs, amounting to forty million, was used by the armed forces and everywhere met with great enthusiasm. Since the war they have been produced by the millions for civilian use and are best known in the form containing 1 pound of solution (fig. 195), although other dispensers hold from 0.33 ounce to 5 pounds. Dispenser systems are used in industry which are supplied from 145-pound cylinders, and agricultural equipment with multiple outlets supplied by one large cylinder are also available.

The aerosol bomb was described and illustrated by Goodhue (368) and, in more detail, by Goodhue and Sullivan (371)[1]. According to the latter, cylindrical tanks commonly used to hold 5 lbs. of dichlorodifluoromethane are obtainable at refrigeration supply stores. These tanks are equipped with a valve having a 1/8 inch female pipe connection (fig. 196).

 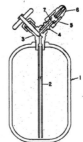

Fig. 195 (at left). Aerosol bomb in action. Roy J. Pence in photo.

Fig. 196 (at right). An aerosol dispenser. After Goodhue and Sullivan (371).

Space Sprays

The basic principle in the use of the aerosol bomb is the release of a solution of insecticide in a liquified gas into the atmosphere through a small orifice. The particles forming the aerosol are extremely small and remain in the air for long periods. It follows that the aerosols as they are prepared for the household "aerosol bomb" are specially adapted for the killing of flying insects. They are consequently known as <u>space sprays</u>. Since the residue left by aerosols is exremely small, they do not have the long lasting or "residual" effect against insects that may be obtained by atomizing the spray liquid on walls, screens, etc., as with the hand atomizer. The convenience and cleanliness of the aerosol bomb, however, have been strong points in the continued favor it is meeting in the household and in industries.

The possibility of using beer cans for the dispensing of aerosols has been considered (735). Such a development would result in a reduced cost of aerosols to the consumer.

---

[1]Goodhue and Sullivan were joint patentees of the aerosol bomb.

Agricultural Aerosols

Many mineral and vegetable oils are used as auxiliary solvents when Freon is used in the production of aerosols, and they are often considered to be important constituents. They influence the insecticidal effectiveness and physical proper- ties of the aerosol, including particle size. Some oils are good solvents for certain insecticides and prevent clogging of the orifice of the aerosol dispenser. The particle size of an aerosol may be changed by any liquid that is added to Freon, including the insecticide. Acetone has been found to be the most satisfactory solvent to use for increasing particle size. With increase in particle size, the aerosol is more suitable for use in field crops, because the deposit of insecticide is increased. Thus particle size constitutes the important difference between aerosols used for field crops and aerosols used as space sprays (370).

Greenhouse Aerosols

There have been many instances of successful use of aerosols in the control of greenhouse pests (369). Nicotine in a liquid aerosol was found to be twice as effective as in a smoke from a burning mixture. Cyclamen mite on snapdragons was controlled by the use of lorol thiocyanate aerosols. Thrips, whiteflies, aphids, sowbugs, ants and crickets were controlled by DDT aerosols. Good control of mush- room flies in mushroom houses was obtained with pyrethrum aerosols. The usual 1- pound aerosol bombs may suffice, although a light container holding 10 pounds of solution has been used. This container is equipped with flexible hose attached to a 2-foot pipe with a nozzle, and can be used like a knapsack sprayer. A formula giving the maximum effectiveness with little or no plant injury consisted of 5% technical DDT, 10% Velsicol AR-60 and 85% Freon-12. The organic phosphates have been used, with suitable precautions to protect the operator.

The cylinder of aerosol solution may be suspended by a bracket above the tops of the plants in the greenhouse. One such apparatus was fitted with a solenoid valve, and was equipped with a multiple nozzle that discharged the aerosol in 4 di- rections. The valve was actuated by an electric connection operated by a switch outside the greenhouse, which obviated, of course, the necessity for wearing a respirator (787).

Type 3. Cold-Air Atomizer:- The cold-air atomizer of the paint gun or De Vil- biss type lends itself well in experimental work for producing small quantities of fine sprays, such as used in Peet-Grady chambers or their more modern equivalents, and in turntable tests. They have not been used on a large scale because it is difficult to rapidly atomize large quantities of liquid with this type of equip- ment, especially liquids of high viscosity, such as oil. However, when using small quantities of oil solutions, volatile solvents of low viscosity can be added and the resulting solution can be broken up into particles of aerosol dimension by means of cold-air atomizers.

Type 4. Fan-or Blower Fog-or Mist Generator:- Various types of fog-or mist generators are available. These are similar in principle to the citrus spray- duster described in the previous section, in that is, an air stream is utilized to break and carry the liquid insecticide, but the liquid may be so finely atomized as to justify placing the equipment in the general category of "fog generators". The equipment can also be adjusted so as to produce larger particles, which may be classified as mists. Atomization of the liquid may be augmented by a system of disks rotating at high speed (1019). Such a machine is shown in fig. 197, A. A high velocity blower with a capacity of about 4500 cu. ft. per minute forces the liquid out horizontally from the disks. The machine will deliver up to about 250 gallons per hour. It is mounted on a base which allows a 20° tipback and a rota- tion of 360°. With one xylene-motor oil solution, particles of 40 microns mass median diameter were obtained, but smaller particles may be obtained with oils of lower viscosity or by adding volatile solvents.

Another type of air-atomizing fog-or mist generator is mounted on a turntable so that it can be revolved in a circle of 360 degrees by the slight pressure of the

Fig. 197. Aerosol-generating equipment. A, fog-or-mist generator utilizing a system of disks to aid in breaking up the
liquids; B, fog-or mist generator that can be revolved by a
slight pressure of the foot; C, machine that can simultaneously
apply a fog or mist and a dust; D, fan-type fog-or mist generator with pusher-type propeller; E, knapsack high-pressure type
of aerosol generator; F, "dry type" of fog or mist generator.
Photos courtesy of: A, Hession Microsol Corp.; B, Lawrence
Sprayer Co.; C, Agricultural Equipment Co.; D, Accurate Tool Co.;
E, Banta & Driscoll Co.; F, Todd Shipyards Corp.

operator's foot (fig. 197, B). It has been used for treating park and shade
trees, orchard trees, truck crops, and fields. Golf courses and city dumps have
been treated for mosquito abatement, and a wide variety of other uses have been
found for the equipment.

In the fog-or mist generator shown in fig. 197, C, the air stream is generated
by powerful blowers of new and revolutionary design. A finely divided spray is
atomized into the air stream as the air leaves the blower. Solutions, emulsions,
or suspensions may be introduced into the air stream. Dust and liquid may be
simultaneously applied. The air velocity may be varied at will from a gentle
breeze to a hurricane velocity of 250 miles per hour and with 8000 or more cu. ft.
of air per minute. An effective spray may be distributed for a distance of over
125 feet when the blower is operated at 3800 r.p.m. with the pattern of spray about
20 feet wide 100 feet from the nozzle. The tops of the tallest park and shade
trees may be reached with the spray. Particle size may vary from 50 to 150 microns.

The unit shown in fig. 197, D is equipped with an aeroplane engine driving a
pusher type propeller creating an air blast which receives the insecticide from an
independently operated insecticide supply system with a supply tank of 550 gallons
capacity. The circulating and feed pump are driven by means of a power take-off

from the truck engine. From 0.5 to 25 gallons of liquid may be dispensed by this machine per minute. By means of a universal mounting, the nacelle can be swung in a full circle horizontally, and from 25 degrees below the horizon to vertically upward. It is operated by remote control from the operator's cab situated above the truck driver's cab.

Type 5. High Pressure Aerosol Generator:- An apparatus has been perfected which consists of a steel pressure chamber or cylinder, called the applicator (fig. 197, E), designed to stand high pressure, and containing a floating piston equipped with seal rings separating a permanent charge of nitrogen gas from the liquid. A high pressure, hydraulic, hand-operated pump, either attached or as an independent unit, depending on the model, is used to fill and refill the applicator. The liquid is lifted from the container and injected into the pressure chamber, thus depressing the separating piston against the nitrogen gas. Without liquid in the applicator, the nitrogen gas exerts a pressure of 350 pounds per square inch. By pumping slightly over 3 quarts of liquid into the chamber, the pressure may be increased to a maximum of 1000 pounds. The piston in the chamber forces the liquid through an orifice 0.020 inch in diameter, thus producing an extremely fine dispersion of the solution in air. The apparatus is so designed that the rate of discharge is uniform at all pressures from the maximum of 1000 lbs. to the minimum of 350 lbs.

The applicator is provided with a 3 ft. flexible pressure tubing at the end of which is flexed a rod and the nozzle. The discharged applicator weighs 23 lbs.

and the hydraulic pump 21 lbs. Somewhat more than 3 quarts of liquid insecticide is discharged in 20 to 22 minutes. The applicator is built as a backpack and can easily be carried in this manner.

This machine has been used for the application of insecticides, plant-growth regulators and weed killers (847). The discharged droplets are so small that they can not be seen, but they may be photographed with the aid of a filter (fig. 198). It will be noted that in this type of aerosol-generating equipment the insecticide is not carried through the trees by a propellant (gas, steam or wind). The pure insecticide is forced through the tree by the momentum afforded by a propellant force separated from the insecticide. In all the previous types of equipment the propellant gases were mixed with and released with the insecticide; the compressed or expanding gases develop the velocity necessary to create the atomization.

A high pressure type of generator, designed in Argentina, has been reported in which a Ford engine is used to compress tin tetrachloride in one cylinder and the insecticide dissolved in acetone or similar volatile solvent in another cylinder. Both solutions are released through nozzles so situated that the solutions will mix and produce a fog (1019).

Fig. 198. Application of 2, 4-D to a citrus tree with "Hi Fog" aerosol generator. Courtesy Banta & Driscoll Co.

High pressure fog generators are practical only for liquids of low viscosity.

Type 6. "Dry Type" Fog or Mist Generator:- In the "dry type" of hot-gas atomizer, the maximum velocity of exhaust gas from an internal combustion engine is obtained by using a venturi tube. An insecticidal oil solution is injected into the gas at the point where the venturi is constricted. Particle size can be reduced by increasing the speed of the engine and by decreasing the flow of liquid. The capacity in gallons per hour is roughly about 1/3 the brake-horsepower of the engine (1019). With oil solutions, particle sizes from 2-3 microns up to coarse sprays can be produced, but difficulty is experienced in breaking up water solutions.

EQUIPMENT FOR APPLICATION OF INSECTICIDES

The machine shown in fig. 197, F, evolved from an oil-fog generator manufactured for military purposes. It is a "dry type" of generator, using no water. An oil solution of the insecticide is sprayed into a mixing chamber where the droplets are picked up by a blast of hot gases (the product of combustion) maintained at a controllable temperature and pressure and are thus dispersed into smaller droplets when discharged at atmospheric pressure. Emulsions or suspensions may also be dispersed. (360) A fog generator of this type was used in experiments on the control of vegetable insects. The machine could produce aerosol particles of the following sizes: small, 10 microns; medium, 20 microns; and large, 30 microns (939).

Fig. 199. Drift of aerosol from a "dry type" of generator. Courtesy Todd Shipyards Corporation.

DDT, Rothane, piperonyl cyclohexenone, and pyrethrum aerosols were successfully used to control garden leafhoppers, and DDT and hexaethyl tetraphosphate for melon aphids. The effective distances of aerosol drift (fig. 199) depended to a certain extent on the type of insecticide used. On windy days a hood for confining the aerosol appeared to be necessary for good control. It appeared that the large particles, 30 microns or more in diameter, were the most effective.

A distinctive type among the "Dry Type" aerosol generators uses high pressure gas like the aerosol bomb, but the gas in this case is air which is generated as required by an air compressor. The air atomizes and propels the insecticide. In addition, exhaust gases are utilized to decrease the viscosity and surface tension of the liquid material to assist atomization and add propelling power.

Type 7. "Wet Type" Fog or Mist Generator:- In the modification of the Army screening-smoke generators for the application of insecticides, the production of larger sizes of particles was one of the essential requirements. Prominent in this investigation was Professor V. K. LaMer of Columbia University, who was instrumental in the development of the Hochberg-LaMer generator, designed at the Columbia University Laboratory. This machine can be adjusted to produce almost any desired particle size. Thermostatically controlled temperatures between 300° and 600°F control particle size. Between 60 to 120 lbs. of pressure is developed by steam in a heater coil. A relief valve in the latter prevents excessive pressure. The maximum temperature generated is only about half as high as that used in producing smoke screens. The Hochberg-LaMer machine is a "wet" type of aerosol generator, employing water and an oil solution of the insecticide. Emulsions or suspensions can also be dispersed by this machine (360).

Much investigational work has been done with this type of generator in the treatment of large areas in the Solomon Islands for the treatment of mosquitoes. It was shown that the fog generator used was 6 times as efficient as standard hand and power spray treatments. In the control of sand fly, Stylocanops albiventris, the results were again very satisfactory. (122, 123, 124)

Another fog generator requiring water for its operation is shown in fig. 200. The principle of the machine is simple. A duplex plunger pump[1] is used for pumping water and insecticide materials. From a 50-gallon tank the water is pumped through a series of coils in which it is heated by a flame from a burner utilizing fuel oil. Superheated steam varying in temperature from 300° to 900°F can be generated at a pressure of 100 lbs. or more. The temperature is regulated by a thermostat. Insecticide material from another 50-gallon tank is pumped into a nozzle

---

[1] Some later models now have 2 pumps, one for the insecticide and one for the water.

Fig. 200. A cloud of aerosol from a "wet type" of generator.
Courtesy Besler Corporation.

where it is atomized by the steam. This produces an insecticidal fog. Particle size depends on the formulations and on the steam temperature.[1] (356)

Selection of Aerosol-Generating Equipment:- Yeomans (1019) has listed the factors to be considered in the selection of a field-model aerosol machine as follows: (1) original cost, (2) simplicity of design, (3) operating cost, (4) possibility of breakdown of insecticide, (5) types of solutions the machine will handle, (6) range and ease of particle-size control. Since there is no machine to date that is outstanding in all these factors, a selection should be based on the suitability of a machine in respect to the most important requirements of the pest control job to be done.

AIRCRAFT FOR THE APPLICATION OF INSECTICIDES

The first use of aircraft for the application of insecticides appears to have been in connection with the dusting of the Catalpa sphinx in woodlands in Ohio in 1921, then by the U.S.D.A.; Bureau of Entomology and Plant Quarantine in the dusting of cotton and in 1923 in the control of anopheline mosquito larvae in Louisiana. Subsequent development of aircraft application always received a large part of its stimulus from mosquito abatement work, and in World War II, airplane application of dusts and liquid concentrates in the mosquito-infested Pacific Islands proved to be the only practicable method of large scale mosquito eradication, especially in combat zones. The opportune discovery of the amazing insecticidal properties of DDT played an important rôle, of course, in the development of airplane application. The extent of interest in aircraft application is indicated by the voluminous bibliography on this and related subjects pertaining to the relation of aviation to economic entomology (413). There are now 75 agricultural aircraft companies in California alone.

---

[1] The following data were supplied by the Besler Corporation:

Discharge of water in gals.per hour	Discharge of insecticide in gals.per.hour	Temp. °F	Particle size in microns
40	40	600	10 - 30
80	80	350	20 - 100
80	120	350	30 - 100

EQUIPMENT FOR APPLICATION OF INSECTICIDES

Crop dusting by means of planes has been conducted for years for the control of cotton insects. It lends itself especially well to the treatment of large, continuous areas of a single crop, such as ordinarily is the case where cotton is grown. Other especially well adapted uses for airplane dusting, as well as for the application of air-atomized sprays and aerosols, is in the control of forest insects and truck and forage crop insects. Airplanes have also been used in treating orchards, however, despite the usually small areas in a cultivated unit (fig. 201).

Fig. 201. Citrus dusting by plane. Courtesy C.F.G.E.; Pest Cont. Bur.

The main advantage of aircraft application is that it is rapid and is a labor-saving device; in addition it makes possible the application of insecticides and fungicides during periods when the soil is too wet for the use of ground equipment. This is an important consideration in certain areas of frequent rains during the pest control season. Aircraft application is also advantageous for the treatment of closely growing and tall crops which may be damaged by ground equipment, even of the high clearance type.

On the other hand, aircraft application has certain definite limitations. The drift of dusts and sprays onto properties where they are objectionable has posed some difficult problems to pest control operators even with ground equipment, and with air-borne equipment this problem is greatly accentuated. The degree of thoroughness of application that can be attained with ground equipment is usually not attainable with air-borne equipment, even with the use of greater amounts of materials per acre. In the control of certain pests, a good deposit of insecticide on the undersides of the leaves is essential, but air-borne equipment has never been able to compete with ground equipment in this respect. In one investigation it was found that, on the average, only 7.2% of the droplets of an air-atomized spray settled on the bottoms of glass slides placed in various locations in the treated swath. Lack of adequate particle-or droplet size control, and more especially, lack of uniformity in distribution over the dusted swath, are also factors limiting the use of aircraft application. As the distance from the center of the treated swath increases, the deposit of insecticide rapidly decreases. In addition the insecticide-depositing properties of the equipment vary greatly from day to day and according to the height of the aircraft above ground. (1009)

A list of crops dusted by airplanes in California in 1946 has been compiled (38). Over a quarter million acres was treated. The greatest acreage of a single crop treated was of tomatoes, mainly for tomato worms and mites (85,476 acres), next was cotton, mainly for lygus bugs and red spiders (56,278 acres) and grapes, for mildew and leafhoppers (44,868 acres). In 1947 the acreage treated by plane had increased to 614,348 acres, the largest items being the control of corn earworm or corn earworm and mites, tarnished plant bugs, leafhoppers, red spiders and other mites, mildew and weeds. The principal materials applied by plane were, in order of acreage treated, DDT and sulfur, sulfur, calcium arsenate and sulfur, DDT, 2,4-D, calcium arsenate, and dinitro compounds (39). In addition to its uses in the application of insecticides and fungicides the plane has been used for applying herbicides on grain fields, rice fields, alfalfa irrigation and drainage ditches. Many thousands of acres have been seeded by plane and fertilizer has been broadcast extensively. These varied uses of agricultural aircraft enable the commercial operators to keep their equipment operating over a greater period with resulting reduced costs for the farmer.

The helicopter (fig. 202) lends itself well to insecticide application, especially for orchard work and for small areas that are not readily accessible with the conventional type of aircraft, and interest in this type of aircraft in this connection is increasing despite the high initial cost of the equipment (742). The following are among the advantages possessed by the helicopter for dusting,

293

Fig. 202. Dusting citrus trees with a helicopter. Courtesy Bell Aircraft Corp.

spraying or seeding, as stated by Kelley (481): "The rotors of the machine provide the motive for diffusing materials, with more than a million cubic feet of air being thrust downward each minute at a velocity of about 12 miles per hour. This force carries the insecticidal or fungicidal material down through the foliage and swirling upwards again to coat the under sides of the leaves and all surface parts of the plant. The helicopter can pass from tree to tree, pausing over each for treatment as long as may be necessary to insure effective coverage; or for large areas of infestation, it can fly at speeds of 15 to 80 m.p.h., leaving behind a swath of toxic material averaging 60 feet in width."

It was found in dusting operations in the Yakima, Washington region that only 7 to 10 minutes of every hour a plane operates was spent actually applying dust, whereas the helicopter, with its ability to make sharp turns and land in any small open space, dusted 40 minutes per hour. Planes usually travel at about 80 miles per hour in applying insecticides. An agricultural model of a helicopter is being built that will carry a 400-pound load, either solid or liquid, at 40 miles per hour, laying down a 60-foot swath, and covering 150 to 200 acres per hour. This model sells for $25,000.

The majority of fixed winged planes carry from 500 to 900 pounds of dust or spray. Smaller ships are coming into favor, however, which sell, converted for dusting, for around $3000. The charge by commercial operators for applying dusts to truck crops may vary from as low as 35 cents per acre (plus materials used) to around $2.00 per acre for small acreages (377).

On citrus, applications of either dust or liquid has had only very limited trial, although Roney and Lewis (753) report that airplane spraying of 50 per cent wettable DDT at 8 pounds per acre gave satisfactory results against citrus thrips in Arizona in 1947. Sulfur dust has also been applied on citrus in thrips control.

Apparatus for Dusting:- French (350, 351) discussed the development of the devices used in aircraft dusting and spraying, and the following discussion is based largely on his account. The first hoppers were built to clamp onto the outside of the fuselage of the plane. Only dusts were applied by plane in those days. The agitator and feeding device was operated by a hand crank. Next the hopper was built in the observer's seat of a two place ship. A circular pipe thrust through the bottom of the fuselage served as an outlet. Then the "air-suction" hopper was built, eliminating all mechanical motions but the opening or closing of the feed gate. The feeding device consisted of an air tube extending down through the center of the hopper. When air passed through this tube, it drew the dust out of the hopper. Air turbulence was depended on to distribute the dust after it left the hopper. The first planes used would carry about 350 pounds of dust.

The venturi spreader, located directly below the hopper and outside of the fuselage, was the next step in advance. Then a reel-type agitator driven by a small propeller blade was added. Practically all aircraft dusters use this system today, with various modifications. The carrying capacity of planes commonly used today is around 800 to 1000 pounds. Another improvement on present types of planes is the location of exhaust pipe outlets so as to eliminate the danger of carbon sparks igniting flammable dusts.

Apparatus for Spraying:- Concentrated insecticides and fungicides have been applied by aircraft for the past 15 years. The types of planes have been about the same as those used for applying dusts, generally biplanes powered by Curtis engines.

EQUIPMENT FOR APPLICATION OF INSECTICIDES

To atomize and distribute liquid sprays, rotating brushes or rotating spinners have been used, these being driven by small auxiliary propellers or fan blades at speeds varying from 1,800 to 2,500 r.p.m. Liquid is pumped in and is thrown from the periphery of these units and is broken up and dispersed by the plane's slip stream.

Fig. 203. A helicopter in action. Above, applying 2 quarts of DDT and 5 gallons of water per acre on young lettuce for control of beet armyworm, _Laphygma exigua_; below, applying DN-Dust-D8 on lemon trees for control of citrus red mite, _Paratetranychus citri_. Courtesy R. M. Boughton.

Booms and nozzles are also used (fig. 203, above). These are much the same as the ones utilized in ground spraying. Various kinds of nozzles are used. Pump pressures of from 20 to 100 pounds are utilized to force the liquid from the nozzles. Single stage centrifugal pumps are generally being used, driven from a power take-off shaft from the engine by means of multiple V-belts. The pump discharge can be directed either into the boom or into a by-pass line leading to the liquid storage tank, where it serves for agitating the liquid. A clutch on the power take-off drive permits the utilization of full engine power in the plane's take-off.

French favors the boom type of dispensing device because it appears to lay down a more uniform spray pattern than the rotating units. He states that for tree crops the boom type of device may not have any advantages.

French considers the width of the spray swath to be influenced by (1) the design and speed of the plane, (2) by the location of the dispersing device with relation to the slip stream, and (3) by the height of the plane above the ground.

Much of the drift problem is avoided if sprays are used instead of dusts. It is of course necessary, for practical purposes, to use concentrated liquids when spraying. Because of the heavier particles, spraying can be done while air current velocities are higher than permissible for dusting. Large droplets have sufficient mass to overcome adverse air currents, but tend to reduce the degree of distribution of the spray, as compared to smaller droplets. They appear to be satisfactory, however, for weed control, where droplets in the range of 300 to 500 microns in diameter are being used. French believes that liquid spraying will soon replace poisonous dusts as a pest control measure.

A discussion of the more practical installations for dispersing insecticides with aircraft has been prepared by the U.S.D.A. Bureau of Entomology and Plant Quarantine (32).

TENT PULLERS

The labor shortage following our entry into World War II posed some difficult problems to pest control operators, not the least of which was covering the trees with tents in orchard fumigation. Here again, necessity was the mother of invention. The idea of the mechanical pulling of tents was not new, for a mechanical tent puller mounted on and operated by a tractor had been assigned to the California Cyanide Company in 1926. The patent eventually came into the hands of the American Cyanamid Company, which released it in 1942 for public use. Thus no restriction exists in the development of tent pullers (979).

The first really practical mechanical outfit for pulling tents was constructed by Ben Aldrich of Santa Paula during the winter of 1941-42 (979). In this machine there were two long iron arms rigidly attached at the base to a long horizontal rotating shaft placed above and parallel to a truck chassis and operated by a power take-off from the truck engine. The tents were attached to the ends of these arms (fig. 204) in much the same manner as they would be attached to the poles pulled in the conventional manner of that period. Within a year, 50 mechanical tent pullers had been built and were operating in California citrus orchards and many more were built in succeeding years until today practically all fumigation is being done with mechanical tent pullers (fig. 205, 132). Tent pulling can also be used for lifting tents when loading (fig. 206). Although the crews used in fumigation are as large as formerly, strong men are no longer required for pulling the tents, resulting in much greater latitude in the hiring of men. Moreover, an orchard can now be more rapidly fumigated.

Fig. 204 (at left). Tents connected to end of **arms** of mechanical tent puller. Courtesy American Cyanamid Company.

Fig. 205 (at right). Pulling a fumigation tent over a tree by means of a mechanical tent puller. Courtesy American Cyanamid Company.

Fig. 206. Tent puller used for lifting rolled tents when loading. Courtesy Woglum.

EQUIPMENT FOR APPLICATION OF INSECTICIDES

Mr. Aldrich built his second tent puller so as to utilize hydraulic power, and other outfits were later built on this principle, although the majority are entirely mechanically operated. A later innovation was the mounting of the mechanical puller on a tractor, with the power take-off from the rear tractor shaft, and many are now mounted in this manner, some with power take-off from the shaft in front of the machine and operating from a planetary type transmission. This transmission appears to be among the simplest to operate (979). The construction of collapsible arms in the tent puller is also a good feature which is incorporated into nearly all mechanical tent pullers (fig. 207). In some outfits the collapsing of the arms must be done by hand, but on others the hinged arms are hydraulically raised or lowered.

Mechanical tent pullers have without a doubt proved their practicability. As many as 80 tents in a string can be handled by these machines and it makes no difference whether they are large or small. Since the tents are more lifted than pulled, they cause less damage through breaking branches, knocking off fruit and abrasion of fruit than was formerly the case.

Fig. 207. Mechanical tent puller: Above, arms extended; Below, arms folded. Courtesy Caterpillar Tractor Company.

CHAPTER XVI
EXPERIMENTAL DESIGN AND EVALUATION OF TREATMENT

Anyone who has had experience in the evaluation of insecticides is aware that a treatment may be repeated in a number of trials against a given number of insects and in accordance with some uniform experimental procedure, but with somewhat different results from each trial. The task of the investigator is to determine whether differences he observes between different insecticides or different types of treatment are consistently greater than those which may be expected from repeated trials of the same treatment. In field experiments, variations in the vitality of the host plant, as affecting the vigor and natural mortality of the insects, variations in population density as it affects the per cent kill, variations in the age distribution of the insects, unavoidable variation in dosage or thoroughness of application, and a host of other sources of error, add to the complications of the evaluation of treatments. These sources of variation may be minimized or to a large extent be eliminated in controlled laboratory experiments.

An example of precision equipment used in laboratory toxicity tests is shown in fig. 208. This is a combination "dust settling" and "mist settling" tower used in the entomology laboratory at the University of California, Los Angeles. Insect-or mite-infested fruits are set on cups attached to springs so that they will be pressed firmly into the holes in the lid of the turntable (B) when it is placed over the fruits and screwed into place, as shown in the figure. The drawer containing the turntable, shown in the open position in fig. 208, is then pushed into place so that the turntable will be enclosed by the settling tower (C). This glass tower is 11.5 inches in diameter and 28 inches high. It was once the gasoline container of an old-fashioned gasoline pump. A weighed amount of dusting powder or liquid is placed in vial A and is blown by means of a compressed air stream through the vent which can be seen in the center of the turntable. A cloud of dust or mist is blown to the top of the tower and then settles slowly and uniformly on the surfaces of the infested fruits. The uniformity of deposit of the insecticide is further increased by the revolution of the turntable holding the infested fruits. In tests of "residual effect" of the insecticide, uninfested fruits are treated and insects or mites are placed on the treated areas of the fruits at various periods after the treatment.

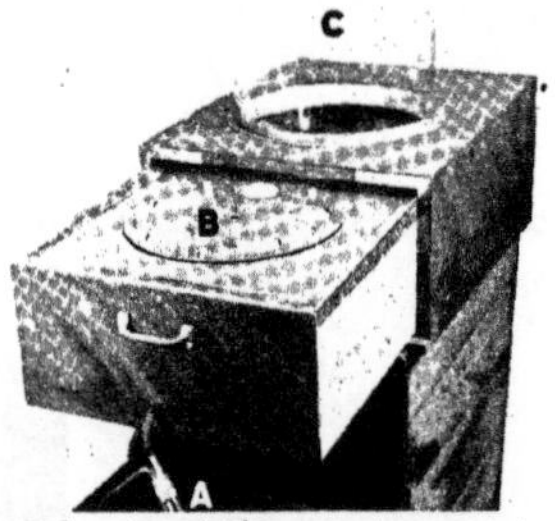

Fig. 208. A settling-dust-or mist tower. A, weighed or measured quantity of dust or liquid; B, turntable with infested fruit in place. The turntable is then slid into place under the glass settling tower (C), the lower portion of which is shown in the photograph. Dusts or mists are blown up through a vent in the center of the turntable.

With this and many other types of apparatus which have been designed for precise laboratory experiments with sprays, dusts, fumigants, and aerosols, the emphasis is on the elimination of sources of variability. Known quantities of insecticide are applied according to techniques that are uniform and reproducible.

Even though the greatest care may be exercised in the design of laboratory insecticide-application equipment to avoid sources of variation in the treatments, some sources of experimental error may still remain. After the investigator has eliminated, as far as is practicable, the avoidable sources of variability in his particular experimental set-up, a statistical evaluation of his preliminary data will provide him with information as to the number of insects he must use, and the number of trials, in a particular series of treatments, before he can be reasonably certain that the differences between treatments which he observes are caused by the intrinsic differences in the insecticidal efficacy of the treatments and not by experimental error and chance variation.

Laboratory tests can be much more rapidly made than field tests, and at much less expense. Often a compound may be procurable only in very small quantities until laboratory tests indicate that it has some potentialities as an insecticide. The laboratory tests are useful to "screen out" the most promising of a group of compounds from the standpoint of their insecticidal efficacy and their effect on potted plants. The more promising compounds must then be tested in the orchard to determine the required dosage and most expedient formulations under practical orchard conditions, where dosages must usually be considerably higher than in the laboratory.

In orchard experiments, the insect substrate may vary from the rough absorptive bark of the branches to the smooth cuticle of the leaves and fruit, and this may have some effect on insecticidal effectiveness. The long-term residual effect of the insecticide on the insect population can be determined by population studies in the orchard. The long-term effect of the insecticide on the vigor of the tree and on production of fruit can also be studied only in orchard experiments. Observation of the effect of the treatment on the work of beneficial insects may comprise an important part of the orchard experiments. Obviously, both laboratory and field work are important in insecticide investigations. The two methods should complement each other in the well-rounded research program.

The importance of the initial "screening out" in the laboratory of compounds of no promise as insecticides is indicated by the experience at the Citrus Experiment Station, where in recent years approximately 1800 organic compounds, exclusive of fumigants, have been tested according to standardized and reproducible techniques. Only about 50, or less than 3%, have shown enough promise to be further studied in field experiments, and it is believed that less than 10 have shown enough promise in the laboratory to justify extensive field investigation. It would not be economically sound, even if it were physically possible, to adequately test each of these 1800 compounds in field plots.

The investigator must bear in mind that there will always exist, in any experiment to determine the effectiveness of a treatment, a minimum and irreducible amount of variability, due to chance alone, which cannot be eliminated by further refinement of technique even in laboratory work. This can be illustrated by tossing 50 coins and recording the number of times they land with heads up. The number of heads among the 50 coins will not be 25 every time, but will _tend_ toward this number as the number of trials increases. The expected _standard deviation_ from the average figure of 25 may be calculated from the formula $\sqrt{p \cdot q / n}$ when p is the true percentage of heads in the group of coins, q is 100-p and n is the number of coins in the group. It can be calculated that the standard deviation is about 7% when the ratio of heads to tails is at or near 50/50. This means that about two-thirds (0.6745) of the times the 50 coins are tossed, the number of heads will be within 7% of the expected mean of 25. No further reduction in variability can be effected without increasing the number of coins. If one were to make a test with 50 insects, using a treatment with which the expectancy of mortality is 50%, the minimum standard deviation would be 7%. It can be seen from the formula that with either an increase or a decrease in mean mortality the standard deviation would decrease. Although the standard deviation, at a given mortality cannot be smaller than the figure calculated from the above formula, it can be considerably larger from sources of variation other than from random sampling. To reduce standard deviation then, all known sources of variability must be eliminated as far as is practicable. Then further reduction can be effected, in a given test, by increasing the number of insects, as can be seen from the formula. Some of the methods employed to reduce or eliminate sources of variability in the testing of insecticides for the control of citrus pests will be described.

THE REARING AND TREATING OF INSECTS AND MITES

One source of variability encountered in field experiments, or in laboratory experiments in which infested material to be treated is brought in from the orchard, is the lack of uniformity in the ages of the insects used in the tests. Even if a certain stage or instar can be satisfactorily isolated from the others in making the mortality or population density counts, the differences in ages of the insects in the stage or instar in question may result in variation in susceptibility to the treatment. In laboratory experiments this source of variability can be eliminated by rearing the insects so that at the time of treatment all those in a given series of trials will be of the same age or of closely similar ages.

California Red Scale:- The California red scale can be conveniently reared on mature banana squashes (Cucurbita maxima) or gourds. At the Citrus Experiment Station squashes are ordinarily used. These are placed on racks in a warm, dry room and they remain in good condition for periods ranging from 6 months to a year, permitting the scales to develop for several generations. They are large and can support a high population of scale. The squashes are elongate and can be oriented so that the red scale crawlers, which are positively phototropic, will congregate on the tip of the squash directed toward the source of light. Red scale crawlers can be readily transferred from squash to other host material by light brushing with a camel's-hair brush. Grapefruit have been used extensively because they will remain in good condition longer than other citrus varieties. A specified number are infested with crawlers each day. Half grown to mature grapefruits may be used, depending on the season of the year the tests are made. The infested fruit is kept at about 76°F, resulting in the scale becoming mature in about 40 days (531). The scales are usually used as test insects when they have reached maturity on the grapefruits; however, for certain experiments insects in some immature stage of development may be desired. In either case, the test insects are all of the same age.

The scale-infested grapefruits lend themselves well to spraying and fumigation experiments. In spraying, they are placed on a turntable 51 inches from the disk of a laboratory precision sprayer generating 300 pounds of pressure (fig. 209), A 3/64 inch disk aperture is used in the nozzle. The fruits are exposed to the spray stream for 4.5 seconds, which allows for 2 complete revolutions of the turntable. Four grapefruits are used per treatment. Usually 400 to 500 insects are counted on these 4 fruits. Thus a series of 10 treatments would require the examination of from 4000 to 5000 scales.

After treatment, the fruits are placed on trays and set aside in a room maintained at a temperature of about 76°F and a relative humidity of 40 to 60%, where they remain for 3 weeks. By that time the dead scales dry out and turn dark in color and are easily differentiated from the live scales in the examinations made to determine per cent mortality.

Fig. 209. Red scale-infested grapefruit being oil-sprayed by means of a precisely determined and accurately reproducible technique at the University of California Citrus Experiment Station. J. P. La Due in photo.

The results obtained from various oil sprays must of course be correlated with oil deposit on the fruit surface. The oil is stripped from the fruit by means of a fine stream of petroleum ether from a wash bottle and is then recovered in Babcock milk-test bottles and weighed according to a standardized procedure (531).

Effect of Crowding

Many data have been published to show that the effectiveness of an oil spray against red scale decreases with increasing population density of the scale (494, 257). The same relationship was found to hold true with adult camphor scales, Pseudoaonidia duplex (Ckll.) (211). In one experiment it was found that before treatment the per cent of insects found to be dead (per cent natural mortality)

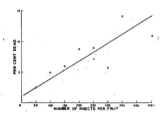 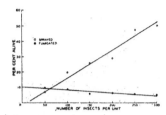

Fig. 210 (left). Increase in natural mortality of red scale on lemons with increase in population density. After Ebeling (257).

Fig. 211 (right). Per cent survival of sprayed and fumigated red scale on
lemons as influenced by population density. After Ebeling (257).

increased with increase in population density (fig. 210). In counts made a month
after treatment with oil spray, the per cent of insects found to be alive increased
with increasing population density. Fig. 211 shows two straight lines depicting
the relation of per cent survival of adult red scale to population density on lemon
fruits in an orchard containing both oil sprayed and fumigated plots. With oil
spray there was about 6 times as high a percentage of insects alive on lemons with
250 to 300 scales per fruit as on lemons with 50 or less scale per fruit (257).

The red scale can, by capillary action, draw oil off the sprayed tree surface
toward its body, where the oil readily penetrates under and through the waxy ventral covering of the scale. The oil can be drawn from a considerable distance
about the body of the scale, and if the scales are close together, they may draw
oil away from one another. There is then less oil available per scale than if they
were farther apart. On the other hand, it can be seen from fig. 211 that crowding
does not have the same effect on the results of HCN fumigation. The gross per cent
survival decreases a little with increase in population density, reflecting the
effect of the increase in natural mortality shown in fig. 210. The fumigated insects all have access to the same amount of gas, regardless of the degree of crowding, so population density can have no effect on the per cent of kill.

In experiments with oil spray, when the relation of survival (or mortality) to
population density is indicated, as in fig. 211, a more accurate comparison between
treatments can be made than if the average per cent survival is given, for the
population density of red scale may vary greatly between trees and between plots.

In laboratory spray experiments the effect of population density is overcome
by counting only those scales separated from one another by at least the width of a
scale (approximately 1.5 mm). In addition, only those scales situated on a delimited equatorial region of the fruit are examined.

Other Subtropical Fruit Insects:- Other scale insects may be reared on banana
squash, including yellow scale, purple scale, and latania scale; and the black
scale can be reared on banana squash from second instar to maturity. The latania
scale, a pest of the avocado, can be reared on potato tubers. Mealybugs and black
scale can be reared on potato sprouts. The method for mass culture of mealybugs
has been described by Smith and Armitage (799). Greenhouse thrips can be easily
reared in enormous numbers on oranges or grapefruit. The fruits are kept in large
cloth-covered battery jars and the propagation is begun by placing a few adult
females, which are parthenogenetic, in the jars. Breeding is then continuous. A
method of rearing citrus thrips in the laboratory, for toxicological tests, was
developed by Munger (630). A cage was devised which was gas-permeable and in which
the same plant tissue could be kept in a healthy condition throughout the incubation period of the egg (7 to 35 days). A compilation of descriptions of many
methods which have been successfully used by various investigators to rear insects

was published by the American Association of Economic Entomologists (142). The
rearing methods were grouped into those pertaining to (1) insects that attack
plants, (2) insects that attack stored products, and (3) insects affecting man and
animals.

A FAIR SAMPLE

While in laboratory investigations techniques are developed to eliminate
sources of variability, in field work the plots must be so arranged that statistical analysis of the data will show the part played by the sources of variability.
In either case the investigator must be able to determine what is a fair sample to
represent an insect population. In laboratory experiments much smaller samples
can be effectively utilized than in the field. Generally it is considered that 25
to 50 insects per treatment is a minimum, except in occasional experiments in which
detailed examination of each insect is involved, as in the determination of the
period required for the death of an insect. Usually several hundred insects are
used per treatment. A greater reduction in error would be attained by using the
minimum number of insects per test, but making many tests. Repetition of tests is
time consuming, however, and since test insects can usually be obtained in large
numbers, the tendency is to use large numbers of insects per test, but to make only
a few tests. However, at least 3 or 4 replications of a test should be made. To
be of greatest advantage, the replications should be made on different days (see
913).

In field experiments, much larger numbers of insects are used per test, and
while it may not be practicable to repeat a test as many times as might be desirable in a given orchard, it is necessary to repeat the test in different localities,
for insecticidal effectiveness and tree reaction may vary conspicuously even in as
limited an area as a single county.

ALLOWANCE FOR NATURAL MORTALITY

In experiments in which there is an accumulation of dead insects in situ on
both treated and check plots as, for example, with scale insects, or in experiments
in which there occurs an appreciable mortality within the period required for
treatment and examination of results, a correction for natural mortality must be
made. A simple formula which appears to have been first used by Abbott (1) is as
follows:

$$\frac{x - y}{x} \times 100 = \text{per cent control}$$

when x is the per cent remaining alive in the untreated plot and y is the per cent
remaining alive in the treated plot.

Sometimes, especially when mortality is high, the per cent of survival may
appear to the investigator to be the more significant figure. For example, there
appears at first glance to be little difference, proportionately, between a 92% and
a 96% kill, but in the latter case only half as many insects have survived the
treatment. The true comparison of the two treatments, from a practical standpoint,
is the relationship of the percentages of survival, namely, 8 to 4. If the effectiveness of the treatment is designated in per cent survival, a correction can be
made for natural mortality merely by dividing the per cent alive in the treated
plot by the per cent alive in the untreated plot. The resulting decimal, multiplied by 100, is the net per cent survival thus:

$$\frac{y}{x} \times 100 = \text{net per cent survival.}$$

MEAN, STANDARD DEVIATION, AND STANDARD ERROR

Let us suppose an investigator has determined the mean per cent mortality of
red scale on 200 fruits from each of 2 plots of orange trees. For each plot the
mean would be the sum of the mortalities on the individual fruits divided by the
number of fruits. The significance of the mean per cent mortality would depend on

the degree of variability between the percentages of mortality on the 200 fruits, or, in other words, the degree of deviation from the arithmetic mean. If the mean per cent mortality is found to be different in the two plots, how is the investigator to know whether or not the difference could be attributable to experimental error or the degree of error inherent in random sampling? To measure this error he may calculate what is known as the standard deviation (σ) by means of the following formula:

$$\sigma = \sqrt{\frac{\Sigma\, d^2}{n-1}}$$

when $\Sigma\, d^2$ is the sum of the squares of the individual deviations from the arithmetic mean, and n is the number of fruits examined per plot. The quantity n-1 represents the number of "degrees of freedom" (see p. 305).

The standard error of the mean may then be calculated as follows:

$$\sigma_m = \frac{\sigma}{\sqrt{n}}$$

The standard error of the difference between the means of the two plots, which is the measure of the significance of the difference, may be calculated as follows:

$$e^2 = \frac{\sigma_1^2}{n_1} + \frac{\sigma_2^2}{n_2}$$

Having calculated the standard error, the next step is to determine the significance of the difference between the true means. The difference is divided by the standard error and the resulting value is called the t value. By consulting a table (822, p. 55), one may determine whether with the calculated t value and number of "degrees of freedom" in the experiment in question, the difference is significant. In general, however, it may be stated that if there is a reasonable number of degrees of freedom, as would certainly be the case in the above experiment, and the value obtained for t is greater than 2, the difference is "significant" in the sense accepted in scientific work. If the value of t is about 3, the difference can be said to be highly significant.

DOSAGE-MORTALITY CURVES

Much experimental work has to be with the relationship of dosage of insecticide to mortality. A complete treatment of this subject is presented by Finney (310). After the per cent kill at different dosages of insecticide has been determined, the data may be plotted on coordinate paper so as to graphically demonstrate the relation of dosage to mortality. The resulting curve is called the dosage-mortality curve. Experiments with a wide variety of organisms have shown that dosage-mortality curves are typically sigmoid (S-shaped) in character (fig. 212). This means that increments in the per cent kill (expressed as ordinate values) per unit of abscissa are smallest near mortalities of 0 and 100% and largest near 50%. It can be seen that the sigmoid curve is merely the cumulative expression of the "normal curve" of frequency distribution.

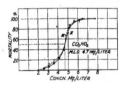

The relative insecticidal efficiency of various insecticides can be more clearly depicted graphically if the sigmoid curves can be transformed to straight lines. This transformation can be effected by changing the abscissa (dosage) values to logarithms and the ordinate (mortality) values to probability units or "probits" (fig. 213). The "probit" value for a given per cent kill may be obtained by referring to a table worked out by Bliss (90). Fig. 214 is a chart obtained from a paper by Lindgren (526)

Fig. 212. Toxicity of chloropicrin to Tribolium. Typical sigmoid dosage-mortality curve. After Strand (853).

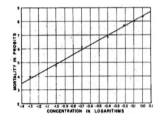
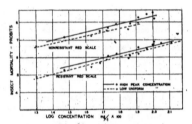

Fig. 213 (left). Toxicity of· quinoline to eggs of Lygaeus kalmii. A .dosage
mortality curve transformed to a straight line by use of logarithms and probits.
Redrawn.from C. F. Smith (784).

Fig. 214 (right). Comparison of mortality of nonresistant and resistant red
scale at high and low peak concentration of HCN gas. Straight lines obtained
by plotting log and "probit" values. After Lindgren (526).

which contains a graphic presentation of dosage mortality relationships which would·
be more difficult to quantitatively interpret by inspection without the log-probit
transformation. Especially such small differences as were found between "high peak
concentration" and "low uniform concentration".(fig. 214) are depicted to best advantage by straight lines.

ANALYSIS OF VARIANCE

In many cases an experiment involves more than two treatments. There may be
as many as 10 or 20. A method of statistical analysis, called analysis of variance,
may be applied in such an experiment. Obviously, since the complexity of the ex-.
periment has increased, the computations involved become more intricate, but nevertheless, with the aid of a mechanical calculator, an involved problem can be solved
with surprising rapidity. Analysis of variance lends itself to many varied types
of problems of evaluation of insecticides, especially in field experiments, and is
probably the most common method of statistical analysis used by economic entomolo-·
gists.

Analysis of variance is a method of statistical analysis worked out by the
noted statistician R. A. Fisher, who says of the method: "The analysis of variance
is a method of arranging arithmetical facts so as to isolate and display the essential features of a body of data with the utmost simplicity" (314). In fig. 215

Table 1.—Period required for 50 per cent mortality (M.L.P.)[1] of potato tuber moth larvae wet with
light-medium oil alone and with added toxicants. Temperature, 85° F.

MATERIAL	Test No.—1	2 .	3	4	5	Mean
1. Oil alone	150	210	170	150	180	172
2. Oil plus 1 per cent dinitro-o-cyclohexylphenol	6	7	6	7	6	6
3. Oil plus 1 per cent Lethane 440[2]	15	15	11	12	12	13
4. Oil plus 0.625 per cent derris extractives[3]	27	24	26	24	22	25
5. Oil plus 1 per cent nicotine	50	41	35	34	35	39
6. Oil plus 1 per cent Thanite[4]	45	45	40	30	40	40
7. Oil plus 0.1 per cent pyrethrins (5 per cent pyrethrum extract)	90	105	90	100	90	95
8. Oil plus 4 per cent DDT[5]	95	125	110	140	110	116
Significant difference at odds of 19:1						14.5

[1] Median lethal period.
[2] A proprietary mixture containing 50 per cent b-butoxy-b'-thiocyanodiethyl ether.
[3] Butyl phthalate solution containing 5 per cent derris extractives (50 per cent rotenone), used at a concentration of 1 part to
7 parts of oil.
[4] Fenchyl thiocynyl acetate.
[5] 2,2-bis-(p-chlorophenyl)-1,1,1-trichloroethane.

Fig. 215. A table in.which experimental data have been arranged for an·
analysis of variance. ·After Ebeling (265).

EXPERIMENTAL DESIGN AND EVALUATION OF TREATMENT

is shown such an arrangement of data obtained in the course of some studies on petroleum oils in which potato tuber moth larvae were used as test insects (265).
The period required for the death of 5 out of 10 larvae (median lethal period),
after immersion in the oil, was recorded, and in fig. 219 each datum is the median
lethal period (M.L.P.) for a single test for 1 treatment. It will be noted that
there were 5 replications of each treatment. This table is typical of the manner
in which data are arranged for an analysis of variance. Each datum could as well
have been any other criterion of insecticidal effectiveness. Again the statistical
analysis involves the squaring of each datum and the summation of the squares.
Analysis of variance is based on the important fact that the sums of squares and
the "degrees of freedom"[1] are additive. The explanation of the calculations can be
found in various text books on statistical analysis, such as Goulden (375),
Snedecor (822), and Fisher (313). Suffice it to state here that the calculations
arrive at an "F-value". By consulting a table (822, pp. 174-177), it can be determined whether the calculated F-value is sufficiently high, with the number of
"degrees of freedom" in the experiment in question, to indicate that there is a
statistically significant difference between the treatments, taking the experiment
as a whole. That is, the value of F tells us whether or not the differences found
between treatments could be attributed to experimental error or error in sampling.
If the F-value is found to be sufficiently high, a further calculation will give us
the "least significant difference" (L.S.D.) at odds of 19:1 or 99:1, between any
two plots in the entire experiment.

In fig. 215 it will be noted that the L.S.D. at odds of 19:1 was found to be
14.5 minutes. Odds of 19:1 (P=0.05) are generally considered as sufficient, in
scientific investigations, to justify the assumption that a difference is significant, although the L.S.D. at odds of 99:1 (P=0.01) is often stated in scientific
papers. If a difference is found to be barely significant at odds of 19:1, it
means that there is only 1 chance in 20 that the difference could be due to chance;
if the difference is barely significant at odds of 99:1 the chance of its being
due to chance is only 1 in 100. In a given problem, the L.S.D. at odds of 99:1
would, of course, necessarily be considerably larger than the L.S.D. at odds of
19:1.

In comparing the mean M.L.P. for treatment 2 with the M.L.P. for treatment 3
(fig. 215), it can be seen that the difference (7 minutes) is not significant at
odds of 19:1. The difference between the mean M.L.P. for treatment 2 and 4 (19
minutes) is more than enough to be significant at odds of 19:1, and we can assume
with considerable confidence that oil plus 1% dinitro-o-cyclohexyphenol is more
toxic to potato tuber moth larvae than oil plus 0.625% derris extractives. In a
similar manner comparisons can be made between any two treatments in the experiment.

In the experiment depicted in fig. 215, if the variability between the 5 tests
with a single treatment had been less, the L.S.D. would have been less. If the
variability had been greater, the L.S.D. would accordingly also have been greater.
If an experimental technique can be refined so as to keep variability at a minimum,
small differences between treatments may be detected and proved to be significant,
but under circumstances in which a greater degree of variability is involved, only
larger differences can be proved to be significant.

ORCHARD EXPERIMENTS WITH RED SCALE

Design of the Experiment:- The design of an orchard experiment will depend on
a number of factors. If not more than about 6 treatments are to be tested, the
experimental design known as the Latin square is well adapted. Let us say there
are 5 treatments to be tested, which can be designated as A, B, C, D, and E. An
orchard may be divided, for example, into 25 squares of 16 trees each, 5 squares in
each direction. In determining the sequence of treatments, a purely random distribution may be followed, or a random distribution with the proviso that no treatment should appear more than once in a single row of squares. If 16 trees were
used in each square, the sampling could be done on the 4 center trees to eliminate

_____
[1]The number of independent observations less 1. In fig. 215 there are 8 plots repeated 5 times.
The number of degrees of freedom (or n-1) is (8 x 5) - 1 = 39.

305

error caused by a dust, for example, drifting from one plot to another. If the pest population has some tendency to migrate, the separation of the trees to be sampled from those of another plot is again of value. The Latin square arrangement of plots was well illustrated in an experiment with codling moth on apples by Hansberry and Richardson (405). Spencer and Osburn (826) have discussed the randomized-block arrangement for insecticide experiments in citrus orchards. In their experiments, if 10 plots were involved (including the standard oil spray and no treatment), there would be 10 blocks, each of which contained 10 trees. In every block, each treatment would be applied to a single tree. The order of treatments is determined entirely by chance.

In work with citrus insects, the Latin square arrangement of plots is often not practicable. The trees are often closely spaced and the grower may object to the excessive driving back and forth required with the Latin square arrangement, with its accompanying destruction of fruit and foliage. There is also usually objection to the needless packing of the soil. In citrus orchards a treatment is usually applied to the trees in two rows which can be sprayed with a tank, or even a half a tank of spray material, depending on the size of the trees. In red scale investigations it has been found, empirically, that 8 or 10 trees are a sufficient number for a treatment, so if 14 or 16 trees are sprayed, 10 trees of average size and vigor may be selected for obtaining data. These are tagged and charted on cross-section paper. Sometimes 2 or 3 replications of a treatment are made in an orchard, although, as will be shown later, if both pre-treatment and post-treatment counts on the same tree are made, as in population density studies, the replication of treatments in a single orchard is less urgent, and the investigator may choose instead to spend his limited time and materials in repeating his experiment in other regions.

In red scale orchard experiments the investigator is interested in the long-term effectiveness of the treatment, as affected not only by the initial per cent kill, but also by the residual effect of the treatment on the settling and development of the "crawlers" of those scales not succumbing to the treatment. Consequently he may not bother to determine the per cent kill, or if he does, it will be only supplemental to more important data which indicate what effect the treatments have on the population density perhaps 6 months or a year after treatment. His first problem, then, is to employ an adequate method of sampling to determine the population density of the scale on the 8 or 10 trees in each plot which were selected as the "data trees".

The Distribution of Red Scale in a Citrus Tree:- The typical density distribution of the red scale population in a citrus tree in southern California is shown by fig. 216. The data from which the chart was developed were accumulated over a period of 2.5 years, in the course of experiments made in 8 orchards in widely different localities. The average percentage of insects in the four quadrants of the tree, based on counts made on green twigs, leaves, and fruits, were found to be as follows: northwest, 28.4; northeast, 43.7; southwest, 12.1; and southeast, 15.8.

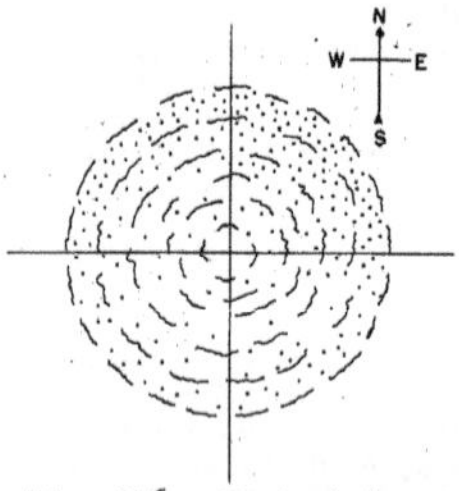

The northeast quadrant of the tree nearly always has more scales per unit than the northwest quadrant. Also the southeast quadrant usually has more scales per unit than the southwest quadrant. Invariably the north half of the tree has more scales than the south half, except when the trees are north of a road and the south half of the tree obtains more dust than the north half. This causes a reversal of the usual population distribution, for the dust favors the settling and development of scale insects.

Fig. 216. Typical density distribution of red scale in a citrus tree in California.

During the summer months, the natural mortality of scales on the south half of the tree is much greater than on the north half. Even adult scales on exposed fruits and leaves are sometimes killed by the extreme temperature. The writer has occasionally noticed all the scales on the exposed portions of some fruits killed by the heat after a prolonged hot spell. On the remainder of the fruit no more than natural mortality could be found. Probably an even more important factor than the mortality of the older scales is the effect of high temperatures on the inhibition of the settling or development of the newly emerged nymphs (crawlers).

In the northern hemisphere the north half of the tree is the cooler half during the summer months. During July, August and September, the hottest months in southern California, the sun strikes the tree from the south, its rays striking the earth at from 21 to 47 degrees from the perpendicular.

The data presented above also show that the east half of the tree has on the average more scale than the west half. On the east halves of the trees were found 59.4% of the red scales. The west half of the tree is exposed to the sun in the afternoon, when the temperature is higher than in the forenoon, and for this reason one might expect a lower population on the west half than on the east.

Despite the adverse effect on the red scale on certain parts of the tree caused by exposure to the direct rays of the sun, the great majority of scale crawlers, because of their positive phototropism, have a tendency to migrate to a position on the part of the tree on which they are born which is exposed to the greatest amount of light. Thus about 95% of the red scale on the leaves are on the upper surface, nearly all the scale on the horizontal branches are on the upper halves of the branches, and a great majority of the scale on the fruits are on the halves facing the outside of the tree. All these facts must be taken into consideration in deciding on an adequate sampling technique.

When plotting the number of insects per unit against the number of examined units (fig 217), it can be seen that the population densities are very asymmetrical in their distribution. This increases the difficulty of sampling as compared to the sampling of populations of normal frequency distribution. Nevertheless the problem still remains of determining the number of units per tree which will constitute a fair sample. Beyond a certain number, a further increase in number of units sampled would yield increasingly diminishing returns. Further accuracy in the evaluation of the treatments might be more efficiently obtained by increasing the number of trees per plot or by replications of the experiment.

Sampling:- To determine the adequacy of the different sized random samples, the deviations of the averages of the mean populations in different sized samples from the "true mean" were determined.[1] The "true mean" was the mean population on a given unit based on counts made on all the units on a tree. For example, let us suppose that a count is made of the live adult red scale on a tree and the mean is found to be 20 insects per fruit. Suppose also that 40 fruits are sampled at random and that in this sample the mean number of insects per fruit is found to be 25. The difference in the "true mean" and the mean of the 40-fruit sample is 5, which is a 25% deviation from the "true mean". That is the basis upon which the data shown in figure 218 were calculated.

Fig. 218 shows the rate of decrease of the deviations of the means of the different sized samples from the "true mean" with increasing number of units sampled. The data are the averages obtained by three workers. The greater the number of units sampled, and the greater the number of insects per unit, the lower the difference in the mean population found by the various samplers from the "true mean". It can be seen from figure 218 that in orchards containing a reasonably dense population, there was little advantage in sampling more than 40 units per tree. Often, in actual practice, 20 units are used, obtained only from the north halves of the trees, where the red scale population is most dense. It can be seen that the lowest variability at any given population density is on the green twigs. The fruits are ordinarily utilized, however, because the scale population builds up in a shorter period following treatment on the fruit than on any other part of the tree.

[1]Ebeling, W. and C. O. Persing. 1945. Evaluation of insecticides used against red scale by estimate of population densities. Manuscript on file at the Division of Entomology, University of California, Los Angeles.

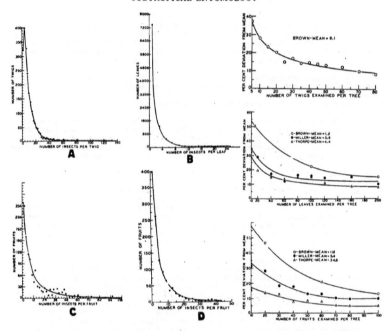

Fig. 217 (left). The asymmetrical distribution of red scale population densities on: A, lemon twigs; B, lemon leaves; C, lemon fruits, and D, orange fruits.

Fig. 218 (right). Rate of decrease of deviation from the "true mean" with increase in size of sample.

In the above experiment, the orchards were sprayed in September, 1943. The fruits and twigs had become sufficiently reinfested with live adult female scales by January, 1944 to make possible a post-treatment population study, but on the leaves no appreciable number of scales were found until December, 1944.

In their work with Florida red scale, Spencer and Osburn (826) kept records of total and living scales from 20 leaves, obtained around the tree at shoulder level, from each of 10 trees per plot. Such counts were made just before spraying, 1 and 3 months after spraying, and at the end of the year.

Having decided on the number of units per tree to be sampled, it was next determined how many insects should be allowed to develop on the treated plots before the post-treatment population density estimates are made. Since the fruit is most commonly used as the unit sampled, this study was confined to the fruit. It was found that the higher populations were less variable, but that if on the average there were between 20 and 30 scales per fruit, the coefficient of variation was close to its minimum value and not much advantage would accrue from allowing the trees to become still more heavily infested before making the counts. Thus after treatment, citrus trees should be allowed to attain a moderate infestation before an estimate of comparative infestations is made, but allowing the trees to become

EXPERIMENTAL DESIGN AND EVALUATION OF TREATMENT

very heavily infested increases the amount of counting to be done without proportionately increasing the accuracy of the estimate. On the other hand, an estimate made when the population of scales averages less than 10 per fruit may be relatively unreliable as compared to an estimate which might be made in the same orchard a month later with a greater population density.

Evaluation of Orchard Treatments for Red Scale:- In field investigations of insecticides used in the control of the red scale, usually 10 trees have been used per plot. These 10 trees are selected for uniformity of size, form and degree of infestation from a considerably larger number of trees treated with each insecticide. Although 2 or 3 subplots have been used in randomized blocks, the estimate of population both before and after treatment on the same tree appears to obviate the necessity for repetition and randomization. However, an experiment is usually repeated in other localities.

With 10 trees per plot and with a pre-treatment and post-treatment estimate of population density, analyses of variance have shown that differences in infestation, at prolonged periods after treatment, which are of economic importance can be readily detected.

The data obtained in orchard experiments may be presented in a variety of ways. The "corrected relative infestation" has been utilized as a direct comparison of effectiveness. The population density of the insects in the plot sprayed with oil of a grade and at the concentration ordinarily used in the district in which the experiment is made, is used as a standard of comparison and is equated to 100. Corrections are made for differences in the degree of infestation of the various plots based on the pre-treatment counts. (268)

ORCHARD EXPERIMENTS WITH CITRUS THRIPS AND CITRICOLA SCALE

With nearly every species of insects a different method of orchard evaluation must be devised. The following are methods which field experience has shown to be both practicable and adequate for the evaluation of treatments used against the citrus thrips and the citricola scale.[1]

Citrus thrips treatment evaluations are based on (1) the amount of thrips-scarred fruit or (2) an index of the thrips population on terminal growth, or (3) an index of the injury by thrips to terminal growth. Citricola scale treatment evaluations are based on counts of the number of scales found on a definite number of leaves and/or twig terminals. With each insect the technique of evaluation must be changed as the season advances to suit the needs of the changing tree and insect conditions.

CONTROL OF CITRUS THRIPS ON ORANGES AND GRAPEFRUIT FOR THE PREVENTION OF SCARRING

Treatments are applied at petalfall, which is late March or early April for the desert sections, and late April or early May in the central and southern interior sections of California. Usually a single, well-timed treatment will prevent fruit scarring.

No pre-treatment counts are made for petalfall applications. Final evaluations of treatments are made after the fruits have colored and the scars can be readily observed by determining the per cent of the outside fruits scarred. All outside fruits in a band 3 to 6 feet from the ground are examined directly on the tree on each of 16 trees in each plot. Depending on the amount of surface affected, the fruits are arbitrarily classified as slightly or severely scarred.

Control of Citrus Thrips on Orange and Grapefruit for Protection of New Growth:-
Applications of insecticides to protect new growth are usually made in August and September, although October treatments are necessary some years. A pre-treatment

---

[1]This section was written by W. H. Ewart and is based on orchard treatment evaluation methods used at the University of California Citrus Experiment Station.

index of the average number of thrips per terminal is made by collecting thrips from 8 new-growth terminals on each of 12 trees in each plot in a specially constructed "thrips cage" (fig. 219).[1] Each terminal is rapped sharply five times against a screen through which the thrips fall onto a removable "Deadline"-covered plate in the bottom of the cage. The thrips from all the terminals in each plot are collected on a single plate which is then removed and placed in a case where it is stored until counts can be made in the laboratory with the aid of a binocular microscope. The average number of thrips per terminal represents the population index. Post-treatment evaluations are made in a similar manner at 1 to 2 week intervals until such time as the thrips population increases to the point where retreatment is necessary or the experiment is completed.

Control of Citrus Thrips on Lemons in Southern California:- Applications of insecticides are started when the population index is 0.5 thrips per terminal as determined by the "thrips-cage" method. This is usually about the middle of May, but some years may be as early as the beginning of April. Because of the continuous growth characteristic of lemons, subsequent treatments are usually necessary throughout the summer and early fall.

.Fig. 219. W. H. Ewart using a "thrips cage" to determine the population density of citrus thrips on a lemon tree.

Pre-treatment and post-treatment counts are made by collecting the thrips in the "thrips-cage" in the manner described above.

Another technique used for evaluating treatments for the control of citrus thrips, particularly on lemons, is the "terminal scarring index" method. Twenty-five new-growth terminals from each of 20 trees per plot are individually classified, on the basis of the degree of injury caused by thrips, into the following six categories:

1. No visible injury.
2. A trace of injury on one or more leaves. No malformations.
3. Appreciable scarring on all leaves. No malformations.
4. Scarring on all leaves with one or more leaves deformed.
      Some stem scarring.
5. Over 50 per cent of leaves deformed. Severe scarring on
      all leaves or stems.
6. All leaves badly deformed or abscissed. Severe scarring
      of all leaves and stems.

The terminal scarring index is obtained by the following formula:

$$I = \frac{a_1b + a_2b - - - - - - - - + a_6b}{N}$$

I = Terminal scarring index.

a_1, a_2 --a_6 = The values of the injury categories, ranging from 1 to 6.

b = Number of terminals in the injury category.

N = Total number of terminals examined.

Control of Citricola in Late Winter and Early Spring in Central California:- This treatment is normally applied between the middle of February and the middle of March.

[1] The "thrips cage" is a modified McGregor thrips collecting trap. It consists of a metal box 7 inches by 10 inches by 2 inches. The top of the box is covered with No. 14 wire screen and the bottom is equipped with a removable metal slide which, when in operation, is covered with a thin film of "Deadline" or similar sticky substance.

310

EXPERIMENTAL DESIGN AND EVALUATION OF TREATMENT

Pre-treatment counts of the number of immature living scales per leaf are made on 25 leaves in a band 3 to 6 feet from the ground from the north side of each of 8 trees per plot. In case of very heavy infestation, 10 leaves per tree are sufficient. The first post-treatment count is made in April on the number of adult living scales per terminal on 10 one-foot terminals sampled at random within an area 3 to 6 feet from the ground on the north side of each count tree. The second post-treatment count is made in early August on immature scales, after all eggs have hatched, in the same manner as the pre-treatment count. The final count of adult scales is made the following April in the same manner as the first post-treatment count.

In cases in which treatments give outstanding results, the experiment may be continued a second year using the same counting techniques as during the first year.

Usually treatments can be satisfactorily evaluated by the above technique when leaf and terminal samples are taken only from the periphery of the north side of the tree. However, in certain types of experiments in which the degree of coverage obtained by different types of application equipment is being studied, the inside as well as the tops of the trees are sampled also.

Control of Citricola Scale in May and June in Central California:- Treatments for the control of citricola scale at this time of year are usually made in conjunction with measures for the control of citrus thrips to prevent fruit scarring.

Pre-treatment counts are made in April of the number of adult scales per terminal, while post-treatment counts of immature scales on leaves are made in August and February as previously described.

Control of Citricola Scale in Late Summer and Fall in Central California:- This treatment is usually made from late August to the middle of October.

Pre-treatment counts are made in August of immature scales on leaves while post-treatment counts of immature scales on leaves in February, adult scales on terminals in April, and immature scales on leaves in August are made as previously described.

SOME METHODS OF APPRAISING CITRUS RED MITE AND BUD MITE POPULATIONS, UNDER FIELD CONDITIONS

Citrus red mite:- In evaluating field populations of citrus red mite it has generally been found impractical to count an adequate sample of mites and their eggs in the field.

Several procedures have been developed in appraising citrus red mite populations under field conditions. These have varied with the particular type of study in progress. Jones and Prendergast (467) described a method used in a critical study of mite and predator populations. Henderson and McBurnie (423) developed a "brush" method of evaluating the egg and mite population on citrus leaves. These methods are time consuming and not adapted for estimating population increase following extensive acaricide applications where limited assistance is available. Under these conditions, a method commonly used is to examine at random the leaves and fruit on various parts of each of twelve trees. Each tree is assigned a population index. The averages of these values are used as the estimate for each plot. Population trends are determined at frequent intervals.

Citrus bud mite:- A method for the field evaluation of citrus bud mite was developed with the early biological and control studies at the Citrus Experiment Station. This technique consists of removing 15 terminals from each of 8 trees per plot. These terminals are brought to the laboratory and 5 buds from each terminal are dissected with the aid of a dissecting microscope of 20 to 30 power magnification. Each bud is assigned a population index according to the following 5 categories:

311 ·

0 - No mites found
1 - 1 mite
2 - 2 to 5 mites
3 - 5 to 15 mites
4 - over 15 mites

The mean from these categories is used as the population index for each plot.

RAPID ORCHARD GRADING

The previously discussed methods of evaluation of population density are used in connection with field experiments in which treatments are being evaluated with the greatest practicable degree of accuracy. The data obtained must be of such a nature as to be amenable to statistical evaluation. More rapid methods of determining degree of infestation are available for those interested in obtaining data upon which to base their seasonal pest control operations.

Packing House Inspection for Red Scale:- The least time-consuming method of arriving at relative infestations is by inspection of the fruit in the packing house as it is going over the conveyor belt in the grading operation. Graders are able to report on per cent of infestation in different orchards, or in different parts of a single orchard, if the fruit is kept separate in the picking and hauling operation. This method is especially valuable in red scale protective districts as organized in California, where it is essential to discover an insect infestation as early as possible so that eradication work can proceed at once. An observer may be stationed at the conveyor belt solely for the purpose of looking for red scale. In the red scale protective districts, if red scale are found, even if only in a single box of fruit, a tree to tree inspection is made of the orchard. The infested trees are charted and a control program is immediately started in accordance with the rules set up in the "district" (see pp. 94 to 96). Picking foremen may also be requested to look for red scale in each box in the orchard.

Orchard Inspection for Red Scale:- In orchard inspections, there have been a variety of methods employed for the appraisal of insect infestation, but the majority of these conform in principle to a method discussed by Worthy and Wilcomb (1017) with special reference to the red scale. The tree is inspected from the outside, with no effort made to push aside branches or to observe the interior of the tree. The inspector walks slowly around the tree and grades it according to the number of scales found. In order to standardize the work as much as possible, the County Agricultural Inspectors in Los Angeles County spend about a half minute per tree. A given tree may be placed on one of 6 grades, from 0 to 5. The grade number is placed on the square on a sheet of cross-section paper which corresponds to the location of the tree in the orchard (fig. 220). It will be noted that in the system represented by fig. 220, every tree in every fifth row is inspected. A dot is placed in the appropriate square when no scales are found on a tree. A 5 represents a very severe infestation. The intervening numbers represent the intervening degrees of infestation. The 1's and 2's are considered to be "lightly infested", the 3's "medium" and the 4's and 5's "heavy". The numbers may be averaged to obtain a figure representing the average degree of infestation in an orchard.

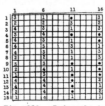

Fig. 220. Red scale infestation grading chart. Explanation in text. After Worthy and Wilcomb (1017).

For a more thorough and accurate picture of the red scale infestation in an orchard, every tree can be graded in the same manner as discussed above. This is called a "tree to tree inspection". Instead of indicating the degrees of infestation numerically, various intensities of stippling, cross-hatching, or shading can be used to indicate progressive degrees of infestation. Such a system is illustrated in fig. 221, which depicts the increasing degree of infestation of a 5-acre lemon orchard from the date of the first inspection on June 1, 1939 (fig. 221, A) to the date of

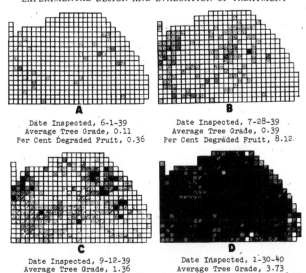

A

Date Inspected, 6-1-39
Average Tree Grade, 0.11
Per Cent Degraded Fruit, 0.36

B

Date Inspected, 7-28-39
Average Tree Grade, 0.39
Per Cent Degraded Fruit, 8.12

C

Date Inspected, 9-12-39
Average Tree Grade, 1.36
Per Cent Degraded Fruit, 5.98

D

Date Inspected, 1-30-40
Average Tree Grade, 3.73
Per Cent Degraded Fruit, 36.05

Fig. 221. Increasing degree of infestation of a 5-acre lemon orchard from 6-1-39 to 1-30-40. Various intensities of stippling, cross-hatching, or shading indicate progressive degrees of infestation. Adapted from Worthy and Wilcomb (1017).

its last inspection on January 30, 1940 (fig. 221, D). In this 7-month period the average infestation, based on the numerical system of grading, increased from 0.11 to 3.73. The amount of degraded fruit caused by red scale alone increased from 0.42 box, in a pick of 117 boxes on June 3, 1939, to 67.2 boxes, from a pick of 364 boxes on May 8, 1940. The entire record obtained from the packing house is shown in table 10.

Orchard grading is thus seen to be valuable as an indication of the time to treat in order to avoid financial loss from degraded fruit. A "tree to tree in-spection" also indicates to the pest control operator what trees might advanta-geously be given a second treatment in the "spot treatment" method of pest control. For example, an entire orchard may be fumigated and the most severely infested trees can be tagged, on the basis of the inspection chart, and can later be sprayed, or vice versa. The lightly infested trees will thus be given a single treatment and the heavily infested trees a "combination treatment", yet the expense is not as great as if the entire orchard were combination treated. Orchard grading also af-fords a method by which the county agricultural inspectors can report to the agri-cultural commissioner the condition of the orchards in their districts. They do their inspecting during the month of April each year and the reports are sent in on special forms as soon as the work is completed. The type of treatment, the brand name of the insecticide, and the dosage are also listed. These reports must reach the agricultural commissioner's office by May 15.

Sometimes instead of inspecting every 5th row, the "hour glass" system of in-spection is employed. The inspector inspects every tree in two diagonals as, for example, along a row running from the northwest to the southeast corner of the

Table 10
Number of boxes of degraded lemons in a 5-acre lemon orchard
in 6 picks made from June 3, 1939 to June 3, 1940.

Number of Field Boxes Reduced in Grade by Red Scale

Date Picked	No. Boxes	Grade	Boxes	Per Cent Degraded Fruit
6-3-39	117	2	0.42	0.36
		3	----	
		Culls	----	
8-8-39	17	2	1.38	8.12
		3	----	
		Culls	----	
9-22-39	609	2	11.5	5.98
		3	15.8	
		Culls	9.1	
2-13-40	405	2	66.4	36.05
		3	35.5	
		Culls	44.1	
3-28-40	685	2	98.6	28.16
		3	56.0	
		Culls	38.3	
5-3-40	364	2	36.7	18.46
		3	22.1	
		Culls	8.4	

orchard and again from the northeast to the southwest corner. An examination of the border rows completes the inspection. If red scale are to be found at all, they should be found most easily along the border rows. The time required for the inspection of an orchard can be greatly reduced by inspecting only every fourth or every fifth tree. This, of course, applies also to the system of inspection in which every fifth row is examined.

Inspection for Black Scale:- In orchard grading for black scale, the twigs must be examined, for the inspector is searching for the conspicuous adult scales, which have long since migrated from the leaves to the twigs (see p. 429). A common method employed by agricultural inspectors is to select 5 twigs, each about 8 inches long, from every third tree. Ordinarily 3 of these twigs are taken from the outside and 2 are selected by reaching through the outer "shell" of the tree. Since the black scale, like the red scale, are located predominantly on the north side of the tree, the twigs are selected from the north side only. According to this system of grading, if no more than an average of 0.1 adult black scale is found per twig in an orchard, the orchard is considered to be in an excellent condition. If from 0.1 to 1.5 scales are found per unit, the orchard is considered to be in a good, "commercially clean" condition. The fruit is not smutted and the tree has suffered no appreciable damage from the scale. The treatment of the previous year has apparently been successful. If from 1.5 to 3 scales are found per twig, the previous year's treatment must have been of questionable effectiveness. The orchard must, of course, be treated at the proper period during the current season to avoid severe damage. If from 3 to 10 scales are found per twig, the orchard is considered to have a heavy infestation and more than 10 scales per twig represent a severe infestation.

CHAPTER XVII
BIOLOGICAL CONTROL

Biological control pertains to the suppression of insect pests or weeds by the use or encouragement of those organisms which in nature tend to maintain the pest or weed populations at low densities. The brief treatment of the subject contained in this chapter, compared to the much greater emphasis that has been placed on insecticides, is not meant to imply a lack of appreciation of its importance in applied entomology. It is rather a recognition of the fact that biological control is a highly specialized field of entomology engaging the efforts of a relatively small percentage of entomologists, to say nothing of general horticulturists.

The grower, the manufacturer of insecticides and his research and sales staff, the pest control operator, thousands of federal, state and county entomologists, and every owner of a home or a garden is either engaged in or must make decisions relating to the application of insecticides. In contrast, the responsibility for the importation and liberation of beneficial insects is wisely placed in the hands of a relatively few highly specialized government and state entomologists. A few well-trained experts suffice for this work and, in fact, incalculable harm could be done by the unwitting introduction of injurious insect species presumed to be beneficial, or hyperparasites that might parasitize introduced or native beneficial species, if this work were not done under the strict control of these experts. The specialized knowledge, and also the facilities, required for the identification, introduction, biological study, mass rearing, and liberation of beneficial insects, precludes the employment of a large number of workers in this field and makes unintegrated and unorganized effort undesirable.

If the introduced beneficial insect is successful in its new habitat, nothing further need be done to facilitate its work; if it is unsuccessful, the environmental factors that result in its failure usually cannot be changed.[1] Occasionally it is feasible to rear a beneficial species which succumbs to the adversity of its new environment during a certain period of the year. It is then possible to make mass liberations of the insects upon the return of favorable conditions. This has been done in California, notably in connection with the predator _Cryptolaemus montrouzieri_, but even this work requires relatively few workers, per unit of acreage affected, compared to the application of insecticides. If the scope of present day biological control is to be broadened, however, it would appear that the greatest likelihood of a great increase in activity would be in the field of mass rearing and liberation of beneficial species requiring a certain degree of aid from man. Likewise the near extermination of beneficial species by the use of modern "super-insecticides" could conceivably lead to mass rearing and liberation of the beneficial species affected so as to reestablish the "balance of nature". In such an event, biological control would no doubt engage the thought and attention of a much greater percentage of potential readers, and the writer would accordingly happily enlarge the scope of the present chapter.

For a general survey of the subject of biological control, Sweetman's "Biological Control of Insects" and Clausen's "Entomophagous Insects" are recommended. For more detailed discussions of specialized aspects of biological control the literature cited in the above books may be consulted or, for the more recent contributions, the usual general indices and bibliographies or abstracting journals may be consulted.

---

[1] At least the climatic factors cannot be changed. However, it is believed that other factors affecting the success of parasites and predators, such as the stage of the host at a given period, the periodic drastic scarcity of the host, alternate host plants, lack of food for adult parasites, etc., are subject to manipulation by man. It is believed, for example, that by manipulation of the environment, a greater degree of biological control of black scale in "single-generation areas", as well as of citrus red mite, citricola scale, California red scale, and other pests, might be effected. Such manipulation may or may not be undertaken in conjunction with mass liberations of the parasites or predators.

THE HOST RELATIONSHIPS OF ENTOMOPHAGOUS INSECTS

Predators and Parasites:- Entomophagous (insect-eating) insects are generally divided into two groups: predators and parasites. The distinction is based on differences in the method of feeding. A predator requires more than a single individual of the host species for completing its development and a parasite passes its entire larval state within or upon a single individual insect. However, there is no definite division between predation and parasitism and it is sometimes difficult to know whether to call an insect a predator or a parasite. (790)

A host insect may be considered to be a meal for a predator and a source of continued larval sustenance for a parasite. It follows that predators feed on their succession of victims from the outside, whereas parasites may feed on their victim from either the outside or inside, usually the latter.

Fig. 222. A predator. Syrphid fly larva feeding on an aphid. Photo by Roy J. Pence.

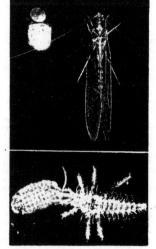

In figs. 222 and 223 are shown examples of predatism. The insects the predators are holding in their jaws represent a single meal in the succession of meals that are required to insure the completion of their development.

Fig. 224 shows an example of parasitism in which the larva of the aphelinid parasite (Aphytis diaspidis How.) feeds under the scale covering of the

Fig. 223. The green lacewing fly, Chrysopa californica. Above, adult and cocoon; below, larva with a young black scale in its jaws. X.4. Photo by Roy J. Pence.

host, Hemiberlesia lataniae Sign., but feeds externally on the eggs and body of the host. A. diaspidis is an example of an external parasite. In fig. 224 the larva has completed its feeding and has pupated. The pupa is surrounded by a ring of meconia (fecal deposits) which are deposited by the larva after it has finished feeding and just before pupation. They are always seen with the pupa.

In fig. 225 is shown an example of parasitism in which the internal parasite, the braconid Lysiphlebus testaceipes (Cresson), completes its development within the body of its aphid host and emerges from the host as an adult. The parasitized aphids live and feed until the parasite has completed its larval development. The vital organs of the host are the last to be consumed. The aphid furtunately ceases reproducing a few days after the parasite lays its eggs. The last larval instar of Lysiphlebus quickly empties the body of the aphid of all edible tissue. The aphid assumes the typical gray, bloated, spherical appearance of the parasitized "mummy" (fig. 225). The parasite larva chews away the ventral integument of the aphid and spins its cocoon. The silk of the cocoon adheres to the inside of the body of the aphid and to the plant tissue at the point of former contact of the removed venter of the aphid and the leaf (863). That is why parasitized aphid "mummies" stay attached to the leaf surface.

Lysiphlebus is about the same size as its host. Parasites, however, may be much smaller than their host, whereas predators are usually larger than their host.

Fig. 226 is a photograph of a microtome section of a California red scale on a lemon fruit. Inside the body of the scale may be seen the pupa of a hymenopterous internal parasite. Successive sections of this scale showed that the internal tissue had been completely removed by the parasite.

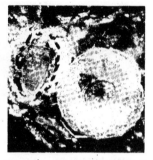

Fig. 224. Aphytis diaspidis, an external parasite on latania scale. Scale covering removed to show pupa of parasite. x 20. Photo by Roy J. Pence.

Fig. 225 (at left). The internal parasite Lysiphlebus testaceipes. Left, adult; right, "mummies" of parasitized aphids, showing exit holes made by emerging parasites. Photo by Roy J. Pence.

Internal parasites do not necessarily need to complete all pre-adult stages of their development within the host, but only the active (larval) stage. The transformation to the adult form may take place outside the body of the host. Again using the aphid as a host example, we may cite the interesting case of internal parasitism by the braconid parasite Praon similans Prov. in which the pupal stage is completed outside the host, but utilizing the host as a "roof". This interesting parasite was illustrated by Howard (447) and later discussed in more detail by Wheeler (942). The mature larva of Praon leaves the body of the aphid through an opening made in the venter of the abdomen, but the aphid is retained as a sort of roof which is held up by the tent-like walls of a light layer of silk (fig. 227). The parasite then spins its cocoon inside this tent. The cocoon is also spun of silk, which is packed tightly and thickly. In fig. 227 may be seen the exit hole made by the adult parasite upon leaving its curious tent.

Parasites usually lay their eggs on or in their hosts in various developmental stages of the latter. Different parasite species may oviposit on or in

Fig. 226. Photomicrograph of California red scale on a lemon fruit, showing the pupa of a parasite within the body of the scale. X35. Microtome section prepared by Miss E. P. Baker. Photo by Roy J. Pence.

317

Fig. 227. The pupal "tent" of an internal parasite Praon similans constructed beneath the body of its host. X10. Photo by Roy J. Pence.

different stages or instars of their host. Some lay their eggs within the eggs of their host (fig. 526). Some lay their eggs on the cocoon of the host or in its burrow or even on the plant surface away from the body of their prospective victims. Those parasites laying their eggs[1] on the body of the host may merely attach them to the outer surface or imbed them in punctures made by the ovipositor in the integument of the host. The extremely varied methods of oviposition used by different parasite species have been discussed in detail by Clausen (167). Fig. 228 shows an ichneumonid parasite ovipositing in the body of a lepidopterous larva. The appendages which helped to guide the ovipositor to the point of insertion of the egg do not enter the larva. The ovipositor is thrust into the body of the larva and the egg is inserted "hypo-dermically". The egg is larger than the diameter of the ovipositor, so it must become very elongate as it is compressed and forced through the ovipositor. This is a remarkable phenomenon when we reflect that no muscles occur throughout the length of the long ovipositor. The egg expands again after it is within the host insect. Some parasites have a remarkable ability to locate the host even when it is some distance removed from the surface. Their ovipositors are modified for the dual function of drilling through such substances as wood or well-packed frass, as well as ovipositing in the unseen host (fig. 229).

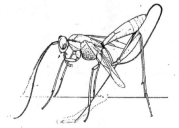

Fig. 228 (left). An internal parasite ovipositing in a larva. Greatly enlarged. Photo by Roy J. Pence.

Fig. 229 (right). A Macrocentrus female in position to inject an egg into the body of an oriental fruit moth larva. The latter is attacked only while in its burrow. From Finney et al (312) after Garman and Brigham.

The examples of parasitic insects given above all belong to the order Hymenoptera, which seems to contain the greatest number of parasites.

A distinction has been made between (1) insects feeding upon the bodies of other insects without killing them and (2) insects which at first avoid the vital organs of the host but later, just before pupation, consume the vital organs and kill the host. The former have been called parasites and the latter parasitoids. If this distinction is made, only a few entomophagous insects (the Strepsiptera and the bee louse, Braula) can be called parasites. Since the word parasite continues

[1]Not all parasites lay eggs; some deposit larvae. This would seem to be a useful adaptation, because if eggs are laid on the host larva and it sheds its skin before the eggs are hatched, the eggs are wasted.

318

to be used in its former inclusive sense in the greater part of the literature and by the entomologists actually engaged in biological control work, and since the distinction in terminology seems to serve no useful purpose, the term will be used in its original sense in this text.

Predatism is found mainly in the orders Neuroptera, Odonata, Plecoptera, Heteroptera, Coleoptera, and Diptera. A few species may be found among other orders, as, for example, the predatory thrips and the coccid-eating Lepidoptera. The parasites are highly specialized insects and might be expected to be found among the highest insect orders. A large number of species occur among the Hymenoptera and Diptera and a few among the Coleoptera and Lepidoptera.

The order Hymenoptera contains the most important group of beneficial insects. Parasites are found among 50 families. The 6 parasite species illustrated above were Hymenoptera. The most important families are the Braconidae, Ichneumonidae, Chalcididae, Trichogrammatidae, Ptermomalidae, Scoliidae, and Tiphiidae. All these, except the family Chalcididae, contain nothing but parasitic species. However, from the standpoint of citrus pests, the most important families of parasitic Hymenoptera are the Aphelinidae (Aphytis, Aspidiotiphagus, Coccophagus, Prospaltella, etc.) and the Encyrtidae (Anarhopus, Anagyrus, Comperiella, Encyrtus, Leptomastidea, Microterys, Metaphycus and Tetracnemus). These two families contain so many species, including all the above genera, attacking mealybugs and scale insects and these coccids are nearly everywhere the most important citrus pests. Among the braconids are found the most important enemies of the fruit flies in the genera [1]Diachasma and Opius. The Trichogrammatidae, Mymaridae, and Scelionidae are egg parasites. The Formicidae are probably the most important of the predaceous groups of Hymenoptera. (167, 863)

Whereas parasites develop in or on their hosts only while they (the parasites) are in the larval stage, predators attack their hosts while in either the immature or adult stages, or in both.

Hyperparasitism:- Hyperparasitism generally refers to parasitism other than the primary parasitism discussed above, although it is sometimes used synonymously with secondary parasitism. A parasite of a predator such as a ladybird beetle, a lacewing fly, or a syrphid fly, is not a hyperparasite because it is not a parasite of a parasite (790). Fig. 230 shows the Australian hyperparasite Myiocnema comperei Ashmead thrusting its ovipositor through the derm of a black scale, Saissetia oleae Bernard, and laying an egg in a primary parasite which is within the body of the scale (802).

Indirect Parasitism:- In this type of hyperparasitism, three insects are involved: first, the host of the primary parasite; second, the primary parasite; and third, the indirect parasite which attacks the host not for the purpose of feeding on it, but for the purpose of feeding on its parasite. The following definition has been suggested: "Indirect parasitism is that type of symbiosis in which the one parasite attacks a living host insect upon which it itself is incapable of feeding, for the sake of the primary parasite which it may harbor" (790). The classic example of indirect parasitism is the interesting chalcidoid parasite, Perilampus hyalinus, whose host relationship was first worked out by Smith (789). Quaylea whittieri, a parasite of the introduced black scale parasite, Metaphycus lounsburyi, is another example.

Fig. 230. Myiocnema comperei, an injurious hyperparasite in act of depositing an egg in a beneficial primary parasite concealed within a black scale. After Smith and Compere (802).

Secondary Parasitism:- A hyperparasite is a secondary parasite when it is a primary parasite of a primary parasite. This type of host relationship is closely

[1]Diachasma is a synonym of Opius.   319

allied to indirect parasitism for it brings about the same result, the destruction of the primary parasite, although the manner of accomplishing this end is quite different. The indirect parasite oviposits on the host of the primary parasite while the secondary parasite oviposits directly on or in the body of the young primary parasite. As might be expected, the life history of the secondary parasite is much simpler than that of the indirect parasite.

Smith (790) suggests the following definition: "Secondary parasitism is that type of symbiosis where a parasite destroys a primary parasite by direct attack, and not through the medium of the host of the primary parasite." He states that this phenomenon is of very common occurrence and that practically all species of primary hymenopterous parasites, especially the cocoon-forming groups, are affected. Great care is being exercised to avoid the introduction, or at least the liberation, of secondary parasites, although secondary parasites already existing in our native fauna can and often do parasitize the introduced beneficial species.

It is known that secondary parasites do not often become as important a factor as was once believed. According to Smith and Flanders (806), "The generally indiscriminate host relations of hyperparasites indicate that only rarely will an introduced species do more than replace a native species through competition. The introduced primary parasites that have brought about control of a pest are species that are equally successful in their native habitat, in spite of attack by their natural enemies. Unlike parasites of introduced pests, the introduced hyperparasite is likely to find all suitable niches occupied."

Tertiary Parasitism:- A hyperparasite in the third degree is a tertiary parasite of the species designated as the host of the series. It is a primary parasite of a secondary parasite; conversely, the latter is its primary host. We find tertiary parasitism defined by Smith (790) as "that type of symbiosis where a parasite is obligatory upon an obligatory secondary." Smith expressed the opinion that "this sort of thing can hardly go on ad infinitum as Burns would have us believe," and stated that it is doubtful whether an obligatory quaternary parasitic insect exists. They have been recorded, but an examination of such cases has revealed that they were only accidentally quaternary, the same species being by nature either secondary or tertiary. Apparently no findings to this date have changed the belief expressed by Smith in this connection. In fact, even tertiary parasitism seems to be a relatively uncommon phenomenon.

Solitary parasitism:- Solitary parasitism "occurs when only one individual parasite develops to maturity feeding on an individual host" (806). A parasite may be solitary in development for one of the following reasons:

(1) "Only one egg is deposited per host.
(2) "The host cannot support more than one individual parasite as with Metaphycus helvolus Comp.
(3) "Its larvae engage in internecine action.
(4) "It is inherently obligatorily solitary in development regardless of the size of the host and the amount of available food as with Coccophagus trifasciatus Comp." (806)

Gregarious Parasitism:- Some species of parasites deposit two or more eggs in or on a host insect that can support more than one individual parasite. These species are habitually gregarious and the occasional solitary individual is not adapted to pupate successfully in the viscous or fluid contents of the host occasioned by the lack of a sufficient number of individuals in the host. (806) A gregarious life in the host is thus obligatory for such a species. On the other hand, so many larvae may develop in a host insect that none can survive.

Superparasitism:- Superparasitism has been defined as follows: "Superparasitism is that form of symbiosis occurring when there is a superabundance of parasites of a single species attacking an individual host insect" (790). It is believed that superparasitism occurs most readily in species unable to distinguish between parasitized and unparasitized hosts.

BIOLOGICAL CONTROL

Scale insects afford a convenient example of superparasitism. Observation of a number of these insects will reveal that some individuals have only one parasite exit-hole, while others may have 2 or more. Those bearing 2 or more exit-holes, exhibit either superparasitism or multiple parasitism, the latter referring to the presence of more than one species of primary parasites.

It is not considered wasteful for a parasite species to lay more than one egg in a host because there is always a superabundance of eggs. The only disadvantage of superparasitism, from the point of view of control of the host, appears to be the loss of effective searching time (331). The time spent in oviposition on a host insect already doomed to destruction could be spent in searching for another insect. In other words, the effective life of the parasite is shortened. Likewise the searching activity of the parasite population is reduced by the loss of parasites by starvation when too many individuals exist in a single host insect.

It is believed, however, that superparasitism is not generally disadvantageous, from the standpoint of biological control, and that under certain conditions it may actually be advantageous to the parasite species, for if the host population becomes extremely low, relative to the parasite population, superparasitism increases, with the result that the chances of the host becoming eradicated over wide areas diminishes, an obvious advantage to the parasite species. This viewpoint is far removed from the one expressed by Fiske (316), who believed that superparasitism is nearly always disadvantageous.

Multiple Parasitism:- This phenomenon has been defined as "that form of symbiosis where the same individual host insect is infested simultaneously with the young of two or more different species of primary parasites" (790). Multiple parasitism is considered to be less frequent than superparasitism.

In multiple parasitism usually only one individual survives. When competing parasite species are equally fitted to survive, which species actually survives depends on (1) "the time of attack, the oldest consuming most or all of the nutrient host tissues, (2) the type of mouthparts and aggressiveness, (3) the secretion of substances toxic to the rival, and (4) the relative length of life cycle" (790).

An example of the possible economic effect of multiple parasitism is offered by Smith and Flanders (806). The parasite Macrocentrus ancylivorus is extrinsically and intrinsically superior to another parasite, Ascogaster carpocapsae, against the oriental fruit moth. Although the first instar larva of A. carpocapsae, when inhabiting the same host individual as the first instar larva of M. ancylivorus, is invariably destroyed, when the two species are reared simultaneously the total number of parasites produced is greater than if only one species is reared.

Some investigators have been led to believe that multiple parasitism may result in less effective biological control of an insect species than would be accomplished by a single effective parasite species. Smith (792), however, after a consideration of the many aspects of the problem, came to the conclusion that "on theoretical grounds as well as on the data so far available, the policy of entomologists in introducing all available primary parasites on an injurious species is justified, and that this policy should be continued." He points out another factor that should not be lost sight of in this connection, namely, "the superior 'balancing' effect that several species of natural enemies have over a single one. Sudden changes in environment often serve to reduce the efficiency of a parasite and cause extreme fluctuation in the abundance of the host, to the detriment of our crops, whereas if there are several species at work they are not all likely to be affected in the same degree, and some will step in and fill the gap left by the delinquent one. This tends strongly toward maintaining a constant population of phytophagous insects, as opposed to a fluctuating one."

Synchronization of Life History of Parasite and Host:- It is obvious that a sine qua non for the success of a parasite is the favorable synchronization of its life history with that of its host. The hymenopterous parasite Metaphycus lounsburyi may be cited as an example in this connection (800). This parasite of the black scale, introduced into California in 1918, spread very rapidly in the coastal areas. In these areas there are two overlapping generations of black scale per year. This results in any instar or stage of the insect being found at any period of the year. Such a condition is favorable to M. lounsburyi, for the developmental stage of the scale suitable for oviposition is the "rubber stage" which could be found at any period of the year. In the hotter inland areas the black scale has only one generation per year, because of retarded development during the hot summer months. There the "rubber stage" insects are found almost exclusively in the late fall and winter months, and conditions are favorable for the development of the black scale only during this limited period. The majority of the parasites succumb without being able to reproduce during the remainder of the year. M. lounsburyi never became established in significant numbers in the inland areas of California because its life history did not synchronize in these areas with that of its host.

After Metaphycus lounsburyi became established throughout the coastal areas for a few years it tended to "even up" the hatch of black scale for, as stated before, it oviposits in scales in the "rubber stage", which occurs just before maturity. "In the course of time only full-grown scales with eggs, or very small scales, were left, and the broods became even. This was to the disadvantage of the parasite, as there were long periods of time when no 'rubber stage' scales were available, and the parasite populations were reduced to the very small number that could find an occasional scale developing unevenly in the brood. Pepper trees, probably because of the more or less continuous production of offshoots which are very favorable for scale development, carry uneven black scale infestations; the scale has therefore been under good biological control on these trees since the establishment of lounsburyi. Formerly, pepper trees were so badly infested that many large trees were destroyed." (800)

Metaphycus helvolus, introduced in 1937, proved to be much more successful in combating black scale than did M. lounsburyi. It is able to attack a greater range of developmental stages of the black scale and greater numbers are able to "carry over" from one favorable period to the next. The life history of M. helvolus is less incompletely synchronized with its host than is that of M. lounsburyi, particularly so in the "uneven hatch" of the black scale in coastal areas of California. In inland areas, M. helvolus is not as effective in controlling the black scale as it is in coastal areas because scales of a suitable age for parasitism are very rare during the summer months. Two insectaries, in certain inland areas, are now engaged in rearing M. helvolus for liberation when the black scale is of the proper size for parasitization in the fall.

THEORETICAL BASIS OF BIOLOGICAL CONTROL

The study of the intricate relationships between the many factors, physical and biological, which influence the density of an animal population has offered a supreme challenge to biologists, dating back to the pioneer work of such men as Buffon, von Humbolt, Malthus, Saint-Hilaire and Haeckel. It was Haeckel, in 1869, who first used the word "oekologie" as the name for this branch of biology, defining it as the "relation of the animal to its organic as well as its inorganic environment, particularly its friendly or hostile relation to those animals or plants with which it comes in contact." Many biologists have expounded the intricacies of ecologic phenomena and have presented much descriptive material and valuable data. Relatively few have attempted an analysis, organization, and simplification, of animal ecology, and it is these few whom biologists will remember with special appreciation. In 1927, Elton in his "Animal Ecology" was among the first to attempt to show that there can be order even in complexity--always a laudable contribution, especially in biological sciences, in which complexity often leads to bewilderment and a sense of futility. In 1937, Chapman, in another book of the same title, "Animal Ecology", again aimed at analysis and organization.

Dr. Chapman's views were well known a few years before the publication of his book, and those of us who were students during those years well remember the profound influence they had on the teaching of the subjects of insect ecology and insect epidemiology, and on the thinking along those lines. This influence lasted for about a decade before it was supplemented by new concepts which greatly de-emphasized certain aspects of the previous points of view and placed others in their true relationships in accordance with a new and more dynamic interpretation of the environmental complex.

Biotic Potential vs. Environmental Resistance:- Chapman considered organisms to have innate biotic characteristics which could be summed up in the term "biotic potential", defined as the "inherent property of an organism to reproduce and to survive." Biotic potential was considered to be the sum of the animal's (a) reproductive potential, as influenced by the number of young produced, the sex ratio, and the number of generations in a given period, and (b) survival potential[1] as based on its ability to obtain adequate food and protection. Pitted against this biotic potential of a species was the "environmental resistance" comprising (a) physical factors, such as are commonly recognized as comprising weather and climate, and (b) biological factors such as competition for food, competition for space, competition for suitable shelter, and predators and parasites. (156)

As a statement of the factors affecting an animal population, there has been no important divergence of opinion from Chapman's presentation, but there has certainly been a divergence of views regarding their relative importance. This divergence from what might be considered as the classic point of view was engendered by the addition of a basically new concept to the study of animal populations which was independently conceived by an Australian entomologist, A. J. Nicholson, and Professor Harry S. Smith of the University of California.

Capacity of the Environment:- Smith (806) has pointed out that the fate of an introduced insect species in its new environment depends on the capacity of the latter for this particular species. The California citrus areas had a high capacity for the cotton-cushion scale, Icerya purchasi, because this species was quite distinct taxonomically from other citrus-inhabiting Homoptera and consequently was not attacked by the native entomophagous fauna. Citrus trees soon became festooned with the white egg masses of this species, and citrus growers, with good reason, anticipated the doom of California's citrus industry. Fortunately, the California environment also had a high capacity for a predaceous beetle introduced from Australia to combat the cottony-cushion scale, namely, the famous vedalia, Rodolia cardinalis. The vedalia, as an enemy of the cottony-cushion scale, filled a niche which had previously been unoccupied; in other words, the California environment had a high capacity for the vedalia. When the dipterous parasite, Cryptochaetum iceryae, was later introduced from Australia to prey on cottony-cushion scale, it merely resulted in the reduction of the vedalia population, because the capacity of the new habitat for enemies of the cottony-cushion scale was limited; only a limited number of insect enemies of this species could be supported, whether they were of one or more than one species.

Harry S. Smith Professor of Entomology and Chairman of the Division of Biological Control, College of Agriculture, University of California.

Often entomophagous insects have been introduced into a country in which related species with similar intrinsic ability to combat a given host insect already existed. The result of such introductions has been merely to reduce the population of the related entomophagous species, with no advantage from the standpoint of the combined effect against the host insect. If, however, the introduced species has intrinsic qualifications making it superior to other enemies of the host species in question, it will not only tend to eliminate the other species, but will become established in larger numbers than the previous combined population of enemies of the host insect. If the searching power of the new entomophagous insect is not re-

[1]Not all biologists agree with the placement of survival potential under biotic potential; some believe it should be placed on the other side of the equation, under environmental resistance.

duced by native parasites it may practically eliminate the native species and its
success against the host insect may be spectacular.

It will be noted that the success of an introduced insect species depends
primarily on the nature of the biotic rather than the physical factors of its new
environment, presuming of course, that the climate of the new environment is favorable. It will now be shown that this is due to the fact that biotic factors,
especially natural enemies, increase in the intensity of their effect with increase
in the population density of the new species.

Density-Independent and Density-Dependent Factors as They Affect Insect Populations:- Howard and Fiske (449) distinguished two categories of natural causes of
mortality among insects. In one category are factors which cause a constant per
cent of mortality regardless of the abundance of the insects; these they called
"catastrophic" factors. In the other category were factors causing an increasing
per cent of mortality as the numbers of the host increased; these were called
"facultative" factors. Smith (794) designated these factors, respectively, as
density-independent and density-dependent mortality factors. It can readily be
understood that the physical factors (climate, weather) offering "environmental resistance" as conceived by Chapman are density-independent,[1] that is, they vary independently of variations in the insect population, and the biological factors
(competition for food, competition for space, competition for shelter, predators
and parasites) are density-dependent, that is, they are affected by a rise or fall
in the population of the host insect.

The economic entomologist is interested in the mortality factors which can
determine the average population density or equilibrium position of a species. If
such factors can succeed in maintaining an equilibrium position at or below the
"economic zero" point they are important, otherwise they are not. Smith (794) has
shown that density-dependent mortality factors can determine the population equilibrium position of a species, and that density-independent factors operating
alone can never do so. In the case of a host insect under biological control, the
per cent mortality caused by biotic factors, particularly entomophagous insects,
increases relative to that caused by physical (climatic) factors when the host
density tends to increase and, conversely, decreases when the host density tends to
decrease. The density-dependent factors are the only ones that are truly regulatory.

Weather may fluctuate and may cause fluctuation in numbers of the host insect
and may thus be of economic importance. A distinction must be made, however, between these fluctuations and the average population densities controlled by biotic
factors.

Injurious insects have often been accidentally moved from one country where
their average population density had been satisfactorily low, from an economic
standpoint, to another country where their average population density rapidly increased to an alarming extent. This increase was not caused by a more favorable
climate, in fact, the climate sometimes was less favorable. The increase was
caused by the absence of density-dependent factors which operated in the old environment, but not in the new. Since such density-dependent factors as competition
for food, competition for space, and competition for suitable shelter presumably
would operate to about the same extent in the new environment as in the old, the
density-dependent factors which were not automatically attained in the new environment, namely, the predators and parasites, could logically be suspicioned as being
the "missing link" in the combination of factors which provided successful control
of the pest in the old environment. It is because the entomophagous insects are
the biotic factor which can be transported and controlled by man that they are the
most important consideration in biological control. Therefore a brief consideration of the factors which determine their success and economic value will be attempted.

[1]Climate is a density-dependent mortality factor only in the special case of insects existing
in protective niches in the environment which are limited in numbers. Individual insects in
excess of the number which can be maintained in these ecological niches are destroyed by the
climate and thus climate can be said to increase per cent of mortality as population density increases. Likewise climate indirectly affects the equilibrium position of an insect population by
favorably or adversely affecting the biotic density-dependent factors. (794)

BIOLOGICAL CONTROL

Reproductive Capacity:- In his "Animal Ecology", Dr. Chapman anticipated a "revision" of ecological concepts as he presented them, as new information is brought to light, but he could hardly have envisioned the complete subservience of certain characteristics of "biotic potential" in the new point of view initiated by Nicholson and Smith. Smith (795) states that "There is no positive correlation between the maximum possible rate of reproduction of a species and its average population." It follows that reproductive capacity may logically be divided into two types: potential and effective, although this differentiation has played no part in the thinking of many biologists.

The species which can reproduce most rapidly is not necessarily the most abundant in numbers of individuals, a point which was strongly emphasized by Darwin in his "Origin of Species". It was pointed out by Smith that the parasite's capacity to lay eggs is far in excess of what is needed to enable it to control its host. It follows that the effect of such a parasite on the host population is not likely to be limited by its potential reproductive capacity. On the other hand, the effective rate of reproduction of a parasite has an effect on the rate of change of the host population which is determined by the difference between the effective rate of increase of the parasite and that of the host.

Searching Ability of the Entomophagous Insect:- Much emphasis has been placed by Smith (795) on the searching ability of entomophagous insects. It is considered to be by far the most important property of an entomophagous insect affecting the population density of the host insect. If the parasite is to be the biotic factor that is to bring about equilibrium in the population of its host, wherein birth rate and death rate reach equality, it must find and destroy that portion of the progeny which, if not destroyed, would result in an increase in the population of the host. However, it is important to bear in mind that this equilibrium condition could be maintained at any density, and in biological control we are concerned with what this density may be. This in turn is determined by the capacity of the entomophagous insect to discover the host in relation to the population density of the latter.

If a phytophagous insect is introduced into a country, it eventually reaches a condition of equilibrium at which point environmental resistance, possibly including some native entomophagous insects, is already destroying as many individuals as are being produced. However, this equilibrium position of the population may be far in excess of "economic zero". It is then essential, in order to effect biological control, to introduce another species or group of species of entomophagous insects which, when added to the other factors, physical and biotic, comprising the environmental resistance, will result in the destruction of as many individuals as are produced, but at a population level which is at or lower than the "economic zero". Under such conditions the host insect will be more widely scattered and the introduced species, to be effective, must have adequate searching ability. Here again Smith was led to an interesting deduction which would be overlooked in a less critical examination of the problem. An effective parasite--one that is able to find and destroy the surplus progeny of a host at a low host density--may destroy no greater percentage of the host than an ineffective parasite which allows the host to be maintained at a much higher density. Per cent of parasitization does not measure the relative effectiveness of two parasites. Searching ability is the paramount consideration.

The searching capacity of a parasite depends on a number of qualities among which the important ones are (1) its power of locomotion, (2) its power of perception of its hosts, (3) its power of survival, (4) its agressiveness and persistence, (5) its power to use its ovipositor, (6) its power of egg deposition, (7) its power to produce females in place of males, (8) its power to develop more rapidly than the host, and (9) its power to occupy host-inhabited areas (795, 331). The black scale parasite Metaphycus helvolus is deficient mainly in the slowness with which it oviposits. When ants are abundant, the female is prevented from ovipositing, for an ant is likely to encounter the parasite before it is through ovipositing. However, other black scale parasites not possessing this weakness have others even more serious. The only attribute of a parasite which can not be determined in the laboratory is its ability to occupy host-inhabited areas.

The above qualities, rather than reproductive capacity, determine a parasite's effectiveness, for reproductive capacity is usually, if not always, more than ample. In fact, parasites which lay the greatest number of eggs are the least effective searchers. Searching capacity tends to be inversely correlated with reproductive potential (331). One would expect a parasite species laying the fewest eggs to be the one most capable of maintaining the host at a low density.

Different parasite species have different host-finding abilities. De Bach and Smith (224) determined the difference in the host-finding ability of 2 pteromalids, Mormoniella vitripennis and Muscidifurax raptor. They used house fly puparia, distributed at random in a container filled with barley, as the hosts. Fig. 231 shows the difference in the rate of increase of the two parasites as the population density of the host is increased. On the assumption that each host discovered is oviposited in and produces a parasite, the curves shown in fig. 231 can also be considered as a measure of searching ability. It will be noted that both species found their hosts at a rate that increased with increasing host density, but there was a marked deceleration in the rate with which Mormoniella found its host as the population density of the latter increased. Since Mormoniella spends as much as 6 or 8 hours on each host it discovers, the percentage of time spent in non-searching activities increased with increase in the population density of the host, and this was considered to be the cause of the deceleration in rate of increase even at low host population densities.

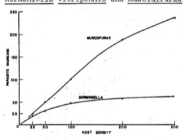

Fig. 231. Difference in the rate of increase of two parasites with increasing population density of their host. Redrawn from De Bach and Smith (224).

The curves shown in fig. 231 agree with a formula worked out by Gause in his "Struggle for Existence" to express the relation between increase of the predator population with increase in the concentration of the population of the prey (355). This relation was mathematically expressed as follows:

$$y = a \left(1 - e^{-kx}\right)$$

In the present experiment, y=parasite increase in terms of number of hosts discovered, a= limit approached asymptotically by y, e= base of natural logarithms, k= a constant determining the rate at which the change of y with host density decelerated, and x= host density.

The Nature of the Distribution of the Host Species:- For effective biological control, the nature of the distribution of the host insect is important for the same reason that the searching ability of the parasite is important, namely, that it affects the population density at the point of equilibrium. Two host species with the same average density of population per given area may yet have an entirely different type of distribution or dispersion within that area. Some species are closely grouped in colonies even though these colonies may be widely scattered. Other species are more evenly scattered throughout the infested area. The distance which an entomophagous insect must travel to reach its prey is much less when the host insects have a colonial type of distribution than when they are uniformly separated. It follows that of two entomophagous species of equal searching ability, the one attacking a host species with a colonial type of distribution will be more effective in keeping its prey at an "economic zero" level than one attacking a host species that is widely scattered. It can be seen that the efficiency of an entomophagous insect depends on the nature of the distribution of its host as well as on its own power of discovery.

BIOLOGICAL CONTROL

An example of the influence of host distribution on the efficiency of an entomophagous species is afforded by the relative effectiveness of 2 species of coccinellid predators introduced into California many years ago. The first to be introduced was the vedalia, Rodolia cardinalis, to control the cottony-cushion scale, Icerya purchasi, an insect with a colonial type of distribution. The vedalia was highly effective in keeping the cottony-cushion scale under control and, in fact, its success was so spectacular that it became one of the classic examples of perfect biological control. The other predator, Rhizobius ventralis, was introduced to attack the black scale, Saissetia oleae. It appeared to have about the same intrinsic ability as Rodolia cardinalis and a similar success was expected in the control of the black scale. It is now well known that R. ventralis failed utterly to live up to expectations. The great difference in the success of these two predators is not to be found in any difference in their biotic potential or searching ability, but in the nature of the spatial distribution of their host. The larva of the vedalia feeds within the large egg-sac of its host and has enough food regardless of the scarcity of its host. The larva of Rhizobius must feed on black scale nymphs that have traveled some distance from the parent scale and may succumb when the host population is sparse (795).

The black scale was finally controlled not by a predator, but by a hymenopterous parasite, Metaphycus helvolus. Parasites are usually more effective against host insects that have a uniformly-scattered type of distribution because they fly in the host seeking stage. Predators must seek the host in the larval stage as well as the adult stage, and since they are not winged in the larval stage they have relatively low powers of locomotion and low searching ability. They are ordinarily effective only against host insects with a colonial type of distribution, such as cottony-cushion scale, mealybugs and aphids.

The greater effectiveness against mealybugs of certain imported hymenopterous parasites, as compared to the already established predator, Cryptolaemus, was prognosticated by Smith and Compere on the basis of their greater searching ability. To quote these investigators: "The citrophilus mealybug normally becomes comparatively scarce in the groves during the period from July on. During this season the mealybugs occur to a considerable extent as isolated individuals scattered over the tree and many of these are able to escape destruction by a predator such as Cryptolaemus. These mealybugs are not sufficiently abundant to do any harm, but they give rise to the heavy spring generation which is the one that is most likely to cause injury. It was thought that an active hymenopterous parasite, which should be most effective during this portion of the season, would be better adapted to the destruction of these isolated individuals than a predator such as Cryptolaemus, and that if an effective parasite of this kind could be found and established, it might serve to reduce very greatly the numbers of mealybugs which produce the spring hatch". (803)

Among the 4 species of mealybugs commonly found on citrus, Baker's mealybug, Pseudococcus maritimus, is less satisfactorily controlled by "crypts" than are P. citri and P. gahani. Among the factors accounting for this is the fact that Baker's mealybug lays its eggs in small masses throughout the tree instead of concentrating them into fewer but larger masses as do P. citri and P. gahani. Baker's mealybug is thus less amenable to successful control by insect enemies with relatively limited searching capacity than are the mealybugs which are less scattered in their distribution. By the same token, the long-tailed mealybug, P. adonidum, is also not as effectively controlled by Cryptolaemus as P. citri and P. gahani because it gives birth to active young, which do not remain concentrated in distribution as long as the eggs in the egg masses of the latter species. (482)

Another factor responsible for the greater effectiveness of the parasites is the fact that they can complete their development in one host insect while predators need more than one and usually many host insects. The explanation of this apparent paradox is that the parasites, being better searchers and needing less food, are able to complete their development when the host population is very low, and they are therefore effective at low host densities (331). From the economic standpoint, the most valuable entomophagous species is the one that can keep a pest at a low density.

Low host density favors the parasites relative to predators, in still another way. The power of hymenopterous parasites to produce females in place of males increases, within limits, as the host density decreases, and as a consequence the relative power of searching is increased. (331)

Flanders (331) has brought out the interesting point that the hymenopterous parasites are such effective searchers that it is fortunate for them that their efforts are not coordinated, or they would exterminate their prey and themselves succumb. As pointed out by Nicholson (653), the action of each parasite is independent, and as the parasites become more numerous their searching areas overlap. The chances increase of one parasite searching in areas already searched over by another, and the searching efficiency per individual decreases. This gives host populations which would otherwise be exterminated a chance to survive. In addition, the fecundity of a hymenopterous parasite may decrease with increasing population density of the parasite population because of the fact that an individual parasite is reluctant to oviposit on hosts upon which others of its species have already oviposited or walked over (225).

Population Oscillations and the Interaction of Host and Parasite:- The rhythmic rise and fall in the population densities of animals results in oscillations characteristic of the species in a given environment. In regions of the earth exhibiting seasonal climatic variations, oscillations are of course affected by the seasonal variations in temperature, humidity, and duration of daylight, which in themselves cause considerable fluctuation in population. Among subtropical pests, the greenhouse thrips, Heliothrips haemorrhoidalis, attacking avocados and oranges out-of-doors in southern California, is a good subject for the study of seasonal population fluctuation because the vitality of the tree usually is not seriously affected by populations allowed to develop for years without pest control treatment. There may be as many as 5 or 6 generations of greenhouse thrips per year. In winter the number of active insects is greatly diminished by low temperatures, and in the more temperate sections the winter may be passed almost exclusively in the egg stage. These eggs hatch in February. Population studies show first a slow increment in population density, but later in the year the population rises abruptly in accord with the well-known sigmoid or logistic curve normally followed in population growth.

One can arrive at the characteristic form of the curve for population increase by applying the well-known Verhulst-Pearl equation:

$$D = \frac{A}{1 + Ce^{-kt}}$$

when D= population density, A= equilibrium position for the species in question, C= initial population, e= base of natural logarithms, k= a constant determining the rate of change of population density with increase in time, and t= time. Smith (794) plotted a curve (fig. 232), based on the preceding formula, showing the increase with time in an adult population, beginning with a population of 10 individuals and equilibrium position (A) of 500, the number of generations (2.72) necessary for the population to reach D=A/2, and a coefficient of increase of 3 at D=0.

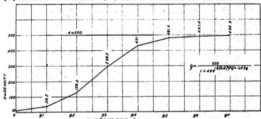

Fig. 232. Population growth curve. After Smith (794).

It was also assumed in plotting the curve that the species consisted entirely of females which produce only females parthenogenetically. This would be in accord with the actual situation with respect to the greenhouse thrips, the seasonal population curve for which is shown in fig. 522. In Smith's curve, an extremely small coefficient of reproduction was purposely used to simplify computations and to keep the graph within reasonable bounds.

The population of the greenhouse thrips reaches a peak every fall before it is reduced by adverse weather conditions. Not only does one observe this annual seasonal fluctuation, however, but in addition the mean population fluctuates from year to year, and this type of fluctuation appears to be influenced mainly by minimum winter temperatures.

With citrus red mite, Paratetranychus citri, another pattern of seasonal fluctuation is noted. Population peaks occur during the 2 seasons of moderate weather conditions, the spring and fall. The citrus aphids reach their peak of population density during the spring. Again, as with all species, the mean population density varies from year to year. Thus all species have their characteristic pattern of population fluctuation.

Biotic factors also affect the nature and the degree of population oscillations, and these are not necessarily seasonal. We shall be concerned here with interactions between the host insect and its entomophagous enemies.

Insect species have characteristic "equilibrium positions" with regard to mean population densities, which may change from year to year, as stated before, because of changing climatic influences. Both the host insects and their parasites oscillate perpetually about their equilibrium positions (655). "When the two factors host population density and parasite population density are considered together, it is apparent that their relative numerical relations alone are sufficient to explain population oscillations about an equilibrium position obtained over successive generations" (225). Since the parasite population is determined by the previous generation of host insects, there is a lag in the oscillation of the parasite in relation to its host, graphically shown in fig. 233. This lag causes the oscillations to tend continually to increase in amplitude. It will be noted from the graph that when the host is at its "equilibrium position" or "steady density" the parasites are either at their maximum or their minimum density, thus causing the host to either decrease or increase.

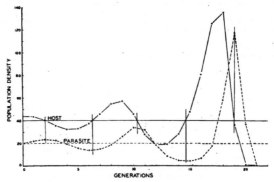

Fig. 233. The lag in the oscillation
of the parasite population in
relation to that of its host.
After Smith (795).

The population oscillations can not, of course, increase in amplitude indefinitely, but, as pointed out by Nicholson and Bailey (655), they can become sufficiently great to result in regional extermination of the host and the formation of small, widely-separated groups of the host species within which the oscillations induced by the host-parasite interaction will continue, but with some of the hosts escaping into surrounding unoccupied country. The foregoing hypothesis may explain the "spotted" distributions of insect populations, not correlated with special "ecological niches", that can often be observed in nature (795).

Smith (795) points out that "if a parasite having a low power of discovery interacts with a host having high powers of dispersal, extreme oscillation at a high average density will ensue. A parasite having high powers of discovery interacting with a host with low powers of dispersal will tend to produce slight oscillation at low average density, etc." The predaceous beetle Rhizobius, which was imported to prey on the black scale, has been used as an example of an entomophagous insect having low powers of discovery attacking a host having high powers of dispersal. If this insect were the only natural enemy of the black scale it would result in extreme oscillation at a high average density.

Advantageous Sequence of Parasitism:- It was once believed by some workers in the field of biological control that no single species of entomophagous insect is able to control an injurious species, but that a sequence of species is necessary to attack the injurious insect in different stages of development. It is now known, of course, that injurious insects have on a number of notable occasions been controlled by a single predaceous or parasitic species when all other species combined had failed. Nevertheless it stands to reason that a number of entomophagous species, neither one of which is capable of controlling the pest alone, might successfully control it by the combined effect of their attacks on successive instars or stages of the pest, provided that their habits lead the various species to attack different instars or stages of the injurious species.

FERTILIZATION AND SEX RATIO IN RELATION TO BIOLOGICAL CONTROL

As previously stated, the order Hymenoptera contains the majority of important parasitic species. Certain features of the interesting biologies of these insects are of interest in connection with biological control. The following is based on a study outline prepared by Smith and Flanders (806).

Among the Hymenoptera, the normal male has only one parent, that is, it is uniparental, and the female may be either uniparental or biparental. Those which are biparental develop only from haploid eggs that become diploid by fertilization. There are species in which all the females and males are produced by unmated females, and these are said to be "uniparental in reproduction", while species in which the majority of the females are produced by females mated with uniparental males are said to be "biparental in reproduction".

As might be expected, the Hymenoptera must have ovaries of different types. Ovaries producing only diploid eggs are called thelyotokous, those producing both haploid and diploid eggs are called amphitokous, and those producing only haploid eggs are called arrhenotokous. The latter type are found in females of all biparental species as, for example, the well-known black scale parasite, Metaphycus helvolus.

Mated females with haploid eggs and with a plentiful supply of sperm usually produce offspring of both sexes, but in the majority of species the sex ratio is highly variable and depends directly or indirectly on external factors. It will be remembered from Chapter 2 that the females of most insect species possess a seminal receptacle or spermatheca which receives the sperm during coition and discharges sperm into the oviducts or the vagina for a considerable period thereafter when eggs pass through. The stored sperm may retain its viability throughout the life of the female and only one fertilization is necessary even though a large number of eggs may be laid. In the case of some hymenopterous parasites, if egg deposition is too rapid, the spermatheca is fatigued or is devoid of spermathecal fluid and

330

consequently does not discharge sperm. The result is the production of males only. The rapid egg deposition can occur only when the population density of the host is high. Therefore an increase in population density leads to an increase in the proportion of males produced. On the other hand, when the population density of the host is low, the proportion of females produced is high. This is an obvious advantage, from the standpoint of biological control, for we are interested in the maximum searching capacity of a species when the population of the host is low, so that it may be retained in a state of low density.

A low host density may also affect the sex ratio by causing an increase in the size of the individual host insect. The male may not be able to consume enough of the host tissue to render the host habitable. Likewise the female parasite is more apt to use her stored sperm when ovipositing on large insects than when ovipositing on small insects, and thus sperm is wasted which could later be used to insure female progeny (examples: Tiphia popilliavora and Metaphycus helvolus).

In the case of gregarious species, superparasitism, which usually takes place when the host individuals are scarce, may cause a selective elimination of females, for the male can attain maturity on less food than the female.

Among intrinsic characteristics of biparental species determining sex ratio, the number of eggs deposited at one insertion of the ovipositor may be a factor, for the greater the number of eggs deposited at a time, the greater the proportion of males, other things being equal. The first eggs to be laid are fertilized and the last are apt to be unfertilized. Trichogramma females when ovipositing in Estigmene acraea deposit 3 eggs at a time. The first two are fertilized and the third unfertilized, resulting in a 2 to 1 sex ratio (2 females and 1 male). (318)

Relation of Sex Phenomena to the Mass Rearing of Parasites:- In the orchard, where the female parasite must search far and wide for the host insects, there is not likely to be a surplus of males. In the insectary, where an abundance of hosts can be supplied, a surplus of males can be produced. The problem in mass rearing is to insure that as small a percentage as possible of the progeny should be males, for every male born, under proper circumstances, could have been a female. A surplus of males means a waste of potential females. The density of the host population in an insectary where mass rearing takes place should be at such a level as to permit of a maximum production of females. This increases the rate of propagation and consequently the volume of production per unit of time. Likewise in the colonization of the parasites in the orchard, they should be released when field conditions are optimum for the production of females, for the possibility of the establishment of the species in an orchard is thus enhanced. An interesting discussion of the relation of sex ratio to the practical problems of biological control is presented in an article on "The Surplus Male" by Flanders. (330).

Among intrinsic factors serving to increase the efficiency of the hymenopterous parasite is its ability to (1) absorb its eggs prior to ovulation (oosorption) and (2) its ability to store ripe eggs after ovulation. Considerable research on these phenomena has been done by Flanders (325). He believes that oosorption and storage of ripe eggs are adaptations for maintaining the reproductive capacity of the species when environmental conditions are unfavorable for oviposition and for permitting the restraint in oviposition which is so essential to host selection. Oogenesis and oosorption may occur simultaneously or, in the prolonged absence of hosts, oogenesis may cease and oosorption become complete. Species in which this occurs are said to be in a condition of "phasic castration", the function or development of gonads being temporarily inhibited. Metaphycus helvolus feeds on the body fluids issuing from the wound on the host insect made by its ovipositor. In this species "the resumption of oogenesis after phasic castration is solely dependent on host-feeding."

HOST FEEDING BY ADULT PARASITES

It is not generally known that many species of hymenopterous parasites feed on their hosts while they (the parasites) are in the adult stage. This is known as host-feeding or host predatism.

Flanders (322) has given us an insight into many peculiar habits of the hymenopterous parasites, among which is host-feeding:

"While M. helvolus is within its host it does not store up food sufficient for the development of its complement of eggs. It is necessary therefore for the adult female to feed on the host's blood as well as on honey dew, the usual food of chalcids.

"To accomplish this, the sting again comes into play. This time no egg is laid but a substance is injected into the scale which keeps the wound open and enables the female to suck up the host's blood. The scale dies as a result of this treatment. Females that lay eggs every day must frequently feed on the host's blood. One female was thus able to lay 740 eggs over a period of 60 days, laying from 3 to 18 eggs a day."

Host-feeding may be as great or even a greater factor in the reduction of the host population than parasitization. Metaphycus helvolus may kill from 70 to 97% of a black-scale infestation over a period of several months. Parasitization accounts for only 20 to 25% of this mortality and host-feeding the remainder. Black scale fed upon by Metaphycus helvolus are flattened and dry. Host-feeding is most important in the early phases of a black scale infestation; as the scales increase in size parasitization becomes relatively greater as a mortality factor. While per cent parasitization is not markedly affected, within limits, by the population density of the parasites, the per cent of scales killed by host-feeding follows a general upward trend as the parasite density increases (220).

An investigation was made of certain citrus orchards in which the red scale population has remained low despite the absence of pest control treatments. The resulting survey encouraged the belief that the principal red scale parasite, the golden chalcid, Aphytis chrysomphali Mercet, is the principal cause of the low red scale populations in the 14 orchards under observation. The main effect of the golden chalcid is believed to be the feeding of the adult on the body fluid of the host. The parasite, with the aid of its ovipositor, forms a straw-like wax tube in the body of the red scale and sucks the body fluid from the latter. It is hoped that with further knowledge on the subject, a biological method of control of the red scale, possibly aided by insectary culture and periodic colonization of the golden chalcid in the orchards, can be effected, at least in the majority of localities.

Techniques have been devised by which the effect of the natural enemies of the California red scale, as well as of other pests, may be assessed in the orchard. Figure 234 shows such a technique in operation. It involves the use or organdy sleeve cages placed over branches infested with red scale. One sleeve is left slightly open at both ends so that natural enemies have ready ingress and egress. The cloth of the other sleeve is impregnated with technical DDT by dipping it in a solution of DDT and allowing it to dry. This sleeve is then placed over a branch so that the foliage does not touch the cloth and so that it is closed at both ends.

Since the red scale is sessile, neither the tree nor the insect host contact the DDT, but the active predators or parasites of the red scale eventually come into contact with the DDT-impregnated cloth of the closed sleeve and are eliminated. The host-insect population in this sleeve then increases as it would if there were no natural enemies.

A comparison of results in the paired sleeves gives a direct evaluation of the effectiveness of the natural enemies, especially when the entire tree is used as a check against results in the open sleeve. (223)

THE RELATION OF INSECTICIDES TO BIOLOGICAL CONTROL

This subject would be too vast to allow of an adequate treatment here. It is intended merely to discuss briefly the relation of insecticides to biological control as it relates to subtropical fruit insects in California. This has posed a

Fig. 234. Field experimentation on the effect of natural enemies on the California red scale. Left, closed DDT-impregnated sleeve which eliminates and excludes natural enemies; right, open, untreated sleeve which permits natural enemies ready ingress and egress. Courtesy P.H. DeBach.

particularly serious problem to entomologists since the development, since World War II, of a number of synthetic insecticides which have a prolonged residual contact effect. Thus some insects may be killed by walking over a surface to which DDT has been applied as long as 2 months previous. This residual action is selectively destructive to parasites, for the pupal stage of these insects may be within their host at the time an insecticide is applied and can usually survive an insecticidal treatment not having residual effect, even though the adult parasites which happen to be on the tree at the time of treatment are destroyed. Upon emerging, these insects are free to oviposit in host insects which did not succumb to the treatment or to fly to nearby untreated orchards. Likewise, in the case of stomach poisons or insecticides not having a residual contact effect, it is possible for parasites to reenter a treated orchard and rapidly reestablish themselves, whereas with such insecticides as DDT, benzene hexachloride, toxaphene, chlordan, parathion, etc., large numbers of parasites entering an orchard are killed before being able to oviposit, merely by walking over the insecticidal residues.

Some examples of Interference with Biological Control by Insecticides:- Previous to the advent of DDT, certain insecticides were known to adversely affect the work of predators and parasites although insecticides which are applied as a gas, such as hydrocyanic acid, or which rapidly decompose when exposed to light, air and moisture, such as nicotine, pyrethrum and rotenone, have not greatly hampered the work of entomophagous insects. Oil spray is reported to have resulted in an increase in citrus blackfly on orange trees over that of untreated trees in a few months after treatment because of the destruction of parasites (166). Oil spray is generally believed to have a detrimental effect against entomophagous insects when used in citrus pest control in California. Sulfur was also found to be inimical to biological control. Sulfur can only exert its injurious effect at temperatures above 70°F.

The first observable reaction of M. helvolus, brought in on dusted citrus foliage and kept in a laboratory at 80°F, was to attempt to remove the sulfur from their tarsi and antennae. The antennae were soon paralyzed and were dragged beneath the body, and the parasites died within 24 hours. Even when, after 3 months, the parasites were confined to the leaves for several hours and then placed on black scales suitable for parasitization, they were unable to recognize them as hosts in which to oviposit and never gained back the ability to do so. (326)

It may be added in this connection that even road dust and other forms of inert residue are inimical to many species of insects, including the hymenopterous parasites (321). On the other hand, the same dust or inert residue may be favorable to the development of the principal citrus-infesting hosts of these parasites, the scale insects.

Insecticides may adversely affect M. helvolus in yet another way. This parasite can thrive best when the black scale is in an "uneven-hatch" condition, that

333

is, when there are various sizes of scales in an orchard at any one time. Pest control treatment tends to bring about an "even-hatch" condition of the black scale, by killing the smallest scales but allowing many of the mature individuals to escape (323).

With the advent of DDT, and later other insecticides with prolonged residual effect, many instances could be noted of adverse effect of pest control treatment on entomophagous insects. Sometimes plots of orange trees sprayed with DDT preparations could be distinguished from a distance from the curling of leaves caused by aphids, which were not killed, while the predators and parasites were eradicated. Nearly always the treated plots were severely infested with red mite if measures were not taken to separately control these pests.

A study was made of the predator populations in three citrus orchards containing plots treated with talc, with talc and DDT, and with plots left untreated as checks. The predators were principally Stethorus picipes, Oligota oviformis, and Conwentzia hageni. The treatments were made on March 7-11, 1946 and the counts of the predators were made on May 24, 1946, about 2.5 months later. It was found that the ratio of mites to predators was 13:1 in the check plots, 57:1 in the plots treated with talc, and 237:1 in the plots treated with talc and DDT. Since in the talc-treated and talc-DDT-treated plots the effect of inert residue, which ordinarily increases a mite population, was approximately the same, the difference in the mite: predator ratio in these plots was attributed to the effect of the DDT in killing the predators (225).

Selective Insecticides:- The selective eradication of pests has been successfully practiced in Great Britain in connection with the control of aphids attacking truck crops. It was found that short exposures of 40 to 60 seconds to nicotine vapor at a concentration of 0.8 mg of nicotine per liter at temperatures of 60° to 80°F results in 80 to 99% kill of certain aphids but does not seriously affect coccinellids, syrphids, or the larvae of the braconid parasite Aphidius (741). According to Ripper (741), "These beneficial insects surviving the nicotine vapor treatment outnumber the surviving aphids and decimate the latter in the weeks following the treatment, so that after an initial high percentage of kill, the infestation is further reduced. Thus, this method gives the best control of aphides hitherto demonstrated on farm and market garden crops in Great Britain, and has therefore been used on large commercial scale during the past four years. It is called field fumigation or gassing."

Dr. Ripper believes that where entomophagous insects are allowed to decimate the survivors of insecticide treatments, resistant races are obviously less apt to develop. He believes that a study of the combined action of chemical and biological control such as practiced in wide scale control measures which he has instituted in Great Britain will determine whether the limitations of chemical control as generally practiced may be overcome.

The limitations on chemical control imposed by the destruction of beneficial insects were discussed by Ripper (741), who suggested the use of "selective insecticides." Selective insecticides form the basis of British Patent 505853 (1939), which relates to the preparation of stomach poisons the individual particles of which are coated with substances digestible only by certain groups of phytophagous insects. The resulting preparation is claimed to be harmless to hymenopterous parasites and certain predators. More recently, DDT particles have been coated with degraded cellulose by an acid precipitation of the degraded cellulose from alkaline solution to which a DDT suspension had been added (743). After the acidification, a hardening process greatly improves the characteristics of the coating. The coating eliminates the contact action of DDT, but when sprayed leaves were eaten by phytophagous insects, these of course had the ability to digest the cellulose coating and were killed by the DDT acting as a stomach poison, while the predators and parasites were unharmed. Fig. 235 shows the differential effect of the coated-DDT preparation, as compared to normal DDT, on the carnivorous Lucilia sericata, as compared to the lack of differential effect on the phytophagous Pieris brassicae.

334

BIOLOGICAL CONTROL

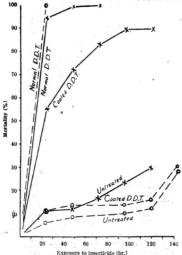

Fig. 235. Effect on *Pieris brassicae* larvae (———x———) and on *Lucilia sericata* adults (———o———) of "coated" and untreated DDT. After Ripper *et al* (743).

Still another development in the field of selective insecticides was brought to light in an announcement by Dr. Ripper at the annual meeting of the A. A. E. E. on December 15, 1949, on the practical utilization of a selective insecticide, octa methyl pyro phosphoramide. This compound was first synthesized by G. Schräder, already well-known for his syntheses of the first insecticidal organic phosphates.

The above compound, when absorbed by the treated plant, will kill such pests as aphids and red spiders feeding on the plant. Contact insecticidal effect is avoided, so parasites and predators are not destroyed. With this new "systemic" insecticide, prolonged control could be effected with one application, whereas plots sprayed with parathion and paraoxon were quickly reinfested, resulting in 10 to 14 days in a higher population of aphids and mites than before treatment.

RELATION OF INSECTICIDES TO POPULATION BALANCE

Nicholson (654) has discussed some factors relating to insect populations which may pertain to some of our pest problems in which it has been noted that increasingly more severe measures must be utilized to keep a pest under "commercial control" as the years go by. According to Nicholson, this may be the case even when the continued application of an insecticide does not result in the development of resistant races or even if it does not result in the destruction of natural enemies. When the treatment is first applied there is a great diminution in the population density of the injurious insect, which disturbs the natural balance between it and its environment. If the treatment is continued, a new state of balance will ultimately be attained, the spray program becoming one of the factors with which the insect population is balanced. It does not necessarily follow, however, that the density at which the new balance is reached will be lower than before; it may even be higher. Nicholson suggests that in field experiments, estimates of the initial results of a treatment be supplemented by detailed ecological studies so that the characteristics and influence of competition in the particular problem in question may be discovered.

Still another point must be taken into consideration in an evaluation of the effect of pest control on population balance, namely, the effect of the pest control measure in causing the starvation of the entomophagous insects. Flanders (320) believes it to be inevitable that a reduction of the host population below that which is necessary to maintain the entomophagous insects will lead to periodic outbreaks of the host species.

THE USE OF ORGANISMS OTHER THAN INSECTS

Microorganisms:- The insects are by far the most important organisms utilized in biological control. Other organisms have to date played a minor rôle, at least as far as their artificial utilization is concerned. Among the non-insect enemies of injurious insects and mites, the various microorganisms (bacteria, fungi, viruses, and protozoa) have appeared to afford the most promise, but not to such an extent as to provoke universal enthusiasm. It appears that the microorganisms that are pathogenic to insects are in many instances already widely distributed and

335

their success depends upon a combination of circumstances which it is often either
not possible or not practical to bring about artificially. The proper weather
conditions, for example, can not be artificially effected. The success of pathogenic microorganisms is often predicated upon a dense host population, the very
condition which, to be successful, the microorganisms would be expected to prevent.

In relation to the control of subtropical fruit insects, the most determined
effort at biological control with microorganisms has been the utilization of entomogenous fungi to control scale insects and whiteflies in Florida (see p. 398).
The "friendly fungi" were grown in large quantities on sterilized media in wide-
mouthed pint bottles and were distributed on the infested trees in the form of a
spray, using a regular spray rig. Belief in the efficacy of the fungi had been
bolstered by the knowledge that when Bordeaux sprays were applied to citrus trees
there was an increase in scale insects and whiteflies. It was assumed that this
was due to a decrease in the fungus population caused by the copper sulfate. The
problem was at long last scientifically studied by Holloway and Young (440) with
special reference to the purple scale. These investigators found that the increase
in scale population was caused by the residue left by the Bordeaux spray, which
produces a favorable environment for scale insects. Other finely divided solids
which have no fungicidal action are known to have a similar effect. They found
that the reduction in fungi did not significantly affect the scale population. The
experience with "friendly fungi" in Florida provides us with an illustration of the
importance of sound, thorough, quantitative investigation of all problems involving
host-parasite relationships, whether the parasites be insects or other organisms.

Sweetman (863) has provided us with a valuable review of the widely scattered
literature on the use of microorganisms to control insect pests, to which the interested reader is referred for the developments on this subject up to 1936. It
will be found that many instances are on record of effective use of bacteria, fungi,
viruses and protozoa, at least on a limited scale and under favorable conditions.
Within the last decade considerable interest has been aroused in Europe over a
French preparation of the dried spores of bacteria pathogenic to lepidopterous
larvae, called Sporein. Many references to successful control with this preparation on a limited scale may be found in the Review of Applied Entomology. Likewise,
in the United States, the control of Japanese beetles has afforded us with the only
practical large-scale example of artificial control of a pest by the dissemination
of microorganisms, in this case, by infecting the soil, where the immature stages
of the beetle are spent. The spore-dust method of colonizing Bacillus popilliae
Dutky, which causes the "milky disease" of the Japanese beetle, was devised by the
U. S. Department of Agriculture about 10 years ago, and has been successfully applied over most of the territory infested with Japanese beetle in the eastern
United States through Federal and State cooperation (945). The spore dust is now
being produced commercially. Two pounds of spore dust, released as 453.6 separate
2-gram quantities, is distributed over an area of 45,360 square feet or 1.04 acre.
The milky disease is slow in acting against a beetle. In southern New York, by increasing the standard spore dust dosage of 2 pounds per acre a thousand fold, the
period required for control was reduced only from 4 to 2 years. Where immediate
control is imperative, chemical control measures must be resorted to. Twenty five
pounds of actual DDT per acre results in almost complete control (2).

The "polyhedral diseases" of which jaundice of the silkworm and the "wilt" of
the gypsy moth and the tent caterpillar are familiar examples, are common among the
Lepidoptera (839). They result in the internal body tissues of the insects becoming liquefied, with many refractive and crystal-like polyhedral bodies appearing
among the disintegrating tissues. These usually range from 0.5 to 15 microns in
diameter. It is generally believed that the polyhedra are by-products of virus
diseases. Some evidence was produced which indicated that the polyhedra served as
a protection for virus held within them. The wilt disease in gypsy moth larvae is
a filterable virus. With the aid of a dark-field microscope, tiny granules may be
seen which appear identical to those seen in infected tissues (358).

In 1947-48, Gernot Bergold, in the Division of Virus Investigations of the
Kaiser-Wilhelm-Institute at Tübingen, Germany, with the aid of the ultracentrifuge
and the electron microscope, was able to demonstrate the serological relationships

of polyhedra protein and polyhedra virus. The virus particles (rods) had a tendency to lie parallel in bundles (90, 91). Later Dr. Bergold, investigating an infectious disease of a spruce budworm, Cacoecia murinana Hb., with an electron microscope, found, in the cell plasma of the larvae, many egg-shaped particles (fig. 236, A) with dimensions of 0.36 X 0.23 micron.[1]
In 0.02 molar Na_2CO_3 solution these bodies were found to have a rod-shaped aperture in each particle (fig. 236, B), giving them the appearance of "coffee beans." From these rod-shaped apertures small rodlets emerged (fig. 236, C) which were very similar to polyhedra rodlets and were the infectious agents of the disease. The rodlets averaged 272 millimicrons in length and 50.6 millimicrons in diameter. These closely resembled the rodlets of the polyhedral diseases of Bombyx mori and Porthetria dispar. As was demonstrated, however, they did not arise from polyhedra, but from the egg-shaped structures. The latter were called "virus capsules:" Bergold's observations and conclusions were confirmed independently and almost simultaneously by Steinhaus (842 a).

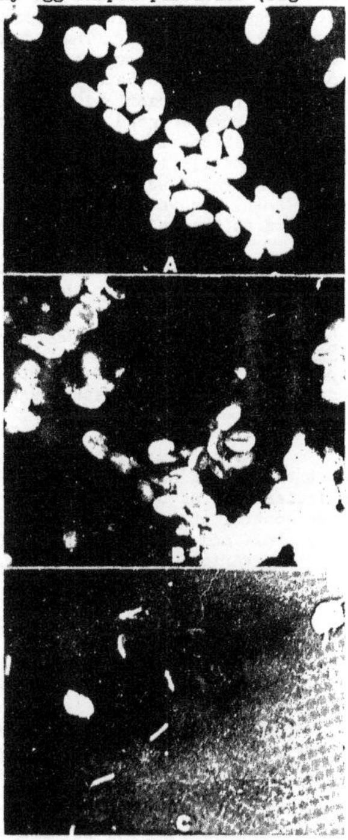

Steinhaus (842) reported the presence in the cytoplasm of diseased cutworms, of many minute discrete granules about 0.4 micron in length and nearly spherical. Steinhaus showed almost simultaneously with Bergold that the virus particle was contained within each of the granules. He recently published electron micrographs of the polyhedra and the virus particles of the alfalfa caterpillar polyhedrosis. The disease in Cacaecia, as well as that in the cutworm are members of a group Steinhaus is calling "granuloses." The first of these diseases was discovered by Paillot and called pseudograsserie. (see 840)

Within recent years a 4-unit course on Insect Pathology has been instituted at the University of California, probably the only course of its kind in the world. This course is taught by Edward A. Steinhaus, author of Insect Microbiology and Principles of Insect Pathology. The course is described in the University's catalogue of courses as follows: "General insect pathology and microbiology, including the biological relationships between all types of microorganisms and insects. Detailed study of bacterial, fungous, virus and protozoan diseases of insects; non-infectuous diseases of insects; histopathology." Dr. Steinhaus is leader of the Laboratory of Insect Pathology of the University's Division of Biological Control, whose Chairman is Professor Harry S. Smith. Experience with entomogenous organisms to date shows sufficient promise to warrant the thorough and systematic attack on the problem which is now being made under Dr. Steinhaus' able direction.

Fig. 236. A, virus capsules from larva of Cacoecia murinana; B, capsules showing apertures from which rodlets (C) have emerged. X 14,500. Original an electro-micrograph X 25,000. Courtesy Dr. Gernot Bergold. See also similar photograph by E. A. Steinhaus (842 a).

In a discussion of the possibilities along the line of biological control of injurious insects with pathogenic microorganisms, Steinhaus (839) made the following statement: "To be a useful member of the sciences employed by entomology,

[1]From a manuscript entitled "Über die Kapselvirus-Krankheit," by Gernot Bergold, kindly sent to the writer, with electromicrographs, by the author on December 11, 1948.

insect pathology should be recognized as a distinct field in its own right. No longer can we make substantial progress by the hit-and-miss methods of the past. What is urgently needed is the sympathetic, moral, and financial support of basic research into the various biological relationships existing between insects and microorganisms, and into the many factors concerned in the spread of diseases among insects in the field. There must also be a systematic search for new diseases occurring naturally among insects and which with the proper techniques may be put to practical use. We have placed the cart before the horse by jumping ahead to obtain as many of the practical benefits from the field as we could without paying for them in terms of good, sound research. The few initial investigations have for some time been yielding diminishing returns. We must now go back and build a scientifically firm foundation of fundamental knowledge to place under the flimsy structure which we have on our hands at present. Having this foundation, then and only then, can we build a structure of broad and practical usefulness to man in his efforts to control the insect pests which plague him and his world."

Nematodes:- Although nematodes have been given very little study with regard to their utilization in the biological control of insects, instances of insect mortality caused by nematodes have been recorded. Sweetman (863) refers to a number of instances of high mortality to grasshoppers, a soil infesting fly maggot, Sciara coprophila, Japanese beetles, and injurious nematodes, brought about by parasitic nematodes. He points out, however, that the life history and ecological relationships of most species are not known and that further studies would be justified.

Fishes:- Vertebrates have long been utilized for the destruction of insects, but there does not appear to be any likelihood of extensive expansion of this particular mode of biological control. Certain fishes have proved to be valuable in the control of mosquitos, but modern chemical control measures are especially practical and highly effective in this particular field of insect control. Nevertheless various species of fish, particularly top minnows, may advantageously be used in pools.

Amphibians:- Examination of the stomach contents of amphibians reveals that enormous numbers of insects are consumed by these animals, but while numbers may in themselves be impressive, it must be borne in mind that they are only relative and give no indication of any practical value in relation to the destruction of a particular pest. However, it appears that the Surinam toad, Bufo marinus, has successfully controlled sugarcane white grubs, Phyllophaga sp. in Puerto Rico, where they were at one time a serious pest (909).

Reptiles:- Knowlton and Janes (498) found that a number of lizards were efficient destroyers of insects and fed on large numbers of beet leafhoppers, Eutettix tenellus, but no data were presented to indicate an appreciable reduction in leafhopper population attributable to these reptiles. The above investigators brought out the point that the lizards did not discriminate between injurious and beneficial insects.

Birds:- It is known that birds consume enormous quantities of insects, and their mobility enables them to quickly concentrate in areas of high food supply, namely, in areas of insect outbreaks. There are, indeed, many authentic instances of reduction of local insect outbreak by birds. The early Mormon settlers in Utah witnessed the wholesale destruction by gulls of white grubs which had become so numerous as to threaten the people with starvation. The reduction of the insects was so spectacular as to lead the Mormons to believe that the appearance of the gulls at a moment so crucial in their history was an act of Providence. Nevertheless, in comparison to the numbers of outbreaks, the percentages of adequate control by birds is small. Besides, the birds have a fault in common with other vertebrate predators in that they are not density dependent and do not differentiate between beneficial and injurious insects. In fact, a large percentage of the insects they consume have within them a parasite.

The birds do not attune their numbers as delicately to the rise and fall of an insect population as the short-lived entomophagous insects. A sharp rise in the

338

population of an injurious pest will not be followed by a sharp rise in the bird population. The birds are much like the density-independent factors of the "environmental resistance" in this respect, and are subject to the same limitations. Moreover, they are predators, and, as such, they find themselves in an environment of limited capacity. It has been pointed out that if all birds were suddenly destroyed, there would be some increase in the population of injurious insects on that particular season, but the following season the population of entomophagous insects would accommodate itself to an environment of a somewhat increased capacity, due to the destruction of the birds, and the population of injurious insects would be maintained at as low a level as before--possibly in some cases even lower (855).

Bird lovers appear to have come to the conclusion, possibly justified, that man is so greedy that only extravagant claims as to the commercial value of birds will incite any general interest in their protection. Unfortunately, these claims cannot, for the most part, be substantiated in an objective analysis of all the factors that enter into the "equation."

Mammals:- The class Mammalia contains many insectivores. Shrews and moles eat many insects, but usually cause more damage than benefit. Bats eat many flying insects and are said to destroy many mosquitoes. Bears consume huge quantities of insects, but can hardly be said to lend themselves satisfactorily to utilization as insect predators in the protection of agricultural crops. The same may be said of skunks. However, the first legislation for the protection of wild animals in New York state was passed at the demand of hop growers who wished to have skunks protected because they considered them to be important enemies of grubs attacking hops. An investigation by the U. S. Biological Survey showed that in New Mexico skunks were the most important natural enemy of the range caterpillar (863). Mice and arboreal squirrels also feed on insects .

SOME ACHIEVEMENTS IN BIOLOGICAL CONTROL, WITH SPECIAL REFERENCE TO CALIFORNIA

The fact that some insects prey upon others has been known for ages, and some systematic study of the subject is known to have been made over 300 years ago. Howard (448) called attention to the fact that Aldrovandi was the first to notice the exit of the larvae of Apanteles glomeratus from the common cabbage caterpillar, and that Vallisnieri (1661-1730) was apparently the first to realize the existence of true parasitic insects. In the last quarter of the 18th and the first half of the 19th century, insect parasites were described by Reamur and DeGeer. The value of biological control was pointed out by Erasmus Darwin and a forest entomologist, Ratzeburg. Hartig, in 1827, recommended the rearing of parasites in large rearing cages. The carabid beetle, Calosoma sycophanta, was collected and placed on poplars in France to destroy caterpillars of the gypsy moth in 1840. The predatory mite Apanteles glomeratus, was sent to the United States from Europe in 1883 and proved to be an important parasite of the imported cabbage worm (448).

In subsequent developments in the field of biological control, California was destined to play an important rôle (see 285, 800). It has already been shown that the cottony cushion scale, a severe pest of citrus trees, was an important factor in bringing about the initial action in quarantine legislation and experimentation with orchard fumigation and oil sprays in California. This insect was also an important influence in the development of biological control as a distinct field of research and practice in entomology, like mechanical, cultural, and chemical control, and, in fact, superseding the latter whenever possible. The outstanding success of the vedalia lady beetle, Rodolia cardinalis (fig. 237), in combating the cottony cushion scale, served to stimulate public confidence in and support for a sound and systematic program of organization, research and foreign exploration in relation to biological control.

Vedalia:- The cottony cushion scale was known to have been introduced from either Australia or New Zealand, so naturally it was concluded that explorations for possible native enemies of the insect might bear fruit in those countries. Albert Koebele, upon the recommendation of C. V. Riley, was sent to Australia by

339

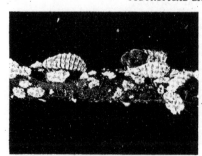

Fig. 237. The vedalia, Rodolia cardinalis, on orange twig infested with cotton-cushion scale. Left, larva; right, adult emerging from pupal case. Photo by Roy J. Pence.

the government, using a fund appropriated by the State Department, as this country's contribution toward the international exposition to be held in Melbourne, Australia. This subterfuge was occasioned by the fact that Congress had passed a bill forbidding foreign travel by employees of the U. S. Department of Agriculture at government expense (448). The importation and successful establishment of the little lady beetle which came to be popularly known as the vedalia was the outstanding accomplishment of Koebele's trip. D. W. Coquillett, of HCN fumigation fame, was the one who received and colonized the beneficial insects sent to California by Koebele. He received his first shipment (28 beetles) on November 23, 1888, and the first beetles to be liberated were placed in the same orchard that figured so prominently in the early investigation of orchard fumigation, that of J. W. Wolfskill (fig. 121). Within a year the dreaded cottony cushion scale had ceased to be a factor to be considered in California citriculture. Furthermore, the vedalia has been taken to all citrus areas of the world where the cottony cushion scale has become established, and everywhere it has been a success.

As previously stated, the spectacular success of the vedalia led to public confidence in and support of biological control, but for some years this work was pursued with apparently insufficient recognition of the dangers inherent in indescriminate insect introductions. This situation was corrected in 1911, when Professor A. J. Cook of Pomona College became State Horticultural Commissioner for California. He appointed a trained entomologist, Harry S. Smith, formerly with the Parasite Laboratory of the U. S. Bureau of Entomology in Massachusetts, to work on the introduction, investigation, and liberation of beneficial insects in this state. The insectary was reorganized and Smith was made superintendent in 1913. It is said by Howard (448) that it was only this fortunate circumstance that prevented the Federal government from prohibiting further introduction of insects from foreign lands. In 1923 Smith joined the staff of the University of California and was stationed at the Citrus Experiment Station at Riverside. Here, under his direction, a large staff of experts on biological control was gradually built up, and a considerable number eventually became located at Berkeley (see p. 86). One success after another rewarded the efforts of this group of investigators in bringing about the complete or partial control by natural means of many important pests, including weeds. Meanwhile entomological literature has been enriched by the results of their basic research on problems relating to insect populations and the complicated biology and taxonomy of the parasitic hymenoptera.

The most prominent successes in biological control in California will later be discussed in some detail in Chapter 18 et. seq. in connection with the various subtropical fruit pests. These include, besides the vedalia, the following beneficial insects now attacking subtropical fruit pests:

1. Cryptolaemus montrouzieri, first introduced by Koebele from Australia in 1892. Shortly after the introduction of the citrophilus mealybug, Pseudococcus gahani, first observed in California in 1913, it became apparent that this pest would rival the erstwhile cottony cushion scale as a pest of citrus, if not successfully combatted. Chemical control measures had proven ineffective. Cryptolaemus, both in the adult and larval stages (fig. 238), successfully attacked mealybugs, but could not survive the winters in California in sufficient numbers to effectively combat the spring build-up of the mealybug population. Investigations led to a

successful method of mass rearing of Cryptolaemus in insectaries, and this became a well-
established method of control (121). Fifteen
insectaries were established by various counties and citrus associations, and by 1928, when
the production of "crypts" reached its peak,
millions of these insects were being reared annually and liberated in infested orchards.

2. The Hymenopterous parasites, Coccophagus gurneyi, Tetracnemus pretiosus and Anarhopus sydneyensis (fig. 239). While Cryptolaemus
appears to have saved the citrus industry from
ruin, and the working out of a successful means
of controlling mealybugs with this insect was
an accomplishment of great importance, this species can hardly be considered as an ideal
natural enemy of mealybugs, for its rearing and
liberation entail considerable expense. The reports that certain effective hymenopterous
parasites had been found which were capable of
maintaining themselves throughout the winter
months without the help of man, was hailed with
great enthusiasm by citrus growers and entomologists in California. C. gurneyi and T.
pretiosus were found in Australia by Harold
Compere in 1927, along with other parasites and
predators, and all were successfully introduced
into California (183). Their success in controlling citrophilus mealybugs proved to be almost as complete as that of vedalia in the control of the cottony cushion scale. Of the two
species, C. gurneyi is dominant. The rearing
of Cryptolaemus montrouzieri has lost its former significance in biological control although
despite this fact, over 20 million were reared
in 1947. This species continues to be reared
merely to maintain a stock to liberate in occasional isolated infestations to speed the
demise of these incipient colonies, which, however, would be doomed in the course of a season
by the aforementioned hymenopterous parasites.

The hymenopterous parasite Anarhopus
sydneyensis was reared from the long-tailed
mealybug, Pseudococcus adonidum, in Australia by S. E. Flanders in 1931 and was successfully introduced into California. This species
was particularly effective in combating the
long-tailed mealybug on the avocado.

3. Metaphycus helvolus:- This chalcid
parasite (fig. 240) is one of 28 species sent
to California from South Africa by Harold
Compere in 1937. Prior to its introduction,
the black scale, Saissetia oleae, was considered almost as important as the red scale as a
citrus pest in California. Within a few years

Fig. 238. Cryptolaemus montrouzieri. Larvae and adults feeding on Baker's mealybug, Pseudococcus maritimus (Ehrh.) and
their eggs. X3. Photo by Roy J.
Pence.

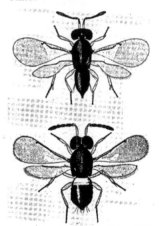

Fig. 239. Adult parasites of the
citrophilus mealybug. Above,
Tetracnemus praetiosus; below,
Coccophagus gurneyi. X40. After
Compere and Smith (183).

after the introduction of M. helvolus, the black scale had been reduced to such low
numbers that in only an occasional orchard, mainly in the interior regions, was
treatment necessary. Many parasites and predators had previously been successfully
introduced for the control of black scale, but the total effect of all these on the
black scale population was not sufficient to obviate the necessity for annual arti-

341

Fig. 240. Metaphycus helvolus. Parasite found highly effective in control of black scale. X50. Photo by Roy J. Pence.

ficial control measures. The much greater success obtained with M. helvolus is due to a number of favorable circumstances pertaining to the biology and habits of this species (see p. 433).

In the interior regions the black scale has only one generation per year and this condition is not as favorable to M. helvolus as the "double brooded" condition of the black scale in coastal regions (see p. 430). The black scale has not been as completely controlled by M. helvolus as the cottony cushion scale and the mealybugs have been controlled by their imported natural enemies, but considering the high degree of control obtained and the former importance of the black scale problem, this instance of biological control should be ranked among the outstanding successes in this field of applied entomology. In interior or semi-interior climatic zones where control for red scale is not required, as in the Piru district in Ventura County, M. helvolus in some years is not completely effective. In such areas the parasite is in some instances being reared in insectaries and liberated when the black scale is in the proper condition, in August and September. In such cases a great saving in cost is effected, as compared to spraying or fumigation, by the mass-rearing and liberation of M. helvolus.

M. helvolus is reared on black scale, which in turn is reared on potato sprouts. The black scale crawlers, as well as those of other insects which can be used as hosts, are collected in large quantities by placing the parent scales in "shadow boxes," so constructed that light, which enters from one direction, causes the positively phototropic crawlers to converge to a point, where they are blocked by a shadow and are forced to pile up. They are then transferred in enormous numbers to the potato sprouts. The parasites oviposit on black scale when they are from 0.6 mm to 2.15 mm long. About a million parasites are reared per year in the cooperative insectary at Fillmore.

The above brief sketch has indicated the outstanding success in biological control in California in relation to subtropical fruit insects, namely, the control of the cottony cushion scale, the control of mealybugs attacking citrus (Pseudococcus citri, P. gahani, P. maritimus and P. adonidum), mealybugs attacking the avocado (P. adonidum) and mealybugs attacking the grape (P. maritimus), and the control, with occasional exceptions in some interior sections, of the black scale. However, the total activity of those engaged in biological control work in California is only in part represented by the above successes. Many subtropical fruit pests which still require the annual or occasional application of insecticides are to some extent held in check by the imported natural enemies, and it is not known how much more severe many of our pest control problems would be without the aid rendered by beneficial insects.

It should also be borne in mind that much effort is expended in attempting to find a biological control for insects which appear practically immune to this form of control. The California red scale is a case in point. Despite the many years of world wide search for effective parasites for this pest, no encouraging results had been obtained. One would expect that the possibility of finding a successful parasite had become about nil, yet the importance of this serious pest demanded that the investigation of the problem be continued (793, 806). Indeed, as the result of recent developments, there is some hope of establishing new natural enemies of the red scale in California which may be more effective than those which are now present in this state. The University has recently successfully established a Chinese form of Aspidiotiphagus citrinus which is being propagated in large numbers on potato tubers infested with California red scale and is being distributed throughout the red scale-infested areas of California. The yellow scale-infesting form of this species is an important parasite in California, especially in the cooler districts, and still another form of this species is the most common parasite of the purple scale.

It is of interest to note that before the days of rapid airplane transportation it was necessary to take California red scale to the Orient to be certain that experiments with parasites made there would be with our strain of red scale and not some other physiologically different, but morphologically indistinguishable strain. Because of the citrus canker quarantine, red scale for this purpose was grown on sago palms instead of citrus. Years later S. E. Flanders discovered, in connection with the rearing of the parasite Habrolepis rouxi, that the red scale, when feeding on sago palm, is not suitable for parasite development. Thus, in the light of this discovery, other work previously done in China had to be repeated. The red scale shipped to the United States from foreign countries are now placed in sealed containers and sent by air mail to the Experiment Station at Riverside. The parasites emerge from the scales in the quarantine rooms in the insectary.

Along with the red scale-infesting form of Aspidiotiphagus citrinus, another Chinese parasitic insect of the genus Casca is being investigated, as well as a number of predators. A resurgence of interest in the possibilities of the golden chalcid, Aphytis chrysomphali, in view of its host-feeding propensities, has already been discussed (see p. 332).

The periodic depletion of the red scale population by insecticides puts any potentially effective parasite at a disadvantage in comparison with the natural enemies of such pests as the cottony cushion scale and the mealybugs, because no insecticides were effective in destroying the latter species (807). The parasites had a chance to increase until they became effective agents of biological control.

It is therefore likely that even a good parasite would not be as spectacular against the red scale as previous natural enemies have been against insecticide-resistant pest species. It is possible, however, that the recently introduced natural enemies of the red scale may lend themselves to mass culture and liberation in infested orchards or that some other means of human manipulation may be designed to overcome the disadvantages they will suffer as the result of present day pest control programs.

An example of present efforts directed toward the biological control of red scale is shown in fig. 241. Potato tubers are used to support heavy infestations of red scale which in turn are used for the propagation of the natural enemies of red scale. The equipment illustrated in fig. 241 is that used for (1) obtaining the parasitization of the red scale by five species of parasitic wasps (Aphytis n.sp., Aspidiotiphagus n.sp., Casca n.sp., Comperiella bifasciata,[1] and Prospaltella n.sp.) recently collected by the University of California, Division of Biological Control, in East Asia, and (2) the daily distribution of such parasites in the citrus orchards of California.

Fig. 241. S. E. Flanders examines parasitized red scale reared on potatoes in the Insectary at the Citrus Experiment Station. The cylindrical cages, which cover the wire baskets, are used for rearing and distribution of red scale parasites. Explanation in text.

[1] Two million individuals of the Chinese race of Comperiella bifasciata were reared and liberated by the University of California, beginning in 1941. By 1948 this species had become well established in the Riverside and Escondido areas, but the parasites became abundant only where the red scale population was abundant. It is believed that of the 2 million C. bifasciata released, a few were so genetically constituted as to be able to adapt themselves to southern California conditions, and that these were the progenitors of those strains now existing in Riverside and Escondido. Individuals of the Riverside "strain" are now being reared for liberation in other areas. It is believed that they may prove to be effective in reducing peak infestations of red scale, but that they will probably be inefficient where the scale infestations are light. (221a)

SUBTROPICAL ENTOMOLOGY

Infested potatoes, with all scale equal in age, are placed on shelves in the glass-topped parasitization cages (in background of photograph) where the scale is immediately attacked by thousands of parasites. Each cage contains only one parasitic species. After an exposure to attack of a day or two, the infested potatoes are placed in wire baskets. A few days before the adult parasites are due to emerge from the scale, the baskets are covered with a cardboard cylinder which is provided with a plastic funnel leading into a short piece of glass tubing. After the parasites emerge from the scale, they ordinarily collect in the glass tube, since at that time they are positively phototropic and negatively geotropic.

These parasitic wasps attack only certain stages of the scale. All the scale on the potatoes are equal in age and the life cycle of the scale is about 38 days at 80°F., so the emergence of the parasites is restricted to a period of a week or two.

The county agricultural commissioners cooperate with the University of California in the release and distribution of newly introduced natural enemies of agricultural pests. To facilitate the distribution and establishment of the new red scale parasites, the counties are supplied with parasitized red scale in completely assembled emergence units as shown in fig. 241. The glass tubes containing the parasites are taken to the field for release.

In South Africa also, the possibilities of control of the California red scale by natural factors are being considered. As in California, it has been observed that certain citrus orchards scattered about through the Union have been commercially free of red scale for prolonged periods. Ullyett (901) points out that natural and chemical control of red scale in South Africa are usually incompatible. For natural control, a moderate population of scales is necessary. This may involve a loss of exportable fruit during the period of establishment of the parasites, but it was observed that orchards left untreated for 2 years did not require treatment afterwards, and some have remained under natural control for 4 to 6 years. Ullyett emphasizes the point that the establishment and maintenance of satisfactory biological control are not as simple as chemical control and the guidance of a specialist in this field of work would be required.

LIST OF SUCCESSFULLY ESTABLISHED INSECTS INTRODUCED
FROM FOREIGN COUNTRIES

Essig (285) has given us an historical account of the introductions of beneficial insects up to the year of the publication of his A History of Entomology. Following is a list of the imported species of predators and parasites which have actually become established in California to date. This list does not include over 50 species of entomophagous insects introduced into California in 1947 and 1948 by the Division of Biological Control of the University of California, mainly from South Africa and China, to prey on red scale, purple scale, black scale, and citrus mealybug.

Beneficial Insects

	Origin	Host
A. Predators		
1. Coccinellidae (lady beetles)		
Cryptolaemus montrouzieri Mulsant	Australia	Mealybugs
Cybocephalus sp.	China	Diaspine scales
Exochomus quadripustulatus (Linn.)	Europe	Woolly apple aphid
Lindorus lophanthae (Blaisdell)	Australia	Red, purple scales, unarmored scales
Orcus chalybeus (Boisduval)	Australia	Red, black and other scales
Rhizobius ventralis (Ehrhorn)	Australia	Black & other scales
Rodolia cardinalis (Mulsant)	Australia	Cottony cushion scale
Scymnus binaevatus Mulsant	South Africa	Mealybugs
Scymnus bipunctatus Kugel	Philippines	Mealybugs

344

BIOLOGICAL CONTROL

B. Parasites

1. Pteromalidae
   Scutellista cyanea Motchulsky — South Africa — Black, soft brown, hemispherical scale Ceroplastes.

2. Aphelinidae
Aphelinus mali (Haldeman)	Eastern U.S.	Woolly apple aphis
Coccophagus gurneyi Compere	Australia	Mealybugs
Coccophagus trifasciatus Compere	South Africa	Black scale
Coccophagus ochraceous Howard (accidentally established)	Africa	Black scale, European fruit lecanium
Coccophagus capensis Compere	Africa	Black scale
Coccophagus rusti Compere	Africa	Black scale

3. Encyrtidae
Anarhopus sydneyensis Timberlake	Australia	Citrophilus mealybug, long-tailed mealybug
Anagyrus aurantifrons Compere	South Africa	Mealybug
Anagyrus nigricornis Timberlake	Hawaii	Long-tailed mealybug
Comperiella bifasciata Howard	Japan	Yellow scale
	China	Red scale
Leptomastidea abnormis (Girault)	Europe	Citrus mealybug
Metaphycus lounsburyi (Howard)	South Africa	Black scale
Metaphycus helvolus (Compere)	South Africa	Black scale
Tetracnemus pretiosus Timberlake	Australia	Citrophilus mealybug
Tetracnemus peregrinus Compere		

4. Eupelmidae
Lecanobius utilis Compere	West Indies South America	Black scale, hemispherical scale
Tetrastichus brevistigma Gahan	Eastern U.S.	Elm beetle

5. Tachinidae
   Erynnia nitida Robineau-Desvoidy — Europe — Elm beetle

6. Ichneumonidae
   Bathyplectes curculionis (Thomson) — Europe (via Utah) — Alfalfa weevil

In 1947 and 1948 over 50 species of entomophagus insects were introduced into California by the Division of Biological Control. Of these the following species are being propagated:

    Enemies of Red Scale:
        Predators:
            Chilocorus distigma Klug from Africa
            Chilocorus wahlbergii Mulsant from Africa
            Chilocorus nigritus Fabricius from China
            Lotis neglecta Mulsant from Africa
            Lotis nigerrima Casey from Africa
            Sticholotis sp. from China
            Pharoscymnus exiguus (Weise) from Africa
        Parasites:
            Aspidiotiphagus sp. from China
    Enemies of Purple Scale:
        Predators:
            Telsimia emarginata Chapin from China
            Aleurodothrips fasciapennis Franklin from China
            Diplosis sp. from Italy
    Enemies of Pseudococcus citri:
        Predators:
            Scymnus quadrivittatus Mulsant from Africa
            Pullus sp. from China
        Parasites:
            Anagyrus sp. from Africa
            Coccophagus sp. from China*
        *Test cultures incomplete

345

SUBTROPICAL ENTOMOLOGY

Enemies of Black Scale:
    Predators:
        Chilocorus angolensis Crotch from Africa

The many successful importations of beneficial insects indicated in the above list, in addition to others which were imported into California from other parts of the United States (see Quayle, 724) include many species which are attacking such pests or potential pests as the red, purple, yellow, black, citricola and soft brown scales, as well as mealybugs, aphids, and mites, and together with native beneficial species, they must certainly exert a considerable effect in mitigating the ravages of injurious species.

Biological Control of the Oriental Fruit Moth:- Biological control in California has, of course, extended its benefits to many crops besides the citrus and other subtropical fruits, although it is in connection with the latter that the greatest successes have been attained. A relatively recent contribution in the field of biological control relates to the oriental fruit moth, Grapholitha molesta (Busck), subsequent to its accidental introduction into California. The work that was done in connection with this problem has not received the wide acclaim and recognition from the general public that was accorded certain other previous accomplishments in biological control because it was done in anticipation of and for the prevention of a serious pest problem rather than, as in previous cases, for the control of an already-established serious pest.

No one seems to know where the oriental fruit moth originated. It was introduced into this country from Japan about 1913 and became established in Washington, D.C., then gradually spread throughout the eastern United States. It attacks a number of deciduous fruits, particularly the peach, and has become a pest of major importance in the East. It is not known to what extent this pest would attack subtropical fruits in California, but probably at least some would be hosts.

These insects are related to the codling moth and are similar in shape, but somewhat smaller. They are about 0.25 inch long and have a wing spread of 0.5 inch. They are colored a nearly uniform dark grey. The flat, whitish eggs of the female are laid on the leaves or, occasionally, on the twigs, soon after the peaches bloom. The larvae of the first generation bore into the tender twigs. They closely resemble codling moth larvae. When they are full grown, they spin cocoons and transform to moths. There are from 1 to 7 generations per year, the later generations attacking the fruit in much the same way as the codling moth. Artificial control has always been rather difficult.

On September 30, 1942, a well-established infestation of oriental fruit moth was discovered by a county agricultural inspector at Yorba Linda, in Orange County, California. The presence of this serious pest in California posed a serious threat to the deciduous fruit industry. The value of the crops grown in California and known to be hosts of the oriental fruit moth was estimated to be, in 1942, well over a hundred million dollars. Surveys were immediately started to delimit the infested area with a view toward possible eradication, and Quarantine No. 7 was issued, establishing a quarantine on interstate shipment of host materials. It soon became evident, however, that the oriental fruit moth had already become too widely distributed to justify any hope for eradication. It was realized that plans had to be made for an eventual control program rather than a program for the prevention of spread.

Research was immediately initiated by the University of California College of Agriculture on the possibilities of chemical and biological control of the oriental fruit moth. The funds for this investigation were appropriated by the California State Legislature to the State Department of Agriculture, which contracted with the University of California to carry on the required research. DDT showed promise of far surpassing previously used insecticides in its effectiveness as a control measure, should the need arise. We are concerned here, however, with the steps that were taken to insure a rapid and effective distribution of parasites to all infested areas. It had already been known from experience in the East that the

346

Fig. 242. Insectary of the Division of Biological Control, University of California Citrus Experiment Station, Riverside, California. Photo by Roy J. Pence.

Fig. 243. A tray of potatoes infested with potato tuber moth larvae to be used as hosts of the parasite, Macrocentrus ancylivorus. Arrow points to "cocooning cards" for the parasites. Courtesy Division of Biological Control, University of California.

hymenopterous parasite Macrocentrus ancylivorus Roh. (fig. 229 was effective in controlling the oriental fruit moth when present in sufficient numbers. Under the direction of Professor H.S. Smith, a program for the rapid mass-rearing and liberation of this parasite was launched which involved a new production technique, to be described below. The original research and development was centered at the modern insect-proof insectary at the Citrus Experiment Station (fig. 242) but mass production finally became centered at a newly established production center built for this purpose at Richmond, California.

In the East, the mass-rearing of Macrocentrus ancylivorus had involved the use of the strawberry leaf roller, Ancylis comptana as the host insect. Obviously, the discovery by S. E. Flanders that the larva of the potato tuber moth, Gnorimoschema operculella, was as satisfactory a host as any for Macrocentrus, paved the way for a far more practical method of rearing this species than any that had previously been employed. The technique involved in the new mass-rearing method has been described by Finney et al (311, 312). The potato tuber moths were reared in movable open boxes inclosing a tray (fig. 243) for supporting a layer of potatoes. The trays were stacked as shown in fig. 244. It was discovered that freshly made punctures through the skin of the potato greatly increased the number of tuber moth per potato. The potatoes were punctured by rolling them under leather mittens studded with brads or by allowing them to fall through a chute containing brad-studded baffle boards. Cloths containing tuber moth eggs were

Fig. 244. Stacked trays of potatoes with tuber moths used as hosts in the rearing of the parasite Macrocentrus ancylivorus. Courtesy Division of Biological Control, University of California.

placed over trays of punctured potatoes, egg side down, and the newly-hatched larvae moved directly onto the potatoes. After one week in the incubation room at 80°F, the egg cloth was pulled out of each unit through a 1.25 inch aperture in the end of the box and 100 females of Macrocentrus, together with a number of males, were injected through the aperture (fig. 244) which was then closed. Thirteen days after the introduction of the parasites, the potatoes were discarded and the Macrocentrus cocoons which had formed in "cocooning cards" were taken to the emergence room or shipping room and the trays were thoroughly cleaned for reloading.

Each tray held about 20 pounds of potatoes which, when well infested, yielded over 4000 mature tuber moth larvae. The maximum production from each unit was about 1500 female and 1300 male Macrocentrus. The "cocooning cards" which held the Macrocentrus cocoons were formed into triangular cases. When the parasites were about to emerge they were shipped in cardboard containers to the points of liberation and suspended in peach trees by the loose ends of wire bands.

The technique for production and liberation of Macrocentrus was a joint contribution of the staff of the University's Division of Biological Control. The parasites were produced by the University and delivered to the various sites of oriental fruit moth infestation where they were colonized by field men of the State Department of Agriculture or, in southern California, by the county agricultural commissioners.

The cost of production of over nineteen million Macrocentrus in 1944 was shown by Smith (798) to be about 78 cents ($0.78) per 1000 parasites with considerable likelihood that research underway at the time would still further increase the efficiency of mass production methods. This should be compared to the cost of $30 per 1000 by the strawberry-leaf-roller method of parasite production which had previously been accepted as the standard method of mass production.

It is considered likely that the potentialities of the oriental fruit moth as a pest in the majority of agricultural regions in California are not as great as in regions of summer rainfall. This does not detract, of course, from the fact that from the standpoint of the soundness of its organization, the imagination and ingenuity displayed in research and development, and the rapid and unqualified success attained, the oriental fruit moth investigation is a classic among projects of this type. In addition to its contribution as a solution of the problem at hand at the time, the principles and techniques worked out in the course of the investigation can be used in connection with the mass culture of other parasites and can be utilized in the event of the occurrence of similar problems in the future.

WEED CONTROL

Like insect pests, weeds can be controlled by legislative, mechanical, cultural, chemical, and biological methods. We are now concerned with biological control, and biological weed control involves a curious reversal of the process thus far discussed in this chapter. The "injurious" insect species is introduced and parasites and predators, if any, must be avoided. Here again, it is obvious that introductions must be made under the strict control of those who understand the dangers inherent in this particular method of biological control. The botanical relationships of the weed to be destroyed and the feeding habits of the potential weed-destroying insect must both be thoroughly understood. If the weed is botanically related to cultivated plants of economic importance, the likelihood of an introduced insect pest of the weed becoming a pest in this country is enhanced and such a possibility must be avoided. Likewise if the potential weed-destroying insect is a polyhagous species, feeding on a wide variety of plants, it must be excluded from consideration. To be considered for weed control purposes, an insect must be highly specialized in its feeding habits, being able to survive on only a few, or preferably only one plant species, the pest plant which it is desired to combat. Such insects are likely to be found among the higher insect orders, from an evolutionary standpoint, in which feeding habits are often highly specialized. Even species in these orders must be carefully studied as to their feeding habits in their new environment before they are finally released. It is almost axiomatic that an intro-

348

duced species, providing climate is suitable, will thrive better than it did in its original habitat. If an insect is imported for the purpose of destroying weeds, it is likely to be a greater success in its new home than in the country of its origin, for in its new environment it will be free of the parasites, predators and diseases that had adapted themselves to this particular species. By the same token, if this introduced weed pest should adapt itself to feeding on cultivated plants in its new habitat, it might become a pest of first magnitude.

A more or less standardized procedure in determining whether an insect is or is not a potential pest of cultivated plants is to form a list of all plants of economic importance in the country of its origin and offer the insect the choice of either feeding on these plants or starving. This test is made in the country of origin. If the insect passes this test, that is, if it feeds on no economic plants even under starvation conditions, the same test is repeated in the country to which the insect is to be introduced and with reference to the plants of economic importance in the new region. This test must be made in an insect-proof building to prevent accidental escape of the tested species.

SOME EXAMPLES OF BIOLOGICAL CONTROL OF WEEDS

The Prickly Pear in Australia:- The prickly pears (Opuntia sp.) are members of the cactus family found in North and South America. A number of separate introductions of Opuntia of various species were made by the early colonists, but the main pest, O. inermis originated from one plant brought to New South Wales in a flower pot in 1839. Originally the cactus was deliberately taken from one locality to another in Australia to be grown as hedges around the homesteads. About 1870, however, the prickly pear got beyond control. By 1900 it covered 10 million acres and by 1920 it covered 60 million acres and was expanding at the rate of a million acres per year, rendering the infested land unsuitable for grazing.

In 1925 a tunnelling caterpillar, Cactoblastis cactorum (Berg.), native to Uruguay and northern Argentina, was introduced into Australia. The original and only shipment consisted of 2750 eggs. In 1926 and 1927, 9 million eggs, reared in insectaries, were distributed to all infested areas and further rearing became unnecessary, for plenty of material could be collected in the field.

The caterpillars eat out the interior of the pear joints. If they are not eaten out entirely they nevertheless succumb to "wet rots" caused by various fungi and bacteria. Even the roots of the plants are invaded and the collapse of great areas of cactus in a single season is indeed spectacular.

Other pests of cactus which have been imported into Australia include the cochineal mealybugs, Dactylopius sp., the red mite, Tetranychus opuntiae, a bug, Chelinidea tabulata, and a beetle, Moneilema ulkei.

Sweetman (965) lists the most important pest plants to have been subjected to biological control as the following: the cacti, (Opuntia spp.), Lantana camara, gorse or furze (Ulex europaeus), blackberry (Rubus fruticosus), ragwort (Senecio jacobaea), St. John's wort (Hypericum perforatum), piri-piri or bidi-bidi (Acaena sanguisorbae), cockle burrs (Xanthium spp.) Clidemia hirta, nut grass (Cyperus rotundatus), Stachytarpheta indica, thistles, skeleton weed (Chondrilla juncea), and certain types of aquatic vegetation. The majority of the successful weed control projects have been in Australia and New Zealand.

The Klamath Weed in California:- The Klamath weed, Hypericum perforatum L., originated in Europe, where it is called St. John's wort, and was first found in Northern California in the vicinity of the Klamath River in 1900. It soon began to take over pasture land, eliminating other plants that had been valuable in supporting grazing animals. In addition it was found to be toxic to sheep and cattle, particularly the latter.

The Klamath weed was introduced into Australia in 1880, taking complete possession of much valuable grazing land. The search for insect enemies was begun by the Australians in 1920. The starvation tests made in Europe and Australia to de-

limit the host range of the insect enemies of the weed were valuable to California
investigators when importations were begun into this state in 1944. The U. S. Department of Agriculture authorized the introduction of three beetles known to be
enemies of the Klamath weed: Chrysolina hyperici (Forst.), C. gemellata (Rossi),
and Agrilus hyperici Cr. with merely a proviso that feeding tests be made on sugar
beet, flax, hemp, sweet potato, tobacco and cotton. A cooperative project was
forthwith set up between the U.S.D.A. Bureau of Entomology and Plant Quarantine and
the University of California (437).

When tests showed no feeding or egg laying on the tested hosts, 4 colonies of
beetles were released late in 1945 and one of these became established. During
1945 and 1946 a total of 10,938 adults of Chrysolina hyperici were liberated, and
liberations continued in increasing numbers.

Feeding tests with Chrysolina gemellata were completed in January, 1946, and
on February 13, 650 adults were released in one location in each of 4 counties.
By the time of emergence of adults in April and May, 1947, the adults could be
found a quarter of a mile from the point of release (fig. 245).

At one test location in Humbolt County where the beetles were released in
January, 1946, a continuous stand of about 20 acres of Klamath weed was destroyed
in less than 3 years. During the winter and spring of 1947-48, about 557,000
beetles, reared in the insectary, were released at 122 locations in 19 counties.
Millions of beetles will be required to control the Klamath weed over its entire
range and it will require several years for present colonies to increase to the
necessary numbers. (439)

Fig. 245. Left, Klamath weed in bloom;
center dark strip, the weed dying from
the attack of Chrysolina beetles; right,
portion of field cleared of the noxious
weed. After Holloway and Huffaker(439),

CHAPTER XVIII
CITRUS PESTS

INTRODUCTION

The Citrus Industry:- The world production of citrus fruits in the 1946-47 crop was estimated[1] to be 342,000,000 boxes, of which 249,000,000 boxes were oranges, 65,000,000 boxes were grapefruit, and 28,000,000 were lemons. The United States led in production with 47% of the oranges and tangerines, 95% of the grapefruit, and 52% of the lemons. This production was 27% over the 1935-39 average and is increasing rapidly. Practically all the increase, however, is in Texas and Florida.

Citrus production is confined to a belt extending around the world, with the equator as its center, and extending roughly 35° latitude north and south of the equator, although proximity to large bodies of water, the influence of warm ocean currents, or the nature of the air drainage, and possibly other factors, may in some parts of the world extend the range of citrus to 40° latitude, north or south, or slightly beyond, as for example, in California, the Mediterranean region, Russia, and New Zealand. Also, in a given locality, the production of citrus fruits is delimited by minimum temperature, which in turn may be influenced by altitude and air drainage.

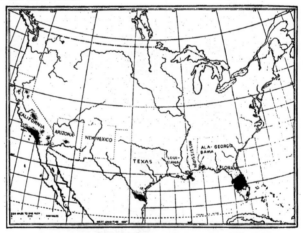

Fig. 246. Citrus areas (in black) of the United States.
After Weber and Batchelor (929).

As can be seen from fig. 246, in the United States commercial citrus production is confined largely to California, Florida and Texas, but some commercial production can be found in Arizona, Alabama, Louisiana, Mississippi, and Georgia. The acreage of bearing citrus trees in the United States in 1945-46 was as follows: in California there were 240,900 acres of oranges, 14,700 acres of grapefruit, and 65,400 acres of lemons; in Florida there were 288,400 acres of oranges and tanger-

---

[1] From United States Department of Agriculture estimate reported in the California Citrograph 32(11):476.

ines and 96,900 acres of grapefruit; and in Texas there were 29,800 acres of oranges and 77,000 acres of grapefruit. Thus the acreage of bearing citrus trees in Florida was about a fifth greater than in California while in Texas it was about a third that of California. In Arizona there were 19,800 acres of citrus, about 65% grapefruit. The remaining states contained only about 6,000 acres of citrus.

Citrus Pests of the United States:- On the accompanying maps are shown the locations of the citrus areas of California, Florida and Texas. With each map is a list of the citrus pests, in order of their importance, in each of the citrus districts. It will be noted that California and Florida both have the purple scale and the citrus red mite among the four most important pests. In Texas the rust mite is by far the most important pest; in Arizona, the citrus thrips is the only pest of importance, and in Louisiana and Alabama the citrus pests are more or less the same species as in Florida. The citrus nematode, although probably of considerable economic importance, is not included on the lists of citrus-infesting species because of the uncertainty as to its true status as a pest. It does not appear to be possible, in the state of our present knowledge, to compare the importance of the citrus nematodes to that of other pests whose injury to citrus trees is more readily evaluated.

CALIFORNIA
CITRUS AREAS
C-COASTAL
Int-INTERMEDIATE
I-INTERIOR
CC-CENTRAL CALIFORNIA
NC-NORTHERN CALIFORNIA
D-DESERT

California Citrus Pests in Order of Their Economic Importance
(see pages 531 to 537 for common names)

Coastal

1. Aonidiella aurantii
2. Paratetranychus citri
3. Aceria sheldoni
4. { Lepidosaphes beckii
     Lepidosaphes gloverii

5. { Aphis spiraecola
     Aphis gossypii
     Myzus persicae
     Toxoptera aurantii

352

CITRUS PESTS

6. { Argyrotaenia citrana / Pyroderces rileyi / Holcocera iceryaella
7. Helix aspersa
8. { Pseudococcus citri / Pseudococcus adonidum / Pseudococcus maritimus / Pseudococcus gahani

9. Saissetia oleae
10. Tetranychus sexmaculatus
11. Tetranychus lewisi
12. Phyllocoptruta oleivora
13. Heliothrips haemorrhoidalis
14. Pantomorus godmani
15. Scirtothrips citri

Intermediate

1. Aonidiella aurantii
2. Paratetranychus citri
3. Scirtothrips citri
4. { Aphis spiraecola / Aphis gossypii / Myzus persicae / Toxoptera aurantii

5. { Argyrotaenia citrana / Holcocera iceryaella
6. Aceria sheldoni
7. Saissetia oleae
8. Lepidosaphes beckii
9. Pantomorus godmani
10. Aonidiella citrina

Interior

1. Aonidiella aurantii
2. Paratetranychus citri
3. Scirtothrips citri
4. Aonidiella citrina
5. Saissetia oleae
6. Coccus pseudomagnoliarum

Central California

1. Scirtothrips citri
2. Coccus pseudomagnoliarum
3. Aonidiella citrina
4. Empoasca fabae
5. Icerya purchasi[1]
6. Aonidiella aurantii[2]

Northern California

1. Scirtothrips citri
2. Aonidiella citrina
3. Coccus pseudomagnoliarum
4. Aonidiella aurantii[2]

Desert

1. Scirtothrips citri

Minor Pests

Archips argyrospila
Aspidiotus hederae
Camnula pellucida
Ceroplastes cerripediformis
Coccus hesperidum
Diabrotica undecimpunctata
Formica cinerea var. neocinerea
Frankliniella gossypii
Frankliniella moultoni
Hercothrips fasciatus
Hypothenemus citri
Iridomyrmex humilis
Leptoglossus zonatus

Marmara salictella
Melanoplus mexicanus
Microcentrum rhombifolium
Myzus persicae
Nysius ericae
Oedaleonatus enigma
Parastichtis pupurea var. crispa
Polycaon confertus
Pseudococcus krauhniae
Serica alternata
Serica fimbriata
Scudderia furcata
Solenopsis xyloni var. maniosa
Xylomyges curiales

---

[1] The cottony-cushion scale is of importance as a pest only in orchards in which the DDT-sulfur thrips control program has eliminated the vedalia ladybird beetle.

[2] The California red scale has gained a foothold in a few orchards and attempts are being made to eradicate it or limit its spread.

Florida
Citrus Regions
1. St. Johns River
2. Central Florida
3. Indian River
4. Lower Indian River
5. Gulf Coast
6. Peace River Flatwoods
7. Ridge

Florida Citrus Pests in Order of Their Economic Importance[1]
(See pages 531 to 537 for common names)

St. Johns River

1. Lepidosaphes beckii
2. Phyllocoptruta oleivora
3. Dialeurodes citri
4. Paratetranychus citri
5. Aphis spiraecola
6. Tetranychus sexmaculatus
7. Pseudococcus citri

Central Florida

1. Lepidosaphes beckii
2. Phyllocoptruta oleivora
3. Chrysomphalus aonidum
4. Dialeurodes citri
5. Paratetranychus citri
6. Aphis spiraecola
7. Pseudococcus citri
8. Tetranychus sexmaculatus

Indian River

1. Lepidosaphes beckii
2. Phyllocoptruta oleivora
3. Paratetranychus citri
4. Chrysomphalus aonidum
5. Dialeurodes citri
6. Aphis spiraecola
7. Pseudococcus citri

Lower Indian River

1. Lepidosaphes beckii
2. Phyllocoptruta oleivora
3. Chrysomphalus aonidum
4. Dialeurodes citri
5. Paratetranychus citri
6. Aphis spiraecola
7. Pseudococcus citri
8. Tetranychus sexmaculatus

[1]Correspondence of Aug. 20, 1948, from W. L. Thompson, Entomologist, Citrus Experiment Station, Lake Alfred, Florida. The map is copied from one prepared by the Department of Agricultural Economics of the Florida Agricultural Experiment Station on Sept. 1, 1945. It is based on aerophotographs, data obtained from the United States census and from technical agricultural workers.

CITRUS PESTS

Gulf Coast

1. Lepidosaphes beckii
2. Phyllocoptruta oleivora
3. Paratetranychus citri
4. Chrysomphalus aonidum
5. Dialeurodes citri
6. Tetranychus sexmaculatus
7. Aphis spiraecola
8. Pseudococcus citri

Peace River Flatwoods

1. Lepidosaphes beckii
2. Phyllocoptruta oleivora
3. Paratetranychus citri
4. Chrysomphalus aonidum
5. Dialeurodes citri
6. Tetranychus sexmaculatus
7. Aphis spiraecola
8. Pseudococcus citri

Ridge

1. Lepidosaphes beckii
2. Phyllocoptruta oleivora
3. Paratetranychus citri
4. Chrysomphalus aonidum
5. Dialeurodes citri
6. Tetranychus sexmaculatus
7. Aphis spiraecola
8. Pseudococcus citri

Minor Pests

Anoplium inerme
Aonidiella orientalis
Ceroplastes floridensis
Chaetoanaphothrips orchidii
Chrysomphalus dictyospermi
Coccus viridis
Cryptocephalus marginocollis
Crytophyllus concavus
Diabrotica vittata
Diabrotica spp.
Dysdercus suturellus
Frankliniella cephalica bispinosa
Lepidosaphes gloverii
Leptoglossus gonagra
Leptoglossus phyllopus
Megalopyge opercularis

Microcentrum retinerve
Myzus persicae
Nezara viridula
Pachneus litus
Papilio cresphontes
Parlatoria pergandii
Phobeton pithecium
Platoeceticus gloverii
Platypus compositus
Pulvinaria psidii
Pulvinaria pyriformis
Schistocera americana
Scolytus rugulosus
Sibine stimulea
Unaspis citri
Wasmannia auropunctata

The Citrus Pests of Texas in Order of Their Economic Importance[1]
(see pages 531 to 537 for common names)

1. Phyllocoptruta oleivora
2. Anychus clarki
3. Aonidiella aurantii
4. Lepidosaphes beckii
5. Lepidosaphes gloverii
6. Anastrepha ludens
7. Parlatoria pergandii
8. Solenopsis geminata
9. Chrysomphalus aonidum
10. Leptoglossus phyllopus
11. Nezara viridula
12. Aphis gossypii
13. Coccus hesperidum

CITRUS AREAS OF TEXAS

[1]The map of the Texas Citrus Areas and the list of the pests, in order of their importance, were obtained through the courtesy of Mr. S. W. Clark, of the Texas Gulf Sulphur Company.

NEMATODES (Phylum Nemathelminthes)

CITRUS NEMATODE

Tylenchulus semipenetrans Cobb

The citrus nematode has been found in all the citrus-growing regions of the United States and in Australia, China, South Africa, Spain, Malta, Israel, and South America, and even as early as 1914 it was stated that this pest probably occurred throughout the world wherever citrus has been grown for a long time (171). It apparently infests all varieties of citrus. J. C. Johnston, University of California Extension Specialist, believes that the citrus nematode is present in practically all California citrus soils.

History:- The citrus nematode was discovered in California in 1912 in soil scraped from roots of citrus trees with badly mottled leaves.[1] It was first found in Florida in 1913 and is believed to be associated with the "spreading decline" of citrus trees in that State (857).

N. A. Cobb of the U. S. Department of Agriculture made a detailed description and illustrations of the citrus nematode (fig. 247) and worked out its life history (170, 171). He was unable to find this species on any host but citrus in an examination of a wide variety of plants. Thomas (872) further investigated the citrus nematode and was also unable to find it in any roots other than those of citrus. However, the citrus nematode was later isolated from olive roots (944). Thomas described the symptoms of nematode attack and experimented with possible control measures, none of which were successful. Foote and Gowans (338) appear to have been the most recent workers to discuss the nematode problem in California, and some of their findings will be presented later.

Description, Life History (171):- The adult females are easily identified (fig. 247), but the larval forms so closely resemble other species that only with the aid of an oil-immersion lens and considerable knowledge of this group of worms can reliable identifications be made. The males

Fig. 247. Citrus root nematode, Tylenchulus semipenetrans. A, male X 300; B, female X 240; C, females with their head ends embedded in root. X 100. Adapted from Cobb (171).

are also identified with difficulty. The males lack an efficient puncturing organ and seldom, if ever, enter sound roots. They mature rapidly and decrease in size and become more slender as they become older, which leads to some doubt as to whether they feed at all. The males vary in length from 280 to 340 microns and measure from 10 to 17 microns at the point of greatest diameter (871). From the illustrations (fig. 247) it appears that the female is of about the same length as the male, but is much broader even before fertilization. The citrus nematodes are yellowish or brownish in color.

The citrus nematode bears a hollow spear with its acute point in the mouth. A narrow tube leads from this spear through the oesophagus to a sucking bulb or pump. The mouthparts are placed in the root surface, the spear is inserted into the tissue and the liquids are drawn out by means of the sucking bulb The anterior portion of the body of the female is thrust into the root tissues (fig. 248)

---

[1] Byars (136) has pointed out that citrus nematodes may also be found on the roots of healthy trees, but Thomas (872) stated that no badly infested tree was found to be in good condition.

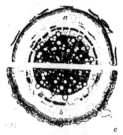

Fig. 248. Left: a, healthy citrus root; b, diseased root, with female nematode shown at (c) with her head end embedded in root X 30. Right: nematode-infested citrus root. a, nematodes; b, separation of outside of root from axial position. X 5. After Cobb (171).

and the posterior egg-bearing portion remains outside. The female can never withdraw from the rootlet, for the body is constricted at the surface of the root, and becomes larger in both directions from this point. The portion of the body outside the root, however, becomes by far the larger.

The eggs are deposited one at a time until a batch of 12 to 20 or more are laid. They are sometimes encased in a mass of "gummy" or "gelatinous" matter. These are thin-shelled, relatively large and are, of course, deposited on the outside of the root. The female larvae insert their heads into citrus rootlets at an early age and, as stated before, they remain in this position. Thomas (872) found 100 nematodes imbedded in the tissue of a piece of citrus root 4 mm long and in another piece of root of the same length he found 108.

The life cycle of the nematode may be completed in from 6 to 8 weeks (872). This accounts for their rapid rate of increase. Foote and Gowans (338) estimated from their samplings that there may be 764,000,000 citrus nematodes in the "tree space" of a mature citrus tree (20 x 22 feet) from a depth of 6 inches to 24 inches. This estimate was based on the average sample in the orchards in which the sampling was done, which contained an average of 1200 citrus nematodes per 40 grams of dry soil. From a depth of 0 to 72 inches the population would be 4,000,000,000 per tree space.

Dissemination:- The citrus nematode is relatively slow and weak, and it is improbable that it ever migrates any great distance through its own muscular exertions (171). However, its small size is in favor of its transportation from place to place in the soil by such agencies as soil water, subterranean insects, worms, and burrowing insects. When brought to the surface by cultural operations, the eggs and free-living younger stages may then be transported from tree to tree, or orchard to orchard, or in fact, even from district to district. It is believed that irrigation water is one of the principal means of distributing this parasite after it has once become established through the planting of infested nursery stock (171). Nursery stock is considered to be another important means of spreading the pest, and is usually the means of initially establishing it in a district previously free of nematodes.

lemon trees that he has seen failing from what is known as "slow decline" have also been heavily infested with citrus nematodes, and that in his opinion, in many cases symptoms attributed to "over-irrigation" are closely associated with citrus nematode infestations.

As its specific name indicates, <u>Tylenchulus semipenetrans</u>, in contrast to some other common nematodes, does not entirely enter the root, as explained previously. Another characteristic of a citrus nematode infestation is that this species does not form knots or enlargements of any kind on the roots, in contrast with some other species, such as the root-knot or garden nematode, <u>Heterodera marioni</u> (Cornu).

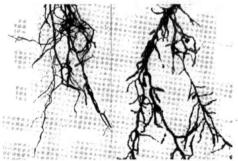

Fig. 249. Left, roots of lemon tree free of nematodes; right, roots of lemon tree from which 26,800 nematodes were taken from 1 gram of root sample. After Foote and Gowans (338).

Soil particles often cling to infested roots (fig. 249) and are not easily removed. This is because of the gelatinous material which covers the eggs. The bark of the rootlet separates from the inner woody portion (fig. 249). Often when the soil is shaken from the roots the outer portion falls away, leaving the white inner woody cylinder.

It is believed that the ability of the tree to grow roots faster than they are killed off by the nematodes is the controlling factor in tree vigor. The margin of safety in root growth when nematodes are present is sometimes so small that adverse conditions which the tree would ordinarily tolerate may upset the delicate balance between root growth and nematode damage and cause a deterioration of the tree.

Cobb (171) mentions the possibility of a secondary invasion of organisms to further increase injury to the roots. Foote and Gowans (338) state that a purplish-colored fungus may be found on citrus roots which are heavily soil-incrusted as the result of nematode infestation. L. J. Klotz has identified this fungus as a species of <u>Fusarium</u>. Foote and Gowans believe that when roots up to 1/4 inch in diameter are rotting, following a preliminary rotting of feeder roots, that a secondary organism may have started on the nematode-weakened rootlets and traveled back to the larger roots.

Control:- Cultural practices appear to have no effect on nematode infestations. Drying of the soil, by irrigating only a portion of the root area at a time, has been tried without success. Soil fumigation has been found to be beneficial when the infested soil is fumigated prior to planting. Carbon bisulfide, ethylene dichloride, chloropicrin and DD (mixture of 1, 2 - dichloropropane and 1, 3 - dichloropropylene) have been used successfully and have resulted in an almost complete elimination of nematodes from infested soil. Young trees planted in treated soil developed almost 50% more rapidly (as based on comparative foliage volume) than trees planted in infested soil.

Treatment of infested soil in which trees are growing has as yet not proved to be successful because the tolerance of the roots for the various fumigants and other compounds which have been used has been no greater than that of the nematodes. A material which could be successfully used and applied in some practical way, such as by addition to irrigation water, would certainly be a boon to the citrus industry of California as well as many other regions.

It is believed that the fumigation of nursery plots would be practical and would result in the newly-planted trees being free of nematodes. This would be of special value if the trees were to be planted in treated soil or soil in which citrus had not previously been grown.

Resistant Rootstock:- In Argentina, where nearly all trees are infested with citrus nematode, one rootstock, Poncirus trifoliata (Linn.), is highly resistant to the attacks of this pest. In orchards in which other rootstocks are heavily infested, only a few male nematodes, and no females, have been found on the above rootstock species. Experiments are being made in Florida to determine whether hybrids between P. trifoliata and other stocks might retain the nematode resistance of this species and yet be superior in other respects. (243a)

SNAILS (Phylum Mollusca)

EUROPEAN BROWN SNAIL

Helix aspersa Müller

The European brown snail has been widely disseminated throughout the world by being accidentally distributed on plants or intentionally as an article of food. In California this snail may be found in the majority of cultivated districts from San Diego County to Sonoma County. It was intentionally introduced from France between 1850 and 1860 and "planted" among the vineyards along Guadalupe Creek in Santa Clara County and later in San Francisco and Los Angeles (72). Although Helix aspersa has some value as an article of food in some European countries, other species are more popular. In California H. aspersa is not eaten, except by an occasional European immigrant, but it has considerable economic importance as a pest in vegetable and flower gardens and on citrus and avocado trees. On the latter they may consume a large percentage of the foliage (fig. 250) and may also devour the bark from green twigs. Likewise holes may be eaten into the peel of the fruit, (fig. 250) and even the smaller abrasions may provide a means of entry for fungus organisms.

Snails are a pest of citrus only in the relatively humid and cool coastal areas in California. They are most active in citrus orchards during warm, damp weather, especially during the late winter and spring months. This is the time when most damage from snails occurs.

Description (72):- The shell of the European brown snail is of a grayish-yellow and brown color (fig. 251). The brown is usually in five interrupted bands, the second and third bands being ordinarily confluent, giving the appearance of a single band. Sometimes all the bands are united. When fully developed, the shells have 5 or 4-1/2 whorls and are an

Fig. 250. Brown snails feeding on an orange. Note also the damage to foliage. Photo by C.O. Persing.

Fig. 251. European brown snail, Helix aspersa. X 1.0. Photo by Roy J. Pence.

inch or more in diameter. The form of the shell is that of a dextral, or right-hand spiral, the sinistral or left hand spiral being very rare.

The body is light to dark gray, and when fully extended it is about two inches long. As seen in fig. 252, the head is provided with two pairs of retractile tentacles. The upper and larger pair are the ocular tentacles, which possess dark pigments at the tips and serve as eyes. The lower pair of tentacles are very sensitive to touch. The mouth is surrounded by fleshy lips. A small, hard chitinous jaw is found on the upper side of the mouth, just inside the lips. Muscles are attached to the upper edge of this jaw. The free sharp edge is toothed, and is capable of cutting off particles of food and conveying them into the oesophagus.

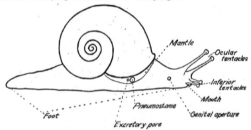

Fig. 252. Diagram of external anatomical features of a snail.
After Basinger (71).

The snail glides along by means of the long, flat, muscular organ known as the foot (fig. 252). Mucus which is constantly secreted by glands in the foot, facilitates this means of locomotion.

Life History (72):- Snails are hermaphrodites and mutual fertilization takes place. Both individuals may lay eggs. Oviposition usually occurs within three to six days after fertilization. The snail then selects a spot of damp soil and prepares in the ground a nest for the eggs.

Fig. 253. Nest and eggs of Helix pisana, which are similar to those of H. aspersa. After Basinger (72).

According to Basinger (72), "The ground is scraped away with the mouth-parts and the loose soil is worked back underneath the foot. The undulations of the muscles of the foot serve as a kind of endless belt in moving the loose soil back out of the way. By this process, requiring several hours for completion, a hole one-quarter inch or more in diameter and one to one and one-half inches deep is finally made, with a rounded cavity at the bottom one-half to three-quarters of an inch in diameter for the egg mass." An egg mass is shown in fig. 253. Although the eggs are laid singly, they adhere to one another to form a loose mass with an average of 86 eggs. They are white, spherical, and about one-eighth of an inch in diameter. After ovipositing, the snail closes the opening of the nest, thus concealing its location. A little mound of earth and excrement indicates the location of the nest, (fig. 253).

The young snails possess a shell of somewhat more than one whorl at the time of hatching They usually remain in the nest from two to four days, then work their way to the surface. Two years are required to reach maturity.

Helix aspersa is distinctly nocturnal. The snails become inactive when dry conditions prevail, sealing themselves to various objects as, for example, the trunk of a tree, or closing the opening of the shell with a parchment-like epiphragm (fig. 254). They then resume activity under more humid conditions. This species can survive a temperature at least as low as 14°F according to E. J. Lowe (72). A certain percentage of the snails bury themselves in loose soil to a depth of one-quarter to one-half of an inch and hibernate during the winter months and on into late summer.

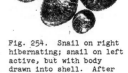

Control:- For the control of the brown snail poison bran mash consisting of one part calcium arsenate and sixteen parts bran was recommended (72). This was scattered in small particles where the snails were most likely to find it. In citrus orchards it was scattered under the tree as far as the foliage extends, but not in the spaces between the trees.

Fig. 254. Snail on right hibernating; snail on left active, but with body drawn into shell. After Basinger (72).

More recently the following developments have changed snail control methods: (1) application of baits to the tree instead of on the ground, (2) fresh orange pulp instead of bran, (3) metaldehyde baits and later (4) tartar emetic sprays (689). It was found that calcium arsenate-bran baits give best results when applied to the tree by means of a power blower, which assured more uniform application of the material than hand application. About one-half pound is used per tree. However, calcium arsenate - fresh orange pulp bait can best be applied by hand, using three-fourths to one pound per tree. Usually two men stand in a truck bed and throw the bait onto the trees while the truck is driven between two rows of trees. Metaldehyde baits are most effective when placed in small piles on the ground around the periphery of the tree using one-third to one-half pound portions. The snails are attracted to and paralyzed by the metaldehyde, but must lie out in the sun to succumb. This places a limitation on the effectiveness of the bait which makes it less effective than the calcium arsenate-bran or calcium arsenate-orange pulp baits.

Under ideal conditions no differences in effectiveness were noted between tartar emetic spray and the calcium arsenate-bran or calcium arsenate-orange pulp bait in the control of either small or large snails. However, under conditions not ideal for snail control, as when it is a little too dry, the tartar emetic spray is the most effective. This was explained by the fact that tartar emetic spray itself provides enough moisture on the tree to stimulate some snail activity. Another factor favoring tartar emetic is that it will stick to the tree better than the bait, especially the calcium arsenate-bran bait, during windstorms. (689)

Tartar emetic sprays can be applied with broom guns, boom sprayer, or spray-duster. When using broom guns or boom sprayer, two pounds of tartar emetic and four pounds of white or brown sugar are recommended per 100 gallons of water. Three to four gallons of spray are applied per average-sized mature orange tree. The spray-duster can be used effectively on lemon and sparsely foliated orange trees. About one gallon of finely atomized spray is then applied per average-sized mature tree. When applied by means of the spray-duster, the concentration is six pounds of tartar emetic and twelve pounds of white or brown sugar per 100 gallons. Spray rigs should be thoroughly cleaned before applying tartar emetic sprays, especially if lime and copper compounds have been used (688).

In comparison with the cost of treatment for some other pests, snail control is relatively inexpensive. The least expensive material is calcium arsenate-orange pulp. The cost of treatment with this bait averages about three cents per tree. The most expensive of the recommended treatments is tartar emetic spray, which costs five to eight cents per tree.

No treatment is effective if the snails are not active, so the treatment should be applied at the right time of the year (late winter and spring months) and during periods of moderately warm weather and when moisture collects on the tree at night. It is advisable to turn under the covercrop before applying the treatments. This permits the snails to leave the covercrop and migrate to the trees. One should wait a week or more after the covercrop is turned under before applying the snail treatments (688).

Handpicking may be resorted to, at least on a small scale. This is best done at night during seasons when the snails are active. When it is too dry for the snails to be active, they may be picked from the trunks of citrus trees, where most of them are sealed fast.

Ducks and geese can destroy large quantities of snails. They are commonly kept, especially in avocado orchards, where no cultivation is practiced, to destroy the European brown snail. Usually no further control measures are necessary.

WHITE SNAIL

Helix pisana Müller

The white snail is widely distributed in Europe and Africa and is of some importance there as a citrus pest. It is eaten by the Sicilians, Italians and French in enormous quantities, and one may see crocks "full of the poor snails already cooked, seasoned with garlic, oil and parsley, or served with tomato sauce and exposed in an artistic way to stimulate the appetite of the gourmands" (233).

The white snail was found at La Jolla, California, in 1918. It was probably introduced by some European resident, in order to propagate it as a table delicacy. An eradication program was immediately inaugurated which led to an almost complete extinction of the potential citrus pest (70). By 1922 the snail had again increased to large numbers. Another eradication campaign was jointly undertaken by the Bureau of Pest Control of the State Department of Agriculture and the Horticultural Commissioner's Office of San Diego County, under the direction of A. J. Basinger. The white snail is somewhat smaller than the previous species, lighter in color, and has a similar life history and habits.

A few minor outbreaks of the white snail have occurred since the 1922-23 eradication campaign near the original infestation and in two other localities in southern California, but all have apparently been successfully eradicated, principally by means of poisoned bait.

THE MITES (Order Acarina)

This economically important group of pests is related to the true spiders; the spider and the mite belong to the same class of arthropods, the Arachnida. The arachnids resemble the insects in being small, predominantly terrestrial animals and in possessing tracheae. They differ from the insects, however, in having usually 2 or 4 pairs of legs, no antennae, no true jaws or compound eyes, only 2 body regions, the head and thorax being fused, in the position of the openings of the reproductive organs, these being located ventrally near the front of the abdomen, and in having no conspicuous metamorphosis. The mites and ticks (Order Acarina) differ from the spiders in having little indication of body regions or segments, and in the newly hatched young having only 3 pairs of legs in all but the family Eriophyidae, which have 2 pairs of legs in all stages. Some adult mites of the Podapolipolidae and Trichadenidae have only 3 pairs of legs, the same as their larvae. The adults of these species, as well as the larvae of the majority of mites, constitute the only exception to the rule that only the insects have 3 pairs of legs. The ticks are larger than the mites and have a tougher cuticle; otherwise there is no important difference.

362

CITRUS PESTS

The mites constitute an important group of citrus pests, especially in the United States. In California the citrus red mite (red spider) may be considered as second only to the California red scale in economic importance as a citrus pest, while the citrus bud mite is also a very important pest, but in a more limited area in the coastal regions. In Texas the citrus rust mite is by far the most important of the citrus pests. In Florida it is always a severe pest, ranking among the first 3 in importance and some years it might be considered to be the number one pest of citrus, surpassing even the purple scale in importance.

The economically important species of mites attacking citrus are found in two families, the Eriophyidae and the Tetranychidae.

The eriophyids are the bud mites and blister mites, various species of which can be found on nearly all varieties of subtropical fruit trees. They are exceedingly small; in fact, a 12-power hand lens must be used to readily discern these mites. Their bodies are narrow and elongate, and they have only 2 pairs of legs. They are pale straw to yellow orange in color. An extensive series of taxonomic papers on the eriophyid mites is being written by H. H. Keifer, and is being published from time to time in the Bulletin of the California State Department of Agriculture. The first paper was published in 1938 and 14 such studies have been published to date. Descriptions are given for hundreds of species which have been collected in California by Mr. Keifer and other entomologists, and excellent illustrations of the species accompany each paper. Many new species have been described in this series.

A taxonomic treatment of the so-called "red spiders" the family of Tetranychidae can be found in a paper by McGregor (553) in which the more common species are described and illustrated. A new and more comprehensive paper, by the same author, will soon be published.

The tetranychids, although they are small, are much larger than the eriophyids. Their bodies are oval in shape and often reddish in color. As previously stated, they have only three pairs of legs when they are first hatched and gain another pair during the first molt.

CITRUS BUD MITE

Aceria sheldoni (Ewing)

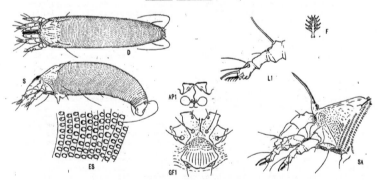

Fig. 255. Citrus bud mite, Aceria sheldoni. D, dorsal side; S, left side, ES, detail of skin on side; AP1, internal female genitalia; GF1, female genitalia and coxae from below; F, featherclaw, L1, left foreleg; SA, side view of anterior section of mite. After Keifer.

363

History:- The citrus bud mite, Aceria sheldoni (Ewing) (fig. 255), came to the attention of entomologists relatively recently, having been discovered on lemon trees near Santa Paula, California on June 17, 1937. It was at once associated with certain abnormalities in the growth and development of lemon foliage and fruit which had previously had no satisfactory explanation. Subsequent investigation disclosed that the bud mite was widely distributed in southern California. This widespread distribution coupled with the fact that the characteristic malformations of foliage and fruit caused by the mite had been periodically noted by citrus growers for at least 20 years previous to its discovery, indicates that it has been in the state for many years. After its discovery in California, the mite was found on citrus trees in Australia and on mandarins in Hawaii (112).

Distribution and Hosts:- Since Citrus is the only known host of the mite in California, it is generally supposed to have originated in some region where the tree is native and to have been introduced into California on one or more of its varieties. In the United States the mite occurs only in California and only in the counties of Santa Barbara, Ventura, Los Angeles, Orange, Riverside, and San Diego. While the mite occurs on lemons, limes, oranges, grapefruit and other citrus varieties, it is principally a pest of lemons.

Description and Life History:- The following description and account of the life history of the eriophyid mite is taken from Boyce et al (113):

"The adult mites are very small in size, averaging about 1/50 of an inch in length. They vary in color and are light yellowish or pinkish. In shape they are elongate and cylindrical. The body is divided into two general portions: the cephalothorax at the fore end of the body is short and bears two pairs of legs, and the mouthparts, which may be considered of the general rasping-sucking type. The abdomen is long and tapering, with many annulations or rings.

"The mites inhabit protected locations within the buds, in the region where the petiole of the leaf oppresses the bud, among the tender tissues of a bud that has just begun to elongate, within a developing blossom, or beneath the button of a fruit. Eggs are found wherever the mites are present. Each female apparently deposits about 50 eggs.

"The eggs are pearly white, nearly spherical in shape and average about 1/500 of an inch in greatest diameter. The incubation period varies from two to six days, depending upon prevailing temperature. There are four developmental forms of the immature mites, namely, larva, moult I, nymph, and moult II. The larva is hatched from the egg, and the adult emerges from the form referred to as moult II. The larva and nymph are motile and feed on the plant tissues, while the two moulting forms are quiescent and do not move about. The total development period from egg to adult under field conditions during the summer season is about 10 days; during the fall season about 15 days, and during the winter season about 20 to 30 days."

Injury:- The citrus bud mites may be seen with the aid of a hand lens under the button of the fruit--as many as 1000 mites have been found under one button. The area on which the mites feed becomes blackened, but it appears that this type of injury is not economically important.

The mites also feed within the leaf-axil buds, in which up to 50 mites are commonly found and as many as 300 have been counted. Boyce et al (112, 113) have described the resulting injury in detail. The most severe symptoms of mite injury may be seen on lemon trees. Sometimes the entire bud is destroyed or it may develop into an abnormal or inferior type of growth. The twigs may develop into a bunched or rosetted type of growth and the leaves may assume irregular shapes (fig. 256). The blossoms may be stunted and misshapen (fig. 257). The most spectacular abnormalities are found among the fruits, which may assume very curious and grotesque shapes (fig. 257). More than 95% of the buds may be infested with the citrus bud mite when the trees have been severely infested for several months.

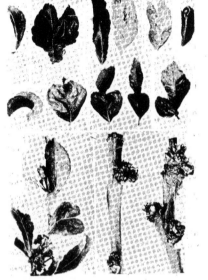

Fig. 256 (at right). Leaves, buds and twigs of lemon deformed by bud mites. After Boyce (106).

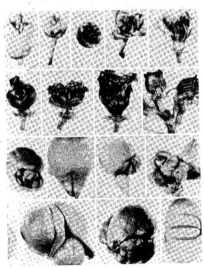

Fig. 257 (at left). Blossoms and fruit of lemon deformed by bud mites. After Boyce (106).

365

Bud mites also attack orange trees, on which the characteristic symptoms of injury are similar to those on the lemon, but the fruit deformities are not as pronounced. The fruit may be flattened vertically and seams or ridges may extend part way or the entire distance from the stem to the stylar end of the fruit. The Valencia orange may be deformed so as to resemble the Navel orange at its stylar end.

Control:- Boyce et al (113) found the mite to be amenable to control with oil or sulfur sprays. They recommended light medium oil at the conventional dosages of 1-2/3 to 2% or lighter dosages containing rotenone if the heavier dosages were found to be undesirable in certain localities or under certain conditions. The sulfur sprays were believed to have limited usefulness because of the likelihood of sulfur injury to the trees under high temperature – low humidity conditions. In addition the sulfur preparations are not effective against red scale and red spider, two important pests that are usually present in the areas where bud mite is a problem. Sulfur preparations may usually be safely used in the spring, however, and a satisfactory formula has been 2 gallons of liquid lime sulfur, 4 pounds of dusting sulfur, and a half pound of casein spreader to 100 gallons of water. The casein is added to the dry sulfur in a bucket and stirred and made into a paste before being added to the spray tank. The lime sulfur is then poured into the tank. Regardless of the material selected for the treatment, a spray in the spring, followed by another in the fall, was recommended.

The control of the bud mite at present is substantially the same as indicated above. Recommendations vary to suit the needs of different districts.

In the coastal areas severe injury from bud mite often occurs when oil spray is omitted even for a single season. In fact, more than one spray a year is ordinarily necessary to prevent losses from bud mite. In lemon districts close to the coast, as in Ventura and Santa Barbara Counties, where the fruit picks are fairly uniformly distributed throughout the year and the effect of oil on maturing lemons is less severe than farther inland, it is possible to spray practically any time of the year following a pick. An oil spray every 7 or 8 months gives excellent bud mite control as well as red mite control. Farther from the coast, where the bulk of the crop matures in the spring, and oil spray during hot weather often causes injury, oil spray should be applied in late fall or late spring. A spray at each of these seasons, the spring spray applied late, after the greater part of the fruit has been picked, provides a good "two-spray program" for red scale and red mite. (982)

For the control of bud mite on oranges a thorough spray with a low dosage of light medium oil, 1-1/4 to 1-1/2%, or a low dosage of oil-rotenone spray, are effective. These are applied in the fall months. (982)

CITRUS RUST MITE

Phyllocoptruta oleivora (Ashmead)

Origin and Distribution:- The russeting of oranges in Florida was first connected with the rust mite, Phyllocoptruta oleivora (Ashm.), in 1878 or 1879, but the original habitat of the rust mite is considered to be southeastern Asia (724). In this country it now occurs in all the citrus-growing states, being the principal pest of oranges and grapefruit in Texas and among the most important of the pests of those fruits in Florida. On page 354 the rust mite is rated second to the purple scale among the citrus pests of Florida. That is because the purple scale does more damage to the trees than the rust mite. On the basis of the amount of money spent in control, however, the rust mite would rank first. In California the rust mite, although it has been in this state for about 50 years, was until comparatively recently almost completely confined to the Chula Vista, Lemon Cove and El Cajon districts in San Diego County. Although it causes severe injury to oranges in Texas and Florida, it has only occasionally been noted on oranges in San Diego County despite its persistence as a pest of lemons. It causes lemons to turn a silver color, and is accordingly called the "silver mite" in the section when it attacks lemons in California.

CITRUS PESTS

There has been a relatively recent spread of the rust mite to a number of areas in California far from its original area of distribution in San Diego County. Although the mite was found on lemons in North Whittier Heights in 1934, it quickly disappeared and has not been seen in that area for the last 10 years. In 1942 it became a pest of oranges in the San Luis Rey Valley and now requires frequent treatments to prevent damage in that district. In 1945 the rust mite was found on oranges at San Juan Capistrano in Orange County and has since spread throughout the San Juan Capistrano orange district. A previous infestation at San Juan Capistrano had subsequently disappeared. In 1945 the rust mite was also found on both oranges and lemons at Santa Paula in Ventura County, and in 1946 infestations were found on oranges at Tustin in Orange County, and at Rancho Santa Fe in San Diego County. (985)

Description and Life History:- Rust mites are parthenogenetic, that is, they reproduce without fertilization. Males, if such exist, have not been found. The eggs are found in pits or depressions on the fruits and leaves and are laid singly, but may occur in groups. The egg is spherical, smooth, and semitransparent or pale translucent yellow. The mites have been observed to lay as many as 29 eggs. The incubation period of the eggs averages about 3 days in summer to about 5-1/2 days in winter. Like the citrus bud mite, these mites are members of the family Eriophyidae, and have the elongate form and two pairs of legs common to species in this family. As in the case of the citrus bud mite, there occurs a larval and nymphal stage, the former lasting 1 to 3 days in summer and 3 to 6 days in winter, and the latter lasting 1 to 3 days in summer and 4 to 13 days in winter. (1020)

In California the rust mites may be found on the trees at any time of the year, but the mite population does not increase from November until the first warm weather in the spring. In Florida a fungus destroys large numbers of rust mites after summer rains, sometimes almost completely wiping out the entire rust mite population in some districts. In California, rains do not occur in summer and apparently no control by fungus diseases occurs.

The adults (fig. 258), along with the citrus bud mites, are the smallest of the citrus pests, being on the average about 0.12 mm in length. They are elongated and wedge-shaped and in general appear superficially much the same as the citrus bud mite, which has already been described. In addition to its two pairs of legs, the rust mite is assisted in moving about and clinging to trees by a pair of lobes or false feet, located on the last abdominal segment. Attached by means of these lobes, the mite can raise its entire body and turn about in various directions.

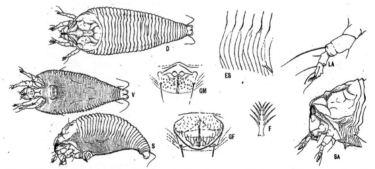

Fig. 258. Citrus rust mite, Phyllocoptruta oleivora. D, dorsal side; S, left side; V, ventral view; GM, male genitalia; GF, female genitalia; ES, detail of skin on side; F, featherclaw; LA, left foreleg; SA, left side of anterior section more or less turned ventrad. (After Keifer).

367

The mites are at first light yellow and straw colored, but some specimens become a darker yellow or nearly brown a few days after reaching maturity. The average period of life for adults reared in confinement was 7.6 days. (1020)

It can be seen that a generation of mites can develop in from 7 to 10 days in summer and this accounts for the enormous numbers that may sometimes be found on citrus trees, and which together with the myriads of cast skins may give the trees a dirty or powdery appearance.

Injury:- The rust mite punctures the epidermal cells of the rind of the fruit with its piercing mouthparts. The outer layer of cells is destroyed. As stated before, injured lemons have a silver color and injured oranges and grapefruits a russet color (fig. 259). The grade of the fruit is, of course, greatly reduced.

A certain type of "tear staining" of grapefruit was once believed to be caused by a fungus disease, but is now known to be the result of rust mite attack (1020). A severe infestation of mites may result in an injury on lemons and grapefruit known as "shark skin". The epidermal cells can then be turned back and pulled off.

Fig. 259. Rust mite injury to oranges. Photo by C. J. Duffy, Coatepec, Mexico.

The rind of fruits affected by rust mite is thicker than normal and the fruits are smaller. Since the epidermal cells are broken, the fruit loses water rapidly and must reach its destination promptly after picking. It has also been reported that affected fruit decays more rapidly than unaffected fruit and is more acid. The writer has seen production records from a Mexican orchard which showed a 30% reduction in production of oranges caused by rust mite infestation when the production of treated blocks of trees was compared to the production of untreated blocks. In a three-year experiment in Florida, it was found that rust mite control increased the annual production of oranges by an average of 0.63 box per tree, besides increasing size, appearance and quality (669).

In Florida there have been indications that a heavy rust mite infestation on the leaves and twigs of the summer and fall growth may contribute to the occurrence of mesophyll collapse during the winter months (883). Mesophyll collapse is a spotted necrosis of the foliage, usually on orange trees. In some orchards where a spotted infestation of rust mite occurred, mesophyll collapse could be found only on trees infested with the mites, but not on those trees which were free of mites. The citrus red mite has similarly been considered a contributing factor in the occurrence of mesophyll collapse on orange trees in California.

Control:- Fortunately, in California the citrus rust mite is easily controlled by sulfur dust or lime sulfur spray. In the warmer areas, sulfur can not be used in the summer and early fall without danger of injuring the citrus foliage and fruit, but temporary relief may be obtained by using oil spray. The sulfur can later be applied during cooler weather. Two dustings are recommended during the winter-spring period. Lime sulfur spray can, of course, also be used, but is more expensive (992). The sulfur dust applied in winter should be effective until the fall application of oil spray.

In Texas, rust mite is controlled by means of from 2 to 4 applications of sulfur dust. The first is applied when the fruits are about the size of a pea, around the middle of May, the second dust is applied from 4 to 6 weeks later, and the remaining dusts are timed to suit weather conditions. About 90% of the dusting for rust mites in Texas is done by means of airplane dusters.

In Florida it may under certain conditions be difficult to control the rust mite, especially if Bordeaux has been applied for fungus diseases. Sulfur in some form is used for control. It may be in the form of a dust, in which 5 to 10 pounds of lime are added to 90 to 95 pounds of sulfur to increase flowability. Sometimes rust mites multiply so rapidly that the greater rapidity of the dust treatment as compared to spraying may save the fruit from much mite injury, and besides, dusting is much less expensive, than spraying. Owners of large orchards usually resort to dusting for mite control. Spraying with lime sulfur, however, has certain advantages. It has considerable effect in controlling scale insects and whiteflies, even though this control is not as good as that of oil spray. In the case of light infestations, the lime sulfur spray may obviate the necessity for a separate oil spray treatment for scales and whiteflies. Lime sulfur was formerly used at the rate of 2 gallons to 100 gallons of water, but in recent years the tendency has been to use less lime sulfur and to add from 5 to 10 pounds of wettable sulfur to each 100 gallons. This makes a safer spray, but fully as effective. Usually three treatments per year are necessary for adequate mite control. (874, 671, 924) If sulfur is used as a spray, the treatment will not be as adversely affected by rain as if dust is used.

CITRUS RED MITE

Paratetranychus citri (McGregor)

History:- The citrus red mite, Paratetranychus citri McG., is the first of the species discussed in this chapter belonging to the family Tetranychidae. It has been known to occur in California since 1890, at which time it is reported to have been introduced from Florida (724). It is very similar in appearance to European red mite, P. pilosus C. & F., of deciduous trees and was considered by various workers to be either synonymous with or a race of this species until McGregor and Newcomer (566) definitely established its present taxonomic status.

In California the citrus red mite is most commonly known as the citrus "red spider" while in citrus areas in this country outside of California, it is known as "purple mite".

In about 10 years after its introduction into California, the red mite became an important pest of citrus in this state. Its life history was recorded and a number of control measures were recommended, (1014), none of which are in use today. Ten years later, Quayle (713) wrote a bulletin on various species of mites attacking citrus trees, in which the biology of the citrus red mite was more intensively studied than ever before and its numerous natural enemies were described and discussed.

Distribution:- The citrus red mite appears to be native to North America. Although it occurs in Texas, Florida, and other Gulf States, it has in the past been a relatively unimportant pest there,[1] while in California it is second only to the red scale as a citrus pest. In California, the red mite was until comparatively recently confined to coastal and subcoastal areas, and it was believed that it had reached geographic limits precisely conditioned by climatic requirements of the species. Rather suddenly the red mite became so abundant in the orchards of the San Jacinto Land Company near Riverside in 1937 and 1938 as to require treatment. The mite had previously been practically unknown in the Riverside area. In 1939 it became a major pest throughout the Riverside area and soon was found abundantly in Redlands. At present it is as important in such dry inland areas as Riverside and Redlands as it is in the coastal areas. The dry north winds add to injury done by the mite and result in its damage being greater in inland areas than nearer the coast. The mite has not yet appeared in significant numbers, however, in the San Joaquin and Sacramento Valley citrus areas[2] nor in the Coachella and Imperial Valleys and Arizona.

[1]It appears that within recent years the citrus red mite (purple mite) has become a major pest in Florida (883), but it has not assumed the importance in comparison to other major pests that it has had for many years in California.

[2]Until recently it was believed that no citrus red mites could survive in the hot and presumably climatically unfavorable citrus areas of the San Joaquin and Sacramento Valleys, but in the spring of 1948 an infestation was found in a Valencia orchard near Woodlake in which heavy dosages of DDT-kerosene had been applied experimentally for yellow scale. Eradication measures were promptly taken. It is likely that the mites had long existed in the orchard but were held to extremely low population density until the DDT destroyed their natural enemies. Another infestation has been reported from a DDT-treated orchard at Hamilton City and still another in a citrus nursery in Kern County. (522)

Description and Life History:- The appearance of the various stages of the citrus red mite are similar to those of the European red mite, which have been well illustrated by Newcomer and Yothers (see fig. 462).

Egg:- The egg of a citrus red mite can be distinguished from those of other mites found on citrus trees because of the vertical stalk attached to the egg, from the tip of which 10 or 12 guy threads radiate to the surface upon which the egg is attached (fig. 260). The bright red eggs are about 0.13 mm in diameter, spherical, but slightly flattened in vertical diameter. Unfertilized females may deposit eggs, but these always develop into males. The efforts of the emerging larva cause the egg to split at its equator, then the upper half of the

Fig. 260. Citrus red mite. Left, male; right, female; center, webbing radiating from stalk of egg. X 18. Photo by Roy J. Pence.

egg shell is pushed back as if it were hinged on one side. After the larva emerges, the upper half springs back into place and the egg again appears normal except for a translucence indicating the absence of the embryo.

From 20 to 50 eggs per female are deposited at the rate of 2 or 3 per day. The eggs are mainly deposited on leaves, fruit and twigs. On the leaves they are deposited on both surfaces, but most abundantly along the midrib. The eggs hatch in from 8 to 30 days, depending on temperature.

Larva. The first instar is called the larva, and in common with all the order Acarina with the exception of the eriophyids, the larva has only 3 pairs of legs. The larva becomes about 0.20 mm in length, and molts in 2 or 3 days. Since metamorphosis is very simple, the larva appears much like the adult mite.

Protonymph. During its first molt, the citrus red mite acquires its fourth pair of legs and is called the protonymph. It is then 0.20 to 0.25 mm long and appears much the same as it did in the previous instar except for its greater length and the extra pair of legs. This stage may be called the first nymphal stage. It lasts 2 or 3 days during the summer months.

Deutonymph. The period between the second and third molt, the second nymphal stage, is called the deutonymph. The deutonymph measures 0.25 to 0.30 mm in length. Two or 3 days are spent in this stage in the warmest weather. The molting of the deutonymph in its transformation to the adult has been described by Quayle (713, 724). The skin is split transversely around the middle of the body between the second and third pairs of legs and the parts of the old cuticle are separated on the dorsal surface by a humping of the middle of the body. The front part of the body is then pulled back out of the old skin and then, with its front legs free, the mite crawls out of the posterior part of the skin. The cast skins may be found in large numbers on the leaf or fruit surface when one is searching for mites or mite eggs with a hand lens or binocular.

The adult female. The adult female is distinctly red in color, often a dark velvety red, or may have the purplish cast that gives it the common name of "purple mite" in the Gulf States. It is from 0.32 to 0.37 mm in length, oval in shape, and well rounded above. Large, white bristles arise from prominent tubercles on the back and sides of the mite (figs. 260 and 261). These distinctive characters, coupled with the stalks and grey threads on the eggs (fig. 260) should serve to make this species easily identified. The adult may live as long as 18 days in summer, making the entire life of the mite from 35 to 40 days. The female is immediately fertilized after molting, and egg laying begins in 2 or 3 days. Thus the period from egg to egg may be as little as 3 weeks under the most favorable temperature conditions to as many as 5 weeks in winter.

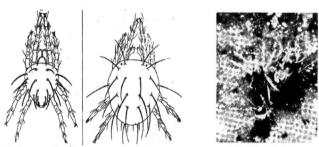

Fig. 261 (at left). Citrus red mite, _Paratetranychus_ _citri_. Left, male; right, female. After Woodworth (1014).

Fig. 262 (at right). Citrus red mite. Left, male, right, molting female; below, an egg. Photo by Roy J. Pence.

The adult male. The male (fig. 261, left) is considerably smaller than the female and is also quite narrowed posteriorly. Its coloration is about the same as that of the female. There are about as many males produced as females. Fig. 262 shows a male standing alongside a molting female, in a commonly assumed position. Mating takes place at the termination of the molting. While the intense light used in photography ordinarily causes the male to move about actively, the presence of the female causes him to become quiet, making photography possible.

Seasonal history. It appears from the life history data presented above that there may be as many as 12 to 15 generations of mites per year. The greatest numbers, however, are found in the spring months, and again in the fall, when the temperature is neither too high nor too low. In the summer, prolonged hot spells may result in the death of all active stages of the mite in the citrus orchards (968). The coldest weather experienced in southern California, however, does not result in an appreciable mortality of the citrus red mite although, of course, the activity of the mite is greatly reduced (964).

The short period required for this species to reach maturity results in a rapid increase in population. The citrus grower, having observed only a few mites on his trees in the late summer, is often surprised to find a heavy infestation within a few weeks after cool weather sets in, and may notice damage from mites before arrangements can be made to treat the orchard.

Injury. The mites feed on the fruit, leaves and green bark of citrus trees. They are most abundant on new growth. The lemon appears to be the preferred host, but oranges and grapefruit may also be severely attacked. Extraction of chlorophyll by the piercing-sucking mouthparts of the mites results in a pale grayish or silvery color (fig. 263).

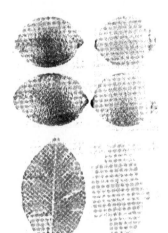

Fig. 263. Red mite injury on lemon fruit and foliage. Left, uninfested fruit and leaf; right, infested. Photo by Roy J. Pence.

371

This adversely affects the appearance of the mature fruit and may result in a lower grade of fruit. Defoliation is often caused by severe infestations of mites and this is most evident on the current or last flush of growth and in the tops of the trees. Red mites cause or accentuate a spotty necrosis of the soft interior tissue of orange leaves known as mesophyll collapse (see fig. 105). This condition is also brought about by dry north winds, saline soil, inadequate irrigation, or a combination of factors, including mite infestation (963). As stated before, hot dry winds accentuate defoliation when mites are present or may result in a drying of the young, tender leaves on the trees. Defoliated twigs, especially on oranges, may die before acquiring a new set of foliage.

The writer once examined a navel orange orchard near Pomona, California, a part of which was protected from north winds by a hill. This orchard had a moderate infestation of citrus red mites. It had been divided into numerous experimental plots treated in August, 1945, with spray oils, which give a good control of red mite, and non-oil sprays having no effect in controlling red mite. The orchard was examined in September after a few days of hot, dry north wind. It was found that in the parts of the orchard protected from the direct force of the dessicating wind, there was little defoliation and no burning of foliage, but in the unprotected part of the orchard, much defoliation had occurred in the upper north sides of non-oil-sprayed trees, and many leaves which had not fallen were dry and brown in color. No damage was found in the oil-sprayed plots in any part of the orchard, whether protected from the winds or not. This orchard afforded a good illustration of the fact that a combination of red mites and hot, dry winds may cause a severe type of injury to citrus trees in the case of a mite infestation which of itself might not result in serious injury, at least if the orchard were to be treated for mites within a reasonable period.

Besides debilitating the tree, the red mite causes economic loss by weakening the fruit stems so that the fruit falls more readily, after it becomes mature, than it would if it were not infested. Navel oranges which mature during the winter months may be left on the trees for months after they have reached maturity. During the spring, orchards which were not oil sprayed in the fall or winter may be infested with mites. It has often been observed, in cases in which only a part of the orchard has been treated for mites, that much less fruit is found on the ground in the treated part than in the part infested with mites. It has been shown that a mite infestation will so weaken an orange tree as to make it more susceptible to injury from oil spray the following year (see p. 202).

It is also quite generally believed, from years of experience and observation, that red mite infestations cause a reduction in the size of the fruit. The tendency has been in recent years to attribute increasingly greater importance to the citrus red mite with respect to its ability to cause damage to citrus trees and fruit. It is now generally believed that light infestations of red mite, such as were formerly considered to be of no importance, can cause considerable damage and should be treated.

Control:- Sulfur dusts were formerly used with a fair degree of success, but gradually became increasingly ineffective. They were largely displaced by the highly refined petroleum oils when these came into use about 1925, and the latter had the advantage of controlling a larger number of the more serious pests than had sulfur. Even to this day, oil spray is the most effective treatment for citrus red mite, and in orchards in which oil sprays are annually applied for scale insects, the red mite is usually not a problem. Unfortunately, hydrocyanic acid gas fumigation has no value in the control of the red mite, and fumigation for scale insects usually entails an additional treatment for red mite.

Under certain conditions it is not considered advisable to use oil spray on citrus trees and consequently much research work was done, especially by the Citrus Experiment Station (102, 110), in an attempt to develop a satisfactory method of controlling red mite that would not involve the use of oil. Until recently, the most useful of the materials which were developed was DN Dust-Dust D 8, used at 1 pound per tree. This should be applied by means of the multivane fan type of power duster (see p. 277), which was developed by University of California entomologists and engineers especially for use in citrus dusting. The dust is applied at night and early morning, to avoid drift caused by movement of air (fig. 264).

Fig. 264.   Rear view of a citrus duster in operation.
After Boyce et al (118).

Two treatments per year are usually necessary to control red mite by means of DN-Dust-D8, and it is advantageous to follow with the second application of dust from 10 to 20 days after the first treatment (110). The first application kills those mites present on the tree at the time of dusting as well as those which may hatch from the eggs within 4 days after the treatment. The second treatment kills the remainder of the mites that were in the egg stage at the time of the first treatment. A hydrocyanic acid gas fumigation results in a good kill of the active mites, and if this is followed in 10 to 14 days by a DN-Dust, a control almost as effective as that from two DN dusts may be obtained.

If for some reason it is more convenient or more practicable to spray than to dust, another formulation containing the same toxic ingredient as DN-Dust-D 8 may be used. This is known as DN-111, and is used at 3/4 to 1 pound to 100 gallons of spray. This is applied only to the exterior of the tree, with no attempt being made to drench the inside branches with spray such as is the case in spraying for scale insects. The spraying may be done with the purely mechanical "boom sprayers" or with ordinary spray rigs equipped with "broom guns" with 6 or 8 nozzles to increase the speed of application.

If the DN-Dust-D8 or DN-111 treatments are followed in a few days by high temperatures, a burn may be produced on foliage and fruit, especially on the south side of the tree and in localized areas in an orchard such as benches or pockets in which temperatures become especially high. In interior or intermediate areas such injury may occur at temperatures from 90° to 95°F or above, while in coastal areas similar injury may be produced at temperatures of from 85° to 90°F. The United States Weather Bureau at Pomona forecasts temperatures two days in advance, and injury can usually be avoided by heeding these forecasts and discontinuing dusting operations when temperatures are predicted that would result in injury in the particular area in question.

Promising new materials:- The DN preparations have not been generally as effective as the regular oil sprays in controlling citrus red mite. Usually improper dosage or method of application or rainfall soon after treatment account for the unsatisfactory results, but sometimes it is not apparent why the results are inferior. This uncertain effectiveness of the DN preparation as well as occasional instances of injury, some of them quite severe, has prompted much investigation on new materials which might prove to be satisfactory. Several of these show considerable promise (463).

373

It appears that the most promising of the new materials in the control of 1
citrus red mite is K-1875 (p-chlorophenoxy) methane. A 40 per cent wettable pow
of this compound is being sold under the trade name "Neotran". Jeppson (464) su
gests that Neotran be applied by means of spray-dusters at 7-1/2 to 10 pounds pe
acre in 100 gallons of water, or with the regular spray equipment, using 1-1/2 t
2 pounds per 100 gallons of water. The addition of 1 to 2 pints of spray oil, a
an adhesive agent, per 100 gallons of spray, has resulted in improved control or
when the treatments have been followed by rains. Dusts have not been as effect:
as the spray-duster treatments or the conventional sprays. K-1875 is effective
against the six-spotted mite, but is not effective against other species of mite
or against insects. It is one of the safest of all insecticides from the standpoint of its toxicity to man and warm-blooded animals.

Thompson (883) recommends to Florida growers that they should apply contro:
measures to citrus red mite (purple mite) while the population is still very lig
DN preparations or oil sprays are recommended. Emphasis is thus placed on preve
tative treatment, applied in September and October. The mites are usually most
abundant in Florida between November and June. DN preparations for control of
purple mite or six-spotted mite may be combined with sulfur sprays or dusts directed against the rust mite.

SIX-SPOTTED MITE

Tetranychus sexmaculatus Riley

The six-spotted or yellow mite was mentioned by Hubbard (454) as a pest of
oranges in Florida in 1885. In Florida it is now considered to be less common o
citrus than the citrus red mite (purple mite) (883), but sometimes it becomes so
numerous on citrus trees as to cause defoliation and drop of fruit. It was intr
duced into California in the late eighties on nursery stock obtained from Floric
In California it is occasionally a pest in the coastal areas.

Life History and Description:- The globular, colorless or transparent or ve
pale greenish-yellow eggs are loosely attached to the web which covers the color
or attached to the surface of the leaf underneath the web. The eggs bear a sta:
(fig. 265, below) like the eggs of the citrus red mite, but have no guy fibrils
radiating out from the stalk. They are 0.11 mm in diameter. From 25 to 40 egg;
are deposited during a period of 10 to 20 days. In June from 5 to 8 days are re
quired for hatching and in winter this period may be prolonged to 3 weeks. The
larval stage in warm weather requires 2 days, the protonymph 2 or 3 days, the
deutonymph another 2 or 3 days and another 2 or 3 days elapse before egg laying
commences. Thus the life history is much the same as for the citrus red mite.(

The body of the adult female is about 0.30 mm long, somewhat smaller than
of the preceding series. The body is oval, lemon-yellow in color with blackish
spots usually grouped in three blotches along each side of the body (fig. 266),
giving the mite its common name. These spots, however, are not always conspicu
and may sometimes be lacking, as in the specimens shown in fig. 265, above. Th
male is smaller than the female, and the body is narrowed posteriorly (fig. 265
Fig. 267 shows the typical sequence of action in the mating of the six-spotted

Injury:- The six-spotted mite causes a very characteristic injury because
feeding is confined to definite areas on the underside of the leaf, usually alc
the midrib and larger veins (fig. 268). A depression in the leaf is formed whe
colony of mites has established itself and the leaf becomes yellow in that area
covered with a web which apparently affords some protection for the mites. The
upper surface of the leaf at the infested area is raised and is yellow or yell
white in color and has a smooth, shiny surface. In Florida, where injury may
come quite severe, especially in the months of February and March, the infeste
areas converge, the leaf becomes entirely yellow and may become distorted and
shapen, and it prematurely dries up and falls.

Control:- With the occasional infestations that justify control measures,
same treatments may be used as are recommended for the control of the citrus r
mite, in fact, usually treatments for red scale or red mite in the districts i
which the six-spotted mite is found prevent the latter from becoming a pest.

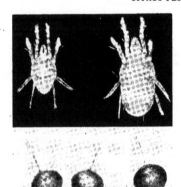

Fig. 265

Fig. 266

Fig. 268

Fig. 265. The immaculate form of the six-spotted mite, Paratetranychus sexmacula-tus. Above, male (left) and female (right). X 50. Below, stalked eggs. X 30. Photo by Roy J. Pence.

Fig. 266. Female of six-spotted mite, showing a characteristic distribution of spots After McGregor (553).

Fig. 267. Mating sequence of the six-spotted mite. The male is behind and below the female. Photo by Roy J. Pence.

Fig. 268. Undersides of lime leaves, showing injury caused by the six-spotted mite. Photo by Roy J. Pence.

SUBTROPICAL ENTOMOLOGY

LEWIS SPIDER MITE

Tetranychus lewisi McGregor

This mite was found on fruits of the Navel orange in Corona, California by
H. C. Lewis in 1942. In recent years it has rapidly spread and may be found on

citrus in Los Angeles, Ventura and Santa Barbara
Counties, but is especially severe in Ventura
County, where at present (1948) it infests at
least 500 acres of citrus. On the other hand, in
the Corona district in Riverside County, where the
mite was originally found, recent attempts to find
it have failed.

Description:- The female (fig. 269) averages
0.36 mm long and 0.17 mm wide. Visible from above
are the 24 strictly dorsal setae, not arising from

Fig. 269. Lewis spider mite,
Tetranychus lewisi. X 110.
After McGregor (563).

tubercles, and a pair of similar setae on the
lateral margin of the body and an inconspicuous
pair of setae at the tip of the abdomen. The oval
body is at first pale greenish-amber, but deepens

to amber and attains a varying number of black spots along the lateral margin.
Usually there is one spot over each third coxa and a pair near the posterior tip
of the abdomen. The male is smaller and narrower and more wedge-shaped. Some
taxonomic characters of both sexes are shown in fig. 270.

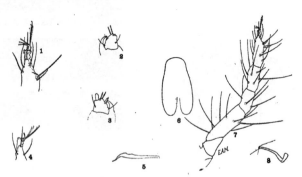

Fig. 270. Tetranychus lewisi. 1, Tips of tarsus of leg I of female; 2, side view
of terminal portion of palpus of male; 3, side view of terminal portion of palpus
of female; 4, tip of tarsus of leg I of male; 5, lateral view of penis; 6, mandibular plate; 7, leg I of female, viewed from outside; 8, collar trachea, viewed
laterally. After McGregor (563).

The almost spherical egg, with its very slender axial stalk, is at first almost colorless, but before hatching it becomes straw-colored. It is 0.12 mm in
diameter.

Injury:- The Lewis spider mite is found only on the fruit, and not until the
fruit has matured to the stage in which color begins to appear. It causes a
"silvering" of lemons and "russetting" of oranges somewhat similar to that of the
silver or rust mite. The greatest infestation appears to be on trees or limbs
that are sparsely foliaged and on outside fruit. Considerable webbing is produced
by this species.

CITRUS PESTS

TARSONEMUS MITE, Tarsonemus bakeri Ewing

During the summer of 1937, a crinkling or pitting of lemon leaves in an orchard near La Habra, California, was called to the attention of the mite specialist, E. A. McGregor. On this foliage McGregor found an extremely small mite, only 1/160 of an inch in length (0.162 mm), or about as long as the citrus bud mite, but about twice as wide as the latter species. The mite was found to be Tarsonemus bakeri (Family Tarsonemidae) (fig. 271).

The mites were first observed feeding within the buds which grow at the base of the leaf stem and were believed to be the cause of the tips of the bud becoming brown and dead (561). As high as 44% of the buds were found infested and frequently between 10% and 30% were injured. Later examinations made in a large number of orchards showed that from 10 to 100% of the small lemons were infested. The mites gain access to the tiny fruits soon after the petals fall and are concentrated in the very restricted space between the inner base of the calyx cup and the outer rim of the "button".

Hosts and Distribution:- Tarsonemus bakeri was observed on lemons, oranges and grapefruit. It was found in many locations in Ventura, Los Angeles, Orange, San Diego, San Bernardino, and Riverside Counties (561).

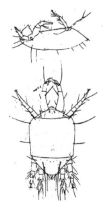

Fig. 271. Tarsonemus bakeri. Male, above; female, below. X 210. After McGregor (561).

Injury: According to McGregor (561) the stunting of bud development results in many branches being leafless on the infested trees. To find the mites, the buds must be dissected, for the mites feed between the bud bracts. Another symptom is the crinkling and dimpling of the leaves.

It was stated that when infestations are heavy one can find many malformed and lopsided lemons. The lemons have a tendency to fall. According to McGregor (561), the malformations of the foliage and fruit caused by the tarsonemid mites are readily distinguished from those caused by the citrus bud mite. He believes that the tarsonemid mites may be instrumental in spreading the Alternaria fungus which is the direct or contributing cause of black rot, center rot, end rot, break down, and other maladies.

It should be mentioned at this point that not all authorities are in agreement that the injury attributed to Tarsonemus bakeri is caused by this mite. H. S. Fawcett believed it was merely the crinkly-leaf strain of the psorosis or "scaly bark" virus.

Two other tarsonemid mites, T. approximatus and T. assimilis are reported by Quayle (724) as occurring on citrus.

OTHER MITES

In Texas, second in importance among the mites is a tetranychid which was formerly believed to be Paratetranychus citri, but which McGregor (558) described as Anychus clarki. This mite feeds principally on the upper surface of the leaves, moves more rapidly than the citrus red mite, and forms practically no webbing. It is far less important in Texas than the rust mite. It is controlled by the sulfur dustings applied for rust mite.

Brevipalpus (Tenuipalpus) californicus (Banks) (Family Pseudoleptidae) (fig. 272) was originally described from individuals taken on citrus in Redlands, California, in 1903, and was described by Quayle (713) as a small, red species, characterized by its flatness, and as the only species of the genus recorded from

377

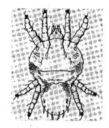

Fig. 272 (at left). Brevipalpus (Tenuipalpus) californicus. After Quayle
(711).

Fig. 273 (at right). Injury caused on lemons by Brevipalpus lewisi. After
Lewis (521).

the United States at that time. The mites are described as being usually motionless and resembling minute scales. Up to the time of Quayle's account of Brevipalpus (Tenuipalpus) californicus in 1912, no damage had been reported. Likewise
these mites had later been found from time to time infesting citrus in various
areas in southern California, also without causing injury, until in 1936 they occurred abundantly on lemons at Corona and were controlled with one sulfur dusting.
In 1938 the mites were found to be causing injury to Valencia oranges at Sanger in
the San Joaquin Valley where an oil spray appears to have eradicated them.

Brevipalpus (Tenuipalpus) californicus is listed as a pest of citrus in New
South Wales where it is called the "silver mite". In Argentina, Tenuipalpus
pseudocuneatus Blanchard is fifth in importance as a citrus pest in the Misiones-
Corrientes region and is considered to be more important than the rust mite (see
p. 543). This species also causes the characteristic reticulated brownish blemish
on grapes in New South Wales (14).

A scabbiness of oranges and mandarins in Spain which appears to be somewhat
similar to Brevipalpus injury in California is reported to be caused by a Tenuipalpus mite thought to be a new species (698). This mite also causes partial defoliation. It is controlled by 2% lime sulfur (28-30° Bé) applied when the mites first
appear on the green fruit.

In the fall of 1942 a mite resembling Brevipalpus inornatus (Banks) (Tenuipalpus bioculatus McG.) was observed by Harold Lewis at Porterville, California,
causing injury to tender wood and fruit stems of lemon and causing a darkened,
scarred area on the surface. The fruit buttons were also darkened. Some leaf injury, of no importance, was also discerned. Damage to fruit, however, resulted in
over 25% being scarred in a characteristic manner as shown in fig. 273. The
orchard in which the damage occurred was not treated for thrips. It is likely that
the usual treatments for scale or thrips are sufficient to keep this potentially
injurious mite in check (521). The same mite subsequently became a serious pest of
pomegranates in the San Joaquin Valley, causing a brownish discoloration and checking of the fruit. It has been described as a new species, Brevipalpus lewisi, by
E. A. McGregor (565) (see p. 677).

A scarlet red mite, 0.238 mm long, somewhat similar to the preceding species
except for its more intense color, was first collected on lemon fruits in San Diego
County by R. S. Woglum in 1943. It may also be found on lemon foliage, but is more
prevalent on the fruits, especially near the calyx. It causes some stippling of
the rind. This mite was also described as a new species, Brevipalpus woglumi, by
E. A. McGregor (565).

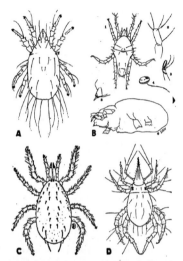

Fig. 274. Some mites that may be found on citrus trees. A, <u>Tyroglyphus americanus</u>;
B, <u>Tydeus ubiquitous</u>; C, <u>Seilus pomi</u>; D, <u>Bdella utilis</u>. C and D are predaceous.
A, after Quayle (711); B, after McGregor (555); C, after Newcomer and Yothers (648);
D, fr. E. O. Essig, INSECTS OF WESTERN NORTH AMERICA. Copyright 1926 by The Mac-
millan Company & used with their permission.

The mites may be found immediately after an HCN fumigation for they have no
tracheae and are not very susceptible to fumigation. In the genus <u>Tyroglyphus</u>
there are four long bristles in a transverse row on the dorsum of the cephalothorax.

<u>Pediculoides ventricosus</u> is a pale yellow mite less than 1 mm in length which
may be found under dead scales, and sometimes this mite becomes a pest in insectaries, where it interferes with the rearing of insects.

<u>Tydeus gloveri</u> Ash. is a pale reddish or yellowish eupopid mite with a sub-
pyriform body and rather short legs. It is stated by Banks (58) to feed on the
young and eggs of purple scale. A California species, <u>T. ubiquitus</u> McG. (fig.
274, B), is a pale species about 0.225 mm long, with a median lavender stripe on
the abdomen, which feeds on the dead bodies and lifeless eggs of insects on citrus
trees. McGregor (555) states that this mite is the most generally abundant of all
animals to be found in citrus foliage in California. It is usually running about
actively. It lays its eggs at the end of a stalk, usually in close proximity to
dead scale insects.

Quayle (724) lists a number of the mites that may be found on citrus, but
cause little or no injury. Among these are <u>Tydeus</u> sp., a white to pale pink mite
which may be recognized by its cast skins being grouped in definite areas, usually
on the underside of the leaf. The mite appears to be a scavenger.

379

Caligonus terminalus Banks is a red mite which has been found on citrus in San Diego County. It may be found in colonies of the six-spotted mite, where it appears to take advantage of the protection of the profuse webbing of the latter species. This mite is not as dark red as the citrus red mite and its body is shiny instead of velvety as in the case of the latter. Its eggs are paler than those of the red mite and they do not have a stalk.

A spinning mite, Tetranychus yumensis McG., can occasionally be found on citrus trees in the Yuma Valley, Arizona, and in the Coachella Valley (556). The female is about the size of the six-spotted mite, is oval in shape and is rusty red, usually with a few small dark spots. The dorsal setae are pale, but do not arise from tubercles as in the citrus red mite. The egg is spherical and has a stalk with a few guy fibrils extending to the leaf from its tip. The mite is found on the lower surface of leaves and spins considerable webbing. It causes a blotchy appearance to the foliage and severe infestations may cause some defoliation. The mite prefers grapefruit and lemons to oranges. It can be controlled with sulfur dust.

NATURAL ENEMIES (713)

The mites have many species of natural enemies, mainly predators, and they may be found abundantly on mite-infested citrus trees, but unfortunately the natural enemies do not obviate the necessity for annual treatment in the majority of orchards. Among the natural enemies of the phytophagous mites may be found other species of mites that prey upon them. The most important of the predaceous mites is Seilus pomi Kalt., a pale, oval species 0.4 mm long, sparsely covered with short curved hairs (fig. 274, C). This mite is found throughout the Pacific Coast States and in the eastern part of the country. Other predators among the mites are the whirligig mite, Anystis agilis Banks (fig. 275), a very active, bright red mite nearly 1 mm long; Bdella utilis Banks (fig. 274, D), one of the "snout mites", also bright red in color and several species of Erythraeus, which are large mites of a reddish to purplish color.

The dusty-wing, Conwentzia hageni Banks (fig. 276), is a neuropteron about 0.20 inch in length and greyish white in color. The name "dusty-wing" comes from the gray powder that covers the wings. The larvae have the alligator-like form common among Neuroptera as well as among the larvae of lady beetles. The larvae are usually red, white and black in color. It is the larvae which prey on mites, piercing their bodies with needle-like mandibles and maxillae enclosed in a cone-shaped beak, then sucking out the body juices of their prey.

A staphylinid beetle, Somatium oviformis (Casey) (fig. 277), has been found feeding on citrus red mite in both the adult and larval stages. The adult is a small black beetle densely clothed with hairs, with short elytra, and with the abdomen strongly curved upwards. The larva is about 2.5 mm long and yellow in color.

Stethorus picipes Casey (fig. 278) is a polished black coccinellid beetle clothed with fine hairs and about 0.04 to 0.05 inch in length. The larva is of about the same length as the adult, dark gray to black in color, and the body is clothed with numerous hairs arising from small tubercles. In addition the entire body is covered with short spines. One larva of Stethorus picipes Casey consumed 189 mites in 30 days, and another larva consumed 110 mites in 13 days.

A predaceous thrips, Scolothrips sexmaculatus Perg. (fig. 469), is one of the few carnivorous thrips. These thrips have repeatedly been observed on red mite, usually the eggs and younger mites, but sometimes the adults. Both the nymphs and adults prey on the mites. The adult thrips is pale yellow and has 3 light brown spots on the fore wings. The specimen shown in fig. 469 was taken on almond, where it was feeding on the two-spotted mite.

Arthrocnodax occidentalis Felt (fig. 279) is a dipterous insect, the larva of which feeds on various species of mites, especially the six-spotted mite, whose protective web appears to offer the best conditions for the larvae. The larva has the typical maggot form, is 2.2 mm long, and yellow in color. One larva fed on 165 mites in 15 days and another fed on 380 in 17 days.

A carnivorous bug, Triphleps tristicolor White (fig. 280), feeds on mites both in the adult and nymph form. This adult bug is black, 1-1/2 mm long, and the

Fig. 275*

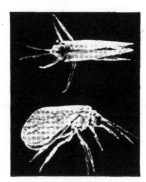

Fig. 276

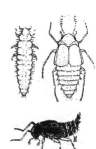

Fig. 277

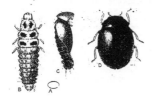

Fig. 278

Fig. 279

Fig. 280

Fig. 275. Whirligig mite <u>Anystis</u> <u>agilis</u>.*
Fig. 276. A dusty wing, <u>Conwentzia</u> <u>hageni</u>. X 10. Photo by Roy J. Pence.
Fig. 277. <u>Somatium</u> <u>oviformis</u>. Larva, adult, and characteristic pose of adult.
                       X 12. After Quayle (711).

nymph is of an amber color. These bugs consume large numbers of mites, but unfortunately do not occur in large numbers.

The brown lacewing, Hemerobius pacificus Banks (fig. 281), is as the common name indicates, brown in color, 10 mm long to the tip of the wings, and is covered with short hairs. It lays its oval, white eggs on the lower leaf surfaces. The larva has the typical alligator-like appearance of the larvae of lacewing flies. It has long blade-like mouthparts, which distinguish it from the coccinellid larvae with which it could possibly be confused. The larva may consume from 500 to nearly 900 mites in the course of its development, averaging from 30 to 45 mites a day. The mites are impaled on the long narrow mandibles and their body fluids are withdrawn.

Fig. 281. Brown lacewing, Hemerobius pacificus. After Quayle (711).

The green lacewing, Chrysopa californica Coq. (fig. 223), is the most common of the lacewings on citrus trees. However, it is primarily a predator of aphids and will be discussed in a later section.

GRASSHOPPERS AND KATYDIDS

Grasshoppers occasionally cause damage to citrus by feeding on fruit and foliage. As usual with foliage feeding insects, damage is apt to be most severe to young trees. Damage to citrus from grasshoppers occurs throughout the world wherever citrus trees are grown. In California, grasshoppers damage citrus mainly in orchards bordering foothills, uncultivated grass lands, or other places affording favorable conditions for a large grasshopper population. In summer the rapid drying of range grasses and the harvesting of alfalfa forces the grasshoppers to search for new sources of succulent food, and migration to citrus orchards is the result. A specially severe attack of grasshoppers was once reported in northern California in which many young orange trees were killed, large, bearing trees were defoliated, and the fruit was badly pitted (902). In central California the grasshoppers may also in some years become injurious to citrus trees and fruit (966, 967). Injury to citrus trees was rather extensive in Florida in 1947. One of the damaged orchards was about 75% defoliated (385).

Since grasshoppers are general feeders, the species attacking citrus are the ones which happen to be common in the region. It has been stated that the most common species in California are probably Camnula pellucida Scudd., Oedaleonotus enigma Scudd., and the different varieties of Melanoplus mexicanus (724). The species principally involved as a pest of citrus in Florida is Schistocerca americana (Drury) (385).

Grasshoppers deposit their eggs in the soil. They drill a hole as deep as possible with the tip of the abdomen, then deposit 20 to 100 eggs in packets, cementing them together with a secretion. A single female may deposit as many as 20 of these egg packets. In cooler temperate regions, the eggs are laid only in late summer and fall and represent the hibernating stage of the insect, but in the subtropical and tropical regions active forms may occur throughout the year (287). As explained in a previous chapter grasshoppers have a simple or incomplete type of metamorphosis, and consequently the nymphs resemble the adults even immediately after hatching. When about half grown the nymphs begin to develop their wing pads. In the case of Schistocerca americana in Florida, it is stated that the eggs hatch in probably 2 to 3 weeks in summer and that it probably requires from 6 to 8 weeks for the nymphs to mature. They become sexually mature within a few weeks after the last molt (385).

Control:- In California, grasshoppers attacking citrus orchards have been controlled first by broadcasting poison bran mash in pasture lands adjoining citrus orchards and later, after the grasshoppers have started moving into the orchards, furnish labor and equipment for spreading the poison bran mash (966). It is possible that new insecticides may replace the poison bran mash in future outbreaks.

Griffiths and King (383) made a comparative study of chlordan, chlorinated camphene, benzene hexachloride, and parathion with respect to their residual effectiveness and compatibility with other insecticides used in the control of citrus pests, using the American grasshopper, Schistocerca americana (Drury) as the test insect. They state, "Under Florida citrus grove conditions benzene hexachloride (0.45 lb. gamma isomer per acre) and parathion (0.45 lb. active ingredient per acre) were found to exhibit little if any residual toxicity after three days. Chlordan (1.5 lbs. per acre) and chlorinated camphene (4.5 lbs. per acre) has some toxicity for more than one week. Dusts or sprays appeared to be about equally satisfactory. All four materials were found to be compatible with wettable sulfur (10 lbs. per 100 gallons), 0.67 lb. 40 per cent dinitro-o-cyclohexyphenol per 100 gallons, neutral copper (3 lbs. 34 per cent metallic copper per 100 gallons), and zinc sulfate (3 lbs. zinc sulfate plus 1 lb. lime per 100 gallons). All such combinations gave comparable mortalities."

Katydids

The katydids, which belong in the family Tettigoniidae, somewhat resemble in form the grasshopper (Locustidae), but are more delicate in structure, are often nocturnal, and can be distinguished from grasshoppers by their very long antennae, which are often longer than the body, and by having four segments in each tarsus. The majority are green in color, and this, along with the characteristic wing venation, makes it difficult to distinguish them from the foliage in trees. Some katydids have broad fore wings, and these are usually found among trees and shrubs. Others have narrow fore wings and are more apt to be found in bushes or tall weeds and grasses. Katydids are well known for the ability of the male to make characteristic stridulatory sounds. These sounds are produced chiefly toward evening and throughout the night. The base of the fore wing is modified so that rubbing the wings together will produce a sound, and this sound varies with the species.

Injury from katydids is similar to that of the grasshoppers, but is not likely to be as severe. Katydids are widely distributed but are not likely to occur in large numbers in any one locality.

Fig. 282. Fork tailed katydid. X 0.8.   Fig. 283. Angular-winged katydid. X 0.7.
Photo by Roy J. Pence.   Photo by Roy J. Pence.

In California the two species of most importance have been found to be the fork-tailed katydid, Scudderia furcata Brunner (fig. 282) and the angular-winged katydid, Microcentrum rhombifolium Sauss. (fig. 283) (443). The first of these

383

species is·so named because of a peculiar forked appendage at the tip of the abdomen of the male. The latter species can be distinguished by its larger size, broader wings, and, in the case of the nymph, by its hunchbacked appearance. It is lesser in importance of the two species because it does not attack the fruit and is less abundant, presumably because of a higher rate of parasitization.

In Florida the katydids Crytophyllus concavus and Microcentrum retinerve are listed as being occasionally injurious to citrus (926).

Injury:- The fork-tailed katydid may be found actively feeding on orange trees about the time of petal fall and at that time the insects may damage blossom buds by gnawing a hole through the petals in order to feed on the pistil and ovary. Later small oranges are attacked, and still later the larger fruits (fig. 284). The latter are attacked on the side facing the inside of the tree, for the katydids will feed in the shade if possible. The injury to these fruits may not be apparent until the fruit is picked. At that time a characteristic circular grayish, depressed and calloused area is seen on the fruit.

Both of the aforementioned species of katydids feed on foliage (fig. 285) usually selecting the new, tender growth. Unless the nymphs are unusually numerous the damage to foliage is not important.

Life History:- Nearly everyone who has done any exploring in the out-of-doors has seen katydid eggs and has remembered them because of the peculiarity of their arrangement. Many species, including Microcentrum rhombifolium, lay their smooth, flat, ovoid eggs in a characteristic fashion in long rows, often along a slender twig or the edge of a leaf, so that the eggs overlap in shingle fashion, each egg overlapping the one below it (fig. 286). The eggs are laid in two rows, each egg in one row alternating with an egg in the other row. They are often highly parasitized. The eggs of Scudderia furcata are hidden in the plant tissues. The majority are inserted into the edges of the old tough leaves, between the upper and lower surfaces (fig. 287). Where the eggs are inserted, the edge is usually chewed away, and with the aid of a hand lens one may often see the protruding edge of the egg. Likewise the outline of the egg may be seen if the leaf is held against the sun. Rarely more than ten eggs are laid in a single leaf, the average being about three eggs. With each of the above species there are six nymphal generations.

Control:- Although the newer insecticides which are being applied to citrus trees in the control of grasshoppers would probably be effective applied in the same manner against katydids, this apparently has not yet been done. Cryolite or barium fluosilicate may be applied to the trees as a dust or as a spray in the rare cases when katydids do sufficient damage to justify control measures (724).

TERMITES

Termites are of world-wide importance because of their damage to wooden structures, but they are seldom pests of living trees. Occasionally, however, they have been known to feed on the bark of the citrus tree both above and below the ground level (fig. 288). Young trees may be completely severed near the soil line. The grapefruit trees shown in fig. 288 were set out in the orchard in May, 1929, and the trunk was completely severed below the soil line, as shown in the photograph, by July of the same year. This injury occurred in the Coachella Valley and was caused by the desert dampwood termite, Kalotermes simplicicornis Banks.

The writer has seen termite damage to the roots of grapevines in the desert valleys of southern California.

The following species have been at one time or another responsible for damage to citrus in California: the desert subterranean termite, Heterotermes aureus Snyder, the western subterranean termite, Reticulitermes hesperus Banks, the desert termite, Amitermes perplexus Banks (A. acutus Light), the desert dampwood termite, Kalotermes simplicicornis Banks, the common dampwood termite, Termopsis angusticollis Hagen, and the common drywood termite Kalotermes minor Hagen (724). All the above species except the last are subterranean, that is, they must have their nests below ground, or, if the nests are above ground, they are formed of thick, hard,

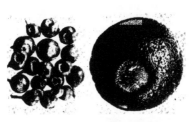

Fig. 284

Fig. 285

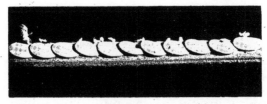

Fig. 286

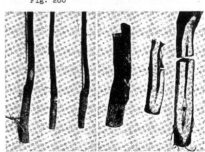

Fig. 287

Fig. 288

Fig. 284. Left, young oranges with holes made by katydids; right, a healed scar of an older orange. After Horton and Pemberton (443).

Fig. 285. Orange leaves injured by katydids. After Horton and Pemberton (443).

Fig. 286. Eggs of angular-winged katydid. Note nymph emerging from last egg on left. Photo by Roy J. Pence.

Fig. 287. Eggs of fork-tailed katydid inserted into the edges of an orange leaf. After Horton and Pemberton (433).

Fig. 288. Termite injury to grapefruit trees. Left, recently planted trees completely severed below the ground by the desert dampwood termite; right, internal injury to grapefruit trees by the same species. Photo by R. H. Smith.

earthen walls and have a direct connection with the ground. Kalotermes minor may feed in the tree above ground and without any ground connections. They usually gain entry through wounds or crevices, or where a large limb has been pruned off, operating in the dead heartwood of old trees and sometimes extending their operation into live tissues.

In Florida, as in California, the subterranean species are the more important, especially Termes flavipes Kollar (925). They are chiefly pests of young trees that are banked with soil during the winter as a protection from cold, especially if pieces of old wood are carelessly left in the soil used for this purpose. In banking trees, only earth free of pieces of wood should be used and the banks should be removed as soon as danger from injuriously cold weather has passed. If a termite nest is located, carbon disulfide or cyanide solution can be poured into the nest as in ant control. In Florida, a dry wood termite, Neotermes castaneus Burm., has caused injury to citrus trees in the same manner as Kalotermes minor in California (926).

THRIPS (Order Thysanoptera)

CITRUS THRIPS

Scirtothrips citri (Moulton)

Distribution:- The citrus thrips is probably third in importance among citrus pests in California as a whole, but is the most serious pest in the San Joaquin Valley, in the desert regions of California, and in Arizona. It is native to southwestern United States and does not occur in the Gulf States. In California it may be found wherever citrus is grown, but is a serious pest only in the hot, arid, interior valleys. In this state, the citrus thrips has been a serious pest of oranges, especially the Navel Orange, for the last 50 years, but did not become a serious pest of the lemon until about 10 years ago. It is also an important pest of the tangerine and grapefruit in the Coachella Valley. The citrus thrips confines its attacks almost entirely to citrus. It occurs on the pepper tree (Schinus molle), which McGregor (564) states is the only other host on which it is known to overwinter in the egg stage. It has also been found sparsely on umbrella trees (Magnolia tripetala), roses, grapes, deciduous trees, and alfalfa.

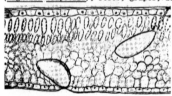

Fig. 289. Semidiagrammatic representation of the position of the eggs of citrus thrips in the citrus leaf tissue. After McGregor (564).

Life History:- The reniform (kidney-shaped) eggs of the citrus thrips are about 0.20 mm in length, and are inserted beneath the cuticle of new leaves (fig. 289), leaf stems, green twigs, fruit stems, and fruit. Many eggs, as in the case of so many small insects and mites, are laid under the sepals of the fruit. As many as 250 eggs may be laid by a single female. It is in the egg stage that the citrus thrips passes the winter. The eggs hatch at about the time of the appearance of the new spring growth on citrus trees in March. Eggs which are laid in the fall may not hatch until the following March, but during the warmest periods of the year they may hatch in 6 to 8 days.

There are two nymphal instars. The first instar nymph is, of course, very small, with a narrow tapering abdomen and prominent bright red eyes. It is colorless when hatched, finally becoming yellowish. This instar averages about 0.41 mm in length. The second-instar nymph is more robust and more deeply colored than the previous instar, becoming rather orange-colored. It measures about 0.90 mm in length. The nymphal instars combined require from 4 to 14 days, depending on temperature (441). The majority of the nymphs drop to the ground and seek concealment for the passing of the prepupal and pupal stages and transformation to the adult.

The third instar is called the prepupal stage. It is somewhat shorter than the second instar nymph, being 0.56 mm in length, but it differs in appearance mainly in the possession of wing pads. These extend to the hind margin of the second abdominal segment or slightly further, depending on the age. The antennae are extended forward.

The fourth instar is called the pupal stage. It is 0.67 mm long. The wing pads are longer than in the preceding stage, extending to the sixth and eventually as far as the ninth or tenth abdominal segment. The antennae lie backward over the head, just the opposite of their position in the previous instar.

The combined prepupal and pupal stages require from 4 to 20 days, depending on temperature (441). Pupation takes place in dead leaves and rubbish under the trees, under clods, or in crevices in the soil. No food is taken during either the prepupal or the pupal stage. This is true even of some species of thrips, such as the greenhouse thrips, which remain on the fruit and foliage during the prepupal and pupal stages.

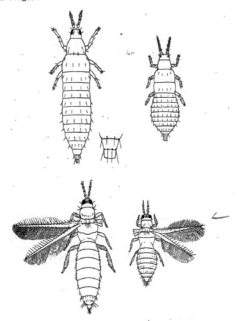

Fig. 290. Comparison of western flower thrips, _Frankliniella moultoni_, with citrus thrips, _Scirtothrips citri_. Above, second instar nymph: left, western flower thrips with enlarged drawing of tip of abdomen, showing row of spines; right, citrus thrips. Below, adult: left, western flower thrips; right, citrus thrips. After Ebeling (249).

The adult female (fig. 290, below) is 0.60 to 0.88 mm in length, the male being somewhat smaller. Both sexes are orange yellow in color and winged. Citrus thrips can reproduce without mating, but the progeny under such circumstances are

The grower, pest control manager, agricultural inspector, or entomolo
want to know how to find the thrips when they first begin to appear in the
in order to gain an idea of the abundance of the insects in a particular c
or locality, or to gain information to be used in timing the first treatme
Many people have become alarmed at finding large numbers of thrips in the
at about the period of the year when the first citrus thrips may be expect
appear. The citrus thrips, however, does not feed on the blossoms; the sp
is apt to find there is the flower thrips, Frankliniella moultoni Hood, or
ally some less common species of Frankliniella. Usually large numbers of
thrips may be obtained by striking a group of orange blossoms against the
the hand. These thrips, however, confine their feeding to the blossoms, p
on the nectar, and do no damage to either fruit or foliage. The flower th
not even be induced to injure fruit or foliage when confined to the blosso
enormous numbers (249). Some flower thrips may wander off onto the small
and new leaves, especially after the petals start to drop, and it behooves
entomologist to be able to identify such individuals in order to establish
or not they are citrus thrips. After all the petals have dropped, flower
are no longer found on citrus trees, and the thrips that are found after t
period are almost certain to be citrus thrips. Before all the petals have
however, one might wish to identify the species he finds on the young frui
foliage. For field identification, it should be borne in mind that the ad
rus thrips run about actively, while the flower thrips are comparatively s
in behavior and are often motionless. The adult citrus thrips are light a
color while the adult flower thrips are straw-colored or dark brown. Ther
readily apparent difference in the color of the nymphs. The differences i
and shape of the nymphs and adults, as well as differences in the arrangem
spines, are shown in fig. 290.

Injury:- The citrus thrips punctures the epidermal cells and withdraw
cell contents. On the fruit, the injury is characterized by a ring of rou
ened scabby, greyish tissue around the stem end and irregular areas of thi
teristic "thrips scarred" tissue over the remainder of the fruit. While t
is small, the thrips feed under and near the sepals. As the fruit enlarge
injured center becomes farther and farther removed from the "button" and u
forms a conspicuous band of "scabby" tissue in a complete ring, (fig, 291)
badly infested orchards as much as 90% of the fruit has been known to be c
because of "thrips scarring," and a 40 or 50% cullage is not uncommon in i
orchards.

The feeding of the thrips on the new terminal growth of foliage resul
the leaves' becoming thick and leathery, with the characteristic grayish
tissue. The leaves are usually dwarfed and distorted (fig. 292, left).
thrips occur principally on the tender young growth, they may be found on
orange most abundantly on the spring and fall flushes of growth, whereas
lemon they may be found continuously throughout the spring, summer and fa
the new growth develops over most of the year.

Besides injuring new twigs, leaves and fruit, the thrips also attack
The buds may be destroyed and adventitious buds later developing in the s
may also be destroyed, or if not, a "multiple bud" condition may develop
rosette type of growth. If the buds fail to develop at all, the twigs ma
without foliage, and this condition is commonly called "rat-tailing" (fig
right).

On oranges, particularly on Valencias, the citrus thrips may produce
kind of injury known as "silvering." This is particularly prevalent in T
County. It is believed that rind oil, released by the thrips' punctures,

388

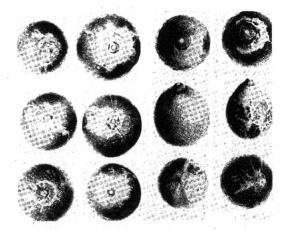

Fig. 291. Injury caused by citrus thrips, _Scirtothrips citri_, on fruits: left, oranges; right, lemons. After Boyce (106).

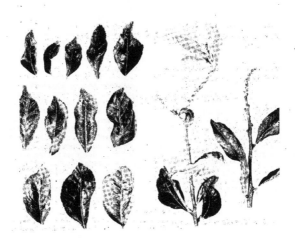

Fig. 292. Injury from citrus thrips. Left, scarred and distorted leaves; right, "rat-tailed" twigs. After Boyce (106):

the fruit. The scarring becomes more pronounced as the fruit matures and is particularly noticeable in late picked Valencias. It sometimes causes a considerable reduction in the grade of the fruit. (975)

The stems of blossoms may also be attacked, causing the blossoms to drop. Throughout the summer and fall the absence of new growth on citrus trees is a reminder of the damage suffered as a result of the attacks of this insect.

Control:- As the result of the first investigations on the control of citrus thrips (468), this insect was treated by means of 3 or 4 sprayings per year with lime sulfur, or lime sulfur and Black Leaf 40, until the value of sulfur dusting was demonstrated in 1927 (554). During the next decade, investigations on citrus thrips control were directed primarily toward the working out of sulfur-dusting programs for the various areas. One of the advantages of the sulfur-dusting program was its lower cost as compared to the previous spraying program.

During 1937 and 1938, many compounds were tested at the University Citrus Experiment Station with regard to their effectiveness against citrus thrips, and among these the most promising was tartar emetic used as a spray (116). The addition of equal parts of sugar to the tartar emetic was found to greatly improve its effectiveness. The spray is applied in the spring on orange trees after the majority of the petals have fallen and before the new fruits have reached pea size. When used with a spray-duster, 1-1/2 lbs. of tartar emetic and 1-1/2 lbs. of sugar in 25 to 50 gallons of water has been the usual amount applied per acre. If a boom sprayer or broom guns are to be used for the application of the spray, 2 lbs. of tartar emetic and 2 lbs. of sugar should be added to from 100 to 200 gallons of water and applied to an acre of trees. Lemons normally suffer little thrips damage to foliage and fruit before the last of May. The first treatment is made at that time or when the thrips first become apparent. For the control of citrus thrips on lemons, using a spray-duster, 3 lbs. of tartar emetic and 3 lbs. of sugar are added to 20 to 30 gallons of water, and if a boom sprayer or broom guns are to be used, the amount of tartar emetic and sugar should likewise be correspondingly increased over that which is used on oranges.

When tartar emetic was first used, it immediately became apparent that previous methods of control had been quite inadequate, for after the trees were sprayed with tartar emetic, a vigorous flush of new growth appeared, far in excess of that which would be found on sulfur-dusted trees. The excellent control afforded by the tartar emetic, to which the trees responded so quickly, showed that even a relatively light population of thrips, such as had been associated with previous methods of treatment, was sufficient to interfere considerably with the growth of the tree and with fruit production. Sulfur dust, however, remained in favor among some growers in central California or the most eastern section of northern California who wished to obtain at least a partial control of citricola scale and black scale along with the thrips treatment. The sulfur dust is effective against the crawlers of these two species of scale when temperatures are high. Likewise lime-sulfur spray is sometimes used in winter in central California as a combined thrips-citricola scale control, usually supplemented with sulfur dust the following spring.

sistant. The orchards in which spraying with tartar emetic had been practiced
he greatest number of times in past years did not necessarily have resistant
rips. A number of tests were made which showed that the development of citrus
rips resistance is not dependent entirely on the previous extent of tartar emetic
eatment. (691)

Tartar emetic is still recommended where experience has shown it to be effecve and it still gives excellent results against nonresistant thrips. As soon as
e first resistance was noted, however, investigations were begun in search of a
bstitute for tartar emetic in the increasingly greater areas in which the latter
s becoming ineffective. Such a substitute was soon found. A spray was recom-
ended consisting of a quart of nicotine sulfate and 2 lbs of sugar per 100 gal-
ns (691). If this spray is applied with a boom sprayer or with broom guns,
out 2-1/2 to 3 gallons of spray should be applied per full-sized tree. Using
e spray-duster, 1 gallon should be applied to the same sized tree. As used to-
y this spray generally contains only 4 lbs. of sugar to 100 gallons. Likewise,
powdered nicotine preparation (nicotine-bentonite) called Black Leaf 155 may be
sed instead of nicotine sulfate. Seven lbs. of Black Leaf 155 and 4 lbs. of
gar are added to 100 gallons of water and the same gallonage per tree is applied
s when the nicotine sulfate is used.

The "nicotine bait" sprays have been fairly satisfactory, but not as effective
s the tartar emetic sprays in nonresistant areas.

Increasing resistance to tartar emetic and unsatisfactory control have re-
ently been noted in several areas in Arizona (753). The recommendation in
rizona, where tartar emetic is still generally effective, is for 3 lbs. of this
aterial and 3 lbs. of sugar in 25 to 50 gallons per acre, using a spray-duster,
r in 200 to 300 gallons per acre with a conventional sprayer, when blossoms have
argely dropped. Where tartar emetic is not effective 8 pounds of 50% wettable
DT per acre is recommended as a spray applied in 100 gallons per acre with a spray
uster or in 200 to 300 gallons per acre with a convention sprayer. Dusting was
ound to be less effective than spraying. However, sprayed orchards should be
atched for the possible appearance of cottony cushion scale, as it has been found
n California after spraying with DDT.

On the basis of promising results obtained with various formulations of DDT
n 1944-45, certain suggestions were made for its use for growers who wished to
ry this new insecticide (109). It was used in the commercial treatment of a con-
iderable acreage in 1946 with satisfactory results, as far as control of thrips
as concerned. In many of the orchards treated with DDT, a heavy infestation of
ottony cushion scale developed, and this was believed to result from the elimina-
ion of the vedalia lady beetles (294, 295). The vedalia returned in these or-
hards in sufficient numbers, however, to again reduce the cottony cushion scale
o their usual low numbers. Nevertheless in view of the risk involved in any pro-
onged return of the cottony cushion scale in injurious numbers, it is recommended
hat DDT should not be used if there is more than a trace of cottony cushion scale
n an orchard, and in no event should more than one application of DDT be made
1thin the period of a year.

Assuming that the above conditions are heeded, the following treatments have
een suggested for the control of citrus thrips on oranges in central California
hen it is not attempted to control citricola scale simultaneously (106):

(1) "For control of citrus thrips in early May to prevent fruit scarring:
    a) Two applications of 2 per cent DDT-sulfur dust containing at least 85
       per cent sulfur at the rate of 100 pounds per acre. The first applica-
       tion should be made when most of the petals have fallen, and second
       application two or three weeks later;

or: b) Four pounds of 50 per cent DDT wettable powder suspended in 100 gallons water per acre applied with a spray duster.

(2) "For control of thrips in late summer, that is, after August 15, to prevent injury to young foliage:

a) Eight pounds of 50 per cent DDT wettable powder suspended in 100 gallons of water per acre applied with a spray duster."

When DDT is to be used on lemons in California in early summer, 4 lbs. of 50% DDT wettable powder to 100 gallons of water per acre, applied with a spray-duster, is recommended. In late summer, the amount of DDT should be doubled, that is, 8 lbs. should be used per 100 gallons. (293)

Combined Citrus Thrips - Citricola Scale Treatment:- If lime sulfur is to be used as a combined control of citrus thrips and citricola scale, it is most effective when applied at 2% concentration immediately after the petals fall, with the addition of 4 lbs. wettable sulfur to 100 gallons of water. Four to 6% lime sulfur, applied in February or March, has been used for the control of citricola scale, but has not proved to be dependable as a control for citrus thrips (106).

Two treatments of 2% DDT-sulfur dust applied in May at 150 lbs. per acre per application proved to be satisfactory in two years of experimental work (1944-45), but in 1946 the control of citricola scale was not satisfactory, for reasons that are not yet understood. A spray after the middle of August with 1 lb. of 50% DDT wettable powder and 1 gallon of kerosene per 100 gallons of water, thoroughly applied with a conventional spray rig, has been an effective treatment for both citrus thrips and citricola scale. (295)

GREENHOUSE THRIPS

Heliothrips haemorrhoidalis (Bouché)

Distribution:- The greenhouse thrips, Heliothrips haemorrhoidalis (Bouché), is a widely distributed tropical and subtropical species which also occurs in temperate climates in greenhouses. In California it attacks avocados and, in limited areas in San Diego and Ventura Counties, it also attacks citrus. It is found on citrus in Florida and Honduras, but without being a pest (877). In New Zealand, however, the greenhouse thrips causes injury to citrus and is sprayed with oil and nicotine sulfate (195). Likewise in the Sukhum region of the U.S.S.R. greenhouse thrips is a pest of avocado, orange and persimmon. It is sprayed with nicotine sulfate or anabasine sulfate in 1% soap solution or with oil emulsion (911). It is also a minor pest in New South Wales and West Australia.

Description:- The greenhouse thrips is a rather large species, about 1.25 mm in length. The adult has a black, reticulated body surface. The nymphal and pupal stages are yellowish white in color. The life history and habits of this species will be discussed in the chapter on avocado pests.

Injury:- The greenhouse thrips was first noticed to be a pest of citrus in the summer of 1936 at Rancho Santa Fe in San Diego County. The injury caused by the insect, its life history, and methods of control were described by Boyce and Mabry (114). The greenhouse thrips, by sucking out the contents of the epidermal cells of the leaves and fruits, including the chlorophyll or pigment, causes these to turn pale in color. The injury is most apt to be found where two fruits are in contact, for that is where the thrips tend to congregate. The "flecked" appearance of the affected areas is also characteristic. This is caused by the drops of liquid excrement which are carried about at the caudal tips of the nymphs until they become quite large, and then are deposited on the surface upon which the thrips are feeding. No scars or leaf deformities, such as result from an infestation of citrus thrips, are caused by the greenhouse thrips. Nevertheless, the injury caused by the latter species results in a degrading of the fruit, and the greenhouse thrips is consequently a pest of major importance in the areas in which it attacks citrus.

CITRUS PESTS

Control:- As a result of their investigations, Boyce and.Mabry (114) recommended either pyrethrum extract or nicotine sulfate at the rate of 1 to 1200 in he regular late summer oil spray for scale insects and the citrus red mite, to be ollowed by a spray in the spring, if necessary, when the pyrethrum extract or nicotine sulfate could be used alone or in combination with 0.5% light medium oil. The removal of "off bloom" fruit was urged prior to treatment in the late summer, for a large percentage of the thrips occur at that time on these fruits.

The present recommendation for the control of greenhouse thrips on oranges is to spray 2 and sometimes 3 times with pyrethrum extract, 3/4 pint, and 2 quarts of oil as a spreader. Pyrethrum is preferred to nicotine.sulfate because the latter is so disagreeable to apply that it results in a less thorough application. The periods between applications vary between 3 and 6 weeks, depending on seasonal temperatures. Either the first or the second application of pyrethrum may be put in the regular oil spray for scale insects. Usually the two applications are sufficient, but the thrips should be carefully observed, for a third application is sometimes necessary (675). In Ventura-County, where about 200 acres of Valencia oranges are infested in years when conditions are favorable, 3/4 pint of pyrethrum extract is added to the 1-1/2 gallons of light medium oil to 100 gallons of water that is regularly used in the spring for the control of bud mite.

OTHER THRIPS

Bean Thrips:- The bean thrips, Hercothrips fasciatus (Perg.), which attacks a wide variety of plants, but especially the prickly wild lettuce (Lactuca sp.), sometimes spends the winter on oranges, particularly in the navels of Navel oranges in Tulare County. It was formerly necessary to fumigate California Navel,oranges destined for Hawaii because of the bean thrips, but the fumigation was not carried out during World War II, when the shipment of oranges to Hawaii was under army control, and finally in 1947 the quarantine was rescinded. The bean thrips is a dark species, like the greenhouse thrips, but possesses two white areas on the fore wings and white bands on the legs and antennae.

Cotton Thrips:- On January 23, 1934, Quayle (723) found the cotton thrips, Frankliniella gossypii Morgan, in great numbers in the vicinity of Hemet, Californ .a, on Valencia oranges, grapefruit and lemons, but most .abundantly on Navel Cranges. The presence of young nymphs indicated that eggs had been deposited on ie infested trees. This species was stated by Dudley Moulton to have occurred nly on cotton before its discovery on citrus by Professor Quayle.

Florida Flower Thrips:- The Florida flower thrips, Frankliniella cephalica ïispinosa (Morgan), was once considered to be a pest of some importance in Florida because it was believed that this species was responsible for considerable scarring resembling that caused by the rubbing of the fruit against the tree during wind storms (926) (see fig. 293). It has been proved by Thompson, however, that the injury formerly attributed to the flower thrips not only resembles wind injury, but is wind injury. The injury done by the flower thrips is a minor scarring around the blossom end which seldom covers enough surface to cause the fruit to be

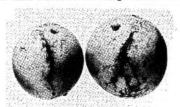

Fig. 293. Scars on oranges caused by wind, but sometimes mistaken
for thrips injury. After Thompson (877).

393

put in a lower grade (fig. 294, A, B). The control of these thrips does not prevent the more conspicuous scarring which was formerly attributed to them, so it is recommended that no control measures be used against this species (877).

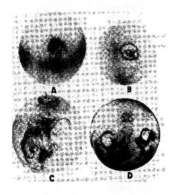

Fig. 294 (at left). Injury to oranges caused by thrips in Florida. A and B, scars on the stem end caused by Florida flower thrips; C, large blotches on mature fruit caused by orchid thrips while the fruit was young; D, scars at points of contact caused by orchid thrips on nearly mature grapefruit hanging in a cluster. Courtesy W. L. Thompson.

Fig. 295 (at right). Orchid thrips, Chaetoanaphothrips orchidii. After Kurosawa (499).

Orchid Thrips:- The orchid thrips, Chaetoanaphothrips orchidii (Moulton) (fig. 295), is another Florida species. It was first collected in a grapefruit orchard on Merritt Island in 1937 by Thompson (876). Since then this thrips has been found in many localities in Florida and has been injurious to grapefruit in Lake and Orange Counties. Its injury is said by Thompson to be more conspicuous than either that of the Florida flower thrips or wind scarring.

The orchid thrips is light yellowish in color, and the nymphs and adults are very active. They are found feeding at points of contact of fruits in a cluster or where leaves come in contact with the fruit (fig. 294, B, C). They prefer green immature fruit.

According to Thompson (877), "The size and appearance of the injury differs according to the age of the fruit at the time it is attacked. When young fruit is attacked the injury appears as a solid area due to a point contact. After the fruit has matured the early injury has the appearance of a silvery to dark brown blotch, sometimes two to three inches wide (fig. 294, C). The discolored blotch resembles the injury caused by a heavy infestation of rust mites on grapefruit called "sharkskin". The injury produced on more mature fruit usually takes the form of a dark brown ring since the mature fruit has flattened to form an area of contact into which the thrips are unable to penetrate. Again the discolored area resembles the injury caused by rust mites on a more mature fruit. The ringed type of injury (fig. 294, D) is much more common than the blotched type and has been observed on grapefruit and to some extent on oranges for a number of years. The ringed type of injury was thought by many to have been the result of an oil burn from the rind of the fruit caused by oil being pressed out of the rind by the weight and rubbing of large fruits hanging in clusters."

The stink bugs (family Pentatomidae) sometimes cause damage in Florida by extracting the juices of ripening citrus fruits and weakening the fruits so as to cause them to drop. There may be as much as a 50% loss of fruit from the attacks of plant bugs (926). Tangerines are preferred by the bugs, with oranges second choice. The most destructive of these bugs in Florida is the southern green stink bug, Nezara viridula L. which breeds particularly in legume cover crops and may attack the fruit if the cover crop is allowed to stand too long. The nymphs may be controlled with oil emulsions, nicotine products, or soap solutions. The adults may be shaken into nets and later destroyed. Excellent control of southern green stink bug on okra was obtained with 10% sabadilla dust. This material proved to be superior to benzene hexachloride or methoxy DDT (1010).

The citron bug, Leptoglossus gonagra, is dark brown to almost black in color and about 0.75 inch long. These bugs breed on citrons which have partially decayed, and if infested citrons are growing in an orchard, the bugs may attack tangerines and oranges, preferring the former (878). The citron bugs cause a drying of the attacked portions of the fruit and may cause the fruit to drop. Thompson recommends the removal of the citrons to prevent the infestations. The adult bugs may be shaken from the trees into nets or pans and destroyed. The collections should be made in the early morning while it is cool and the bugs are not flying about. Chlordan dust or spray as used in grasshopper control (p. 383) will also control the citron bug (884).

In Florida, citrus grown near cotton may be attacked by the cotton stainer, Dysdercus suturellus (Her.-Sch.), a dark red bug which feeds on cotton balls, but sometimes invades citrus orchards. The injury is similar to that of the green stink bug and the control is also similar.

The leaf-footed bug, Leptoglossus phyllopus (L.), occurs in Florida and as far westward as Arizona. It appears somewhat like the citron bug, but has a yellow stripe across the wings and a wider fourth segment on the hind legs, giving it the common name of "leaf-footed bug". These bugs also prefer the tangerines. They may be removed by jarring the trees. In the Imperial Valley, California, the Western leaf-footed plant bug, Leptoglossus zonatus (Dallas) (fig. 296), occasionally causes damage to citrus grown near pomegranate trees, upon which they breed. As is typical of the attacks of bugs, the fruits become dry and drop. This pest was at one time so serious as to lead to tent-fumigation with HCN gas. One grower jarred the bugs from the trees early in the morning and led a flock of turkeys through the orchard to feed on the bugs.(721)

Fig. 296. Western leaf-footed plant bug, Leptoglossus zonatus. X 2.5. Photo by Roy J. Pence.

395

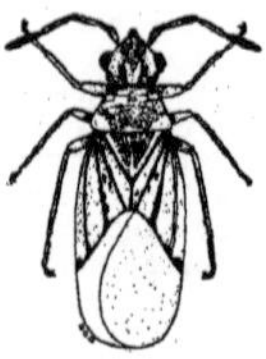

The false chinch bugs, **Nysius ericae** (Schilling) (fig. 297), have occasionally been injurious to citrus in California. The adults are small, light or dark gray lygaeids, 3 to 4 mm long. The nymphs are pale gray, with a reddish brown abdomen. This species multiplies in great numbers in native grasslands, and when the grasses or weeds dry up or are destroyed, the bugs sometimes move to citrus trees and may cause injury, especially to young trees. In large numbers they may cause a drying of the leaves.

Fig. 297. False chinch bug, **Nysius ericae**. X 9. After Essig.

Among the newer insecticides DDT and benzene hexachloride dusts have been used with fair success. In June, 1948, a heavy infestation of false chinch bugs occurred in a peach orchard near Palmdale, California. A 5% chlordan dust applied with a hand duster appeared to satisfactorily control the insects, so 20 acres were dusted by airplane, using 50 lbs. of 5% chlordan dust on 10 acres and 100 lbs. per acre on the other 10 acres. After 12 hours the chinch bugs began to leave the trees and after 3 days practically none were left and control seemed to be complete. It is hoped that chlordan will continue to give effective control of the false chinch bugs.

Suborder Homoptera (leafhoppers, whiteflies, aphids, mealybugs, scales)

The second of the two suborders, the Homoptera, includes the most important groups of citrus pests, either from a world or a national standpoint. The scale insects, for example, are probably more important, from the standpoint of their destructiveness to citrus, than all other pests combined. The aphids, mealybugs and whiteflies are also of world-wide importance as citrus pests.

LEAFHOPPERS

The leafhoppers (family Cicadellidae) are small slender insects, usually about a fourth of an inch in length. They are characterized by the bristle-like antennae inserted in front of and between the eyes and the two parallel rows of spines along the hind tibiae. In California, the potato leafhopper, **Empoasca fabae** Harris, or "green leafhopper," as it is known locally, breeds in large numbers on wild plants and cultivated crops and swarms into citrus orchards in Tulare County during the fall months (965). The leafhoppers spend the winter in the shelter of the citrus trees. They insert their sucking mouthparts into the ripening oranges, causing the release of oils from the oil cells in the fruit rind, which results in a spotting of the fruit. The damage is greater on Valencia oranges than on Navels, because the former remain on the tree longer and are therefore subject to attack over a longer period. Also the spotting is more prominent on Valencias than on Navels. Damage to oranges starts after the fruit begins to break color. The attacked portion of the fruit remains green while the remainder becomes the natural orange color.

Control:- Whitewash acts as a repellent, and when leafhoppers are noticed in an orchard in appreciable numbers, or if past experience in a particular orchard indicates control measures to be desirable, this spray can be used to good advantage. Usually it is desirable to use a zinc-copper-lime spray for correction of zinc deficiency and the control of septoria fungus and brown rot. This formula calls for 5 lbs. of zinc sulfate, 1 lb. of copper sulfate and 4 lbs. of hydrated lime to 100 gallons. If, in addition, leafhopper control is desired, the amount of lime can be increased to 15 or 20 lbs. per 100 gallons on Navels and 20 to 30 lbs. per 100 gallons on Valencias. An "outside coverage" spray is adequate for the control of leafhoppers, but should be applied before they have had a chance to injure the fruit.

WHITEFLIES

The whiteflies (family Aleyrodidae) are tiny insects from 1 to 3 mm long, the adults of which have their wings covered with a white, powdery wax which gives the insects their common name. The nymphs, after the first instar, are flattened, oval, and are similar in appearance to the early instars of the unarmored scale insects. Like the scale insects, the whiteflies lose their legs and antennae after the first molt, but unlike the scale insects, the females gain them back in the adult stage. The nymphs are frequently characterized by a fringe of conspicuous, white, waxy plates or rods extending out from the margin of the body. The most characteristic structural feature of the nymphs is a vase-shaped orifice within which the anus opens. This is called the vasiform orifice.

The whiteflies injure citrus trees in the same manner as aphids and scales, by withdrawing large quantities of sap from the trees and by being responsible for the growth of sooty mold fungus, Meliola camelliae (Catt.), which develops on the sugary excrement of the whiteflies known as "honey dew."

CITRUS WHITEFLY

Dialeurodes citri (Riley and Howard)

The citrus whitefly was at one time the most important citrus pest in Florida, but today ranks below the purple scale, rust mite, and possibly others, in this respect. It is nevertheless a serious pest in that state. It appears to have been introduced into Florida some time between 1858 and 1885 from Asia (615). It is now found throughout Florida and the other Gulf States.

In California the citrus whitefly was discovered in May, 1907, at Marysville, California, and an eradication campaign was begun with HCN gas which, however, was unsuccessful. The citrus whitefly was again discovered in a number of localities, including several in southern California, but always the infestations were either eradicated or the whitefly population was kept at a very low ebb and further spread from the initial foci of infestation was thus prevented (285). Since the initial infestation was found, the citrus whitefly has been discovered in over three dozen widely separated areas in California, and the combined cost of the various eradication programs was approximately $375,000, but prior to the recent outbreak in Fullerton (in 1947) no citrus whiteflies have been seen in this state since its reported eradication in 1942 (893).

The treatments which have been used in later years have been light medium oil sprays, such as used for scale insects on citrus, or HCN fumigation. Not only citrus trees, but all other host plants, such as a number of ornamental shrubs, have been treated. The oil spray, of course, is the most practical treatment for dooryard trees and ornamental shrubs.

Life History (615):- The pale yellow eggs are scattered singly on the undersides of the leaves, especially young leaves. They are attached to the leaf by a stalk. They hatch in from 8 to 24 days, depending on the season. Infertile eggs develop into males only. After hatching, the young nymph (crawler) moves about for several hours and settles on the underside of the leaf. After the first molt the legs become vestigial. The pupal stage is reached after the third molt. Nymphal life averages from 23 to 30 days. The pupa resembles the last nymphal instar (fig. 298, right) and requires from 13 to 304 days for development. The adult whitefly (fig. 298, left) lives on an average, about 10 days, but has been known to live for as long as 27 days. The adult female lays about 150 eggs under outdoor conditions. The entire life cycle from egg to adult requires from 41 to 333 days, a great variation being noted even among eggs laid in the same leaf on the same day.

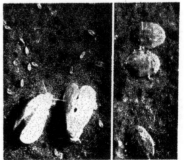

Fig. 298. Left, adults and eggs; right, nymphs (or pupae) of the citrus whitefly. X 12. After English and Turnipseed (277)

397

Injury:- The citrus whitefly can be found on a large number of host plants, but it prefers the various citrus varieties, chinaberry and umbrella trees, gardenia, privet, prickly ash and Japanese persimmons. The destruction of wild or useless plants growing near citrus orchards is recommended, especially the chinaberry and umbrella trees (926).

The whiteflies weaken citrus trees by the extraction of large quantities of sap. In addition, as stated before, they are responsible for the development of the unsightly sooty mold fungus developing on their excretions. Although whiteflies do not attack the fruit, their honey dew drops on the fruit, and the resulting sooty mold fungus must be removed by washing. Some investigators believe that the sooty mold fungus on the foliage reduces the amount of light reaching the leaves and thus interferes with the production of starch. It is believed that the sooty mold fungus is conducive to an increase in the purple scale and Glover's scale populations. (926)

CLOUDY-WINGED WHITEFLY

Dialeurodes citrifolii (Morgan)

The cloudy-winged whitefly, Dialeurodes citrifolii (Morg.), is very similar in appearance to the citrus whitefly, but while the eggs of the latter are pale yellowish in color, those of the former species are black. Likewise the species can be distinguished by the fungi which attack them. The yellow aschersonia fungus attacks only the cloudy-winged whitefly. The name "cloudy-winged" comes from the darkened area in the middle of each wing. The summer brood of the cloudy-winged whitefly lags a couple of weeks behind that of the citrus whitefly. On the whole, the cloudy-winged whitefly is less abundant in Florida than the citrus whitefly, but in parts of southern Florida it is the more abundant of the two species, and in recent years its importance has been increasing.

WOOLY WHITEFLY

Aleurothrixus floccosus (Maskell)

The wooly whitefly was first observed infesting citrus in Florida in 1909 by E. A. Back, having previously been known as a pest of citrus in Cuba. It is believed, however, that this species is probably a native of Florida, where it had been known to occur on the sea grape, Coccolobus for many years, and that a strain has comparatively recently developed on citrus which has spread from the 1909 infestation on citrus at Tampa. This strain, however, may have developed elsewhere, to be later introduced into Florida. (925)

The common name of the species is derived from the curled waxy filaments, having the appearance of wool, which cover the pupae. The eggs are brown in color and shaped somewhat like a short sausage and are laid mostly in circles. The adults are more yellow than the previously described species, are very sluggish, and seldom fly.

FLORIDA OR GUAVA WHITEFLY

Trialeurodes floridensis (Quaintance)

This species is quite abundant on guavas and avocados in Florida and occasionally found on citrus. It resembles the citrus whitefly, Dialeurodes citri, but the larvae are smaller, a deeper yellow, and somewhat thicker. (926) Recently an extensive systematic study of the North American species of the genus Trialeurodes has been made by Russell (758).

Control of Whiteflies

The Florida Agricultural Experiment Station recommends that in the fall the flight of the adult whiteflies should be watched, and after they have stopped flying, treatment should be deferred another 10 days to insure the hatching of practically all the eggs. The easily-killed nymphs should then be sprayed. The period for spraying is thus in late September or early October. In case the cloudy-winged whitefly is more abundant than the citrus whitefly, it is well to wait a little longer to spray, because egg laying occurs later in the year with this species. (921)

Spray oils or oil emulsions should be used in the control of whiteflies at a concentration insuring 1% actual oil in the spray mixture. Particular attention should be given to the undersides of the leaves, for the larvae are found there exclusively. A fall spray for scales will, of course, also control whiteflies. (921)

Entomogenous Fungi:- A measure which was at one time widely employed as a specific treatment for whiteflies was the use of entomogenous "friendly fungi." The best known and most widely distributed of the entomogenous fungi in the Gulf States is the red aschersonia, Aschersonia aleyrodis Webber (fig. 299), so-called because it produces orange-red pustular growths on the bottoms of white-fly-infested leaves, which range in size from 1/32 to 1/8 inch. Only the immature stages of the whitefly are infected.

Fig. 299. Red aschersonia fungus, Aschersonia aleyrodis Webber, parasitizing whitefly larvae on citrus leaves in Florida. Photo by Roy J. Pence.

A method of growing red aschersonia in large quantities on sterilized media in wide-mouthed pint bottles and distributing it on trees infested with whitefly larvae was perfected by Berger (84, 85, 86) following the discovery by Fawcett (300) that this and other Aschersoniae could be readily grown in pure culture on sweet potato and other media containing sugar. The cultures were sold to growers for one dollar a bottle. Each bottle contained enough of the culture to treat an acre of whitefly-infested trees (86). The contents of a bottle were mixed with 70-100 gallons of water in a spray rig. Only portions of the tree were sprayed with this mixture, averaging about a gallon per tree. The spray was directed against the undersides of the leaves. Applications were made preceding a period of rain, if possible. Cultures were prepared by the Florida Plant Board, beginning in 1915, but the Board discontinued this practice in 1943.

The yellow aschersonia, A. goldiana Lace and Ellis, was grown and distributed for the treatment of the cloudy-winged whitefly, and the brown whitefly fungus, Aegerita webberi Faw., was grown for the treatment of either the citrus whitefly or the cloudy-winged whitefly.

It has been known for many years that Bordeaux spray used on citrus trees in Florida is followed by an increase in scales and whiteflies. It was assumed that this was caused by the reduction in entomogenous fungi. Holloway and Young (440) investigated this problem with special reference to the purple scale, and came to the conclusion that the reduction in fungi did not significantly affect the scale population, but that the residue left by the spray, whether copper was present or not, was the factor resulting in the increase in purple scale. Low residue copper sprays are now used in Florida instead of the conventional Bordeaux spray formulas.

In the above experiments, no increase in the percentage of infection of purple scales by entomogenous fungi resulted from applying a spray of fungus spores. The same conclusion with reference to whiteflies had been reached by Morrill and Back (616).

The friendly fungi are present in all orchards and increase in numbers under proper weather conditions. However, they cannot be depended on for adequate control of scales and whiteflies or for the production of clean fruit. Since it has been shown that nothing is gained by artificially adding more fungi to those already present, the "friendly fungi" are no longer even mentioned in the official cooperative spray schedules which are annually issued in Florida.

Insect enemies:- A number of predators and parasites are listed by Quayle (724) as enemies of whiteflies in the United States, but neither singly nor in the aggregate do these species effect an adequate control.

APHIDS

The aphids (family Aphididae) comprise a large and economically very important family of small, soft-bodied, fragile, phytophagous insects. They may be winged or wingless, naked, like the citrus aphids, or covered with white, waxy secretions, like the wooly apple aphid. A single species may change from sexual to parthenogenetic forms, depending on the season. The antennae are 3-to 6-segmented and bear sense organs or sensoria, which in the case of citrus-infesting aphids are circular. The rostrum, a modification of the labium, is 3- or 4-segmented and well developed, and the lancets (maxillae) which issue from the beak and penetrate the plant tissue are very long (fig. 300). Wings are present on migrating or sexual forms, and are normally as long as or longer than the body, thin and transparent, having few veins, and usually folded roof-like over the back. Aphids bear oil- or wax-secreting organs called cornicles, which are tubes of variable size and shape extending from the posterior part of the dorsal portion of the abdomen.

Fig. 300. Head and rostrum of Macrosiphum solanifolii.

The biologies of the aphids are very complex because of the number of forms of a species throughout the year, including both sexually and parthenogenetically reproducing forms, and because of the strong influence of climatic factors on the biology. The habit of seasonal alternation of host plants also complicates the study of their biologies. The kinds of forms that may be found among some species in temperate zones are listed by Essig (287, p. 333) and these should be studied by anyone having an interest in the technical terminology pertaining to these forms. Briefly, and in non-technical terms, it may be stated that the characteristic seasonal life history of aphids in temperate zones and requiring alternate host plants is as follows: Overwintering eggs hatch in the spring, giving rise to apterous females that reproduce by parthenogenesis. One or more generations of apterous parthenogenetic females follow, then alate (winged) parthenogenetic females appear. These alate individuals disperse and start new colonies, often on a different species of host plant. There follow several generations composed of either apterous or alate females, or both, all of which are parthenogenetic. In the fall the sexual individuals are produced on the original host, the sexual or oviparous females being apterous and the males being alate. These mate, and the females lay the overwintering eggs.

There are numerous exceptions to the typical life history. For example, the most important aphid attacking citrus in the United States, Aphis spiraecola, has a life history which follows the above pattern in Maine, but in Florida and in California it occurs very sparingly on its secondary host, and many individuals pass the winter in constantly reproducing colonies of parthenogenetic females on citrus. The overwintering egg stage is consequently of no importance in the maintenance of the aphid population. Several of the species attacking citrus live only in tropical or subtropical regions and never produce eggs. Some species which may be found on citrus, such as the melon aphid, Aphis gossypii, and the green peach aphid, Myzus persicae, are general feeders and do not require a single host or combination of hosts to insure their continued development throughout the year, but will disperse from plant to plant within the range of the winged or crawling migrants.

CITRUS PESTS

R. C. Dickson, who has made a study of citrus aphids, informs us that the species infesting citrus in California are, in order of their importance:

Spirea aphid, _Aphis spiraecola_ Patch
Cotton or melon aphid, _Aphis gossypii_ Glover
Green peach aphid, _Myzus persicae_ (Sulz.)
Black citrus aphid, _Toxoptera aurantii_ (Fonsc.)
Potato aphid, _Macrosiphum solanifolii_ (Ashm.)
Foxglove aphid, _Myzus convolvuli_ (Kalt.)
Cowpea aphid, _Aphis medicaginis_ Koch
Dock aphid, _Aphis rumicis_ Linn.

Of the above 8 species, only the first 4 are ever serious, and most of the damage is done by the first two. In California, _Aphis spiraecola_ and _Toxoptera aurantii_ are practically confined to citrus. Infestations in the spring may be found on just a few terminals on an occasional tree in the orchard, but after the terminals become crowded, winged forms appear which start new colonies on previously uninfested terminals on the same or a different tree. Finally the entire orchard may become infested. The other species of aphids, however, ordinarily attain their high population densities in citrus orchards or cover crops and other nearby vegetation and then move to citrus. Usually there is a mass migration of aphids to citrus trees at the time the cover crop is being disked under.

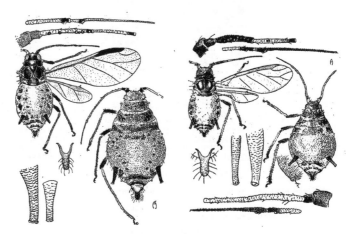

Fig. 300a. Left, _Toxoptera aurantii_; right, _Aphis citricidus_. Both of these aphids have a practically world-wide distribution on citrus except that the latter species is not found in the United States. After Essig (287a).

The brown citrus aphid, _Aphis citricidus_ (Kirkaldy) (fig. 300a, right), which may also be found in the literature under the names _Myzus citricidus_ Kirkaldy, _Aphis tavaresi_ Del Guercio, and _Aphis citricola_ van der Goot, is widely distributed throughout the world except in the United States. Along with _Toxoptera aurantii_, this species is believed to be a vector of the virus disease of citrus trees in Brazil known as _tristeza_. The above species have been described and illustrated by Essig (287a).

401

SUBTROPICAL ENTOMOLOGY

KEY TO APHIDS COMMONLY INFESTING CITRUS

IN CALIFORNIA - WINGED FORMS

A. Frontal tubercles prominent, exceeding vertex (fig. 301, A, B).

    B. Abdomen pale, without dark patch or cross-markings, cornicles reticulated
        at tip (fig. 301, R).

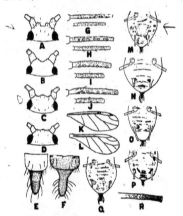

Fig. 301. Taxonomic characters for citrus-infesting aphids. A, head of Myzus
persicae; B, head of M. convolvuli; C, head of Toxoptera aurantii; D, head of Aphis
spiraecola; E, cauda of A. gossypii; F, cauda of A. spiraecola; G, 3rd and 4th antennal segments of A. gossypii; H, 3rd and 4th antennal segments of A. medicaginis;
I, 3rd and 4th antennal segments of A. spiraecola; J, 3rd and 4th antennal segments
of A. rumicis; K, wing of T. aurantii; L, wing of A. gossypii; M, dorsum of M. persicae; N, dorsum of A. spiraecola; O, dorsum of M. convolvuli; P, dorsum of A. gossypii; Q, dorsum of A. rumicis; R, tip of cornicle of Macrosiphum solanifolii.

        ————————————— Macrosiphum solanifolii (Ashm.)

    BB. Abdomen with dark markings, cornicles not reticulated at tip.

        C. Cornicles clavate, frontal tubercles gibbous, converging, abdomen with
           a dark dorsal patch (fig. 301, M).

        ————————————— Myzus persicae (Sulz.)

        CC. Cornicles cylindrical, frontal tubercles not converging, abdomen with
           dark transverse dashes (fig. 301, O).

        ————————————— Myzus convolvuli Kalt.

AA. Frontal tubercles not prominent (fig. 301, C, D).

    B. Median vein once-branched (fig. 301, K).

        ————————————— Toxoptera aurantii (Fonsc.)

    BB. Median vein twice-branched (fig. 301, L).

C. Cauda with fewer than 5 hairs on each side, sensoria absent from 4th antennal segment (fig. 301, E, G, H).

    D. Third antennal segment with 7 to 10 sensoria, abdomen with a few broken dark dorsal dashes (fig. 301, G, P).

        _____ Aphis gossypii Glover

    DD. Third antennal segment with 3 to 7 sensoria, dorsal surface of abdomen with black transverse bars (fig. 301, H).

        _____ Aphis medicaginis Koch

CC. Cauda with more than 5 hairs on each side, sensoria present on 4th antennal segment (fig. 301, F, I, J).

    D. Third antennal segment with 7 to 12 sensoria arranged in a single row, dorsal surface of abdomen without markings anterior to cornicles (fig. 301, I, N).

        _____ Aphis spiraecola Patch

    DD. Third antennal segment with 12 to 20 sensoria, dorsal surface of abdomen with dark, broken transverse dashes (fig. 301, J, Q).

        _____ Aphis rumicis Linn.

SPIREA APHID

Aphis spiraecola Patch

The spirea aphid is considered by R. C. Dickson to be the most important species on citrus in California. Likewise Goff and Tissot (363) and Watson and Berger (926) consider it to be the most important aphid on citrus in Florida and the latter rank this species fourth in importance among the citrus pests of Florida, ahead of the Florida red scale. In California, even in the coastal areas where their attacks are the most severe, the aphids are of less economic importance than the scale insects and mites.

Aphis spiraecola is a small aphid, 1.8 mm long,[1] and apple-green, almost identical in color with the young tender citrus leaf. In the winged forms, when the wing pads begin to form the thorax changes color, first becoming pale pink and finally dark brown or almost black when the wings are fully developed. The abdomen, however, usually remains green.

The spirea aphid is closely related to the green apple aphid, Aphis pomi De Geer, and, in fact, not all authorities are agreed that it is anything more than a physiological race of the latter species. That it is at least a physiological race of A. pomi was proved by Tissot (891), who obtained A. pomi from apples in Pennsylvania and tried to transfer them to citrus in Florida, but succeeded in getting only 2 out of 47 to live to maturity and reproduce, while of a similar number transferred to apple in Florida at the same time and under similar conditions, all developed and reproduced abundantly.

Injury:- As in the case of the other citrus-infesting species, the most noticeable damage is that of leaf curling (fig. 302). The aphids attack the new, tender terminal growth and prefer the undersides of the leaves. One or two aphids can cause the leaf to curl slightly and a heavy infestation causes it to curl strongly, with the underside of the leaf and the attached aphids inside the leaf

---

[1] The dimensions given for the citrus-infesting aphids are the average of from 20 to 30 measurements of each species made by the writer.

roll that is formed by the curling. The infested leaves are dwarfed and lose their oily appearance, and the twig which bears them is also stunted and twisted. Young trees that are annually attacked by aphids grow very slowly and 5- or 6-year-old trees may have attained the size of a normally-developed 2- or 3-year-old tree. In searching for aphids, one is more apt to find them on young trees than old ones, for the young trees have new growth on them at practically any season of the year.

The most immediate economic loss from aphids occurs if the population is abundant when the trees are in bloom, for the flower buds, flowers and very young fruits may be injured. The flowers

Fig. 302. Curling of terminal leaves of orange caused by aphids. Photo by Roy J. Pence.

may drop, and the feeding on the tiny fruit causes a bumpiness which remains even on the mature fruit. Another economic loss results from the fruit being blackened with sooty mold fungus which develops on the "honey dew" excreted by the aphids. The honey dew drops onto all parts of the tree, including the fruit, and the resulting sooty mold fungus must be removed by washing.

Life History:- Experiments on the life history of Aphis spiraecola were made out of doors by Miller (609) at Lake Alfred, Florida. As far as he was able to ascertain, no males or oviparous forms appeared throughout the year. Although eggs have been found at Gainsville, Florida, on spiraea in winter, none have ever been found on citrus. According to Miller (609), Tissot found in 1925-26 that the eggs did not hatch, but in 1926-27 many eggs hatched. The nymphal period in the alate forms was longer and the rate of reproduction was less than in the apterous forms.

In Florida alate aphids were found to be present at all times of the year, but they were most abundant in April and May. Winged forms developed as a response to aging or hardening of the substrate upon which the aphids were feeding, or to crowding (609). As will be shown later, it has been proved in the case of the cotton or melon aphid that the condition of the food plant, and not crowding, causes the production of winged forms (363). It is under these conditions that aphids are forced to move to new terminals on a tree or to new trees, and winged forms become necessary for distribution of the population.

COTTON OR MELON APHID

Aphis gossypii Glover

The cotton or melon aphid, while not considered to be as important as the green citrus aphid, may nevertheless cause serious injury. It is widely distributed throughout the world and occurs on a large number of host plants. It is a serious pest of cotton and melons and a number of vegetables. It probably attacks citrus wherever it is grown. This species on citrus is always a dark gray or dull black in color, but on some other hosts it may be black, gray, green, or yellow. It is the smallest of the citrus-infesting aphids, being on the average about 1.5 mm long.

Life History:- In the northern parts of its range, this species passes through the typical aphid life history, but in the citrus areas of the United States the sexual forms are omitted, reproduction being entirely by parthenogenesis. Experiments were made on melon in Florida in an insectary open on all sides except for a wire screen (363). In one year a minimum of 16 and a maximum of 47 generations completed their life cycle. The nymphal period varies from 3 to 20 days, average 7.3 days; reproductive period, 2 to 31 days, average 15.6 days; post reproductive

CITRUS PESTS

period, 0 to 21 days, average 5.3 days; length of life, 9 to 64 days, average 28.4 days; number of young born per day, 1 to 14, average 4.3; average number of young per female, 67.

No winged forms were seen by Goff and Tissot (363) on the plants used for rearing the aphids except on some plants which were set aside and became crowded with aphids. These investigators made an experiment to determine whether the winged forms developed as a result of crowding or because of changes in their food plant which took place as a result of the heavy aphid population. "A young plant was allowed to become overcrowded until practically all forms produced were of the winged type. All aphids were then removed from the plant and a half dozen newly born young from one of the aphids used in the life history study were scattered on the plant so that they were not in contact. However, they developed as winged forms, while sisters of these on a plant in good condition did not develop wings, showing that in this case it may have been due to the condition of the plant and the food it furnished."

GREEN PEACH APHID

Myzus persicae (Sulzer)

The green peach aphid, Myzus persicae (Sulzer), is widely distributed throughout the world on a very large number of host plants. It has been recorded on citrus in most of the citrus-growing regions of the world and probably occurs on this host wherever it is grown. In this species on citrus the apterous individuals are pale green while the alates, when mature, are quite dark with a black dorsal patch on the abdomen. The color of the green peach aphid is similar to that of the green citrus aphid. The green peach aphid may also be distinguished by the converging frontal tubercles of the head and the slightly swollen cornicles. This species measures 2.1 mm in length.

BLACK CITRUS APHID

Toxoptera aurantii (Fonscolombe)

The black citrus aphid, Toxoptera aurantii (Fonsc.), is another cosmopolitan species, occurring wherever citrus is grown. It is a small aphid, 2.1 mm long, which may be dull or shiny black or a mahogany brown color, sometimes with a purplish hue. The most reliable distinguishing character, however, is the black stigma on the leading edge of the fore wing and the once-branched media vein in the fore wings. According to Dickson, one of the reasons this species is of minor importance is because it reaches high population densities too late in the year to cause economic damage to citrus, at least in California.

OTHER APHIDS

The potato aphid, Macrosiphum solanifolii (Ashm.), is a large aphid, 3.3 mm long, with a pale green and a pink variety. A recognition character for this species is the reticulation at the tip of the cornicles which may be found on both sexes and in all forms. In cold regions it spends the winter on roses, while in summer it is apt to become a serious pest of potatoes, and it has been found on 20 other hosts (679). On citrus the potato aphid seldom, if ever, increases in number to the point where it is injurious.

The convolvulus aphid, Myzus convolvuli (Kalt.), occasionally may be found on citrus, but it never causes injury. The apterous individuals are pale green and the alates quite dark. This species is 2.8 mm long.

The black clover aphid, Aphis medicaginis Koch, is a shiny black species when mature, and the legs and basal half of the antennae are white or pale yellow. This species averages 2.1 mm in length. It occurs on a wide variety of plants, but chiefly on legumes. No sex forms have been found in the West, and the winter is passed on clover, melilotus, alfalfa, and optuntia (282). Occasionally small colonies may be found on citrus in California and Arizona.

405

The dock or bean aphid, Aphis rumicis L., is one of the most common of the dull black species of aphids and it occurs throughout the Western States. It is a small species, 1.8 mm in length. This species winters on euonymus and dock, then migrates to a large number of plants, particularly beans, on which they may do considerable damage. After the beans have been harvested and the plants are plowed under, this aphid returns to its winter hosts (861). Dr. Dickson informs us that only occasionally and in small colonies is this species established on citrus.

Natural Enemies

Aphids are attacked by many species of insect enemies, and these destroy large numbers, but while the aphid population is on the increase, they usually do not result in control, and artificial treatment must therefore be utilized. Once the population is diminishing because of adverse weather conditions or the gradual hardening of the foliage, the predators and parasites can quickly wipe out an infestation. Therefore it is usually inadvisable to treat an aphid infestation if it is rather late in the season of spring growth and the natural enemies are abundant.

The great majority of the predators of the aphids belong to one or the other of three orders: Coleoptera, Diptera, and Neuroptera. Aphid predators found in Florida have been discussed by Cole (175).

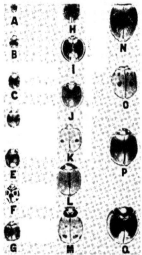

Fig. 303. Some predaceous lady beetles. X 2.5. A, Scymnus bipunctatus Kug.; B, Lindorus lophanthae (Blaisd); C, Rhizobius ventralis (Er.); D, Exochomus fasciatus Csy.; E, Hyperaspis lateralis Muls.; F, Hyperaspis c-nigrum Muls.; G, Rodolia cardinalis (Muls.); H, Cryptolaemus montrouzieri Muls.; I, Orcus chalybeus (Boisd.); J, Chilocorus stigma (Say); K, Adalia bipunctata (L.); L, Coccinella californica Mann.; M, Hippodamia convergens Guer.; N, H. ambigua Lec.; O, Olla abdominalis Say; P, Cycloneda polita Csy.; Q, Axion plagiatum (Oliv.). Photo by Roy J. Pence.

Coleoptera:- Apparently Coleoptera is the most important of the orders of predators attacking aphids, but primarily a single family is concerned, the family Coccinellidae (lady beetles) (fig. 303). The lady beetles, with the exception of

the genus Epilachna, are practically all beneficial insects, feeding in both the
larval and adult stages on injurious insects, especially the aphids and scales.
The lady beetles lay their yellow oval eggs in clusters on various parts of the
plant. The long axes of the eggs are at right angles to the surfaces on which they
are deposited. Among the other two groups of aphid predators, the long axes of
syrphid fly eggs are parallel with the surface upon which they are laid and those
of the lacewing flies are laid at the ends of stalks.

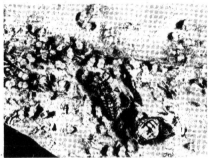

Fig. 304. Convergent lady beetle, Hippodamia convergens. Larva and pupa in a
colony of green peach aphids on an orange leaf. X 2. Photo by Roy J. Pence.

The larvae of the coccinellids are carabidoid (alligator-like) (fig. 304),
somber in color or sometimes with bright markings. The adults are oval or circular, convex or hemispherical, with the head small, partially withdrawn into the
prothorax or concealed under the pronotum (fig. 303). The smooth shiny elytra
are of many colors and usually have distinct patterns of spotting or marking by
which the species may be differentiated. Enormous numbers of the adults may be
found in the mountains in the fall, congregated among leaves and plants for their
winter hibernation. In winter they become deeply covered with snow. Some may
also be found in large numbers in the valleys, aestivating at the bases of wide-
spreading plants or in secluded places in trees, as in heavily foliaged young citrus trees, during summer. Clausen (162) made a biological investigation of what he
considered to be the 8 most important species of lady beetles predaceous on aphids
in California. The adult of each species will be described below. It should be
borne in mind that the colors and markings are quite variable, and the descriptions
given are descriptions of the most common forms ordinarily seen.

The California lady beetle, Coccinella californica Mann. (fig. 303, L), is 6
or 7 mm long, has elytra of yellow or red color or with intermediate shades. The
pronotum is black and has yellow spots on the anterior edge. A black spot is
located at the anterior juncture of the elytra.

Coccinella trifasciata Linn. is 5 or 6 mm long, has yellow and red elytra with
three black transverse bars (fascia) on each elytra, the first pair of fascia meeting at the anterior juncture of the elytra. As in the previous species, the prothorax is black, with two yellow spots on the anterior margin of the pronotum. In
California a variety of this species named C. trifasciata var. juliana Muls. is
more abundant than C. trifasciata. It differs in having only one transverse sub-
basal fascia.

The convergent lady beetle, Hippodamia convergens Guer. (fig. 303, M), is from
6 to 7 mm long. The elytra are yellow, orange, or red and with 6 black spots on
each elytron. The specific name comes from the two oblique converging white lines.
A black spot is also located at the anterior juncture of the elytra.

407

Hippodamia ambigua Lec. is in size, shape, and coloration similar to *H. convergens*, except that it has no spots on its elytra except for the black spot at their anterior juncture. The pronotum has the two oblique converging white lines as in *H. convergens*. Some authorities, in fact, believe it to be a spotless variety of the latter species.

The ashy gray lady beetle, *Olla abdominalis* (Say) (fig. 303, 0), is 4 to 5.5 mm long and is nearly circular in shape as contrasted to the more elongate shape of the previous species. The pale yellow prothorax has 7 black spots. The elytra are of a greyish yellow color, and each elytron has 7 or 8 black spots arranged in transverse rows, four in front, then 2 or 3, and finally 1 spot near the posterior margin.

Olla oculata Fabr. is similar in size and shape to the previous species, but is black, with two large red spots on each elytron and with a yellow spot on each side of the pronotum with one in between in the anterior margin. The lateral spots each have a black dot in the center which resembles a pupil and gives the beetles their common name, the eyed lady beetle. This species could easily be confused with the twice-stabbed lady beetle, *Chilocorus stigma* (Say) (fig. 303,J), which it resembles in size, shape, and color and markings of the elytra. *C. stigma*, however, has no markings on the pronotum. Also, its legs and undersides are black, while those of *Olla oculata* are reddish.

The two-spotted lady beetle, *Adalia bipunctata* L. (fig. 303, K), is 4 to 5.5 mm long, the elytra are yellow to red in color with a circular black spot in the center of each elytron. The pronotum is black, with a large yellow spot covering each lateral margin, two very small yellow spots at the anterior margin of the pronotum and with a small basal yellow fascia at the posterior margin.

The western blood-red lady beetle, *Cycloneda munda* (Say), is 4 or 5 mm long, and the elytra are pale to very bright red, with no markings. A more or less circular yellow spot occurs on each side of the pronotum, in the center of which is a round black spot, giving the appearance of a pair of eyes. Proceeding backward from the median anterior margin of the pronotum, is a yellow stripe extending half way back to the posterior margin.

Clausen (162) worked out the life histories of the 8 species of lady beetles mentioned above and presented in tabular form the number of days spent by the various species in the egg stage, each of the four larval instars, and in the pupal stage. The total number of days from egg to adult varied from 21.0 for *Olla abdominalis* to 33.2 for *Hippodamia ambigua*. The number of eggs deposited under normal field conditions varied from 200 to 500 and occasionally more, and extended over a period of 4 to 8 weeks in case the female lived the full adult life. The average number of aphids eaten per individual larva varied from 216 for *Cycloneda munda* to 475 for *Coccinella californica*. The number of aphids eaten per day varied from 11.4 to 24.9 for the larvae and from 15.6 to 56.1 for the adults. *Hippodamia convergens* proved to be the most voracious feeder among the adults.

Goff and Tissot (363) mentioned *Hippodamia convergens* as the most important of the predators of melon aphids in Florida. The second most abundant they believed to be the blood-red lady beetle, *Cycloneda sanguinea* Fabr. This is very similar to the Pacific Coast species *C. munda*, but *C. sanguinea* extends westward only to Arizona. Other lady beetles common among melon aphids were listed as several species of *Scymnus* and the two-spotted lady beetle, *Olla abdominalis* var. *sobrina* Csy (363). Miller (609) in listing the predators of the green citrus aphid, mentions the above species, but gives first place to *Cycloneda sanguinea*, and in addition lists a number of minor species.

Leis conformis Boisd., a lady beetle introduced from China by the University of California Citrus Experiment Station, became established in Florida in at least two counties. It has been reported that this large beetle eats more aphids per day than a dozen native ones (923). In addition, the Chinese beetle is not as susceptible to attacks by hymenopterous parasites and entomogenous fungi as the native lady beetle.

CITRUS PESTS

Diptera:- Just as the most important aphis-feeding Coleoptera belonged to a single family, so the majority of the important aphis-feeding Diptera belong to the family Syrphidae (syrphid flies). The adults are small, medium sized or large flies, very active on the wing, often gaudily striped with yellow, and wasp-like or bee-like in appearance, and commonly mistaken for wasps or bees by the layman. One of the common names, "hover fly" comes from the habit of hovering in mid-air with the wings in rapid motion, then suddenly darting to another place to again remain poised in a fixed position.

The eggs are laid singly on the plant surfaces. Those found among aphid infestations on citrus trees are usually elongated, white, and reticulated. They are laid horizontally on the leaf surface as contrasted to the vertical position of lady beetle eggs. Likewise the latter are found in groups rather than single.

The larvae of the species attacking aphids represent one of the four general types of syrphid fly larvae called the aphidivorous type (287). They can be seen almost anywhere where aphids are abundant. They are elongated, somewhat flattened, pointed anteriorly and truncate posteriorly like the house fly maggot. In color they are green, brown, gray, or mottled. The larvae can often be seen groping amongst the aphids, their anterior portion waving about in the incessant search for prey. The aphids are seized and held aloft until sucked dry (fig. 222). The syrphid fly larva, while holding the aphid aloft, keeps only its most posterior position in contact with the plant surface, apparently by means of four tentacle-like pseudopods which may be seen protruding from about the truncated posterior portion of the body. Unlike the lady beetles, the syrphid flies feed on aphids only in the larval stage, the adults frequenting flowers and feeding on nectar.

The following have been mentioned as being the principal aphidivorous species of California (215):

The large syrphid, Scaeva pyrastri (Linn.), was considered to be by far the most important of the syrphid flies attacking aphids in California. It is probably the best known of the syrphids because of its abundance and its large size. It is 11 to 14 mm long, has a yellow face, reddish brown eyes and antennae, dark metallic blue thorax covered with soft pile, and velvety or shiny black abdomen. The pale-green, white-striped larvae are said to destroy about a thousand aphids in the 2 or 3 weeks of their larval existence. They may first be seen among the aphid colonies in March and start preying on the aphids somewhat earlier than the lady beetles.

The genus Syrphus includes some of the more important species of the family. The American syrphid, Syrphus americanus Wiedemann, is 9-10 mm long, metallic greenish in color, and has yellow transverse bands across the abdomen. The arenate syrphid, Syrphus arenatus (Fallen) is 9-12 mm long, metallic green or black, with the three principal yellow bands interrupted and those on the third and fourth segment arenated. The larvae are yellowish, brownish or purplish and attack many species of aphids. S. opinator O. S. is a western species 9-11 mm long, with an unbroken yellow transverse band on the second and third abdominal segments. S. ribesi is 7-12 mm long, has the first yellow band interrupted in the middle, the second entire, and the third entire but with a median incision. The venter is alternately marked with yellow and black bands. S. torvus is similar to S. ribesi, but has hairy eyes.

Allograpta obliqua (Say) is small and slender, 6-7 mm long, with narrow yellow transverse bands anteriorly and oblique and longitudinal lines at the tip of the abdomen. The larvae are light green, with white markings. It has been stated that they consume an average of 34 aphids per day (609).

The four-spotted aphis fly, Baccha clavata (Fabr.), is 9-11 mm long, and has a slender, wasp-like body. It is black with clear wings. The larvae are dull green and have a red dorsal line. B. lugens differs from the previous species in having a smoky film on its wings and abdomen. Both of the latter species are important predators of citrus aphids in Florida (609).

Fig. 305. Eggs of Chrysopa californica. Above, hatched eggs on
stalks attached to an orange, X 3.5;
below, egg a short period before
hatching, X 28. Photo by Roy J.
Pence.

Fig. 306. Larva of the green lacewing emerging from a molted skin.
X 6. Photo by Roy J. Pence.

Neuroptera:- In the order Neuroptera we
again find but a single family of importance
among the aphidiphagous predators. This
family, however, contains the well-known
green lacewing, Chrysopa californica Coq.
(fig. 223). The life history of this species
has been worked out by Wildermuth (950). The
adult is a beautiful insect, 9-14 mm long,
with delicate, iridescent, gauzy wings, free
of hairs, and folded roof-like over the body,
and with large golden eyes. The males are
slightly smaller than the females and neither
take food in the adult stage. The adults
give off an offensive odor when crushed.

The egg is placed in a long hair-like
stalk about a half inch in length (fig. 305).
It is oblong and very small. The eggs are
always a source of wonder to those who discover them for the first time. The larvae
are cannibalistic and it is supposed that
the eggs are laid at the ends of stalks so
as to be out of reach of the larvae as they
are hatched.

The larvae (fig. 306) are thysanuriform
(alligator-like) in shape, hairy, and with
large sickle-like mandibles. The larval
habits are described by Wildermuth (950) as
follows:

"The hatching process requires but a
few minutes, but the larva rests on the empty
eggshell for some time after emergence. When
the eggshell becomes dry and hardened, the
larva lazily crawls down the supporting egg
stalk and eagerly begins searching for food.
If small aphids or thrips nymphs are present,
it quickly seizes one of these and begins
feeding. If only full-grown and large aphids
are present, it is more cautious, running in
a circle around the tempting and monstrous
meal or following the aphid, ever and anon stopping as if to consider whether or
not it could safely attack a creature so many times larger than itself. Finally,
however, its increasing hunger apparently overcomes all fear, and it pounces on its
prey. The aphid is lifted bodily off its feet, the lacewing larva all the time
crushing, piercing, and sucking its prey. The larvae of all lacewing flies extract their food from their host by piercing it with their long, powerful mandibles,
which are hollow, the internal fluids of the host being rapidly absorbed through
them. With abundant food present the larva grows rapidly and quickly takes on a
robust appearance."

Wildermuth (950) found that the total larval period required 21 days in
February and 12 days in October, 1915. The average number of aphids eaten per
larva was 88 in February and 143 in October.

Parasites:- The only important parasite of citrus aphids in California is the
hymenopteron, Lysiphlebus testaceipes (Cresson), but large numbers of aphids are
destroyed by this insect. L. testaceipes is 2 mm long, with black head and thorax,
brown or dusky abdomen (except for the anterior part of the second segment, which
is pale yellow), and with dull yellow-brown (testaceous) legs. Its work is indicated by the swollen, spherical, gray bodies ("mummies") of its victims, each with
a circular exit hole in the dorsum, where the parasite escaped after completing its

immature stages in the aphid. Sometimes a "cap or "lid" remains attached to the edge of the emergence hole (fig. 225).

This parasite lays from 400 to 500 eggs, depositing only one egg in a host. It is small wonder then that the entire aphid population on a leaf may be destroyed. However, since aphids can thrive at a lower temperature than the parasite, they can increase in large numbers in the spring before Lysiphlebus becomes abundant, (724). By that time injury to the foliage has already occurred.

Another parasite of aphids, Praon similans Prov. (fig. 227), is worthy of mention more from the standpoint of its curious habits than because of its economic importance, for it is not often found parasitizing citrus aphids. This interesting braconid is discussed on page 317.

When to Treat:- In California the control of aphids on mature trees is usually not considered except in the spring, for it is in the spring that damage can be done to blossoms and fruit. It has been the observation of those familiar with the problem that heavy infestations of aphids usually follow cold winters. The aphids may begin to appear in February and treatment may be necessary in March, but infestations demanding immediate control measures are more apt to occur during the month of April. The aphid population declines with the approach of hot weather and increases again in the fall, so as far as citrus is concerned, the most favorable periods of the year for aphids happen to coincide with the spring and fall flushes of tender new growth.

The stunting of new growth is not considered as important on mature trees as on young trees on which the full stature of the tree is yet to be attained. Since young trees have a certain amount of new growth at almost any period of the year, and since protection of new growth is especially important on young trees, treatment for aphids might well be undertaken at any period of the year. Also, because of the continuous growth on young trees, they may have to be treated several times a year.

On mature trees, if many colonies of aphids have become established, it is best to treat while the new leaves are still small, for after the leaves have become cupped or curled, thus protecting the aphids within, it is difficult to contact them with the spray; besides, the damage by that time has already been done. Moreover, when the leaves harden, the aphids can no longer feed on them, and those which have not migrated must die whether they are sprayed or not. Another factor making late treatment inadvisable is the abundance of beneficial insects which, if unmolested, will quickly decimate an aphid population, especially if the population is already waning.

The rapid and effective action of natural enemies under certain conditions was once demonstrated in a spectacular manner in the experimental orchard of the University of California, Los Angeles. On April 4, 1947, a heavy infestation of aphids was being attacked by large numbers of natural enemies, especially the lady beetle Coccinella californica and the hymenopterous parasite Lysiphlebus testaceipes. Six Valencia orange trees were sprayed with a very effective non-oil spray containing 2.5% rotenone, which was used at 1 quart to 100 gallons. This treatment resulted in practically complete control of aphids. Within a week, however, there were no more aphids on the untreated trees than on the sprayed trees; it was difficult to find a live aphid anywhere in the orchard. If the entire orchard had been sprayed, disappearance of the aphids would no doubt have been ascribed to the spray treatment.

Florida citrus growers are advised to control aphids in the winter to prevent a possible infestation in the spring. The aphids usually begin to develop winged forms about the middle of March, when the foliage hardens. Nine-tenths of the aphids may then develop wings. They are then more difficult to control. (923)

Control

The standard control measure for aphids for many years has been nicotine sulfate, and it is still as effective as any known insecticide or combination, but is more difficult to procure than in pre-war years. Lime sulfur may be added for mite control, and this, of course, is the usual procedure, for citrus red mite or six-spotted mite are nearly always an annual problem. Light medium oil at 0.5 to 0.75% with rotenone preparations may be used if it is desired to control or contribute to the control of black scale or bud mite.

In the Exchange Pest Control Circulars, until nicotine preparations became difficult to obtain, the following recommendations were made for the aphid-red mite or aphid-red mite-bud mite combinations:

Standard Sprays:-

(1) "Nicotine sulfate 3/4 pint; lime sulfur--coast 1 gallon, interior 1-1/2 to 2 gallons;[1] spreader 1/2 lb. calcium caseinate spreader or 3 oz. albumen spreader; water to make 100 gallons.

(2) "Nicotine sulfate 1 pint, soap powder 3 lbs., water to make 100 gallons. (This formula may be used in districts where lime sulfur is unsafe).

(3) "Light medium oil 1/2 to 3/4 gallons, rotenone powder 3/4 to 3 lbs. (depending on brand), water to make 100 gallons. (Used where black scale[2] or bud mite is also a problem)."

Dusts:- "In dusting for aphis, use Commercial No. 8 to No. 10, or self-mixed. The standard formula for self-mixed dust is: nicotine sulfate 2 quarts, hydrated lime 50 lbs., sulfate of ammonia 2 lbs. (This equals a No. 9 dust.) Forty to 50 lbs. of dust per acre are adequate.

"Note: Where mottle-leaf correction is desired, add 1 lb. of zinc oxide to 100 gallons of spray in formula No. 1."

In 1947 nicotine was not mentioned in the Pest Control Circular in the aphis control recommendations because it was practically unavailable. It was stated that "Rotenone powders are used with light medium oil, sometimes with light oil, at 0.5 to 1.5%, the higher dosage where other pests are present as spider, bud mite, etc. Zinc oxide for mottle-leaf may be used with this spray or cryolite for orangeworms, but it is not advisable to add both cryolite and zinc at the same time." (989) Various brands of rotenone powders were commercially available to be used at various dosages varying from 1/3 - 1/2 to 2 - 4 lbs. per 100 gallons, depending on the concentration of rotenone. Likewise a light oil containing 0.25% rotenone to be used at 1.25 to 1.5% was mentioned as being available for aphis spraying.

Non-oil spray-materials are also available containing 2.5% rotenone. These are used at 1 pint to 100 gallons, and appear to be about as effective against aphids as nicotine sulfate. HETP (hexaethyl-tetraphosphate) or TEPP (tetraethyl pyrophosphate) both as a dust and a spray, are being used to control citrus aphids. The dust should be used at 90 lbs. per acre. The liquid is used at 1/2 pint to 1 quart to 100 gallons, depending on the type of material and equipment (whether spray-duster, boom, or conventional equipment). Several lethane dusts may be used. These, being non-volatile, work by contact action only, and should be used at 90 lbs. per acre (989).

In the treatment of young trees in Florida, the tips of the infested twigs are sometimes swished about in buckets of any kind of effective aphicide, or they may be placed in bags of nicotine sulfate-lime dust and shaken to insure contact of dust with all the aphids (923).

---

[1]In the springtime lime sulfur can be used with greater safety in the interior than along the coast, and therefore higher dosages are possible (Author's Note).

[2]In the coastal regions some "off hatch" black scale can be found in the spring about the time treatments are ordinarily applied for aphids. Oil-rotenone is much more effective against black scale than straight oil, and even at the low dosages recommended it has been remarkably successful (Author's Note).

CITRUS PESTS

COTTONY CUSHION SCALE, Icerya purchasi (Maskell)

Family Margarodidae

The introduction of the cottony cushion scale into California and its subsequent impact on the development of entomology in California have been discussed by Essig (285) and Quayle (724). It is a species native to Australia and appears to have been introduced into California on acacia at Menlo Park in 1868 or 1869, and in about 10 years it was causing great damage to citrus orchards in southern California. The energetic measures taken in the fight against this extremely serious pest led to several innovations in pest control practice and legislation. The cottony cushion scale was an important stimulus, directly or indirectly in the development of the following important phases of economic entomology in California: biological control, hydrocyanic acid fumigation, oil sprays, and quarantine.

The cottony cushion scale is widely distributed throughout the citrus-growing regions of the world and occurs on many host plants. The introduction of the cottony cushion scale into Florida came about as a result of the shipping of a box of vedalia lady beetles from California to Florida with some cottony cushion scales enclosed with the beetles to serve as a source of food on the journey! The box was left near some grapefruit trees, which became infested with cottony cushion scales, and this is believed to have started the original infestation (372).

Description and Life History (372).
The Female:- The cottony cushion scale can be easily distinguished from other insects likely to be found on citrus trees. The mature females have bright orange red, yellow, or brown bodies, often partially or entirely covered by yellow or whitish wax. The most conspicuous and characteristic feature about these insects, however, is the large elongated, white egg sac (fig. 307). The sac has a cottony appearance, whence the names "cottony cushion scale" and the Spanish equivalent "escama algodonosa" are derived; likewise the egg sac is fluted (grooved), which leads to the common name, "fluted scale," often used in Australia. The egg sac becomes 2 to 2-1/2 times as long as the body of the adult female, and makes for an overall length of 10 to 15 mm. Inside these egg sacs may be found hundreds of bright red, oblong eggs. Gossard (372) found from 600 to 800 eggs per ovisac and states that he counted over 1,000 eggs from the ovisac of a single female.

Fig. 307. Adult female cottony cushion scale, showing nature of fluted egg sac. X 10. Photo by Roy J. Pence.

Within a few days, in summer, the eggs will hatch, but the incubation period may be extended to two months in the winter. The newly hatched nymphs are bright red, with darker antennae and thin brown legs. The antennae are 6-segmented. In the second nymphal instar, the eyes can no longer be seen from the dorsal aspect. The dorsum is more hairy, the hairs occurring in irregular tufts, and the waxy secretion on the back is more abundant. In the third instar the body is broadly oval in shape and reddish brown in color, but the body is largely obscured by the cottony, waxy secretion. The antennae have become 9-jointed. Immediately after the third molt the insect is freed from its secretion and is still reddish brown in color, with black legs and antennae. The antennae are now 11-jointed. The formation of the ovisac begins as soon as egg laying commences and continues as long as oviposition lasts.

The Male:- Sexual differentiation takes place in the second nymphal instar, at which time the male becomes more slender than the female and has longer and stouter legs and antennae; also the male is more hairy than the female in the corresponding instar. In the third instar the male is narrower, more elongated, and more flattened than the female and the body does not have the black hairs seen on the female.

413

The third instar nymph, prior to transformation to the adult stage, conceals itself under a bit of bark, under leaves, in the crotch of a tree, etc., or may descend to the ground and conceal itself under clods, leaves, or any suitable shelter. In these sheltered places it spins a flimsy cocoon and transforms to the adult stage. The adult male is winged, has a dark red body and dark-colored antennae. Dark whorls of light hairs extend from each segment of the antennae except for the first.

The males are rare. According to Quayle (724), Hughes-Schroder found that the California race of Icerya purchasi consists only of protandric hermaphrodites and males, and that there are no females. Parthenogenetic development does not take place, but the hermaphrodites are self-fertilizing and give rise to hermaphrodites only. Hermaphrodites that have been fertilized by males may produce hermaphrodites only or hermaphrodites along with a small proportion of males.

The Vedalia:- As previously stated, the cottony cushion scale was once a very serious pest, and in fact the very existence of the citrus industry of California was at one time threatened. Insecticides used in the control of this insect proved to be ineffective. Growers were making preparations to have their trees removed. It was known, however, that the cottony cushion scale was being held in check in Australia by its natural enemies, and Albert Koebele was commissioned by the U. S. Department of Agriculture to make the trip to Australia in 1888 to search for these natural enemies. He found several insect enemies of the cottony cushion scale, but by far the most important was the lady beetle, Rodolia (Vedalia) cardinalis (Muls.) fig. 308. Soon after the first shipment of vedalias to California, which, according to Gossard (372) consisted of only 127 beetles, it was realized that the cottony cushion scale had met its Nemesis. The California orchards were practically cleared of cottony cushion scale within 18 months. This marked the first successful introduction of a beneficial insect into any country to destroy another insect, and that is one of the reasons it has been so often referred to as the classic example of biological control of an insect pest. A number of equally spectacular successes have since been attained in the biological control of insects.

Fig. 308. Vedalia lady beetle, adults and larvae, feeding on a colony of cottony cushion scale on a lemon twig. X-3.5. Photo by Roy J. Pence.

Description and Habits of Vedalia:- The adult vedalia lady beetles are nearly hemispherical, 2.5-3.5 mm long, and are irregularly marked dorsally with red and black (fig. 303, G). Red predominates in the females and black in the males. However, the color pattern is often considerably obscured by the many fine body hairs, resulting in a grayish appearance.

The vedalia thrusts its oblong, bright red eggs under the cottony cushion scale or attaches them to the cottony egg sac. As many as 800 eggs are laid by a female. The orange-red larvae enter the egg mass from beneath and feed on the eggs and later on all stages of the host. The adults also feed on the cottony cushion scales in the same manner, but neither larvae nor adults feed on other species of insects.

Cryptochaetum (801):- A small fly known to be parasitic on the cottony cushion scale was Koebele's original objective on the trip to Australia which led to the introduction of the vedalia. This parasite, now known as Cryptochaetum iceryae (Will.), is about 1.5 mm long and of a metallic blue luster. As seen in fig. 309,

Fig. 309. Cryptochaetum iceryae.
After Smith and Compere (801).

it has a broad head and tapering body. Upon emerging from the parasitized scale, the adult pushes back a circular portion of the body wall of the latter, which usually remains attached as if on a hinge (fig. 310). This parasite is more erratic than the vedalia, but at certain times and in certain localities it is a more important check on the cottony cushion scale than its more famous rival.

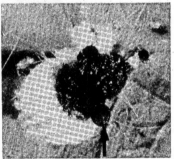

Fig. 310. Cottony cushion scale with circular portion of body wall (arrow) pushed outward by escaping parasite, Cryptochaetum iceryae. X 15. Photo by Roy J. Pence.

MEALYBUGS

The mealybugs (family Pseudococcidae), especially those of the genus Pseudococcus, are widely distributed throughout the world on citrus. They occur on a large number of hosts, but in temperate regions they are found mainly in greenhouses. They are everywhere at least potentially serious, but in most citrus regions they are rather well held in check by beneficial insects. If it were not for this fact, the mealybugs would probably be our most destructive insects, for they do not yield as readily to oil spray, fumigation, and other insecticide treatments as do the other citrus pests.

The mealybugs have soft, oval, flattened, distinctly segmented bodies, although the divisions between head, thorax, and abdomen are not distinct. They are covered with white, mealy wax, and usually with lateral waxy filaments of variable

Fig. 311. Citrophilus mealybug. Cocoons of the males. X 7. Photo by Roy J. Pence.

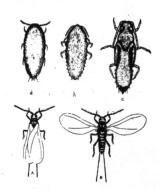

thickness and length, and anal filaments longer than those on the sides. These lateral and anal filaments are useful in the differentiation of the citrus-infesting species. Ferris' The California Species of Mealybugs should be consulted for aid in the identification of slide mounts. Also see Essig (279) for a general discussion of California species.

The female has 3 nymphal instars, then transforms to an adult. The male passes through 3 nymphal instars, then forms a flimsy cottony cocoon about 1/8 inch long (fig. 311) in which the transformation to the adult takes place. The adult male (fig. 312, 313) has a pair of wings and a pair of halteres provided with hooks. It bears two long, white, waxy anal filaments. With the exception of the male pupa, all stages of the mealybugs are motile, and although these insects are sluggish in their movements, they may often be seen moving from place to place.

Fig. 312 (at left). Male of citrus mealybug. a, nymph just emerged from egg; b, nymph immediately after cocoon is made; c, nearly mature; A, adult in normal attitude; B, adult with wings spread. After Essig. (279).

Fig. 313 (at right). Citrus mealybugs in coitu. Left, female; right, male. Courtesy R. S. Woglum.

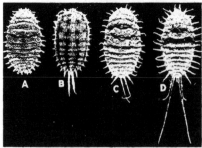

Fig. 314. The 4 species of mealybugs most common in citrus trees in California. A, Pseudococcus citri; B, P. gahani; C, P. maritimus; D, P. adonidum. X 10. Photo by Roy J. Pence.

Clausen (162) considered 4 species of mealybugs as being generally distributed on citrus trees in California. These 4 species are the citrus mealybug, Pseudococcus citri (Risso), the citrophilus mealybug, P. gahani Green, Baker's mealybug, P. maritimus (Ehr.), and the long-tailed mealybug, P. adonidum (L.) (fig. 314). The presence of occasional individuals of other species of mealybugs on citrus was considered by Clausen to be accidental. Clausen was unaware, however, of an infestation of the Japanese mealybug, Pseudococcus krauhniae (Kuwana), infesting orange trees at Ojai, Ventura County, which first came to the attention to entomologists in 1918 (799). This insect is quite similar in appearance to the citrus mealybug except that its body fluid is dark-colored. As far as is known, the Japanese mealybug has not spread beyond the originally infested area.

The citrus mealybug is the principal mealybug pest of citrus in Florida and the long-tailed mealybug can also be occasionally seen on citrus, as well as on avocados and mangos. An entomogenous fungus effectively reduces the citrus mealybug population during the period of summer rains in Florida. (926)

Among the four widely distributed species mentioned above, only one, the long-tailed mealybug, is viviparous. The other three species deposit eggs in fairly compact ovisacs, which when grouped together, form great cottony masses (fig. 315) which are the most conspicuous characteristic of mealybug infestations.

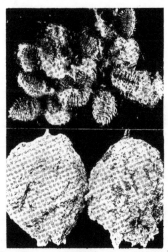

Fig. 315. Above, a colony of citrus mealybugs, show-ing cottony egg masses; X 4 (original); below, egg masses on oranges with a severe infestation (after Quayle).

The mealybugs cause injury by extracting plant sap and by excreting honey dew, which forms a medium for the growth of sooty-mold fungus (fig. 316). They may cause citrus fruits to drop when they cluster about the fruit stems, as they often do. Mealybugs are troublesome only in the more humid coastal districts.

When they are not hidden by their egg masses, many mealybugs can be quite satisfactorily determined as to species by external appearance. It should be borne in mind, however, that a number of individuals of each species may have to be examined before one can be reasonably certain as to the

Fig. 316. A citrophilus mealybug infestation on oranges. Courtesy R. S. Woglum.

identity of the species. Properly mounted species can be determined with greater exactness by means of well-defined taxonomic characters not discernable on un-mounted specimens. Fig. 317 shows certain taxonomic characters that may be util-ized in the determination of stained and properly mounted specimens of the mealy-bugs infesting citrus in California. These consist of penultimate anal lobe cerarii[1] and the ventral side of the anal lobe.

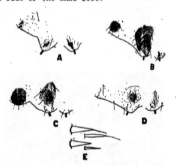

Fig. 317. Penultimate and anal lobe cerarii (left) and ventral side of anal lobe (right) of (A), Pseudococcus citri, (B), P. gahani, (C), P. adonidum and (D) P. maritimus. E, cerarian spine (left) and dorsal body seta (right) of P. kraunhiae (above) and P. citri (below). Except for these spines, the taxonomic characters shown in (A) will serve equally well for P. citri and P. kraunhiae. After Ferris (306).

As an aid in rapid field-identification of species, the following key is adapted from a more extensive key worked out by Basinger (73), so as to pertain only to citrus infesting species of mealybugs. (Also see fig. 314).

KEY TO CITRUS-INFESTING SPECIES OF MEALYBUGS

1. Anal filaments short or not more than 1/2 as long as the body .... 2.
   Anal filaments very long, usually as long as or longer than
   the body .... Pseudococcus adonidum (Linn.).
2. Wax evenly distributed over the dorsum, or, if uneven, arranged
   to form a median longitudinal stripe .... 3.
   Wax unevenly distributed over the dorsum so as to give the appearance
   of 4 longitudinal rows of darker impressed dots, the 2 median rows
   being the most conspicuous. Body fluids dark red or purple ....:...
   .... Pseudococcus gahani Green.
3. Wax usually arranged to form a definite, median, dorsal stripe when
   the purple body color shows through owing to the scarcity of wax.
   Lateral filaments short, tapering from base to tips. Anal filaments not more than 1/4 the length of the body....Pseudococcus citri (Risso).
   Wax distributed over the dorsum in a uniform thin, white, powdery
   covering. Lateral filaments short, slender and of more or less
   uniform thickness. Anal filaments 1/4 to 1/2 the length of
   the body .... Pseudococcus maritimus (Ehr.).

---

[1]A typical cerarius is said by Ferris (306) to consist of "a pair of spines.....set close to-gether at the margin of the body and usually accompanied by a more or less distinct group of tri-angular pores and slender setae."

CITRUS PESTS

CITRUS MEALYBUG

Pseudococcus citri (Risso)

In California, the citrus mealybug was once a serious pest of citrus, but now is only occasionally troublesome in widely separated sporadic attacks in the coastal citrus districts from San Diego to Santa Barbara. The decline in the importance of the citrus mealybug as a pest is due to the successful introduction of a number of natural enemies which will be discussed later. This insect has been known in the citrus orchards of Florida since 1879 and was introduced into California, via San Diego County in 1880 (724).

Description and Life History (162):- The body of the adult female (fig. 314,A) is about 3 mm long, pale yellow to brownish orange, and the lateral filaments are short, stout, and irregular in shape although they gradually lengthen posteriorly and terminate in an elongated pair of filaments often 1/4 the length of the body. The thinness of the waxy covering along a median dorsal band, allowing the darker body color to show through, gives the appearance of a brownish-gray mid-dorsal longitudinal stripe. When these mealybugs are irritated or slightly pressed, they give off a yellowish or orange liquid, in contrast to the claret-colored fluid given off by the citrophilus mealybugs.

The eggs of the citrus mealybug are deposited on the fruit, twigs and foliage, and under loose pieces of bark. The fairly compact network of waxy, interwoven fibers gives the general appearance of a light, cottony mass (fig. 315). A single individual may deposit from 300 to 587 eggs in a period ranging from 6 to 14 days in summer to considerably longer during the cooler portions of the year. The period required for hatching of the eggs likewise varies with the temperature, and ranges from 6 to 10 days to several weeks.

The newly hatched larvae are light yellow and free of wax. In summer, each of the 3 nymphal instars requires slightly over 2 weeks, and oviposition takes place approximately 2 weeks after the third and final molt. The females die as soon as oviposition is completed. Although development in the spring is quite uniform, the variation in the rate of development of different individuals causes an overlapping of succeeding generations so that during the summer and fall all stages may be found at one time. However, it has been determined that there are 2 or 3 generations a year. The winter is passed upon the tree in all stages, although preponderantly in the egg stage. Infestations, as in the case of certain other citrus pests, such as aphids and mites, are the most severe in the spring and fall months, neither the summer nor the winter being favorable.

CITROPHILUS MEALYBUG

Pseudococcus gahani Green

The citrophilus mealybug was found in a restricted locality near Upland, California, during the fall of 1913. After an unsuccessful attempt to eradicate the mealybug in this area, it spread toward the coast, where, by 1928, it was infesting 50,000 acres of citrus in Orange, Los Angeles, and Ventura Counties. It was surpassing the citrus mealybug in the severity of its attacks. The annual rearing and liberation of the beetle Cryptolaemus montrouzieri, beginning in 1916, resulted in fairly satisfactory control of the citrophilus mealybug, but in some years this pest still caused damage. Not until 1929, two years after the introduction of certain hymenopterous parasites from Australia, was the citrophilus mealybug brought under practically complete control.

Description and Life History (162):- Like the preceding species, the citrophilus mealybug is about 3 mm long. The lateral wax filaments of this species (fig. 314, B) are stout and very short, but the pair of anal filaments are long, about 1/3 the length of the body of the insect, and they gradually taper from a rather thick base. The granular waxy covering of the body is not evenly distributed. There is a scarcity of wax in four places on each segment, causing the

419

insect to appear to have four parallel, dark, impressed lines running longitudinally down its dorsum. The body fluid of the citrophilus mealybug is dark in color as contrasted to the light color of the body fluid of the other 3 species found on citrus.

Claret-colored, bead-like drops of liquid are exuded from dorsal glandular pits of this insect when it is slightly pressed or irritated. Two of these pits are close to the anterior and two are close to the posterior end of the insect. The liquid is sometimes very conspicuous, even in immature forms, and aids in identifying the live insect. (799)

The adult females used to congregate on the trunks of infested trees in large masses and deposit their eggs in irregular cottony sacs. Now they can no longer be found in such abundance in California.

Forty individuals examined by Clausen oviposited an average of 533 eggs, ranging from 394 to 679 per individual, and averaging 62 eggs per day. The females lay eggs from 7 to 14 days. The eggs are laid in an egg sac similar to that of Pseudococcus citri.

Nymphs are yellow when first hatched, but change to a deep amber in later instars. The 3 nymphal instars require 15, 13, and 12 days, respectively. After the second molt, the nymphs are similar to the adults in all but size. A. J. Basinger found that the citrophilus mealybug completes its life cycle in from 57 days in summer to 168 days in winter and spring, and that there are 4 generations per year (799). Overwintering is confined largely to intermediate stages, not many adults being found until March and early April.

BAKER'S MEALYBUG[1]

Pseudococcus maritimus (Ehrhorn)

Pseudococcus maritimus is a widely distributed species, and occurs further north than the other 3 species under consideration, attacking grapes as far north as Michigan. Its greatest importance in California is as a pest of grapes, but it also attacks walnuts, pears, and apples, in addition to citrus, and attacks a wide range of ornamental plants both in the greenhouse and outside. It is believed that there exist a number of forms of Pseudococcus maritimus and that different forms attack different host plants or groups of host plants. It is believed, for example, that different forms of this species occur on citrus, grape, and pear. A chalcid parasite, Acerophagus nototiventris, attacks the grape form of P. maritimus, but will not attack the citrus form, nor will it attack any other mealybug on citrus.

Baker's mealybug was quite well controlled by native parasites and predators even before the importation of natural enemies, and has never been a serious pest of citrus. Some years, however, this mealybug may get under the buttons of Valencia oranges and so weaken the stems that an abnormally heavy drop of fruit takes place late in the season in some orchards (976). The presence of Baker's mealybug on oranges prevents the shipment of this fruit to the Hawaiian Islands, which maintains a quarantine against this species (973).

Description and Life History (162):- Baker's mealybug while having the general mealybug shape, is somewhat more elongated than the previously described species (fig. 314, C). The body color is usually pearlish, but sometimes shading into grey. The waxy secretion is powdery in consistency, but often is so thin as to appear to be lacking. The lateral filaments are short, slender, and do not taper. The anal pair of filaments are 1/4 to 1/2 the length of the body. The body fluids given off by the adult females, when irritated, are lemon-yellow to orange.

_____
[1]The common name adopted for this species by the American Association of Economic Entomologists is "grape mealybug," but the writer is reserving this term for the grape form of the species (see p. 571).

The females of this species were found to deposit from 432 to 621 eggs in egg sacs 1/5 to 1/3 inch in length, and the egg-laying period was found to vary from 7 to 15 days. These eggs hatch in about 8 days under favorable conditions or incubation may be considerably prolonged in colder weather. The period required for the three instars averaged 16.3, 13, and 13 days, respectively. Oviposition takes place several weeks after the final molt, after considerable additional growth has taken place.

LONG-TAILED MEALYBUG

Pseudococcus adonidum (Linnaeus)

This mealybug is also a widely distributed species and occurs in many hosts in the greenhouse and out-of-doors. In California, its favored host out-of-doors is the dracaena, on which it may be found in large numbers. In California the long-tailed mealybug has never occurred in serious infestations over a large area of citrus, but has on several occasions occurred in serious outbreaks in certain limited districts. It was a serious pest of avocados in San Diego County before successful species of parasites were liberated to control it in 1941.

Description and Life History (162):- The long-tailed mealybugs, as their common name implies, have very long anal filaments, usually as long as or longer than the body (fig. 314, D). This is the character that most readily separates long-tailed mealybugs from other citrus-infesting species. The marginal filaments, however, are also longer than those of the previously discussed species. The body fluid exuded when these mealybugs are irritated is slightly yellow, but almost colorless. The body proper is about 3 mm long.

With regard to its biology also, this species is the most readily distinguished among the citrus-infesting mealybugs, for unlike the other species, the long-tailed mealybug is viviparous. The young are born alive under a thinly woven, cottony network of waxy threads which the female weaves about her body. This affords some protection for the young for a while before they start feeding. The production of young was found to take place for from 10 to 21 days. Of 15 individuals investigated an average of 206 young were produced, or 13 per day. It is thus seen that the long-tailed mealybug does not produce as many young as the previously discussed species; likewise, there is a much heavier mortality among the first-instar nymphs. It was found that the first-instar nymphs completed their development in from 10 to 20 days, the second in from 8 to 22, and the third in from 7 to 20 days, averaging 16.5, 15, and 12.5 days, respectively. Fertilization was found to take place largely during the third nymphal period, and the production of young began within 10 to 15 days after the casting of the third nymphal skin.

Biological Control

As stated before, the various artificial measures brought to bear against the mealybugs on citrus had, from a practical standpoint, been failures. Such measures as water-washing and oil spray provided temporary relief, but the mealybug population increased very rapidly after treatment. It was observed that in orchards regularly fumigated with HCN gas, the long-tailed mealybug was kept under control, but the fumigations were ineffective against the other three citrus-infesting species. Fortunately for the future of the citrus industry in the United States as well as many other parts of the world, biological control has been highly effective against the citrus-infesting mealybugs. The mealybugs happen to be especially amenable to biological control. Their soft bodies, lack of protection, slow movement, and more particularly their habit of feeding in clusters, makes them an easy prey even for entomophagous insects of relatively low searching ability. Some of the beneficial insect species which were either indigenous or had been introduced before the introduction of the mealybugs into California were destroying large numbers of mealybugs before the more effective specific enemies of mealybugs were introduced, but these were obviously unable to keep the mealybugs under commercial control. They included lady beetles, especially various species of the genus, Scymnus, and a number of lacewings, particularly Sympherobius angustus (Banks) (fig. 318), syrphid flies, and other Diptera among the predators, as well as a number of hymenopterous parasites. These were usually effective in controlling the mealybugs if ants were kept off the trees, but sometimes they were not present in sufficiently large numbers to effect a satisfactory control even when ants were not present to interfere with their activities.

Fig. 318 (at left). Cocoon and adults of the slender brown lacewing, Symperobius augustus, a mealybug predator. X 6.3. Photo by Roy J. Pence.

Fig. 319 (at right). Larvae of Cryptolaemus montrouzieri. X 1.4. Photo by Roy Pence.

The Mealybug Destroyer, Cryptolaemus montrouzieri (Muls.):- This lady beetle was first introduced into California by Albert Koebele in 1892. Koebele procured this predator under the joint auspices of the United States Department of Agriculture and California's old State Board of Horticulture.

The mealybug destroyers, more commonly known in California as "crypts," are shiny black, with the head, prothorax, tips of the elytra, and abdomen reddish (fig. 303, H). They are 3 to 3.5 mm long. The small, oval, yellow eggs are laid singly in the cottony egg sac of the mealybug or in the vicinity of the mealybug clusters. The larvae (fig. 319) are over twice as long as the adults, being from 7 to 10 mm in length. They are covered with a white, cottony secretion of wax, including long waxy filaments, which causes them to be confused, especially before they have reached full size, with the mealybugs among which they are found (fig. 238). The arrangement of the wax filaments, however, is much different from that of any citrus-infesting mealybugs.

The larvae are depended upon to do the major part of the control work, despite the fact that the adult beetles feed upon mealybugs almost exclusively. Smith and Armitage (799) list the following factors as favoring the use of "crypts" for the biological control of mealybugs: (1) they have no known parasites, (2) they do not tend to disperse widely when colonized, (3) they are easily propagated under insectary conditions, and (4) they feed on all the important citrus infesting mealybugs.

Artificial Production and Distribution of "Crypts":- After their introduction into California, the "crypts" became widely distributed throughout the regions where mealybugs are found, but they subsequently disappeared from all citrus sections except for San Diego and Santa Barbara Counties. It soon became apparent that in order that the "crypts" might occur on the citrus trees in sufficient numbers to control the mealybugs, they would have to be reared and distributed by man. A practical method of rearing "crypts" was immediately indicated when it was found by H. S. Smith that mealybugs develop readily on potato sprouts (fig. 320). The artificial mass production of "crypts" in insectaries is described in detail by Smith and Armitage (799). Some of these insectaries are under the supervision of the county agriculture commissioners' offices and some are privately operated.

Idaho russet potato tubers are planted in trays the standard dimensions of which are 16 x 18 x 4 inches. They are planted in a soil medium composed of 4 parts of light sandy loam and 1 part of screened dairy manure. This is covered with a 1/2 inch layer of coarse plaster sand. Fifteen to 18 potatoes are planted in a tray. The trays are placed on racks in insectaries where 360 trays are stored in a standard room. They are kept in subdued light so as to promote vertical growth and to limit the formation of chlorophyll which inhibits the settling of the mealybug nymphs. A temperature averaging 65°F and a humidity of about 70% is maintained. The sprouts are watered every 10 days except for the top trays and the trays exposed to the ventilators, to which water must be added every 5 days.

Fig. 320. Potato sprouts infested with mealybugs which are reared as food for Cryptolaemus. Photo by R. S. Woglum.

As food for the "crypts," the citrus mealybug, Pseudococcus citri, has been used because it does not have as great a tendency to wander about as other mealybug species, it has a shorter life cycle than citrophilus, and it infests the potato sprouts very heavily.

For the infestation of the sprouts with mealybugs, portions of sunflower or a variety of mallow grown for the purpose are placed on trays in which mealybugs are hatching. The mealybugs infest these hosts, and the latter are then placed on trays of potato sprouts which are to be infested. The mealybugs transfer to the potato sprouts as their temporary hosts begin to wilt. During this period the temperature is kept at 80° F with a relative humidity of 70%.

The "crypts" are introduced not later than two weeks after the date the potato sprouts are infested with mealybugs. Twenty-five adult "crypts" are required to "sting" each tray of infested potato sprouts. The beetles are allowed to remain on the sprouts about 18 days, after which they are removed, mixed with other beetles, and used to infest other sprouts. The beetle larvae pupate after attaching themselves to strips of burlap sacks used for that purpose. The adults, as they develop are attracted to the light of the cloth-covered windows of the insectaries, from which they are collected by hand. The average minimum production of "crypts" per tray is 400, although often 700 to 1,000 adults are reared per tray.

When collected, the "crypts" are placed in small gelatin capsules, 10 to each capsule. The beetles are scooped directly into the capsule, and an average of 2,000 beetles per hour can be collected by one man. It is possible for the beetles to live in these capsules for 5 to 7 days with little or no mortality. However, they are generally liberated in the orchards in from 48 to 72 hours after being collected. The beetles are liberated in the orchards by the grower, after written permission from the agricultural inspector of the district, or they may be liberated by the insectary officials. The contents of a capsule (10 beetles) are thrown through an opening into the center of the tree. The beetles begin to fly immediately after leaving the capsule and land on the inside branches. One man can liberate as many beetles per hour as two men can collect.

There were at one time 16 insectaries in which "crypts" were being reared (799), but the number of insectaries in operation gradually diminished as parasites were later introduced which did not require artificial rearing; also the remaining insectaries cut down on their production of "crypts" and are now sometimes rearing other beneficial insects besides the "crypts". Nevertheless, the latter for many years played an important rôle in the battle against what were at the time the most important insect enemies of citrus, and many are still being reared to control

occasional outbreaks of mealybugs which are not being controlled sufficiently
rapidly by existing natural enemies.

The Sicilian Mealybug Parasite, Leptomastidea abnormis (Gir.):- The mass production of Cryptolaemus was carried on mainly because of the seriousness of the
citrophilus mealybug which had taken the place of the citrus mealybug as Enemy
Number One of citrus. The decrease in the importance of the latter species was
largely due to the introduction in 1914 of Leptomastidea abnormis from Sicily and
its establishment in California by the California State Department of Agriculture
(791). This hymenopterous parasite spread rapidly, aided by artificial propagation
and distribution by the State Insectary. One shipment of these parasites was sent
to Florida. Soon it was practically impossible to find a colony of citrus mealybugs which was free from attack by the Sicilian mealybug parasite. Unfortunately,
it did not attack other mealybug species in California.

The Sicilian mealybug parasite can quite readily be distinguished from other
mealybug parasites. It holds its characteristically banded wings aloft in a
peculiar position. The insect is 0.75 to 1.50 mm long and is of a golden or dusky
yellow color.

Hymenopterous Parasites Effective Against the Citrophilus Mealybug:- The
Sicilian mealybug parasite and other natural enemies had successfully maintained a
natural control of the citrus mealybug and Cryptolaemus was fairly satisfactory in
the control of the citrophilus mealybug. The rearing of the beetles was an expensive process, however, and in addition control was not always completely satisfactory, for sometimes the beetles did not increase sufficiently rapidly in the
orchards in which they were liberated to prevent damage. The number of orchards
in which control by the liberation of "crypts" was unsatisfactory increased each
year. Despite the biological control in effect at the time, the citrophilus mealybug was considered one of the major pests of citrus. It was also an important pest
of ornamentals and in the same localities caused serious damage to deciduous fruits,
particularly to pears and apples.

It was believed that the biological control of the citrophilus mealybug could
be further improved by the introduction of more natural enemies of this species,
especially hymenopterous parasites. The conclusion that the citrophilus mealybug
must have originated in some country in the Pacific maintaining direct steamship
communication with San Francisco, plus the inability of investigators to find it on
the Asiatic mainland or in Japan, Formosa and the Philippines, led to the conclusion that this species, and its natural enemies, might be found in Australia, New
Zealand, or some of the South Pacific Islands. Accordingly, Harold Compere de-
parted on August 27, 1927, and soon after his arrival in Sydney, Australia, he
found the citrophilus mealybug sparsely distributed on choisya and oleander. He
found a number of hymenopterous parasites attacking the mealybugs, and two of
these, Coccophagus gurneyi Compere and Tetracnemus pretiosus Timberlake were
destined to become, along with the vedalia lady beetle, classics in the history of
biological control.

The parasites found by Compere were immediately successfully introduced into
California, and in the summer of 1928, Coccophagus and Tetracnemus became widely
disseminated by natural means (183). In the spring of 1929 an appreciable reduction of mealybugs in the districts where the parasites had been thoroughly established was noticed. In 1929 the work of the parasites was aided by production and
distribution in certain insectaries. By the spring of 1930 the parasites had become established throughout the entire acreage in southern California infested with
citrophilus mealybugs. The "peak-hatch" of mealybugs has been kept under commercial control and, in fact, it is difficult to find specimens for classroom demonstration.

The descriptions and biologies of the natural enemies of the citrophilus
mealybug shipped to California by Compere in 1928 are contained in a report by
Compere and Smith (183). This paper also contains the story of the search for the
parasites, their introduction into California, some interesting theory on biologi-

cal interrelationships between hosts and parasites, and the story of the spectacular demise of the citrophilus mealybug in California. This paper is recommended for entertaining as well as informative reading.

Coccophagus gurneyi Compere

The female of this parasite (fig. 239) averages about 1 mm long and is colored black, with a conspicuous band of yellow across the base of the abdomen by which it can be quite readily distinguished. The body of the male is entirely black and consequently is more difficult to distinguish from other parasites (179).

The females were observed to be long lived and to continue oviposition over a considerable period. Normally only one egg is deposited in a mealybug, but the females do not distinguish between parasitized and unparasitized hosts and as a consequence several females may oviposit on the same mealybug. In summer the eggs hatch in about 4 days and approximately 27 days are required for development from the egg to the adult stage. The body of a small or partly grown mealybug is completely filled by the fully mature larva of Coccophagus, but a mature mealybug is not fully occupied. The mummified bodies of the dead mealybugs are usually grayish or fuscous (dark brown, approaching black) owing to the dark colored pupal remains of the parasite seen through the derm. Near the posterior end may be seen the exit holes through which the parasites issue. The mummified mealybugs may be seen most abundantly in concealed places such as under tree bands, in dried leaves or under loose bark, which seems to indicate that parasitized mealybugs seek concealment.

Tetracnemus pretiosus Timberlake

This parasite (fig. 239) is 0.9 to 1.5 mm long. It can be distinguished from C. gurneyi by the absence of a yellow band at the base of the abdomen and by its greater activity. The males have branched antennae (889).

The adults are relatively short lived, and deposit their eggs within a few days after emergence. Oviposition is very rapid, compared to C. gurneyi. The eggs are extremely small (0.03 mm long). A period of 23 days was found to elapse between oviposition and the beginning of the emergence of adults of one particular group of these parasites. As in the case of C. gurneyi, only one parasite issues from a mealybug even though several eggs may have been laid, the supernumerary eggs and larvae being destroyed. The "mummies" of mealybugs parasitized by Tetracnemus are paler in color than those parasitized by Coccophagus.

The Relative Value of Coccophagus and Tetracnemus:- It is believed that Coccophagus and Tetracnemus working together are more effective than either would be alone (183). Coccophagus happens to be the more effective of the two parasites in California. If both are confined in a cage with an abundance of citrophilus mealybugs, Tetracnemus gradually becomes eliminated. However, in the field, when the host population becomes low, which it usually is, overlapping of parasitism which results when the host is abundant, no longer occurs, and the effect of one parasite on the other is reduced to a minimum. Under such conditions, since the two species have slightly different habits and habitats, each species destroys some host individuals which would not have been destroyed by the other species.

Tetracnemus is very scarce during the winter months, while Coccophagus is abundant and active. Thus the latter species is very important for the purpose of destroying the overwintering mealybugs which start the incipient spring infestations which were so troublesome in past years. During the summer, however, Tetracnemus destroys many mealybugs. Since it oviposits very quickly, ants have less opportunity to interfere with its activities, and dissections have shown that the ratio of mealybugs destroyed by Tetracnemus, as compared to those destroyed by Coccophagus increases with increasing ant infestation.

Parasites of the Long-Tailed Mealybug:- An outbreak of long-tailed mealybugs on citrus near Downey, California, in 1933 resulted in the importation of Anarhopus sydneyensis Timb. (fig. 321) from Australia in 1933, Tetracnemus peregrinus Comp.

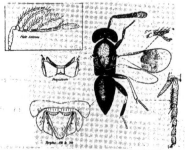

Fig. 321. Anarhopus sydneyensis.
After Compere and Flanders (181).

in the winter of 1934-35, Leptomastix dactylopii How. in 1934, and Anarhopus
fusciventrus (Gir.) in 1936. Likewise several other species of parasites already
existing in California were distributed among long-tailed mealybug infestations.
A. sydneyensis and T. peregrinus, especially the former, proved to be effective in
keeping the long-tailed mealybug under control (319).

    Long-tailed mealybugs still remained a pest on avocados in parts of San Diego
County. Cryptolaemus proved to be ineffective against mealybug species, like
Pseudococcus adonidum , which do not produce egg masses. In the spring of 1941,
the San Diego County Agricultural Commissioner's Office obtained parasitized long-
tailed mealybugs from various areas where parasites were known to be effective and
succeeded in establishing A. sydneyensis and T. peregrinus in the avocado-growing
areas inhabited by the mealybugs. In a year and a half after these parasites were
liberated, it was difficult to find unparasitized mealybugs in avocado orchards
(327). Today the long-tailed mealybug is not a pest of avocados except on scions
of grafted trees, where even an incipient infestation of mealybugs can be harmful
before the parasites have a chance to do their work.

    In 1943 there occurred another outbreak of the long-tailed mealybug on citrus,
this time in the Anaheim area of Orange County. The infested area gradually increased until in 1945 about 1,000 acres were infested to an economically important
degree. During the same period this species increased to a similar extent in
Ventura County, also on citrus. In 1946, entomologists from the Citrus Experiment
made a study of the natural enemies of the long-tailed mealybug and presented the
following data as to the relative numbers of the various species attacking this
pest (222).

    Parasites: Anarhopus sydneyensis, 85%; Tetracnemus pretiosus, 9%; Coccopha-
gus gurneyi, 4%; and T. peregrinus, 1%.

    Predators: The California brown lacewing, Sympherobius californicus, 74%;
Cryptolaemus montrouzieri, 16%; and the California green lacewing, Chrysopa californica, 10%.

    It was found that the mealybug populations were building up early in the
spring, but were reduced to low levels by their natural enemies by June or July.
The predators were given principal credit for checking the mealybugs, and experiments are now projected for the study of the effects of mass liberations of larvae
and adults of the brown lacewing, "crypts" and the green lacewing. The Orange
County Agricultural Commissioner's Office plans to cooperate in this work. (222)

CITRUS PESTS

THE UNARMORED SCALE INSECTS

Family Coccidae

This family of scale insects is represented by a number of species on citrus wherever it is grown, but apparently because of generally effective parasitization, soft scales are almost everywhere far less important as citrus pests than the armored scales.

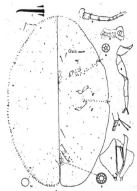

Fig. 322. Dorsal (left) and ventral (right) aspects of the adult female of the soft scale, Coccus hesperidum. 1, Stigmatic setae; 2, antenna; 3, spiracle; 4, quinquelocular pore; 5, leg; 6, claw; 7, anal plates; 8, multilocular pore; 9, general outline; 10, simple pore. After Steinweden (845).

The soft scales, as their name implies, do not possess the scale covering or "armor", separate from the body of the insect, which is secreted by the armored scales. Their bodies are relatively soft in the younger instars, but the mature females may possess a derm which, together with its waxy secretions, may be of considerable toughness, as for example in the case of the black scale, Saissetia oleae Bern. The Coccidae (fig. 322) are oval to circular, flat to hemispherical, or nearly globular in form; segmentation is indistinct, antennae and legs are usually present after the first instar as contrasted to the armored scales, in which they are absent; the posterior extremity is characterized by an anal cleft and a pair of dorsal triangular anal plates anterior to which is an anal ring with 6 to 8 large setae. The soft scales are more or less active in all stages except that some are sessile in the adult stage. As in the related cottony-cushion scales and mealybugs, the males differ greatly in structure from the females after the first molt. They usually possess wings, whereas the females are always wingless. Also like the previously mentioned Homoptera, the soft scales excrete "honey dew" which supports the growth of the sooty mold fungus.

In California the only soft scales of more or less importance are the black scale, citricola scale, and the soft scale. In Florida the soft scale is about the only unarmored scale found on citrus, but is so heavily parasitized that it seldom causes injury. The black scale rarely attacks citrus in Florida although it is abundant on other hosts. This situation may also be observed in northern California and in some other countries, as for example, Mexico, Chile, and South Africa. Apparently under somewhat adverse conditions the black scale can thrive on certain favored hosts, such as the olive, but not on citrus. Three species of wax scale (Ceroplastes) occasionally occur on citrus in Florida, but never become injurious. In Texas the soft scales are not even of minor importance on citrus.

427

The black scale occurs in nearly all citrus regions of the world. It was for many years one of the major citrus pests in California, where it was found abundantly before 1880 (724). Essig stated in 1926 that the black scale "is by far the most important member of the entire order from an economic viewpoint, being responsible for losses of more than two million dollars annually to fruit growers in California alone" (282). Smith and Compere, in 1928, stated that the black scale at the time was "generally recognized in California as the pest of first importance in both the citrus and the olive industry" (802).

The black scale was the most widespread of the scale insects and usually demanded annual treatment with HCN fumigation or oil spray, although in the majority of orchards the same treatments were needed anyway for red scale. The black scale occurs on an enormous number of host plants. Some of these hosts, notably the pepper tree, Schinus molle, once constituted a menace to citrus orchards because they served as a source of reinfestation after the trees in the orchard had been treated for black scale This was true, however, only as long as there was no effective black scale parasite in California. Now that we have the parasite Metaphycus helvolus, the presence near the orchard of such hosts as the pepper tree, oleander and nightshade are believed to be advantageous. On the nightshade and the vigorous new shoots low down on the pepper tree and the oleander, the black scale have several broods per year, even in interior areas. M. helvolus can find black scales in the right stage of development for parasitization during periods of the year when in the citrus orchard suitable hosts cannot be found. Border rows of pepper or oleander around citrus orchards thus serve as a source of parasites for the orchard under conditions in which they would not survive if there were nothing but orange trees in the vicinity. Thus the parasites are present to oviposit on black scale on the citrus trees when the scale is in the stage of development permitting parasitization.

Some growers have in recent years planted oleanders as borders for their citrus orchards to insure the continuance of a parasite population in the orchards throughout the year. In such cases the oleanders are called "foster hosts". (333) Likewise investigators at the Citrus Experiment Station are experimentally "dusting" black scale crawlers onto the trees at appropriate periods of the year to bring about the "uneven hatch" conditions favorable to parasitization by helvolus.

The black scale has a large number of natural enemies, but all these combined seldom sufficed to bring the black scale under control until the introduction of Metaphycus helvolus (Compere) from Africa by the University of California Citrus Experiment Station in 1937 started the black scale on its decline to its present status as a minor pest. The potentialities of M. helvolus as a parasite of the black scale were first noticed in a few orchards in 1939 and by 1941 the black scale population had been reduced to an unprecedented low ebb nearly everywhere (797). To this day only occasional orchards have needed treatment for black scale

Description of Adults:- The full-grown females are 3 to 5 mm long, brown or black in color, nearly circular, hemispherical, with a very tough derm, and with two transverse and one longitudinal ridge forming a letter "H", on the dorsum of the majority of individuals (fig. 323). The males are usually very scarce even when the females are abundant. They have the usual two wings of the coccid males, are about 1 mm long and honey yellow in color (709).

Life History (709). The Egg:- The eggs are oval, 0.3 mm long, and pearly white, changing to cream or pinkish. A few days before hatching they assume a reddish-orange hue and the eye spots of the embryo may be discerned. The females will lay on the average about 2,000 eggs. In late spring and early summer these may be seen by turning the black scale over and looking inside the "capsule" (fig. 324).

Fig. 323 (at left). Black scale, Saissetia oleae, on avocado twig. Above, X 2; below, X 0.7. Photo by Roy J. Pence.

Fig. 324 (at right). Two female black scales turned over. Left, eggs removed; right, scale "capsule" filled with eggs. X 8. Photo by Roy J. Pence.

Incubation may require as little as 16 days, or may in the case of second brood scale in the coastal areas, be prolonged for a month or six weeks in winter. Nearly all the eggs hatch, but a large percentage of the "crawlers" fail to settle on the tree and a great mortality of the young scales after they have settled may occur during periods of hot, dry weather.

The First-Instar Nymph:- The "crawlers" are 0.34 mm long, light brown; with black eyes, and six-segmented antennae. The body is flat and oval. There exists an arch at the posterior tip of the black scale from which the crawlers issue. They travel about considerably before settling on the leaves or green twigs. It is during this period that the black scale can be widely disseminated by wind, birds and insects, and by man and his equipment. This is also true of all other scale insects. Quayle (714) proved that an orchard apparently completely free of black scale can be reinfested in a single season if it is so situated that the prevailing winds can carry crawlers from an adjoining infested orchard.

The nymphs double in size before the first molt. In summer the first molt occurs in from 4 to 6 weeks after birth, but may be prolonged to 2 months or more in winter in the coastal "second brood areas".

The Second-Instar Nymph:- Except for its greater size, the second instar black scale appears much the same as the first instar. A longitudinal ridge begins to take shape, however, along the median line of the dorsum. A gradual obliteration of the two ends of this ridge leaves a central portion which later forms the bar of the letter "H" seen on older individuals. The second molt occurs on an average of about 2-1/2 to 3 months after the birth of the scales, while in the winter this period is, of course, much longer. The molting process is similar to that of other unarmored scales. The old skin is split at the anterior end and pushed backward until it is free from the insect. The integument of the antennae, beak, and legs are sloughed off with the molted skin. After the second molt the scale has reached the adult stage.

Migration from Leaves to Twigs:- After they have once settled, the scales remain stationary until some stimulus induces them to migrate. This stimulus may be provided at any time by merely detaching a leaf, causing it to become withered. The nymphs will wander off in search of a more suitable substance on which to continue development. This is true of all the unarmored scales, and provides a means by which the investigator can transfer them from one host to another. Under ordinary field conditions, however, the black scales will leave the foliage and migrate to the leaves sometime in the fall or winter months. Probably the majority migrate to the twigs and branches after the second molt. This migration is apparently a provision for obtaining a permanent food supply, for in the case of deciduous trees, all the scales would be destroyed if they remained on the leaves. In the case of citrus trees only a portion of the leaves fall each year, but the migrating instinct still prevails.

The Adult Female:- After the second molt the female increases rapidly in size and changes its shape, becoming nearly circular and hemispherical. The letter "H" becomes distinctly outlined on the dorsum. As the egg laying stage is approached, many of the scales become a dark mottled gray. The insects have then reached the "rubber stage". This term is often seen in pest control literature because it denotes the period when ordinary pest control treatments for black scale are no longer effective. Black scale should be treated before the "rubber stage" occurs.

When egg laying begins, the scales become more leathery, acquire a smoother surface and also become much darker in color, finally becoming a dark brown or black. The size of the adults is quite variable; they are larger on young, vigorous trees than on old trees and are larger on succulent twigs than on older twigs. They vary also with the host plant, being, for example, larger on citrus than on olive. They are also smaller when crowded against one another than when they have plenty of space.

As oviposition takes place, the body of the female, at first distended with its huge egg content, gradually recedes toward the roof of the "capsule" formed by the thickened derm. In the spring one may examine the venter of the scale and see at first only a mass of pinkish eggs (fig. 324, right), or eggs and hatched nymphs. These can be shaken out by tapping the "capsule" and the venter of the female body may then be seen flattened against the roof of the capsule (fig. 324, left). The appendages, including the considerably protruded vagina, can be seen with the aid of a good hand lens. It will be seen that while the female has retained its antennae and legs, these appendages have developed very little after the first instar, and have become nonfunctional after the migration away from the leaves. The legs and antennae of the black scale, as well as other soft scales, can be seen from the dorsal aspect only during the crawler stage; later the lateral margins of the scale greatly overreach their extremities (fig. 324).

Development of the Male:- No difference between the sexes can be noted until after the first molt. Then the male becomes more elongate, its length being 1.5 mm and its width 0.64 mm. Near the end of the second instar the eyes, which the females never possess, become visible as small dark areas on the front margin of the body. About 4 weeks are spent in the second instar, at the termination of which a puparium is formed, which completely covers the insect, although it is not readily discernible because of its transparency.

The next stage of development is called the prepupa, and this stage lasts from 5 to 8 days in warm weather. The color is light brown, the eyes being dark red or brown. On the ventral side the antennae and legs are plainly visible in close contact with the body.

The next stage, the so-called pupa, is 1.2 mm long and 0.4 mm wide. It is colored the same as the prepupa except that it has a larger amount of pigment at the anterior end and has an entirely red head with black eyes. The antennae, legs and wing pads have become more conspicuous and, while they lie closely oppressed to the body, they are distinct from it and are readily lifted away. Eight to 12 days are spent in the pupal stage, but the fully developed male remains beneath the puparium for 1 to 3 days before emerging. Its long, white caudal filaments can be seen extending out beyond the posterior tip of the puparium.

Seasonal History of the Black Scale (709):- Except in certain regions close to the coast, the black scale has only one generation per year. Most of the eggs are deposited by midsummer and the majority of the crawlers have already emerged from under the "shell" or "capsule" of the adult female by the middle of July. While in the "single-brood" areas the main egg-laying period of the black scale is during May and June, the limits of egg laying will be from April to September, inclusive. Near the coast in the "double-brood" area, egg-laying may occur again in the late fall or early winter months, and in this area there may be two generations a year, or at least a partial second generation. The "single brood" areas are also called areas of "normal hatch", and the "double brood" areas are called areas of "irregular hatch" or "off hatch". In pest control recommendations, for instance, one often reads the terms "normal hatch black scale" or "off hatch black scale".

CITRUS PESTS

<u>Injury</u>:- In common with other sucking insects, the black scale withdraws large amounts of sap from the tree and probably causes injury in this way. Defoliation, fruit drop and dead wood result from severe infestations. The sooty mold fungus which develops on black scale-infested trees must be removed from the fruit by washing. It is believed by some that the sooty mold fungus may also reduce photosynthesis by reducing the amount of light reaching the leaf surface.

Biological Control

<u>Predators</u>:- In contrast to insects with a colonial type of distribution, such as the cottony-cushion scales and the mealybugs, the black scales, which are widely scattered over the tree and rather uniformly distributed, are not as amenable to control by predators. The most effective stage of the predators, the larval stage, does not have the searching ability of the winged parasites.

<u>Rhizobius ventralis</u> (Ehr.) (fig. 303, C) was found to be the lady beetle which most commonly feeds on black scale, feeding on the eggs and newly hatched young before they have come out from under the "shell" of the adult and also on the younger scales after they have settled on the tree. The adult is 3 mm long, broadly oval, and covered with gray hairs. The larva is 5 or 6 mm long, the dorsal surface black and the ventral surface gray, and with dark-colored legs. (709)

Other coccinellids, chrysopids and syrphids will feed on black scale, but are primarily predators of other species of insects.

<u>Parasites</u>:- The black scale is attacked by many parasites, the majority of which were introduced from Australia or Africa, but only after the introduction of <u>Metaphycus helvolus</u> could good control of black scale over a large area be attributed to natural enemies. The majority of the black scale parasites were described and discussed by Smith and Compere (802), and Compere (180). The species which will be discussed are all in the order Hymenoptera, and include only species likely to be seen by the casual observer and which destroy large numbers of black scale.

<u>Scutellista cyanea</u> Mots. is a common parasite species in California and in many other citrus regions. It is an African species which was introduced into California via Australia in 1902. In a sense it may be considered to be a predator, for its larvae feed on the eggs of the black scale deposited in the brood chamber under the scale body.

The adult <u>Scutellista</u> can often be seen walking slowly about among black scales. It is a "hump-backed", blue insect with light brown antennae (fig. 325). The exit holes in the scale shells are larger than those of other parasites. Usually there is only one exit hole, but sometimes 2 or 3 or rarely 4. The white, grub-like larva may often be seen feeding on the black scale eggs.

Quayle (709) found the percentage of scales parasitized by <u>Scutellista cyanea</u> to be as high as 75%. While <u>S. cyanea</u> undoubtedly destroys many scales, a fundamental weakness of the parasite, as pointed out by Quayle, is that it does not consume all the eggs within a black scale "capsule", and as long as even a

Fig. 325. <u>Scutellista cyanea</u> X 18. Photo by Roy J. Pence.

small percentage of the eggs are allowed to hatch, the end result may not be much different than if all the eggs had hatched, for only a limited number of scales, far below the reproductive capacity of the black scale, can become established and develop within a given area of the surface. Quayle counted as many as 400 to 700 young scales on each of the leaves arising from twigs that contained adults 75% of which had been parasitized.

<u>Metaphycus lounsburyi</u> (Howard):- This internal parasite was sent from Africa to the State Insectary and was reared and distributed under the direction of

431

H. S. Smith. In 1919 the first colonizations were made, one in Santa Paula, Ventura County, typical of "uneven-hatch" areas; and one in Alhambra, Los Angeles County, in an "even-hatch" district. In both areas the parasites proved to be very successful when first introduced, and their success led to further colonizations. In certain uneven-hatch areas the black scale population was so reduced that fumigation was discontinued.

After its initial successes, Metaphycus lounsburyi became a relatively minor factor as a biological control measure for the black scale. M. lounsburyi oviposits only in the "rubber stage" scales, and it is only a matter of time until only full-grown scales with eggs or very small scales remain, resulting in an "even-hatch". Since under such circumstances there are prolonged periods when no "rubber stage" scales are present, the parasite populations are greatly reduced. M. lounsburyi thus became ineffective after an initial period of considerable reduction in the black scale population. However, M. lounsburyi has continued to control black scale on pepper trees, where the continuous production of offshoots, favorable to black scale development, results in an "uneven-hatch" condition despite the continued presence of the parasite (800).

A hyperparasite, Quaylea whittieri (Gir.), which parasitizes M. lounsburyi, was intentionally introduced from Australia in the year 1900 in the belief that it was a primary parasite of the black scale. During the period when the black scale was abundant and the primary parasite population was high, this hyperparasite resulted in a considerable reduction in M. lounsburyi, but after M. lounsburyi reduced the scale population, thereby reducing its own population, Q. whittieri was no longer an important factor. Q. whittieri is evidently not able to exist in destructive numbers when the population density of Metaphycus is relatively low, and this being the case, the hyperparasite could not be of any great importance.

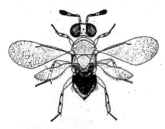

Metaphycus lounsburyi is from 0.7 to 1.4 mm long, with the dorsum lemon or orange-yellow in color, with distinct black markings (fig. 326). The venter is uniformly light gray. The duration of the life cycle is from 2 weeks to several months. Another parasite, M. stanleyi Compere is similar to M. lounsburyi in general appearance, but is distinguished by the darker venter of the abdomen, which contrasts sharply with the light-colored venter of the thorax.

Fig. 326. Metaphycus lounsburyi, parasite of black scale. After Smith and Compere (802).

Metaphycus helvolus (Compere):- This parasite (fig. 240) is one of 28 species shipped to California from South Africa by Harold Compere in 1937. In South Africa it was not considered to be an important parasite because on black scale attended by many ants, which affect this species more than they do other black scale parasites, M. helvolus was rare (323). This is because of the fact that M. helvolus oviposits slowly and feeds for a time on the fluids issuing from the wound made by its ovipositer. It is therefore more interfered with by ants than rapidly moving and rapidly ovipositing species.

The field evidence of the successful work of M. helvolus in 1939 stimulated considerable mass production of the parasite by county and other cooperating insectaries after propagation methods had first been worked out at the insectary of the Citrus Experiment Station. This aided in the rapid dispersal of the parasite, with the excellent results which have already been described. M. helvolus is now keeping the black scale population down to below the point of economic injury in all but a small percentage of the orchards in some of the "even-hatch" areas. In addition to this good control of the black scale, M. helvolus has brought about an almost complete annihilation of Saissetia nigra, once the most important pest of ornamental shrubs and vines in the Los Angeles area. It also attacks the European fruit lecanium, Lecanium corni, the citricola scale, Coccus pseudomagnoliarum, and the soft brown scale, C. hesperidum.

Description and Life History:- Metaphycus helvolus is small, the females being 1 mm long and the males somewhat smaller. The females are orange-yellow in color and the males are dark brown (323).

Usually only one egg is deposited in a host, and it is deposited near either the anterior or posterior margins of the host after the ovipositor has pierced the dorsum. M. helvolus attacks black scale nymphs varying from 0.6 mm to 2.5 mm in length. The nymphs when they are attacked may thus be in the first, second, or early third instars. The method of oviposition employed by M. helvolus, and peculiar to many hymenopterous parasites, is described by Flanders (321) as follows:

"On finding a suitable scale, the female places the tip of its sting against the top surface of the scale and drills through it. Thereupon the female explores with its flexible sting the body cavity of the scale. Finding the place suitable as a home for its offspring, the female deposits an egg within the scale hypodermically by means of the sting. This is done by forcing the egg into the sting so that it flows through like sand through an hour glass. Only in this manner is oviposition possible, since the diameter of the sting is much smaller than the diameter of the egg."

An average of about 400 eggs are deposited per female. The latter are very long lived. Their long life and huge egg laying capacity are due in part to their habit of consuming, between egg depositions, the body fluids of their hosts.

The factors which result in Metaphycus helvolus being a successful parasite may be summarized as follows:

1. It has a short life cycle (about 15 days). It may have as many as 8 generations to one of its host.
2. It has a long egg laying period because of its habit of feeding on host fluids.
3. It deposits only one egg per host and thus spends a larger percentage of its time in search of hosts than some other species of parasites.
4. None of the scales parasitized by M. helvolus produce offspring.
5. The adult destroys many host insects by feeding on their body fluids.
6. Because of its high searching ability it is effective at low host densities.
7. The host density is ordinarily kept very low and the parasites become so scattered that they are not easily found by hyperparasites.

Other Parasites:- There are over 80 parasites of the black scale. The genus Coccophagus is especially well represented among these parasites. The species discussed above, however, especially Metaphycus helvolus, account for practically all the parasitization of black scale now taking place in California. Metaphycus stanleyi is now considered to be next to M. helvolus in importance as a black scale parasite.

Artificial Control

Resistance to HCN:- Difficulty in the control of black scale by means of HCN fumigation near Charter Oak, Los Angeles County, was first called to the attention of Professor H. J. Quayle in 1915 (718). Woglum (960) pointed out that an 18 to 20 cc fumigation schedule was failing to give a good control of black scale over an extended area in the Charter Oak district where once a 13-1/2 cc ("75 per cent") schedule resulted in effective control. Black scale resistance was further experimentally corroborated by Gray and Kirkpatrick (381). The "resistant area", in the case of black scale, is relatively limited compared to the extent of the areas in which red scale has become resistant.

Fumigation:- In areas in which the black scale is known to be resistant to HCN gas, oil spray is generally recommended rather than fumigation, but early

433

fumigation, as soon as the hatch is complete, may be quite effective. If possible without undue injury to trees, a 20 cc schedule should be used. In areas in which the black scale has not developed resistance, the dosage of gas will depend on the experience of past years in the particular area involved. The tendency is to use as high a dosage of gas as can be used without undue risk of injury to the trees. This will vary from a 16 cc to 22 cc schedule, depending on the severity of the infestation and on the previous experience in the district, or even the particular orchard in question.

Since the introduction of _Metaphycus helvolus_, the number of orchards treated for black scale has decreased to a small fraction of the number treated in former years. For example, the Los Angeles County Agricultural Commissioner's Office reported that as early as 1941 only 24 orchards out of the 5239 citrus orchards in the County had heavy infestations of black scale. Even before the introduction of M. helvolus, however, the majority of citrus orchards were fumigated primarily for red scale even when black scale was present. Dosages adequate for a good kill of red scales are even more effective against the black scale if the treatment is properly timed for the latter. The black scale is in such cases the species determining the date of treatment. While in the case of the black scale there are periods of the year when results are definitely unsatisfactory, in the case of the red scale the date of fumigation makes little difference, although somewhat better kill is obtained in winter and a higher dosage may be safely used at that season. However, at this period of the year the black scale in the "even-hatch" areas has long since reached a developmental stage not susceptible to treatment either with fumigation or oil spray.

The date of treatment of an orchard in which black scale is either the primary pest or the pest which determines the date of treatment, will depend on whether the orchard is in an area of "even-hatch" or an area of "uneven-hatch". In areas of "even-hatch", fumigation begins about the first of August and may continue through September with good results. The best results are obtained when fumigation is done as soon as possible after the last eggs have hatched, but before the oldest nymphs have reached the "rubber stage". In "uneven-hatch" ("double-brooded") areas the black scale begin to hatch sooner than in the "even-hatch" areas and are too far advanced in development ("out of condition") for satisfactory treatment by fumigation 2 or 3 weeks before they are out of condition in the "even-hatch" areas. In the "uneven-hatch" areas, however, the following generation of black scale are ready to treat in late November or early December.

Oil Spray:- If an oil spray is thoroughly applied and properly timed it is very effective in controlling black scale. A light medium oil is used: emulsions at 1-3/4 or 2% concentration, emulsives at 1-2/3 or 1-3/4%.

In timing oil sprays, the dates for most effective spraying are much the same as for fumigation except that oil spray will give a good kill of black scale in a somewhat more advanced stage of development, that is, spraying can be done with good results somewhat later than fumigation. The addition of rotenone to the oil further increases the effectiveness of oil spray against the more advanced instars of the black scale. If spraying is deferred until October, however, the best period, from the standpoint of tree reaction, has been passed. Orange trees are not ordinarily sprayed for black scale after the second hatch in the "two-brooded" areas because of adverse tree reaction. Fumigation in that case would be the preferred treatment.

SOFT SCALE

Coccus hesperidum Linnaeus

The soft scale, more commonly known in California as the soft brown scale, occurs throughout the tropical and subtropical regions of the world and is common in greenhouses in temperate regions. In California it is a common pest of ornamental shrubs and sometimes attacks citrus trees, especially the young trees. A characteristic of a soft scale infestation is that the insects are nearly always found in large numbers in isolated spots on a tree, usually on a few adjoining twigs. Likewise quite commonly only a few trees in the orchard are infested. Ants are always found in great abundance where the soft scale colonies occur, yet the parasites soon reduce the colony despite the ants, and artificial control measures are seldom necessary. The control of the ants, however, hastens the disappearance of the scales.

Description and Life History:- The adult females (fig. 327) are flat, oval, brown to pale yellow, and 2.5 to 4 mm long.

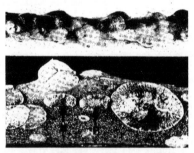

Fig. 327. Soft scale, Coccus hesperidum. Above, an orange twig heavily infested with various stages, X 4.8; below, adult (A), exit hole of a parasite in a nymph (H), and parasite seen within the body of another nymph (P), X 12. Photo by Pence.

The soft scale lays no eggs, the young being born alive. Otherwise the life history follows the usual pattern of the unarmored scales. The female has two molts and matures and produces young in about 60 days in summer. No males have been observed in California. There are 3 to 5 generations per year in southern California, and all stages of the insect may be seen in the same colony at one time.

Parasites:- The soft scale was said to be one of the worst of the scale pests in the early days of the citrus industry in California, comparable in its destructiveness with the black scale. Now it is of practically no importance, and the vastly improved condition with respect to this insect is attributable to the attacks of a number of hymenopterous parasites. Timberlake (887) listed five species as attacking the soft scale in order of their probable effectiveness as follows: Metaphycus luteolus Timb., Microterys flavus (How.), Coccophagus lecanii (Fitch), C. scutellaris (Dal.), and Aphycus alberti How. Since Timberlake's paper a number of parasites introduced for black scale control have attacked the soft scale, but it is still believed that Metaphycus luteolus (fig. 328) is the most effective of the parasites attacking this insect. If the soft scale are attended by ants, however, M. stanleyi becomes a more effective parasite than M. luteolus, for the former oviposits so rapidly that it is less interfered with by the ants.

Metaphycus luteolus is a small species, 0.7 to 0.8 mm long. It is pale yellow with a few brown and black markings. It is widely distributed and destroys its host in the early

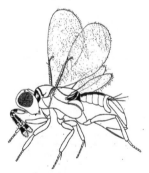

Fig. 328. Metaphycus luteolus. Drawing by Harold Compere.

instars before the scale has had a chance to produce offspring and also before the scale has had a chance to cause serious injury to the tree (887).

Control:- Artificial control, if necessary, as for black scale.

CITRICOLA SCALE

Coccus pseudomagnoliarum (Kuwana)

The citricola scale (fig. 325) is known only in California and Japan. The fact that it does not occur on one of its hosts, the hackberry (Celtis) in Japan indicates that it is not likely that Japan is the native home of the insect (324). In California it occurs on hackberry in a few places, but not more than 20 miles from the nearest citrus plantings.

The citricola scale had for years been confused with the soft scale, which it closely resembles, but it was recognized as being a different species in 1909 by E. O. Essig, then a student inspector from Pomona College, who found it on orange trees in southern California. In 1914 it was described as a new species by Roy E. Campbell, who named it Coccus citricola (143), whence it derived its present common name. Since it was later found to be morphologically identical with a scale insect attacking citrus in Japan which had already been named two months previously by the Japanese coccidologist Kuwana, the scientific name given it by the latter was adopted by all entomologists.

The citricola scale is now an important citrus pest only in the San Joaquin Valley. In southern California this insect was once widely distributed in the inland districts, but now occurs sporadically in serious infestation only in Riverside and San Bernardino Counties. The insect ceased being an important pest in southern California about 1934-35 and has not as yet re-established its former importance in this area.

In California the citricola scale is primarily a pest of citrus, but can be found on a few other plants growing near citrus trees. California hosts, other than citrus, are the hackberry, Celtis occidentalis; buckthorn, Rhamnus crocia; nightshade, Solanum douglassi; pomegranate, Punica granatum; walnut, Juglans regia, and elm, Ulmus americana. (715)

Description and Life History:- The life history may be briefly summarized as follows (724): Between 1,000 and 1,500 eggs are deposited in an egg laying period of 30 to 60 days which may occur over a period of 3 or 4 months in the spring and summer. The first appearance of young may be noted in the latter part of April in Tulare County and about a month later at Riverside, California. The first molt occurs about a month after the eggs are hatched, and the second molt about a month after the first molt. The scales are very flat and transparent while on the leaves during the summer, and these characteristics aid in distinguishing them from the nymphs of the black and the soft brown scales. In November the scales become much darker. They begin migrating to the twigs and continue to do so throughout the winter until, during February and March, the peak of migration is reached. They rarely move to branches more than 1/2 inch in diameter. After they have migrated to the twigs, the scales develop rapidly and are mature by the latter part of April or in May. There is only one generation per year.

The male, which is rare, develops in a manner similar to that of the male of other scale insects (see black scale).

In view of the possibility of confusion of this species with the soft brown scale in field determinations, the following differences in appearance, habits, and life history between the two species, after Quayle (724), have been tabulated.

SOFT SCALE	CITRICOLA SCALE
Distinctly brown as it approaches maturity.	Becomes gray as it approaches maturity.

Gives birth to active young.	Lays eggs.
Occurs in large numbers in colonies usually occupying small portions of a tree and on only a few trees in the orchard.	More or less uniformly distributed over the tree and throughout the orchard.
Different instars and stages on the tree at the same time because there are 3 to 5 overlapping generations per year (fig. 327).	At any period of the year the scales are in about the same stage of development and are uniform in size because there is only one generation per year (fig. 329).
May mature on leaves as well as twigs.	Invariably migrates to twigs before becoming mature.

Fig. 329, Citricola scale, Coccus pseudomagnoliarum on orange twig. X 2. Photo by Roy J. Pence.

For microscopic differentiation of mounted specimens a number of structural differences have been suggested, among them the difference in the antennae. The soft scale have 7-segmented antennae while the citricola scale nearly always have 8-segmented antennae. Despite the greater number of antennal segments, the citricola scale, on citrus, have on the average somewhat shorter antennae than the soft brown scale. (715)

Parasites:- In general, the species of parasites which attack the soft brown scale also attack the citricola scale. The most important species are Metaphycus luteolus Comp., M. stanleyi Comp., M. helvolus Comp., Coccophagus lycimnia (Walk.), and C. scutellaris (Dal.) (324). While M. luteolus is considered to be the most effective of the citricola scale parasites, its effectiveness is reduced by the fact that the adult citricola scale developing from overwintering stages are practically immune to attack. Since in general there is only one generation of citricola scale per year, there is a considerable period in the spring when M. luteolus would not be able to maintain its numbers in a pure infestation of citricola. The parasite therefore is much more effective against citricola when the latter is mixed with soft scale or the second generation of "double-brooded" black scale, which afford sustenance for the parasites during the season when no immature citricola scale can be found. During the past several years in central California, a portion of the spring hatch of citricola scale developed rapidly and matured during the summer. These adults were highly parasitized by M. luteolus and M. helvolus.

S. E. Flanders believes that low temperatures during the winter in central California is the most likely factor preventing control of citricola scale by M. luteolus.

Resistance to HCN Gas:- Resistance of the citricola scale to HCN fumigation was first observed near Riverside, California in 1925 (725). In 3 or 4 years this species became resistant over the greater part of the area in which it occurred in southern California. From December, 1928 to July 1932, A. F. Swain gathered citricola scale from "resistant areas" and "nonresistant areas" and obtained in 8 separate tests of simultaneously fumigated resistant and nonresistant scale an average kill of 47.6% of the resistant scale and 90.4% of the nonresistant scale. (724) Within the last few years citricola scale has become resistant to HCN gas in the Ivanhoe area of Tulare County, California.

437

Fortunately, in the season of 1933-34 the citricola scale was greatly diminished in numbers. in the greater part of its range in southern California through natural causes the exact nature of which have to this day remained a mystery. Only occasionally is the citricola scale sufficiently abundant in southern California to justify treatment. In the Hemet and East Highlands areas, however, citricola scale was abundant until Metaphycus helvolus was introduced.

Artificial Control:- The citricola scale may be controlled by oil spray and fumigation. Since its life history is similar to that of black scale, the recommended period for treatment is also similar, that is, July, August and the first half of September. Recommended dosages of spray oil and HCN gas are also about the same as for black scale. Where the citricola is resistant to HCN gas, as in certain areas in Riverside and San Bernardino Counties, only oil spray is recommended.

In Tulare County, where sulfur dusting was usually employed to combat the citrus thrips, the thrips dusting, together with one or two more applications later in the season, ordinarily served as a satisfactory control for the citricola scale. This treatment is still employed by some growers, DDT for thrips being used, along with the sulfur, in the first dusting.

Lime sulfur in winter is occasionally used instead of the sulfur dusts, but it often results in an injury to the rind of the fruit known as "scratching".

An effective spray for citricola scale which is now being used in Tulare County is composed of 1-1/2 pounds of 50% wettable DDT and 1-1/2 gallons of kerosene plus 2 ounces of blood albumin spreader per 100 gallons of water. This spray is even more adverse to the development of vedalia, the cottony-cushion scale predator, than the DDT-sulfur dust used in thrips control. It is suggested that the spray should not be used if there exists more than a trace of cottony-cushion scale in the orchard. (296)

OTHER UNARMORED SCALES

HEMISPHERICAL SCALE, Saissetia hemisphaerica (Targ.):- This insect (fig. 330) is more nearly hemispherical than the black scale. It is a shiny brown species, 3 mm in diameter. It is a cosmopolitan species, often a pest in greenhouses. Though widely distributed on citrus, it is seldom injurious. Control, if necessary as for the black scale.

Fig. 330. Hemispherical scale, Saissetia hemisphaerica, on a grapefruit twig. X 6. Photo by Roy J. Pence.

NIGRA SCALE, Saissetia nigra (Nietner):- This shiny black tropical species is commonly called "black scale" throughout the world except that in the United States it is called "nigra scale", in order to differentiate it from the more important Saissetia oleae, which was called "black scale" in this country long before S. nigra was introduced. A thorough study of the bionomics of S. nigra was made by R. H. Smith (819). The nigra scale is said to have been seen in California as early as 1900. It became abundant and widely distributed during the decade 1930-1940, and was especially abundant in the southern coastal districts from Santa Barbara to San Diego. It was found on 161 species of plants in California.

CITRUS PESTS

In California there was some apprehension concerning the potentialities of the nigra scale as a pest of citrus, avocado and other subtropical fruits, but the introduction of the parasite Metaphycus helvolus in 1937 marked the beginning of a rapid decline of this pest. M. helvolus was even more effective in combating nigra scale than it has been against the black scale, although it was imported primarily to combat the latter species. The nigra scale, once a serious pest of ornamental shrubs in California, is now rarely seen in this region.

GREEN SCALE, Coccus viridis (Green):- This common pest of the tropics is also known as the green coffee scale, soft green scale, and green bug. It was discovered in Florida in 1942 by inspectors of the State Plant Board, but it appeared that the insect must have already been in the State for several years.

As the name indicates, the adult female is a bright pale green, being somewhat transparent. The early instars are easily confused with corresponding instars of the green shield scale, to be discussed later, but the adult does not have the exterior egg sac like the latter species. The adult female is elongate oval, with a length of from 2.35 to 3.3 mm and a width of from 1.35 mm to 1.65 mm on citrus trees. On their favored host, the groundsel tree, Baccharis halimifolia, the green scale are only about 3/4 as large as on citrus. On the dorsum, the adults have a black U-shaped or irregular internal marking, the closed end of which is located anteriorly. Dead scales are light brown or buff-colored and the black markings disappear

On citrus the green scales are found on the undersides of the leaves, especially along the midrib. The pest has also been found on 71 other species of plants in Florida, but only along the extreme lower east coast (in 1942).

The green scale is parthenogenetic and oviparous. The whitish-green eggs are laid singly and remain under the adult until hatched. This insect is mobile during all nymphal instars as well as the adult stage. The entire life cycle, from egg to egg, required from 59 to 62 days during the late summer months in southern Florida.

WAX SCALES, Ceroplastes sp.:- The wax scales are large unarmored scales with a thick layer of wax covering a soft and delicate derm. They are usually angular, rather than regular in outline. They give off honey dew abundantly, and the sooty mold fungus developing in this honey dew often obscures the beautiful patterns and colors of these insects.

The Florida wax scale, C. floridensis Comst., is a beautiful snow-white scale insect, often with a pinkish tint caused by the red color of the insect showing through the wax covering the derm. The length is about 3 mm. This is the most abundant species of wax scale in Florida. In both Florida and California the barnacle scale, C. cerripediformis (Comst.). (fig. 331), may occasionally be found sparingly in citrus trees. The writer found it to be abundant on limes and oranges near Veracruz, Mexico in 1944. The thick coat of wax of this species is dirty-white in color and mottled with brown, and is divided into distinct plates, one on top and six on the sides. The white wax scale, Ceroplastes destructor Newst.,

Fig. 331. Barnacle scale, Ceroplastes cerripediformis, X 9.
Photo by Roy J. Pence.

439

occurs in Florida and Mexico, but not as a pest of citrus. In Australia, however, it is the third insect in importance in New South Wales and coastal North Queensland, and second in importance in Western Australia.[1]

PYRIFORM SCALE, Pulvinaria pyriformis (Ckll.), is a pear-shaped soft scale, brown in color but nearly surrounded by a white, cottony wax. The full-grown female may be from 2 to 4 mm in length. It is found occasionally on citrus in Florida but not as abundantly as on avocado.

GREEN SHIELD SCALE, Pulvinaria psidii Mask., is related to the pyriform scale and forms even more of a cottony covering for its body than the latter. This scale is about the size of the pyriform scale, greenish brown in color, and the mass of waxy threads covering the body becomes several times the size of the body. The early instars are easily confused with those of the green scale, Coccus viridis (Green). This species is widely distributed in southern Florida, but is only occasionally found on citrus. This is one of the serious scale pests of southeastern Asia.

THE ARMORED SCALE INSECTS

Family Diaspididae

The armored scales constitute the predominant citrus-infesting insect fauna in the majority of the important citrus-growing regions of the world, and from an economic standpoint may be considered to be more important than all other citrus pests combined. Not only are they widely distributed, but it is difficult to control them. Generally the armored scales also cause more injury to citrus trees than unarmored scales. For example, despite its smaller size, the California red scale can cause greater injury, in the form of defoliation and dying back of twigs, than the black scale, even with fewer individuals per unit of tree surface.

The armored scales are smaller, on the average, than the unarmored scales. They may be circular, subcircular, elongate, thread-like, or may assume a characteristic oyster-shell shape such as typified by the purple scale. The segmentation of the body of the adult female is obscure. The head and thorax are fused, and segments 2 to 8, inclusive, of the abdomen are fused into a much constricted area called the pygidium. The pygidium bears the anus dorsally and the vagina ventrally, as well as wax pores and tubes leading to wax glands which open on the dorsal and ventral surface of its posterior tip. The taxonomy of this group is based largely on the structures and markings found on the pygidium, and especially the pygidial fringe, which is strongly chitinized and consists of lobes, plates, and setae of various numbers and forms in the different species. The antennae are absent or vestigial and the eyes and legs are absent. Two pairs of thoracic spiracles are present, but no abdominal spiracles.

After the first molt one can begin to distinguish between the sexes. The male scale becomes more elongate. Finally emerging from under the scale is a minute and fragile two-winged insect with well-developed antennae, eyes, and with anal filaments or a stylus.

The typically diaspine life history may be synoptically presented as follows:

Female: (a) crawler, (b) first instar settled stage, (c) second instar, (d) third instar (adult).
Male: (a) crawler, (b) first instar settled stage, (c) second instar, (d) prepupa, (e) pupa, (f) adult.

One of the conspicuous differences between the unarmored and the armored scales is that the latter, as implied by the term "armored" are hidden under an "armor" or "scale" covering or, more often, they are encased between a relatively thick and tough upper scale covering and a much thinner one below (fig. 332).

---

[1]Correspondence from Dr. A. J. Nicholson, August 15, 1947.

Fig. 332. Cross section of the California red scale
showing thick dorsal scale covering or "armor" and a
thinner one below. X 100. After Ebeling (251).

These coverings consist of wax secreted by the insect plus the exuviae (cast skins).
The wax is secreted by the aforementioned wax glands of the pygidium. The exuviae
are added to the armor during the molt. The circular scales then add threads of
wax, secreted from the pygidium, to the outer edge of the armor, revolving all the
while their bodies, which are not attached to the armors between molts. An outer
fringe of wax is thus added to the armor, and this thin fringe of wax offers protection to the growing insect, which eventually fills its new covering and molts,
thereby adding another exuvia to its armor. The process is diagrammatically illustrated in fig. 333, taken from a paper on "The nature and formation of scale insect
shells" by Metcalf and Hockenyos (601). The elongate or oystershell-type scales
merely move their bodies from one side to the other, as they add on the threads of
wax to the posterior margin of the armor. They do not completely revolve. The
armor thus becomes elongate instead of circular.

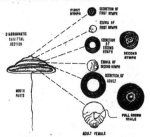

Fig. 333. Development of the scale covering or "armor" of a
circular diaspine scale. After Metcalf and Hockenyos (601).

The mouthparts of the Homoptera, as we have seen from chapter 2, are modified
for piercing and sucking. Those of the superfamily Coccoidea (whiteflies, cottony-
cushion scales, mealybugs, and scale insects) are of special interest to us because
this superfamily includes the majority of the principal citrus insect pests. We
have waited until now to discuss the mouthparts of the Coccoidea so that the subject could be discussed in connection with the study of the economically most important coccoid family, the Diaspididae.

The chief characteristics of the mouthparts of the Coccoidea are the pairs of
mandibles and maxillae, which are very long and bristle-like. Each part is, in
fact, called a bristle. It is difficult to differentiate the mandibles from the
maxillae unless one traces them back and determines the position of their articula-
tion to the head.

The maxillae form a hollow tube (lumen), through which the plant fluids are
drawn (823). The salivary canal also accompanies the food channel. Surrounding
the maxillae are the mandibles, and the entire bristle fascicle is called the
rostralis. When the rostralis is only partway extended into the host plant, the
remaining portion is looped into a long internal pouch called the crumena. This

loop of the rostralis can be seen in cleared specimens or in specimens to which oil has been added to the ventral surface to make the body wall translucent.

The four bristles of the pairs of mandibles and maxillae are often forced apart in the process of clearing, staining, and mounting, in the preparation of mounted specimens, so that one often sees the bristles considerably removed from one another except near the base of the rostralis. Normally, however, the four bristles are attached together by some kind of mechanism which, while preventing them from separating, allows them to slide past one another. This must be understood in order that the mechanics of the penetration of the plant tissue by the rostralis may be visualized.

The mouth bristles are not moved by simultaneous contractions of their muscles. The piercing of the host tissue is left to the outer bristles, the mandibles, but these alternate in their action. The successive stages of the process were worked out by Weber and are described by Snodgrass (823)[1] as follows:

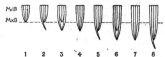

Fig. 334. Successive stages in the insertion of the feeding bristles of Hemiptera. From Snodgrass (823) after Weber.

"When the insect begins an insertion of its bristle bundle (fig. 334, 1), one mandibular bristle is thrust out a short distance in advance of the other to puncture the food tissue (2), and then the opposite mandibular bristle is protracted until its tip meets that of the first (3). Now the two maxillary bristles are lowered together until their tips lie between those of the two mandibular bristles (4). At a single thrust a bristle is extruded no farther than the maximum distance the short protractor muscle can drive it with one contraction. This distance at best is insignificant compared with the depth to which the bristle bundle can finally be sunken into the food tissue. Repeated thrusts, therefore, are necessary (5 to 8)." The penetration of woody material by the long, delicate thread-like rostralis has long been an entomological mystery. The mechanism of the musculature involved in this process has been described by Snodgrass (823).

The liquid food of the coccoid insects is drawn up through the lumen formed by the maxillae by means of the contraction and expansion of the pharynx, which is motivated by a series of powerful divaricator muscles (fig. 335). In the red and yellow scales, the movement of this "pharyngial pump" can be clearly seen with the aid of a microscope. Nel (640) proved with the red and yellow scales that the saliva is passed through the rostalis, through a different passage, while the liquid food is being imbibed. In the case of the red and yellow scales, the toxic action of the saliva is indicated by the yellow streaks marking the passage of the rostralis through a citrus leaf (fig. 336). The citrus leaf is so thin in relation to the length of the rostralis that the latter must be extended laterally through the leaf tissue. If the rostralis of a live red or yellow scale is examined with a microscope, a small bead of clear liquid can be seen being formed at its tip, indicating the extent to which saliva is forced through the rostralis.

Red scale crawlers have a tendency to settle in minute depressions or pits on their substrate, if such pits can be found. On citrus fruits, pits are numerous. Beneath the bottoms of these pits are located the oil glands. The red scale inserts its rostalis near the bottom of the pits, but avoids the oil gland, as Nelson (641) demonstrated by cross sections of the rostralis in situ. Nelson found that although successive instars inserted their rostralis to a greater depth in the fruit,

---

[1] From FUNDAMENTALS OF INSECT LIFE by Metcalf & Flint. 1932. Courtesy of McGraw-Hill Book Co.

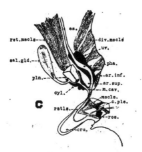
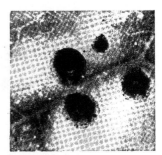

Fig. 335 (at left). Longitudinal section of mouthparts of Aonidiella aurantii or A. citrina. After Nel (640).

Fig. 336 (at right). Red scale on lemon leaf after immersion in clearing solution. Chlorotic areas indicate the path of the rostralis. X 18. Photo by Roy J. Pence.

the ratio of depth to length of the rostralis became less, for the rostrali proceeded more horizontally than vertically. After each molt, the old rostralis remained in the fruit and the rostralis of the next stage was inserted in a new location. The tip of the rostralis always ended in a cell, and such cells were injured by feeding of the red scale females in the second and third instars, but not by individuals in the first instar. The dimensions of the rostralis in the first, second and third (adult) instars, as well as the depth of penetration into the fruit of the rostralis, as determined by Nelson (641) are given below:

Instar	Diameter in mm	Length in mm	Depth of tip Below Surface in mm
First	0.0020	0.324	0.25
Second	.0024	1.037	0.61
Third	.0042	2.690	0.98

Fig. 337 (see p. 444) shows 2 photomicrographs of the rostralis of the red scale in situ in a lemon leaf. The leaf and scale were treated with KOH, bleached in alcohols and cleared in xylene. These photomicrographs show the irregular path followed by the rostralis through the leaf tissue and also the great distance penetrated.

CALIFORNIA RED SCALE

Aonidiella aurantii (Maskell)

Distribution and Economic Importance:- The California red scale is probably the most important citrus pest in the world. It is not as widely distributed as the purple scale, but causes greater injury and is more difficult to control. In California, South Africa, Australia and New Zealand the California red scale is the principal pest of citrus, but in many other countries it is also very important and may be the most important citrus pest in some areas of the country. Among these countries or regions are Mexico, Chile, Argentina, Brazil, Palestine, and islands of the eastern Mediterranean. In Argentina the red scale occurs in all citrus districts and is considered second only to the fruit fly, Anastrepha fraterculus Wied., as a pest of citrus.

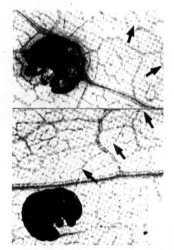

Fig. 337. California red scale. Path of the stained rostralis (shown by arrows) through the cleared tissue of a lemon leaf. X 28. Photo by Roy J. Pence.

In California the red scale is widely distributed, although the vigilant quarantine and control and eradication measures which have been taken in certain protective districts has resulted in a few red scale-free areas in regions in which climatic conditions are favorable for its development, notably in Ventura County. Occasional infestations of red scale have occurred on ornamentals of citrus, in the Coachella Valley, the San Joaquin Valley, and as far north as Butte and Glenn Counties. It is probable that nowhere in the citrus-growing areas of California or Arizona is the red scale limited by climate, and any further limitation of its spread must be by quarantine and the establishment of eradication districts if and when they are needed in the areas now free or nearly free of red scale.

In Texas the California red scale was once an important pest and caused serious damage (161). The first treatments for California red scale started in 1925, and by 1928 about 75% of the orchards were heavily infested. In 1933 two hurricanes blew in across the Gulf of Mexico from Florida. The one in September continued for 25 hours and brought with it 25 inches of rain. In the following years much pink scale fungus, Nectria diploa B and C, could be seen in the citrus districts of Texas, although previously it had been seen in only one locality. Although it is now known that the pink scale fungus is not parasitic, nevertheless some observers believe that some truly parasitic fungus may have been similarly blown in and distributed by the hurricanes and that this resulted in the control of red scale by natural means. Others believe that the hurricanes wrought such destruction to citrus trees and loss of fruit that oil spraying was discontinued for several years for economic reasons and that this may have resulted in the restoration of a natural balance between the red scale and its natural enemies that has maintained the red scale at a satisfactorily low density to this day.

It is now believed that what has been identified as California red scale in Florida is yellow scale, Aonidiella citrina, which is sparsely distributed in that state.

History:- The red scale is thought to be indigenous to southeastern Asia, although the usual means of tracing the origin of introduced insects have not led to

definite conclusions as to the native home of this species (793). It was undoubtedly introduced into Australia whence, between 1868 and 1875, it was introduced
into Los Angeles, probably on nursery stock (724). For many years red scale was
not as widespread in California as the black scale, but now it can be found nearly
everywhere, although it is not yet a pest in citrus orchards in the San Joaquin
Valley and the desert regions and, as stated before, in limited areas from which
it has been excluded by protective programs. The red scale was not a specially
severe problem after the beginning of commercial fumigation and up to the time of
the establishment of resistant strains, a period which varied according to the
region. It was the development of the resistant strains which made the red scale
problem the number one citrus insect problem in California. Both resistant and
nonresistant red scale can be found in all districts (236), but it is reasonable
to believe that the resistant race will continue to spread as it has in past years
and finally become the dominant race everywhere.

Description of Adult Female:- Nel (640) gave the average dimensions of the
scale covering (armor) of the adult female California red scale as being 1.9 mm. in
width and 1.77 mm in length. He did not state, however, from which host the insects were taken. Great differences may be found on the average size of the red
scale not only on different host species, but also on different morphological parts
of the same host. The body of the female is somewhat smaller than the scale covering. The writer measured the widths of the bodies of an average of 185 adult female red scales on each of 14 host species or varieties and found them to vary from
1585 ± 4.1 microns on the fruit of the grapefruit, Citrus decumana to 1080 ± 7.8
microns on the leaves of the eucalyptus, Eucalyptus globulus. The shape (ratio of
width to length) also varied on different host species. On citrus the insects
were larger on the fruit than on the leaves and larger on succulent "sucker" twigs
than on the slow-growing twigs or on the branches.[1]

The apparent reddish color of the scale covering of the insect in situ is not
the color of this structure itself, but rather the color of the body of the female
as seen through the scale covering. The ventral scale covering consists of the
lower half of the molted skins plus a mass of many threads supplied by the wax
glands of the pygidium.

The body is reniform in shape because the lateral margins of the body extend
posteriorly as far as the tip of the pygidium. On the pygidial fringe there are
three pairs of well developed lobes, the two inner pairs being distinctly notched.
Between the lobes and laterad of the outer lobes are deeply fringed plates. Two
internal sclerotized structures (structures A and B, figures 338 and 339, p. 447)
are visible anteriomedially on the ventral side of the pygidium of the red scale.
These structures were found by McKenzie (570) to be the only taxonomic characters
by which the red and the yellow scale (Aonidiella citrina (Coq.)) could be reliably
differentiated. These two insects, now considered as distinct species, were formerly considered to be varieties of the same species. The yellow scale possesses
structure A, but the latter differs in shape from that of the corresponding structure on the red scale. Structure B does not occur on the yellow scale.

It will be noted that the adult female red scale has no legs and the well-
developed antennae of the first nymphal stage have degenerated into the simple, inconspicuous appendages shown in figure 338. Taking Nelson's (641) average figure
for the length of the rostralis of the adult female red scale, namely, 2.69 mm, it
can be seen that the rostralis is about twice as long as the greatest diameter of
the body.

Life History:- As with all the principal citrus-infesting scale insects in
California, the life history of the red scale was worked out by Professor H. J.

SUBTROPICAL ENTOMOLOGY

Henry J. Quayle
Professor Emeritus of Entomology, University of California Citrus Experiment Station; pioneer in research
on the biology and control of subtropical fruit pests.

First Nymphal Instar:- The California red scale does not lay eggs, but gives
birth to active young. After birth, the young remain a day or two under the
scale covering in the region of the pygidium. After emerging from under the scale
covering they crawl about actively and are at this period commonly called
"crawlers." (fig. 340, A) The majority emerge before noon and settle within six
hours. A few of the "crawlers" may crawl about for a day or two but 95% settle
within one day (724). On the leaves of citrus the crawlers usually settle beside
the midrib or prominent veins; on the fruit they usually settle in the tiny depressions marking the location of the oil glands. The females never change their
location. Mortality is higher at this stage than any other. The mortality of females was found to vary from around 30% for those individuals from crawlers
settling in the spring to 100% for those individuals from crawlers settling during
the winter months (91). These figures include mortality in any stage of development, but the greatest mortality is among the crawlers. Quayle found only 41% were
able to settle and continue development under favorable conditions and when protected from natural enemies. During the winter months, the mortality is greatly
increased.

The host species has a great effect on the percentage of red scale crawlers
which will settle and develop to maturity. The writer found that out of 75 red
scale crawlers transferred to the leaves of eight citrus varieties, the following
number of insects settled and developed to the second instar:

Host	Number of Scales
Citrus limonia Osb. (Eureka lemon)	56
Citrus grandis Osb. (Imperial grapefruit)	48
Citrus grandis Osb. (Frizzelle pummelo)	31
Citrus sinensis Sw. (Trovita orange)	26
Citrus sinensis Sw. (Ruby orange)	25
Willow-leaf mandarin x Imperial grapefruit	14
Citrus deliciosa Sw. (Willow-leaf mandarin)	13
Citrus deliciosa Sw. (Dancy tangerine)	2

In another experiment the writer placed 20 red scale crawlers on each of six
oranges and six lemons on July 31, 1931, and these were examined on September 4,
1931. The crawlers were transferred by means of a camel's-hair brush and the
fruit was kept in a moist chamber at a temperature of 68°F. A repetition of the
experiment was made in which 45 crawlers were transferred to each of five Valencia

446

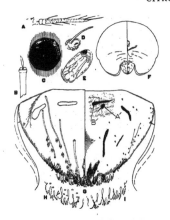

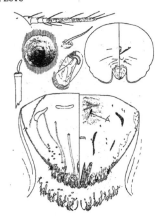

Fig. 338

Fig. 339

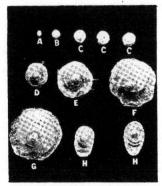

Fig. 340

Fig. 338. California red scale, Aonidiella aurantii (female). A, antenna of first instar; B, enlargement of free end of tubular duct; C, scale covering; D, antenna of adult; E, spiracle; F, general features of adult; G, pygidium, showing sclerotized structures A^1 and B^1; H, dorsal aspect of pygidial margin; I, ventral aspect of same. Compare G, A^1 and B^1 with fig. 339. After McKenzie (570).

Fig. 339. Yellow scale, Aonidiella citrina (female). A^1, sclerotized structure used to distinguish A. citrina from A. aurantii. Compare with fig. 338. After McKenzie (570).

Fig. 340. Life stages of the California red scale: A, crawler; B, whitecap; C, nipple stage; D, first molt; E, second molt; F, gray adult; G, adult; H, male scale covering. X 11. Photo by Roy J. Pence.

447

oranges and five Eureka lemons on August 4, 1931 and were examined on September 4, 1931. The results of the experiments·are shown in Table 11. Nearly equal numbers of crawlers settled and reached the whitecap stage on the two hosts, but a far greater percentage reached maturity on the lemons than on the oranges.

Table 11

Relative Percentage of Crawlers of the California Red Scale
Developing to Maturity on Fruit of Orange and Lemon.

A. Crawlers transferred July 31, 1931

Fruit	Males	Females	Average per Fruit
Orange	1	1	
	3	9	
	3	1	1.83 ± 0.34 males
	1	6	
	3	3	3.83 ± 0.79 females
	0	3	
Total	11	23	
Lemon	1	8	
	4	6	
	10	11	4.50 ± 0.76 males
	3	13	
	4	4	7.83 ± 0.89 females
	5	5	
Total	27	47	

B. Crawlers transferred August 4, 1931

Fruit	Males	Females	Average per Fruit
Orange	4	2	
	2	12	
	0	4	3.00 ± 0.41 males
	3	8	
	6	17	8.60 ± 1.64 females
Total	15	43	
Lemon	3	11	
	7	16	
	6	27	6.25 ± 0.72 males
	9	25	
		Fruit destroyed	19.75 ± 2.18 females
Total	25	79	

From the figures in table 11, it can be calculated that only 92 insects out of 345 (26.6%) reached maturity among the 345 crawlers placed on the orange fruits and 178 out of 300 (59.3%) reached maturity on the lemon fruits. An additional effect of the host species was found to be the length of the life cycle. The first emergence of young was noted seven days earlier on the lemon fruits than on the orange fruits.

Within an hour or two after the crawler settles, a cottony secretion is secreted over the body. Soon white threads of wax envelop the body and extend down the sides to the substrate. The insect becomes circular in form. After a couple of days the mass of waxy threads becomes more compact and the insects are then called "whitecaps" (fig. 340, B). Heavily infested fruits may at times be nearly completely covered with whitecaps except where the more mature stages occupy the surface (fig. 341). In a few more days the wax settles down still further except for a central prominence or nipple, and the insect is then said to be in the "nipple" stage. The scale covering is spread out thinly beyond the body margin of the insect and loses its cottony appearance. After 12 or 15 days the insect molts (fig. 340, D), adding its exuvia or cast skin to the already secreted scale covering. The scale covering cannot be lifted from the body of the insect during the period it is molting.

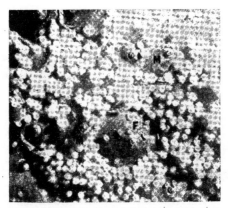

Fig. 341. California red·scale. F, female; M, male. Note crawlers on scale covering of female and the many whitecaps nearly completely covering the surface of
the fruit. X 13. Photo by Roy J. Pence.

Second Instar:- After the completion of the first molt, a rim of wax is extended beyond the margin of the former scale covering. The covering may again be
readily lifted. The female rotates in order that the wax threads secreted by the
pygidium can be added uniformly to the outer margin of the scale covering so as to
form a circular armor. Quoting from Nel (640): "Some interesting observations as
to the secretion of this wax were made by placing a·second-stage female, deprived
of its wax covering, under a microscope and studying the flow of wax filaments
from the abdominal pores. These filaments were at first excreted vertically, and
upon attaining maximum length they fell over as a result of their own weight. They
then became entangled in the pygidial fringe, because of· the up and down movement
of the pygidium. It appeared that the pygidial lobes and plates acted as the
teeth of a comb, causing the tangled mass of wax filaments to be more closely
united and formed into a ribbon which then passed out to form the waxy portion of
the scale covering." The elongate shape of the scale. covering of the male is due
to the fact that the male, while it at first rotates like the female, later ceases
to rotate completely, with the result that the scale covering increases in one direction only, thus becoming elongate (fig. 341, M). However, it never reaches the
length of the female scale covering, despite its elongate character.

The second molt (fig. 340, F) is similar to the first. At the termination of
the second molt the insect has reached the adult stage.

Adult Female:- After the second molt, the female again rotates, extending another rim of wax beyond the margin of the second instar scale covering. This wax
rim is soft and light gray in color, and gives the name "gray adult" to this particular phase of the adult development. The scale covering may again be easily detached.· It is during this "gray adult" stage (fig. 340, F) that fertilization
takes place. The insect then enters the second phase of its adult development
(fig. 340, G) characterized by the following features: rotation of the body and
secretion of wax ceases, the dorsal scale covering is attached to the body, and
the thorax extends backward in a large rounded lobe on each side of the pygidium,
giving the fully mature insect its reniform shape.

Not only is the fully mature female protected by its dorsal "armor," but it is
also protected below by a ventral membrane. A portion of the ventral membrane

449

usually remains on the substrate when the scale is lifted away (fig. 342). The dorsal covering or "armor" is the thicker, measuring about 65 microns in thickness, while the ventral membrane measures about 20 microns (251).

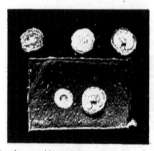 

Fig. 342 (at left). Ventral surface of fully mature female red scale. Upper row: left, scale sprayed with oil to make waxy membrane translucent; middle, scale with ventral membrane intact; right, scale with ventral membrane removed. Lower: adult female on a section of leaf; showing ventral membrane torn off and adhering to leaf when insect is lifted.

Fig. 343 (at right). Tracheal system of a California red scale (gray adult). After deOng, Knight and Chamberlin (231).

If the mature female is placed on its back on a microscope slide and a drop of oil is placed on the ventral surface, making it translucent, large numbers of embryos may be seen within the body. It appears as if the insect were completely filled with embryos. If a number of females are examined, one may be found giving birth to an active nymph. The tracheal system can also be plainly seen, and because of the fact that oil, upon entering the tracheae, changes the refractive index of their contents, the progress of spray oil through the tracheal system can easily be studied (fig. 343).

Duration of the Various Instars:- Nel (640) made a series of eight life history studies of the California red scale reared on young seedling Valencia orange trees. The experiments were made in a lath house. The mean temperature during the course of the experiment varied from a minimum of 58.3° and maximum of 95.1° F in August, when the experiment was begun, to a minimum of 39.8° and maximum of 74.3° F in November, when the experiment was concluded. The crawlers with which each of the eight experiments was begun were transferred to the orange seedlings from August 10 to September 29, 1928. The mean number of days spent by the females in the various instars and molts were as follows: first instar, 13.00; first molt 3.38; second instar, 8.63; second molt, 3.63; third instar up to the time of reproduction, 32.50; total period in days from free moving crawlers to start of reproduction, 60.26. According to Yust (1022), females produced young for periods varying from an average of 42.0 days, for those whose reproduction began in April or May, to 197.8 days for those whose reproduction began in November, 1935. In another investigation with red scale reared on lemon fruits, the number of crawlers produced per day per female varied from 0 to 14 (average 4.0), and the number of crawlers produced per female varied from 18 to 176 (average 72.6) (91). Young are produced only by fertilized females.

The Development of the Male:- The male develops according to the typical coccoid metamorphosis. No distinguishable difference can be seen between the two sexes until after the first molt, when the male becomes elongate and two pairs of conspicuous purplish eyes appear. The second molt is followed by the prepupal stage, during which the sheaths of the antennae and wings become visible (figs. 344 & 345). The next molt is followed by the pupal stage. The sheaths of all appendages are

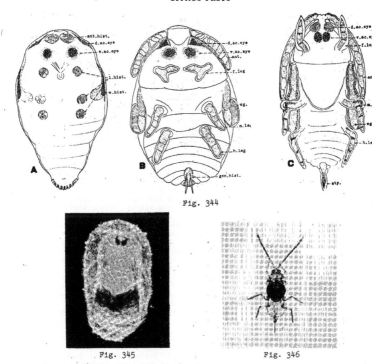

Fig. 344

Fig. 345   Fig. 346

Fig. 344. Métamorphosis of the male Aonidiella aurantii or A. citrina. A, second instar; B, prepupa; C, pupa. After Nel (640).

Fig. 345. Male pupa of California red scale. Scale turned over to show body of insect beneath. Photo by Roy J. Pence.

Fig. 346. Adult male of Aonidiella aurantii. X 22. Photo by Roy J. Pence.

then quite conspicuous, and the style, which can be seen protruding from the posterior margin, appears in this stage. The next molt brings the male to the adult stage. It makes its way from under the scale covering (fig. 340, H) and emerges as a frail, weak-flying, orange-yellow insect (fig. 346) with the usual insectan appendages: a pair of wings, the second pair being replaced by a pair of hooked halteres; eyes, antennae, legs, and a long style - in sharp contrast to the morphologically degenerate female. When mature, the male is 0.6 mm long (exclusive of the style) and has a wing expanse of 1.5 mm. It is orange-yellow in color.

Sex Ratio:- Quayle (708) found the ratio of females to males developing from crawlers liberated from January 2 to July 24, 1910, to be 1.8:1, but the ratio of females to males among those liberated from August 18 to November 29, 1910, was practically 1:1. Of 52 red scale adults reared on orange seedlings by Nel (640) during the second half of the year 1930, 319 were females and 202 were males. Of 65 red scales which reached maturity from 70 crawlers transferred to an orange seedling in the early spring, 19 were females and 46 males. The writer, from a transfer of 645 red scale crawlers, found a sex ratio of females to males among the insects developing to maturity, of 2.60:1 on mature Valencia orange fruits and 2.57:1 on mature Eureka lemons.

Seasonal History (236):- In November, 1939, experiments were begun by R. C. Dickson with a view of determining the number of generations per year of the red scale in various localities in southern California. These experiments were continued for two years. Small screen cages were hung in various kinds of trees which would not be subject to pest control treatment, and in each cage were placed two grapefruits infested with red scale crawlers of the non-resistant race. The grapefruits were partly covered with wax, which reduces transpiration and prolongs the period during which the fruits can be used as satisfactory hosts, without sufficient desiccation to interfere with the development of the scales. As soon as the young began to appear, they were transferred to new grapefruits.

Under these field conditions, Dickson found the number of days required for red scale crawlers to reach maturity (the beginning of the production of young) varied from 65 to 74 days in Riverside County and the eastern extremity of the citrus district of San Bernardino County among insects starting their development during the summer months and prior to August 6, to 246 to 262 days among insects starting their development in September and October in two localities in Santa Barbara County. He found that the greatest number of generations which would develop per year varied in 1939-40 from 2.1 and 2.2 in Santa Barbara County, to 3.4 to 3.6 in Riverside County. In 1940-41 the greatest number of generations varied from 2.0 in Santa Barbara County to 3.1 in Mentone, Elsinore, Corona, Riverside, San Fernando and El Cajon. By "greatest number of generations per year" is meant the number produced when records are kept of the first crawlers to emerge from each generation of scale. If a record were kept of average individuals instead of the earliest-emerging individuals, the numbers of generations per year would be found to be somewhat less than those stated above. It is of interest to note, however, that a greater number of generations per year develop in the inland areas than in the coastal areas.

Injury:- The red scale is not only the most difficult of the citrus-infesting insects in California to control, but it also is the most injurious. Injurious effects include the yellowing of foliage and defoliation; the dying back of small twigs and finally, if the insect remains unchecked, the large branches; the dropping of the current season's fruit and a reduction of production for one or more years after a severe infestation, even though the infestation is controlled by artificial control measures; and the culling of the infested fruit. Different degrees of damage will result on trees of different degrees of vitality, and, in the writer's opinion, a given number of red scale per unit of tree area will cause more damage in the inland areas than near the coast. A light infestation of red scale will often cause a gumming of the branches of orange trees in the easternmost citrus districts, but a much heavier infestation will fail to cause this type of damage in coastal areas.

One of the reasons that the red scale is such an injurious pest is that it attacks all parts of the tree: the trunk, branches, twigs, leaves and fruit. It is also likely that the red scale injects a toxic substance into the tree tissue with its saliva, for a yellowish discoloration of the leaf tissue may be seen where the mouthparts of the insect have been inserted (fig. 336). If this is true, it may account for the greater injury caused by this species as compared with other scale insects.

The red scale occurs on a large number of hosts, and some of these, such as the castor bean, rose, nightshade, eucalyptus, carob, walnut, and laural sumac, serve as sources of reinfestation of citrus orchards which have been recently treated. Only on citrus, however, among the fruit trees, is the red scale a serious pest.

Distribution of the Red Scale Population in Citrus Trees:- Comparisons were made by Dickson (236) of the per cent of female insects reaching maturity on fruit exposed to the sun on the south side of the tree as compared to fruit in the shade on the north side of the tree. The experiments were begun on May 12, July 30, and October 18, 1941. The average per cent of transferred crawlers (females) reaching maturity on the fruits in the shade on the north side of the tree was 19.3% while for those on the fruits exposed to the sun on the south side of the tree it was 3.3%. No account was made of males which might have developed, but they are on the average only about half as numerous as the females.

Ebeling and Persing[1] found the average population density of adult female red scales in 8 citrus orchards located in various areas of southern California to vary on the different quadrants of the tree in the following ratio: northeast, 3.60; northwest, 2.34; southeast, 1.31; southwest, 1.00 (see p. 306).

Dissemination:- Like the other scale insects, the red scale is distributed from tree to tree and orchard to orchard primarily at the "crawler" stages of development. The crawlers are small and are constantly moving about, thus being especially amenable to dislocation from the tree surface and subsequent dissemination over a wide area by means of the wind. The blowing of the older, weakened leaves off the trees, if these leaves should happen to harbor an infestation of red scale, would also appear to constitute a means of dispersal for this species. The crawlers, being continuously active prior to their permanent settling on the tree surface, are apt to crawl upon the feet or bodies of birds and insects. It is known that dispersal of red scale takes place in this manner, for almost invariably these insects can be found adjacent to birds' nests, if an infestation occurs anywhere within a reasonable distance, even though no scale may be found on the remainder of the tree or even in the remainder of the orchard. Likewise red scale is conspicuously abundant on susceptible host plants growing around bird baths.

Of special interest are the methods of red scale dispersal that are subject to the control of man. The problem is of special interest to those who are concerned with the management of pest control districts where an attempt is made to keep certain uninfested areas entirely free of red scale. Of these, the distribution of various stages of the insect on nursery stock and bud wood are probably the most important, but if quarantine laws are properly enforced, this should not prove to be an important hazard. The distribution of crawlers from one orchard to another by means of horses, tractors and trucks, pickers' sacks, clothing, and ladders may be of some importance in the dispersal of the species, but it is hardly conceivable that this could be of practical importance unless the infested articles were transferred within a day from a heavily infested orchard to one free or nearly free of scale. If red scale is quite generally distributed throughout an orchard it is not likely that the possible addition of a few more insects from another orchard could be of practical importance.

The question is often asked as to the advisability of treating picking boxes after their use for the picking and hauling of infested fruit. In answer to this question it should be pointed out that unless the boxes are to be moved into an infested orchard within a couple of days there is no chance of their being a vehicle for the transportation of red scale crawlers into another orchard, for red scale crawlers have never been known to survive longer than two days without a suitable host upon which to settle and feed.

Even if picking boxes were to be transported before the end of the two-day period, the disinfestation of the boxes could hardly be justified unless they were to be taken to an isolated orchard known to be free of scale. Quayle (726) has pointed out that past experiences in box sterilization for such insects as mealybugs has given little evidence that this procedure is of value in preventing the dispersal of insects. It is at best but one factor among many that might account for the dissemination of scale insects, and the most important factors are beyond man's control.

[1]Ebeling, W. and C. O. Persing, 1946. Evaluation of insecticides used against red scales by estimates of population densities. Unpublished manuscript on file at the University of California, Los Angeles.

Natural Enemies:- Although every citrus region in the world has been searched for parasites of the California red scale in parasite explorations dating back as far as 1891 (793) all attempts at control by means of parasites have been a failure. There is at present only one parasite, the golden chalcid, Aphytis chrysomphali (Mercet) (fig. 347), which is likely to be found widely distributed, though in limited numbers, attacking the red scale. This parasite has been considered to be of little importance in its control, although further studies in recent years have led to the belief that through its host-feeding propensities, the adult parasite may be of some value in limiting the increase of a scale population. It is believed that some of the cases of unexplained red scale mortality which have been noted in occasional orchards may be due to host feeding by the golden chalcid.

The golden chalcid is a very small parasite, yellow in color, and 0.78 mm long. It deposits its egg on or under the body of the red scale, but beneath the scale covering. The larva feeds externally on the body of the red scale and the adult emerges either from under the scale armor or through it via an exit hole gnawed through the armor. (708)

A red scale-attacking race of a hymenopterous parasite from South Africa, Habrolepis rouxi Comp., was liberated on red scale-infested trees throughout the citrus belt in California in 1937 and 1938. This insect was recovered on untreated lemon trees in San Diego in October, 1943. It was not found, however, in oil-sprayed orchards. H. rouxi appeared to have much promise in the laboratory but not in the orchards after it was liberated. It does not seem to like the citrus tree as a habitat and leaves it at the first opportunity (796). The female of this species (fig. 348) deposits eggs in the "blood" of the scale. The larvae develop and pupate inside the host.

Among the predators, Lindorus lophanthae (Blaisd.) (fig. 349) is commonly associated with red scale as well as the purple scale. This beetle is 2 to 3 mm long with metallic black elytra etched with grayish or light brown hairs and with a brown prothorax with a faint darker horizontal band across the middle. It has a brown ventral surface and legs. Quayle found that this species lays one or two eggs beneath the scale. The larva consumes the scale and eats out an irregular hole in the scale covering, then proceeds to consume other scales. The adult also feeds on red scales. (708) S. E. Flanders informs us that the twice-stabbed lady beetle, Chilocorus stigma (Say) (fig. 350), will greedily feed on red scale, but apparently it does not occur in sufficient numbers in the field to effectively combat the scale. Another species of this genus, C. distigma (fig. 351), which was introduced from South Africa in 1948, attacks the red scale. The female lays its eggs under the scale. It is not yet known how effective this predator will be under field conditions. A considerable number of predatory insects and mites feed on red scale crawlers or various stages of the settled insects, but they appear to have little effect in controlling this pest.

Sokoloff and Klotz (824) found a soil-inhabiting microorganism, which they called Bacillus "C", to be able to invade and destroy the adult red scale under certain laboratory conditions. Attempts at mass innoculation of red scale in a lemon orchard failed to result in any significant reduction in the population of this insect.[1]

Control

Despite the fact that the control of the California red scale is more difficult than that of any other citrus pest, and the combined research activity in southern California directed against this insect has been greater than that directed against any other, the methods of control are substantially the same as they were 20 years ago, namely, fumigation and oil spray. In view of the widespread occurrence of resistant red scale in nearly all citrus districts, it can be said that the pest control situation, with regard to the red scale, is not as satisfactory as it was several decades ago. Present control measures not only are not sufficiently effective against the red scale, but they also are inimical to

---

[1] Notes on file at the Citrus Experiment Station, Riverside.

Fig. 347

Fig. 348

Fig. 349

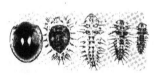

Fig. 350

Fig. 351

Fig. 347. The golden chalcid, <u>Aphytis</u> <u>chrysomphali</u>. X 55. After Quayle (724). Courtesy Comstock Publishing Company.

Fig. 348. <u>Habrolepis</u> <u>rouxi</u>. X 30. Photo by Roy J. Pence.

Fig. 349. <u>Lindorus</u> <u>lophanthae</u>. X 12. After Essig (279).

Fig. 350. Adult, pupa, and 3 larval instars of the twice-stabbed lady beetle, <u>Chilocoris</u> <u>stigma</u>. X 2.5. Photo by Roy J. Pence.

Fig. 351. <u>Chilocorus</u> <u>distigma</u>, a recently introduced predator of the red scale. Larva, pupa and adult. X 3.0. Photo by Roy J. Pence.

the welfare of the tree, ranging in degree of injury from acute damage to trees and fruit to a more insiduous and accumulative debilitation of the tree. Much research continues to be directed toward the working out of a safer and more effective treatment for the control of the red scale.

Resistant and Nonresistant Races and Protective Stupefaction

The red scale was first found to be resistant to HCN in the Corona area about the year 1914. Subsequently the areas in which red scale was found to be resistant became more numerous and greater in extent. In districts in which an 18 to 20 cc fumigation schedule was once so effective against red scale that a single treatment would keep the orchards commercially clean for 2 to 4 years, it was finally necessary to use 22 to 24 cc to keep the orchards in a commercially clean condition for a single year. In an experiment in which 5 ml of HCN was introduced into 100 cu. ft. of space in an air tight fumatorium (approximately equivalent in concentration to a 16 cc schedule under a tent under field conditions), an 89% kill of resistant red scale and a 99.8% kill of nonresistant red scale was obtained (236). In this experiment 55 times as many scales survived the treatment among the resistant scales than among the nonresistant scales.

Further increase in dosage in orchard fumigation became impossible because the limit of the tolerance of the tree to the gas had been reached. It was often necessary to use oil spray in addition to fumigation to satisfactorily combat the scale.

Laboratory Experiments:- The resistance of red scale to HCN gas has been the subject of considerable investigation. It has been shown that much more HCN can be recovered from nonresistant red scales than from resistant red scales, and the same relation held true with regard to nonresistant and resistant black scales (534). Red scales taken from mature lemons from areas in which the red scale was known to be nonresistant had only 60.4% as much wax as scales taken from mature lemons from areas in which this insect was known to be resistant. Differences in chemical composition between the two races also were found. (397).

The experimental evidence for the resistance of the red scale consisted in the variation in kill of insects collected from a number of districts and fumigated together, on the same host, in an airtight fumigation chamber. An experiment made by A. F. Swain in this manner with scales taken from areas in which red scale was known to be resistant, and fumigated together with red scale taken from areas in which the insect was known to be nonresistant, involved 15 separate tests. Using various dosages of HCN gas, the average kill of resistant red scale for the 15 tests was 37.7% and for the nonresistant red scale it was 77.6%, a difference of 44.9%.

According to Lindgren (526), "The resistant red scales are better able to survive a dosage of 28 cc than the nonresistant red scales an 8 cc dosage (or even a 4 cc dosage, as a few preliminary experiments with the dosage indicate). A kill of 98.36% of the nonresistant red scale was obtained with a dosage of 8 cc and an exposure of 15 minutes, whereas a kill of 97.62% of the resistant red scale was obtained with 28 cc of HCN per unit and an exposure of 45 minutes."

Repeated fumigations will increase the resistance of a red scale population which is a mixture of nonresistant and resistant individuals (1022, 528). The latter investigators found in one test that resistance progressively increased with 11 fumigations, but that 5 additional fumigations had little or no further effect on resistance. It is assumed that in many orchards in which red scale is considered to be resistant, average resistance will increase still further with successive fumigations before the ultimate degree of resistance is reached by the red scale populations in these orchards.

Research on the subject of resistant races at the University of California Citrus Experiment Station was greatly facilitated by the rearing of a nonresistant race and a resistant race in adjoining rooms in the insectary. The original stocks of scale were collected August, 1936. The resistant, race was obtained from Corona, California, and the nonresistant race from an isolated orchard in the foothills east of Glendora, California. These strains have been reared on squash under identical conditions ever since, and still exhibit the original difference in their susceptibility to HCN gas. The crawlers of these scales, being positively phototropic, will crawl toward the ends of the squashes facing the source of light. There they are brushed off into beakers and later brushed onto grapefruit, and are then allowed to develop on the grapefruit until they reach the stage of development required for the experiment. Experiments can be made with pure cultures of resistant or nonresistant scales of uniform ages. (525)

Experimenting with his resistant and nonresistant races of red scale, Lindgren (526) was able to show that among resistant red scale, other instars besides the adult, and including the motile first instar nymphs, also are resistant to HCN gas. Interesting differences in the two races in their response to HCN gas were discovered. It was found that when resistant red scale were exposed to low concentrations of HCN gas (0.121 mg HCN per liter), and then immediately exposed to an ordinarily lethal concentration, the "stupefaction" of the insects at the lower concentration resulted in a reduced kill. When the same procedure was carried out with the nonresistant scales, however, no decrease in kill resulted. Likewise when nonresistant red scales were exposed to the low concentration for 10 minutes, no effect on the susceptibility of the scales to the higher concentration was noted. Only after an exposure of 20 minutes to the sublethal dosage did the nonresistant scales show evidence of "protective stupefaction" with the resulting reduced kill when subjected to the high gas concentrations. In the resistant race the effect of the sublethal dose of gas disappears after 2 hours, while it requires 3 hours or longer for this "protective stupefaction" to disappear in the nonresistant strain.

Hardman and Craig (408) made the discovery that in both the resistant and nonresistant races of the red scale the spiracles close within 3 to 5 minutes after the cyanide reaches them. In the resistant race the spiracles remain closed for at least 30 minutes, if the cyanide gas remains, while in the nonresistant race the spiracles stay closed for only about 1 minute. It was shown by Quayle (730), however, that the interesting physiological difference found by Hardman and Craig could not account for the difference in the resistance of the two races for the following reasons: (1) the difference in the tolerance of the two races is noted when they are fumigated for only 2.5 minutes, before the individuals of either race close their spiracles (as well as for prolonged periods, up to and including 2 hours), (2) the difference in tolerance occurs independently of stupefaction, and (3) the difference in tolerance occurs during the molting process, when the spiracles are not functioning.

Mature female red scales of the race resistant to HCN are also resistant to methyl bromide. This was also true of the "early gray" adults in 40 minute fumigations, but in 120- to 180-minute fumigations no differences in susceptibility between the two races in the early gray adult stage could be found (1023). There is no difference in the susceptibility of the resistant and nonresistant races of red scale to oil spray (527). Little difference was found in kill between resistant and nonresistant races of California red scale when ethylene dibromide was used as the fumigant. There was likewise very little difference in the susceptibility of different stages of the insect. A low dosage of HCN, sufficient to "stupefy" the red scale, will cause the scale to be more difficult to kill with ethylene dibromide than if no stupefying charge of HCN had been applied.(529).

Genetic Difference Between the Resistant and the Nonresistant Races of the Red Scale

The fact that Lindgren's two stocks of red scale, originally varying greatly in their resistance to HCN gas, maintained this difference from generation to generation for the past 12 years, under identical conditions, proves that resistance is inherited. Dickson (235) undertook the task of determining the genetic mechanism by which this resistance is inherited.

Dickson removed the males from red-scale-infested grapefruit and this grapefruit was placed in a cage with grapefruit bearing red scale of the type desired for the male parents. Since red scale do not reproduce without fertilization, the only females producing young were those fertilized by males of the opposite race as regards resistance. When the young from this cross were born, they were brushed onto clean fruit and the fruit was held until the female scales reached maturity, and was then fumigated with HCN for 40 minutes at 75° F in a 100 cubic-foot fumatorium. Some fruits were retained for making further crosses. Dosages of 1 cc or 2 cc of HCN per 100 cu. ft. were used in each fumigation.

Counts of the percentage of adult red scale females surviving the fumigation treatments showed that the F_1 females were intermediate in per cent survival between the 2 parental strains. The per cent survival of the F_2 females was intermediate between that of the F_1 females and that strain to which their paternal grandmothers belonged. Dickson's conclusion was that resistance, in the case of the California red scale, depends on a single gene (or group of closely linked genes) in the X-chromosome and is therefore sex-linked.

Red scale was once easily killed by HCN fumigation in California, but the resistance factor is believed to have been present in some individuals. The strain of red scale possessing the resistance factor would, of course, increase in relation to the nonresistant population with successive fumigations. Wherever fumigation is continued, it is believed that the red scale population will gradually contain an increasingly greater percentage of resistant individuals until eventually practically all the population will be of the resistant strain, after which no further increase in resistance is to be expected.

Fumigation

Although fumigation under tents was first conceived and commercially applied in connection with the control of the cottony cushion scale, it was soon apparent that it afforded an excellent means of controlling other scale insects. Of all the hundreds of gases tested, hydrocyanic acid gas to this day has been found to be by far the most effective and most practical for orchard fumigation. Until resistant strains of red scale began to develop, HCN fumigation was so effective that one fumigation was usually sufficient to control red scale for a period of 2 to 4 years, but, as stated before, HCN fumigation against the resistant race of red scale is far less effective. Until the resistant strain of red scale is universally distributed, however, recommendations as to whether or not HCN should be used, or as to the proper dosage of gas, will depend on whether past experience in a particular orchard has indicated the red scale to be resistant or nonresistant.

HCN fumigation is the most effective single treatment for nonresistant red scale. An 18 cc schedule or more should be used on oranges and a 20 cc schedule or more should be used on lemons. The fumigation schedule is usually increased from 4 to 8 cc if the scales are resistant, but even this increased concentration of gas is not as effective against resistant red scale as the lower concentration is in the control of nonresistant red scale. It represents the upper limit, however, of the gas concentration which can be tolerated with reasonable safety to citrus trees when the usual fumigation precautions are observed (see pages 244 to 246). Despite these precautions, serious injury to citrus trees sometimes results from the high gas concentration. The actual fumigation schedule employed depends on previous experience in the orchard or the district in question as to the amount of gas the trees can tolerate without being injured.

From the standpoint of the life history of the red scale, it probably makes little difference at what period of the year fumigation is done, for all stages of the insect can be found on the tree at any period of the year. It is not possible to time the treatment for a particular susceptible stage of the insect, as it is possible to do for black, citricola, or purple scales. Nevertheless, red scale is more susceptible to HCN fumigation when it has been "preconditioned" at low temperatures prior to fumigation (732, 526, 1024) so best results at a given concentration of gas are obtained during the coldest months of the year. In addition, a higher

concentration of gas can be used with safety to the trees in winter than during
the warmer seasons of the year. Dosages of from 24 to 28 cc are used for resistant
red scale on both oranges and lemons in the coastal districts and from 20 to 24 cc
in the interior. The winter fumigation season usually extends from the middle of
December to the end of February for oranges or to the end of April for lemons.
This season cannot be extended further because of injury to the spring flush of
growth, the blossoms, and later the young fruit.

The period actually chosen by the grower or pest control operator for red
scale fumigation may depend on a number of practical considerations. If the scale
infestation is so severe that further delay in fumigation will result in injury to
the tree or a reduction in the market value of the crop, the tendency is to fumigate at the earliest opportunity. Likewise experience has shown that it is sometimes difficult to get the work done in California in the winter months because it
is often impossible to do any orchard work for a week or two at a time because of
unfavorable weather or the inability to move equipment through the soft, rain-
soaked soil.

Late summer and fall fumigation may be desirable if black or purple scales
are a problem, for the fumigation must then be timed for the latter insects at
their most vulnerable stage of development.

The summer fumigation dosages for resistant red scale are 16-20 cc in the
coastal areas to 20-24 cc in the interior, in the fumigation of oranges; for
lemons, the dosages are 22-24 cc in coastal areas to 24-28 cc in the interior (728).

In the case of oranges, fumigation may be begun in the latter part of July and
continued until October. The orange tree is rather susceptible to injury for the
remainder of the fall season, especially during the month of October when fumigation often causes pitting of fruit and leaf drop, particularly in the case of
Valencias. If fumigation cannot be completed before the first of October, it
should be deferred until winter. Fumigation may start a week or two earlier on
Navel oranges than on Valencias because of the larger size of the fruit. The old
crop of Valencias should be picked before fumigation because of the drop of fruit
caused by pulling the tents over the tree.

The red scale problem is usually more acute on lemons than on oranges. The
lemon is a better host, and a larger percentage of scales are able to successfully
settle on the tree and reach maturity. Also the life cycle is somewhat shorter,
at least on the fruit. Lemons are, on the average, grown at higher elevations
than oranges, because of their greater susceptibility to injury from low temperatures. This results in a higher night temperature and more rapid development of
the scales. (248) Mature fruit, which is apt to be infested with red scale, is
found on the tree at all seasons of the year. The worst results from fumigation
are on the fruit, which becomes pressed against the tent, thus preventing the gas
from gaining access to some of the insects. In four field experiments in which
orchards with both orange and lemon trees were fumigated, it was found that fumigation results were better on orange trees than on lemon trees despite an average
of 4.5 cc higher dosage schedule in the fumigation of the latter. In laboratory
experiments about twice as great a per cent of survival was found on fumigated
lemon fruits, leaves and green twigs than on the same parts of the orange. (246)

Precautions:- The precautions ordinarily observed in the fumigation of citrus
trees, to avoid as much as possible adverse tree and fruit reaction, are discussed
in Chapter 14.

Double fumigation:- Some growers who have experienced difficulty in controlling red scale with fumigation have fumigated twice in a single year without any
definite interval between the two fumigations. Others have timed the fumigations
to take advantage of certain facts relating to the life history and differential
susceptibility of life stages. The latter procedure has been called "double fumigation" (727, 728). It is most extensively used in the Red Scale Protective Districts of Ventura County where fruit reaction to oil spray appears to be especially

severe, or at least the growers are generally more apprehensive concerning the adverse effects of oil spray than in the majority of citrus districts. Usually when two treatments must be used the grower prefers the "combination treatment" (oil spray followed by fumigation) because it is less expensive than two fumigations and is effective against citrus red spider and bud mite as well as against the red scale.

In the "double fumigation" the assumption is made that all insects except those in the second molt or those in the second molt and early gray adult stage (the two most resistant life stages of either resistant or nonresistant red scale) will be killed by the first fumigation. In actual practice this goal is not entirely realized for a number of reasons. The second fumigation is then timed so as to kill the survivors of the resistant stages after they have developed into stages less resistant to HCN, but before they have a chance to give birth to crawlers, for the latter may spread the infestation. It can be calculated that under average summer conditions if only second molt insects survive the fumigation, a second treatment from 12 to 44 days after the first would find all the scales in vulnerable stages and no crawlers present. If, however, some early gray adults survive the first fumigation, the ideal time for the second fumigation is from 28 to 32 days after the first, an all too short period for practical purposes. If still other stages escape the first fumigation, the second fumigation can no longer be directed against a preponderance of susceptible stages. (727)

Oil Spray

As stated in Chapter 12, the refinement of petroleum fractions of the proper distillation range, plus the discovery of "quick-breaking" emulsions, paved the way for the relatively safe and effective use of oil spray for the control of citrus-infesting scale insects. Only the increase and spread of the resistant race of the red scale, however, brought about the condition which obtains in California today, in which about 3/4 of the treatment for scale insects is with oil spray. Never again was the pest control situation as satisfactory as in the early days of HCN fumigation. Nevertheless, oil sprays must be credited with having played an important rôle in the pest control program in the "resistant areas". Fortunately the race of the California red scale which is resistant to HCN is just as readily killed by oil spray as is the nonresistant race (208, 527).

Even though the per cent kill from fumigation may still be higher than from oil spray in the majority of "resistant areas" it does not follow that the long term results will be superior. Spray oil, besides its initial effect in killing a certain percentage of the insects, has a "residual effect" by virtue of the film of oil left on the tree to inhibit the settling and development of red scale crawlers (see page 198). The combined result of the initial kill and the residual effect is in many cases superior to that of fumigation in the areas where resistant scale predominates. In addition the oil spray has the following advantages: it is less expensive, an advantage that increases with decreasing size of the trees; it is effective against citrus red mite[1] and citrus bud mite; it may be used as a carrier for other insecticides, such as cryolite for orangeworms or rotenone for aphids; and it may be used as a carrier for minor element deficiency correctives such as zinc oxide.

The difference in susceptibility of citrus trees to injury in the interior as compared to the coastal districts is just the opposite in the case of HCN fumigation and oil spray. Whereas HCN is apt to cause more injury to trees and fruit in the damper coastal areas than farther inland, the opposite is true in the case of oil spray. Accordingly, the dosages of oil must be adapted to the requirements for each district. In general in the inland districts a more adverse tree reaction will result from the use of oil spray than in the coastal districts. Within a given district, however, tree reaction is more severe in some areas than in others.

---

[1] Possibly against this advantage should be balanced the possibility that annual treatment by oil sprays resulted in the development of a strain of red mites more prolific and more injurious to citrus trees than that which existed in the pre-oil spray era, as some authorities believe to be the case.

Likewise a grower's experience may lead him to the conclusion that his trees are more susceptible to oil spray injury, or less susceptible, as the case may be, than those of his neighbor's or in comparison with the average condition for his district.

Even an oil dosage that is near the upper limit of tree tolerance comes far from giving a 100% kill of the red scale actually hit by the spray, and careless application may leave many insects unwet by the spray. The per cent kill is especially low on the "gray bark" just below the green twigs, and also on the rough bark of the larger limbs. The oil is absorbed so rapidly by the porous bark that the quantity remaining on the surface is not sufficient for penetration under the body of the scale and into the spiracles. In a lemon orchard sprayed with 1-2/3% "heavy" tank mix oil, it was found that 24 hours after treatment, 52.4% of the red scales showed by the translucent appearance of the waxy ventral membrane (fig.342), that oil had penetrated beneath their bodies and presumably had entered their tracheae. Counts made 6 weeks later showed a net mortality from the spray of 86.3%, indicating that 33.9% of the insects were killed without tracheal penetration, that is, without suffocation. Some of the insects which do not have their spiracles plugged with oil eventually die, presumably from a prolonged impairment of their physiological processes caused by the presence of oil on and in their bodies. (251) Nevertheless, a survival of 13.7% on the branches (and often the survival is much higher than in this particular instance) may result, before the approach of the following year's spray season, in a heavy reinfestation in the orchard and considerable damage to the trees and to the crop of fruit. In such cases a fumigation in addition to the spray may be indicated in the pest control program of a single year.

Precautions:- The precautions ordinarily observed in the spraying of citrus trees, to avoid as much as possible adverse tree or fruit reaction, are discussed in Chapter 12.

Recommendations for the Spraying of Oranges:- August and September are the preferred months for the spraying of orange trees. Mature Valencia oranges should be picked before the spray is applied. Oil spray applied in October is deleterious, particularly with regard to its effect on the following year's crop. It may be that the embryonic fruit buds are in a stage of development in which they are particularly susceptible to injury from oil spray. Also, desert winds are quite common in October, and these produce an adverse weather condition for both spraying and fumigation. Subsoils are apt to be drier than at any other time of the year, and this may serve to weaken the resistance of the trees to the adverse effects of pest control treatments. Later spraying, as in November, may cause an impaired coloration of the following year's crop, especially in the case of Navel oranges.

Navel orange trees may be sprayed during the winter months, after the fruit is picked and before the period of bloom. This period is so limited and weather conditions are often so adverse, that only a small percentage of the acreage could be treated at this time of the year. The most important advantage of spraying after the fruit is picked is that injurious effects of oil spray on the fruit are avoided. Of special economic importance is the avoidance of excessive water spot, a malady brought about by prolonged rainy or wet periods and which is greatly accentuated by oil spray (see pp. 207-210). Not all growers agree that the spraying of Navel orange trees after the pick is desirable, even when practicable. Some have observed an excessive defoliation and formation of dead wood and reduced production in these winter-sprayed orchards. Others seem to have carried out a winter spray program on Navels with good success.

Valencias have also been sprayed during the winter and spring months, particularly in the Orange County - Whittier area. As with the Navels, adverse tree reaction has been noted. With both varieties, damage has been more noticeable from the later work, after the blossom growth was advanced, than in February. This has been especially true when sudden warm spells have followed the treatments. (1006)

According to the entomologists of the Exchange Pest Control Bureau, in addition to causing an adverse tree reaction, the spring oil spraying has been less

461

effective in the control of scale insects and citrus red mites than the same treatment applied during the usual pest control season. It is their opinion that if oil spray is used in the winter-spring season, it should be only in cases of extreme necessity and when fumigation is not possible.

A light medium oil (emulsives, 1-3/4%, emulsions, 2%) is generally recommended for the spraying of orange trees in the interior and intermediate districts. For coastal districts, a medium oil at the same concentration may be used if the infestation of scale is especially heavy.

Recommendations for the Spraying of Lemons:- In the case of lemon trees, early spraying is not especially desirable, in fact, it is often avoided to prevent the dropping of ripe or nearly ripe fruit which often occurs during periods of hot weather following an oil spray. In any event, it is usually desirable to complete all spraying for oranges first, before spraying of lemons is commenced. If black or purple scale is an important consideration, or if citrus red mite requires early attention, lemon trees may be sprayed in the same period as the oranges. For the major part of the lemon acreage, however, oil spraying commences in October and extends through November and often considerably later. Pickable sizes of fruit should be removed before an oil spray is applied.

The writer does not wish to imply that all authorities are in agreement as to the best time for the spraying of lemons. Newcomb (643), for example, states: "In the fall it is advantageous under our conditions (in Corona) to finish spraying lemons before October first in order that the oil may have time to leave the fruit before it is picked. Green fruit sprayed later in the fall is often difficult to color with ethylene in the packing house. Late fall spraying has also resulted in bronzing of the yellow fruit held on the trees."

Obviously, the spraying schedule must be arranged to accommodate the picking schedule, or vice versa.

For the spraying of lemons, a medium or heavy medium oil (emulsives 1-3/4% and emulsions 2%), should be used. The heavy grade of oil was once widely used for the spraying of lemons, but it is now rarely used except in the Chula Vista section. There it is used chiefly because of the difficulty of controlling citrus red mite from one year to the next with any lighter grade of oil.

The generally unfavorable results from winter and spring spraying in the case of oranges does not apply to the spraying of lemons. Much spraying of lemons is done in the spring, March and April being the preferred months. Medium or heavy medium oil may be used. An increasing number of growers are spraying lemon trees in March or April with light medium oil as a part of an annual two-spray red scale and citrus red mite control program. The other light medium spray is applied in the fall. The two-spray program has resulted in improved red scale and mite control with apparently no greater tree reaction or effect on production than would be obtained from a heavy medium oil applied once a year.

Miscellaneous Varieties:- For the spraying of red scale on grapefruit, a medium oil may be used in areas where only a light medium oil would be recommended for the spraying of oranges. The grapefruit tree is more tolerant to spray oil than the orange. On the other hand, tangerine trees are more susceptible to oil spray injury than orange trees.

Although there are only 755 acres of limes in California out of a total citrus acreage of approximately 330,000 acres, the severity of the damage sometimes resulting from the use of oil sprays on limes justifies a few words of warning. Spraying done in November and December has been particularly injurious. Injury has been caused by both light medium and medium oil. When injury has occurred, it has not become evident until several months after the trees were sprayed. Leaves begin to wilt, then dry so fast that they remain attached to the tree (fig. 352). Dead bark beginning on smaller branches extends downward, sometimes as far as the trunk, and in some cases the entire tree may be killed. It is recommended that lime trees be

fumigated instead of sprayed, but if
they are sprayed, the treatment should
be made before the end of September. .
(983)

Combination Treatment:- When resistant red scale infestations become
so severe that neither oil spray nor
fumigation will keep the trees "commercially clean" for a year - that is,
until the normal pest control season of
the following year - an effective control can still be obtained by means of
the "combination treatment" or "spray-
fumigation treatment". According to
this procedure, the oil spray is applied as usual in the late summer or
fall months and is followed by a fumigation. In field experiments, some ad-

Fig. 352. Lime tree killed by late fall
spraying with oil. After Woglum (983).

vantage has been found in fumigating from 10 to 15 days after spraying, if both
treatments are completed before winter (731). Perhaps the action of the oil spray
in loosening or "lifting" the margins of the scales tends to make fumigation more
effective, at least this can be shown to be the case if the scale margins are
lifted by means of a knife blade, to simulate the typical action of the oil, and
then fumigated. Lindgren and Dickson (527) in laboratory experiments found that
scales surviving oil spray were no more easily killed by fumigation than unsprayed
scales, but they do not record whether the sprayed insects had the margins of their
armor raised above their substrate as so often occurs in the field. However, regardless of the interval between spraying and fumigating, the combination treatment
is highly effective. In laboratory experiments, no difference was found in effectiveness of the treatment whether the fumigation followed or preceded the spray.
Neither was there any significant advantage in any particular time interval between spraying and fumigation among the time intervals that were tested (from 1 to
14 days). (527)

The progeny of resistant scale surviving an oil spray are no less resistant
to HCN than their progenitors (1027).

Oil spray results in a much better kill on the green twigs, leaves and fruits
than on the rough bark. Fumigation results in much better kills on the rough bark
than on the fruit. This is due in part to the fact that when the tent is pressed
against the periphery of the tree, such parts of the tree as come in contact with
the tent fabric do not obtain the full gas concentration and as a consequence the
survival of red scale is high in these areas. In addition, it is reasonable to
suppose that the scales located on the larger branches, which are smaller in size
and appear to be less vigorous than those on the fruit, green twigs and young
leaves, would also succumb more readily to HCN. When oil spray and fumigation are
both applied within a single pest control season they are thus seen to have a
complementary effect. The oil spray is most effective on the parts of the tree
where HCN is least effective, and vice versa. The two treatments are also complementary in another way. The HCN is least effective against the second molt and the
early gray adults (527) but oil spray is very effective against these very stages
(251). Oil spray, on the other hand, is least effective against the adult stage,
which is relatively easily killed by fumigation.

The oil spray usually precedes the fumigation in actual field practice for
several practical reasons. In the treatment of orange trees it is desirable to
spray before the first of October. The grower wishes to make sure of getting the
spray applied during this period, but he may not be able to arrange for fumigation
as well as spraying. Pest control operators concentrate on the spraying of oranges
in August and September because they know that the spraying of lemons and the fumigating can be successfully done later in the year. Sometimes the citrus red mite
infestations become severe early in the pest control season, especially on lemons,
and this again favors oil spray before the fumigation.

The fumigation may be deferred until the winter months. Probably the complementary effect of the two treatments is then partially lost because of a certain amount of dispersal of the progeny of the survivors of the oil spray from the branches, where survival is high, back to the periphery of the tree. On the other hand, the red scale should more readily succumb to the fumigation in the winter months, because of the low temperatures, and in addition the trees can tolerate a greater HCN gas dosage. The results of combination treatments, when the fumigation was deferred until winter, were found on the average to be about as good as when the fumigation followed rather soon after the oil spray. (731)

Possible New Insecticides for Certain Armored Scales:- A satisfactory substitute for the present types of oil spray, or the use of lighter petroleum fractions made effective by the addition of toxicants, has long been the goal of investigators working on the red scale problem. The possibilities of the use of DDT with 3% kerosene have been extensively investigated (260, 268, 147). Although no general recommendations have as yet been made, some growers are using a two-spray program with 3% DDT-kerosene in areas where a substitute for the regular oil sprays is especially desirable, such as in the Navel orange areas where much loss of fruit has been suffered from water spot. However, the difficulty experienced in the control of citrus mite infestations that follow the application of DDT to citrus trees, and some evidence that successive years of DDT application to citrus may provoke certain symptoms of chronic phytotoxicity, have indicated that recommendations for the use of DDT in red scale control would not be justified.

In experiments made with DDT and spray oil or kerosene used against the yellow scale, Aonidiella citrina, the scale population was increased over that found in the plots sprayed with oil alone (268). The same thing was found to be true when DDT was added to oil in the control of the Florida red scale, Chrysomphalus aonidum (384). Parasite and predator populations were reduced by the oil-DDT sprays, but not materially affected by oil spray alone. There was an irreducible minimum of DDT remaining on or inside the leaf for many months, but deposits of 0.5 mcg DDT/cm^2 of leaf surface have no effect on Florida red scale or its natural enemies. (384)

With purple scale, the addition of DDT to petroleum sprays appears to increase the long-term effectiveness of the latter even more than in the case of the California red scale (533). Here again, however, not enough information is available on the various ramifications of the problem to justify recommendations at this time.

Experiments have indicated promise for the new organic phosphate insecticide, parathion, in formulations containing no oil whatever.[1]

YELLOW SCALE

Aonidiella citrina (Coquillett)

The yellow scale is not as widely distributed over the world as the red scale, having been recorded only from Australia, Japan, India, California and Texas. It was accidentally introduced into the San Gabriel Valley, California on small orange trees imported from Australia.

The yellow scale was described as a variety of the red scale by Coquillett in 1891, and as such it was considered until Nel (640), as a result of an intensive morphological investigation of the red scale and yellow scale, came to the conclusion that they were distinct species. The morphological characters used by Nel to separate the species, however, were found to be unreliable by other workers because of their great variability. This led to further investigation by McKenzie (570), who found certain internal sclerotized structures on the ventral side of the pygidium by which red scale and yellow scale could be distinguished (figs. 338 and 339). Otherwise, the two species are practically identical in structure and their life history is also very similar.

---

[1] G. E. Carman. 1948. Unpublished data on file at the Citrus Experiment Station, Riverside.

It has always been possible to distinguish red scale infestations from yellow scale infestations in the orchard because of the difference in color of the live insects and also on the basis of the difference in their habits. The difference in color is implied in the common names of the two species. S. E. Flanders has found that the color difference can be accentuated by rearing the insects on banana squash. The red scale then becomes grayish brown or chocolate-colored and the yellow scale is of a more pronounced yellow color than when it occurs on citrus. In addition to the difference in color, the somewhat greater convexity of the red scale aids in the separation of the species.

Differences in ecology and habit further serve to distinguish yellow from red scale in the field. Although yellow scale can be found sparsely distributed in coastal and intermediate citrus districts, it is abundant only in such interior areas as the citrus districts of the San Joaquin and Sacramento Valleys and in the Redlands and Highlands districts in southern California. On the other hand the red scale has demonstrated its ability to thrive in both coastal and interior districts. It is now found abundantly in the Redlands-Highlands yellow scale area and apparently only prompt and vigorous eradication measures have prevented widespread and serious infestations in the San Joaquin and Sacramento Valleys.

There is also a rather distinct difference in the habits of the two species with reference to the parts of the citrus tree they infest. The red scale indiscriminately infests all the morphological parts of the tree, while the yellow scale predominantly infests the fruit and leaves and is only sparingly found on the twigs and branches.

Injury:- The yellow scale may cause defoliation, deadwood, and dropping of fruit, in addition to the economic loss occasioned by the infested fruit, which is culled. It probably causes as much damage per individual as the red scale, but since it is not as widely distributed throughout the tree, the total effect on the tree of a severe yellow scale infestation is not as great as that of a red scale infestation. As in the case of the red scale, the yellow scale occurs on a wide variety of hosts besides citrus.

Parasites:- It appears that the most important parasite of the yellow scale is Comperiella bifasciata How. (fig. 353). C. bifasciata is of a general metallic color, but with the front and vertex of the head a creamy white and bearing a dark longitudinal median stripe. The most striking characters of this species are more conspicuous in live than dead specimens, for their brilliant metallic

Fig. 353. Comperiella bifasciata depositing its eggs in a yellow scale. X 20. After Compere and Smith (182).

reflections are then seen to best advantage. Also in life the apical portions of the wings are held semi-erect, but the wings straighten out after death. (178, 182)

It was once believed that this hymenopterous insect parasitized the California red scale in Japan, but later investigations showed that the latter insect was not the California red scale, but another species, which is now called the Asiatic red scale, Aonidiella taxus Leonardi.

Superficially, A. aurantii and A. taxus appear identical, but properly cleared and stained specimens reveal that the pygidium of the latter species "entirely lacks the prevular scleroses and apophyses" (571).

Comperiella bifasciata imported into California oviposited in the California red scale, but only one male was ever recovered. This species successfully developed in the yellow scale, however, and the first liberations in citrus orchards were made in 1931. The amazing specificity of some species of parasites is illustrated by C. bifasciata which can develop successfully in Chrysomphalus aonidum, A. taxus and A. citrina but not in A. aurantii which is so similar in appearance to

A. citrina that it required considerable taxonomic research to find the microscopic structural characters by which the species may be differentiated.

In the Redlands-Highlands area, where the yellow scale was once a major pest, C. bifasciata did excellent work in reducing the yellow scale population to the few minor outbreaks that have occurred in this district in the last decade. In Tulare County, however, it generally has not been able to keep the yellow scale population sufficiently low to obviate the necessity for treatment. Insecticide treatments greatly deplete the parasite population, and the high degree of control obtained results in the starvation of remaining individuals. Comperiella nevertheless tends to retard the build-up of yellow scale after treatment and it is stated to have reduced the cost of yellow scale treatment even in Tulare County. (332)

A Chinese race of C. bifasciata has been introduced into California which is capable of developing in the red scale, and while this insect can be recovered in the field, it has not given evidence of having practical value. It is planned, however, to give this species a further test by the use of the mass culture and periodic colonization methods.[1]

Fig. 354. Aspidiotiphagus
citrinus.
After Quayle (708).

Second in importance to Comperiella bifasciata is another hymenopterous parasite, the yellow scale-infesting form of Aspidiotiphagus citrinus Craw. (fig. 354). This species is of a general yellowish black color, but has black eyes and red ocelli. Like C. bifasciata the yellow scale-infesting form of A. citrinus is not able to develop in the California red scale. This insect was apparently introduced along with the yellow scale and became abundant along with the increase of the latter species. It seems to have been the factor which practically eliminated the yellow scale from the San Gabriel Valley. (332)

A third parasite, Prospaltella aurantii (Howard) occasionally destroys a high percentage of yellow scale in southern California, but it is not known to occur north of Tehachapi.

Control:- Fumigation and oil spray as for red scale are recommended for the control of this insect. Data obtained by Lindgren from fumigations of yellow scale, nonresistant red scale, and resistant red scale together in the same fumatorium, indicate that yellow scale are more resistant to HCN than nonresistant red scale, but they are much less resistant than resistant red scale (236). Fumigation has generally been a rather successful method of treatment.

Oil spray is more effective against yellow scale than against red scale because the former do not occur on the rough bark, where oil spray is less effective than on any other part of the citrus tree.

OTHER SPECIES OF AONIDIELLA

McKenzie (571) included in a single paper descriptions of all of the then known 17 species of Aonidiella, including excellent illustrations to aid in the identification of 13 of these species. He later added two new species to this Aonidiella complex (572). This clarification of the taxonomic status of the many species involved is of great benefit as a basic step in the formulation of systematic and equitable quarantine, survey, and control and eradication policies by state and federal officials. It is also valuable in connection with parasite exploration work. Insects which were once believed to be the California red scale are now known to be closely related, but distinct species. This explains certain failures of parasites which were taken from what appeared at the time to be California red scale but now are known to be other species. These parasites were not

---

[1] Correspondence from Professor H. S. Smith, December 10, 1947.

successful against the true California red scale. Parasites are often so specific in their host preferences that they will develop successfully in only a single insect species. For example, two parasites which successfully attack the yellow scale in California cannot develop in the red scale.

FLORIDA RED SCALE

Chrysomphalus aonidum (Linnaeus)

The Florida red scale is the only important citrus pest to be described by Linnaeus, who described it in his tenth edition of Systema Naturae (1758) from specimens collected in India under the name of Coccus aonidum. Among the relatively minor citrus pests, Linnaeus described the soft brown scale, Coccus hesperidum, and certain fruit piercing moths.

The Florida red scale is one of the two most important scale insects of Florida, where it was apparently introduced from Cuba in 1874 (596). It is the most important citrus pest in Egypt, where it is known as the black scale, the most important citrus pest of some sections of Mexico, and it also occurs on citrus in the Gulf States and in many other citrus producing areas throughout the world. In California it is a pest only in greenhouses. It appears to be unable to tolerate climates as arid as those in which California red scale thrives. In Florida this pest occurs most abundantly along the lower east coast, where the climate is especially favorable for its development, but in the northern part of Florida, where freezing temperatures sometimes occur, it is not so abundant. When a tree is defoliated, as by a freeze, the Florida red scale may be practically eliminated. Mathis (596) lists 191 host species of the Florida red scale.

Description of Adult (596):- The dorsal scale covering is circular and convex, and is made up of 3 distinct rings. The innermost ring is nearly central. It is light brown, while the second ring is reddish brown. The third, which is wider

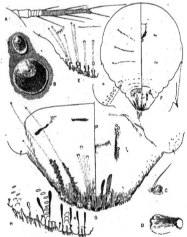

Fig. 355. Florida red scale, Chrysomphalus aonidum. A, antenna and cephalic margin of crawler; B, armor of male and female; C, antenna of adult female; D, anterior spiracle of adult female; E, pygidium of second-instar female; F, general features of adult female; G, pygidium of adult female; H, dorsal aspect of detail of pygidial margin. After Ferris (318).

than the other two combined, is reddish brown to black, with a thin gray margin. The armor of the adult female Florida red scale can be lifted away from the body of the insect, in contrast to that of the adult female California red scale, which

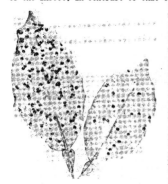

is attached to the body. This is a convenient means of differentiating the two species in areas in which they occur side by side. Another difference is that whereas the body of the adult female California red scale, as seen when lifting the armor of the grey adult, is somewhat reddish and "kidney shaped" or "heart shaped", that of the adult female Florida red scale is bright yellow in color and "the shape of a wide short top" (926), (fig. 355, p. 467). It is a little over 1 mm in length. Another difference in the two species is that the California red scale gives birth to active young while the Florida red scale lays eggs. As the eggs are laid, the body of the female becomes progressively smaller. When oviposition is complete, the body is very much shrivelled and practically clear in color. The armor of the Florida red scale is much darker (fig. 356) than that of the California red scale and it is also more convex. Taxonomic characters used in the determination of stained and mounted specimens are shown in fig. 355.

Fig. 356. The Florida red scale, Chrysomphalus aonidum, X 0.7. Photo by Roy J. Pence.

Life History (596):- The oval, lemon-yellow eggs are deposited beneath the dorsal scale covering, where they remain until they hatch. Females continued to lay eggs for a time after the scale covering was removed. It was found that in warm weather a female may lay 2 eggs per hour, and 334 of the bright yellow, active crawlers were removed from an isolated female on a fruit in a 51-day period. The crawlers could live without food on slightly moistened filter paper from 6 to 13 days and on dry filter paper for 3 or 4 days, showing a much greater ability to survive than the California red scale. The crawlers will settle on leaves or fruits of any age, with the exception of very immature fruits. Scales were rarely found on green wood unless the infestation was extremely heavy. None were found on gray wood.

Upon settling, the first instar nymph continues its development to the period of the first molt in 46 days at 59°F to 15 days at 82°F. The second instar female completes its development in from 36 days at 61°F to 11 days at 81°F. As in the case of the first instar, the optimum temperature for development was between 78° and 83°F.

The Florida red scale cannot reproduce without fertilization. It is believed that fertilization occurs at night shortly after the female completes the second molt. The period from completion of the second molt to the first egg deposition ranged from 2 to 4 weeks and the oviposition period ranged from 1 to 8 weeks. Thirty females on orange fruits produced from 32 to 334 crawlers, with an average of 145. Twenty-five females on leaves produced from 21 to 156 crawlers, with an average of 80.

The Male:- The development of the male follows the usual pattern of coccoid male development (see page 450). Winged adults were found 15 days after the first molt was complete. The total period for development ranged from 78 days at 61°F to 28 days at 83°F, about the same period as required for the female to complete the second molt. The male takes no food, and 4 days was the longest period that an adult male was kept alive. Of 15,738 scales examined over a 12-month period, 59% were females, the range being from 52 to 68%. Ninety-six per cent of the males and 13% of the females were on the upper side of the leaves. Experiments indicated that light is one of the most important factors influencing the distribution of the females. In the absence of light the females were evenly distributed between the

two surfaces of the leaves. Both light and gravity appeared to affect the distribution of the male, although other factors appeared to exert an influence.

Seasonal History (596):- Population studies showed a peak in living scales and
ovipositing females in August, September and October in Florida. It is believed
that 5 or 6 generations of Florida red scale would occur per year at a mean temperature of 74°F. Three or 4 generations per year have been reported in Palestine at a
mean yearly temperature of 67°F and 5 generations at a mean yearly temperature of
73°F. (766) In Florida there are usually 4 generations per year (Fla. Bul. 462).

Injury and Economic Importance (596):- Yellowish areas form on the leaves beneath the scales and eventually cover the entire leaf as the infestation increases.
Trees may become nearly completely defoliated, with a resulting decrease in their
vitality and productivity. Infested fruits are unattractive and besides they do
not attain their color uniformly, and the grade of the fruit is consequently reduced. Sometimes the canning plants refuse to accept heavily infested fruit because of the difficulty of removing the scales and because of the possibility of
their becoming incorporated into the finished product.

Natural Enemies:- The hymenopterous parasite Aspidiotiphagus lounsburyi (Berl.
and Paoli) is the only parasite of any importance attacking the Florida red scale
in Florida (596). It attacks mainly the immature stages. In 1942, the percentage
of parasitized scales ranged from 7.4 in August to 20.0 in November. Prospaltella
aurantii (How.) and Pseudohomalapoda prima (Gir.) have been recorded as being parasites of the Florida red scale. The former is found as a late-instar larva and as
a pupa in second-instar female and male scales, while the latter species is found
as a late-instar larva or pupa in third-instar female scale. (384) Comperiella
bifasciata How., introduced from the Orient in 1924, became successfully established in California as a parasite of the Florida red scale on an ornamental plant,
Aspidistra, to which the Florida red scale is practically confined in this state.
(804)

The twice-stabbed ladybird beetle, Chilocorus stigma (Say.) (fig. 303; J), is
numerous in some orchards in the latter part of winter and both the larvae and
adults feed on all stages of the Florida red scale. If it fails to pull off the
covering of the scale, this predator chews a hole through the dorsum of the scale
covering and devours the body. The larva of Chrysopa lateralis Guerin is said to
be able to devour as many as 12 mature Florida red scales in an hour. This predator pulls the dorsal scale covering until it is loose, then inserts its mandibles
under this covering and feeds on the live insect body. The predaceous mite,
Hemisarcoptes malus (Shimer), is occasionally found under ovipositing females and
seems to cause some mortality of crawlers. In general, however, parasites and
predators are not very important factors in the control of the Florida red scale.
(596)

Certain fungi once believed to be important in the biological control of
Florida red scale have proved to be of little or no value.

Control:- The treatment for both Florida red scale and purple scale in Florida
is oil spray, and since the recommendations as to season and dosage are practically
the same for these two pests, they will be discussed jointly in the section on
purple scale (see p. 474).

DICTYOSPERMUM SCALE

Chrysomphalus dictyospermi (Morgan)

The dictyospermum scale is widely distributed on citrus and is the most important citrus pest in the western Mediterranean basin. In California, where the
dictyospermum scale was first reported in 1909, it has occasionally been found on
lemons and avocados, especially the latter, but has never become abundant or widespread. In Florida, dictyospermum scale is widely distributed, and is a major
pest of the avocado. Serious infestations of dictyospermum scale are sometimes

found on tangerines, oranges, and grapefruit in Florida (879). A survey showed the insect to be attacking citrus in Lake, Orange, and southwest Seminole counties.

The insect occurs on a large number of host plants and is common in greenhouses. The injury to citrus is much the same as that of other armored scale insects whose potentialities for injury are somewhat reduced by their scarcity on the twigs and branches.

Life History and Habits:- This insect, like the Florida red scale, deposits eggs, and like the Florida red scale, also, the armor can at all times be lifted from the live body of the insect. This species might be confused with the California red scale in the field, but the lifting of the armor of the adult female dictyospermum scale is a convenient way of distinguishing the insect. The armor is lighter in color than that of the California red scale. The body of the female found beneath the armor resembles the pear-shaped body of the Florida red scale more than the kidney-shaped body of the female California red scale. The dictyospermum scale is less likely to occur on the twigs and branches than the California red scale, but does not avoid these parts of the tree to the same extent as the Florida red or the yellow scale. Stained and mounted specimens may be identified on the basis of taxonomic characters shown in fig. 357).

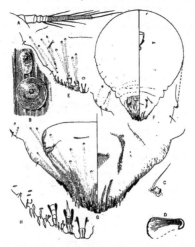

Fig. 357. Dictyospermum scale, Chrysomphalus dictyospermi. A, antenna and cephalic margin of crawler; B, armor of male and female; C, antenna of adult female; D, anterior spiracle of adult female; E, pygidium of second-instar female; F, general features of adult female; G, pygidium of adult female; and H, dorsal aspect of detail of pygidial margin. After Ferris (318).

The life history of the dictyospermum scale was worked out by Cressman (206). All stages of the insect are of a lemon-yellow color. The color of the scale covering varies from a yellow-brown to brown among immature insects, but among older insects all variations from gray to dark brown may be observed. Although parthenogenesis in this species has been reported, Cressman found fertilization to be necessary for reproduction in the New Orleans district. All the scales he studied were oviparous. The life history of the dictyospermum scale is similar to that of the California red scale. In the insectary Cressman reared 5 to 6 generations a year. There may be 3 or 4 generations per year out of doors in California (724).

Cressman (206) recovered the following parasites from the dictyospermum scale: Aphytis chrysomphali (Mercet), Prospaltella aurantii (How.), Signiphora merceti (Mal.), and S. flavopalliata (Ashm.). The effectiveness of these parasites was reduced because they did not kill the scale before it had a chance to lay a certain number of eggs. The coccinellid beetles Rhizobius debilis Blackburn and Lindorus lophanthae Blais. were believed to be more effective as natural enemies.

Control of dictyospermum scale was found to be easily accomplished by means of oil sprays (206).

PURPLE SCALE
Lepidosaphes beckii (Newman)

From a world standpoint, the purple scale vies in importance with the California red scale, for while it is generally somewhat more easily controlled by existing artificial control measures than the red scale, it is more widely distributed throughout the citrus regions of the world than the latter species. It is the most widely distributed of the important citrus scale insects. It appears to thrive in a wider range of climatic conditions than many other scale insects, but prefers the more humid climates.

In California, the purple scale does not occur in the interior districts, but in the coastal districts it is an important pest, being overshadowed in importance only by the red scale, among the insects. It appears to have been introduced on orange trees received from Florida in 1889 (724). In Florida and some of the other Gulf States, the purple scale is the most important citrus pest. In parts of Mexico and the West Indies, in Central America, Peru, Chile, parts of Paraguay and Brazil, and parts of southeast Asia, it is also the most important citrus pest.

Unlike some of the other citrus-infesting scale insects, the purple scale confines its attacks largely to citrus and is not found on many other host plants.

Fig. 358. Purple scale on orange. M, males; F, females. X 6.

Description of Adult:- The armor of the female, which is 2-3 mm long, is shaped somewhat like an oyster shell (fig. 358). It is often curved into the shape of a comma, which accounts for the various Spanish equivalents for the name "comma scale" by which it is known in Spanish-speaking countries of the Western Hemisphere. In color it is purplish or purplish brown. Beneath this armor may be found the elongate body of the female, which averages 1.2 mm in length and 0.7 mm in width (710). The pygidium of the adult female is shown in fig. 359, p. 472, along with that of the closely related Glover scale.

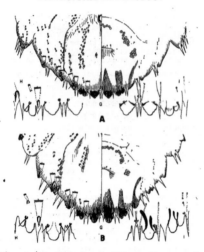

Fig. 359. A, purple scale, *Lepidosaphes beckii*; B, Glover scale, *Lepidosaphes gloverii*. G, pygidium of female; H, dorsal aspect and, I, ventral aspect, of detail of pygidial margin. After Ferris (318).

Life History (710):- The life history of the purple scale was worked out in California by Quayle, and a similar investigation is now being undertaken by the entomologists of the Subtropical Fruit Insects Laboratory at Fort Pierce, Florida.

The Egg:- The pearly white eggs are oval, 0.25 mm long and 0.15 mm broad, and have a very minutely granulate surface. The number of eggs per female was found to be from 40 to 80, arranged somewhat irregularly and with their long axes at an angle with the substrate. They are deposited in 3 weeks or more and hatching occurs in from 2 weeks in summer to 2 months in winter. While depositing eggs, the body of the female contracts toward the narrow anterior end, and the space thus provided is taken up by eggs.

The First Stage Nymph:- The motile crawler is of a whitish color, with the posterior tip brown. It has the usual flat and oval shape of scale crawlers. The length of the body is 0.78 mm and the width is 0.64 mm. The distinctly annulated antennae are 6-segmented. The crawler wanders about for a period varying from a few hours to a day or two, then settles on a branch, twig, leaf or fruit and begins to secrete two coarse, cottony threads of wax from just under the margin of the anterior end. These eventually form a cottony mat covering the insect. The threads of wax are believed to form a protective function while the more compact scale covering is being formed, and young coccinelid larvae, upon encountering the threads, will immediately back away or turn in another direction without molesting the protected insect. The mass of threads remain on the insects until they are about half grown. When the scales are numerous they form a white, fuzzy coating over the surface. During the period the scales are covered with the masses of waxy threads, they are said to be in the "fuzz stage". This term is familiar to entomologists and pest control operators, for it is only in the "fuzz stage" that the purple scale can be successfully treated with either oil spray or HCN fumigation. In its more advanced stages of development the purple scale is killed with even greater difficulty than the red scale.

armor as in the case of the red scale, but since the pygidium of the purple scale is moved only from side to side, the armor is extended in only one direction, becoming elongate, but progressively wider, as the insect increases in size. The ventral covering consists of waxy secretions only, with no cast skins.

Second Instar Nymph:- The first molt occurs about 3 weeks after birth in summer. The molting process differs considerably from that of the red or black scales. The red scale splits its old skin so that half is incorporated into the dorsal and half in the ventral scale covering. The unarmored scales molt in such a manner that the skin is split at the anterior end and is pushed back off the insect. In the case of the purple scale, the rent in the old skin occurs in an irregular line a short distance from the margin on the sides, between the mouthparts and the antennae in the front and just anterior to the pygidium. After emergence of the anterior end, the rest of the body is pulled forward and downward, leaving the exuvia attached to the dorsal covering. The period required for the second molt is about the same as that required for the first molt, or about 3 weeks. Soon after the second molt, fertilization takes place and eggs are deposited about 15 days later.

Life history studies of the female purple scale conducted in a screened, roofless insectary at Auburn, Alabama, by English and Turnipseed (277) resulted in the following data:

Period	Number of days		
	Minimum	Maximum	Average
Birth to first molt	12	44	21.3
Birth to second molt	21	62	34.9
Birth to first egg	33	148	62.0
Incubation	6	50	15.3
Complete development	42	198	77.3

The Male:- As usual with scale insects, the sexes are indistinguishable until after the first molt. The male scale covering is shorter and much narrower than that of the female (fig. 358, M). Beneath this scale covering the development of the male takes place in accordance with the usual diaspine pattern for the males. The male emerges from under the scale covering as an adult about 50 days after birth.

Seasonal History (710):- At Whittier, California, the purple scale was found to pass through its life history in 3 months or less during the warmer periods of the year, and 2 generations develop from May to October, allowing for another generation to get fairly well started before the coldest weather of winter, and in the coldest winters the individuals of this generation will not deposit eggs until spring. Mild winters allow for the development of more or less of a fourth generation.

Injury:- The purple scale, like the California red scale, attacks all parts of the tree--the fruit, leaves, green twigs, and branches--and injury can be very severe. In California the damage from purple scale, such as defoliation and deadwood, is most severe in limited patches scattered about over the lower north side of the tree. These rather sharply delimited injured areas may spread and eventually envelop the entire tree, but treatment usually prevents this most extreme condition of purple scale injury. West of the Andes Mountains in Peru, where heavy deposits of dust and inadequate methods of oil spray application prevent effective treatment for the pest, it is killing trees and ruining entire orchards.

On the leaves, the area surrounding each scale or group of scales turns yellow and remains discolored after the scales are removed. Similar areas on the fruit do not attain their normal color, but remain green. The purple scale is negatively phototropic, and in Florida the crawlers seek to avoid direct sunlight by settling under slightly loose sooty mold or the secretions of the wooly whitefly or in the leaves curled by citrus aphids. All these conditions favor an increase in the scale population. Heavy infestations of purple scale also occur around and under the calyx, and Fawcett (301) showed that the scales increase the incidence of stem-end rot in citrus fruits in Florida.

Natural Enemies:- In California the parasite most commonly attacking the purple scale is the purple scale-infesting form of Aspidiotiphagus citrinus Craw. (fig. 354). The round holes in the posterior one-third of the second instar scale coverings show where these parasites have emerged. It is never of great importance in the control of the purple scale, especially in orchards where spraying or fumigation is regularly practiced. The most common predator appears to be the coccinellid beetle Lindorus lophanthae Blaisd., which has already been described in the discussion of red scale predators (p. 454). Scymnus marginicollis Mann. is another coccinellid commonly found attacking the purple scale. This beetle is of about the same size as L. lophanthae, but is of a dull black color instead of the metallic bronze luster of the latter species. Other coccinellids, as well as lacewing flies and mites, occasionally feed on purple scales. In Florida the twice-stabbed beetle, Chilocorus stigma Say, is considered to be of some importance as a predator. It is now known that certain fungi previously believed to be of value in the control of scale insects in Florida are of little or no value. However, according to Miss Fran E. Fisher of the Citrus Experiment Station at Lake Alfred, the endoparasitic chytrid, Myiophagus sp., is a major factor in the biological control of purple scale.

Control in California:- There is usually a hatch of the purple scale in late July and August. This is indicated by the "fuzziness" of the scales caused by the mass of entangled waxy threads secreted by the young nympha. The purple scale is then in the "fuzz stage" and is considered to be especially vulnerable to either fumigation or oil spray. Another hatch occurs in September and October. These are not distinct hatches, and there is a considerable overlapping of generations, but during the two periods referred to there is a preponderance of individuals in the vulnerable "fuzz stage".

Years of experience in California has shown that fumigation in late July and early August has resulted in better control than at other periods of the year. The fumigation schedule should not be less than 18 cc. Heavily infested trees should be sprayed with oil before or after fumigation. Adverse tree and fruit reaction to fumigation will be less at this period of the year than at any other. If treatment is deferred until the next hatch, the purple scale will have spread to the fruit. Likewise conditions for fumigation are less favorable than in August. (971)

The purple scale is again predominantly in immature stages in September, October and November, the exact period varying from year to year, and oil spray is the recommended treatment during this period, using a light medium oil; emulsion 2%, emulsives 1-3/4%. Medium to heavily-infested trees should later be fumigated as soon as the trees are in a condition in which they will tolerate the gas, which is usually in late December and early January. Spraying should be thorough, with emphasis on good coverage of the inside of the tree.

Control in Florida:- For the control of scale insects in Florida, oil spray is used almost exclusively. Refined petroleum of from 70 to 85 seconds Saybolt viscosity is commonly used. Thompson (881) recommends from 1-1/2 to 1-2/3% actual oil applied between June 15 and July 15, or at least before August 1. Injury to the trees sometimes occurs from the late May or early June sprays, and spraying after the recommended period tends to delay coloring of the fruit in the fall and prolongs the period required to color or "degreen" the fruit in the coloring room. Likewise in the fall months if the trees should have both a heavy crop of fruit and a high scale population, the shock to the tree caused by oil spray is particularly severe.

The most effective spray coverage of the tree can be obtained early in the summer before the weight of the fruit has bent the limbs and compacted the foliage. In the fall the clusters of fruit also prevent the turning of foliage by the spray

CITRUS PESTS

William L. Thompson
Entomologist at the University of Florida Citrus
Experiment Station, Lake Alfred, Florida.

stream, thus reducing coverage efficiency. It is believed in Florida that bad results in purple scale control are more apt to result from inadequate spray coverage
than from any other factor. It is especially important in the case of the purple
scale to wet the limbs and twigs because they serve as a source of reinfestation of
the sprayed leaves. (881)

It is important in Florida to prevent purple scales from becoming established
on the fruit because Florida oranges must be artificially colored by means of
ethylene gas, and wherever purple scales occur they affect the fruit adjacent to
their bodies in such a way that it remains green after the remainder of the fruit
has colored. This accounts for the "green spots" in the coloring room. It is more
difficult to obtain a good coverage of the larger fruit in the fall than the smaller
summer fruit.

It was observed in Florida that oil sprays applied after August 1 either delayed or prevented the formation of the maximum quantity of solids in the fruit.
The early application of oil spray is of special importance if the fruit is of a
variety which normally has a low per cent of solids or one which is to be picked
early in the season. (885)

GLOVER SCALE

Lepidosaphes gloverii (Packard)

The common name "Glover scale" has been accepted by the Committee on Common
Names of the American Association of Economic Entomologists (628), and it is the
name that has commonly been used in California, although the name "long scale"
appears to be more commonly used in Florida, and the Spanish equivalent, escama
larga, is used in Mexico. This insect appears much the same as the purple scale
in color and also in shape except that it is even more narrow, but it is of about
the same length (fig. 360, p. 476). The Glover scale is not as apt to be curved
or "comma-shaped" as the purple scale. Often infestations considered to be purple
scale are found, after more careful identification, to be composed of Glover scale
or to include many of the latter (fig. 361, p. 477).

Fig. 360. Purple and Glover scale. The two large specimens in the center are purple scales. After Woglum (999).

The following structural comparisons will aid in differentiating between
purple scale and Glover scale (999):

Purple scale	Glover scale
Adult scale broadly trumpet- or musselshell-shaped. Sides never parallel. Length 2 to 3 mm.	Adult scale long and narrow with sides almost parallel; straight or sometimes curved in shape. Length 2.5 to 3 mm.
Ventral covering whitish, completely covering the insect; entire, without a median division. Edge attached some little distance from scale margin (fig. 362).	Ventral covering whitish and divided lengthwise at the middle. Covering extends almost to the edge of the scale (fig. 362).
Eggs deposited irregularly beneath the scale (fig. 362).	Eggs in two rows beneath scale. (fig. 362)

In this state Glover scale is of little importance, being confined to a small
area, San Juan Capistrano Valley in northern San Diego County. The insect is found
in many citrus regions throughout the world, but apparently is not as widely distributed as the purple scale.

The life history of Glover scale is similar to that of the purple scale and
the two species have about the same natural enemies. Control measures are also the
same as for purple scale.

OTHER ARMORED SCALES

GREEDY SCALE, Aspidiotus camelliae Sign., attacks a great number of ornamental
plants the world over, and if it were not for the practically universal use of pest
control measures of one kind or another, this insect would be found attacking citrus in California more frequently than is now the case. However, it is readily controlled by oil spray or fumigation. The scale coverings of the females are circular, 1 to 1.5 mm in diameter, very convex, light gray, with subcentral yellow or
dark brown exuvia.

OLEANDER SCALE, Aspidiotus hederae (Vall.):- This insect is similar in appearance to the greedy scale. The female scale coverings are circular, 1 to 2 mm in
diameter, flatter than those of the greedy scale, light gray, with yellow or light
brown exuviae which are central or subcentral, but are not as often to one side of
center and not as far away from the center as in the case of the greedy scale. The
pygidial characters of properly cleared and stained specimens must be examined before Aspidiotus camelliae and A. hederae can be differentiated with certainty. The
former has one pair of lobes on the pygidium while the latter has 3 pairs. These
two species are thus easily distinguishable by microscopic examination (fig. 363).

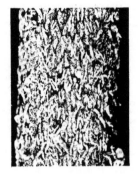
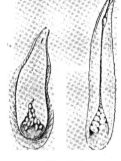

Fig. 361   Fig. 362

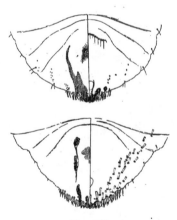

Fig. 363

Fig. 361. Glover scale on orange twig. X 4.5. Photo by Roy J. Pence.

Fig. 362. Ventral aspect of two Lepidosaphes scales with eggs. Left, purple scale; right, Glover's scale. After Woglum (999).

Fig. 363. Above, greedy scale, Aspidiotus camelliae; below, oleander scale, A. hederae. After Dietz and Morrison.

Like the greedy scale, the oleander scale attacks many species of host plants and is widely distributed throughout the world. It can become numerous on citrus in California on untreated trees. The oleander scale is an important pest of ripe lemons in Italy during the spring and early summer. It causes a distortion of the fruit (712).

CHAFF SCALE, _Parlatoria pergandii_ Comst.:- The chaff scale is not found on citrus in California, but in Florida and other Gulf States it is widely distributed. It is seldom a serious pest, probably because oil sprays used for other scale insects and for whiteflies are so effective in the control of this species. Parasites and entomogenous fungi are also effective in keeping down the numbers of this pest; nevertheless, in orchards in which pest control is not practiced, the chaff scale may sometimes be found in injurious infestations in some districts. It infests all parts of the tree, especially the twigs and branches, and since the scales overlap one another, the tree is given the appearance of being covered with chaff, hence the common name "chaff scale".

The scale covering of the female is variable in shape from circular to elongate, rather thin, more or less transparent, and of a brownish or grayish color, and the yellowish exuvium is marginal (fig. 364). It measures 1 mm in shortest diameter. The adult female under the scale covering is dark purple. The female lays an average of about 16 large eggs. The scale covering of the male is oblong and whitish. It is about half the length of the elongate females, and proportionately more narrow.

There are 4 generations of this insect per year.

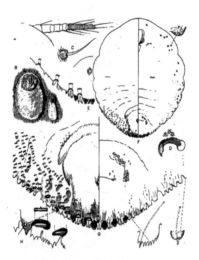

Fig. 364. Chaff scale, _Parlatoria pergandii_. A, antenna and cephalic margin of crawler; B, armor of male and female; C, antenna of adult female; D, anterior spiracle of adult female; E, pygidium of second-instar female; F, general features of adult female; G, pygidium of adult female; H, dorsal aspect of detail of pygidial margin. After Ferris (318).

CAMPHOR SCALE, Pseudoaonidia duplex (Ckll.):- This species was found on camphor trees, Cinnamomum camphora, in New Orleans, Louisiana in 1920, apparently having been introduced from Japan. It is now found in all the Gulf States except Florida. Besides being a pest of the camphor and the satsuma orange, Citrus nobilus unshiu, it has also been found in serious infestations on Seville orange, Citrus aurantium, Kaki persimmon, fig, and pecan and in lesser number on other citrus varieties and other subtropical fruit trees, as well as many other plants. (207)

The oval, purple-colored eggs are found beneath the scale covering of the female. The newly hatched nymph secretes a white waxy covering similar to that of other armored scales and goes through the typical diaspine life cycle. The covering of the adult female, when not crowded by other insects, is nearly circular, distinctly convex, with a sub-central nipple (fig. 365). At maximum size it is 3 to 4 mm in diameter. A ventral layer of wax usually adheres to the bark when a scale is removed, much the same as with the California red scale. The body of the adult is first pink, then purple. The covering of the male insect is oblong-elliptical. (212)

Fig. 365. Camphor scale, Pseudoaonidia duplex. X 3.0. After Cressman and Plank (212).

In summer the periods required for the completion of the various stages of the females are as follows: first instar, 10-11 days; second instar, 15-17 days; preoviposition period, 17-20 days; first egg to emergence of crawlers, 9-11 days; total 51-59 days. In New Orleans there are three generations per year and a partial fourth if the weather is warm.

The most important predator of the camphor scale is the orange bagworm, which feeds on these insects and uses the scale coverings to aid in the construction of the larval case. As many as 60% of the scales may sometimes be destroyed. (699)

The most important treatment is a winter spray, using 2.5 to 3% of oil. This, however, must be supplemented by other sprays used during the spring or summer. The latter should be used at 1-3/4 to 2% concentration. The camphor scale has been successfully treated in greenhouses with hydrocyanic acid gas. (212)

CITRUS SNOW SCALE, Unaspis (Chionaspis, Prontaspis) citri (Comst.):- This species has a world wide distribution in the citrus belt, but is not found on citrus in the more temperate of the citrus regions, as for example, in California. It occurs on citrus in Florida, but is not widely distributed there. The writer has found severe infestations of this scale insect on citrus on the coastal plain of Veracruz State, Mexico, and in the Cauca Valley, Colombia, where it is the most important pest of citrus. In the Concordia region, Argentina, Unaspis citri is the most important of the scale insects, and after a rapid build-up of the pest in Tucuman Province, it was declared a "national pest" in 1938, making control measures compulsory (418). In New South Wales, Australia, Unaspis citri, called the white louse scale, is second only to the California red scale in importance.

The female scale covering is oyster-shell shaped, but much broader at the base than the purple scale. It is 1.5 to 2.25 mm long and dark brown in color, with a lighter margin. It has a prominent longitudinal ridge (fig. 366, p. 480) and this, together with its color and broader outline, distinguish the insect from the purple scale (fig. 358).

This insect gets the name "snow scale" and the Spanish name piojo blanco (Mexico and Colombia) and cochinilla blanca (Argentina) from the white color of the male scales. The male is elongate and has a prominent longitudinal ridge and a fainter one on each side (fig. 366, p. 480). This insect heavily infests the larger branches and trunks of citrus trees, and since the males are more numerous and more conspicuous than the females, they often give the trunk and branches the

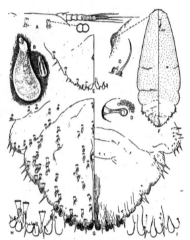

Fig. 366. Citrus snow scale, Unaspis citri. A, antenna and cephalic margin of crawler; B, armor of male and female; C, antenna of adult female; D, anterior spiracles of adult female; E, pygidium of second-instar female; F, general features of adult female; G, pygidium of adult female; H, dorsal aspect and I, ventral aspect, of detail of pygidial margin. After Ferris (318).

appearance of being covered with snow. The infestation begins on the trunk and branches and reaches the leaves and fruit only after the tree is very severely infested. Injury to the tree may be very serious, including a splitting of the bark, allowing for the entrance of fungi, and severe defoliation may occur.

Where Unaspis citri occurs, another scale insect, Pinnaspis aspidistrae (Sign), the males of which are similar in superficial appearance to those of U. citri, may also be abundant, but it confines its attacks to the foliage. This insect occurs in greenhouses, but not on citrus, in the United States.

Control:- In Colombia lime sulfur or oil sprays are used in the control of the snow scale (541). In Argentina also these two insecticides are used, but a scorching of the leaves frequently occurs if an effective dosage (4%, 32°Bé) of lime sulfur is used. If lower dosages are used, the treatment must be repeated. Oil spray can be applied to the trunk and branches at 2 to 2.5% and then sprayed on the entire tree at a reduced dosage 15 to 20 days later. High concentrations of either insecticide can be safely applied to the trunk and larger branches with a brush, to prevent the spread of the insect to the twigs, leaves and fruit. (418)

In New South Wales, Australia, the snow scale (white louse) can be controlled by HCN fumigation, or by lime sulfur spray (1.5 to 1.6% calcium polysulfide) applied at any time of the year. The lime sulfur hinders the formation of the scale covering and prevents the emergence from under the scale covering of crawlers and adult males. The population thus gradually declines. Oil sprays often give poor results because the scales are located chiefly on the older bark, where too much of the available oil is absorbed. (420)

RUFOUS SCALE, Selenaspidus articulatus (Morg.):- The rufous or West Indian red scale is a flat, circular, pale brown species, similar to the dictyospermum scale

Fig. 367. Rufous scale, _Selenaspidus articulatus_. A, antenna and cephalic margin of crawler; B, armor of male and female; C, antenna of adult female; D, anterior of spiracle of adult female; E, pygidium of second-instar female; F, general features of adult female; G, pygidium of adult female; H, dorsal aspect of detail of pygidial margin. After Ferris (318).

This species has a wide distribution in tropical and subtropical regions of the world. It occurs on many hosts, but seems to prefer citrus. H. L. McKenzie believes the species to be of African origin. It is established in southern Florida on citrus, fig, palms, and a number of other plants. It has been intercepted in quarantine in California, on citrus fruits from Florida, and several times on bananas. The rufous scale is a serious pest of citrus in some districts in the West Indies and Mexico.

In Mexico and Peru, the writer has found _Selenaspidus articulatus_ and _Lepidosaphes beckii_ occurring together on citrus trees. In Peru, _L. beckii_ is more important than _S. articulatus_ near the coast, while the latter species is the more important of the two pests a little farther inland where the climate is less humid. From this observation it may be inferred that the rufous scale could probably become a serious pest in much of the citrus-growing area in California, if it were introduced.

Experience in Peru shows that the rufous scale can be satisfactorily controlled by oil spray as used for purple scale. It is, in fact, more easily controlled than the latter species because it occurs only on fruit, foliage, and green twigs, but not on the larger twigs and the branches, where oil spray is the least effective.

BUTTERFLIES AND MOTHS (Order Lepidoptera)

ORANGEWORMS

The larvae of four species of small moths attack oranges in certain citrus districts of California. They are the only lepidopterous insects attacking citrus

481

in California that are important pests. As a group they are known as the orange-worms. Damage from orangeworms may some years be negligible, but some years these insects used to destroy 40% or more of the crop before satisfactory control measures were worked out (76). The orangeworms will be discussed in order of their present importance as pests of the orange in California.

ORANGE TORTRIX

Argyrotaenia (Tortrix) citrana (Fernald)

Although it has been known to be present in California at least since 1885, the orange tortrix was not an important pest until about the last 20 years. This species is a serious pest on citrus only in the coastal and certain intermediate areas, but is only rarely injurious in such districts as Riverside and San Bernardino. It is never of economic importance in the San Joaquin, Sacramento, Coachella and Imperial Valleys. Basinger (75) found the orange tortrix in a great variety of host plants. It is reported that the damage done by the orange tortrix to prunes and apples in some California districts in recent years makes it of considerable economic importance as a deciduous fruit pest. (96) The orange tortrix is thought to be native to southwestern United States.

Description of the Adult:- The adult female moths are about 10 mm long and have a wing spread of about 16 mm. The males are somewhat smaller (fig. 368, A, B). The adult moth is brownish or buff-colored, with a saddle of a darker shade across the folded wings. When the moth is at rest, the folded wings flare out a little at the tip like a bell. There are about an equal number of males and females.

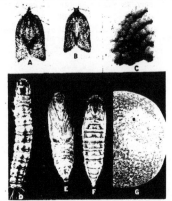

Fig. 368. Orange tortrix, Argyrotaenia citrana. A, adult female; B, adult male; C, egg mass; D, larva; E, ventral view of pupa; F, dorsal view of pupa; G, hole in fruit formed by larva. A and B, X 3. C, greatly enlarged; D, E and F, X 4; G, X 0.60. Arranged from Basinger.

Life History (505, 75). Egg:- The eggs of the orange tortrix are laid on either surface of green orange leaves, smooth green twigs, or on the fruit. They are deposited in masses and in such a manner that they overlap like shingles (fig. 368, C). When first laid, the eggs are pale green to cream-colored, but they turn darker as the embryos mature. After the eggs hatch they look like a silvery patch. The individual eggs are flat and oval and have a finely reticulated surface. An individual female may lay about 200 eggs in several masses.

Larva:- When first hatched, the larvae are about 1.5 mm long. When full-grown they are usually about 12 to 14 mm long (fig. 368, D). The head and prothoracic plate, and usually the body, are straw-colored, although the body may sometimes be greenish, dark gray or smoky-colored. They are very active and will wriggle away backwards or sideways when disturbed, or they may drop to the ground or remain suspended from the leaf on a silken thread which they can ascend again.

The larvae are solitary. They usually become located in the young tips of twigs, among the infolding leaves, which they web together lightly for a temporary nest. They may also make nests among the buds and blossoms, thus exhibiting both the leaf-rolling and nest-making tendencies of members of the family Tortricidae (the leaf-rollers), to which this species belongs. The larvae feed on the plant parts enclosed by their nests. Later in the season dried petals and pistils of the blossoms which have fallen and collected in various parts of the tree are webbed together and fastened to a leaf or fruit cluster to form a nest (fig. 369). This nest is oc-

Fig. 369. Nests of the orange tortrix on orange leaves. Photo by A. J. Basinger.

cupied all summer or longer and is sometimes utilized by several generations. Other larvae may locate under the sepals of young orange fruits or in old, curled leaves, leaves webbed together, or they may locate between a leaf and a fruit or between two fruits.

There may be from 5 to 7 larval instars per year or even more. At room temperature (58° to 78°F) it was observed that the average period in the various stages, using 50 individuals, were: egg, 9 days; larva, 40 days; pupa, 10 days; total 59 days. At constant temperatures ranging from 35° to 85°F the number of days required from egg to adult varied from 236 to 44, respectively.

Pupa:- The pupa (fig. 368, E and F) is 8 mm long, light brown, with the posterior end pointed and bearing a cremaster with 8 small hooklets. The insect pupates in its last larval nest or location, although by the time the adult is ready to emerge, the pupa has often worked itself to the outside or partly outside the nest.

Seasonal History (75):- In localities considerably removed from the coast, as for example in Corona, two generations per year have been noted, one extending from about the middle of February to the middle of May, and the other from the middle of May to the middle of February. During the hottest period of the year a condition approaching aestivation sets in and development is very slow. In the coastal regions various stages of development may be observed at any time of the year-- that is, the generations overlap.

Injury:- According to Basinger (75), the larvae of the orange tortrix feed on the tender foliage of citrus trees and frequently on the petals and ovaries of the buds and blossoms. Their greatest economic damage, however, is to the fruit. They may scar the fruit by the superficial feeding of very young larvae, while the latter are under the sepals or buttons of newly formed fruits, or larvae of all sizes, especially those beyond the second instar, may bore holes in the fruit (fig. 368, G). Injured fruits are likely to drop. Those which do not drop are likely to be overlooked during the sorting and grading operations in the packing house. Since the holes are an avenue for the entry of fungus organisms, infested fruits are much more likely to decay in transit to the consumer.

Fruits growing in clusters are more apt to be attacked by orange tortrix than single fruits. Most commonly the larvae feed at the point of contact of two adjoining fruits.

The larvae of the orange tortrix feed more extensively on the peel and also into the pulp of the orange than any of the other species. As a consequence, fruits attacked by the species are apt to drop sooner than those attacked by the other species of orangeworms. The principal damage on Navels occurs during the late fall and winter period. This was formerly true of the Valencias, but during the last decade the winter damage has generally been light on this variety, with the exception of occasional orchards, and the late spring and summer damage has been the more important. (115)

Occasionally the orange tortrix attacks lemons. Important injury was caused to lemons in a few orchards in the coastal area of Ventura County in the winter season of 1946-47 (115).

Natural Enemies:- The natural enemies of the orange tortrix often are able to control this pest. Of the 12 parasites found attacking this insect by Basinger (75), Apanteles aristoteliae Viereck, Hormius basilis (Prov.), Exochus sp. and Campoplex n. sp. were the most important, especially the first two. A. aristoteliae is about 3 mm long, black, and with long, filiform antennas. Except for the light brown stigma, the wings are clear. It attacks all instars of larvae, but is most frequently found in larvae which are less than half grown. H. basilis is slightly less than 3 mm long, has a yellowish body, dark eyes, and clear wings except for the clouded stigma.

PYRODERCES

Pyroderces rileyi (Walsingham)

Pyroderces was found on oranges in Sweetwater Valley, San Diego County, in 1926. Prior to 1939, this insect was not known to occur on oranges in Orange County, but has recently become prevalent. In the spring of 1948, pyroderces caused severe damage to Valencia oranges in Florida.

Pyroderces has for years been known as a scavenger insect commonly found on cotton wherever this plant is grown in North and South America, the West Indies, and the Hawaiian Islands. It is of interest to the cotton industry because of the necessity for distinguishing the small reddish larvae of this species from those of the pink bollworm, Pectinophora gossypiella, in areas in which an attempt is being made to keep out the latter species by quarantine. For this reason a complete description of Pyroderces rileyi was made by August Busck (133).

Description (133):- The moth (fig. 370, above) has a wing expanse of 9 to 12 mm. The fore wings are chestnut brown with whitish or straw-colored streaks, edged by irregular black scales. The head is a light chestnut-brown, the abdomen reddish brown, and the legs light reddish, with black annulations on the tarsi and tibiae. The full-grown

Fig. 370. Pyroderces rileyi. A, adult; B, side view of head; C, larva. X 8. After Busck (133).

larva (fig. 370, below) is 7 to 8 mm long, and the abdomen is a deep wine red. The head is light brown with mouthparts, including the labrum, black. The thoracic shield is broad, strongly chitinized, and dark brown. The pupa is on the average as long as the larva. It is light yellowish-brown in color and has a smooth surface.

Habits and Injury:- On orange trees pyroderces is also predominantly a scavenging insect, feeding on dry or decayed fruits after they are first injured by other insects, but it also feeds to some extent on sound orange fruits. Only a small percentage of the larvae feed on the fruits and the feeding may be accidental, for not much of the peel is consumed. Ten or 12 pyroderces have been observed on a single fruit without any signs of feeding on the fruit. Nevertheless, the larvae may be so abundant that a certain amount of damage may be done to the peel. Sometimes the larvae have eaten as far as the albedo of ripe Valencia oranges. However, the larvae cannot complete their development on oranges alone; they evidently need dead plant or animal organic material in their diet. In the field the principal source of the dead material eaten by pyroderces larvae appears to be the dead floral parts of orange blossoms. These are webbed together in small or relatively large quantities on the surface of fruits and leaves. (115)

Observations to date indicate that during the winter and spring months only a relatively small number of pyroderces larvae remain on the trees. They are most commonly found during this period inside mummified fruits on the ground or still hanging to the tree, and generally begin to appear in numbers on the trees during June and July. (115)

HOLCOCERA

Holcocera iceryaella (Riley)

The genus Holcocera is composed of species of more or less scavenging habits, being only secondarily feeders on live plant or animal material. The species H. iceryaella was once supposed to be a beneficial insect. Albert Koebele found this insect feeding on living black scale and dead cottony-cushion scale in the vicinity of Los Angeles in 1886. E. O. Essig in 1916 listed other coccid hosts of this insect, and I. R. Horton recorded it as a mealybug predator in 1918 (71). Although primarily a scavenger, holcocera does not have as pronounced scavenger tendencies as pyroderces.

Basinger (71) was the first to record this species as doing damage to oranges similar to that of the orange tortrix, but he stated that the holes and channels they made in the oranges were not quite so deep and pronounced as those of the latter species. Basinger (71) stated that at the time of his observations there were more holcocera in Orange County than tortrix. During recent years, however, holcocera has not been of much importance except in a few Valencia orchards (115).

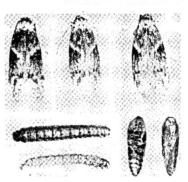

Fig. 371. Holcocera iceryaella. Above. adult moths; below: left, larvae; right, pupae. X 5. Arranged from Basinger.

Description of Stages (71):- The adults (fig. 371) are small gray moths, 6 to 8 mm long and with a wing expanse of about 15 mm. When at rest, they have a linear appearance, for the wings are folded straight back and are only a little wider toward the posterior end than at the head. The V-shaped mark on the back, when the wings are folded, is caused by the joining of the two oblique

SUBTROPICAL ENTOMOLOGY

Fig. 372. Holcocera iceryaella.
Nests constructed between two
oranges. Photo by A. J. Basinger.

lines of lighter color on each forewing.
The forewings are fringed around their
entire margin, but with the fringe on the
hind margin extra long.

The larvae (fig. 371) are 7 to 9 mm
long when full grown. They are brownish
gray with longitudinal stripes of a
lighter color. The head and prothoracic
plate are brown to shiny black. The
larvae may occupy nests made by webbing
together bits of trash, such as old
fallen blossoms, or may occupy old
abandoned tortrix nests or old curled
leaves. They often locate at the spot
where two oranges come together, and here
they make a nest (fig. 372) and feed on the oranges. The larvae are moderately
active and irritable, but less so than tortrix larvae.

The pupae (fig. 371) of holcocera are about .5 mm long, dark brown, and with
the abdominal segments closely oppressed. They may be found on the larval nests,
appearing like dark brown grains of wheat.

PLATYNOTA

Platynota stultana Barnes and Busck

Playtnota stultana is a small tortricid moth which has been found attacking a
wide variety of flowers and subtropical fruits in California and Mexico and was

Fig. 373. Platynota stultana.
X 4. Photo by Roy J. Pence.

reported as a pest of roses in a greenhouse in
Virginia (642). The wing expanse of the adult
(fig. 373) varies from 12 to 16 mm. The fore
wings are dark brown with the outer half a lighter
yellowish brown. The labial palpi and antennae
are long in proportion to the size of the insect.
The eggs are laid in flat clusters and overlap
like shingles. The egg contents are pale green,
giving the egg cluster a color similar to that of
the leaves. There are five instars of larvae.
They are yellowish white when newly hatched, with
head and prothoracic shield brown. As they mature
they become yellowish to brownish green, but the head and prothoracic shield remain
brown. A ragged median dorsal stripe runs the full length of the body. This
represents a clear area in the body wall through which the brownish dorsal blood
vessel can be seen. The pulsations of this vessel can be readily observed. The
average period from egg to adult is 48.6 days. (642)

Various entomologists have from time to time reported platynota to be of some
importance as a pest of citrus, but later investigations have tended to cast some
doubt as to whether this insect ever was of commercial importance. During 1934
and 1935, Basinger (74) made a study of orangeworm populations, and found that
platynota comprised a very small percentage. He found that insects pointed out to
him as being platynota were in reality the then little-known Holcocera iceryaella.
Boyce and Ortega (115) state that platynota is the least common of the four orangeworms and that it is doubtful that this insect is responsible for important economic damage on citrus.

COMPARISON OF THE FOUR SPECIES OF ORANGEWORMS

A resume of the descriptions of the full grown larvae and the adults of the
four species of orangeworms is presented here in tabular form, so as to facilitate
a comparison of the species.

486

Species	Full Grown Larva	Adult Female
Tortrix	Length, 12-14 mm. Head, prothoracic plate, and usually body straw-colored, although body may sometimes be greenish dark gray or smoky-colored. Actively wriggle away backwards and sideways when disturbed.	Length 10 mm, wing expanse about 16 mm. Outline of wings bell-shaped when at rest. Brownish or buff-colored.
Pyroderces	Length 7-8 mm. Abdomen deep wine red; head light brown with black mouthparts; thoracic shield dark brown.	Wing expanse 9-12 mm. Fore wings chestnut brown with whitish or straw-colored streaks, edged with irregular black scales. Head light chestnut brown; abdomen reddish brown.
Holcocera	Length 7-9 mm. Brownish gray with longitudinal stripes of a lighter color. Head and prothoracic plate brown to shiny black.	Length 6-8 mm; wing expanse about 15 mm. Linear appearance when at rest. Color gray, with V-shaped mark on back of lighter color. Wings fringed.
Platynota	Average length, 13 mm. Yellowish to brownish green; head and prothoracic shield brown. Ragged median dorsal stripe runs full length of body. Represents clear area through which brownish dorsal vessel can be seen.	Wing expanse 12-16 mm. Fore wings dark brown with outer half lighter yellowish brown. Labial palpi and antennae long.

The Control of Orangeworms

Investigations by A. J. Basinger as early as 1925 presaged the success of the fluorine compounds in the control of orangeworms. A report on experiments leading to the use of cryolite as the standard remedy for orangeworms was made by Basinger and Boyce (76). Research and field observations on orangeworm control have been continued to the present time by Dr. Boyce and his assistants. The following recommendations are based on a report on experiments made by Boyce and Ortega (115).

Among the many materials tested for the control of orange tortrix and holcocera, cryolite is still the most effective. None have been effective against pyroderces.

For control of orange tortrix on Navels, cryolite applied from May through August is usually highly satisfactory. On Valencias, a fall treatment is necessary to avoid drop of fruit during the winter season. Another treatment is required during May or June. The fall treatment should be a spray - either cryolite with regular oil spray for scale insects or a separate treatment with cryolite after the oil spray. Cryolite should not be used with oil, without the approval of the manufacturer of the particular brand of oil being used, because of the possible adverse effects of the cryolite on the performance of some of the spray oils.

A 50% cryolite dust mixture at 1 lb. per tree is most commonly used in the spring, although evidence seems to indicate that a spray, using 3 lbs. of cryolite to 100 gallons of water, will afford a somewhat higher degree of control.

If a dust is used, the cryolite may be incorporated with DN-Dust for the combined control of orange tortrix and citrus red mite. A proprietary product

487

containing this combination of materials is on the market. Cryolite may also be added to a DN-111 spray in the spring season when the latter is being used for the control of red mite. In this mixture also, 3 lbs. of cryolite is used per 100 gallongs.

It is difficult to decide whether treatment will or will not be justified in any particular season. Unless an orchard is known to have a history of annual tortrix infestation, or if the infestation is sufficiently severe to be evident at the proper period for treatment, the grower must consider his treatments merely as insurance against possible losses from the pest.

FRUIT TREE LEAF ROLLER

Archips argyrospila Walker

Two of the previously discussed species of orangeworms which attack orange fruits in the coastal and intermediate areas in southern California, namely, the orange tortrix and platynota, are "leaf rollers" belonging to the family Tortricidae. Another tortricid has been found in limited, but serious infestations on navel oranges in Tulare County, California (1003). It is the fruit tree leaf roller, Archips argyrospila Walker, a common and widely distributed pest of deciduous fruits, especially the apple.

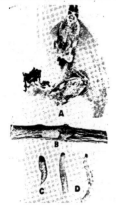

The adult moth (fig. 374, A) has a wing expanse of 20 to 25 mm of a fawn or rusty brown color, and possessing a prominent light spot near the middle of the fore wings, as well as other irregularly placed spots (281). From May to August the adults lay their eggs in irregular flat masses, containing from 50 to 100 eggs which are covered over with a whitish or greyish "cement". These egg masses occur chiefly on the trunk and branches, and are familiar to anyone who has worked extensively with deciduous fruit trees, as in the winter pruning. In March, April, or May, the eggs hatch and the cement-covered egg masses are perforated with the many holes left by the hatching larvae (fig. 374, B).

Fig. 374. Fruit tree leaf roller, Archips argyrospila. A, adults and injury to leaf, X 1.7; B, egg mass on twig, X 1.7; C, pupa, X 1.7; D, larvae, X 1.3. Courtesy R. S. Woglum.

According to Woglum and Lewis (1003), hatching on orange trees east of Lindsay, California commenced about March 15. The larvae fed on the young spring growth and rolled or folded the leaves by means of their webs. As they became longer the larvae fed within the rolled leaves. On some trees nearly all the terminal growth was infested.

The mature larvae (fig. 374, D) on citrus were found to be 18 to 20 mm long, and had colorless or pale green bodies with light or dark brown heads and thoracic shields. The larvae fed through late March and early April and pupated in late April in rolled leaves on the trees. Adult moths emerged in early May and deposited their egg masses on the gray wood of the larger branches. These eggs do not hatch until the following spring, there being only one generation per year.

Control:- In late March, 1947, about 600 acres of orange trees were dusted with DDT dusts for the control of fruit tree leaf roller. On Navels, representing the greater part of the acreage, a dust containing 2% DDT and 80% sulfur was used; on Valencias, a 5% DDT-talc dust was used. On Valencias it was necessary to avoid the use of sulfur because of the irritation to the fruit pickers. The Navel oranges had, of course, already been picked. The dusts resulted in very satisfactory control and were made at a period when they fitted into the citrus thrips control program with little or no additional cost (1003).

CUTWORMS

Larvae of a number of species of night flying moths known as "cutworms" (family Noctuidae) may occasionally cause injury to citrus.

In 1934 and 1935 a noctuid determined as Xylomyges curiales Grote (fig. 375) was found to be especially abundant in a limited area in Tulare County, California, and another species, Parastichtis purpurea var. crispa, was also present in large numbers (724).

Cutworms spend the day in the soil and feed on roots or the stems of small plants at night, thus "cutting" them off near the ground. They sometimes climb citrus trees, especially in the early spring after the covercrop on which they have been feeding is plowed under, suddenly depriving them of their accustomed source of food.

Fig. 375. A cutworm, Xylomyges curialis. X 1.3. Photo by Roy J. Pence.

The cutworms make holes extending through the rind of the fruit. The cutworms can also cause damage to the bloom, and this damage can be done in a short time. If the worms are present in large enough numbers to warrant control, dusting with 50 to 80% cryolite or barium fluosilicate has been recommended, the latter strength if the worms are doing serious damage. (962)

ORANGE DOG

Papilio cresphontes Cramer

This large yellow and black swallowtail butterfly is a minor pest of citrus in Florida and other Gulf States as well as in the Antilles and South America. Another species, P. thoas, is very similar in appearance and habits. Two or three larvae may defoliate a tree in a few days (926). The caterpillar may attain the length of 2-1/2 inches, and is then a dark brown, with light yellow patches. When not feeding, the caterpillar pulls back its head into the enlarged front part of the body, causing the whole anterior to resemble the head of a dog. When disturbed, the caterpillar gives off a very strong and disagreeable protective odor. The usual control on small trees is by handpicking. P. zelicaon Boisd. is a yellow or orange and black species with a wing expanse of 80 mm which may feed on tender leaves of citrus in California. It has been known to defoliate citrus trees. (442)

OTHER MOTHS

NAVEL ORANGEWORM, Myelois venipars Dyer:- This is a snout moth (Pyralidae) which has never been of economic importance on oranges, but in recent years has worked over into the walnuts, on which it is becoming a major pest. The insect will be described and discussed in the chapter on walnut insects.

LEAF MINERS:- The work of a leaf miner can occasionally be seen on citrus, avocado, sapote, and other subtropical fruits in California. The small, reddish larvae (fig. 376, A, p. 491) mine the fruit and green twigs, and since these mines are formed just below the epidermis, they may be easily seen. The mines wind about for a considerable distance and become wider with the increasing size of their occupant (fig. 376, C, p. 491). The insect and its work are of considerable interest to laymen, but of no economic importance. The adult is a very small tineid moth (fig. 376, B, p. 491). R. S. Woglum reared one of the larvae to maturity and it was found to be a western form of Marmara salictella Clemens. Its primary host is

489

willow, and occasionally oranges and other cultivated hosts are attacked when growing near willow trees. (1000)

SLUG CATERPILLARS:- Some interesting, but economically unimportant insects of the families Megalopygidae and Eucleidae, known as "slug caterpillars", occur in Florida. The larvae are short and slug-like, and have stinging hairs which, when touched, result in a momentary itching sensation for some people and for others a very painful swelling. They feed on citrus foliage but are too few in numbers to be injurious. The adults have stout, hairy bodies, and are night-flying moths. Three species may occasionally be found on citrus in Florida and other Gulf States. These are the puss moth, Megalopyge opercularis S. & A.; the saddleback caterpillar, Sibine stimulea Clem., and the bag moth, Phobetron pithecium S. & A.

BAGWORMS:- The bagworms, Platoeceticus gloverii Packard, weave themselves a conical bag of silk which is then covered with bits of leaves, bark, moss, and other fragments, and often with dead scale insects. Then thorns and sticks, with their sharp tips to the outside, are fastened to the outside of the case. Only the head and legs extend from the bag. During the winter the bag is fastened to the tree by means of silken threads (fig. 377). The female moth is wingless and does not leave the bag. The male has a wing expanse of 1-1/2 inches and is pale dull brown in color.

The larvae feed on orange leaves and the epidermis of green twigs and may also feed on the orange peel, making wounds somewhat similar to those made by grasshoppers and katydids, but less deep (926).

BEETLES (Order Coleoptera)
WESTERN SPOTTED CUCUMBER BEETLE
Diabrotica undecimpunctata Mannerheim

The western spotted cucumber beetle (fig. 378) is about 5 mm long, has green elytra with 12 black spots, and the antennae, legs and body are black. It occurs abundantly west of the Rockies on native plants. Sometimes when the native vegetation dries up in summer in California, the cucumber beetles or "diabroticas" move onto cultivated vegetation, including citrus trees, on which they occasionally do considerable damage to the tender new growth and young fruits. On the leaves the beetles eat the epidermis, leaving the translucent framework of the leaves. Small scars are made on the fruit.

All stages but the adult are spent in the soil, the larvae feeding on the roots of the host plants. The life history required 107 days at 60°F and 27 days at 85°F. (605)

Effective control of the western spotted cucumber beetle in deciduous fruit orchards was obtained with a dust containing 0.15 or 0.2% pyrethrin and 2.0% Lethane 384 in talc, applied at 50 lbs. per acre. More recent experience indicates that good control may also be obtained with DDT preparations. (605)

A closely related species, the southern corn rootworm or spotted cucumber beetle, Diabrotica duodecimpunctata (F.), attacks the satsuma oranges of southern Alabama, but confines its injury to the fruit, eating small holes through the skin and into the pulp, upon which it feeds (243). The striped cucumber beetle, D. vittata (F.), attacks citrus, particularly young satsumas, in Florida. It attacks both leaves and fruit. It cannot breed in sandy soil, and is scarce in the main citrus belt of Florida, but sometimes occurs in immense numbers in the Everglades, and occasionally does considerable damage in satsuma orchards in west Florida. (926)

FULLER ROSE BEETLE
Pantomorus godmani (Crotch)

This insect is a snout beetle occurring in all citrus regions of the United States and quite widely distributed throughout the world. It is of a grayish-brown color and 1/3 inch in length (fig. 379). The body narrows anteriorly toward the head and ends in a short, broad snout. The beetles lay their eggs in crevices

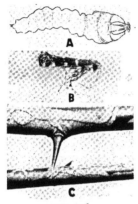

Fig. 376   Fig. 377

Fig. 378

Fig. 379

Fig. 376. A leaf miner, Marmara salictella. A, larva; B, adult; C, mines of larva
           in orange twigs. A and B greatly enlarged, after Woglum; C, original.
Fig. 377. Overwintering bagworm in case attached to twig. After Watson & Berger(926).
Fig. 378. Diabrotica undecimpunctata. X 5. Photo by Roy J. Pence.
Fig. 379. Fuller rose beetle. Above, adults; below, larva. X 4. Photo by Roy Pence.

Fig. 380. Injury to orange foliage caused by the Fuller rose beetle. X 0.35. Photo by Roy J. Pence.

in the tree, under loose bark, or under the buttons of the fruits. Upon hatching, the larvae drop to the ground and may feed on the roots of citrus trees, if other food is not available, but where weeds and Bermuda grass occur it appears that the beetles feed on the roots of these plants, for it is under such conditions that they breed up in their greatest numbers and cause the greatest damage to the citrus foliage. There appears to be two broods of beetles, one in the spring and one in the fall, the individuals of each brood requiring a year for the completion of their life history. (969)

The beetles feed on the foliage, attacking the leaves at their margins and working toward the middle. This gives the leaves a characteristic ragged appearance (fig. 380). This type of injury is commonly observed on the lower parts of the tree, and in the case of large trees, it is not considered to be of importance. On small trees the leaf area consumed is greater in relation to the total leaf area of the tree, and the damage done by the beetles assumes somewhat greater importance. The most serious damage, however, seems to be the destruction of buds and the suppression of new growth on young trees. Usually the worst injury occurs on re-buds and replants in old orchards.

Control:- A dust of 50% cryolite and 50% talc has proved satisfactory in the control of the Fuller rose beetle, or the cryolite may be combined with sulfur. More recently it has been found that DDT dust is very satisfactory in the control of this insect. The adult rose beetle does not fly, and young trees have been protected by cotton bands, applied to the trunk so as to lap over. These bands keep the beetles from crawling from the ground to the tree. This is a specially convenient method for use in the control of the beetles on small dooryard trees when materials or equipment for spraying or dusting are not available.

CITRUS ROOT WEEVIL

Pachnaeus litus (Germar)

Another curculionid, the citrus root weevil, feeds on the blossoms and leaves of citrus in Florida (926). It causes considerable damage to lime trees in the Florida Keys, but less on the mainland. The adults are greenish-blue beetles, 1/3 to 1/2 inch long. The larvae feed on the roots of citrus trees and cause even more damage than the adults. This and related curculionids are among the worst pests of citrus in the West Indies (1007). In Cuba and Jamaica, P. litus is a serious pest and is called the verde-azul (green-blue). Young oranges, as well as the foliage, are attacked, and the scars on the fruit become more noticeable as the fruit becomes larger. Another smaller green-blue weevil, P. citri, occurs in Jamaica. Eight other species of weevils are listed as citrus-feeders in the West Indies by Wolcott. (1007)

A chrysomellid beetle, Cryptocephalus marginicollis Suffr., which is very common in Cuba, but which had not previously attacked citrus, was reported by one grower as a serious pest of oranges in 1944 (922). The beetles feed in great numbers on recently-set oranges and have caused as much as 25% of the fruit to fall.

BRANCH AND TWIG BORERS

A number of species of coleopterous larvae injure citrus trees by boring into the trunk, the branches, or the twigs. In southeastern Asia they are sometimes

serious pests (165). In the United States they only occasionally cause injury.
Among the measures that might aid in combating these insects is pruning off infested twigs and branches and destroying them in
order to prevent the spread of the insects to adjoining trees. The larvae may sometimes be destroyed in their burrows by means of a short wire
or a knife blade.

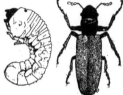

The branch and twig borer, Polycaon confertus
Lec. (fig. 381), is 7 to 15 mm long, mostly black,
but with brown elytra. The eggs are laid in
crevices of the bark and upon hatching the larvae
bore in, completely mining the heartwood. Essig
(278) listed the orange among the host plants and
stated that the live oak appears to be the native
host. The Western twig borer, Amphicerus cornutus   Fig. 381. Branch and twig
(Pallas), and the apple twig borer, A. bicaudatus   borer, Polycaon confertus.
(Say), are listed as pests of citrus (282). They   After Essig.
are dark brown or black bostrichid beetles, 11 to
13 mm long. The orange sawyer, Anoplium inerme (Newm.), and the twig girdler,
Oncideres cingulatus (Say), are listed by Hubbard (454) as pests of citrus. The
former is 11 to 15 mm long, dark brown, and hairy. The latter is 16 mm long,
brownish, and with a broad transverse band on the elytra caused by gray pubescence.
The larva bore beneath the bark, girdling the twig. Pecan, walnut, and persimmon
are also attacked.

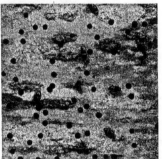

Fig. 382. Fruit tree bark beetle, Scolytus rugulosus. Adults and
exit holes on surface of bark. X 1.7. After L. M. Smith (808).

Some damage to citrus trees in Florida is caused by shot-hole borers (family
Scolytidae) (fig. 382) particularly in the lowlands where the trees have been injured by high water. The flow of sap in these trees is not sufficient to keep the
beetles from feeding. They may cause the death of trees which would otherwise
probably recover their vigor. It was found that these beetles can successfully be
controlled by means of DDT. Forty per cent wettable DDT at 2-1/2 ounces per gallon
of water, and applied by means of a knapsack sprayer to infested trunks of the citrus trees, resulted in good control. Usually only the trunks were infested. The
DDT kills the beetles walking over the surface of the bark. (882) Probably a DDT
spray would be equally effective against other species of barkbeetles and borers.

HYMENOPTERA

Next to the parasites, the ants[1] are the most important Hymenoptera in relation to citrus. In tropical regions the leaf cutting ants (Atta sp., especially

[1] See Eckert and Mallis (272) for a general discussion of the principal species of ants in
California as well as a key to aid in their identification.

A. sexdens and Acromyrmex sp.) as well as other phytophagus species, may do considerable damage to citrus. In Texas and Florida leaf cutting ants also do some damage.

FIRE ANT
Solenopsis geminata (Fabricius)

In California, Arizona, Texas, Puerto Rico and Haiti, the fire ant (fig. 383, above) causes injury by feeding on the tender twigs, bark, and leaves of citrus trees. In Texas thousands of young citrus trees are damaged by these ants every fall. In California these ants seem to prefer young avocado to young citrus trees.

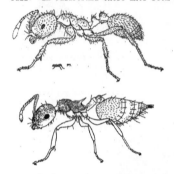

Fig. 383 (at left). Above, the fire ant, Solenopsis geminata; below, the acrobat ant, Cremastogaster lineolata. After Woodworth.

Fig. 384 (at right). Nest of the fire ant, Solenopsis geminata, at the base of a citrus tree. After Clark (160).

This species varies in length from 1 mm in the workers to 6 mm in some of the winged forms. The workers are pale yellowish or reddish with blackish abdomen. The winged forms may be of the same color as the workers or they may be entirely shiny black or entirely reddish. They are capable of stinging severely. They construct large nests (fig. 384) at the bases of citrus trees in Texas and often completely girdle the tree above or below the ground and cause its death (160).

Solenopsis geminata has been responsible for much damage to citrus trees in Haiti, Trinidad and Grenada. Citrus trees in regions of high rainfall are planted on mounds, in order to avoid "foot-rot". The area between the trees is often kept entirely bare of weeds and the ants which normally feed on these weeds take up their abode in the mounds and feed on the bark of the citrus trees, sometimes girdling them completely and sometimes bringing about their deaths more slowly. In the north of Haiti, 20% of the grapefruit trees in young orchards were said to have been killed by fire ants (635).

Control:- Benzene hexachloride is proving to be superior to insecticides formerly used against this insect in Texas. It appears to be non-toxic to citrus trees when used in quantities sufficient to control the ants and is toxic and repellent for a considerable period. A 0.5% (gamma isomer) dust can be blown into the nests with small, hand-powered dust blowers (351). The control now being recommended in California is 5% chlordan dust, which has proved to be very effective.

THE LITTLE FIRE ANT
Wasmannia auropunctata (Roger)

In some sections along the east coast of Florida, the little fire ant is a serious pest to the men who must work in citrus trees, such as pickers, pruners,

494

CITRUS PESTS

and sprayers. The ants nest in the soil or under debris, and climb the trees in search of honey dew. They occur in great numbers and climb onto the men who work in the trees and sting them. A severe itching or burning sensation lasts for a half hour or more and the itching may recur at intervals for several days. DDT proved to be more toxic to this ant and provided a longer lasting control than any other material tried (672). Preliminary trials indicated the most effective means of application of the DDT was in fuel oil applied as an emulsion. Sprays containing 0.5 to 1 pound of DDT per 100 gallons, applied to the entire inside of the tree, have controlled the ant and prevented infestation for an entire year. Other formulations, including a 10% DDT dust, were effective. The spray applied only to the trunk will control the ants for 1 or 2 months and would have less adverse effect against beneficial insects.

DDT was found to have somewhat longer residual effect against the little fire ant than did chlordan. Both had considerably longer residual effect than benzene hexachloride (1010).

THE ACROBAT ANTS, Cremastogaster sp.:- These ants (fig. 383, below) have caused injury to citrus in Texas by honeycombing the branches and twigs. The workers also attend aphids and other honey dew-producing insects. They are small, sluggish ants with short, wide, sharply pointed abdomen, often held over the rest of the body, resulting in the name "acrobat ant". These ants can be controlled by injecting small quantities of carbolineum into the tunnels or cavities which they have made in the tree (352).

ARGENTINE ANT

Iridomyrmex humilis Mayr

This species (fig. 385) was first observed in this country in New Orleans in 1891. It was supposed to have been introduced in coffee ships arriving from Brazil. In California it was first collected in 1905, probably near Ontario. By 1916, the Argentine ant had become established in 16 separate localities in California. Now it is generally established throughout the State at the lower altitudes in moist situations around residences and in orchards (272).

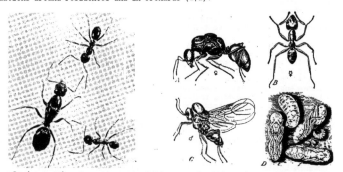

Fig. 385 (at left). Argentine ant, Iridomyrmex humilis. Queen with two workers in attendance. X 6.0. Photo by Roy J. Pence.

Fig. 386 (at right). Argentine ant, Iridomyrmex humilis. A, wingless female; B, worker; C, male; D, immature stages; a, eggs, b, young larva; c, full-grown larva; d, pupa, lateral view; e, pupa, ventral view. X 8. After Back.

Description:- The workers (fig. 386, B) are 2.2 to 2.6 mm long, brown, with the thorax scapes and legs somewhat lighter. The females (fig. 386, A) are brown, with darker abdomen. They are 6 mm long, and when seen they may have their wings

495

or may have lost their wings and possess only wing pads. However, they rarely fly. The males (fig. 386, C) are winged, shiny brownish black and 5 to 6 mm long. The antennal pedicel of the Argentine ants consists of only a single segment.

No perceptible odor is noticed when a single individual is crushed, but a characteristic musty or "greasy" odor may be noticed when large numbers are crushed. A good diagnostic characteristic of the species is its tendency to travel in definite trails. The Argentine ant as a house pest, however, may be confused with the odorous house ant, _Tapinoma sessile_ (Say), which also travels in trails. The latter species, however, is blacker, more squat, has a broader abdomen, moves more slowly, and when it is crushed it has a pungent odor.

Life History:- The elliptical, pearly white eggs (fig. 386, D-a) are 0.3 mm long and hatch from 12 days to 2 months, depending on temperature. At 81°F the larval stage (fig. 386, b and c) may require only 11 days, but this period may increase to 2 months at a mean daily temperature of 52°F. The pupal stage (fig. 386, e and f) requires from 10 to 25 days. Thus the minimum period from egg to adult is 33 days and the maximum about 140 days, with an average of 74.

The Colony:- According to Woglum (958) the Argentine ant, like most other ant species, forms nests under the surface of the ground. The nest is an elaborate system of closely connected galleries and cells, with usually only 2 or 3 openings to the surface. The nest may cover an area of over 10 square feet, but is seldom more than 8 or 10 inches in depth. In winter the nests are on the sunny south side of the trees, usually just outside the "drip" of the branches, but in summer they are more apt to be on the east or north margin. Often there is a colony for each tree in the orchard.

Around residences the nests of the Argentine ant are located in moist and dark situations. In California they begin to increase in numbers toward the end of February and reach their maximum numbers during July, August, and September, then decrease about the middle of October. During the winter they congregate and form large colonies in limited areas such as in leaf accumulations under hedges, in the ground, or in manholes or basements kept warm by steam pipes. In the spring and summer the ants gradually spread over larger areas. Their nests may then be found along the edges of walks, under shrubbery and in other situations kept well moistened. Although the winter nests in the ground are usually around 6 inches deep, in summer they are only 1/2 to 1-1/2 inches below the surface of the soil (272).

The summer colony consists of 3 forms - workers, males, and queens. The workers gather the food and attend to the work of the community. They may be seen on the trails both day and night. They are imperfect females which cannot reproduce. They are the most numerous form and the ones usually seen, although queens can sometimes be seen out on the trails with the workers. In summer new nests are established and a fertilized queen must then accompany the workers. The queens are winged for a short time during the mating period. The males retain their wings. They have no other function aside from the fertilization of the queen. Mating takes place in the nest which may have many queens showing no antagonism to one another. The queens lay enormous numbers of eggs.

The Argentine ants drive out nearly all native species in the areas they infest. If the Argentine ant is controlled, the native species will soon reappear.

Feeding Habits:- In the orchard, the preferred food of the Argentine ant is the honeydew excreted by aphids, mealybugs (fig. 387), whiteflies, and soft scales. Ants in appreciable numbers in a citrus tree are a certain sign of honeydew-excreting insects. The ants may also feed on nectar from blossoms and the bodies of certain insects. In the home, the Argentine ant feeds on a great variety of foods such as meats, sugar, sirups, and fruit, seeming to prefer the sweet substances. However, ants may also remove small seeds from garden seedbeds and may kill newly-hatched chicks to obtain their blood. Argentine ants sometimes invade bee hives for the honey and this may lead to the destruction of the bee colony (813).

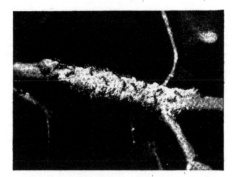

Fig. 387. Argentine ants attending citrus mealybugs. Slightly enlarged. Courtesy R. S. Woglum.

Why Ant Control is Desirable:- The constant attendance of honeydew-excreting
citrus pests by the Argentine ant protects these pests from their natural enemies.
Some entomophagous insects are more deterred by ants than others. Parasites which
move about slowly and which require relatively long periods to oviposit are more
adversely affected by ants than the more active species. The black scale parasite
Metaphycus helvolus is more adversely affected by ants than some other species.
The extent to which ants can protect the honeydew-excreting insects can best be
demonstrated by controlling the ants and noticing the results on the pest in
question. The soft brown scale, which is ordinarily heavily parasitized, may
nevertheless build up its population into sizable colonies if attended by ants.
Control of these ants results in the annihilation of the colony by predators and
parasites, particularly the latter.

It is not necessarily the honey dew-producing insects alone that are favored
by the presence of ants. Armored scales, which do not produce honey dew, may also
be protected from normal parasitization by the presence of the ants tending the
honey dew-producing species. Argentine ants tending soft brown scale on orange
trees interfere with the activities of Comperiella bifasciata, the most important
of the yellow scale parasites. Yellow scale was found to be most abundant on
orange trees on which soft scale (and ants) were present, and the greatest density
of yellow scale was found near the soft scale colonies. Ant control resulted in a
reduction of both soft scale and yellow scale. (329)

Control:- It appears that barriers of various types placed on the trunk are
not practical. They are too expensive, and if the "skirts" of the tree touch the
ground, as is usually the case, the ants will use them as a means of gaining access
to the tree. Control by means of "ant syrup" is recommended as being inexpensive
and also very successful if properly done. A syrup prepared according to the
"Government formula" is recommended. This contains sugar, honey, and arsenic. It
may be purchased at from 1 to 2 dollars per gallon. However, some well-known ant
syrups which are slight modifications of the Government formula have been widely
and successfully used.

The ant syrup is placed in containers which are attached to the trunks of the
tree. A variety of types of containers have been used, including those made of
glass, aluminum, tin and paraffined paper. The latter are the least expensive and
are effective. The containers are mostly of 1 or 2 ounces capacity and usually
have traps and entrances and exits for the ants. Paraffined cups partly filled
with ant syrup can be purchased. The cups should be attached to the shady side of
the trunk and up off the ground. Every tree should have a cup.

SUBTROPICAL ENTOMOLOGY

The early spring months are the best time to start ant control. Results during the summer, when food is plentiful and temperatures are high, are usually disappointing, especially in the interior areas. The late fall, after the pest control work has been done, is also a favorable period for ant control. About 2 weeks after the poison has been distributed, the cups should be examined. Cups which are empty or with crystallized syrup, or which are contaminated with dirt or with dead ants or mold, should be replaced with freshly-filled containers. This reservicing is important for effective ant control, and should be continued until the ants are eliminated. Ants usually persist longest in marginal rows, where they enter from adjacent properties or roadsides. The servicing of marginal rows is advisable even after the interior of the orchard is found to be free of ants. (974).

For the control of ants around dwellings the ant syrup is also effective. It should be borne in mind that sodium arsenite is a deadly poison and should be kept away from children and domestic animals. The syrup is placed in suitable receptacles, which may be purchased, and may be placed near the foundation of the house or anywhere near trails, so as to attract the ants to the poison. The ants not only feed on the syrup, but also carry some to the queens and larvae in their nests. The receptacles should be refilled with ant syrup every 5 to 10 days (272).

It has been found that the new chlorinated hydrocarbons, such as DDT, benzene hexachloride, and chlordan, are effective when applied over the trails in the form of a spray or dust. Their residual effect makes them more effective than insecticides which were formerly used in this manner, such as pyrethrum and sodium fluoride. Repeated applications over a sufficiently wide area are considered by some people to be more convenient than the placing of ant cups. They also have the advantage of promptly getting rid of the ants in places where their presence is objectionable and rapid action is desired.

Recently Argentine ants have been controlled in citrus orchards by using either 2% of a 40% chlordan emulsion, or a 5% chlordan dust. These are applied to the ground immediately surrounding the trunks of the trees. If the "skirts" of the trees touch the ground, the material must be applied in such areas also.

THE GRAY ANT

Formica cinerea Mayr var. neocinerea Wheeler

This native California ant is gray in color and is a relatively large species, being about 6 mm long. It was once the principal ant species in citrus orchards. In some areas where the Argentine ant has been exterminated, the gray ant has reappeared. Its effect in protecting the soft scales, aphids, and mealybugs from the attacks of their natural enemies is somewhat similar to that of the gray Argentine ant. In the control of this species, a syrup similar to that used for the Argentine ant, but with double the concentration of sodium arsenite, is recommended.

CHAPTER XIX
CITRUS PESTS OF FOREIGN COUNTRIES

In this chapter the principal citrus pests of practically all citrus-growing countries of the world, except the United States, will be discussed. Some of these probably would become pests in this country if introduced, so they are of interest to us not only in connection with the study of the problems of citrus pests and pest control methods in other countries, but also in connection with our own plant quarantine problems. Citrus pests which occur in foreign countries, but also in the United States, have been discussed in the preceding chapter. They are not discussed in this chapter, but are included in the lists accompanying the maps which are to be found on pages 530 to 537. The maps show the locations of the citrus pests in the country discussed and are accompanied by lists of the pests in the country as a whole or in the different citrus areas within the country. In these lists, the pests are usually arranged in approximate order of their economic importance. They will contain some species which are not discussed in the remainder of the text because of their slight importance as pests. The scientific names and their authors and the common names of all pests discussed in Chapters 18 and 19, and included in the lists of pests in the various states and countries, are to be found on pages 530 to 537. For this reason the pests are referred to only by their scientific names on the lists accompanying the maps.

The citrus nematode is not included in the lists of pests because of the uncertainty as to its real status as a pest in comparison with that of the others. It is likely, however, that the citrus nematode is of considerable economic importance in the majority of the citrus areas in which it occurs.

WEBBING SPIDER, *Lampona obscoena* Koch (172):- The webbing spider is not a mite, but a true spider (Araneida) which has been in evidence as a pest of citrus trees in Australia for at least 50 years. It is not a major pest only because infestations are not widespread. Whenever it builds up its population in localized areas, however, it is a pest of first importance.

These spiders are of a light gray color and are about 0.25 inch long. They spin their webs in citrus trees, first webbing together one or two leaves, then gradually increasing their nest until many leaves and even branches are included, making a large unsightly mass, sometimes two square feet in extent, and sometimes with a dozen or more per tree. Fruit production is reduced and in severe cases the tree may be killed by strangulation. Citrus trees are killed in the same way by an unidentified true spider in Trinidad (see p. 540).

Control of the webbing spider has consisted of pyrethrum-kerosene emulsion applied to the nests by means of knapsack sprayers. Best results on the larger nests are obtained if these are torn apart before being sprayed. Escaping spiders are sprayed as they are spinning their way to the ground. A spray towards the end of October, followed by another about three weeks later, is recommended.

CITRUS GREEN TREE HOPPER, *Caedicia strenua* Walker (422):- This species and *Caedicia simplex* Wlk. are more or less permanent pests of citrus in New South Wales. A third species, *C. olivaceae*, occurs in citrus orchards along the coast, but causes little damage. *Caedicia* spp. belong to the family Tettigoniidae, and in the United States they would be called "katydids". These insects occur on eucalyptus, which may be a native host. Blackberry is also an important host.

Description:- The adults of *C. strenua* are 1-3/4 inches from head to tip of folded wings. They are grass-green in color and slender. The fore wings are long, narrow, opaque and grass-green, while the hind wings are very pale green, transparent and fan-like. The posterior edges of the fore wings have a black margin.

The insects gnaw off patches of the rind and also gouge out holes in the fruit, almost to the center. When the fruit becomes mature, the white eaten surface con-

499

trasts sharply in appearance with the remainder of the fruit. Thus the injury is similar to that of the katydids attacking citrus fruits in California. When mild, cool, wet weather occurs up to midsummer, causing a soft, succulent growth of foliage and maintaining the fruit in an attractive condition for prolonged periods, the insects become very abundant and cause much damage.

Control:- The citrus green tree hopper is sometimes hand-picked, but spraying is a preferable method of control. Lead arsenate is used at 1-1/2 lbs. to 40 gallons, but is not recommended for application to citrus trees if other insecticides are available. Cryolite at 1-1/2 lbs. to 40 gallons, plus 1 quart of spray oil as a spreader, is effective and in addition is of considerable value against the dicky rice weevil. Only the outer foliage and fruitlets need to be sprayed. According to limited tests, cryolite, barium fluosilicate and lead arsenate dusts have effectively controlled the "tree-hoppers".

SOUTH AFRICAN CITRUS THRIPS (400, 82)

Scirtothrips aurantii (Faure)

The South African citrus thrips is second only to the scale insects as a pest of citrus in South Africa and in Southern Rhodesia. This species probably worked over onto citrus from native host plants and has been known as a pest of citrus for only about 30 years. It is the only thrips known to damage citrus foliage and fruit in South Africa and Southern Rhodesia. In the Veldt, this insect occurs chiefly on wild species of acacia, but it is found on a wide variety of other wild and cultivated plants, including the eucalyptus.

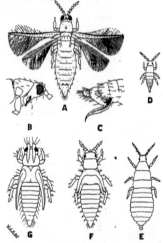

Fig. 388. South African citrus thrips, Scirtothrips aurantii. A, adult female; B, side view of head; C, side view of tip of abdomen with ovipositor; D, first instar nymph just after emergence from egg; E, second instar nymph just before pupating; F, prepupa; G, pupa. After Hall (400).

The South African citrus thrips (fig. 388) is similar in appearance and in life history to the citrus thrips of California (see pp. 386-390). The major types of injury caused by this species are also similar, although a "tear staining" and "late thrips marking" (fig. 389) are also of importance.

500

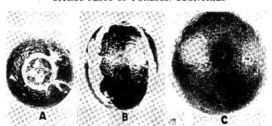

Fig. 389. Injury caused by the South African citrus thrips, Scirtothrips aurantii. A, typical ring marking around calyx, X 1; B, "tear stain" X 0.75; C, russety marking X 0.6. After Hall (400).

A predaceous bug, Orius (Triphleps) thripoborus Hesse attacks the South African citrus thrips in the Transvaal (426a).

Control:- Tartar emetic has proved to be effective in the control of the South African citrus thrips. Alcoholic suspensions of DDT as low as 0.001% have been found to be promising in laboratory work, but 5% DDT dust was less effective. (638)

BANANA RUST THRIPS, Scirtothrips signipennis Bagnall:- This thrips is a serious pest of bananas in various parts of the world. In Queensland, citrus trees growing near bananas are attacked, and even after the bananas are removed the thrips will persist on citrus. A 2% DDT dust preparation has been used in the control of this insect.

LARGER HORNED CITRUS BUG

Biprorulus bibax Breddin

The larger horned citrus bug (spined lemon bug) is indigenous to Australia, where it is a major pest of citrus over a wide area. In the citrus district of subcoastal southern Queensland it is second only to the California red scale as a citrus pest. The native hosts of this pentatomid bug are the native kumquat, Eremocitrus glauca and the finger lime, Citrus australasica. All varieties of cultivated citrus are attacked, but lemons and certain varieties of mandarins are preferred. Citrus varieties other than lemon do not appear to be attacked unless lemon trees are in or near the orchard.

Fig. 390 (at left). Larger horned citrus bug, Biprorulus bibax. X 1.4. After Summerville (859).

Fig. 391 (at right). Larger horned citrus bug on an orange. X 0.45. After Summerville (859).

Description (859):- The adult female (fig. 390) is 14.4 to 22 mm long and 12. to 16.5 mm wide. The male is somewhat smaller. The general color is a shiny lemon-green somewhat lighter on the venter.. After death, this bug changes to a dull yellow color. The pronotum has a large sharp spine at each side, giving the insect its common name.

Injury (859):- If fruit is available, feeding is confined to the fruit (fig. 391), if not, young tender twig growth is attacked to a limited extent. The proboscis appears to barely penetrate the rind and the fruit juices are sucked out. Lemon fruits less than 1-1/2 inches in length fall quickly after a premature yellowing. Larger lemons do not fall so quickly and the premature yellowing does not always take place. When oranges or mandarins are attacked, the premature coloring always occurs, and unless the fruit is quite large, it soon drops. With all varieties of citrus there is a drying out of the tissues of infested fruits. Also there may be a formation of gummy substance inside the fruit, particularly in the case of the lemon. Losses in the case of the lemon may be as much as 90% of the crop.

Control (597):- DDT is very effective against the larger horned citrus bug, thus affording an efficient and easily applied control measure as contrasted to the laborious and exacting technique of HCN fumigation. Likewise, since the eggs are resistant to fumigation, a second treatment is nearly always necessary unless, as rarely happens, the fumigation is perfectly timed. The residual effect of the DDT spray against the individuals that happen to be in the egg stage at the time the spray is applied obviates the necessity for retreating. The DDT spray may be applied whenever the bugs are present in sufficient numbers to be injurious.

The residual effect of the DDT makes a thorough application unnecessary. For a full-bearing tree, if a fine nozzle giving a mist spray is used, only about 1 gallon of 0.2% DDT spray per tree is necessary. Treatment will probably have to be repeated, however, with each of the well-defined bug migration periods in spring, early summer, and midsummer.

The Maori mite (rust mite), Phyllocoptruta oleivora (Ashmead), is not controlled by DDT as it is recommended for the larger horned citrus bug, and the mite population may increase considerably following the summer application of the spray. An early summer lime-sulfur spray, at a concentration of 1 gallon to 35 gallons of water, becomes an essential part of the orchard spraying program. If temperatures are too high for the safe use of lime sulfur, wettable sulfur may be used. If the DDT preparation available is one that can be used with the sulfur preparations, an extra application is not involved in the treatment of mites.

Five species of hymenopterous egg parasites are considered to be important enemies of the larger horned citrus bug.

BRONZE ORANGE BUG

Rhoecocoris sulciventris Stål

The bronze orange bug is another native pentatomid which is a common pest of citrus in Queensland and the north coast of New South Wales. It attacks all varieties of citrus, sucking the sap from young and tender shoots and from leaf, fruit and flower stalks. A secondary form of injury is a fruit and leaf spotting caused by the secretion of a caustic substance which, like that which is secreted by the aforementioned larger horned citrus bug, causes inflammation to tender skin and is particularly painful if it reaches the eyes. The bronze orange bug can squirt its fluid much farther than the larger horned citrus bug, in fact, the slightest irritation of the bug will cause it to squirt this secretion for a distance of 2 feet or more. The secretion is troublesome to persons working in the infested trees. (859)

Description (421):- The adult bugs (fig. 392, left) are nearly 1 inch long and range in color from a light brown with a bronze tinge to a deep brown (almost black). The venter is somewhat reddish.

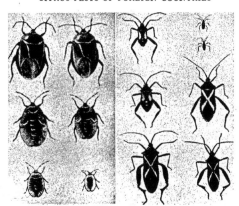

Fig. 392. Left, the bronze orange bug, Rhoecocoris sulciventris, after Hely (421); right, the crusader bug, Mictis profana, after Anonymous (19).

Control:- In limited experiments it has been shown that 0.1% DDT emulsions are very effective against all stages of the bronze orange bug.

THE CRUSADER BUG, Mictis profana Fabricius:- The crusader bug (fig. 392, right) obtains its name from the pale yellowish St. Andrew's cross on the hemelytra. The upper surface of the body is of a general dark brown or greyish-brown color. The under surface of the body, legs and antennae are brown. This species is a general feeder and sometimes attacks citrus. The bugs usually feed on the young shoots, a few inches below the tips. The growth is checked and the shoots often die back to the larger twigs or branches (19, p. 37).

Fig. 393. Rutherglen bug, Nysius vinitor. X 6. After Anonymous (19).

RUTHERGLEN BUG, Nysius vinitor Bergroth:- This bug (fig. 393) is a serious pest of many fruit and vegetable crops in Australia, but it is a minor pest of citrus. The adult female is of a general greyish-yellow color, with darker brown markings. The upper surface of the head and thorax is covered with large pits. The eyes are black and prominent. The adult is about 1/5 inch long. The insect is controlled in Australia by pyrethrum dusts and sprays.

SUBTROPICAL ENTOMOLOGY

CITRUS GREEN STINKBUG (433)

Rhynchocoris humeralis Thumberg

This pentatomid is a serious pest of citrus in China, India, and other regions of southeast Asia. In China, the citrus green stinkbug, together with another stinkbug, Cappoea taprobanensis (Dallas) (fig. 394, C-2), has at times entirely ruined the citrus crop in some localities.

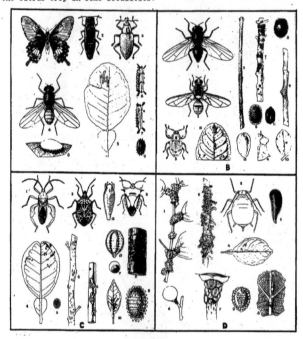

Fig. 394. Some citrus pests of southeast Asia and Indonesia. A: 1, Papilio polytes; 2, Agrilus occipitalis; 3, Hypomeces squamosus; 4, Drosophila punctipennis; 5, Phyllocnistis citrella; 6 and 7, larva of Setora nitens; 8, larva of Altha castaneipars; 9, larva of Cania bilinea.. B: 1, Lonchaea gibbosa; 2, Dacus dorsalis; 3 and 4, Maleuterpes dentipes; 5 and 6, Coccus viridis; 7 and 8, Saissetia nigra; 9, 10 and 11, Parlatoria ziziphus; 12 and 13, Pulvinaria psidii. C: 1, Anoplocnemis phasiana; 2, Cappea taprobanensis; 3, Rhynchocoris serrata; 4 and 5, Aleurocanthus spiniferus; 6 and 7, Aonidiella aurantii; 8 and 9, Pseudococcus hispidus; 10, Aphis citricidus; 11 and 12, Asterolecanium sp.; 13, Eriophyes sp. D: 1, Pseudococcus filamentosus; 2, Icerya purchasi; 3, Aphis citricidus; 4 and 5, Lepidosaphes beckii; 6 and 7, Pseudococcus citri; 8 and 9, Icerya pulcher. After Voute (912).

SERRATE CITRUS STINKBUG, Rhynchocoris serratus Donovan:- This species, (fig. 394, C-3) is a serious pest of citrus fruits in Indonesia, where it causes injury similar to that of the preceding species.

504

CITRUS PESTS OF FOREIGN COUNTRIES

CITRUS PSYLLA, Trioza (Spanioza) erythreae Del Guercio (907):- The citrus psylla is one of the minor citrus pests in South Africa Nyassaland and Kenya. A small pit forms where the insect settles, the nymphs being sessile, and this causes a corresponding swelling on the opposite side of the leaf. Severely infested leaves tend to curl up. The tree does not appear to suffer greatly even from severe attacks.

Diaphorina citri Kuway is another psyllid species which attacks citrus in India Burma, Malaya, Indonesia, China and the Philippine Islands. Diaphorina citri is a serious pest in the Punjab of India where a nicotine sulfate dust at a strength of 3.5 to 4% nicotine sulfate by weight may be used for control. At least 2 applications are necessary, using about 2 to 4 pounds of dust per tree per application. Treatment by means of a hand-duster required from 5 to 8 minutes per tree. (474)

CITRUS BLACKFLY

Aleurocanthus woglumi Ashby

The citrus blackfly was first found by Maxwell-Lefroy in India, George Compere in the Philippine Islands, and R. S. Woglum in Ceylon, all in the year 1910 (239). A few years later the blackfly was reported from various localities in the tropical and subtropical areas of the Western Hemisphere, including the West Indies, and eventually it was found in Mexico. It is consequently of interest to the citrus industry of the United States because of the possibility of its accidental introduction into this country despite the prevailing quarantines.

Description and Life History (239):- When the adult blackfly first emerges, the head and thorax are a bright brick-red color, the front of the head is a pale yellow, the antennae and legs whitish, and the eyes are a deep red or reddish-brown. Within 24 hours after emergence, the insects become covered with a heavy pulverulence, which gives them a general slaty-blue appearance; in fact, they are called the "bluefly" in the Bahamas. Colorless spots on the wings, when these are at rest, form what appears to be a white band across the middle of the dorsum. The adults are a little over 1 mm long (fig. 395, C).

A characteristic thing about the citrus blackfly is the way the eggs are laid. They are, in about 50% of the cases, laid in a spiral path on the undersides of the leaves (fig. 395, A). The female starts laying at a point which becomes the center of the spiral. She then moves forward a little and lays another egg, and thus keeps on moving forward and laying between 35 and 50 eggs per spiral or mass and may lay considerably more than 100 eggs in a lifetime.

The eggs when first laid are creamy white, becoming brown and finally black between the eighth and tenth day after being deposited. They are oblong, with rounded ends, and are attached to the leaf by a short pedicel situated near the posterior end.

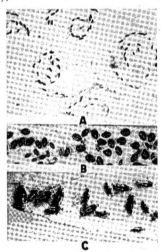

Fig. 395 (at right). Citrus blackfly: A, egg clusters in characteristic spiral arrangement; B, pupae; C, adults. X 5. Arranged from Woglum (997).

The nymphs and pupae are characteristically aleyrodid in shape, but quite spiny (fig. 396). The nymphs are dark brown and the pupae are black.

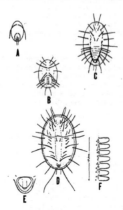

Fig. 396. The citrus blackfly, *Aleurocanthus woglumi*. A, nymph, first instar; B, nymph, second instar; C, nymph, third instar; D, pupa case; E, vasiform orifice of pupa case; E, margin of pupa case. After dietz and Zetek (239).

There are a possible 3 to 6 generations per year, with much overlapping of generations even from the same colony of individuals.

The Citrus Blackfly in Florida:- The citrus blackfly had been a threat to the citrus industry of Florida since its establishment in Jamaica in 1913, but the successful introduction of natural enemies of the blackfly in Cuba (168) reduced the danger of introductions of the insect into Florida by greatly reducing its population density in the West Indies.

When an infestation of citrus blackfly was found in the Island of Key West, Florida, on August 10, 1934, eradication measures were immediately undertaken (649). Oil sprays ranging in actual oil content from 2/3 to 1% were applied at 20-day intervals from September, 1934, to June, 1937, except for about 2 months during the spring of 1936, yet no apparent damage was done to the sprayed trees (citrus and mango). After four complete, intensive inspections were made from February to June, 1937, without the blackfly being found, all eradication activities, with the exception of quarantine enforcement, were discontinued on June 30, 1937, and no reappearance of the pest in Florida has been reported to date.

The Citrus Blackfly in Mexico:- The citrus blackfly was found at El Dorado, Sinaloa, Mexico by A. C. Baker in 1935, and the pest subsequently spread southeastward as far as Mexico City. It also spread up the east and west coasts of Mexico and may now be found in Tamaulipas State in the east, and along the west coast it was observed by Woglum and Smith (1005) as far north as Empalme, near Guaymas, which is about 270 miles south of the Arizona border and only 80 miles south of the major citrus producing area of Sonora State, which is located near Hermosillo. In the course of their observations in the Los Mochis area, in Sinaloa State, Woglum and Smith found no citrus trees which were not heavily infested, and the infestations were among the most severe and destructive they had ever seen of any insect on citrus trees (fig. 397). Crops have failed in this section and the trees have rapidly deteriorated because of the attacks of the blackfly. The infested trees are, of course, heavily coated with sooty mold fungus. All varieties of citrus were equally attacked.

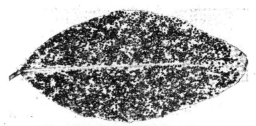

Fig. 397. Undersides of an orange leaf infested with citrus
blackfly, mostly in pupal stage. X 0.67. After Woglum (997).

Although the introduction of the hymenopterous parasite Eretmocerus serius
Silvestri into Cuba resulted in control of the citrus blackfly in from 8 to 12
months, with a resulting reduction in the likelihood of the spread of this pest to
Florida, no such success accompanied its introduction into western Mexico; in fact,
the parasite appears to have no appreciable effect on the blackfly population. It
is believed that it works well only during the rainy season and that the rainy
season in western Mexico is not long enough to enable the parasite to reduce the
blackfly population sufficiently to effect a commercial control. It is believed
that for the same reason Eretmocerus serius might not be effective under the climatic conditions of California and Arizona, and this, together with the nearness
of the Mexican blackfly infestations to our citrus areas in this country, has
caused increasing concern in regard to the citrus blackfly problem.

In 1947 a citrus blackfly control and eradication campaign was begun as a cooperative project between the Mexican Department of Agriculture, the State of
Sonora, Mexico, the citrus growers of Sonora, and the citrus growers of California
and Arizona (994). Ingeniero Salvador Sanchez Colín, Director General of the Mexican Department of Agriculture, was director of this cooperative Mexican-American
project and Mr. R. S. Woglum, at the time Chief Entomologist of the Pest Control
Bureau of the California Fruit Growers Exchange, was Field Manager. The oil spraying operations were under the direction of Mr. Cyril Gammon of the California State
Department of Agriculture, who has had much experience in oil-spraying for the
eradication of citrus whitefly in California. This was an emergency program pending an expected Congressional appropriation. In this program the California-
Arizona citrus growers set a precedent in direct cooperation with a foreign country.

An emulsive oil with rotenone powder was used against the blackfly. In the
preparation of the oil-toxicant, 35 grams of derris root powder, having a minimum
rotenone content of 5%, was mixed with a liter of light medium emulsive spray oil.
This was used at 1-2/3%. Later, in cooperative experiments with the University of
California Citrus Experiment Station, it was found that a kerosene-DDT formulation
was the most effective of all insecticides that were investigated, and it had a
good residual effect. Two pounds of technical DDT are added to 3 gallons of kerosene for each 100 gallons of spray. Blood albumin spreader is used at 4 ounces to
100 gallons. This formulation is now used on all dooryard trees, but the continued
use of light medium oil and rotenone is recommended in commercial orchards until
further experience is gained with kerosene-DDT. (997)

Two new spray rigs and a nurse rig used in this campaign were driven down from
southern California. The citrus growers in the infested districts paid for the
spray oil used in the spraying of their orchards and for the labor. A few other
hosts besides citrus, and including mango, sapote, and myrtle, were sprayed. This
program was terminated on May 1, 1948, when the U. S. Bureau of Entomology and
Plant Quarantine took over the blackfly eradication work in northern Mexico after
an emergency appropriation of $100,000 by Congress for the blackfly campaign for
the 1948-49 fiscal year (998). This work is under the direction of A. C. Baker,

507

who is in charge of the Bureau's fruit fly investigations at Mexico City. The stop-gap emergency campaign initiated by the California-Arizona growers succeeded in bringing the blackfly infestation to a very low point by the time the government was able to take over the direction of the work.

ORANGE SPINY WHITEFLY
Aleurocanthus spiniferus (Quaintance)

This is considered to be the most important of the 4 or more species of white-flies attacking citrus in Japan, and is also recorded from the Philippine Islands, Java, Malay States, India, East Africa and Jamaica. The adult (fig. 394, C-5) is of an orange-yellow color shaded with brownish purple. It is lightly covered with a white waxy powder. (724)

MARLATT WHITEFLY, Aleurolobus marlatti (Quaintance), somewhat resembles the former species. It did not become generally established as soon as Aleurocanthus spiniferus in Japan, but was considered to have potentialities as a pest. (163)

BLACK CITRUS APHID
Aphis citricidus (Kirkaldy)

This mahogany-brown aphid (fig. 394, D-3) with pale, slender antennae, may be found in practically all citrus regions of the world except in the United States. In the literature it is often referred to as Aphis tavaresi Del Guercio or A. citricola van der Goot. This species is often an important pest of citrus and may be controlled by the same measures used for the citrus aphids in the United States (p. 411).

Fig. 398. Comstock mealybug, Pseudococcus comstocki.
After Klein and Perzelan (490).

COMSTOCK MEALYBUG, Pseudococcus comstocki (Kuwana) (745, 490):- Comstock mealybug (fig. 398), originally described from Japan, occurs also in Australia, New Zealand, Madeira Islands, Puerto Rico, Ceylon, Israel, South America, eastern United States, and Louisiana. It has not been found in California. It was not until 1937 that Comstock mealybug was discovered in Israel, but within a couple of years it had already caused much damage to citrus in that country and investigations were immediately undertaken with the view of finding some method of control. In Israel this mealybug has attacked only citrus. After at first causing considerable damage, this species gradually diminished in importance.

CITRUS PESTS OF FOREIGN COUNTRIES

WHITE WAX SCALE

Ceroplastes destructor Newstead

The white wax scale is the third in importance among citrus pests in New South Wales, the most important citrus producing state in Australia.

Fig. 399. White wax scale, Ceroplastes destructor. X 0.56.
After Anonymous (15).

The body of the adult female beneath the copious covering of white wax (fig. 399) is shiny and light-red to dark-brown. The posterior end narrows into a slender, elongate caudal process. The reddish eggs are deposited beneath the body of the female from about November to February. About 3000 eggs are deposited per female. The "crawlers" which hatch from these eggs are reddish and move about actively. After settling, the nymphs secrete a fringe of conspicuous white wax plates which appear much like those secreted by whitefly nymphs. Later the body becomes heavily coated with the white wax, as shown in fig. 394 and are then in the young "peak" stage, and do not lay their eggs until about 10 months later. At maturity the female may measure as much as a half inch in length.

While in the "peak" stage, the white wax scale can be controlled with a "soda spray" made according to one of the following formulas:

    (1) Fresh washing soda - - - - - - - 8 pounds
        Soft soap - - - - - - - - - - - 2 pounds
        Water - - - - - - - - - - - - - 40 gallons
                        or
    (2) Soda ash - - - - - - - - - - - - 3 pounds
        Soft soap - - - - - - - - - - - 2 pounds
        Water - - - - - - - - - - - - - 40 gallons

If California red scale is also present, as is often the case, 1 gallon of oil to 40 gallons of spray should be added instead of the soft soap.

PINK WAX SCALE

Ceroplastes rubens Maskell

This species takes the place of the white wax scale in coastal North Queensland, where it is more important than such well-known pests as the California red scale, citrus white louse (snow scale) and the citrus bud mite. In Japan it is a serious citrus pest, second only to the Yanone scale in importance. The pink wax scale is covered with a waxy covering which is at first a deep rose red but later becomes a grayish pink. The eggs and young are purplish red. This species may be controlled in the same manner as the white wax scale. C. rubens also attacks avocado, cinnamon, mango, eugenia, palm, tea, and other plants, and it occurs in India, Ceylon, and the Hawaiian Islands, as well as in Australia and Japan (724).

Fig. 400. Ceroplastes sinensis. X 1.4. After Gonzales-Regueral y Bailly (366).

CAPARETTA BLANCA or CHINESE WAX SCALE, Ceroplastes sinensis Del Guercio:- This wax scale is found on citrus in Italy and in Spain, principally in the Province of Castellon, but it is a minor pest. The female (fig. 400) may become as much as 7 mm long, 5 mm wide, and 4 mm high. The creamy-white wax of the adult female is divided into 7 plates, 6 lateral and 1 central, each with a darker center. (366)

509

THE PULVINARIA SCALES

The pulvinarias are soft scales which secrete a mass of waxy threads about their bodies in which to conceal the eggs. The orange pulvinaria, Pulvinaria aurantii Cockerell, was once considered to be a very serious pest in Japan, but it was found that even such a relatively ineffective insecticide as kerosene emulsion resulted in reasonably good control (163). Pulvinaria polygonata Cockerell (cellulosa Green) (fig. 401) is considered to be one of the most important of the coccid pests of citrus in the Malaysian region (165). The adult female is 3.5 mm long but, with ovisac, it measures 4.5 to 5 mm in length. It is pale olivaceous above, with the dorsum minutely and closely studded with brown dots, so as to give the appearance of a dark brown or even black to the unaided eye. The extreme margin is unspotted. This insect closely resembles Pulvinaria psidii Maskell, another coccid pest which inhabits the same general region. (382).

The nymphal instars of Pulvinaria polygonata are found largely on the twigs and shoots (fig. 401, A) but many females move to the leaves just before pupation. The white egg masses on the foliage are very conspicuous, and in heavy infestations the leaves may be completely covered with them. Sooty mold fungus is also abundant. (165).

Pulvinaria floccifera Westwood (fig. 402) is a pale yellow insect with a white ovisac 5 to 8 times the length of the body, resulting in the overall dimensions being 8 to 12 mm in length and 2 to 3 mm in width. This species attacks a number of ornamental plants. It has at times increased to alarming numbers on orange trees in localized areas in Spain. Liberation of large numbers of predaceous beetles (Exochomus 4-pustulatus and Cryptolaemus montrouzieri) resulted in a great reduction in the pulvinaria infestation within a few months. (365)

DELTA SCALE, Lecanium deltae Lizer (416):- The delta scale has become increasingly important as a pest of citrus in Argentina in the last decade. These soft bodied scales, which feed on the undersides of the leaves, result in small depressions being formed on the leaves wherever the insects are located. The resulting deformity of the leaf causes considerable difficulty in properly wetting it with oil spray, which is used as the control measure for the delta scale.

BLACK PARLATORIA

Parlatoria zizyphus (Lucas)

The black parlatoria is stated to be one of the most common scale insects of the lemon and orange in Italy, Spain and Northern Africa (724). It is very abundant on citrus in Malaya, the Philippine Islands, Burma, and India. In Siam the insect was observed to be so abundant on lime as to cause severe defoliation (165).

Description (651):- The dorsal scale covering of the black parlatoria (fig. 403) is 1.25 to 2 mm long, broadly oval, approaching rectangular in shape, the black portion being the opaque black exuvia of the second instar female. The latter is rectangular, but with the angles rounded. Two or 3 longitudinal ridges appear on the dorsum. Projecting in front of this is the first exuvia, which is also black and oval in shape. In addition a fringe of wax extends beyond the exuviae, especially posteriorly. The ventral scale covering is composed of waxy secretion, and is pale brown or white in color.

The scale covering of the male is 1 mm long, white, often stained brown or yellow, very elongate, with the black nymphal exuvia at the anterior end.

Biology and Habits:- The black parlatoria female lays only 10 or 12 eggs and has 4 or 5 generations per year. Despite being unprolific, this insect is a serious pest because of its great resistance to insecticides, which makes it probably the most difficult to control of all the scale insects. The black parlatoria infests principally the leaves of lemons, on both surfaces, and sometimes the branches and fruits (366). This species is one of the most common scales occurring on the fruit in the markets in Italy. This may in part be due to the fact that it adheres so firmly to the fruit that it is not easily removed by rubbing. (712).

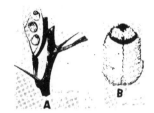

Fig. 401. <u>Pulvinaria polygonata</u>. A, nymphs and adults <u>in situ</u>, X 1; B, adult, X 4. After Green (382).

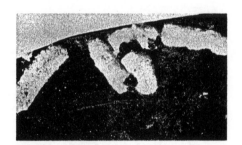

Fig. 402. <u>Pulvinaria floccifera</u>. X 3. After Essig.

Fig. 403. The black Parlatoria, <u>Parlatoria zizyphus</u>. X 10. After Gonzalez-Regueral y Bailly (366).

SUBTROPICAL ENTOMOLOGY

YANONE SCALE

Chionaspis (Prontaspis) yanonensis (Kuwana)

The yanone scale is considered to be indigenous to China, but was introduced to Japan over 42 years ago, having been first found near Nagasaki in 1907. In Japan the insect became the most serious of all the citrus pests. (163) It was considered to be what is known to us as the snow scale, Unaspis citri (Comstock), until it was described as a new species by Kuwana in 1923. U. citri does not occur in Japan. The superficial appearance of both sexes of this species is so similar to that of U. citri that the reader is referred to the description of the latter species (p. 379). The character of the infestation is also similar to that of U. citri, including the predilection for the twigs and branches. As with the latter species, the fruit may become infested in the more severe infestations. Treatments that have been found to be successful in the control of snow scale should also be effective against the orange pulvinaria.

CITRUS FLOWER MOTH OR RIND BORER

Prays citri Milliere

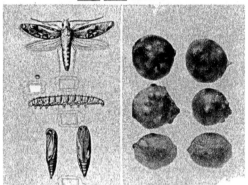

Fig. 403a. Left, the four developmental stages of the citrus flower moth, Prays citri; right, grapefruits, oranges and lemons with swellings of the rind caused by the larvae of either Prays citri or Prays endocarpa. After Garcia (354).

The citrus flower moth (fig. 403a) is a small tineid which is a rather serious pest to the blossoms and fruit of citrus in southern Europe (Italy, Sicily, Corsica), Palestine, India, Ceylon, Malaya, New South Wales, and the Philippines. The moth was described by San Juan (762). The adult is of a general grayish brown color. The fore wings bear numerous irregular markings, and a conspicuous marginal fringe of hairs extends over the distal half of the anal margin of these wings. The moth is from 3.6 to 4.5 mm long and with a wing expanse of from 8.4 to 10.2 mm. The full grown larva is from 4.2 to 5.5 mm long, semi-transparent, and is sparsely provided with fine hairs which are not visible to the unaided eye. The pupa is enclosed in a loose web.

Injury:- In the Mediterranean region the larvae feed on all the flower parts and web them together. In the Philippines and India, the larvae feed on the rind of citrus fruits, next to the pulp (354). This causes the growth of galls which appear externally as swellings.[1] In Palestine the grafted buds on young citrus plantings are injured (92).

---

[1] The rind-boring form is considered by H. T. Padgen to be a distinct species, P. endocarpa Meyrick.

Control:- Promising results were obtained in the control of this insect by means of a spray of 120 g derris powder and 300 g soap in 5 gallons of water applied weekly from petal fall until the fruits are about an inch in diameter. Bait-sprays are made of 20-25 g sodium arsenite and 400-500 g brown sugar in 5 gallons of water. Fallen fruits should be collected and destroyed before the larvae leave to pupate. A hymenopterous parasite attacks the pupae and an ant, Oecophylla smaragdina, prevents the infestation of fruits on trees in which it has made its nests. (354)

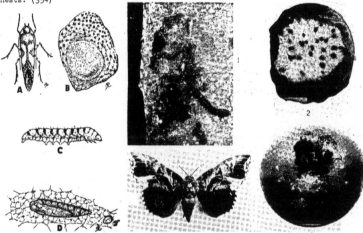

Fig. 404 (at left). Prays endocarpa Meyr. A, adult X 2.5; B, egg on rind of a lime, X 35; C, larva, X 2; D, pupa, X 2. After Padgen (673).

Fig. 405 (at right). Some citrus pests of southeast Asia and the East Indies. 1, injury by Indarbela tetraonis; 2, injury by Prays endocarpa; 3, fruit-piercing moth, Othreis fullonia; 4, injury by the moth borer, Citripestis sagittiferella. X 0.67. After Voute (912).

CITRUS RIND BORER, Prays endocarpa Meyrick (673):- Closely related to the citrus flower moth is the citrus rind borer (fig. 404). It feeds strictly on the fruit (fig. 405, 2), however, and does not attack the blossoms. The eggs are laid singly on the surface of the fruit. Upon hatching, the larvae bore through the egg and into the rind. Gall-like swellings are caused by the feeding and lignified tissue is formed around the entrance burrows. This lignified tissue may extend down into the fruit pulp and ruin it for eating purposes, but the juice can still be used. The larvae, however, remain in the rind. As many as 84 moths have been bred from a single small lemon.

BARK-EATING BORERS

Indarbela quadrinotata Walker and I. tetraonis Moore

Indarbela (Arbela) quadrinotata Wlk. is one of the most injurious pests of citrus in the Central Provinces and Berar in India (825a). The eggs of this lepidopterous borer are laid on the bark in May-June in groups, usually 15 to 25. These hatch in 8 to 11 days, and the larvae feed superficially on the bark, under shelters of wood, silk, and excreta, until September. Then they bore into the bark and feed during the day, but spend the night in their shelters, becoming full grown

513

in December. In late April they pupate in the tunnels. The pupal stage lasts 21 to 31 days and the moths emerge in May and June. Mango, pomegranate, guava and mulberry are also attacked.

In September or October the shelters of the larvae may be located and a wad of cotton wool, dipped in a fumigant such as carbon bisulfide, may be put into the tunnels. The entrances are then closed with clay.

Indarbella (Arbela) tetraonis is at times a serious pest in various regions throughout southeast Asia. Its habits and the injury it causes (fig. 405) are similar to that of the previous species. A mixture of kerosene and petrol, injected into the holes, is stated to be effective against this pest. The larvae may be killed by probing their holes or galleries with pointed wires (157).

FALSE OR CITRUS CODLING MOTH (388, 339, 665)

Argyroploce leucotreta Meyrick

The false codling moth is indigenous to South Africa. This pest is of some importance in all the principal South African citrus regions, but chiefly around Bathurst, Cape Colony. Besides citrus, it attacks acorns, apricots, guava, olive, peach, persimmon, plum, pomegranate, walnut, and a number of wild fruits. The false codling moth is sporadic in its outbreaks, and a season of severe infestation may be followed by one of comparative scarcity of the moths. It is referred to in the literature as a pest of cotton, and occasionally maize, in East Africa, although Omer-Cooper (665) believes the cotton-infesting insect is probably a distinct species or variety.

The false codling moth somewhat resembles the true codling moth, Carpocapsa pomonella (Linnaeus), in all stages. The adult is brown in color, with a semi-oval dark reddish patch mixed with black and with a white center on the fore wings. The hind wings are lighter, greyish brown but darker towards the outer margin. The male has a pale greyish genital tuft and a large, dense bunch of greyish white hairs on the hind legs.

Likewise in habits and the nature of its injury to fruit, the false codling moth is quite similar to the true codling moth.

Control:- Orchard sanitation is practiced for control of this insect, but this is somewhat counteracted in areas in which the false codling moth invades the citrus orchards from nearby native vegetation.

In certain experiments, DDT was found to be superior to benzene hexachloride or nicotine bentonite as a control for false codling moth. The concentration was 1 lb. of actual DDT per 100 gallons of spray. Four applications were made with hand pressure pumps at intervals of 10, 11, and 13 days.

In 1945 and 1946 the false codling moth was identified with fungus infection of avocado fruits in South Africa. DDT suspensions have given satisfactory results in the control of the insects in this connection, but it has proved to be less expensive to cover individual fruits with paper bags, and this has given better protection against the insect than spraying.

LIGHT BROWN APPLE MOTH, Tortrix postvittana Walker:- This species is indigenous to Australia, where it is widely distributed. Besides a wide range of native plants, this leaf roller attacks apples, pears, oranges, and grapes. The larvae feed on the foliage but may also feed on the fruit, causing blemishes. Occasionally they tunnel more deeply into the fruit.

The moth has a wing expanse of about 3/4 inch and is of a general light brown color with darker markings. The flat, oval eggs are laid in groups on the leaves, and the larvae, slightly more than a half inch long when full-grown, are green in color. The larvae spin cocoons within rolled leaves. This pest may be controlled by 1-1/2 lb lead arsenate to 50 gallons of water or, in apple orchards, it is controlled by the usual codling moth treatments (12).

BOLLWORM

Heliothis armigera (Hübner)

Fig. 406. Boilworm, Heliothis armigera. X 1.1.
Photo by Roy J. Pence.

This well-known pest of cotton, corn, and tomatoes (fig. 406) which may often be found in the literature as Heliothis obsoleta Fabricius, is under certain conditions an important citrus pest in the Union of South Africa and Southern Rhodesia. As much as 20 or 30% of the citrus crop has been lost in some years in certain areas because of the ravages of this noctuid, and in some orchards it has been known to cause a total loss of fruit. In addition it is a pest of deciduous fruits.

The bollworm on citrus can be controlled by means of cryolite or sodium fluosilicate dust, which appear to act as contact as well as stomach insecticides against this species (490).

FRUIT PIERCING MOTHS, Family Noctuidae (194, 364, 782, 733, 99):- Certain noctuid moths, comprising 16 or more genera, cause serious damage to citrus fruits by piercing the rind by means of their probosces, a type of injury which, as far as the Lepidoptera are concerned, is practically confined to this group of large, night-flying moths (fig. 407), although a fruit-piercing satyrid has been reported (364). Among the noctuids, the genera Othreis, Eumaenas, Achaea, Serrodes, Sphingomorpha, Anomis, Heliophisma, and Ercheia appear to be the most injurious.

Fig. 407. Fruit-piercing moths. Above, Othreis salaminia; below, Othreis fullonia, female. After Frogatt.

The specially modified mouthparts of the fruit-piercing moths are capable of producing not only punctures, but distinct holes in the rind. The fruits may then

515

drop because of fungus infection (194). When not in use, the proboscis is coiled up under the head like a small watch spring, as is usual for the Lepidoptera, but when they are extended they are about an inch long. They have a sharp tip with serrate cutting edges along the sides, enabling the moths to bore holes into fruits that are quite hard.

The most practical method of avoiding loss is to pick the fruit as soon as possible. Poison baits have not been very effective in reducing loss.

MOTH BORER

Citripestis sagittiferella Moore

Among all the citrus areas of the world, the Malay States and Indonesia are unique in having their most important pests among the Lepidoptera. Probably the most serious of these is the moth borer, Citripestis sagittiferella (fig. 408), or this moth in conjunction with another moth, Prays citri. The moth borer is known to occur only in the region mentioned above. It is particularly destructive at the lower elevations. Grapefruit is the preferred host, and 100% losses of fruit may occur, although orange, lemon and lime are also attacked. The damage done by the moth borer is comparable to that of the Mediterranean fruit fly in the regions most suited to the development of the latter. (165)

Fig. 408 (at left). Moth borer, Citripestis sagittiferella, X 3. After Clausen (164).

Fig. 409 (at right). Moth borer, Citripestis sagittiferella. A, egg X 12; B, pupa X 2.5; C, larva X 28. After Padgen (673).

Life History (673, 165):- The scale-like eggs (fig. 409, A) are laid singly or in clusters on the surface of the fruit. Shortly after hatching, the larvae penetrate the rind and, after the first molt, attack the pulp. The burrows made by the larvae are lined with silken web. From time to time apertures are made to the surface by the larvae, through which frass and excrement is expelled. A large amount of drying sap and excrement may gather at this point, giving the fruit an unsightly appearance (fig. 405, 4). The fruit eventually becomes discolored and may drop in bad infestations. The pupal stage is spent in a cell in the soil beneath the tree. The adults are predominantly a wood brown speckled with drab gray. The wing expanse is 27 to 29 mm. A considerable number of generations may be produced each year.

Control:- Grapefruit have been bagged to prevent oviposition, but this is not considered practical for smaller fruits. Destruction of infested fruits has been recommended, but this measure is not entirely successful because many larvae drop to the ground before the fruit falls (673).

CITRUS LEAF MINER

Phyllocnistis citrella Stainton

Leaf miners may be members of four different orders: Coleoptera, Lepidoptera, Hymenoptera, and Diptera. They have in common the habit of their larvae of living and feeding for a part or all of their existence immediately beneath the epidermal

516

layer of a leaf, twig, or fruit. Leaf miners attract attention because of the curious nature of their injury. A winding mine, visible because of the transparent epidermal layer beneath which the larva feeds, often reveals the entire life history of this insect. From the point where the egg was inserted and the larva was hatched, the mine gradually becomes wider as it increases in length, concomitant with the increase in the size of the larva. The cast larval and pupal skins can often be seen in the mine.

Fig. 410. Citrus leaf miner, Phyllocnistis citrella. Greatly enlarged. After Clausen (164).

Leaf miners may often be seen on citrus trees the world over, but in tropical Asia certain lepidopterous leaf miners are among the major citrus pests. The lyonetid moth Phyllocnistis citrella, a minute, grayish insect (fig. 410) is often a serious pest of nursery and young orchard trees. Young nursery stock is often killed by this pest. The curling of the leaves caused by this pest appears much like the injury caused by citrus aphids. Another lepidopterous leaf miner, Psorostica (Tonica) zizyphi Stainton, is associated with the above species in Ceylon.

PAPILIOS

A number of species of swallowtail butterflies (Papilio) attack citrus throughout the world. In China, Papilio antiphates (Cramer), P. demoleus Linnaeus, P. helenus Linnaeus, P. memnon Linnaeus, P. polytes Linnaeus (fig. 389, A-1), and P. xuthus Linnaeus attack citrus (188). In India, P. demoleus Linnaeus, P. polytes Linnaeus, and P. polymnestor Cramer, as well as other species of Papilio, are serious pests. They are especially serious in nurseries and young orchards, but sometimes attack large trees. As many as a thousand caterpillars have been found on a single medium-sized tree, and sometimes entire orchards are defoliated. Hayes (415) considers the papilios to be the most important citrus pests in India. Handpicking is ordinarily employed in combating the pests, but lead arsenate has been used. In Australia, P. aegeus Donovan, and P. anactus W. S. Macleay occur on citrus; in Palestine, P. machaon Linnaeus; in South Africa, P. demoleus Linnaeus; in Central and South America, P. anchisiades Esper, P. crassus Cramer, P. thoas Linnaeus, and P. idaeus Fabricius.

In southeast Asia the lemon butterfly, Papilio demoleus Linnaeus, is a well-known and easily recognized species. It has a wing expanse of about 3 inches. It is a bluish-green in color, with yellow markings and two eye-like spots on the hind wings.

CITRUS TRUNK BORER (434, 523)

Melanauster chinensis Förster

This cerambycid beetle is one of the most destructive of all the citrus pests of China. It attacks citrus trees in both its larval and adult stages. It is described as being shining black with somewhat rounded spots of white pubescence on the elytra (fig. 411, p. 518). The females are 24 to 40 mm long and the males only slightly smaller.

The adults appear in the spring and lay their eggs in slits which they cut into the bark on the basal portion of the trunk. The adults feed on the leaves, petioles, and bark of twigs and branches. In the case of young trees, they also feed on the bark of the trunk. The larva feeds about 2 months in the inner bark of the basal portions of the trunk. After it has reached a length of 16 mm to 20 mm it begins to bore into the wood, mining the trunk, root, and rootlets. A single larva is able to kill a citrus tree 5 or 6 years old. There is only 1 generation per year.

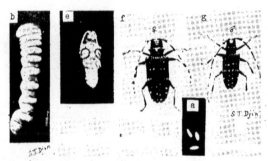

Fig. 411. Citrus trunk borer, Melanauster chinensis. a, eggs; b, larva; e, pupa; f, adult female; g, adult male. X 0.7. After Lieu (523).

Except for ants taking away the eggs, Melanauster chinensis appears to have no natural enemies.

Control:- A lime whitewash applied to any part of a citrus tree will repel the adults and prevent oviposition and feeding. If the beetles are forced to feed on whitewashed material, they are killed by the lime. Pressing the egg slits with the thumb will destroy the eggs. Young larvae can be probed out of their burrows.

CITRUS BARK BORER, Agrilus occipitalis Eschscholtz:- This buprestid beetle (fig. 389, A-2) is a serious pest of citrus in the Philippines, often killing young trees. It is 5 mm long, greenish or purplish bronze, and has grayish markings on the elytra. The larvae mine under the bark and may mine the large roots. The adults feed on the margins of the leaves. Jarring the beetles from the tree or dusting them with cryolite or barium fluosilicate has been used for control. (724)

CITRUS SNOUT BEETLE, Sciobius granosus Fabricius (13):- The citrus snout beetle was first found on citrus in the Muden Valley, Natal, South Africa in 1931, although the insect has been known to commonly occur in that state.

The adult is 7/16 inch long, oval, dull brown in color. It has a rostrum of medium length and breadth. The adult feigns death when disturbed. It cannot fly, the second pair of wings being absent. Although the insect is not strictly nocturnal, it does most of its feeding in the late afternoon and at night. During the day it remains quiet on the tree or under debris near the trunk.

Injury:- Injury consists of the girdling of the roots of the citrus trees by the larvae as well as feeding on the foliage by the adults. The larvae first eat the fine rootlets, then eat grooves into the larger roots. Weevil injury can be found as deep as 18 inches below the surface of the ground. The roots are not only completely girdled in many cases, but are made subject to decay organisms. The adults prefer young leaves, but in severe infestations all types of leaves are eaten. The tree may be so weakened as to be unable to yield a crop.

Control:- The adult beetles have been jarred from the tree and caught on large sheets, then destroyed. Another control measure is the poisoning of the adults by spraying with 20 lbs. of sodium fluosilicate and 20 lbs. of white sugar per 240 gallons of spray, at the rate of 6 gallons per tree. The same mixture may be used to spray clumps of weeds or covercrop growth used as trapcrops. Since the adults do not fly, they may be prevented from ascending the trees by means of suitable collars placed around the trunks. The destruction of vegetation on which the snout beetle breeds in and around the orchard helps to keep down the numbers of the insect. Dry leaves and other debris, in which a certain number of beetles hibernate, should be raked from under the trees.

The fuller rose beetle, <u>Pantomorus godmani</u>, accounts for some damage to citrus trees in the Muden Valley, but it is not expected to ever become as severe a pest as the citrus snout beetle.

DICKY RICE WEEVIL

<u>Maleuterpes spinipes</u> Blackburn

Fig. 412. Dicky rice weevil, <u>Maleuterpes spinipes</u>. X 7.
After Anonymous (15).

The dicky rice weevil (fig. 412) is fourth in importance among citrus pests of New South Wales. It is 2-1/2 to 3 mm long, rather robust, brownish with grayish white markings on the prothorax, elytra, and legs. There appear to be 2 generations a year. The larvae feed on the roots of citrus trees, especially of nursery stock. The adults feed on the foliage and on the skin of the fruit of citrus. As much as 70% of the fruit may be badly disfigured and a further 20% slightly marked when the beetles are numerous. Damage may also be done to young or grafted trees by the feeding on the foliage. (1012)

Control:- Banding the trunks with sticky banding material in July is recommended for the control of these weevils. The banding material should be freshened by rubbing, or more material should be added, from time to time. Where the sun strikes the trunks, the banding material should be applied over grease-proof paper to prevent injury. The branches touching the ground must be cut back so that they are at least 6 inches clear of the ground.

THE CITRUS SNOUT BEETLE, <u>Maleuterpes dentipes</u> Heller:- This snout beetle or weevil (fig. 394, B-3 & 4) is a small brownish species, somewhat related to the dicky weevil, but it can be distinguished by the spine arising from its thorax (fig. 394, B-4). It is a minor citrus pest in the East Indies. Damage from this species is similar to that of the dicky rice weevil, but it is not as important a pest.

THE FRUIT TREE ROOT WEEVIL, <u>Baryopadus</u> (<u>Leptops</u>) <u>squalidus</u> (Boheman):- This Australian species is 1/4 inch long and with yellow body and legs (fig. 413).

Fig. 413. Larva and adult of the fruit tree root weevil, <u>Baryopadus squalidus</u>. X 1.0. After Anonymous (15).

The larvae feed on the roots and the adults on the tender foliage of apple trees, chiefly, but may also attack citrus. Sickly-appearing trees with a general "dieback" condition and with dropping leaves should be suspected as being victims of this weevil, and if so, the beetles should be controlled, for they are capable of destroying an orchard. Banding, as for dicky rice weevil, should be used as a control measure. This should be begun in July. The beetles which collect below the bands should be gathered and destroyed once a week. (15).

THE GREEN SNOUT BEETLE, Hypomeces squamosus Fabricius:- This snout beetle (fig. 394, A-3) is 14 mm long and shiny black flecked with green or yellowish green. The adult feeds on the foliage of citrus trees in the East Indies. (912)

THE YELLOW MONOLEPTA BEETLE, Monolepta australis Jacoby:- This is an Australian species, 1/4 inch long and with yellow body and legs. These beetles sometimes destroy the blossoms of citrus trees in New South Wales, causing crop failure. They may be controlled by means of pyrethrum dusts. (15)

LEAF-CUTTING, FUNGUS-GROWING or PARASOL ANTS

Atta sp. and Acromyrmex sp.

These ants derive their common names from their habit of cutting leaves and carrying them off to their nests, the leaf portions being carried parasol-like over their bodies and eventually titurated and used as a spongy substrate for fungi which are grown in subterranean chambers. Certain bodies produced by the fungi serve as food for the ants. Various species of Atta, Acromyrmex, and other related genera belong to this group of ants, and their range extends throughout the Western Hemisphere from as far north as New Jersey to as far south as Argentina, or roughly from 40° north to 40° south of the equator. Wheeler (943) wrote an extensive account of North American leaf-cutting ants. He states that Atta texana Buckley, in Texas, is the only species in the United States which is sufficiently numerous to be of some economic importance.

In many regions the leaf-cutting ants are the major agricultural pests, and their importance is increasing as new tropical areas are being cleared for cultivation. In Cuba the species Atta insularis Guerin (fig. 414), known locally as

Fig. 414 (at left). Above, the bibijagua, Atta insularis, a Cuban leaf-cutting ant; below, mound of the bibijagua. After Bruner and Valdes Barry (129).

Fig. 415 (at right). Worker of the leaf-cutting ant, Atta sexdens. X 2.5. After Wolcott.

"bibijagua", is a serious pest, and has been known to completely defoliate citrus trees in a single night (1007). In tropical South America, Atta sexdens Linnaeus (fig. 415) and A. cephalotes Linnaeus are the most important species, and are serious pests of citrus and avocado trees growing in or near rain forest areas. One species in Trinidad, A. octospinosa Reich has its colonies, which are relatively small, within the citrus orchards.

The excavations made by the ants may extend over an area of 14,000 square yards and to a depth of as much as 25 feet. The queen, in starting a new colony, makes a food chamber in the ground, which has a single opening. She carries into the posterior portion of the oral chamber a culture of the fungus Rozites gungylophora, and bleached leaf fragments and other matter from the old nest, and places this in the chamber. In a pocket of this spongy mass, called a "fungus garden" (fig. 416), she lays her first eggs. She fertilizes the fungus with her own faeces (fig. 417).

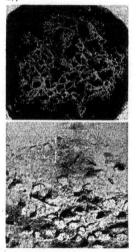 

Fig. 416 (at left). Above, fungus garden of Atta texana. After Wheeler (943). Below, cross section of a colony of A. insularis, showing chambers and fungus gardens. After Bruner and Ochoa (128).

Fig. 417 (at right). Atta sexdens queen fertilizes, with her own faeces, a fungus cultured for food. Profile from photograph by J. Huber. From Wheeler (943).

The smallest workers soon[1] take over the function of manuring the fungus garden and feeding the larvae with their mother's eggs. Other workers cut leaves

_____

[1] Autore (42) states that in the case of Atta sexdens rubropilosa Forel in Brazil, the various castes of workers are produced in 4 to 10 months after the nest is established, soldiers after 22 months, and the sexual forms after 38 months. The nests have then reached their maximum development.

Fig. 418. Lime leaves cut by the leaf-cutting ant, Atta insularis. After Bruner and Valdes Barry (129).

from trees and smaller plants (fig. 418). Worker ants of still another type carry the leaf portions to the hole, where they are taken by other ants which knead the pieces of leaves into minute wads and insert them in the fungus garden. Other workers are engaged in enlarging the subterranean labyrinth to make way for new fungus gardens. The fungus gives rise to transparent pyriform globules which Moeller gave the name of "kohlrabi" and for which Wheeler (943) suggests the term "bromatia". These serve as food for the ants. The bromatia eventually become so abundant that they are used as food for the larvae as well as the workers.

Weyrauch (941), who made a study of the nests of Atta sexdens and A. cephalotes, the two species which cause great damage to fruit and shade trees in the Chanchomayo Valley of the Peruvian Andes, calculated that a large nest may contain about three million ants. He found that the fungus gardens and nests are sometimes destroyed by saprophytic bacteria and is investigating the possibility of culturing these for artificial introduction into ant colonies.

Control:- Citrus trees are among the preferred hosts, but usually only trees 5 years or less of age are seriously damaged. Bruner and Valdes Barry believe that the bibijagua can be controlled by destroying nests, before they become too large, by some other practicable method, such as pouring a little carbon bisulfide into the nest. The initial work involved might be considerable, but in later years only a few new nests established by queens flying in from other areas need be destroyed. (129)

A fumigation method has been recommended for the destruction of the well established bibijagua colony. Fumes are produced by burning sulfur and white arsenic, using 2 parts of the sulfur to one of the arsenic. Machines can be purchased for the generation of the vapors. These are applied through perforations of the ant nests made from the surface by forcing a barreta (bar) through the soil to the large chambers in which the fungus gardens are maintained. By blowing the vapors only into the natural entrances of the nests, only 60 to 70% of the colony is destroyed at best, whereas by using the barreta to make additional holes to all parts of the subterranean labyrinth, 95 to 100% of the colony can be destroyed. Carbon bisulfide and calcium cyanide dust can be used, but both are more expensive than the recommended method, and the latter insecticide involves considerable hazard, especially to unskilled operators. (128)

In South America, fumes of sulfur and arsenic, carbon bisulfide, or hydrocyanic acid have been extensively used to combat leaf-cutting ants. Usually the fumigation must be repeated several times to be effective.

Atta sexdens not only does not cut banana foliage, but is repelled by it. As a protection against damage from leaf-cutting ants, the natives of the Amazon basin sometimes tie fresh strips of banana foliage to citrus and avocado trees every other day while they are in danger of attack. (1007)

FLIES (Order Diptera)

FRUIT FLIES, Family Trypetidae (Tephritidae)

This is a large family of chiefly tropical or subtropical flies, but with a number of species in temperate regions. The adults are usually more or less the size of a house fly. They have vari-colored bodies and have the habit of slowly elevating and lowering their variously mottled wings. Many species insert their eggs under the skin of various fruits and berries, into the husks of seeds and nuts, and in vegetables. These are mined and rendered unfit for food by the resulting larvae.

The larvae of many species have been described by Phillips (693). They are small and yellowish-white in color. A minute black dot, representing the mouthhooks, may be seen on each side of the anterior and smaller end of these typically-shaped maggots. They average from 7 to 9 mm in length, but may vary from 3 to 15 mm. Small spinules lie on the outer surface of the integument and these offer anchorage for the forward or backward movements of the larvae in their tunnels in the fruit and other host material.

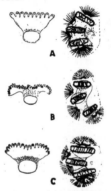

Fig. 419. Anterior respiratory organs (left) and posterior respiratory organs (right) of, (A) Ceratitis capitata, (B) Dacus dorsalis, and (C) Anastrepha ludens. Arranged from Phillips (693).

The larvae possess one pair of anterior respiratory organs (fig. 419) projecting like ears from the dorsal part of the first thoracic segment, close to its posterior margin. Another characteristic feature is seen from a posterior view of the last segment, namely, a pair of posterior stigmata each made up of a group of 3 slits (fig. 419). The position of the stigmata varies with the species. An enlarged view of the posterior respiratory organs shows details which are of prime importance in larval taxonomy. Each dorsal trachea opens into an atrial or stigmatic chamber much larger in diameter than the trachea. Distally the chamber divides into 3 stalked lobes. A "trough-shaped" structure is located at the end of each lobe.

In the United States there are a number of fruit flies that are important pests of deciduous fruits, such as the apple maggot and the current fly. An important pest of walnuts, the walnut husk fly, will be discussed in Chapter 21. On citrus, the fruit flies are of importance in this country only as potential pests and in connection with quarantine legislation. The many fruit and vegetable hosts of the fruit flies are excluded from the United States by quarantines. With the exception of a few Mexican fruit flies in the Rio Grande Valley of Texas, well under control, there are no fruit flies attacking citrus in the United States.

A general discussion of the fruit flies of interest in connection with subtropical fruits is given by Quayle (720, 724), while Smith et al (805) also present a rather comprehensive discussion, but with special reference to the quarantine aspects of the problem.

Life History:- Eggs are deposited in a puncture made by the ovipositor just below the surface of the fruit (fig. 420). From 2 or 3 to as many as 15 eggs may be deposited at a time, and a single female may lay as many as 800 eggs during her normal oviposition period. The maggots burrow through the fruit (fig. 421). When full-grown they drop to the ground and pupate in the soil, emerging later as adult flies. There may be as few as a single generation to as many as 15 or 16 in the warmest regions. Aside from temperature, the sequence of hosts is important in determining the success of a species.

Fig. 420 (at left). Eggs of Mediterranean fruit fly in cavity beneath surface of fruit. After Back and Pemberton.

Fig. 421 (at right). Section of orange infested with Mediterranean fruit fly. Courtesy H. J. Quayle.

Control:- An important aspect of the control of the fruit flies is the "host-free period", which is maintained by the elimination of certain economically unimportant host plants at a period of the year when the main commercial crops are off the trees or in a condition not suitable for oviposition or development of the flies. Artificial control measures are also used. Various types of poison bait syrups, usually with some arsenical as the toxic ingredient, are sprayed on the host trees, and the adult flies alighting on these trees and feeding on the syrup are poisoned. Experience in South Africa against the Mediterranean and Natal fruit flies has shown the sodium fluosilicate is 16 times as toxic as lead arsenate as a bait spray and has certain other advantages (783). DDT dust has been found to be very toxic to the Mexican fruit fly (701).

MEDITERRANEAN FRUIT FLY

Ceratitis capitata (Wiedemann)

The Mediterranean fruit fly may be found throughout the subtropical regions of the world with the exception of southeastern Asia and North America. Its wide distribution and wide range of host plants makes it the most important of the fruit flies. Most deciduous and subtropical fruits are attacked, at least to some extent. The peach appears to be the favorite host. All citrus except the lemon and sour lime are attacked. The Mediterranean fruit fly will oviposit in avocados that have been allowed to soften, but since avocados should be picked while still firm, the Mediterranean fruit fly is not an important threat to this fruit. Some varieties of banana are practically immune. Pineapples have never been found infested by this insect.

Description:- The Mediterranean fruit fly is a little smaller than the house fly and has the colors brown, yellow, black and white arranged in a characteristic pattern (fig. 422). The wings are held in a drooping position. The black spots on the thorax, the two white bands on the yellowish abdomen, and the black and yellow markings on the wings, are distinguishing characteristics for this species.

Quarantine:- Federal Quarantine No. 13 which pertains to Hawaii, excludes all fruits and vegetables except the noncooking type of bananas, pineapples, taros, and cocoanuts. These may be admitted after inspection and certification by quarantine officials. Quarantine No. 56, which pertains to all foreign countries except

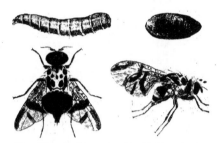

Fig. 422. Mediterranean fruit fly, showing
larva, pupa, and adult. X 6. After Quayle.

Canada, excludes all fruits and vegetables and plants or portions of plants used
for packing for fruits and vegetables, except that certain fruits may be imported
under permit when it has been determined that they are not carriers of fruit flies.
These quarantines pertain to the melon fly, oriental fruit fly, and other species
as well as the Mediterranean fruit fly.

The history of the successful eradication campaign against the Mediterranean
fruit fly in Florida in 1929 is presented on pages 54 and 55.

MEXICAN FRUIT FLY

Anastrepha ludens (Loew)

Although more restricted in its range than the Mediterranean fruit fly, and
more restricted also in the number of hosts it attacks, the Mexican fruit fly is
nevertheless a serious pest in many of the regions where it occurs. It is indigenous to Mexico, and is also found in Nicaragua, northern South America, and with
perhaps a few flies in the Rio Grande Valley of Texas. For a discussion of its
successful control in the latter area, see page 52. Grapefruit is a favored host,
but all varieties of citrus except lemons and sour limes are attacked. In the
native home of the Mexican fruit fly, in northeastern Mexico, it infests the fruit
of the yellow chapote, Sargentia greggii, an indigenous tree. Infestation in the
field has been confirmed for the following hosts: cherimoya, Annona cherimola,
custard apple, A. reticulata, white sapote, Casimiroa edulis, sour orange, grapefruit and shaddock, kid-glove oranges of the mandarin and satsuma types, sweet
oranges, varieties of sweet lime, rose apple, Eugenia jambos, jinicuil, Inga
jinicuil, mammee apple, Mammea americana, mango, peach, guava, pomegranate, pear,
apple, quince, and the yellow chapote. A number of other fruits and vegetables
have been infested under laboratory conditions. (51)

The importance which has been attached to the Mexican fruit fly is attested
by the fact that in 1928 a laboratory of the U. S. Department of Agriculture,
Bureau of Entomology and Plant Quarantine, was located in Mexico City for the
purpose of investigating all phases of the Mexican fruit fly problem. This was a
cooperative undertaking, the Secretaria de Agricultura y Fomento of Mexico providing the buildings and grounds, and the U. S. Department of Agriculture assign-
ing the personnel and providing the equipment. By far the greater part of our
technical knowledge concerning the Mexican fruit fly has been supplied by the investigators connected with the laboratory, headed by A. C. Baker.

Description (51):- The adults (fig. 423, p. 526) are about the size of house
flies and are of a general yellowish brown color. Longitudinal markings of a somewhat lighter color are found in the thorax, especially on newly-emerged flies, and
on the posterior part of the mesothorax there is a small median spot of dark brown.

The eggs are green. The wings are transparent where they are not mottled and striped with yellowish brown bands. The inverted V on the lower part of the outer half of the wing is not connected at the tip and is not connected with the main pattern, thus distinguishing this species from certain other closely related flies. The ovipositor sheath of the female is very long, slender, and tube-like; it is longer than the remainder of the abdomen.

The full-grown larvae are white, dirty white, or yellowish white, and pointed at one end where the black mouthhooks are plainly visible. At the larger, truncate end of the larvae may be seen the posterior spiracles (fig. 419, C) and these constitute one of the diagnostic characters to aid in the separation of Anastrepha ludens larvae from certain closely allied species which may be associated with A. ludens but which are of no economic importance.

Fig. 423. The Mexican fruit fly, Anastrepha ludens, on an orange. X 1.4. Photo by Roy Pence.

Quarantine:- Oranges, sweet limes, grapefruit, mangos, ochras, sapotes, peaches, guavas, and plums are denied entry from Mexico to the United States by Federal Quarantine No. 5. Federal Quarantine No. 64 was enacted to prevent certain fruits from being sent from certain counties in Texas to other parts of the country except under certification by the U. S. Department of Agriculture, and citrus fruits are regularly shipped from these counties under this certification. The fruits affected by Quarantine No. 64 are mangos, sapotes, peaches, guavas, apples, pears, plums, quince, apricots, mameys, ciruelas, and all citrus fruits excepting lemons and sour limes.

ORIENTAL FRUIT FLY

Dacus dorsalis Hendel

The oriental (mango, Malayan, Formosan) fruit fly had long been known as a pest of tropical and subtropical fruits in the Pacific Islands, but before its introduction into the Hawaiian Islands it appeared to be a serious pest only in the island of Formosa (Taiwan). The flies were first collected in Honolulu on May 9, 1946, but surveys showed that they were at that time already widely distributed over the islands of Oahu, Hawaii and Maui. They appear to have been brought to the island of Hawaii by Marines returning from Saipan as early as June, 1945. In 1947 this species was found in Guam (579).

This species had previously been called Dacus ferrugineus Fabricius and D. ferrugineus var. dorsalis Hendel, but the specific name Musca ferrugineus Fabricius 1794 was recently found by Alan Stone to be preoccupied by Musca ferruginea Scopoli 1763 and is, therefore, a primary homonym and consequently invalid. Dr. Stone believes Dacus dorsalis Hendel to be the next available valid name as far as can be determined.

Economic Importance:- This fruit fly occurs on a wide variety of hosts. The avocado, mango and papaya are the most commonly attacked. Citrus fruits, figs, persimmons, and loquats, among the subtropical fruits, also are attacked. The flies have been found in some regions with a climate similar to that of southern California and there is great concern about the possibilities of the pest becoming established in California and becoming a major pest here. The oriental fruit fly is a more injurious species than the Mediterranean fruit fly and melon fly. It has been found in mature pineapples and bananas, not attacked by the latter species, and so the shipment of these fruits to the mainland has been restricted (40).

Appearance:- The oriental fruit fly (fig. 424) averages slightly larger than the house fly. The predominant color is yellow, with dark markings on the thorax.

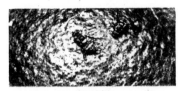

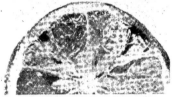

Fig. 424. The oriental fruit fly, Dacus dorsalis. Upper, adults on orange; lower, larvae in pulp of fruit. Courtesy F.G. Holdaway.

A "T" formed by the juncture of a median stripe and a transverse band, is located on the abdomen. Superficially it is similar in appearance to the melon fly, Dacus cucurbitae (Coquillett) except that it has clear wings, while those of the melon fly have a characteristic pattern of darkened areas. (579)

THE SOUTH AMERICAN FRUIT FLY

Anastrepha fraterculus (Wiedemann)

Anastrepha fraterculus, often referred to in the past as the West Indian fruit fly, is distributed through Central and South America and Mexico. It is a serious pest in a large part of the citrus acreage in South America. In Argentina, for example, it is considered to be the most important pest in all citrus areas (see p. 542). Fruit flies were declared a "national pest" in 1937, making their control compulsory. It is stated that all varieties of fruits are attacked, and sometimes vegetables and nuts. (417) A combination of bait-traps, bait-sprays, the destruction of infested fruits, and liberation of parasites is practiced in all orchards of a district. For the bait-traps, wine vinegar and water (1:3) is recommended, but some other combinations are used. For the bait-spray, 5% molasses and 1.5% sodium fluosilicate or sodium fluoride is used. The spray is applied lightly to the trees once a week, but not to branches bearing fruit. Infested fruits are collected daily and buried every 10-15 days under at least 20 inches of well-compressed soil.

Anastrepha fraterculus as it occurs in northern Mexico and the Rio Grande Valley of Texas is considered by Dr. A. C. Baker as a distinct form of the species, based on morphological differences and different host responses (51). The Mexican form attacks guavas and peaches but does not appear to attack citrus, while in South America, citrus is severely attacked. Laboratory experiments confirmed the field observations concerning the host preferences of the Mexican Anastrepha fraterculus (54).

Description:- The adult fly (fig. 425, p. 528) is about 12 mm long, (not including the ovipositor of the female) and it has a wing expanse of about 25 mm. The body is rust-yellow or brownish-yellow, with 3 sulfur-yellow longitudinal stripes on the thorax. The wings are clear except for a characteristic but variable pattern of yellow-brown color. The ovipositor is stout and shorter than the abdomen. It is tapered regularly toward the tip and covered with coarse black hairs. (695)

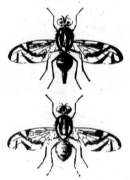

Fig. 425. South American fruit fly, Anastrepha fraterculus. Above, female; below, male. X 18. After Wille (952).

The wing markings of the Mexican form of Anastrepha fraterculus differ from those of the South American form. The inverted V of the wing of A. fraterculus is separated from the main pattern (fig. 426, A) while in the Mexican A. fraterculus the inverted V is connected with the main pattern (fig. 426, B). Although occasional exceptions may occur, this character suffices to differentiate populations as a whole. (51).

A   B

Fig. 426. A, wing of Anastrepha fraterculus; B, wing of Mexican form of A. fraterculus, showing how the "inverted V" is connected with the main pattern. After Baker et al (51).

NATAL FRUIT FLY

Pterandrus rosa Karsch

The Natal Fruit fly, along with the Mediterranean fruit fly, which it closely resembles both in appearance and habits, is a pest of citrus in all the principal citrus districts of South Africa. It is the most important pest of deciduous fruits and attacks citrus, particularly Navel oranges, late in the year. It is an indigenous insect, formerly found only in Natal.

The Natal and Mediterranean fruit flies, which may for convenience be discussed together, are worst where a large variety of fruits are grown and where fruits ripen irregularly over a long period. Both species breed in large numbers in the "bush" at certain times of the year and fly considerable distances to infest orchards. In order to live, the flies must have a regular supply of food and water. The food consists of sweet liquids such as honey dew, plant sap, and fruit juice, and water is obtained from drops of dew on trees.

Control measures include orchard sanitation - the picking of all infested fruit and raking up of dropped fruit. Fruit should be placed in an old oil drum and covered with water and a little oil. The infested fruit should be disposed of once a week. (782)

Excellent results have been obtained from the use of sweet liquid poison baits. The trees are lightly "spot-sprayed" with the following solution:

CITRUS PESTS OF FOREIGN COUNTRIES

```
Water - - - - - - - - - - - - 4 gallons
White sugar - - - - - - - - - 2 pounds
Sodium fluosilicate - - - - - 1 ounce
```

ead arsenate was used in bait sprays, but the sodium fluosilicate
antages. It is soluble in water and does not require as much stir-
times as toxic to fruit flies as lead arsenate. (782) Lead arsenate
" for fruit fly and false codling moth control resulted in a reduc-
the juices of oranges (588).

should be sprayed from at least two sides. Usually knapsack
ed. A course spray leaving large drops is used, and an attempt is
oliage instead of fruit. Bordering native growth or hedges should
. The adult fly population should be determined by means of bait
ourse, spraying should be done only when the flies are present.
ost abundant after hot, dry weather when no dew has fallen during
ght, and that is also a good time to spray, because the flies are
r. Bait spraying should be repeated, if the rain washes the bait
Cooperation of growers for the control of the fruit flies over a
ery desirable.

nzene hexachloride, although very toxic to the flies, have not so
e effective in the control of fruit flies, but it is suggested that
thod of using these insecticides successfully may be developed.(782)

QUEENSLAND FRUIT FLY

Strumeta tryoni Froggatt

land fruit fly is an important pest in Queensland and New South
a, attacking citrus and a large variety of other tropical and sub-
, as well as many deciduous fruits and
d vegetables. This pest is now more
the Mediterranean fruit fly in Austra-
ittle larger than the common house fly,
pearance, and reddish-brown in color.
arked by a number of distinct lemon-
which are very pronounced while the
The abdomen has two transverse bands
n. The wings are clear except for a
d along the front margins and a trans-
ar their base (fig. 427).

Fig. 427. Queensland fruit
fly, Strumeta tryoni. After
Anonymous (15).

1):- Tartar emetic was found to be
ium fluosilicate as a bait spray for
fruit fly. Sodium fluosilicate
njury to foliage and a roughening of
ng lateral branches. The leaf injury
p burn and small circular necrotic patches. When used with sugar,
ic resulted in only a limited amount of tip burn and no burning of
molasses instead of sugar was used with tartar emetic, there was a
ount of tip burn and some slight bark injury.

a for the tartar emetic bait spray that is being used in New South
lows:

```
Tartar emetic - - - - - - - - - - 2 ounces
White sugar - - - - - - - - - - - 2-1/2 pounds
Water - - - - - - - - - - - - - - 4 gallons
```

fluosilicate is to be used, the formula is as follows:

```
Sodium fluosilicate - - - - - - - 2 ounces
White sugar - - - - - - - - - - - 2-1/2 pounds
Water - - - - - - - - - - - - - - 4 gallons
```

529

SUBTROPICAL ENTOMOLOGY

The above formulas are prescribed in the State Regulations. In Australia, measures for the control of fruit flies are compulsory under the Plant Diseases Act. These compulsory measures include the use of foliage poison baits and the regular destruction of all infested and fallen fruits.

JAPANESE ORANGE FRUIT FLY

Dacus tsuneonis Miyake (163)

This species is the only fruit fly attacking oranges in Japan, where it is believed to be indigenous. At times it becomes a serious pest and may occasionally cause the loss of 40 to 50% of the fruit in some localities. The fruit has usually dropped from the tree by the time the larvae have matured (in October or November). The larvae then enter the soil and pupate. The adult flies appear the following June or July and oviposit a few weeks later.

CITRUS FRUIT FLY

Mellesis citri, Chen (156a)

The citrus fruit fly is a serious pest of sweet oranges in China. It has been estimated that near Kiangtsin, Szechwan Province, over 50% of the fruit is infested. The infested fruit usually falls to the ground.

This species, which was described in 1940, is brownish or reddish-yellow and somewhat larger than the average fruit fly. The length of the male or the female without ovipositor is 13 mm. The ovipositor is 6.5 mm long and the wings are about 11 mm long. The hyaline wings have a rather broad costal stripe which is yellow in the middle and blackish brown at the base and apex.

Scientific and Common Names of the Citrus Pests of the World

Aceria sheldoni (Ewing) - - - - - - - - - - - - - - - - citrus bud mite (Eriophyidae)
Achaea janata Linneaus - - - - - - - - - - - - - - -fruit piercing moth (Noctuidae)
Actinodes auronotata Castelnau & Gory - - - - - - - - - - - - - - - Buprestidae
Aethalion reticulatum Linnaeus - - - - - - - - - - - - - tree hopper (Membracidae)
Agonopterix sp. - bud moth (Oecophoridae)
Agrilus auriventris Saunders - - - - - - - - - - - - citrus agrilus (Buprestidae)
Agrilis occipitalis Eschscholtz - - - - - - - - - - - - - - - citrus bark borer
Aleurocanthus cheni B. Young - - - - - - - - - - - - - - - - - Chen's whitefly
Aleurocanthus citriperdus Quaintance & Baker - - - - - - - - - - - - - whitefly
Aleurocanthus husaini Corbett - - - - - - - - - - - - - - - - Husian's whitefly
Aleurocanthus spiniferus (Quaintance) - - - - - - - - - - Orange spiny whitefly
Aleurocanthus woglumi Ashby - - - - - - - - - - - - - - - - citrus blackfly
Aleurolobus citrifolii Corbett - - - - - - - - - - - - whitefly (Aleyrodidae)
Aleurolobus marlatti (Quaintance) - - - - - - - - - - - - - - - Marlatt whitefly
Aleurolobus szechwanensis B. Young - - - - - - - - - - - - Szechwan whitefly
Aleurothrixus floccosus (Maskell) - - - - - - - - - - - - - - - wooly whitefly
Aleurothrixus howardii (Quaintance) - - - - - - - - - - - - - Howard's whitefly
Amsacta lactinea Cramer - - - - - - - - - - - - - - - - tiger moth (Arctiidae)
Anacridium aegyptium Linnaeus - - - - - - - - - - - - - - - - - - grasshopper
Anastrepha fraterculus (Wiedemann) - - - - - South American fruit fly (Trypetidae)
Anastrepha ludens (Loew) - - - - - - - - - - - - - - - - - Mexican fruit fly
Anastrepha mombinpraeoptans Seín - - - - - - - - - - West Indian fruit fly
Anastrepha serpentina Wiedemann - - - - - - - - - - - - - - - - - - fruit fly
Anastrepha suspensa Loew - fruit fly
Ancylocera cardinalis Dalman - - - - - - - - - - - - - - - - Cerambycidae
Anomala undulata Melsheimer - - - - - - - - - - - - - - - - blossom anomala
Anoplium inerme (Newman) - - - - - - - - - - - orange sawyer (Cerambycidae)
Anoplocnemis phasiana Fabricius - - - - - - - - - - - - - - - bug (Coreidae)
Anychus clarkii McGregor - - - - - - - - - - - - - - - Texas citrus mite
Anychus orientalis Zacher - - - - - - - - - - - - - - - - - oriental mite
Anychus ricini Rahman & Sapra - mite

Anychus verganii Blanchard - mite
Aonidiella aurantii (Maskell) - - - - - - - - - - - - - - - California red scale
Aonidiella citrina (Coquillett) - - - - - - - - - - - - - - - - - yellow scale
Aonidiella comperei McKenzie - - - - - - - - - - - - - - - false yellow scale
Aonidiella orientalis Newstead - - - - - - - - - - - - - oriental yellow scale
Aphis citricidus (Kirkaldy) - - - - - - - - - - - - - - - - black citrus aphid
Aphis gossypii Glover - - - - - - - - - - - - - - - - - - - cotton or melon aphid
Aphis nigricans Van der Goot - aphid
Aphis spiraecola Patch - spirea aphid
Aplonobia oxalis Womersley - sour-sob mite
Archips argyrospila (Walker) - - - - - - - fruit tree leaf roller (Tortricidae)
Argyroploce leucotreta Meyrick - - - - - - - - false codling moth (Olethreutidae)
Argyrotaenia citrina (Fernald) - - - - - - - - - - - - - - - - orange tortrix
Aserica (Serica) castanea Arrow - - - - - - - small brown beetle (Melolonthidae)
Aspidiotus camelliae Signoret - greedy scale
(=Hemiberlesia rapax Comstock of European authors)
Aspidiotus hederae (Vallot) - oleander scale
Aspidiotus scutiformis Cockerell - - - - - - - - - - - armored scale (Diaspididae)
Astilus variegatus (Germar) - - - - - - - - - - - - - - - - beetle (Dasytidae)
Atta cephalotes Linnaeus - leaf cutting ant
Atta insularis Guérin - - - - - - - - - - - Cuban leaf cutting ant or bibijaqua
Atta octospinosa Reich - - - - - - - - - - - - - - - - - - - leaf cutting ant
Atta sexdens Linnaeus - leaf cutting ant
Atta texana Buckley - - - - - - - - - - - - - - - - - - Texas leaf cutting ant
Attacus atlas Linnaeus - - - - - - - - - - - - - - - - atlas moth (Saturniidae)
Baryopadus (Leptops) squalidus (Bohemann) - - - - - - - fruit tree root weevil
Bemisia giffardi Kotinsky - whitefly
Biprorulus bibax Breddin - - - - - - - - - larger horned citrus bug (Pentatomidae)
Boarmia crepuscularia Hübner - - - - - - - - - - - - - - - a looper (Boarmiidae)
Boarmia selenaria dianaria Hübner - - - - - - - - - - looper or measuring worm
Bostrychopsis jesuita Fabricius - - - - - - - fig and orange borer (Bostrichidae)
Brachycerus citriperda A. K. Marshall - - - - - - - - - - - - - - - - a weevil
Brachyrhinus cribicollis Gyllenhal - - - - - - - - snout beetle (Curculionidae)
Brachytrupes portentosus Lichtenstein - - - - - - - - - the large brown cricket
Brevipalpus lewisi McGregor - - - - - - - - - - - - - - - - - - Lewis spider mite
Cacoecia podana Scudder - a tortricid
Caedicia simplex Walker - - - - - - - - inland green tree hopper (Tettigoniidae)
Caedicia strenua Walker - - - - - - - - - - - - - - - citrus green tree hopper
Calpe capucina Esper - - - - - - - - - - - - - - - fruit-piercing moth (Noctuidae)
Calpe excavata Butler - - - - - - - - - - - - - - - - - - fruit-piercing moth
Camnula pellucida Scudder - - - - - - - - - - - - - - - clear-winged grasshopper
Cania bilinea Walker - moth (Limacodidae)
Cappaea taprobanensis (Dallas) - - - - - - - - - - - - - - - - pomelo stink bug
Ceratitis capitata (Wiedemann) - - - - - - - - - - - - - Mediterranean fruit fly
Ceroplastes ceriferus Anderson - - - - - - - - - - - - - - - wax scale (Coccidae)
Ceroplastes cerripediformis (Comstock) - - - - - - - - - - - - barnacle scale
Ceroplastes destructor Newstead - - - - - - - - - - - - - - white wax scale
Ceroplastes floridensis (Comstock) - - - - - - - - - - - - Florida wax scale
Ceroplastes sinensis Del Guercio - - - - - - - - - - - - - - Chinese wax scale
Ceroplastes rubens Maskell - red wax scale
Ceroplastes rusci Linnaeus - wax scale
Chaetoanophothrips orchidii (Moulton) - - - - - - - - - - - - - - orchid thrips
Charagia virescens Duble - - - - - - - - - - - - - - - - - - a hepialid borer
Chelidonium cinctum Guerin - - - - - - - - - - - - - lime tree borer (Cerambycidae)
Chelidonium gibbicolle White - - - - - - - - - - - - - - - green citrus longhorn
Chilades laius Cramer - a lycaenid
Chionaspis (Prontaspis) yanonensis Kuwana - - - - - - - - - - - - yanone scale
Chloridolium alemene Thomson - - - - - - - - - - - - - - - - - Cerambycidae
Chondacris rosea De Geer - - - - - - - - - - - - - - - - citrus locust (Locustidae)
Chortoicetes terminifera Walker - - - - - - - - - - - - Australian plague locust
Chrysochroa fulminans Fabricius - - - - - - - - flat-headed borer (Buprestidae)
Chrysobothris lepida Castelnau & Gory - - - - - - - - - - - - - flat-headed borer
Chrysomphalus aonidum (Linnaeus) - - - - - - - - - - - - - - Florida red scale

531

Chrysomphalus dictyospermi (Morgan) - - - - - - - - - - - - dictyospermum scale
Chrysomphalus rossi Maskell - - - - - - - - - - - - - - scale insect (Diaspididae)
Citripestis sagittiferella Moore - - - - - - - - - - - - - moth borer (Pyralidae)
Citriphaga mixta Lea - - - - - - - - - - - branch and trunk borer (Cerambycidae)
Clania minuscula Butler - - - - - - - - - - - - - - - - - - bagworm (Psychidae)
Clitea metallica Chen - - - - - - - - - - - - - - - - - - metallic flea beetle
Coccus hesperidum Linnaeus - soft scale
Coccus pseudomagnoliarum (Kuwana) - - - - - - - - - - - - - citricola scale
Coccus viridis (Green) - green scale
Colaspis flavipes Olivier - - - - - - - - - - - - - - - - - - - Cerambycidae
Colasposoma fulgidum Lefevre - - - - - - blue-green citrus nibbler (Eumolpidae)
Coleoxestia spinipennis Serville - - - - - - - - - - - - - - - - Cerambycidae
Colgar peracuta Walker - Hemiptera
Cotinis mutabilis Gory & Percheron - - - - - - - - - - - - - - - a scarabaeid
Corythuca gossypii Fabricius - - - - - - - - - - - - - - - - - bean lace bug
Cratosomus punctulatus Gyllenhal - - - - - - - - - - - - - - a large weevil
Cryptoblabes gnidiella Millière - - - - - - rind-boring orange moth (Pyralidae)
Cryptorrhynchus corticolis Boheman - - - - - - - - - - - - - - - - - weevil
Cryptothelea variegata Snellen - - - - - - - - - - - - - - - bagworm (Psychidae)
Ctenopseustis obliquana Walker - - - - - - - - - - - - - leaf roller (Tortricidae)
Cyrtophyllus concavus Scudder - - - - - - - - - - - - - - - - - - - katydid
(=Pterophylla camellifolia (Fabricius))
Dacus dorsalis Hendel - oriental fruit fly
Dacus tsuneonis Miyake - fruit fly
Dasychira misana Moore - Lepidoptera
Dendrobias mandibularis Serville - - - - - - - - - - - - - trunk borer (Coleoptera)
Depressaria culcitella Hering - - - - - - - - - - - - - - leaf roller (Oecophoridae)
Depressaria lizyphi S. Hons. - - - - - - - - - - - - - - - - - - leaf roller
Dexicrates robustus (Blanchard) - - - - - - - - - - - - - - - - - Bostrichidae
Diabrotica decolor Erichson - - - - - - - - - - - - - - - - - - Chrysomellidae
Diabrotica duodecimpunctata (Fabricius) - - - - - - - - - spotted cucumber beetle
Diabrotica speciosa (Germar) - - - - - - - - - - - - - - - - diabrotica beetle
Diabrotica undecimpunctata Mannerheim - - - - - western striped cucumber beetle
Diabrotica vittata (Fabricius) - - - - - - - - - - - striped cucumber beetle
Dialeurodes citri (Ashmmead) - - - - - - - - - - - - - - - - - citrus whitefly
Dialeurodes citricola B. Young - - - - - - - - - - - citrus whitefly (China)
Dialeurodes citrifolii (Morgan) - - - - - - - - - - - - cloudy-winged whitefly
Dialeurodes elongata Dozier - - - - - - - - - - - - - - - - elongate whitefly
Diaphorina citri Kuwayana - - - - - - - - - - - - - - citrus psylla (Chermidae)
Diaprepes abbreviatus Linnaeus - - - - - - - - - - - - - - - - - - - weevil
Diaprepes esuriens Gyllenhal - weevil
Diaprepes excavatus Rosenschöld - - - - - - - - - - - - - - - - - - - weevil
Diaprepes scalaris Boheman - weevil
Diaprepes sprengleri Linneaus - weevil
Dichocrocis punctiferalis Guenée - - - - - - - - - yellow peach moth (Pyralidae)
Dionconotus cruentatus Brullé - - - - - - - - - - - citrus flower moth (Miridae)
Diploschema rotundicolle Serville - - - - - - - - - - - trunk borer (Cerambycidae)
Doticus pestilens Olliff - - - - - - - - - - - - dried apple beetle (Anthribidae)
Drosicha contrahens Walker - mealybug
Drosicha stebbingi Green - mealybug
Drosophila punctipennis van der Wulp - - - - - - - - - - - - - black citrus fly
Dysdercus longirostris Stål - - - - - - - - - - - - cotton stainer (Pyrrhocoridae)
Dysdercus suturellus Herrich-Schaeffer - - - - - - - - - - - - - cotton stainer
Eantis palida Watson - - - - - - - - - - - - - - - - - - skipper (Hesperiidae)
Eantis thraso Hübner - - - - - - - - - - - - - - - - - - - citrus skipper
Eburia ocoguttata Germar - - - - - - - - - - - - - - - - - - Cerambycidae
Ectatomma ruidum Roger - ant
Ectinohoplia rufipes Motschulsky - - - - - - - - - - - - - - - - Scarabaeidae
Edessa quadridens Fabricius - - - - - - - - - - - - - - - - - - Pentatomidae
Empoasca fabae Harris - - - - - - - - - - - - - - - - - potato leafhopper
Ephestia vapidella Mannerheim - - - - - - - - - - - - - stub moth (Pyralidae)
Epicometis (Tropinota) hirtella Poda - - - - - - - blossom feeder (Cetoniidae)
Epitetranychus altheae Hanstein - - - - - - - - - - - - - - - - - - - a mite

Epitrix cucumeris Harris - - - - - - - - - - - - - - - - - - - potato flea beetle
Euphoria basilis Burmeister - - - - - - - - - - - - - - - - - mayate (Scarabaeidae)
Euproctis flavata Cramer - - - - - - - - - - - - - - - - tussock moth (Lymantriidae)
Eurytoma fellis Girault - - - - - - - - - - - - - citrus gall wasp (Eurytomidae)
Eutinophae bicristata Lea - - - - - - - - - - - - - - - - - - leaf eating weevil
Fiorinia theae Green - tea scale
Formica cinerea Mayr var. neocinerea Wheeler - - - - - - - - - - - - - - - gray ant
Frankliniella cephalica bispinosa Morgan - - - - - - - - - -- Florida flower thrips
Frankliniella gossypii Morgan - cotton thrips
Frankliniella insularis Franklin - thrips
Frankliniella moultoni Hood - - - - - - - - - - - - - - - - Western flower thrips
Frankliniella rodeos Moulton - thrips
Frankliniella varipes Moulton - thrips
Geisha distinctissima Walker - a fulgorid
Geloptera porosa Lea - - - - - - - - - - - - pitted apple beetle (Chrysomellidae)
Gnatholea eburifera Thomson - Cerambycidae
Gonodonta incurva Sepp - - - - - - - - - - - - - - fruit-piercing moth (Noctuidae)
Gymnandrosoma aurantianum Costa Lima - - - - - - - - the orange moth (Tortricidae)
Gymnandrosoma punctidiscana Dyar - - - - - - - - - - - - - - - - - - - Tortricidae
Heliothis armigera Hübner - - - - - - - - - - - - - - - - - bollworm (Noctuidae)
Heliothrips fasciatus Pergande - thrips
Heliothrips haemorrhoidalis (Bouché) - - - - - - - - - - - - - greenhouse thrips
Helix aspersa Müller - - - - - - - - - - - - - - - - - - European brown snail
Helix pisana Müller - white snail
Helopeltis antonii Waterhouse - - - - - - - - - - - - - - - - plant bug (Miridae)
Helopeltis collaris Stål - plant bug
Hemiberlesia rapax Comstock - greedy scale
=Aspidiotus camelliae Signoret
Hemitarsonemus latus (Banks) - broad mite
Hercothrips fasciatus (Pergande) - - - - - - - - - - - - - - - - - - - bean thrips
Heterodera marioni (Cornu) - - - - - - - - - - - - - - - - - root-knot nematode
Heteronychus sanctae-helanae Blanchard - - - - - - - - black beetle (Dynastidae)
Heterorrhina elegans Fabricius - - - - - - - - - - - a flower beetle (Cetoniidae)
Heterothrips moreirai Moulton - thrips
Holcocera iceryaeella (Riley) - - - - - - - - - - - - - holcocera (Blastobasidae)
Hoplophorion (Metcalfiella) pertusa Germar - - - - - - tree hopper (Membracidae)
Hoplophorion (M.) vicina Fairmaire (fimbriata Stål) - - - - - - - - tree hopper
Hyalarota hübneri Westwood - - - - - - - - - - - - orange case moth (Psychidae)
Hypomeces squamosus Fabricius - - - - - - - - - - - - - - - - green snout beetle
Hypothenemus aspericolis (Wollaston) - - - - - - - - - - bark beetle (Scolytidae)
Hypothenemus citri Ebeling - - - - - - - - - - - - - - - - - - citrus bark beetle
Icerya floccosa Hemphill - Monophlebidae
Icerya montserratensis Riley & Howard - - - - - - - - - - - - - - - monophlebidae
Icerya pulcher Leonardi - giant coccid
Icerya purchasi Maskell - - - - - - - - - - - - - - - - - - - cottony cushion scale
Icerya seychellarum (Westwood) - - - - - - - - - - - - - - - - - - Monophlebidae
Indarbella (Arbela) quadrinotata Walker - - - - - - bark-eating borer (Cossidae)
Indarbella (Arbela) tetraonis Moore - - - - - - - - - - - - - bark-eating borer
Iridomyrmex detectus F. Smith - ant
Iridomyrmex humilis Mayr - Argentine ant
Isochaetothrips striatus Hood - thrips
Isotenes miserana (Walker) - - - - - - - - - - - - - - orangeworm (Tortricidae)
Lachnopus coffeae Marshall - weevil
Lachnopus curvipes Fabricius - weevil
Lachnopus hispidis Gyllenhal - weevil
Lachnopus sparsimguttatus Perroud - - - - - - - - - - - - - - - - - - - weevil
Lampona obscoena Koch - - - - - - - - - - - - - - - webbing spider (Araneida)
Laspeyresia sp. - - - - - - - - - - - - - - - - fruit boring moth (Olethreutidae)
Lecanium deltae Lizer - delta scale
Lepidosaphes beckii (Newman) - - - - - - - - - - - - - - - - - - - purple scale
Lepidosaphes gloverii (Packard) - - - - - - - - - - - - - - - - - - Glover scale
Lepidosaphes lasianthi Green - - - - - - - - - - - - - - - - - - - armored scale
Lepidosaphes pallida Green - - - - - - - - - - - - - - - - - - - pale long scale

glossus gonagra Fabricius - - - - - - - - - - - - - - - - - - - bug (Coreidae)
glossus membranaceus Fabricius - - - - - - - - - - - - - - - - bug (Coreidae)
glossus phyllopus Linnaeus - -· - - - - - - - - - - - - - leaf footed bug
glossus stigma Herbst - - - -·- - - - - - - - - - - - - - - - bug (Coreidae)
glossus zonatus (Dallas) - - - - - - - - - - -· Western leaf-footed plant bug
stylus pleurostictus Bates - - - - - - - - - - - - - - - - - - Cerambycidae
spis japonica Cockerell - - - - - - - - - - - - - - - - - - Japanese long scale
ngaspis rossi Maskell - armored scale
hiidae - colonial spiders (Araneida)
aea gibbosa de Meijere - - - - - - - - - - citrus blossom fly (Lonchaeidae)
aea plagiata Walker - - - - - - - - - - - - - - - - - - fly (Sapromyzidae)
iactylus affinis Castelnau - - - - - - - - - - cockchafer (Melolonthidae)
dactylus pumilio Burmeister - - - - - - - - - - - - - - - - - cockchafer
dactylus suavis Bates - - - - - - - - - - - - - - - - - rose chafer, ahogapollo
dactylus suturalis Mannerheim - - - - - - - - - - - -· - - - - - cockchafer
phora accentifer Olivier - - - - - - - - - - - - - - - - - - Cerambycidae
siphum solanifolii (Ashmead) - - - - - - - - - - - - - - - - potato aphid
terpes dentipes Heller - - - - - - - - - - -· - - - citrus snout beetle
terpes spinipes Blackburn (=phytolymus Oliff) - - - - - - - dicky rice weevil
ra salictella Clemens - - - - - - - - - - - - - leaf miner· (Gracilariidae)
a unipuncta Donovan - - - - - - - - - - cherry tree borer (Xyloryctidae)
termes darwiniensis Froggatt - - - - - - - - - - - - - - - - - - - termite
opygae lunata Cramer - - - - - - - - - - - - - - - - moth (Megalopygidae)
opygae opercularis Abbott & Smith - - - - - - - - - - - - - - - puss moth
auster chinensis Förster - - - - black and white citrus borer (Cerambycidae)
oplus mexicanus devastator Scudder - - - - - - - - - devastating grasshopper
oplus mexicanus mexicanus (Saussure) - - - - - lesser migratory grasshopper
oplus mexicanus spretus (Walsh) - - - - - - - - - Rocky Mountain grasshopper
ona ruficrus Latreille - - - - - - - - - - - - - - - frapuan (Meliponid bee)·.
ona testacea var. cupera Smith - - - - - - - - - - - - - - leaf-cutting bee
centrum lanceolatum Burmeister - - - - - - - - - - - - - - - - - katydid
centrum retinerve Burmeister - - - - - - - - - - - angular-winged katydid
centrum rhombifolium Saussure - - - - - - - - - - - - - broad-winged katydid
s profana Fabricius - - - - - - - - - - - - - -· - - - Crusader bug (Coreidae)
hrosticus citricola Bezzi - - - - - - - - - - - - - fruit fly (Trypetidae)
epta australis Jacoby - - - - - - - yellow monolepta beetle (Chrysomelidae)
hamus versteegi Ritsema - - - - - - - - - - - citrus longhorn (Cerambycidae).
hlebus dalbergiae Green - mealybug
nella maskelli Cockerell - - - - - - - - - - - - - - - - - Maskell scale
ntia munda Stål -· - a pentatomid
is venipars Dyar - - - - - - - - - - - - - - Navel .orangeworm (Pyralidae)
caria brunhea Saunders - - - - - - - - - - - - - - - - sap-feeding ant
persicae (Sulzer) - - - - - - - - - - - - - - - - - green peach aphid
ctus disimulator Boheman - weevil
ytes cordifer Klug - - - - - - - - - - - - - - - - - - - Cerambycidae
a viridula Linnaeus - - - - - - - - - - - - - - - southern green stink bug
s ericae (Schilling) - - - - - - - - - - - - false chinch bug (Lygaeidae)
s vinitor Bergroth - - - - - - - - - - - - - - - -· - Rutherglen bug
eonotus enigma Scudder - - - - - - - - - - - - - - - · the valley grasshopper
icus abbot1 Grote - - - - - - - - - - - - - - - - common bagworm (Psychidae)
icus kirbyi Guilding - - - - - - - - - - - - - - - - - bicho de cesto
toma saundersi Westwood - - - - - - - - - - - - -· - - - - - an embiid
eres dejeani Thomson - - - - - - - - - - - - - - - - - - - Cerambycidae
eres tyrannus Guenée - - - - - - - - - - - - fruit-piercing moth (Noctuidae)
a postica Walker - - - - - - - - - - - - - - - -· - - moth (Lymantriidae)
zia insignis Douglass - - - - - - - - - - - - - - - - greenhouse orthezia
zia praelonga Douglass - - - - - - - - - - - - - - - - citrus orthezia
rrhinus cylindrorostris Fabricius - - - - - - elephant beetle (Curculionidae)
is (Ophideres) ancilla Cramer - - - - - - - - fruit-piercing moth (Noctuidae)
is apta Walker - fruit-piercing moth
is fullonia Clemens (fullonica L.) - - - - - - - - - fruit-piercing moth
is (Ophideres) materna Linnaeus - - - - - - - - - - - fruit-piercing moth
is (Ophideres) salaminia Cramer - - - - - - - - - - - fruit-piercing moth

Oxycetonia jucunda Faldermann - - - - - - - - - - - - - - blossom feeding scarab
Pachneus citri Marshall - - - - - - - - - - - - - - - - blue-green orange weevil
Pachneus citri litus Germar - - - - - - - - - - - - - blue-green orange weevil
Pachneus citri opalus Germar - - - - - - - - - - - - - blue-green orange weevil
Pachnaeus litus (Germar) - - - - - - - - - - - - - - - - - citrus root weevil
Pantomorus glaucus Perty - weevil
Pantomorus godmani (Crotch) - - - - - - - - - - - - - - - - fuller rose weevil
Pantomorus parsevali Costa Lima - weevil
Papilio aegeus Donovan - - - - - - - - - - - - - - - - - - citrus butterfly
Papilio anactus W. S. Macleay - - - - - - - - - - - - - - - citrus butterfly
Papilio anchysiades capys Hübner - - - - - - - - - - - - orange dog (Brazil)
Papilio andraemon Hübner - orange dog
Papilio androgeus Cramer - orange dog
Papilio cresphontes Cramer - - - - - - - - - - - - orange dog (swallowtail)
Papilio demodocus Esper - orange dog
Papilio demoleus Linnaeus - - - - - - - - - - - - - - - - - - lemon butterfly
Papilio memnon Linnaeus - orange dog
Papilio polycaon (Cramer) - - - - - - - - - - - - - - - - perro del naranjo
Papilio polytes Linnaeus - - - - - - - - - - - - - - citrus leaf caterpillar
Papilio thoas melonius Cramer - - - - - - - - - - - - - - - - - - orange dog
Papilio thoas thoas Linnaeus - - - - - - - - - - - - - - - - - - orange dog
Papilio xanthus Linnaeus - orange dog
Papilio xuthus Linnaeus - orange dog
Paralecanium expansum Green - flat scale
Paraleyrodes naranjae Dozier - - - - - - - - - - - - - - - - orange whitefly
Paraleyrodes perseae (Quaintance) - - - - - - - - - - - - - bay whitefly
Parastichtis purpurea var. crispa Harvey - - - - - - - - - - - - cutworm
Paratetranychus citri (McGregor) - - - - - - - - - - - - - citrus red mite
Parlatoria longispina Morgan - - - - - - - - - - - - - long-tailed parlatoria
Parlatoria morrisoni Bodenheimer - - - - - - - - - - - - Morrison's parlatoria
Parlatoria pergandii Comstock - - - - - - - - - - - - - - - - - chaff scale
Parlatoria proteus Curtis - - - - - - - - - - - - - - - - - parlatoria scale
Parlatoria sinensis Maskell - - - - - - - - - - - - - - - - Chinese parlatoria
Parlatoria zizyphus (Lucas) - - - - - - - - - - - - - black parlatoria scale
Patanga succincta Linnaeus - - - - - - - - - - - - - - - - - - grasshopper
Peltotrachelus pubes Förster - - - - - - - - - - - - - - - - - Cerambycidae
Perigea cupentia Cramer - - - - - - - - - - - - - - - - fruit-piercing moth
Perperus insularis Boheman - - - - - - - - - - - - - white striped weevil
Phenacoccus hirsutus Green - - - - - - - - - - - - - - hibiscus mealybug
Phobetron pithecium Abbot & Smith - - - - - - - - - - - hag moth (Eucleidae)
Phthia lunata Fabricius - coreid bug
Phyllocnistis citrella Stainton - - - - - - - - - citrus leaf miner (Lyonetidae)
Phyllocnistis saligna[1]Zeller - - - - - - - - - - - - - - - - - - leaf miner
Phyllocoptruta oleivora (Ashmead) - - - - - - - - - - - - - - citrus rust mite
Phyllophaga (Lachnosterna) citri (Smyth) - - - - - - June beetle (Scarabaeidae)
Phyllophaga (Lachnosterna) puberula (J. Duval) - - - - - - - - - - June beetle
Phyllophaga (Lachnosterna) subsericans (J. Duval) - - - - - - - - - June beetle
Phymateus leprosus Fabricius - - - - - - - - - - - - - - - - - - bush locust
Pinnaspis aspidistrae (Signoret) - - - - - - - - - - - - - - aspidistra scale
Pinnaspis minor Maskell - - - - - - - - - - - - - - - - - lesser snow scale
Platynota stultana (Walsingham) - - - - - - - - - - - platynota (Tortricidae)
Platypus compositus Say - - - - - - - - - - - - - - - ambrosia beetle (Scolytidae)
Platypus sulcatus Dejean - - - - - - - - - - - - - - - - - - ambrosia beetle
Platoecetious gloverii Packard - - - - - - - - - - - - - - - orange bagworm
Polycaon confertus LeConte - - - - - - - - - - - - - - branch and twig borer
Polyrachis semiaurata Mayr - ant
Prays citri Millière - - - - - - - - - - - - - citrus flower moth or rind borer
Prays endocarpa Meyrick - - - - - - - - - - - - - - - - - citrus rind borer
Prepodes quadrivittatus Olivier - - - - - - - - - - - - - - - - - - - weevil
Prepodes roseipes Chevrolat - weevil
Prepodes similis var. amabilis Waterhouse - - - - - - - - - - - - - - weevil
Prepodes vittatus Linnaeus - weevil
Prodenia ornithogalli Guenée - - - - - - - - - - - - - - - - - cotton cutworm

[1]Phyllocnistis saligna is not a pest of citrus.
(See U.S.D.A. Tech. Bull. 252, p. 2).
535

aspis yanonensis Kuwana - - - - - - - - - - - - - - - - - - - yanone scale
pulvinaria pyriformis Cockerell - - - - - - - - - - - - - - - pyriform scale
strophus avidus Marshall - weevil
oaonidia duplex (Cockerell) - - - - - - - - - - - - - - - camphor scale
oaonidia trilobitiformis Green - - - - - - - - - - - - - - trilobite scale
ococcus adonidum (Linnaeus) - - - - - - - - - - - - long-tailed mealybug
ococcus citri (Risso) - - - - - - - - - - - - - - - - - - citrus mealybug
ococcus citriculus Green - - - - - - - - - - - - - - - - Green's mealybug
ococcus comstocki (Kuwana) - - - - - - - - - - - - - - Comstock mealybug
ococcus cryptus Hemphill - - - - - - - - - - - - - - mealybug (on roots)
ococcus filamentosus Cockerell - - - - - - - - - - - - - - - - - mealybug
ococcus gahani Green - - - - - - - - - - - - - - - - citrophilus mealybug
ococcus hispidus Morrison - - - - - - - - - - - - - - - - - - cocoa louse
ococcus krauhniae (Kuwana) - - - - - - - - - - - - - - Japanese mealybug
ococcus maritimus (Ehrhorn) - - - - - - - - - - - - - - - grape mealybug
ococcus perniciosus Newstead - - - - - - - - - - - - - - Libbekh mealybug
ococcus pseudofilamentosus Betrem - - - - - - - - - - - - - - - mealybug
ococcus virgatus (Cockerell) - - - - - - - - - - - - - - - - - - mealybug
omydaus citriperda Tryon - - - - - - - - root bark channeler (Curculionidae)
ptera torquata Dalman - Buprestidae
sticha (Tonica) zizyphi Stainton - - - leaf miner and roller (Oecophoridae)
ndrus rosa (Karsch) - Natal fruit fly
phylla camellifolia (Fabricius) (Cyrtophyllus concavus Scudder) - - - katydid
naria aurantii Cockerell - - - - - - - - - - - - - orange pulvinaria scale
naria cellulosa Green - - - - - - - - - - - - - pulvinaria scale (Coccidae)
naria citricola Kuwana - - - - - - - - - - - - - - - - - pulvinaria scale
naria flavescens Brethes - - - - - - - - - - - - - - - - pulvinaria scale
naria floccifera Westwood - - - - - - - - - - - - - - - - pulvinaria scale
naria horii Kuwana - - - - - - - - - - - - - - - - - - - pulvinaria scale
naria polygonata Cockerell - - - - - - - - - - - - - - - pulvinaria scale
naria psidii Maskell - - - - - - - - - - - - - - - - green shield scale
lerces rileyi (Walsingham) - - - - - - - - - pyroderces (Cosmopterygidae)
ecus kondonis Kuwana - - - - - - - - - - - - - - - - - - ground mealybug
ocoris sulciventris Stål - - - - - - - - - - - - - - - bronze orange bug
hocoris humeralis Thunberg - - - - - - - - - - citrus green stink bug
hocoris longirostris Stål - - - - - - - - - - - - - - - - - stink bug
hocoris serratus Donovan - - - - - - - - - - - - - serrate citrus stink bug
musae Froggatt - Island fruit fly
lophora collaris Germar - - - - - - - - - - - - - - - - - - Cerambycidae
etia coffeae Walker - - - - - - - - - - - hemispherical scale (Australia)
etia hemisphaerica (Targioni) - - - - - - - - hemispherical scale (U.S.A.)
etia nigra Nietner - nigra scale
etia oleae (Bernard) - black scale
tocerca americana (Drury) - - - - - - - - - - - - - - American grasshopper
tocerca flavo-fasciata (DeGeer) - - - - - - - - - - - - - - citrus locust
cotetranychus (Tet.) hindustanicus Hirst - - - - - - - - - - - - - - mite
ius granosus Fahrer - - - - - - - - - - - - - - - - - - citrus snout beetle
cothrips aurantii Faure - - - - - - - - - - - South African citrus thrips
cothrips citri (Moulton) - - - - - - - - - - - - - - - - - citrus thrips
cothrips signipennis Bagnall - - - - - - - - - - - - - - banana rust thrips
rpopa australis Walker - - - - - - - - - - - - - - - - passion vine hopper
rtus rugulosus (Ratzeburg) - - - - - - - - - - - - - - - - shot-hole borer
leria furcata Brunner - - - - - - - - - - - - - fork-tailed bush katydid
leria mexicana (Saussure) - - - - - - - - - - - - - - - Mexican katydid
llera perplexa Westwood - - - - - - - - - - - - - - - - - - - stink bug
aspidus articulatus (Morgan) - - - - - - - - - - - - - - - - rufous scale
othrips rubrocinctus (Giard) - - - - - - - - - - - - - - red-banded thrips
a alternata Le Conte - - - - - - - - - - - - - June beetle (Melolonthidae)
a fimbriata Le Conte - June beetle
orpha nitella Zeller - - - - - - - - - - - - - - - - - - - Setomorphidae
a nitens Walker - Lunacochidae
e stimulea Clemens - saddleback
e trimacula Sepp - Cochlidiidae

```
Solenopsis geminata (Fabricius)- - - - - - - - - - - - - - - - - - - - - - fire ant
Solenopsis xyloni var. maniosa Wheeler - - - - - - - - - - - - - - - - - gray ant
Sparganothis (Platynota) pilleriana Schiffermüller - - - leaf roller (Tortricidae)
Spilosoma metarhoda Walker - - - - - - - - - - - - - - - - - - tiger moth (Arctiidae
Stenozygum personatum Walker - - - - - - - - - painted capparis bug (Pentatomidae)
Strumeta tryoni Froggatt - - - - - - - - - - - - - - - - - Queensland fruit fly
Stromatium barbatum Fabricius - - - - - - - - orange beetle-borer (Cerambycidae)
Taeniothrips frisi Uzel - - - - - - - - - - - - - - - - - - - - - - - thrips
Takahashia citricola Kuwana - - - - - - - - - - - - - - - - - - an unarmored scale
Takahashia japonica (Cockerell) - - - - - - - - - - - - - - an unarmored scale
Taragama repanda (Hübner) - - - - - - - - - - - - tent caterpillar (Lasiocampidae)
Tarucus theophrastus Fabricius - - - - - - - - - - orange hairstreak (Lycaenidae)
Tarsonemus bakeri Ewing - - - - - - - - - - - - - - - - - - - tarsonemus mite
Telenchulus semipenetrans Cobb - - - - - - - - - - - - - - - - citrus nematode
Tenuipalpus californicus Banks - - - - - - - - - - - - - - - - - silver mite
Tenuipalpus obovatus Donnadieu - - - - - - - - - - - - - - - - - - - mite
Tenuipalpus pseudocuneatus Blanchard - - - - - - - - - - - - - tenuipalpus mite
Tetraleurodes mori (Quaintance) - - - - - - - - - - - - - - - - mulberry whitefly
Tetranychus lewisi McGregor - - - - - - - - - - - - - - - - - Lewis spider mite
Tetranychus sexmaculatus Riley - - - - - - - - - - - - - - - - - six-spotted mite
Tetranychus telarius Linnaeus - - - - - - - - - - - - - - - - common red spider
Tetranychus yumensis McGregor - - - - - - - - - - - - - - - Yuma spinning mite
Throscoryssa citri Maulik - - - - - - - - - black and red leaf miner (Halticidae)
Tortrix excessana Walker - - - - - - - - - - - - - - - - - leaf roller (Tortricidae)
Tortrix postvittana Walker - - - - - - - - - - - - - - - light brown apple moth
Toumeyella sp. - - - - - - - - - - - - - - - - - - - - - turtle back scale
Toxoptera aurantii (Fonscolombe) - - - - - - - - - - - - - - black citrus aphid
Trachyderes succinctus Linnaeus - - - - - - - - - - - - - - - - - Cerambycidae
Trachyderes thoraxicus Olivier - - - - - - - - - - - - - - - - - Cerambycidae
Trialeurodes floridensis Quaintance - - - - - - - - - - - - - - avocado whitefly
Trialeurodes vaporariorum (Westwood) - - - - - - - - - - - - greenhouse whitefly
Trigona amalthea Fabricius - - - - - - - - - - - - - - - - - bee (Meliponidae)
Trigona trinidadensis Provancher - - - - - - - - - - - - - - citrus branch girdler
Trioza (Spanioza) erythreae Del Guercio (T. merwei Pettey) - - - - citrus psylla
Unaspis (Chionaspis, Prontaspis) citri (Comstock) - - - - - - - - - snow scale
Uracanthus cryptophagus Olliff - - - - - - - branch and trunk borer (Cerambycidae)
Valanga nigricornis Burmeister - - - - - - - - - - - - - - - - - grasshopper
Vinsonia stellifera Westwood - - - - - - - - - - - - - - - - - an unarmored scale
Wasmannia auropunctata (Roger) - - - - - - - - - - - - - - - - little fire ant
Xylomyges curialis Grote - - - - - - - - - - - - - - - - - - - - - cutworm
Xyleborus affinis Eichhorn - - - - - - - - - - - - - - - - bark beetle (Scolytidae)
Xyleborus testaceus Walker - - - - - - - - - - - - - - - - - - - bark beetle
Zale fictilis Guenée - - - - - - - - - - - - - - - - fruit-piercing moth (Noctuidae)
```

MEXICO (29, 622, 258)

There are probably between 50,000 and 75,000 acres of citrus in Mexico. About two thirds of this acreage is planted to oranges and the remainder to limes. The leading states in the cultivation of the orange are, in order of their importance, Nuevo Leon, Veracruz, San Luis Potosi and Jalisco. Michoacan, Veracruz, and Colima are the most important lime-producing states.

The citrus pest situation in Mexico differs from that of the majority of areas where the citrus blackfly occurs in that the parasite, Eretmocerus serius, which is highly effective in the majority of the citrus-growing regions of the world, is not keeping the citrus blackfly under control under the semiarid conditions of many of the citrus-growing regions of Mexico. As a result it is a serious pest in that country. Scale insects are also a serious problem. As in so many parts of the world, existing equipment and technique of application are inadequate to meet the exacting demands of scale control on citrus trees. The Mexican fruit fly is sporadic in its attacks and is of chief importance as a pest of late-picked Valencia oranges. It does not attack the lime.

Principal Citrus Pests in Approximate Order of Importance

1. Aleurocanthus woglumi
2. Chrysomphalus aonidum
3. Lepidosaphes beckii and/or L. gloverii
4. Aonidiella aurantii
5. Phyllocoptruta oleivora
6. Aphis gossypii
7. Selanaspidus articulatus
8. Unaspis citri
9. Anastrepha ludens
10. Parlatoria pergandii
11. Atta sp.

Minor Pests

Aonidiella orientalis
Ceroplastes cerripediformis
Coccus hesperidum
Coccus pseudomagnoliarum
Dendrobias mandibularis
Dialeurodes citri
Epitrix cucumeris

Euphoria basilis
Icerya purchasi
Grasshoppers
Macrodactylus sp.
Murgantia munda
Noctuids
Papilio cresphontes

Pseudococcus citri
Saissetia oleae
Scudderia mexicana
Scirtothrips citri
Selenopsis geminata
Termites
Tetranychus sp.

CUBA AND PUERTO RICO[1]

Citrus is widely scattered throughout all the six provinces of Cuba and the Isle of Pines. The only citrus grown for export, however, is grapefruit, principally produced on the Isle of Pines. Some citrus, mainly grapefruit, is grown in

---

[1]Information by correspondence of April 14, 1947, from S. C. Bruner, Chief of the Department of Plant Pathology, Agricultural Experiment Station, Cuban Ministry of Agriculture.

CITRUS PESTS OF FOREIGN COUNTRIES

the north central part of Puerto Rico, but the commercial grapefruit industry collapsed after the large acreages of Texas came into production. The cottony cushion scale and the citrus blackfly ceased to be important after the introduction of suitable natural enemies. The rust mite is abundant, but chiefly of importance on fruit grown for export. The spirea or citrus green aphid is of special importance on young trees, and the same may be said of the leaf-cutting ant.

1.	Lepidosaphes beckii	2. Phyllocoptruta oleivora
	Coccus viridis	3. Aphis spiraecola
	Chrysomphalus aonidum	4. Pachnaeus litus
	Toumeyella sp.	5. Atta insularis
	Lepidosaphes gloverii	6. Gonodonta nutrix
	Chionaspis citri	7. Solenopsis geminata
	Icerya purchasi	8. Aleurocanthus woglumi

9. Cryptocephalus marginocollis

The above species are widely distributed over the West Indies. In addition the following pests attack citrus in the West Indies (1007 et al):

Actinodes auronotata	Eanthis thraso	Phyllocoptruta oleivora
Aleurothrixus howardi	Frankliniella insularis	Phyllophaga citri
Aleurocanthus sp.	Gonodonta incurva[1]	Phyllophaga puberula
Aleurocanthus woglumi	Heliothrips haemorrhoidalis	Phyllophaga subsericans
Anastrepha mombinpraeoptans	Lachnopus coffeae	Prepodes quadrivittatus
Anastrepha suspensa	Lachnopus curvipes	Prepodes roseipes
Aonidiella aurantii	Lachnopus hispidis	Prepodes similis var.
Aonidiella orientalis	Lachnopus sparsimguttatus	amabilis
Atta insularis	Neoclytes cordifer	Pseudococcus citri
Ceroplastes ceriferus	Oiketicus abboti	Psiloptera torquata
Ceroplastes floridensis	Othreis apta	Pulvinaria psidii
Chrysobothris lepida	Pachneus citri	Saissetia hemisphaerica
Coccus hesperidum	Papilio andraemon	Saissetia nigra
Corythuca gossypii	Papilio androgeus	Saissetia oleae
Cryptorrhynchus corticalis	Papilio cresphontes	Selenaspidus articulatus
Dialeurodes citrifolii	Papilio thoas melonius	Solenopsis geminata
Diaprepes abbreviatus	Paraleurodes naranjae	Tetranychus sexmaculatus
Diaprepes esuriens	Paraleurodes perseae	Toxoptera aurantii
Diaprepes excavatus	Paratetranychus citri	Trialeurodes floridensis
Diaprepes scalaris	Perigea cupentia	Unaspis citri
Diaprepes sprengleri		Zale fictilis

CENTRAL AMERICA

The number of citrus trees in Central America, including the dooryard trees and those scattered about over the plantations, would no doubt reach an impressive number, but there are few commercial orchards. Except in British Honduras, which exports about 20,000 boxes of grapefruit annually, all the citrus fruit is consumed locally.

The citrus pests in many of the citrus areas of Central America are probably similar, in their order of importance, to the following list submitted for Costa Rica:[2]

1.	Atta cephalotes, A. sexdens	7.	Saissetia oleae
2.	Lepidosaphes beckii	8.	Saissetia hemisphaerica
3.	Pseudococcus citri	9.	Papilio polycaon
4.	Toxoptera aurantii	10.	Anastrepha ludens
5.	Aleurocanthus woglumi	11.	Anastrepha fraterculus
6.	Icerya purchasi	12.	Macrodactylus suavis

13. Oiketicus kirbyi

---

[1] The fruit-piercing moths are pests in the Windward Islands.

[2] Correspondence of February 23, 1948, from Ing. Jorge E. Umana, Jefe Servicio de Entomología, San Jose, Costa Rica.

SUBTROPICAL ENTOMOLOGY

In addition to the above insects, the following are reported as pests of citrus in Guatemala:[1] Phyllocoptruta oleivora, Frankliniella insularis, Frankliniella tritici, Trigona silvestriana, T. amalthea, Solenopsis geminata, Cotinis mutabilis, Anastrepha unipuncta, Dialeurodes citri, Papilio anchisiades, P. thoas, Eantis palida, and Epanteria icasia.

T. D. A. Cockerell informed[2] the writer that Selanaspidus articulatus, Chrysomphalus aonidum, Lepidosaphes beckii and Parlatoria pergandii are the most important scale pests of Honduras.

TRINIDAD[3]

According to a recent estimate, there are on the island of Trinidad 3,000 acres of limes, 3,000 acres of grapefruit, and 1,000 acres of oranges. In contrast to the situation in the majority of citrus areas, the scale insects are of little economic importance. The most important pests of citrus are several species of Atta, which may defoliate the trees. Trigona trinidadensis, a Meliponine bee, attacks the bark and fruit of citrus, often girdling entire branches and causing premature shedding of the fruit. As in the West Indies, the rust mite is of importance only if the crop is to be exported as whole fruit. The larvae of Laspeyresia sp., probably a new species, puncture the rind of citrus fruits, but the majority die after puncturing the rind and before they can proceed further. Orthezia praelonga sporadically causes severe injury to citrus. Cratosomus is a large, attractive weevil which burrows into branches of citrus trees, quickly killing them, and may kill the entire tree by burrowing into the trunk.

Serious trouble has been caused in Trinidad by unidentified colonial spiders. They cover the entire tree with a dense web which causes shading and "strangulation of the growing stems." The fruit drops and the tree stops growing and may even die as a result of an infestation by these spiders. Usually a considerable acreage of citrus is attacked at a time. Dusting with DDT or benzene hexachloride with a high-velocity blower, or tent fumigation with HCN, can be used for controlling this pest.

It can be seen that the citrus pest problems of Trinidad are interesting and in many ways unique. The pests are, in order of their importance, as follows:

1. Atta cephalotes
2. Atta octospinosa
3. Trigona trinidadensis
4. Azteca sp.
5. Phyllocoptruta oleivora
6. Laspeyresia sp.
7. Linyphid spiders
8. Chrysomphalus aonidum
9. Selenaspidus articulatus
10. Lepidosaphes gloverii
11. Saissetia oleae
12. Coccus viridis
13. Orthezia praelonga
14. Chionaspis citri
15. Pulvinaria pyriformis
16. Lepidosaphes beckii
17. Anastrepha serpentina
18. Ceroplastes floridensis
19. Cratosomus punctulatus
20. Pseudococcus citri

COLOMBIA AND VENEZUELA (541, 266)

A comparatively small acreage of citrus is grown in either of these countries, but there is considerable interest in the expansion of this industry. The pests most commonly encountered, in approximate order of their importance, are as follows:

1. Unaspis citri
2. Lepidosaphes beckii
3. Chrysomphalus dictyospermi
4. Toxoptera aurantii
5. Paratetranychus citri
6. Phyllocoptruta oleivora
7. Anastrepha spp.
8. Atta sp.
9. Pinnaspis aspidistrae
10. Pseudococcus citri
11. Chrysomphalus aonidum
12. Coccus viridus
13. Platynota sp.
14. Saissetia haemisphaerica
15. Icerya purchasi

---

[1]Correspondence of February 26, 1948, from Ing. A. Fuentes Novella, Jefe del Departamento de Defensa Agricola, Guatemala, Guatemala.

[2]Correspondence of November 11, 1946, from the Pan-American Agricultural School at Tequcigalpa, Honduras.

[3]Information by correspondence of April 23, 1947, from Allan Pickles, Entomologist for the Trinidad Department of Agriculture.

CITRUS PESTS OF FOREIGN COUNTRIES

ECUADOR[1]

No figures are available on the number of citrus trees in Ecuador, but there is reported to be a considerable acreage of oranges in some of the principal river valleys, especially immediately to the north of Guayaquil and in the Chone-Tosagua section. There are not only enough oranges grown for local consumption, but some are exported to Chile and Peru.

The citrus pests in order of their importance are as follows:

CITRUS AREAS OF ECUADOR

1. Lepidosaphes beckii and L. gloveri
2. Selenaspidus articulatus
3. Icerya purchasi
4. Atta cephalotes and Ectatomma ruidum
5. Gymnandrosoma aurantianum
6. Toxoptera aurantii
7. Aleurothrixus howardi
8. Pinnaspis minor
9. Ceroplastes sinensis

Other species: Coccus hesperidum, Lecanium deltae, Orthezia insignis, O. praelonga, Saissetia oleae, Leptoglossus phyllopus, Dysdercus sp., Papilio epenetus, Papilio zenxis, Megalopyge lanata, Phobetron hipparchia, Lonchaea pendula, Phyllocoptruta oleivora, and Icerya montserratensis.

PERU[2]

There does not appear to be more than 2,000 acres of cultivated citrus, practically all oranges, in Peru. Like many of the Latin-American countries, Peru has intentions of expanding its citrus industry. There is some room for expansion in the irrigated sections west of the Andes, but more particularly in some of the fertile regions of high rainfall in the montaña of the interior, such as Chanchamayo and Huánuco.

CITRUS AREAS

PERU

Particularly in the rainless areas west of the Andes, where dust accumulations on the foliage tend to reduce the effectiveness of oil sprays and favor the increase of scales and mites, the purple scale, Lepidosaphes beckii, and the rufous scale, Selanaspidus articulatus, are extremely severe pests and must be constantly combatted. The citrus pests, in order of their importance, are as follows: (952, 263)

1. Lepidosaphes beckii, 2. Selenaspidus articulatus, 3. Chrysomphalus dictyospermi (At Camaná), 4. Toxoptera aurantii, 5. Parlatoria pergandii, 6. Unaspis citri, 7. Atta sp., 8. Phyllocoptruta oleivora, 9. Pseudococcus sp., 10. Melipona testacea var. cupera, 11. Diabrotica decolor. Other species as indicated for Ecuador.

[1]Both the map and the list of pests were obtained through the courtesy of Ing. Antonio Garcia S., Director Técnico de Agricultura, Dirección Técnica de Agricultura, Quito, Ecuador.

[2]The map is redrawn from one prepared for the writer by Dr. J. E. Wille, Government Entomologist at La Molina.

SUBTROPICAL ENTOMOLOGY

CHILE [1] (264)

In Chile there are about 10,000 acres of lemons and
8,000 acres of oranges. The bulk of the citrus acreage lies
within a hundred miles north and south of the capitol city,
Santiago. The citrus pests, in order of their importance,
with exceptions in certain localities to be named later, are
as follows:

1. Lepidosaphes beckii
2. Paratetranychus citri
3. Pantomorus godmani
4. Pseudococcus citri
5. Aphids
6. Saissetia oleae
7. Heliothrips haemorrhoidalis

The above order of importance would not vary much throughout Chile except that in the Quillota and La Cruz citrus sections Aonidiella aurantii is the most important pest. It does
not occur in other areas. Chrysomphalus dictyospermi is probably about third in importance in Vallenar and Azapa, but does
not occur in the main citrus areas. The mealybugs, Pseudococcus citri, P. gahani, P. longispinus, and P. maritimus,
were once injurious to citrus, but are now controlled by parasites imported from California. Other pests found on citrus
are Ceroplastes cerripediformis, Chrysomphalus dictyospermi,
Coccus hesperidum, Dexicrates robustus, Dialeurodes citri,
Helix aspersa, Icerya purchasi, Saissetia hemisphaerica,
Selenaspidus articulatus, and Tetranychus sp.

ARGENTINA, PARAGUAY AND

URUGUAY [2] (620)

In Argentina there are about 100,000
acres of sweet oranges, 40,000 acres of
Mandarins, and 7,500 acres of lemons. Paraguay and Uruguay also have sizeable acreages
of citrus, the latter with 15,000 acres. The
citrus industry of Argentina is fostered by
the government, and control of certain pests
is compulsory.

The lists of citrus pests, arranged in
order of their importance, for the various
areas shown on the map are as follows:

Paraná delta

Anastrepha fraterculus	Icerya purchasi
Aonidiella aurantii	Pseudococcus citri
Lecanium deltae	Coccus hesperidum
Lepidosaphes beckii	Astilus sp.
Diabrotica speciosa	Papilio thoas thoas
Toxoptera aurantii	

---

[1] Map copied from one prepared for the writer by Ing. Gregorio Rosenberg, Ministerio de Agricultura, Santiago, Chile.

[2] The list of citrus pests, for Argentina, was prepared for the writer by Ing. Aldo Vergani,
the expert on citrus pests in the Argentine Ministry of Agriculture.

Anastrepha fraterculus	Lepidosaphes beckii	Lecanium deltae
Unaspis citri	Chrysomphalus aonidum	Pseudococcus citri
Aonidiella aurantii	Icerya purchasi	Coccus hesperidum
Chrysomphalus dictyospermi	Saissetia oleae	Aleurothrixus howardi
Phyllocoptruta oleivora		Oiketicus kirbyi

Missiones, Corrientes, and Paraguay

Anastrepha fraterculus	Icerya purchasi
Aonidiella aurantii	Lecanium deltae
Chrysomphalus aonidum	Toxoptera aurantii
Pinnaspis aspidistrae	Coccus hesperidum
Tenuipalpus pseudocuneatus[1]	Gymnandrosoma punctidiscana
Phyllocoptruta oleivora	Diabrotica sp.
Aphis citricidus (tavaresi)	Papilio spp.
var. argentinensis	Naupactus disimulator
Chrysomphalus dictyospermi	Sibine trimacula
Lepidosaphes beckii	Anychus verganii

Tucuman and Jujuy

Anastrepha fraterculus	Toxoptera aurantii	Pseudococcus citri
Aonidiella aurantii	Macrosiphum solanifolii	Hemiberlesia rapax
Unaspis citri (418)	Prodenia ornithogalli	Papilio sp.
Chrysomphalus aonidum	Aleurothrixus howardii	Platypus sulcatus
Lepidosaphes beckii	Ceroplastes grandis	Diabrotica speciosa
Frankliniella sp.	Lecanium deltae	Astilus variegatus
Chrysomphalus dictyospermi	Edessa quadridens	Phobetron sp.
Gymnandrosoma aurantianum	Icerya purchasi	Sibine trimacula
Gymnandrosoma punctidiscana		Microcentrum lanceolatum

BRAZIL (621, 432)

Brazil is among the leading citrus-producing countries of the world, being exceeded only by the United States and possibly Spain in the quantity of fruit produced.

Over 125,000 acres of citrus is grown in São Paulo State, of which about 50% consists of Navel oranges. The State of Rio de Janeiro has about 70,000 acres of citrus, mostly of the Pera variety, which resembles the Valencia and is the most popular variety for export. Minas Geraes and Bahia also have substantial acreages of oranges, and in fact, they are produced to some extent in all parts of Brazil. There is also a considerable acreage of grapefruit and lemons in Brazil.

Citrus does not require irrigation, because of the heavy rainfall. Some orchards, especially in São Paulo State, are well cared for, but the majority are badly neglected. In some cases the only work done on the

BRAZIL: LOCATION OF CITRUS INDUSTRY

---

[1] This mite injects a toxin which causes the disease of orange trees known as "lepra explosiva", formerly believed to be caused by a virus. (R.A.E., Ser. A. 34:298, 1946)

543

orchard consists of cutting down the annual growth of vegetation to "discover" the citrus trees prior to harvesting.

Scale insects often cause great damage and, if they are sprayed, the inadequate equipment used for spraying and the careless manner of application of the spray preclude the possibility of adequate control. The rust mite reduces the exportable volume of fruit, and although it is easily controlled, control measures are usually not practiced. The leaf-cutting ants take the usual toll characteristic of tropical and subtropical regions of high rainfall.

Fruit flies are sometimes serious pests, especially in São Paulo, where coffee, their favored host, is extensively grown. They are especially severe on late-maturing varieties.

The insect pests of Brazil, in approximate order of their importance, are as follows:

1. Lepidosaphes beckii
   Chrysomphalus aonidum
   Aonidiella aurantii
   Unaspis citri

2. Phyllocoptruta oleivora
   Thrips

3. Ceratitis capitata
   Anastrepha fraterculus

4. Aphis citricidus[1]
   Toxoptera aurantii

5. Atta spp.

Minor Pests (697a, 734a)

Aethalion reticulatum	Macrodactylus suturalis
Aleurothrixus floccosus	Macropophaga accentifer
Ancylocera cardinalis	Melipona ruficrus
Anomala undulata	Morganella maskelli
Aspidiotus scutiformis	Naupactus paulanus
Ceratitis capitata	Oncideres dejeani
Chrysomphalus aonidum	Pantomorus glaucus
Coccus hesperidum	Pantomorus persevali
Coleoxestia spinipennis	Papilio anchysiades capys
Cratosomus fasciatus	Papilio thoas brasiliensis
Cratosomus reidii	Parlatoria longispina
Diploschema rotundicolle	Parlatoria pergendii
Eburia ocoguttata	Parlatoria proteus
Frankliniella insularis	Phthia lunata (611)
Frankliniella rodeos	Phyllocoptruta oleivora
Frankliniella varipes	Pinnaspis aspidistrae
Gymnandrosoma aurantianum	Pinnaspis minor
Heliothrips haemorrhoidalis	Pseudaonidia trilobitiformis
Heterothrips moreiri	Pseudococcus citri
Hoplophorion vicina	Pseudococcus comstocki
Icerya floccosa	Pseudococcus cryptus
Icerya purchasi	Pulvinaria flavescens
Isochaetothrips striatus	Rophalophora collaris
Lepidosaphes beckii	Saissetia oleae
Leptoglossus gonagra	Schistocerca flavo-fasciata
Leptoglossus pleurosticus	Toxoptera aurantii
Macrodactylus affinis	Trachyderes succinctus
Macrodactylus pumilio	Trachyderes thoraxicus

---

[1] These aphids are vectors of the virus of tristeza, a serious disease of sweet orange on sour root stock.

THE MEDITERRANEAN REGION

Spain

Oranges comprise one of Spain's principal agricultural resources. They were first introduced by the Arabs into Andalusia in the ninth century, spreading from that district into Murcia, Valencia and Castille. The greater part of the production of citrus is now along the Mediterranean coast, particularly in the Valencia district (see map).[1] Spain is exceeded only by the United States, and possibly Brazil, in the production of citrus fruits, but far exceeds all other countries in exports, for 85% of the citrus fruit produced in Spain is exported (17, 27).

Sweet oranges make up the bulk of the citrus production, there being few lemons or grapefruit. Although many varieties of oranges are grown in Spain, the fruit is largely classified for marketing as "whites", "bloods", and "blood ovals." In the region of Seville, a considerable quantity of sour or Seville oranges are grown.

As in the majority of the Mediterranean citrus areas, by the time oranges mature in the fall and are susceptible to infestation by the Mediterranean fruit fly, the weather is too cold for these insects and losses are therefore negligible. Before the return of warm weather in the spring, the greater part of the fruit is harvested.

Despite the fact that citrus fruits have been cultivated in the Mediterranean countries for centuries, none of the indigenous insects have become serious pests of citrus. The principal pests in Spain are the introduced coccids. In addition to the 12 species of coccids that attack citrus in Spain, there are 3 species of Lepidoptera, 3 of Coleoptera and 1 of Diptera, all of minor importance. The lasiocampid *Taragama repanda* may be a serious pest at times in Zone VII (see p.546). The coccids, however, especially the dictyospermum and the purple scales, are important pests everywhere, and much spraying and fumigating is done to control them.

Following are lists of the coccid insects in the various citrus-growing zones indicated on the map (365). The scale insects are listed in order of their importance, with no indication of the relative importance of the citrus mealybug. However, all insects of great economic importance in the zone in question are preceeded by an asterisk (*). The citrus mealybug, *Pseudococcus citri*, is thus considered to be among the serious citrus pests only in Zone VI.

CITRUS AREAS
OF THE
SPANISH LEVANT

---

[1]The map and the information as the relative importance of the insects, are from Gomez Clemente (365).

SUBTROPICAL ENTOMOLOGY

Zone I
Castellón and north to
Tarragona Province.

*Chrysomphalus dictyospermi
*Lepidosaphes beckii
*Lepidosaphes gloverii
Saissetia oleae
Parlatoria pergandii
Icerya purchasi
Pseudococcus citri

Zone II
Burriana and Villareal
(Castellón Province).

*Chrysomphalus dictyospermi
*Lepidosaphes beckii
*Lepidosaphes gloverii
Saissetia oleae
Ceroplastes sinensis
Icerya purchasi
Pseudococcus citri

Zone III
Nules, Vall de Uxó, Onda
and Moncofar to Almenara
(Castellón Province).

*Chrysomphalus dictyospermi
*Lepidosaphes beckii
*Lepidosaphes gloverii
Parlatoria pergandii
Saissetia oleae
Icerya purchasi
Pseudococcus citri

Zone IV
Sagunto, Los Valles, Puzol,
Algimia to the Valencia
region.

*Chrysomphalus dictyospermi
*Lepidosaphes beckii
*Lepidosaphes gloverii
Saissetia oleae
Icerya purchasi
Pseudococcus citri

Zone V
Valencia, Moncada, Bétera,
Torrente, Albal, to Silla.

*Chrysomphalus dictyospermi
*Lepidosaphes beckii
*Lepidosaphes gloverii
Parlatoria pergandii
Icerya purchasi
Pseudococcus citri

Zone VI
High and low riviera of
Júcar, Carlet, Alberique,
Alcira, and Játiva.

*Chrysomphalus dictyospermi
*Lepidosaphes gloverii
Lepidosaphes beckii
Parlatoria zizyphi
Parlatoria pergandii
Saissetia oleae
Icerya purchasi
*Pseudococcus citri

Zone VII
Sueca, Cullera, Tabernes,
Gandía, Oliva (Valencia),
Pego, Vergel, and Denia
(Alicante).

*Chrysomphalus dictyospermi
*Lepidosaphes gloverii
Lepidosaphes beckii
Parlatoria pergandii
Pseudococcus citri
Icerya purchasi

Zone VIII
Oribuela, Dolores (Alicante),
Murcia, Blanca, Ulea, Abaran,
and Totana (Murcia).

*Chrysomphalus dictyospermi
*Parlatoria pergandii
*Lepidosaphes gloverii
Icerya purchasi
Pseudococcus citri

Italy and North Africa

Italy is among the five most important citrus-producing countries of the world, being exceeded only by the United States, Spain, Brazil and possibly China, in quantity produced. Italy possesses the world's largest acreage of lemons, but in total production it is now probably exceeded by California. The most important citrus areas are in Sicily and the southern part of the peninsula (see map). A narrow coastal belt of citrus extending from Genoa to the French boundary lies between 43°50' and 44°30' north latitude, four degrees farther north than the upper limits of the California citrus area (929). Lemons make up about half of the citrus acreage, oranges about a third, and the remainder is devoted to an amazing variety of citrus fruits.

In Algeria about 20,000 acres are planted to oranges and mandarins. Some citrus is also grown in Tunisia and Morocco (see map).

CITRUS PESTS OF FOREIGN COUNTRIES

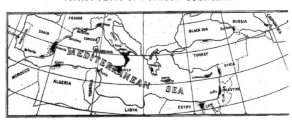

The citrus areas of the Mediterranean basin (929).

In Italy, as well as Algeria, Tunisia and Morocco, about the same group of citrus pests are found as in Spain. Throughout the western Mediterranean basin <u>Chrysomphalus dictyospermi</u>, which in that area is called poll roig (Spanish), <u>bianca rossa</u> (Italian), or pou rouge (French) is the predominant citrus pest. The citrus pests in Italy, in what appears to be approximately their order of importance, are as follows:

1. <u>Chrysomphalus dictyospermi</u>
2. <u>Lepidosaphes beckii</u>
3. <u>Pseudococcus citri</u>
4. <u>Parlatoria zizyphus</u>
5. <u>Aspidiotus hederae</u>
6. <u>Saissetia oleae</u> (North Africa)

7. <u>Lepidosaphes gloverii</u> (Corsica and Morocco)
8. <u>Aonidiella aurantii</u>
9. <u>Ceroplastes sinensis</u>
10. <u>Icerya purchasi</u>
11. <u>Parlatoria pergandii</u>

12. <u>Ceratitis capitata</u>
13. <u>Prays cltri</u>
14. <u>Chrysomphalus aonidum</u>
15. <u>Tetranychus telarius</u>[1]
16. <u>Heliothrips fasciatus</u>
17. <u>Boarmia selenaria dianaria</u>

Crete (43)

1. <u>Aonidiella aurantii</u>
2. <u>Parlatoria zizyphus</u>
3. <u>Saissetia oleae</u>

4. <u>Chrysomphalus dictyospermi</u>[2]
5. <u>Pseudococcus citri</u>
6. <u>Lepidosaphes beckii</u>

7. <u>Coccus hesperidum</u>
8. <u>Ceroplastes rusci</u>
9. <u>Icerya purchasi</u>

Israel[3] (488)

There were about 75,000 acres planted to citrus in Israel (Palestine) before World War II, of which 85% were planted to oranges, 12% to grapefruit, and 3% to lemons, limes, mandarins etc. Citriculture is considered to be the mainstay of the national economy of Israel, accounting for more than 50% of the total export from the country before the war. The Israeli hope to further expand this industry. The shamuti, an orange of high quality, is the predominant variety. There are very few Valencias, for the Mediterranean fruit fly infestations are a serious handicap in the production of varieties ripening in spring and summer.

The California red scale is especially injurious in the interior valleys and the purple scale in the coastal plain. The Florida red scale is primarily a pest in the coastal plain, but may be found in some inland areas.

CITRUS
AREAS

---

[1] Quayle (712) observed a species of red spider, identified by Italian entomologists as <u>Tetranychus telarius</u> Linnaeus, in all citrus sections of Spain and Italy. With few exceptions, they were not sufficiently abundant to cause serious injury. Roberti (751), however, reported serious damage to lemon trees on the Sorrento Coast of Italy in July, 1944, caused by <u>T. telarius</u>. In many orchards there was complete defoliation.

[2] A recent introduction (in 1940) and confined to the Department of Canea.

[3] Correspondence from F. S. Bodenheimer dated January 31, 1948.

Considerable pest control is practiced in Israel. Oil spray or HCN fumigation are used each year against the California and Florida red scales, and the purple scales. The Florida wax scale and the California black scale are combatted with oil spray, and 1 to 1-1/2% oil emulsion is sometimes applied with a brush to the trunks and large branches for the control of the chaff scale. Insectaries are operated for the production of predators and parasites. The citrus pests, in approximate order of their importance are as follows:

1. Aonidiella aurantii
2. Chrysomphalus aonidum
3. Lepidosaphes beckii
4. Pseudococcus citri
5. Pseudococcus comstocki
6. Ceratitis capitata
7. Icerya purchasi
8. Phyllocoptruta oleivora[1]

9. Anychus orientalis
10. Epitetranychus altheae
11. Ceroplastes floridensis
12. Parlatoria pergandii
13. Dioconotus cruentatus
14. Heliothrips haemorrhoidalis
15. Prays citri
16. Nezara viridula

17. Cryptoblabes gnidiella

Syria

The citrus pest situation is much the same as in Israel except that in Syria Chrysomphalus aonidum is the principal pest species.

Turkey (119a)

Citrus fruits, predominantly oranges, are grown mainly in the warm, low-lying coastal districts. There are about 15,000 acres of citrus in the country. The greatest acreage, as well as the most rapid expansion, is in the Icel (Mersin), Seyhan (Adana) and Dörtyöl districts (see map). Except for some of the newer plantings, the trees are seedlings. In almost all the citrus districts there is constant danger from freezing.

CITRUS AREAS

By far the most important pest of citrus is the California red scale, Aonidiella aurantii (Maskell). Parlatoria pergandii, Coccus hesperidum, Pseudococcus citri and Toxoptera aurantii are also serious pests at times. Icerya purchasi was an important pest before the introduction of the vedalia lady beetle around 1934. Other pests of citrus to be found in Turkey are Ceroplastes floridensis, C. rusci, C. capitala, C. sinensis, Lepidosaphes beckii, Saissetia oleae, Cryptoblabes gnidiella, and Epicometis hirta.

Egypt

Although citrus fruits have been grown in Egypt for about 3400 years, it was not until about 30 years ago that a commercial industry began to develop. Nevertheless it was estimated that by 1935 there were not less than

---

[1] The severe attacks by the rust mite in recent years will probably raise this pest to a higher position in relation to certain other pests whose importance has been decreasing (489).

CITRUS PESTS OF FOREIGN COUNTRIES

35,000 acres of citrus in Egypt, mostly concentrated in the upper portion of the Nile River delta. A large part of the citrus area is planted to limes, in the production of which Egypt leads the world.

The Florida red scale is a widespread and serious citrus pest in Egypt, where it is called the "black scale". The hibiscus mealybug is also a serious pest, especially from the standpoint of the difficulty of its control. The citrus pests of Egypt, in what appears to be their approximate order of importance, are as follows:

1. Chrysomphalus aonidum
2. Phenacoccus hirsutus
3. Lepidosaphes beckii
4. Ceratitis capitata
5. Ceroplastes floridensis
6. Aonidiella aurantii
7. Parlatoria pergandii
8. Pseudococcus perniciosus
9. Tenuipalpus obovatus

RUSSIA

Some small citrus orchards could be found in the Transcaucasian region of southern Russia as early as the 17th century, yet by 1928 there were only 1500 acres of citrus in that country. Beginning in 1936 the citrus industry expanded rapidly, with government aid, so that by 1940 there were 42,805 acres, about 75% of which was planted to mandarins, 21% to lemons, and the remainder to oranges (18). In 1947 Russian agricultural scientists visiting in California stated that there were about 50,000 acres of citrus which it was hoped would increase by 50% in a 5-year program being inaugurated at that time.

The greater part of the citrus acreage is situated along the shore of the Black Sea, between the Turkish border and the town of Sotchi. Some citrus is grown, however, on the western shore of the Caspian Sea (see map). All citrus areas are north of 40° N. latitude. The breeding of cold-resistant sweet and acid types of citrus is being undertaken at the government experiment station at Sukhum.

There has been considerable treatment for citrus pests, especially the scale insects, using oil spray and cyanide fumigation. The following list of citrus pests, given in order of their importance, was prepared for the writer in 1947 by Dr. A. A. Gogiberidze, director of the experiment station at Sukhum. It refers to the citrus areas along the Black Sea Coast.

Pests of Principal Economic Importance

Acarina

1. Phyllocoptruta oleivora   2. Paratetranychus citri

Coccoidea

1. Pseudococcus gahani
2. Pulvinaria aurantii
3. Ceroplastes sinensis
4. Leucaspis japonica
5. Lepidosaphes beckii
6. Chrysomphalus dictyospermi
7. Aonidiella citrina
8. Lepidosaphes gloverii
9. Pseudococcus maritimus
10. Coccus pseudomagnoliarum

Pests of Little or No Economic Importance

Orthoptera

1. Anacridium aegyptium   2. Locustidae (4 species)

Thysanoptera

1. Heliothrips haemorrhoidalis

Aphididae

1. Toxoptera aurantii

Coccoidea

1. Icerya purchasi
2. Pulvinaria floccifera
3. Parlatoria zizyphus
4. Coccus hesperidum
5. Parlatoria pergandii
6. Ceroplastes rusci

    7. Saissetia oleae   9. Aspidiotus camelliae
    8. Pulvinaria horii   10. Aspidiotus hederae

Lepidoptera

1. Sparganothis pilleriana

Coleoptera

1. Aserica japonica   2. Hypothenemus aspericolis   3. Aserica castanea

IRAN[1]

    There are about 22,000 acres of citrus in Iran, grown in various regions, but principally along the shore of the Caspian Sea. The production is estimated to be between 58,000 and 60,000 tons per year. Nearly all the pests of importance are scale insects. These are combatted with oil spray, oil plus DDT, kerosene plus DDT, and calcium cyanide fumigation. The cottony cushion scale was introduced into Iran in 1925. Two insectaries are now engaged in rearing the vedalia, Rodolia cardinalis, to combat the cottony cushion scale in newly-infested areas. The California red scale, Aonidiella aurantii, is said to be heavily attacked by the hymenopterous parasite Aphytis chrysomphali.

    The citrus pests, in order of their importance, are as follows:

    1. Chrysomphalus dictyospermi   9. Aonidiella aurantii
    2. Icerya purchasi   Aonidiella citrina
    3. Parlatoria zizyphus   10. Hemiberlesia camelliae
    4. Pulvinaria floccifera   11. Aspidiotus hederae
    5. Pulvinaria aurantii   12. Aspidiotus orientalis
    6. Parlatoria pergandii   13. Pseudococcus filamentosus
    7. Coccus hesperidum   Pseudococcus citri
    8. Lepidosaphes beckii   14. Ceroplastes sinensis
       Lepidosaphes gloveri   15. Taeniothrips frisi
             16. Parlatoria morrisoni

[1]Information by correspondence of May 13, 1948, from A. Davatchi, Professor of Agricultural Entomology, College of Agriculture of Karadj, Iran.

CITRUS PESTS OF FOREIGN COUNTRIES

UNION OF SOUTH AFRICA

As can be seen from the accompanying map[1] there are about 4,825,000 citrus trees in the Union of South Africa, comprising about 50,000 acres. Of this acreage, 81.3% is planted to oranges, 7.8% to grapefruit, 6.7% to naarjes, and 4.2% to lemons. For present purposes 7 areas are designated by numbers on the map and the citrus pests, or groups of pests, in each of these areas are listed in order of their importance.[2]

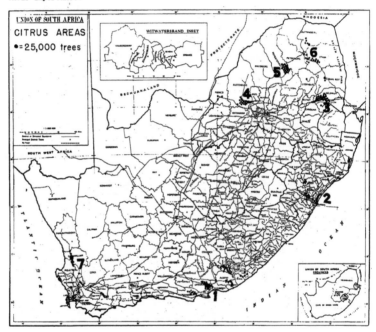

Area No. 1 (see map). This is composed of 3 districts in the Eastern Cape Province situated along the Sunday's River Valley near Port Elizabeth, the Kat River Valley near Fort Beaufort, and the third and smallest area near Grahamstown.

1. Aonidiella aurantii
   Lepidosaphes beckii

2. Pterandrus rosa
   Ceratitis capitata

3. Argyroploce leucotreta

4. Scirtothrips aurantii

5. Pseudococcus citri

[1]Through the courtesy of Dr. F. J. Stofberg, Entomologist in the Subtropical Horticultural Research Station at Nelspruit, E. Transvaal, contact was made with the Field Staff of the South African Citrus Exchange, who provided the map and included data on the location and extent of the citrus industry as of 1948.

[2]Information by correspondence of November 20, 1947, from Dr. F. J. Stofberg, Nelspruit, E. Transvaal.

Area No. 2. This area, known as Muden, is near the town of Greytown, and is situated in the drier inland part of Natal.

1. Aonidiella aurantii	3. Argyroploce leucotreta
Lepidosaphes beckii	4. Pterandrus rosa
Coccus hesperidum	Ceratitis capitata
2. Scirtothrips aurantii	5. Sciobius granosus

Area No. 3. Eastern Transvaal Lowveld, situated along the Crocodile River with Nelspruit as its center.

1. Aonidiella aurantii	2. Scirtothrips aurantii
Lepidosaphes beckii	3. Argyroploce leucotreta
Lepidosaphes gloveri	4. Pterandrus rosa
Chrysomphalus aonidum	5. Pseudococcus citri
Coccus hesperidum	6. Colasposoma fulgidum

Area No. 4. Situated around the town of Rustenburg in the Western Transvaal.

1. Aonidiella aurantii	3. Argyroploce leucotreta
Chrysomphalus aonidum	4. Pterandrus rosa
Coccus hesperidum	Ceratitis capitata
2. Scirtothrips aurantii	

Area No. 5. This area, known as Zebedela, contains the largest single private citrus planting in the Union, with a half a million trees. It is situated in the Northern Transvaal.

1. Aonidiella aurantii	3. Argyroploce leucotreta
Chrysomphalus aonidum	4. Pterandrus rosa
Coccus hesperidum	Ceratitis capitata
2. Scirtothrips citri	

Area No. 6. This area, also in the Northern Transvaal, is situated around the town of Tzaneen and includes the Letaba Citrus Estate.

1. Aonidiella aurantii	2. Argyroploce leucotreta
Lepidosaphes beckii	3. Scirtothrips aurantii
Chrysomphalus aonidum	4. Pterandrus rosa
Coccus hesperidum	Ceratitis capitata

Area No. 7. Clanwilliam, about 150 miles north of Capetown, is a citrus area of some importance. By far the most important of the citrus pests is the California red scale, Aonidiella aurantii. It is more important than all other citrus pests combined.

Citrus in South Africa appears to be free of important mite pests. Occasionally, especially in the Transvaal, severe outbreaks of the cotton boll worm, Heliothis armigera, occur. This pest may cause losses up to 50% of young fruit as well as mature Valencias just after petal drop. The weevil Protostrophus avidus feeds on young leaves of citrus in the Transvaal. Brachycerus citriperda attacks citrus in the Muden area. Leptoglossus membranaceous attacks ripe fruit in some districts. The bush katydid, Phymateus leprosus is a sporadic pest of citrus in the Eastern Cape Province.

SOUTHERN RHODESIA

The principal citrus pests of Southern Rhodesia, in order of their importance, are reported to be as follows:[1]

Major Pests

1. Aonidiella aurantii   2. Scirtothrips aurantii

---

[1] Information through courtesy of Dr. E. Parry Jones, Pest Control Ltd., Salisbury, Southern Rhodesia.

CITRUS PESTS OF FOREIGN COUNTRIES

Minor Pests

1. Heliothis armigera
2. Coccus hesperidum
3. Aphis citricidus
4. Argroploce leucotreta
5. Ceratitis capitata

6. Papilio demodocus
7. Fruit piercing moths
   (Othreis, Achaea,
   Sphingomorpha, Calpe
   and Serodes)

Pests Which Are Present and Might Become Economically
Important Under Certain Climatic Conditions

1. Pseudococcus citri,
   P. filamentosus
2. Lepidosaphes beckii
3. Ceroplastes destructor
   var. brevicauda Hall

4. Icerya purchasi
5. Aphis gossypii
6. Trioza erythreae
7. Chrysomphalus aonidum

INDIA AND PAKISTAN (415, 955)

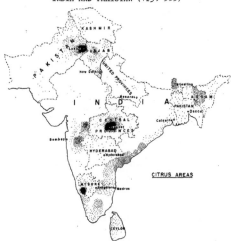

The citrus acreage in India-Pakistan was estimated to be about 130,000 acres
in 1943; about 6.5% of the total fruit acreage (21). The mandarins or loose-skinned
oranges make up over half of this acreage. There is not even enough citrus fruit
produced to supply the domestic demand.

In the majority of the citrus areas of India, as well as in other citrus growing areas of southeast Asia, the Lepidoptera are the most important pests, in contrast to the rest of the world, where the coccids or fruit flies, usually the former, have the greatest economic importance. The principal citrus pests of India, in approximate order of importance, are as follows:

1. Phyllocnistis citrella
   Psorosticha (Tonica)zizphi
2. Papilio demoleus, P. polytes
   and P. polymnester
3. Othreis (Ophideres) materna,
   O. fullonia, O. ancilla,
   O. salaminia, Achaea janatah,
   and Phyllodes consobrina
4. Chrysomphalus aonidum
   Erium sp.
   Aonidiella aurantii, A.
   citrina, A. orientalis and
   A. comperei
   Monophlebus dalbergiae
   Pseudococcus citri
   Aspidiotus destructor
   Lepidosaphes beckii
   Coccus hesperidum
   Fiorinia theae
   Vinsonia stellifera
   Drosicha stebbingi

5. Dialeurodes citri
   Dialeurodes elongata
   Aleurolobus citrifolii
   Aleurocanthus husaini
   Aleurocanthus woglumi
   Aleurocanthus spiniferus
6. Chelidonium cinctum
   Monochamus versteegi
   Stromatium barbatum
   Indarbela quadrinotata
   Indarbela tetraonis
   Peltotrachelus pubes
   Chloridolium alemene,
   Gnatholea eburifera
   Heterorrhina elegans
7. Aphis citricidus
   Toxoptera aurantii
8. Rhinchocoris humeralis
   (in Assam)
   Scutellera perplexa
9. Diaphorina citri

10. Tonica zizyphi
    Tarcus theophrastis
11. Agonopterix sp.
12. Myrmicaria brunnea (Ceylon)
    Dacus dorsalis (Ceylon)
13. Anychus ricini
    Schizotetranychus hindustanicus

INDONESIA[1] (912, 165)

By far the greater part of the citrus of Indonesia is represented by the dooryard trees growing around the native homes. Regular plantings are scarce, and are restricted to a few districts. There are 6000 acres of citrus in the hill sections about 50 miles west of Surabaya. In the Garoet area, in the west central part of Java, there are about 650 acres, and at Pasar Mingoe, grapefruit is quite extensively grown. On the island of Madura, just off the eastern end of Java, there is also considerable citrus. Despite the scarcity of extensive plantings, it has been estimated that there are at least ten million citrus trees in the Dutch East Indies alone. There are many insect pests, but the scattered condition of the trees, which are growing among many other species of trees and shrubs, results in the pests rarely occurring in abnormally large numbers despite the almost complete absence of pest control measures. In the orchards the most important pests are the scale insects and mealybugs. The fruits of some citrus varieties are often seriously damaged by Citripestis sagittiferella and Prays endocarpa, and a weevil, Maleuterpes dentipes, may cause much damage to the foliage. The growth of nursery trees is often seriously hampered by the leafminer Phyllocnistis citrella.

Homoptera

Coccus viridus, Lepidosaphes beckii, Asterlecanium sp., Aonidiella aurantii, Pulvinaria psidii, Parlatoria zizphus, Saissetia nigra, Pseudococcus citri, P. hispidus, P. filamentosus, P. pseudofilamentosus, Icerya purchasi, I. pulcher

Aleurocanthus spiniferus, A. woglumi, A. citriperdus, Aphis citricidus (A. tavaresi), Toxoptera aurantii, Aphis gossypii, A. nigricans, Diaphorina citri

---

[1]Correspondence of June 23, 1947, from Dr. J. van der Vecht, Acting Chief of the Institute for Plant Diseases, Buitenzorg, Java.

CITRUS PESTS OF FOREIGN COUNTRIES

Lepidoptera

Citripestis sagittiferella, Prays endocarpa, P. citri, Phyllocnistis citrella, Othreis (Ophideres) fullonica, Papilio demoleus, P. memnon, P. polytes, Indarbela tetraonis, Amsacta lactinea, Attacus atlas, Boarmia crepuscularia, Dasychira misana, Orgyia postica, Setomorpha nitella, Depressaria lizyphi, Setora nitens, Cania bilinea, Altha castaneipars, Psychidae

Coleoptera

Maleuterpes dentipes, Hypomeces squamosus, Agrilus occipetalis

Hemiptera

Helopeltis antonii, Cappea taprobanensis, Anoplocnemis phasiana, Rhynchocoris serratus

Diptera

Dacus dorsalis, Lonchaea gibbosa, Drosophila punctipennis

Acarina

Eriophyes sp., Tenuipalpus sp.

AUSTRALIA[1] (26)

According to a survey made in 1945, there were 47,490 acres of citrus in Australia at that time. About 60% of the acreage was in the state of New South Wales, with Victoria second in acreage. In New South Wales, 41.8% of the citrus acreage was planted to Valencia oranges, 28.2% to Washington Navels, 11.4% to lemons and limes, and 10.6% to mandarins, the remaining 7% being planted to other orange varieties and to grapefruit. In Victoria there was a relatively larger acreage of lemons and limes (24.5%). In Queensland the mandarins were the most extensively planted citrus variety and there was a larger acreage of commons and other orange varieties than of either Valencias or Navels. In South Australia 57.1% of the citrus acreage was in Washington Navels and 24.1% in Valencias. In Western Australia there was about an equal acreage of Navels and Valencias, and these two varieties made up 76.9% of the citrus acreage of the state.

[1]Correspondence of August 15, 1947 from Dr. A. J. Nicholson, Chief of the Division of Economic Entomology, Council for Scientific and Industrial Research, Commonwealth of Australia, including data from entomologists in the various States of the Commonwealth.

SUBTROPICAL ENTOMOLOGY

Some of the citrus pests of Australia, such as the California red scale, yellow scale, purple scale, black scale, citricola scale, and the citrus bud mite are among the principal citrus pests in California. Others, such as the Florida red scale (circular black scale), the rust mite (Maori mite), and the aforementioned purple scale, are relatively more important in Florida than in California. Still other species, such as the snow scale (white louse scale) (<u>Unaspis citri</u>), white wax scale (<u>Ceroplastes destructor</u>), pink wax scale (<u>Ceroplastes rubens</u>), dicky rice weevil (<u>Maleuterpes spinipes</u>), fruit tree root weevil (<u>Baryopadus squalidus</u>), Bronze orange bug (<u>Rhoecocoris sulciventris</u>), larger horned citrus bug (<u>Biprorulus bibax</u>), Queensland fruit fly (<u>Strumeta tryoni</u>), Mediterranean fruit fly (<u>Ceratitis capitata</u>) and the fruit piercing moth (<u>Othreis fullonia</u>) are either not present or are not sufficiently abundant to be considered as pests in the United States.

The following lists were prepared by entomologists in the various States of the Australian Commonwealth. The approximate order of importance is given only for the more important pests. The others are merely listed alphabetically.

I NEW SOUTH WALES

Major Pests

1. Aonidiella aurantii
2. Unaspis citri
3. Ceroplastes destructor
4. Maleuterpes spinipes
5. Baryopadus squalidus
6. Rhoecocorus sulciventrus
7. Lepidosaphes beckii
8. Strumeta tryoni
9. Aceria sheldoni
10. Phyllocoptruta oleivora
11. Saissetia oleae
12. Coccus hesperidum
13. Helix spp.
14. Toxoptera aurantii
15. Pantomorus godmani
16. Biprorulus bibax
17. Ceroplastes rubens

Minor Pests

Aonidiella citrina
Aplonobia oxalis
Caedicia simplex
Caedicia strenua
Carpophilus sp.
Ceratitis capitata
Chortoicetes terminifera
Citriphaga mixta
Doticus pestilens
Eurytoma fellis
Geloptera porosa
Harmologa miserana
Heliothis armigera
Heliothrips haemorrhoidalis
Heteronychus sanctae-helenae
Hyalarcta huebneri
Icerya purchasi
Iridomyrmex detectus
Lampona obscoena
Lepidosaphes gloverii

Maroga unipuncta
Mictis profana
Monolepta australis
Nezara viridula
Nysius vinitor
Orthorrhinus cylindrirostris
Othreis spp.
Papilio aegeus
Papilio anactus
Perperus insularis
Polyrachis semi-aurata
Prays sp.
Pseudococcus sp.
Rioxa musae
Scolypopa australis
Stenozygum personatum
Tenuipalpus californicus
Tortrix postvittana
Trialeurodes vaporariorum
Uracanthus cyptophaga

II VICTORIA

Major Pests

1. Aonidiella aurantii
2. Aonidiella citrina
3. Saissetia oleae
4. Coccus hesperidum
5. Tortrix postvittana
6. Orthorrhinus cylindrirostris
7. Brachyrhinus cribricollis

Minor Pests

Bostrychopsis jesuita
Coccus pseudomagnoliarum
Caedicia olivacea
Helix aspersa
Icerya purchasi
Toxoptera aurantii

CITRUS PESTS OF FOREIGN COUNTRIES

III QUEENSLAND

The four areas in which the greater part of the citrus in Queensland is grown differ so much climatically that a separate list of pests is given for each. Pests of about equal importance have been grouped together. These different groups, separated by spaces, are arranged in order of their importance.

I Nambour-Brisbane

1.
Ceroplastes destructor
Ceroplastes rubens
Lepidosaphes beckii
Phyllocoptruta oleivora
Strumeta tryoni

2.
Aphis citricidus
Eutinophaea bicristata
Pseudomydaus citriperda
Rhoecocoris sulciventris
Unaspis citri

3.
Chrysomphalus aonidum
Pulvinaria polygonata

4.
Aceria sheldoni
Aonidiella aurantii
Biprorulus bibax
Coccus hesperidum
Dichocrocis punctiferalis
Eurytoma fellis
Othreis fullonia
Scirtothrips signipennis

II Gayndah-Howard

1.
Aonidiella aurantii
Biprorulus bibax
Phyllocoptes oleivorus
Strumeta tryoni

2.
Unaspis citri
Chrysomphalus aonidum

3.
Ceroplastes rubens
Lepidosaphes beckii
Coccus hesperidum

4.
Aceria sheldoni
Aphis citricidus
Othreis fullonia
Scirtothrips signipennis

III Cooktown-Cardwell

1.
Ceroplastes rubens
Lepidosaphes beckii
Phyllocoptruta oleivora
Strumeta tryoni

2.
Aonidiella aurantii
Othreis fullonia

3.
Aphis citricidus
Unaspis citri

4.
Aceria sheldoni
Coccus hesperidum
Paralecanium expansum
Saissetia coffeae

IV Charters Towers

1.
Aonidiella aurantii
Biprorulus bibax
Mastotermes darwiniensis
Phyllocoptes oleivorus
Strumeta tryoni

2.
Aphis citricidus
Chrysomphalus aonidum

Lepidosaphes beckii
Unaspis citri

3.
Aceria sheldoni
Ceroplastes rubens
Coccus hesperidum
Colgar peracuta
Othreis fullonia

IV SOUTH AUSTRALIA

1. Aonidiella aurantii
2. Aonidiella citrina
3. Saissetia oleae
4. Coccus hesperidum
5. Pseudococcus adonidum
6. Othreis spp.
7. Caedicia spp.
8. Orthorhynus cylindrirostris
9. Icerya purchasi
10. Mictis profana
11. Nezara viridula

557

SUBTROPICAL ENTOMOLOGY

V WESTERN AUSTRALIA

1. Aonidiella aurantii
2. Ceroplastes destructor
3. Ceratitis capitata
4. Saissetia oleae

5. Lecanium hesperidum
6. Mictis profana
7. Phyllocoptruta oleivora
8. Heliothrips haemorrhoidalis

NEW ZEALAND

There were about 1500 acres of citrus in New Zealand in 1936, some of which, on the north end of the South Island, extends below 40°S. Longitude. About two thirds of this acreage is planted to lemons; the remainder is about equally divided between sweet oranges and grapefruit. The principal citrus pests are the scale insects. Oil sprays are used to control these pests, and a concentration of 3% summer oil is used against the more resistant species. Two applications, 4-6 weeks apart, are made for the control of Aonidiella aurantii and Ceroplastes destructor. The scale insects, in order of their importance, are as follows: (291)

1. Aonidiella aurantii
2. Saissetia oleae
3. Saissetia coffeae

4. Ceroplastes sinensis
5. Ceroplastes destructor[1]
6. Chrysomphalus rossi

Other Citrus Insects

Aphis citricidus
Charagia virescens
Ctenopseustis obliquana
Heliothrips haemorrhoidalis
Icerya purchasi
Maleuterpes spinipes

Paratetranychus citri
Pseudococcus adonidum
Pseudococcus maritimus
Tortrix excessana
Tortrix postvittana

CITRUS AREAS OF CHINA

CHINA (453)

A large quantity of citrus fruit is grown in China between the latitudes of 23° and 30° north latitude, or, roughly, south of the Yangtze River. The greatest production is in Kwantung province, with Fukien second and Chekiang third. Citrus is also grown in Kwangsi, Kiangsi, Hupei, Kweichow, Yunnan, and Szechwan provinces. China is fourth or fifth among the citrus producing countries of the world being exceeded only by the United States, Brazil, Spain, and possibly Italy. The percentages of the various types of citrus fruit are as follows: loose-skinned oranges, 86.7; sweet oranges, 4.0; and pomelo, 9.3.

Unlike the situation in the majority of the citrus regions of the world, the most serious citrus pests in China are considered to be not the scale insects but the wood borers and leaf miners. Condit et al (188) state that the black and white citrus borer, Melanauster chinensis, "is undoubtedly the worst citrus insect" of Kwangtung Province, the principal citrus-producing area of China. Young trees are frequently killed before coming into production. A leaf-mining beetle known as the "black and red leaf miner", Throscoryssa citri, annually skeletonizes nearly all the foliage of large areas of citrus during March and April. The scale insects also are important. Silvestri (773) considered four armored scales (first 4 given below) to be the most important scale insects. The approximate order of importance of the principal groups of pests seems to be as follows: (188)

[1] In 1945 this species was believed to have been in New Zealand only a short time and was found in only a few orchards.

CITRUS PESTS OF FOREIGN COUNTRIES

1. Melanauster chinensis
   Chelidonium gibbicolle
   Agrilus auriventris
   Chrysobothris sp.
   Nadezhdiella cantori
2. Throscoryssa citri
   Phyllocnistis citrella
   Podagricomela nigricollis
3. Chrysomphalus aonidum
   Aonidiella aurantii
   Chrysomphalus dictyospermi
   Parlatoria zizyphus
   Ceroplastes floridensis
   Ceroplastes rubens
   Icerya purchasi
   Pseudococcus kraunhiae
   Prontaspis yanonensis
   Aspidiotus duplex
   Unaspis citri
   Lepidosaphes beckii
   Drosicha contrahens
4. Mellesis citri (156a)
   Dacus dorsalis (Formosa)
5. Clitea metallica
   Ectinohoplia rufipes
   Oxycetonia jucunda

6. Rhynchocoris humeralis
   Cappaea taprobanensis
7. Aphis citricidus
8. Aleurocanthus citriperdus
   Aleurocanthus spiniferus
   Aleurocanthus woglumi
   Aleurocanthus cheni
   Bemisia giffardi
   Dialeurodes citri
   Dialeurodes citricola
   Aleurolobus szechwanensis
9. Clania minuscula
   Papilio antiphates
   Papilio demoleus
   Papilio helenus
   Papilio memnon
   Papilio polytes
   Papilio xuthus
10. Othreis fullonia
    Othreis salaminia
11. Phyllocoptruta oleivora
12. Brachytrupes portentosus
    Chondacris rosea
13. Oligotoma saundersi [1]

JAPAN (163)

In Japan the citrus-growing areas extend south from Tokyo, at about 35.5° north latitude, to the island of Kyushiu. The most extensive plantings are said to be located in the prefectures of Wakayama and Shizuoka. The greater part of the acreage is planted to the so-called Satsuma or Unshu orange, but grapefruit and Navel oranges are also grown, as well as a few lemons. As is usually the case throughout the world, the most injurious pests have been introduced from elsewhere. The major citrus pests, in order of their economic importance, are as follows:

Scale Insects and Whiteflies

Major Pests

1. Prontaspis yanonensis
2. Ceroplastes rubens
3. Pulvinaria aurantii
4. Parlatoria pergandii
5. Pinnaspis aspidistrae
6. Lepidosaphes gloverii
7. Aleurocanthus spiniferus
8. Aleurolobus marlatti

[1] An embiid that carries a fungus in Formosa.

559

Minor Pests

Dialeurodes citri
Icerya purchasi
Icerya seychellarum
Pseudaonidia duplex
Ceroplastes floridensis
Coccus pseudomagnoliarum
Saissetia oleae
Lepidosaphes beckii
Aonidiella aurantii
Aonidiella citrina
Pseudococcus citri

Coccus hesperidum
Takahashia citricola
Takahashia japonica
Pulvinaria citricola
Aspidiotus perniciosus
Chrysomphalus aonidum
Leucaspis japonica
Bemisia giffardi
Aleurocanthus sp.
Rhizoecus kondonis

Fruit Feeders

1. Ophideres tyrannus
   Calpe excavata
   Calpe capucina

2. Dichocrocis punctiferalis
3. Cacoecia podana

Leaf Miners, Leaf Rollers, and General Feeders

1. Phyllocnistis saligna[1]   2. Depressaria culcitella   3. Papilio xanthus

Fruit Flies

Dacus tsuneonis Miyake

Trunk Borer

Melanauster chinensis

Miscellaneous

1. Aphids
2. Thrips

3. Geisha distinctissima

4. Eriophyes oleivorus
5. Oxycetonia jucunda

PHILIPPINES (165)

Citrus fruits are grown on scattered trees in nearly all parts of the Philippines, but there are very few orchards. The limited commercial production is nearly all confined to the Province of Batangas, on the island of Luzon, from which the Manila market is supplied. The citrus pests are as follows:

Insects Attacking the Fruit

Hemiptera: Rhynchocoris serrata, R. longirostris
Lepidoptera: Dichocrocis punctiferalis, Othreis fullonica, Prays citri
Diptera: Dacus dorsalis, Monacrostichus citricola

Insects Attacking the Foliage

Acrididae: Chondracris rosea, Valanga nigricornis
Lepidoptera: Phyllocnistis citrella, Spilosoma metarhoda, Euproctis flavata, Orgyia postica

Pests Feeding on the Sap

Eriophyidae: Phyllocoptruta oleivora
Miridae: Helopeltis collaris
Pentatomidae: Rhynchocoris longirostris
Chermidae: Diaphorina citri
Aphidae: Aphis gossypii, Myzus persicae, Toxoptera aurantii
Coccidae: Ceroplastes rubens, Chrysomphalus aonidum, Chrysomphalus dictyospermi, Coccus viridis, Lepidosaphes beckii, Lepidosaphes gloverii, Parlatoria pergandii, Parlatoria ziziphus, Selenaspidus articulatus, Pinnaspis aspidistrae, Pseudococcus citriculus, Pseudococcus virgatus, Pulvinaria polygonata, Pulvinaria psidii, Saissetia hemisphaerica
Aleyrodidae: Aleurocanthus spiniferus, Aleurocanthus woglumi

[1]Phyllocnistis saligna is not a pest of citrus.
(See U.S.D.A. Tech. Bull. 252, p. 2).

CITRUS PESTS OF FOREIGN COUNTRIES

Tree Borers

Buprestidae: <u>Agrilus occipitalis</u> (serious pest), <u>Chrysochroa fulminans</u>.

HAWAII (703)

Citrus fruits have been cultivated in Hawaii for well over a century. At one time citrus plantings were more extensive than at present, but were replaced by more remunerative crops. There is not now enough citrus fruit produced to meet local demands.

1. <u>Dacus dorsalis</u>
   <u>Ceratitis capitata</u>
2. <u>Lepidosaphes beckii</u>
   <u>Chrysomphalus aonidum</u>
   <u>Aonidiella aurantii</u>
   <u>Coccus viridis</u>
   <u>Parlatoria pergandii</u>
   <u>Parlatoria ziziphus</u>

<u>Ceroplastes rubens</u>
<u>Icerya purchasi</u>
3. <u>Pseudococcus filamentosus</u>
4. Aphids
5. <u>Phyllocoptruta oleivora</u>
6. <u>Paratetranychus citri</u>
7. <u>Tetranychus sexmaculatus</u>
8. <u>Tenuipalpus irritans</u>

CHAPTER XX
GRAPE PESTS

The grape is considered to be probably the oldest of domesticated fruits, and it appears that it has been cultivated since the dawn of history. The greatest acreages today are in France, Italy and Spain. California, however, has 557,838 acres, and it will be noted that this is over twice as great as the entire citrus acreage of the state, and comprises about a third of the entire fruit and nut acreage, according to the 1947 estimates (137). California has about 80% of the total grape acreage of the United States. Only about 2% of the world's wine is produced in California, but 15% of the table grapes and 30% of the raisins are grown in this state (462). Thus it can be seen that the more serious pests of the grape are of considerable economic importance in California.

Aside from those in California and Arizona, most of the important varieties of grapes grown in the United States have been derived from American wild vines or from crosses between them and Vitis vinifera, the "European" grape, a native of western Asia. The latter variety makes up the commercial acreage in California and Arizona and is the variety that may properly be included among the subtropical fruits. Here it is grown extensively in the hot dry interior valleys and deserts.

The vinifera grape has many pests, of which the grape leafhopper and the Pacific mite appear to be the most important in California at present. Like the pecan, the grape in the United States is attacked almost exclusively by native pests.

ROOT-KNOT NEMATODE

Heterodera marioni (Cornu)

The root-knot nematode or garden nematode is the most important of several species of nematodes that attack grape roots. The biology of this species will be discussed in the section on fig pests (p. 620). On grapes swellings and distortions may be caused by root-knot nematode infestations. In very sandy soils containing a high nematode population, nematode-resistant rootstocks may be necessary if grapes are to be grown. On the other hand, nematodes are usually not a serious problem on loam and clay-loam soils. Solonis x Othello 1613 is a good rootstock that is highly resistant to nematodes and also moderately resistant to grape phylloxera.

GRAPE ERINEUM MITE

Eriophyes vitis Pagenstecher

This microscopically small eriophyid mite overwinters in the grape buds and feeds on the undersides of the leaves in the spring. The leaf reacts by producing a superfluous growth of leaf hairs in the infested area, called an erineum. The erinea, as seen on the lower leaf surface, are at first white but turn brown in color as they become older. Apparently little if any injury is done by these mites, and where sulfur dusting is practiced, as for mildew control, they are seldom in evidence.

GRAPE BUD MITE (811)

Eriophyes vitis Pgst. (bud mite strain)

The grape bud mite is believed to be a physiological strain of the erineum mite. No structural difference has been found between the two strains. The grape bud mite differs from the erineum strain in the following ways: (1) it cannot produce erinea; (2) it lays eggs and breeds under the dormant bud scales; (3) it spends the entire year in the buds and axils of the leaves; (4) it may wander over the open leaf surface in migrating from bud to bud, but probably never lays eggs on the leaves by choice; and (5) it produces certain deformities not produced by the erineum strain.

The bud mite is widely distributed in both coastal and interior districts, and it appears that no grape variety is immune to injury from this pest.

Seasonal History:- The winter is passed by the mites under the bud scales and as deep in the buds as they are able to penetrate. Even after the vines start to grow in the spring, the majority of mites remain under the bud scales, which form a whorl around the base of the new shoot. With the advancing season the bud mites can be found progressively farther out on the canes. It is believed that the bud mites spend their entire life cycle under the bud scales, since eggs, young, and adult mites are found there and all attempts to rear them in small cages attached to leaf surfaces have failed.

Injury:- The grape bud mite causes a shortening of the first 5 or 6 internodes of the new canes of the grapevine and thereby results in a dwarfing of the vine (fig. 428). "Witches brooms" may be produced when the mites cause the death of the terminal bud and the new canes, with the result that 5 or 6 lateral buds are forced out (fig. 429). When the vines are severely infested by this mite, all the winter buds on all spurs may be killed and nothing but suckers are produced in such cases. Several years of such severe injury may result in the death of the vine, or in less severe cases the vine may become deformed. There is sometimes a

Fig. 428 (left). Left, mild injury caused by bud mite on Muscat grape. Basal nodes are somewhat shortened and inflorescences are absent. Right, a normal cane with inflorescences. After Smith and Stafford (811).

Fig. 429 (right). Zigzagged stem and "witches broom" effect caused by the grape bud mite. After Smith and Stafford (811).

40% reduction in grape tonnage in vineyards where mite damage occurs and there have been cases of total loss of crop. Fortunately, outbreaks have so far been sporadic and localized and often occur only on a few vines scattered through the vineyard. Several species of predaceous mites, particularly Seiulus sp. frequently nearly eradicate the bud mite in some vineyards.

Control:- Good control was obtained in one test by applying an oil spray in the fall after the crop was off but before the leaves dropped.

GRAPE RUST MITE
Calepitrimerus vitis (Nalepa)

This is another eriophyid mite which hibernates in the grape buds in the winter and lives on the leaves in the summer, causing those of the white grapes to turn yellow and those of the dark wine grapes to turn red. Sulfur dust as applied for the control of mildew will hold this species in check.

PACIFIC MITE

Tetranychus pacificus McGregor

This mite has come to the fore as a pest on deciduous trees and grapevines in California more or less within the last 2 decades. Before it was described as a new species in 1915, the Pacific mite was not distinguished from the common red spider, Tetranychus telarius (L.)[1] or the two-spotted mite, T. bimaculatus. These species are, in fact, difficult to distinguish except on the basis of microscopic characters – mainly differences in the female tarsi and male genitalia. The Pacific mite is generally pale amber in color, with 2 or more large black areas on each side of the abdomen (fig. 430). The body of the female is larger and more nearly rounded than that of the male. The average length of the female is somewhat less than 0.5 mm.

The Pacific mite was first noticed on grapes in 1928 and within a few years became a grape pest of major importance. It is now a serious pest in northern and central California as far south as Fresno. Around Fresno and farther south the Willamette mite, Tetranychus willamettei McG, is more important as a grape pest.

Fig. 430. Pacific mite, Tetranychus pacificus. A, lateral view; B, dorsal view. X 100. After McGregor.

Life History:- Winter hibernation of orange-colored females without dark spots on their abdomens takes place beneath the bark of the vine above and below the soil line. The first active mites are noticed during the first 2 weeks of March, depending on the season. If there are weeds growing about the vines, the mites may feed on these even before the first grape foliage has appeared. A day or two of feeding causes a change of color from orange to amber and in the appearance of the lateral dark-spots, for these spots are the fecal contents of the viscera. By the time the new grape shoots have 4 to 6 leaves, all the mites have emerged from their winter quarters and are actively feeding and reproducing on the undersides of the leaves. Each female deposits from 50 to 100 eggs over a period of 2 weeks to a month.

The eggs hatch in from 3 to 6 days, depending on the temperature. The newly hatched 6-legged larva is almost colorless but begins to assume a greenish color after feeding a while. It has a round body at this stage. After the first molt the mite has 8 legs. Development from egg to adult requires only 5 to 10 days in midsummer. The adults move about over the leaf surface and the webbing very actively.

Injury:- By midsummer the mites have increased enormously. As many as 1000 mites on a leaf are commonly found. The foliage of infested vines turns color prematurely, that of the black grapes changing to red and that of the white grapes changing to yellow. Defoliation takes place if the vines are heavily infested, and as a consequence the sugar content of the grapes is greatly reduced. If infestations continue in a vineyard for 3 or 4 years, the vines are greatly weakened and are less productive. Occasionally vines are killed by the mites. Factors which cause a weakening of the vines, as for example, a sudden lowering of a high water table, such as occurs in some sections, will be followed by a great increase in the mite population. Late hibernation tends to reduce injury because a longer period is required for the mites to increase to injurious numbers.

---

[1]McGregor (562) believes that Tetranychus telarius does not occur in the United States and that the term "common red spider" has been applied indiscriminately to 2 species, T. althaeae Von Hanst. and T. bimaculatus Harv.

GRAPE PESTS

Natural Enemies:- Although mites have many natural enemies, only two were mentioned by Lamiman (504) as occasionally having some important effect on the Pacific mite population, namely, the six-spotted thrips, Scolothrips sexmaculatus (Perg.), and the predaceous mite, Seilus pomi (Par.)

Control:- Three methods of control are now used against the Pacific mite on grapes (833):

1. Banding. Since the mites do not overwinter under the bark of the spurs, sticky banding material applied at the bases of the spurs will keep them from reaching the new foliage until such time as the canes are long enough to contact the ground. Weeds must be removed and sticky banding material must be applied to stakes and wires of trellised vines. In the meantime the "suckers" must be removed as they appear, in order to prevent them from acting as a "bridge" from the ground to the canes. If the suckers are removed from the vineyard and destroyed, a large percentage of the overwintering mites are destroyed. Banding results in a delay of the peak of infestation. The banding material should be applied before the last of February.

2. Spraying. When the new shoots have 6 leaves, spray with 1.5% light medium summer oil emulsion. Special care should be exercised to wet the undersides of the leaves. A spreader increases the effectiveness of the spray. Repeat in 10 days. Oil will cause severe injury to foliage if used when there is sulfur on the vine. A period of 10 days should elapse after sulfur-dusting before the oil spray is applied.

3. Dusting. DN-Dust D-4 or DN-Dust D-8 may be applied when the first yellowish spots, indicating mite injury, are noticed on the foliage. Special dusting machines, built for dusting vines, should be used. On medium sized vines 30 to 35 lbs. of dust is used per acre, and if many live mites are seen on the vines after dusting, the treatment should be repeated. DN Dust should not be applied at temperatures of 95°F or higher, or injury to foliage may result.

New Insecticides:- Preliminary results have shown great promise for parathion in the control of Pacific mite. Used as a spray at 1 lb. of 15% wettable powder to 100 gallons of water, a 97.1% reduction of mites was obtained, and with 2 lbs., a 99.95% reduction was obtained. The addition of 899 (di-2-ethylphenylpthalate) increased the control efficiency to 100%. A control efficiency of less than 95% in spraying for the Pacific mite is considered usually of little value. (810)

WILLAMETTE MITE

Tetranychus willamettei McGregor

As previously stated, in the San Joaquin Valley around Fresno and farther south, the Willamette mite (fig. 431) is found on grape more abundantly than the Pacific mite. This species may also be found on deciduous trees, and in Western Oregon it is more important than the Pacific mite. According to E. A. McGregor (by correspondence) there is no dependable way to distinguish this mite from the Pacific mite with only the aid of a hand lens, but in general it may be said

Fig. 431. Willamette mites, Tetranychus willamettei McG. and stalked egg. X 8. Photo by Roy J. Perice.

that the Pacific mite usually has definite color patterns on the dorsum of salmon, orange red, greenish yellow, or otherwise, while the Willamette mite is more apt to be lacking in color patterns. It is characteristically pale, lemon amber, often with a few dusky blotches along the sides.

GRASSHOPPERS

The most common species of grasshoppers on grapes is said to be the devastating grasshopper, _Melanoplus_ _devastator_ Scud., although other species also infest grapes (833). The devastating grasshopper is a small, yellowish-brown species with a row of elongated black spots along the middle of the fore pair of wings (282).

In the early summer, when the native vegetation dries, the grasshoppers migrate to the vineyards. Often the vines are completely defoliated and the bark is chewed off the new canes. The winter is spent by the egg stage, in the soil, and the eggs hatch in the spring. The grasshoppers reach maturity by mid-summer and have 1 generation per year.

Control:- Poison bran mash may be broadcast in the vineyard. The following formula is recommended: bran 100 lbs., white arsenic 5 lbs or sodium fluosilicate 4 lbs; and water, to make a mash, about 12 gallons. Half of the bran can be replaced by sawdust. (833)

GRAPE THRIPS (49)

Drepanothrips reuteri Uzel

This species was first noticed on this continent in 1926 and has not yet proved to be an important grape pest when compared to such pests as the grape leafhoppers and the Pacific mite. However, some years certain vineyards may suffer scarring of about 50% of the bunches of grapes, which represents considerable loss in the case of table grapes. Damage to foliage may also be severe.

Description and Life History:- The adult females (fig. 432) are very small, about 0.75 mm long, amber to orange in color, and resemble the citrus thrips both in appearance and habits. The males are still smaller, lighter in color, and possess 2 black, sickle-shaped claspers projecting from the tip of the abdomen. It is believed that the grape thrips hibernate in the soil, in California, chiefly as adults, and survive only in very small numbers. They emerge from hibernation at the time the buds open in late March. The life cycle was found to average 21.9 days in midsummer at Davis, California, with 5 or 6 generations developing annually.

Fig. 432. Grape thrips, _Drepanothrips reuteri_. After Bailey (49

Injury:- Early in the season, when the fruit is about 1/3 grown, the feeding of the thrips causes a scarring of the berries. The skin eventually becomes russeted and checked (fig. 433). Later in the year practically all feeding is on the tender canes and unfolding leaves. When the thrips are abundant the young canes may be killed and the new leaves at the periphery of the vine are curled and appear to be scorched. Young vines may be stunted and their vitality reduced.

Control:- Treatments made for leafhoppers usually control this pest, but if not, a spray of 2 lbs. of tartar emetic and 4 lbs. of sugar to 100 gallons of water, applied in May or June, is recommended (833).

Fig. 433 (at right). A russetting and checking of grapes caused by the grape thrips. After Bailey(49).

566

GRAPE PESTS

FALSE CHINCH BUG

Nysius ericae (Shilling)

The false chinch bug has already been discussed among the citrus pests (p.396). After their migration to vineyards from the drying grasslands in the spring they may become so abundant as to cause the wilting of the foliage in a single day. The grassland from which the bugs are migrating should be burned if possible. The vines can be dusted with calcium cyanide or 20% sabadilla dust. (833)

CICADA

Platypedia areolata (Uhler)

This cicada is a common Western species about 18 mm long and bronzy-blackish in color with yellow and amber markings and a greenish-yellow band between the prothorax and mesothorax (282). It may sometimes cause damage to grapevines by its oviposition in the canes. After the eggs hatch the nymphs drop to the ground. They have been known to feed on the roots of orchard trees for 5 years before reaching maturity, but whether they feed on the roots of grapevines is not known. This pest is controlled if the canes are pruned to a spur and the prunings are burned. (833)

GRAPE LEAFHOPPER

Erythroneura elegantula Osborn

This native American insect is the most widespread and the most important of the grape pests in California and Arizona, besides being a pest of the American vines of the Eastern States. The most severe infestations occur in the dry, hot interior valleys, where the largest grape acreages occur. In the cooler and more humid regions closer to the coast, the leafhopper occurs, but is not a serious pest. Thus in California the grape leafhopper is a serious pest mainly in the San Joaquin and Sacramento Valleys and in the desert grape-growing regions in the Coachella and Imperial Valleys.

The grape leafhopper populations appear to fluctuate from periods of abundance to periods of very low populations. Usually cold and wet weather in winter and early spring tends to reduce the numbers of this pest.

In nearly all the past literature, it will be found that the scientific name Erythroneura comes (Say) has been used for the grape leafhopper in California. According to N. W. Frazier,[1] however, north of the Tehachapi Mountains the one species of leafhopper on grapes is E. elegantula, while in southern California, E. elegantula and E. variabilis Beamer are of about equal importance. No distinction will be made here, however, because of the similarity of habits and control. What was once called E. comes is now considered to be a group of about 50 species, including elegantula and variabilis. The E. comes of the present classification is not reported in the Western States.

Description. Eggs:- The eggs are laid singly in the epidermal tissue of the upper and lower surfaces of the leaves. They are white, bean-shaped, and about 1 mm long. The external indication of the location of the egg is the small, bean-shaped blister, which can be identified with the aid of a hand lens (fig. 434).

Nymphs:- In common with insects having simple metamorphosis, the nymphs resemble the adult, except in size and the absence of wings. There are 5 nymphal instars (fig. 435). When first hatched, the nymphs are semi-transparent and have red eyes. The succeeding stages are of a whitish color. Wing pads begin to be apparent in the third-stage nymphs and become increasingly prominent in the succeeding stages (fig. 428). Although the adults lay about an equal number of eggs on the upper and lower leaf surfaces, the nymphs confine their feeding almost exclusively to the latter.

[1] Correspondence of September 20, 1948.

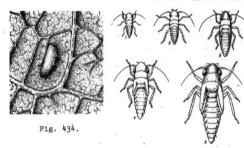

Fig. 434.

Fig. 435.

Fig. 436.

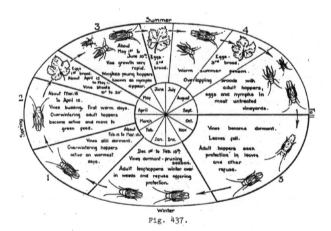

Fig. 437.

Fig. 434.   Egg of grape leafhopper in tissue of leaf. Greatly enlarged. After Lamiman (503).

Fig. 435.   The 5 nymphal instars of the grape leafhopper. Greatly enlarged. After Lamiman (503).

Fig. 436.   Grape leafhopper, _Erythroneura elegantula_ on a grape leaf. X 12. Photo by Roy J. Pence.

Fig. 437.   Seasonal life history of the grape leafhopper. After Lamiman (503).

GRAPE PESTS

Adult:- The adults (fig. 436) are 2.75 to 3 mm long, pale yellow, and irregularly marked dorsally with red. The reddish markings are most conspicuous on overwintering individuals. The hoppers are usually found on the undersides of the leaves and may be seen if the leaves are turned over carefully. However, adult leafhoppers are easily disturbed, and will fly out in swarms if the foliage is violently disturbed.

Life History:- The life history of the grape leafhopper (707) is synoptically presented in fig. 437. The winter is passed by the adults, which hibernate under leaves, dead grass, or in brush, straw piles, alfalfa fields, old cotton fields, etc., within a half mile of vineyards. The hoppers move to green vegetation in February and March, in order to feed, and they attack grape foliage as soon as it appears. They then continue to feed on the vines until the last leaves drop in the fall.

About 2 weeks after the overwintering hoppers attack the vines, oviposition begins. This is usually about April 7 to 15 in the Fresno area. Each female lays about 100 eggs. In the spring the incubation period is 17 to 20 days and in the summer it may be as short as 8 days. About the first of May the young of the first generation begin to appear in the Fresno area, while in the Lodi-Stockton area they appear 1 or 2 weeks later. Maturity is reached in about 18 days. The approximate chronology of the life history of the grape leafhopper in the Fresno area is indicated by the following dates (503):

Generation	Eggs	Hatching	Adult
First	April 7-15	April 24-May 1	May 20-25
Second	June 5-10	June 25-30	July 1-8
Third	July 25-30	August 10-15	August 25-30

The maximum numbers of each stage would be found about 2 weeks later than the dates given above.

Injury:- The leafhoppers make innumerable feeding punctures in the leaves, through which the sap is extracted. Small white spots, caused by the withdrawal of chlorophyll, are the first signs of injury. Finally these coalesce and form a pale-yellowish blotching. Injury is first noticed among the leaves around the crown of the vine. Finally the infested leaves dry up and fall to the ground, often exposing the grapes and resulting in sunburn injury. The great drain of sap and the defoliation causes a reduction in the sugar content of the grapes and in the following year's production. The leafhopper excrement, and the dust and fungi lodged thereon, are another objectionable feature of grape leafhopper infestations, especially in the case of table grapes. The production of raisin and table grapes has been reduced 25 to 30% in some seasons by leafhoppers, and in addition the quality of the grapes is lowered and the following year's production is reduced. (503)

Natural Enemies:- Predators and parasites do not appear to have any decisive effect on the leafhopper population. Apparently the most important natural enemy is an egg parasite, *Anagrus epos* Gir., which has been found to heavily parasitize the later broods of eggs in some seasons.

Control:- Calcium cyanide dust and pyrethrum-oil sprays had for years been used in the control of the grape leafhopper before the advent of DDT. The latter insecticide, however, has now largely replaced them. When applied before the bloom, DDT kills the overwintered adults and the few early nymphs that may be present. Nymphs hatching later and adults flying into the vineyard are killed by the DDT residue left by the spray. It is important that the DDT be applied to the undersides of the leaves, where the leafhoppers feed. DDT should not be applied after the grape berries are formed; some other insecticide must be substituted. However, the early DDT treatment may suffice to control the leafhoppers for an entire season.

Suggestions have been made for the use of DDT in vapo-sprays, fog generators, regular sprays and dusts. All these methods have given excellent control when properly employed. Dusting is the least expensive way of treating for leafhoppers and is much more rapidly done than spraying. The DDT dust, with sulfur, can be applied as a part of the mildew control program. (832, 343)

GRAPE WHITEFLY

Trialeurodes vittatus (Quaintance)

The tiny nymphs of this species sometimes appear on the undersides of grape leaves in large numbers and cause the smutting of fruit and foliage. They are lemon-yellow in color and are surrounded by the narrow fringe of white wax, as is typical of the whiteflies. The naked, dark brown pupa case also possesses a narrow, white, marginal fringe. The adults have immaculate white wings. Sprays applied for leafhoppers keep this pest in control, but 1-1/2% light summer oil plus 1 pint of nicotine sulfate to 100 gallons of spray will give control, if control is necessary.

GRAPE PHYLLOXERA

Phylloxera vitifoliae Fitch

The grape phylloxera (fig. 438) is an insect related to the aphids which attacks the roots of grapevines. It is indigenous to the United States east of the Rocky Mountains, where it occurred on wild species of grapes and whence it was introduced into France in 1863, and into California in 1873. In both regions it caused alarming destruction before being controlled by the use of resistant rootstock. An infested area remains infested as long as any grapevines survive and in such an area ungrafted vinifera varieties cannot be grown except in soil so sandy that it will not crack when dried after a thorough wetting. In sandy loam and heavier soils the phylloxera occurs in greater numbers and spreads more rapidly.

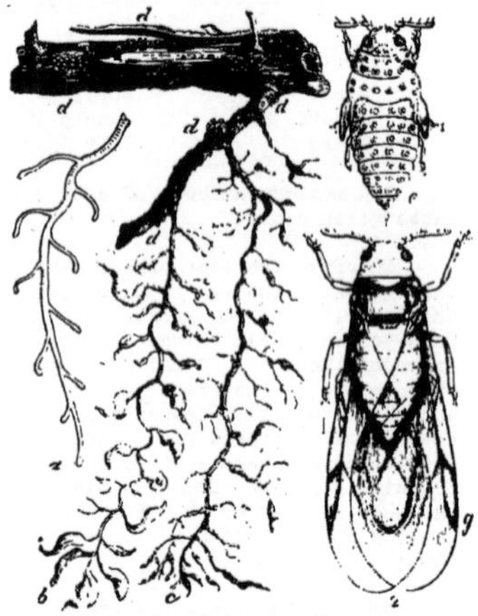

Although isolated infestations of grape phylloxera have been discovered in southern California from time to time, it has not become established as a serious or widespread pest in this part of the state. It is reported that Stanislaus, Merced, and Tehama counties are free of phylloxera, and that Kern, Kings, and Madera counties have only localized infestations (462). In and adjacent to these infested areas phylloxera resistant stocks are required. In San Joaquin

Fig. 438. Grape phylloxera. a, healthy root; b, infested root showing swellings; c, root deserted by phylloxera and in process of decay; d, phylloxera in larger roots; e, nymph; g, winged female. After Riley.

County and the Sacramento Valley, with the exception of Tehama County, infested areas are numerous and resistant rootstocks are advised for loam and heavier soils in the older vineyard areas. Fresno and Tulare counties are reported to have large infested areas as well as many smaller infestations. It is suggested that if no phylloxera is present within a half mile the grower will probably get the best results from rootings of the desired fruiting variety, but the phylloxera situation should first be thoroughly investigated. Information can be obtained from the farm advisor or the agricultural commissioners. With the exception of parts of San

GRAPE PESTS

Benito County, the entire north coast region is quite generally infested with
phylloxera and resistant rootstocks are recommended.

Life History:- Since control measures are not based on a knowledge of the
life history, the details of the complicated life history of this pest will be
omitted here. Briefly, however, it may be said that this insect overwinters on
the roots of the vine and reproduction begins in mid-April. During the following
6 months, several generations are produced, including winged forms. Dissemination
takes place by means of the winged forms as well as wingless individuals which ascend to the surface and wander about seeking new vines upon which to establish new
colonies. All forms are oviparous. The color of the wingless form varies from
yellow to yellowish-green or yellowish-brown and the winged forms are orange with
grayish-black head and thorax. (216)

Injury:- An infested vineyard can at first be distinguished by small "spots"
of dead or dying vines which gain in extent as the infestation becomes older. The
attacks of the phylloxera cause swellings (nodosities and tuberosities) to form on
the roots. It is believed that injury is not so much caused by withdrawal of sap
from the roots as by the decay that occurs in the lesions caused by the phylloxera.
There is little apparent damage to the vine the first year or two it is infested.
Sometimes the first reaction is an unusually heavy crop of grapes, owing to the
"root pruning". The vine gradually declines over a period of years. (216)

Control:- Quarantine, to prevent the spread of infested rooted vines is an
important factor in the control of phylloxera. Artificial means of control, as by
soil fumigants, has not been successful.

For wine-grape varieties in the nonirrigated soils in the coastal valleys of
California, the Rupestris St. George is the recommended phylloxera-resistant stock.
It is not resistant to nematodes. In irrigated soils that are free from nematodes,
Aramon x Rupestris No. 1 is said to surpass Rupestris St. George in respect to the
growth and productivity of the grafted vines. It is even more susceptible to
nematodes, however, than the majority of the fruiting varieties. Solonis X Othello
1613 is a moderately phylloxera-resistant vine which is also highly resistant to
the root-knot nematode. It is considered to be the best stock available in fertile,
irrigated, sandy-loam soils in the San Joaquin Valley. In very sandy soils of low
fertility where vines on Solonis X Othello 1613 are not satisfactory the Dogridge
and Salt Creek rootstocks have been suggested because of their extreme vigor. These
rootstocks, however, are now only in the experimental stage. (462)

SCALE INSECTS

There are about 13 species of scale insects attacking grapes, but rarely in
numbers justifying control measures. Normal pruning practice annually removes the
majority of the scales, but if control is necessary, 4% dormant oil emulsion for
unarmored scales and 6% for armored scales, applied in mid-winter, is recommended.

GRAPE MEALYBUG

Pseudococcus maritimus (Ehrhorn) (grape form)

This mealybug has been described in connection with the discussion of citrus
pests, although the citrus-infesting and the grape-infesting forms are considered
to be distinct. The sooty mold fungus growing on the "honey dew" excreted by this
species, along with the egg masses and bodies of the mealybugs, mars the appearance
of the fruit. If the mealybug population becomes too large the grapes may crack
and be infested by secondary molds.

The grape mealybug is normally controlled by the hot weather of summer. Likewise ant control tends to check the numbers of this species by allowing an increase
in the beneficial insect population. If artificial control is necessary, however,
the vines may be sprayed with a mixture of 3 gallons of oil emulsion and 4 gallons
of lime sulfur solution in mid-winter. Spraying must be done very thoroughly in
order to reach the mealybugs, which are situated deep in the crevasses of the bark.

571

SUBTROPICAL ENTOMOLOGY

The possibility of grape mealybug control with the proper use of parathion dust is indicated in preliminary tests. A 100% mortality was obtained with dust of 1% to 5% parathion and the treated vines remained free of mealybugs for the remainder of the season. (342)

ACHEMON SPHINX MOTH (706)

Pholus achemon (Drury)

These large, nocturnal sphingid moths are widely distributed throughout this country, and in California their main economic importance is as pests of the grape. Outbreaks of this species occur periodically. H. J. Quayle once observed 75 acres of vines so completely stripped of foliage by sphinx moth larvae that scarcely an entire leaf was left in the area.

Appearance:- The mature larvae are about 3 inches long, green, and with 6 to 8 pale yellowish, oblique bars extending across the spiracles (fig. 439). The early instars bear a horn on the posterior end of the body, but this disappears with the last molt and is absent on fully-grown larvae. The adult moth has a wing expanse of 3 to 4 inches. Its general color is a marbled brownish gray, with maroon patches on the body at the base of the wings as well as on the fore wings. The hind wings are rosy pink with a brown border and dark spots (fig. 42).

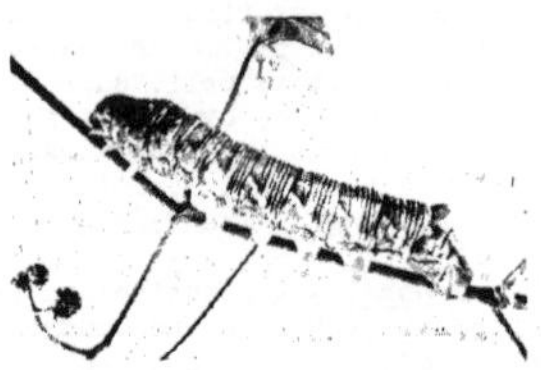

Fig. 439. Larva of achemon sphinx on grape. X 0.4. Photo by Roy J. Pence.

Life History:- The large, green spherical eggs are laid singly on the upper surfaces of the leaves. The 3 larval instars require about 24 days in mid-summer. Pupation takes place in the soil. The adults appear in April and May and again in July and August, indicating 2 broods per year.

Control:- When the larvae of the first brood begin to appear, the vines may be sprayed with 4 lbs. of standard lead arsenate per 100 gallons of water, or they may be dusted with powdered lead arsenate at 1 part to 4 parts of hydrated lime or dusting sulfur. If not too many are present, hand-picking of the larvae may be feasible.

WHITE-LINED SPHINX MOTH

Celerio lineata (Fabricius)

This is the most common of the Western sphinx moths. The species is somewhat smaller than the preceding. The adults are brown with white lines on the head and thorax. The fore wings have white veins and a broad buff stripe extending from the base to the tip (fig. 440). The hind wings are dark brown with a wide rosy band extending across the middle. The larvae are somewhat less than 3 inches long when full-grown and are generally green, with black lines and with yellow heads and yellow anal horns. Occasionally a black form with yellow lines may be found.

There are two broods a year and the life cycle is much like that of the achemon sphinx moth. The type of injury and the control measures are also the same as for the latter species.

Fig. 440. White lined sphinx, Celerio lineata. X 0.40. Photo by Roy J. Pence.

GRAPE PESTS

GRAPE LEAF FOLDER (61)

Desmia funeralis (Hübner)

This pyralid moth (fig. 441) is common in the east and may be found in California and British Columbia in the West. Although the rolled grape leaves which characterize the work of this insect are quite commonly seen, the insect is not often of economic importance.

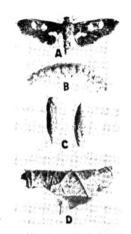

Fig. 441. Grape leaf folder, Desmia funeralis. A, adult; B, larva; C, pupae; D, over-wintering chrysalid in folded leaf. X 2.5. Photo by Roy J. Pence.

Description:- This moth has a wing expanse of about 1 inch, and is dark brown in color with 2 white spots on the fore wings. The female has 2 white spots on the hind wings also, but the male has only one. The larvae are pale green with brown head and brown spot on each side of the first 2 segments. They are about an inch in length.

Life History:- In the spring the flat, iridescent, reticulated eggs are laid singly or in small groups on the undersides of the leaves or on water-sprouts in sheltered parts of the vine. The newly hatched larvae feed in groups on leaves which they have webbed together. Later they roll and tie the leaves in a characteristic manner (fig. 437, D), feed on the edge of the leaf inside the roll and pupate in the rolls. There are 5 larval instars. The life cycle, from egg to adult, requires from 6.5 to 7.5 weeks. There are 3 broods a year. The pupae of the last brood pass the winter in the rolls which are scattered about in the debris on the ground. It has been noted that in the spring the adult females seek protected places in which to rest during daylight hours and their location not only in the vine but also in the vineyard is influenced by this habit. Infestation is usually heaviest near weedy ditch banks, shrubbery, citrus trees, or other places where protection may be found.

Control:- Before the flower inflorescences are formed in the spring, the vines may be sprayed with a mixture of 4 to 6 lbs. of lead arsenate and 5 to 6 lbs. of wettable sulfur per 100 gallons of water, or dusted with 40 to 50% cryolite and sulfur between the time the bloom begins in the early grape varieties and the time it ends in the late varieties. Cryolite dust in July controls the second brood. The removal of sprouts from the vine ("suckering") in the spring may remove as many as a third of the eggs and reduce the insects' potentialities for damage.

CUTWORMS

Several species of gray-brown noctuid moths are occasionally pests of grape-vines because their larvae climb up into the vine in the spring and destroy the buds. These larvae are in the soil or under trash in the daytime and feed at night, thus they are seldom seen. In the late spring, after they are full-grown, they pupate in the soil and the moths emerge and lay eggs in mid-summer. The resulting larvae feed on roots and crowns of low growing vegetation until the following spring, then feed on the grape buds again as did the previous generation.

Some of the common species of noctuids responsible for damage to grape buds are the greasy cutworm, Agrotis ypsilon (Rot.), the variegated cutworm, Lycophotia margiritosa (Haw.), and the brassy cutworm, Eriopyga rufula (Grote).

Control:- Sticky banding material applied to the bases of the canes in spring results in good control of cutworms. Cutworm baits are also effective. Usually the bait is scattered around the base of the vine late in the day.

573

SUBTROPICAL ENTOMOLOGY

WESTERN GRAPE LEAF SKELETONIZER (506, 37)

Harrisina brillians Barnes & McDunnough

This diurnal moth has been known to attack grapes in the southeastern United States for over a century, then worked its way west, being reported in Arizona in 1893. It is believed that this species may have been introduced from Europe, where it had been injurious to cultivated grapes in parts of Italy. Specimens collected in Alpine Valley, San Diego County, in 1941, were identified by H. H. Keifer, systematic entomologist for the State Department of Agriculture, as the above species, and the identification was later confirmed at the U.S. National Museum. It was recalled by Professor E. O. Essig, however, that a number of moths were received by mail from San Diego County at the University of California in 1937, which were definitely determined as being Harrisina brillians. The moths were so badly damaged that no specimens were mounted, and there was no written record of the collection or origin of the moths.

The moth year by year increased its range concentrically from Alpine Valley. In California it caused much greater injury to both cultivated and wild grapes than it had throughout the many years of its occurrence in other regions. Some untreated vineyards in the Alpine area suffered losses of 50 to 100% from defoliation by the larvae of the grapeleaf skeletonizer. When vines were excessively defoliated, a secondary flush of foliage was forced out, but this was killed back by the early frosts and the vines were so weakened that production was greatly reduced the following season. In addition there was considerable feeding on the maturing fruit by the larvae.

The larvae from a single egg mass are able to defoliate an entire vine. The colonies of larvae feed on the undersides of the leaves and it is difficult to detect their presence until the advanced stages of injury have been reached.

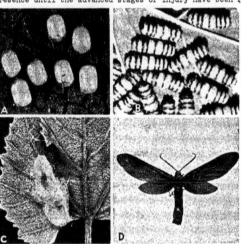

Fig. 442. Eggs, larvae, cocoons, and adult (X2) of the grape leaf skeletonizer. Larvae from San Diego Co., Comm. Off., others from Lange (506).

Description and Life History:- The eggs (fig. 442, A) are laid on the undersides of grape leaves. The larvae (fig. 442, B) at first are white to yellow-white specked with black setae, but after the second instar they change to bright yellow with 2 blue to purple transverse bands and 10 black bands of verrucae (dense tufts of upright hairs). The gregarious larvae feed in large colonies in early instars and split up into smaller groups or occur singly in later instars. There are 5 larval instars.[1] The larvae spin dirty white cocoons (fig. 442, C) in which they pupate. The pupae are yellowish to reddish brown and 7 to 10 mm long. Pupation takes place under the rough bark of mature vines and never on the runners, which is a fortunate circumstance, because it indicates that spread cannot take place by means of the dormant cuttings which are used to propagate grapevines.

The adults (fig. 442, D) are metallic bluish-black or greenish-black, the legs, body and wings being of the same color. They have a long abdomen and a wing expanse of about an inch. There are at least 2 and possibly 3 generations per year.

Eradication Project:- After the area of infestation of the grapeleaf skeletonizer had been delimited, control measures were applied, using in the first campaign a 50% cryolite dust in June, at the rate of 25 lbs. per acre followed by another application with 15 lbs. per acre in July-August, and a third application of 15 lbs. per acre in September, wherever live larvae were still found. In 1947, 74,295 lbs. of cryolite dust was used in 2 applications to 1,411 acres of cultivated grapes and to 63,525 backyard vines in the several small communities in the El Cajon Valley where the pest now exists. Simultaneously, an aggressive campaign of eradication of the wild grapevines in and around the infested region was undertaken. By killing the wild grapevines around the infested area, a host-free protective zone has been maintained. The 1947 inspection failed to show any outward spread of the pest.

The present control area is about 3,000 square miles in extent. In the campaign against the wild grapevine in 1947 alone, 394,300 square yards of vines were cut, 1,206,589 square yards were given the first treatment with 2,4-D and 1,164,880 square yards of regrowth was treated. (39) It is believed that at least 3 applications of 2,4-D will be necessary to eliminate the wild vines. The majority of treated areas were rather inaccessible and far from water. Aerosol generators which can be carried knapsack-like on the back of the operator (see p. 290) proved to be a boon under these circumstances, and gave good results. In a cooperative project with the University of California, it was determined that an aerosol containing 3.5% of 2,4-D would suffice to kill the vines.

At the fall meeting of the California Association of County Agricultural Commissioners in 1947, Cyril Gammon, an entomologist with the State Bureau of Entomology and Plant Quarantine, stated that he believed the chances of eradicating the grapeleaf skeletonizer in San Diego County were good. If the pest would be allowed to spread to the main grape-growing districts of California, it is believed that the cost of control would be over a million dollars annually.

THE RAISIN MOTH (240)

Ephestia figulilella Gregson

The raisin moth occurs in this country in the states of California and Arizona, infesting dried fruits, especially raisins, peaches, apricots, pears and figs.

Description:- The adult moth is about 3/8 inch long when at rest with wings folded and has a wing expanse of 15 mm. The fore wings are grayish, the hind wings satiny white. Both pairs have a prominent marginal fringe. The full-grown larvae are a half inch in length, white and with 4 rows of purple spots along the back.

Life History:- The moths are active chiefly in the early evening. During the day they remain in shaded, protected places. During the 2 weeks of their life the females lay about 350 minute white eggs. These are scattered over the surface of the fruit. They hatch in about 4 days. The larvae feed on the fruit about a month, then enter the soil or some other dark retreat to pupate. The winter is

[1]From a typewritten report on the life history of the grapeleaf skeletonizer by D. W. Robinson, a student at San Diego Junior College.

spent in the topsoil by well-grown larvae. Feeding begins at the time of the ripening of the earliest fruits, such as mulberries, and so breeding may continue from April or May to November.

Injury:- Damage consists largely of the feeding of the larvae on the surface of the dried fruit and the resulting fecal pellets and webbing. Sound grapes may be attacked (fig. 443), but usually the larvae infest only grapes which have been injured from other causes, chiefly the exposure of pulp by the cracking and crushing of the berries in tight bunches.

Fig. 443 (left). Larva of raisin moth feeding on a grape. X.2. After Donohoe et al (240).

Fig. 444 (right). Rolling paper tray of raisins in "biscuit roll". Photo by Perez Simmons.

Control:- Infestation of the earlier varieties of raisin grapes such as Zante and Thompson Seedless, can be reduced by the use of paper trays rolled in the "biscuit roll" (fig. 444), if these rolls are made carefully and if it is not attempted to put too much fruit into them. The raisin moth lays its eggs at night and many are killed by the hot sun, so the trays should be rolled in the afternoon, if possible, to avoid enclosing a living infestation in the rolled packages. Raisins dried on wooden trays are more exposed to attack by the raisin moth, but if protection is desired after completion of drying on spread trays after they have been stacked, this can be accomplished by covering the stack with covers of shade cloth.

Electrically driven raisin cleaners have proved to be effective for removing the eggs and larvae of the raisin moth from Zante and Thompson Seedless raisins but they are not as effective for the Muscat raisins because these do not shatter so well. If the screen is properly operated, about 90% of the eggs and larvae can be screened out. An added advantage in using this machine is that it removes the dust and sand from raisins before they are stored away.

Although the period that boxed raisins are stored on the ranch is influenced by market conditions, it should be borne in mind that as long as they remain at the ranch the infestation of raisin moths is increasing, providing temperatures are favorable. During storage on the ranch, if the boxes are covered with shade cloth much of the raisin moth infestation is prevented.

In the packaging of raisins, as well as other dried fruits, a small quantity of ethyl formate or isopropyl formate may be pumped into the package just before the fruit is put in. Insects or their eggs which may have escaped removal during cleaning and washing operations can be killed by these gases. In this "individual-package" fumigation for a 25-pound box of raisins (0.41 cu. ft.), 4 ml of ethyl formate in hot weather or 7 ml in cold weather will suffice. Isopropyl formate is about as insecticidally active per unit volume as ethyl formate. (775)

GRAPE PESTS

Sanitation, such as raking peach pits, fallen mulberries, or other waste fruits into the sun, is desirable to reduce as much as possible the sources of raisin moth infestations.

GRAPE BUD BEETLE

Glyptoscelis squamulata Crotch

This chrysomelid beetle, indigenous to the Western States, was first noted as a pest of grapevines in the Las Vegas Valley, Nevada, in 1922, and in the Coachella Valley, California, in 1923, whence it spread over the entire valley and became the major grape pest, sometimes destroying as much as 90% of the crop. In 1936 this species was first reported on grapes in the Kings River bottoms near Sanger, California, and is now well established in parts of Fresno County.

Fig. 445. Grape bud. beetle, Glyptoscelis squamulata, feeding on a grape bud. X 2.8. Photo by Roy J. Pence.

Life History:- The adults (fig.445) are about 1/4 inch long and covered with light to dark gray pubescence. They emerge from the ground in the latter half of February and crawl up the vines to feed on the opening buds. The eggs are deposited under the bark and the larvae drop to the ground upon hatching and go down to the roots of the vines as far as 2 or 3 feet below the surface. The winter is passed in the larval and pupal stages in the soil. (253)

Injury:- The grape bud beetles feed on the developing buds at night. During the day they hide under the loose bark of the vines, cracks in the stakes, or under the dry grass in the ground. Since the buds swell and burst open gradually over a period of several weeks, the beetles have ample opportunity to visit each bud, if they are sufficiently numerous. If some buds burst open and grow despite the beetle infestation, the small foliage may be eaten and the development of the new shoot may be so impaired that it cannot yield fruit. If and when the shoots become an inch or two long, the feeding of the beetles is of negligible importance.

Control (261):- Treatment consists of either dusting with 5% DDT or spraying with 1 pound of actual DDT (2 lbs. of 50% wettable powder) or 3 gallons of kerosene containing 1 lb. of DDT (4.7 wt/%) per 100 gallons of water. Between 30 and 40 lbs. of dust or 300 gallons of spray should be used per acre. The dusts should be applied with hand dusters so that the material may be directed against the trunk and crown of the vine and the stakes. The spraying has been done with regular power sprayers equipped with booms, so as to hold the hoses above the vines and allow for the spraying of both sides of several rows of vines at a time. The grape bud beetle population in Coachella Valley has greatly diminished since DDT treatments were begun in 1945.

CALIFORNIA GRAPE ROOTWORM (706)

Adoxus obscurus (Linnaeus)

This chrysomelid beetle is a pest of grapes in Algeria, France, Italy, Germany, and California. In California it has been known since 1880 and occurs on a native plant, the fireweed (Epilobium argustifolium), as well as on the grape.

Life History:- The yellowish-white oval eggs are 1/25 inch long. They are deposited in clusters of 75 to 100 on old wood and may be found beneath the inner layers of bark. The eggs hatch in 8 to 12 days and the larvae burrow down to the roots of the vine. The pupae are found at a depth of 4 to 8 inches in the early spring and the adults appear about May 1. It can be seen that the life cycle is much the same as the previously discussed grape bud beetle, which belongs to the same family of beetles.

The adult beetles (fig. 446) are 1/5 inch long, black, with a pubescence of short gray hairs. There is a brown form with the elytra, tibiae, and basal half of the antennae brown.

Injury:- The larvae may completely destroy smaller roots and gouge out linear grooves or even completely girdle the larger roots. The adult beetles eat out characteristic chain-like slits in the leaves (fig. 446) and gouge out the berries.

Control:- The grape rootworm may be controlled with lead arsenate spray, 3 lbs. per
100 with 1/4 lb. casein spreader, or with a   Fig. 446. The grape root worm,
20% lead arsenate dust, or with 3 lbs. of   Adoxus obscurus, and character-
derris powder (4% rotenone) per 100 gallons of   istic injury to foliage. Arranged
spray. These treatments should be made as soon   from Quayle (706).
as the beetles appear in the spring. Arsenicals
should not be applied to grapes after the flower inflorescences have formed.

It is likely that some of the recently developed insecticides, like DDT, may be effective against the grape rootworm.

DARKLING GROUND BEETLES

The greater part of the damage from darkling ground beetles (family Tene-brionidae) is done by certain species of Blapstinus. The adults are about 1/4 inch long, black or bluish black, and have reddish legs. Young vines are sometimes girdled by these beetles just above the ground. It is believed that the beetles may start feeding only where wounds have been made by cultivating tools. Poison bran mash may be used to control these beetles and 3% DDT dust has been found to be effective.

Control:- Dust consisting of 0.75% gamma isomer of benzene hexachloride, 50% sulfur, and the remainder an inert diluent, was found in 1947 and 1948 to give ex-cellent control of hoplia beetles. About 30 lbs. per acre was applied in the ex-perimental trials. (461a) Previously no satisfactory control for hoplia beetles had ever been available.

THE LITTLE BEAR

Pocalta ursina (Horn)

This scarabaeid beetle is 23 mm long, robust in form, metallic steel blue or bluish or greenish black, with reddish brown elytra, and the body clothed with long, yellowish hairs (282). It is sometimes a minor pest of grapes in the south-ern San Joaquin Valley. The adult beetles attack the tender shoots, usually on 1- and 2-year-old vines. In limited tests, a 1% parathion dust has resulted in the death of all beetles on the treated vines (342).

CLICK BEETLE

Limonius canus (Le Conte)

The adult of this elongate, dark brown elaterid is about 3/8 inch long. It feeds on the developing buds of the grape in the spring, causing injury much like that of the grape bud beetle. If it occurs in sufficiently large numbers it is capable of doing considerable damage.

GRAPE PESTS

The eggs are laid in the soil in the spring and hatch in 15 to 33 days. The larvae, called wireworms, live in the soil for about 3 years before transforming to adults.

Control:- The same as for grape bud beetle.

HOPLIA BEETLES

Hoplia callipyge Le Conte and H. pubicollis Le Conte

These scarabaeid beetles emerge from the ground in the spring and feed on the blossoms of many native plants in the Sierra foothills. They also attack the grape. Hoplia callipyge (fig. 447), known as the grapevine hoplia, is also a serious pest of roses at Fresno during April and May, injuring buds and open rose flowers.

The two species are rather similar in appearance, having reddish-brown elytra, with a darker prothorax, and with an iridescent greenish to gold center. The underside is silvery and shiny. These beetles are rather robust in form and from 1/4 to 1/2 inch long.

The adult hoplia beetles appear about the middle of April, feed on the foliage and forms and sometimes the buds and flowers, of the grape, oviposit, and disappear, all within a period of 3 or 4 weeks. They are not seen again until the following spring. Severe infestations are localized, but much injury may occur in the severely infested vineyards.

BRANCH AND TWIG BORER

Polycaon confertus Le Conte

Fig. 447. Grapevine hoplia, Hoplia callipyge. Above, on grape, X 3; below, attacking rose, X 1/2. Photo by Roy J. Pence.

The adults are 7 to 15 mm long, cylindrical, black with brown elytra, and with the pronotum narrowest at the base (fig. 381). The robust, curved, white larvae taper off in size posteriorly. They are covered with fine hair. The females lay eggs on the vines and the larvae or "grubs" bore into the trunk and arms of the vine. In the spring the adults feed at the bases of new shoots and bore holes into the spurs. Vines may be killed outright as the result of these borers, or as many as a half of the spurs may be killed.

Control:- Spraying is ineffective, but outbreaks can be prevented by annually burning dead and dying vines and brush in and around the vineyard before March. The dead arms found on some vines are favorite breeding places and should be sawed off and burned.

LEAD CABLE BORER

Scobicia declivis (Le Conte)

This beetle is about 1/4 inch long, dark brown or black, with antennae and portions of the legs of an amber color. The lead cable borers normally mine and destroy dead and seasoned oak, and obtain their common name from their habit of occasionally boring through lead telephone cables. With respect to the grape industry, these beetles are of importance because of the fact that they bore into wine casks, especially those made of oak. They are attracted by the alcohol.

Control:- The attacks of these beetles can be prevented if a hot, strong solution of alum is applied to the outside of the casks, followed, after the casks have dried, with raw linseed oil paint.

CHAPTER XXI
WALNUT PESTS

Persian walnuts are grown commercially in the United States only in areas on the Pacific Slope, mainly in California. Batchelor et al (78) list the 8 counties producing more than two-thirds of the State's crop in 1943-1944, in order of their importance, as Ventura, San Joaquin, Contra Costa, Santa Clara, Los Angeles, Tulare, Stanislaus, and Riverside. There were 126,788 acres of walnuts in California in 1947. The center of walnut production is continuously moving northward in this state.

Barrett (64) listed 286 species of insects and mites as occurring on walnut in the United States, but only a few of these are of economic importance, principally the codling moth, the walnut husk fly, the walnut aphid, and within recent years, the navel orangeworm. The codling moth has been a pest of some consequence in California for the last 30 years, the husk fly for only about 20 years, and the orangeworm first appeared as an important pest in 1947. Thus nearly all the important walnut pests are relatively late arrivals.

WALNUT BLISTER MITE

Eriophyes tristriatus (Nalepa)

Often yellow or brown felt-like growths (erinea) may be seen in distinct depressions on the undersides of walnut leaves with corresponding rises on the upper leaf surface (fig. 448, C). Microscopic examination of the erineum before it gets

Fig. 448. The walnut blister mite, Eriophyes tristriatus. A, adult X 100; B, mites in an erineum of a walnut leaf, X 20; C, lower surface (left) and upper surface (right) of walnut leaves, showing erinea produced as a result of mite infestation, X 0.6. Photo by Roy J. Pence.

too old and dry, will reveal the presence of many tiny white mites (fig. 448, A, and B), very elongate and possessing 2 pairs of legs in common with all other members of the family Eriophyidae. These apparently do not seriously affect the walnut tree even when they are abundant, but if control is desired, lime-sulfur 1-10 applied in the spring, when the buds are swelling, will be found to be effective. The mites spend the winter under the bud scales.

RED SPIDERS OR RED MITES

Two-spotted mite, Tetranychus bimaculatus Harvey

The two-spotted mite is a pest on a large variety of deciduous trees and other cultivated crops. In southern California this appears to be the most important of several species of mites that may occur on walnuts. Although its attacks are sporadic and not generally widespread, it can at times do serious damage, especially in the dry interior valleys in mid- and late season.

Description:- The two-spotted mite (fig. 449) is 0.42 mm long. It is variable in color, being usually brick red or ferruginous red, but at times rusty green, greenish amber, yellowish, or occasionally almost black. There are almost always pigmented blotches on the sides of the body which usually coalesce to form two large dark spots, one on each side. Sometimes these spots are divided into a large anterior and a small posterior spot on each side. (553)

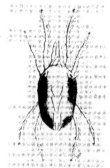

Life History:- The mites occur on both surfaces of the leaves of the walnut although they are not commonly abundant on the lower surfaces. They also occur on the green twigs and on the nuts. An infestation is characterized by large quantities of fine silken webbing on the leaf surfaces. The females deposit their round, pale yellowish eggs beneath and among these webs. In from 3 to 6 days the eggs are hatched and the entire life cycle requires only about 2 weeks. The newly-hatched larvae are nearly colorless, and in common with other tetranychid species, they possess only 6 legs until after the first molt, when they have their full complement of 8 legs. (107)

Fig. 449. Tetranychus bimaculatus McG. X 84. Courtesy E. A. McGregor.

Injury:- Injury is caused by the piercing of the leaf tissue and withdrawal of sap. This causes the leaves to become pale in color. Later the leaves become brown and dry and defoliation occurs. In some cases the majority of the leaves may fall prematurely, causing injury to the developing walnuts and a weakening of the trees. The tendency is for the infestation to begin in a few spots on the lower part of the tree and progress upward, finally affecting the whole tree in severe infestations. Trees already weakened from other causes are the most severely injured, and unlike the citrus red mite, the common red spider thrives best when the soil is deficient in moisture and the trees are wilting. Dust on the trees increases the severity of an infestation. (107)

Control:- Sulfur dust is the insecticide generally used in the control of the two-spotted mite on the wide variety of crops on which it occurs. On walnuts also, it has resulted in good control under satisfactory conditions, but the risk of injury is too great. On the other hand, DN-Dust D-8 affords excellent control when applied with equipment with which a large volume of air moving at a relatively low velocity can be applied. The "fish tail" type of discharge is not used, but instead the "fish tail" assembly is replaced by an adapter suitable for walnut dusting. The recommended dosage is 3 lbs. of DN-Dust D-8 per mature walnut tree or about 50 lbs. per acre. The dust should be applied when injury from red spider is first noticed. It appears to be safe for use on walnut trees even under extreme weather conditions. (107) The use of DDT and other new synthetic organic insecticides in codling moth control increases the red spider population, and, in fact, DN-D8 gives rather poor control in walnut orchards that have been treated with DDT.

OTHER SPECIES

The Pacific mite, Tetranychus pacificus McG., (see p.564) appears to be the most injurious mite species on walnut in central and northern California. When walnuts are interplanted with citrus in southern California the citrus red mite, Paratetranychus citri McG., may become a pest on the walnuts. The closely related European red mite, P. pilosus (C. & F.), may occasionally be a pest on walnuts in central and northern California. In southern California it is not a pest unless its natural enemies are decimated by some of the new insecticides. All these mite species may be controlled as recommended for the common red spider. Treatment should begin when the first signs of injury are noted.

SUBTROPICAL ENTOMOLOGY

FALSE CHINCH BUG

Nysius ericae Schilling

The false chinch bug has already been discussed on page 396 as a pest of citrus trees. It also can be a serious pest on walnuts, and has been known to kill young walnut trees. A 10% DDT dust thoroughly applied both to the trees and the ground, has been found to be a good control for this pest on walnuts.

WALNUT APHID

Chromaphis juglandicola (Kaltenbach)

The walnut aphid is found in Europe and in Western United States. In California it may be an important walnut pest, particularly in coastal areas.

Life History:- (214) This species (fig. 450) is lemon-yellow in color and 1/16 inch long. It occurs sporadically on the undersides of walnut leaves and on young nuts. Only in severe infestations is it found on the upper leaf surfaces. The stem mothers hatch from the overwintering eggs in the early spring and feed on the buds and new leaves as they appear. All have emerged by the time the leaves have fully opened out. When mature, the walnut aphids possess wings and deposit between 25 and 35 young. They are viviparous and parthenogenetic, and many generations of such individuals are produced. Altogether there may be 10 or 11 generations per year.

The true sexes, a winged male and wingless oviparous female, appear in the fall months and the overwintering eggs are deposited on the twigs. The oviparous female may be distinguished by 2 dark areas on the dorsum. When infestations are heavy, they may be seen as early as July, but otherwise not before September or October.

Injury:- The infested leaves may turn brown and drop prematurely and the vitality of the tree may be reduced. Infested trees may become black with sooty mold fungus. Walnut leaves coated with this fungus are more susceptible to sunburn than clean leaves, and defoliation may subject the nuts to sunburn also. There is an increase in the percentage of perforated shells and shrivelled and dark-colored kernels.

Fig. 450. Above: walnut aphid, Chromaphis juglandicola, feeding on walnut leaf. Arrow indicates point of insertion of mouthparts. Below: nymph and winged adult. X 32. Photo by Roy J. Pence.

Natural Control:- Periods of high temperatures and low humidity sometimes greatly reduce the walnut aphid populations during the summer months. Lady beetles, syrphid flies and chrysopids take an annual toll of the aphid population. As high as 88% mortality has been caused by a fungus Entomophthora chromaphidis Burger and Swain (132). As a result, the walnut aphid population fluctuates greatly, not only within a given year, but also from one year to the next.

WALNUT PESTS

Artificial Control:- The least expensive control of the walnut aphid is by dusting, and it is quite effective. From 3 to 5 pounds of nicotine sulfate (40% nicotine) is mixed with 100 lbs. of a good grade of dusting lime (hydrated lime) using the higher concentrations in the heavier infestations. Usually 25 to 30 lbs. of dust per acre is sufficient. From 5 to 8 aphids per leaf are enough to justify treatment. If the trees are to be sprayed with lead arsenate, from 1/3 to 1/2 pint of nicotine sulfate may be added to the spray to control the aphids. Promising results are reported, even under adverse weather conditions, from the use of a speed sprayer to control walnut aphids, using "Black Leaf 40", "Black Leaf Dry Concentrate," benzene hexachloride, and TEPP. (666, 667)

A number of aphicides were tested in 1948 in conjunction with standard lead arsenate or DDT in codling moth control. These materials and the rates per 100 gallons of water were as follows: 14% dry nicotine concentrate, 1 lb.; benzene hexachloride containing 10% gamma isomer, 2/3 lb.; tetraethyl pyrophosphate, 1/8 pint of 40% or 1/4 pint of 20%; and 25% wettable parathion, 1/3 lb. About 55 gallons of spray were applied per tree, using 25-foot towers. All combinations resulted in excellent control of aphids. (609)

Dusts of parathion, Neotran + nicotine, and DN-D8 + nicotine were found to be highly effective against the walnut aphid. The treatments, however, were not effective against the common red spider. (666, 667)

SCALE INSECTS (282, 724)

The scale insects are relatively unimportant pests of the walnut, but at least 4 species may occasionally be troublesome.

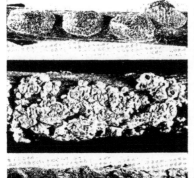

Frosted scale, Lecanium pruinosum Coq. (fig. 451, above), is a large unarmored scale appearing much like the European fruit lecanium L. corni, but is considerably larger. It is 8 to 9 mm long, oval, brown and covered with a powdery white wax which gives it its common name. In past years it seldom caused serious injury on walnut, but since the use of DDT for codling moth control the frosted scale has become an important pest. It may be controlled by a properly timed spray of 1-1/2 gallons of kerosene and 1 lb. of 50% wettable DDT powder to 100 gallons of water.

Calico scale, Lecanium cerasorum Ckll., is another lecanium scale, occasionally occuring on walnuts but not south of the Tehachapi mountains. The females are 6 to 9 mm in diameter, nearly globular, and dark brown with white or yellowish white areas on the dorsum. Dormant oil emulsion 3% can be used to control this pest.

Fig. 451. Above, frosted scale, Lecanium pruinosum, X 3.5; center, walnut scale, Quadraspidiotus juglans-regiae, X 2; below, obscure scale, Chrysomphalus obscurus, X 2. All on Persian walnut. Pence photo.

Walnut scale, Quadraspidiotus juglans-regiae (Comst.) (fig. 451, center), is an armored scale which is known to occur in the United States, Canada and Switzerland on a number of deciduous fruit and shade trees. The female is circular, flat, 2.25 to 3 mm in diameter, gray to reddish brown and with a brown subventral exuviae. It occurs on the twigs. The walnut scale is not considered to be sufficiently important to warrant control measures.

SUBTROPICAL ENTOMOLOGY

Italian pear scale, Epidiaspis leperii (Sign.) (E. piricola Del G.), is another armored scale which may be found on walnuts. The female has a circular, dark gray armor hiding the dark red or purplish body. The scale covering of the male is elongate and white. There appears to be 1 generation per year. This species may be controlled by spraying the trunks and larger branches of the trees with oil spray or lime sulfur. However, if the trees are kept free of lichens, this scale cannot survive. The most effective spray mixture with which to eliminate the lichens is 12 lbs. of hydrated lime and 1/2 gallon of spray oil to 100 gallons of water.

The obscure scale, Chrysomphalus obscurus (Comst.), is a small armored scale of a grayish color which closely sim1lates the bark of the walnut tree (fig. 451, below). This species will be discussed more fully in the chapter on pecan insects, for it is relatively more important on pecan than on walnut. It is seldom seen on the latter host.

California red scale, Aonidiella aurantii Mask (p.443), can sometimes be found on walnuts, but seldom to such an extent as to be injurious. The greatest economic importance which can be attached to red scale infestations on walnuts is in relation to red scale eradication campaigns, as in areas under the control of "protective associations" or "protective leagues." In this connection many walnut trees have been fumigated by the Ventura County Citrus Protective League. The difficulties involved in the fumigation of large walnut trees can well be imagined. Tents as large as 112 x 105 feet and using 1250 square yards of material have been employed to cover the trees, and as much as 7 pounds or more of HCN gas per tree has been used. In southern California no oil spray should be used on walnut trees when they are in foliage, but such sprays could be used in the dormant season.

CODLING MOTH

Carpocapsa pomonella Linnaeus

Although the codling moth is a universal and age old pest of the apple and was introduced into California about 75 years ago, its appearance as a pest of walnuts was not noted in California until 1909 (340). The codling moth had become a serious pest in a limited area in southern California by 1918. At present it is widely distributed on walnuts and annual treatment must be resorted to if serious damage is to be avoided. It is considered to be the major pest of walnuts.

Description. Eggs:- The eggs are flat, disc-like, and are only about half the diameter of an ordinary pin head (1/25 inch). They are white when first laid, but become darker as the embryo develops. They are laid singly on foliage or nuts.

Larva:- The newly hatched larvae are whitish and have large black heads (fig. 452). The full-grown larvae are about 3/4 inch long, are white to yellow or pinkish, and with the head and thoracic shield brown.

Pupa:- Pupation takes place in any hidden retreat affording protection, in a white felty cocoon. The pupa is yellow to dark brown, and about 1/2 inch in length.

Adult:- The adults (fig. 453) have a wing expanse of about 3/4 inch. The general grayish cast of the wings blends the adult moths with the usual color of the bark upon which they are usually resting, so they are rather inconspicuous. The fore wings are brownish with several gray or paler cross lines and with a dark golden and coppery patch at the apex. The hind wings are paler in color. The outer margins of the wings are fringed with hairs.

Life History (718a):- The winter is passed in the larval stage in cocoons which may be found under the bark, in old pruning cuts, crotches, in litter found on the ground, or other protected places. Pupation takes place in early spring, and the first brood of moths appear in April. This spring brood of moths will continue laying eggs until July. The eggs are laid predominantly on the twigs and leaves in the first generation and predominantly on the nuts in later generations.

Fig. 452 (left). Codling moth larvae (arrows) entering through the blossom ends of walnuts. Courtesy A. E. Michelbacher.

Fig. 453 (right). Codling moth, *Carpocapsa pomonella*; X 3.5. Photo by Roy J. Pence.

The moths may be seen flying about at dusk, and if the temperature is 60° F or higher, with little or no air movement, the females deposit their eggs at that time.

The first brood of larvae appear during May and June. These at first enter, or attempt to enter, the calyx end of the nut, but generally the largest portion of this brood of larvae hatch after the nuts have attained sufficient size to contact one another in clusters. Then the larvae commonly enter the nuts at these points of contact. (fig. 454) Still later in the season, when the shell of the nut becomes hard, the larvae enter through the fibrous tissue at the stem end. From 35 to 45 days are required for the larvae to become fully developed.

Seasonal History:- Depending on seasonal temperatures, there may be a complete second generation of codling moth and even in some years a partial third generation.

Injury:- The majority of the nuts infested by the first brood of larvae which have gained entry into the calyx will drop. The most important injury, however, results from the feeding of the larvae in the walnut kernels. To the loss of nuts must be added the extra packing costs caused by the sorting out of infested nuts. All varieties of Persian walnuts are attacked, but the Seedling, Placentia, Chase and Payne are the most susceptible and the Eureka and Franquette appear to be somewhat less susceptible to attack. A 25% loss of nuts in untreated orchards is not uncommon, but infestations as high as 50 to 75% are rare. (101)

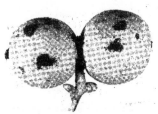

Fig. 454. Walnuts showing points of entry of the codling moth, *Carpocapsa pomonella*. X 1. Photo by Roy J. Pence.

Natural Enemies:- A number of parasites and predators attack the codling moth, the commonest being the egg parasite *Trichogramma minutum* Ashm. Parasitized eggs may be distinguished by their darker color. Larvae and pupae are attacked by the parasite Aenoplex carpocapsa Cushman and others, as well as a number of predators, particularly certain carabid and clerid beetles. The total effect of natural enemies in the codling moth population, however, is small. (718a)

Control:- In northern California, basic lead arsenate has become increasingly less satisfactory for codling moth control and it has been found that standard lead arsenate, with a safener added, can be safely used and is much more effective. The following formula is the most satisfactory that can be recommended for general use (606):

Standard lead arsenate........................	3 pounds
Safener (Delmo Z, a commercial basic zinc sulfate product containing 50% zinc expressed as metallic, or a similar product)........................	1 pound
14% nicotine dry concentrate.................	1 pound
Light medium summer oil emulsion containing 80% to 83% oil.................	1/3 gallon
Water..	100 gallons

The nicotine, which is added for aphid control, is slurried in a pail, after which the lead arsenate and safener is added while continuing to slurry. The slurried mixture is added to the spray tank while filling the tank with water and with the agitator in motion, when the tank is 1/3 to 1/2 full. The oil is added when the tank is 3/4 or more full. One-fourth to 1/3 lb. of depositor can be added to increase deposit and sticking ability. This also should be slurried along with the dry ingredients. Benzene hexachloride, parathion and tetraethyl pyrophosphate have shown promise as aphicides, but the nicotine preparations would have the least effect on natural enemies and are therefore recommended pending further experience with the newer insecticides.

DDT has been highly effective in codling moth control, but tends to greatly increase the mite population by destroying natural enemies. Some growers, however, have used DDT on a limited acreage, and for these growers the following formula has been suggested (606):

50% DDT wettable powder......................	1 pound
DDT depositor................................	1/3 pound
14% dry nicotine concentrate.................	1 pound
Light medium summer oil emulsion containing 80% to 83% oil.................	1/3 gallon
Water..	100 gallons

The DDT and depositor are slurried in one pail and the dry nicotine concentrate in another. The first slurry is added to the tank, with agitator in motion, while it is only 1/3 to 1/2 full of water. Then the nicotine slurry is added and finally the oil. It is recommended that if DDT is used, standard lead arsenate should be used in alternate years.

The use of DDT sprays has been investigated in southern California with regular spray equipment, speed sprayer, and aerosol-generating equipment. DDT dusts were applied with the ordinary walnut duster and by airplane. Two applications of 2 lbs. 50% wettable DDT powder per 100 gallons, applied with a regular spray rig or with a speed sprayer resulted in excellent control--0.5% and 0.8% wormy nuts, respectively, compared to 21.7% wormy nuts where 2 sprays of basic lead arsenate were applied with a regular spray rig, Likewise 3 applications of 10% DDT dust or 2 applications of 20% DDT dust, each application at the rate of 5 lbs. of dust per tree, gave very good results, leaving only 1% wormy nuts. Significantly inferior control was obtained when the dust was applied by plane. With the aerosol generator some burn of foliage nearest the nozzle of the applicator was obtained; 2% wormy nuts were found. Results with DDT spray or dust proved to be superior to those obtained from the regular spray program of two treatments of basic lead arsenate. (666, 667)

Baits consisting of malt, yeast and water may be used to attract moths and determine the extent of moth activity. The bait pans are placed in the trees when the walnuts begin to set, about a dozen to an orchard. Treatment can be timed on the basis of information obtained from the catch of adult moths.

NAVEL ORANGEWORM

Myelois venipars Dyar

The navel orangeworm is a rather recently established pest on walnuts, but was known for many years, as a scavenger in cracks or wounds of Navel oranges. Essig (282) points out that what appears to have been the first observation on this pyralid moth was made in 1899 by T. D. A. Cockerell, who found the larvae in oranges which had colored prematurely, near Phoenix, Arizona (173).

R. S. Woglum found the navel orangeworm in fallen oranges in the State of Sonora, Mexico, in 1914. Don C. Mote found the larva of this species in fallen Navel oranges in Arizona in 1921 "feeding either in, or near, the bast fibre (core) of the orange with a mass of frass near by" (619). At the time some apprehension was felt concerning the possibility that this insect might have something to do with the dropping of oranges and the spreading of Alternaria citri, the causal organism in black end rot of Navel oranges. California established a quarantine against the Navel orangeworm. A survey made in 1930 revealed that with the exception of a few rare instances, this insect attacks only split or wounded oranges, or those infected by "black rot," and is not a primary pest but is more in the nature of a scavenger. The quarantine against Arizona Navel oranges was then discontinued. (539)

A few Navel orangeworms were found in mummified oranges in Orange County in late 1942 in the oriental fruit moth survey (480). It has also been found infesting almond, jujube, loquat, peach (mummified), prune, quince, and Persian (English) walnut. A survey made in December, 1945, revealed light infestations of the Arizona navel orangeworm in split Navel oranges in eastern San Bernardino and Riverside counties (984). It has not yet been reported from northern California.

The Navel Orangeworm in Walnuts:- The navel orangeworm was found in Orange County in 1943 and there were several records of it occurring in walnuts in the field in 1945. In 1946, cull walnuts infested with codling moth or injured in other ways, were infested with Myelois larvae in several walnut packinghouses in southern California, and, after the main harvest was over, they were found in nuts left in the orchards. Early in 1947 the navel orangeworm was found in walnut trees. The eggs are deposited on green nuts. The larvae can enter any hole in the nut, and since the newly-hatched larvae are so small, they could probably enter very small holes and possibly they may even bore through the tender tissue of the stem end of the young nuts. The writer found a navel orangeworm in an almond which had bored a hole directly through the husk on the side of the almond. Often several larvae are found in a walnut and as many as 56 in a nut have been reported. In 1947, infestations in walnuts were found in Los Angeles, Orange, Riverside, San Bernardino, Santa Barbara and Ventura Counties, and the navel orangeworm had become a major pest of walnuts.

It is of interest to note that a related species, Myelois ceratoniae Zell. is reported to have been reared from stored English walnuts in South Africa (64).

Description and Habits. Egg:- The eggs are creamy white when first deposited, but later become pink. They are "diamond striated" and are almost microscopic in

587

size. They are laid in the navel end of injured oranges or on green walnuts. After the walnut hull splits, the adults will oviposit on the walnut as well as the hull.

Larva:- When full grown, the larva is 0.5 to 0.75 inch long and pale pink to deep pink in color. In October and November the larval stage requires 30 to 50 days and is apparently divided into 6 instars. The larva overwinters in mummified and rotting fruit hanging in trees, or in nuts, a further indication of its scavenger habits.

A description and figures to aid in the separation of the larvae of the navel orangeworm from the larvae of the codling moth, which they somewhat resemble superficially, may be found in a mimeographed announcement (E-27, November 28, 1947) of the California State Department of Agriculture.

Pupa:- At first pinkish, later turning to dark brown. It is formed in a silken cocoon within the orange or nut (fig. 455).

Adult:- Wing expanse 16-19 mm; color pale grayish marbled with brown and black (fig. 456).

Control:- Before DDT was used for codling moth, the standard practice was to fumigate the walnuts with methyl bromide while they were in the sacks

Fig. 455 (left). Myelois venipars. Pupa in hull of an almond. Photo by Roy J. Pence.

Fig. 456 (right). Arizona navel orangeworm, Myelois venipars. X 3.8. Photo by Roy J. Pence.

in which they were gathered. Ortega (666) reports that in experimental work in 1947, when excellent control of codling moth was obtained with DDT, the navel orangeworm was of minor consequence. During the summer of 1948, this observation was substantiated by commercial experience, for where DDT was used in walnut codling moth control, the navel orangeworm was of no commercial importance.

CATALINA CHERRY MOTH (46)

Melissopus latiferreanus (Walsingham)

The Catalina cherry moth was formerly known to infest only the Catalina cherry in southern California and the large green galls on blue oak in central California. It was found on the latter host by Essig (282). Investigations conducted in 1944 showed that this insect was an important pest of walnuts in the Sacramento Valley. Although some varieties, such as the Payne, appeared to be more susceptible than others, nuts of the majority of varieties were attacked. The pest is now known to occur throughout nearly all of California. Besides the hosts already mentioned, it attacks acorns, hazelnuts, and filberts, but is of economic importance on walnuts only in portions of the Sacramento and Napa Valleys. It is likely to become destructive only in areas where its preferred host, the oak apple gall[1] (fig. 457, above) is produced in abundance. The moths migrate from the oaks to walnuts.

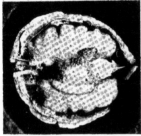

Fig. 457. Above, the "oak apple," gall of the gallfly Andricus californicus, on oak, X 0.35. Below, larva of the California cherry moth infesting a walnut, X 1.5. Photo by Roy J. Pence.

Description:- The adult moth (fig. 458) has a wing expanse of 12 to 15 mm. It is pale or dusky bronze in general coloration and has 2 coppery areas at the tips of the fore wings. The whitish, disc-like eggs are laid on the fruit of the host plants or on oak galls. The larvae (fig. 457, below) are 18 to 20 mm long when full grown. They are similar to those of the codling moth, which belongs to the same family of moths (Olethreutidae), but in the later stages of larval development the two species can be distinguished by superficial examination. The Catalina Cherry moth larvae are then a rather uniform creamy white while those of the codling moth are somewhat pinkish.

Taxonomically, the larvae of the two species are similar, having similar plates around the abdominal setae and similar proleg hooks. However, the head of the Catalina cherry moth larva is of a clear amber color, while the head of the codling moth has black stripes. The lateral tubercle on the ninth abdominal segment has 3 setae and is rarely divided, while on the codling moth larva the lateral 3 setae on the ninth abdominal segment are usually on a bisetose upper tubercle and a unisetose lower tubercle.

Seasonal History:- Since two crops of oak galls are produced each year, a highly satisfactory source of food is maintained for two and a partial third brood of moths which are produced each year. Catalina cherry moth infestations on walnuts usually do not increase markedly until late August or early September. From then on the infestation tends to increase until the crop is harvested.

Injury:- Bait trap catches have shown that while the moths are present throughout the summer, they do not damage the nuts until late August or September. The larvae cannot develop on the green husks of walnuts. Likewise they are not able to penetrate sound husks until they begin to crack at the period of maturity. Having gotten through the husk, entrance into the nut proper is made at the stem end. Although the larvae could feed on the walnut kernel at any stage of its development, they cannot gain access to it until cracks appear in the husks.

Fig. 458. Catalina cherry moth, Melissopus latiferreanus. Left, male; right, female. X 2. Courtesy Oscar Bacon.

---

[1] The galls are caused by species of Cynipidae, small, wasp-like insects which insert their eggs in the twigs of oak trees. The galls begin to develop soon after the eggs hatch, and are believed to be caused by the larvae. The galls are of different shapes and colors, depending on the species of cinipid involved.

589

Control (47):- It has been shown that the Catalina cherry moth is most effectively controlled by spraying just before the husks of the nuts begin to crack. In the case of the Payne variety, this is usually during the last half of August.

DDT, chlordan, methoxychlor, and standard lead arsenate have been tested, but the latter is the only one of the group that has shown any promise in the control of the Catalina cherry moth. No control program has been as yet developed, however, that would warrant a recommendation.

Harvesting at the earliest possible date is known to greatly reduce the infestation. Once the husks start to crack, the infestation increases rapidly, and a delay of a week or two in harvesting may result in great loss. Along with rapid harvest, prompt drying of the nuts should be practiced, for the larvae cannot develop in dried walnut kernels. Very small larvae that may have already gained entry to the nut are then killed before doing detectable damage.

OTHER LEPIDOPTERA

The fruit tree leaf roller, Archips argyrospila Walker, occurs on many varieties of fruit trees and has been discussed among the citrus pests (p. 488). The larvae may also be occasionally found in walnuts, where they may be mistaken for those of the codling moth. They are about the same size as the codling moth, but are cream-colored, while the codling moth is often pink. The larvae feed on the foliage causing the ragged appearance shown in fig. 459. The fruit tree leaf roller is readily controlled with a 5% DDT dust, and this treatment is being used on walnuts.

The Indian meal moth, Plodia interpunctella (Hbn.) (fig. 502), attacks fruits and nuts in storage. The larvae may be found in walnuts and may be confused with those of the codling moth, although they are somewhat shorter, being about a half inch long, and are not colored as distinctly pink. They can be controlled by heat or fumigation treatments.

The red-humped caterpillar, Schizura concinna (A. & S.) occurs on a variety of fruit and nut trees, including the walnut (724). The larvae first feed in colonies on the lower leaf surface, then, as they become older, on the edges of the leaves. The full grown larva, about 1-1/4 inches in length, is characterized by a red hump on the 4th body segment. The adult is dark brown, has a wing expanse of 1-1/4 inches, and lays white

Fig. 459. Injury to walnut leaves caused by fruit tree leafroller. Photo by Roy J. Pence.

globular eggs in clusters of from 50 to 100 on the under side of the leaf.

The walnut span worm, Coniodes plumigeraria (Hulst) has a wingless female and a gray winged male. The caterpillars may become 20 mm long, have 3 pairs of prolegs and are "pinkish gray varied with darker gray or purplish, or black and yellow" (282). Pupation occurs in the soil. In February or March the females lay oval, flattened eggs in masses on the limbs and small twigs. Except for one severe infestation noted by Essig (282) and a few other limited outbreaks, this species has been of no economic importance on walnuts. Arsenical or fluorine sprays or dusts can be used for control, or tanglefoot bands may be used to keep the wingless females from ascending the trees (724). Another span worm, Sabulodes nubilata (Pack.), the caterpillar of which is similar to that of the former species, is occasionally injurious to walnuts in southern California. Both sexes are winged. (724)

WALNUT PESTS

BEETLES

A number of coleopterous wood borers may be found attacking the Persian walnut. They are principally flat-headed borers (Buprestidae) and round-headed borers (Cerambycidae). The larvae of the buprestids have greatly enlarged and flattened thoracic segments and their tunnels and exit holes are oval-shaped. On the other hand, the larvae of the cerambycids have cylindrical thoracic segments and leave cylindrical tunnels and exit holes in the twigs and branches of their host trees, as do also the branch and twig borers (Bostrichidae) that may be found attacking walnuts.

Buprestidae

Dicerca horni Crotch (fig. 460) appears to be the flat-headed borer most often causing injury to walnut trees in California. In common with the majority of coleopterous wood borers, this species is most apt to attack trees somewhat weakened from disease or neglect, but may hasten the death of a tree which might other-

 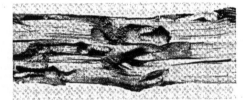

Fig. 460 (left). Dicerca horni. X 2.5. Photo by Roy J. Pence.

Fig. 461 (right). A walnut branch infested by Pacific flatheaded borer, Chrysobothris mali. X 1.4. Photo by Roy J. Pence.

wise recover. The adult is about 12 mm long, of a bronze color, and with small, black, narrow broken ridges on the dorsum. The larvae are about 37 mm long. (282) Another borer apt to be found on walnuts is the well-known Pacific flatheaded borer, Chrysobothris mali Horn (fig. 461).

Cerambycidae

Among the round-headed borers, the hairy borer, Ipochus fasciatus Lec. (fig. 462), appears to be the most common on walnuts. It is currently abundant on Persian walnut trees in subdivided areas where often the trees are not irrigated and the present drought (1947-48) has tended to weaken the trees and make them especially susceptible to attacks by wood borers. The adult hairy borer is small for a cerambycid, being only 5 to 8 mm long. It is a dark mahogany brown throughout with the exception of 2 transverse broken and irregular bands of white on the elytra. The entire body is sparsely covered with long white hairs. The suture where the elytra come together is not visible to the unaided eye. With the exemption of the long antennae

Fig. 462. The hairy borer, Ipochus fasciatus. X 5. Photo by Roy J. Pence.

typical of the cerambycids, this beetle is rather ant-like in appearance. The larvae are commonly found in the small twigs of the Persian walnut, leaving circular exit holes about a millimeter in diameter. Two other round-headed borers, Synaphoeta guexi (Lec.) and Xylotrechus nauticus (Mann.), have been reported taken from the limbs and trunks of dead or dying trees by Essig (282).

591

SUBTROPICAL ENTOMOLOGY

Bostrichidae

Polycaon confertus Lec. and P. stouti Lec. are sometimes found in dead and dying Persian walnut trees (64). The adults sometimes burrow into living twigs at a crotch or next to a bud axil and cause the limbs to break. This type of damage may be found on all kinds of deciduous fruit, nut, citrus, avocado, and olive trees. P. confertus (fig. 381) has been discussed in previous chapters. P. stouti is 17 - 20 mm long, entirely black, with large pits over the surface of the head and pronotum and smaller pits on the elytra.

Control for Wood Borers

Infestation by wood borers can to a large extent be prevented by keeping the trees in a thrifty condition. If infested twigs or branches can be found, they may be pruned off and burned to prevent further spread of the emerging adults. Whitewashing and tree wrappers or protectors may prevent egg laying.

WALNUT HUSK FLY (100)

Rhagoletis completa Cresson

The walnut husk fly, the first species of the economically important family Trypetidae[1] to become a major pest in California, was found by S. E. Flanders in several varieties of Persian walnut in the Chino section of San Bernardino County in October, 1926. In 1927, walnuts from the same area were found to have larvae feeding in their green husks, and a trypetid species emerged from these nuts the following June (1928). Thus another major walnut pest had become established in California, presumably brought in by auto tourists who had gathered infested black walnuts when passing through Kansas, Oklahoma, Texas, or New Mexico.

The walnut husk fly now infests several thousand acres of walnuts in southern California. Ten years after it was known to occur in California it had spread as far east as Yucaipa and Moreno, southward as far as Elsinore, and Westward as far as El Monte.

The varieties most susceptible to attack are the Eureka, Franquette, and Payne, while the Placentia, Ehrhardt, and Neff varieties and the majority of seed-lings are much less susceptible to attack, but may suffer considerable injury in the oldest infested areas.

A comprehensive and detailed four-year study of the bionomics of the walnut husk fly was made by A. M. Boyce, beginning 1928, and the following data are ob-tained largely from his report on his investigations (100).

Description. Egg:- The somewhat curved egg, which is at first pearly white in color, becomes somewhat darker before hatching. Fine reticulations, more dense at the posterior end, cover the surface of the shell. They average about 1 mm in length and 0.22 mm in width. The tracheal system is plainly visible.

Larva:- The first-instar larvae are from 1.8 to 2.9 mm long and are nearly transparent. The second-instar larvae are 4.0 to 4.5 mm long, whitish, and semi-opaque, although the dark contents of the alimentary canal can be easily seen from the exterior. The yellowish-white third-instar larvae average about 9.0 mm long and 2 mm wide.

Pupa:- The barrel-shaped, straw-colored pupae are about 5 mm long and 3 mm wide. They resemble a grain of wheat in shape and color. The anterior spiracles project conspicuously and are dark brown.

Adult:- The adult flies are about the size of the common house fly, being 4 to 7 mm long. They vary in color from tawny to blackish, with a faint, narrow, lemon-yellow stripe extending lengthwise on the thorax and a small triangular area of

[1]Family Tephritidae according to the latest classification.

similar color on the posterior end of the thorax (the scutellum). Three dark bands extend transversely across the wings (fig. 463).

Fig. 463. Female walnut husk fly, dorsal aspect, showing characteristic position of wings. X 7. After Boyce (100).

Life History:- In early July the adults begin to emerge from the soil beneath infested trees and emergence continues until about October 1. About 10 to 15 days after they emerge from the soil they begin to lay eggs. One female may deposit from 200 to 400 eggs in her lifetime. These are deposited in groups of 15 in a small cavity made by the female in the green husk tissue. In 5 to 10 days these eggs hatch and the tiny maggots begin to tunnel through the inner husk. During the period of about a month that is required for them to pass through their 3 instars, the larvae consume a large part of the husk (fig. 464). When they become full grown they emerge from the husk and drop to the soil. They burrow down into the soil to a depth of several inches. A puparium is formed within 24 hours after the larvae enter the soil, but 36 hours after the formation of this puparium the larvae molt once more (their fourth molt) before entering the true pupal stage. There is only one generation of husk flies per year, but not all flies emerge the year after the larvae enter the soil. Only 71% of the flies emerge the year after the larvae drop to the ground and begin their pupation. Twenty seven per cent do not emerge from the soil until the second year and a few remain in the soil until the third or fourth year after the beginning of pupation.

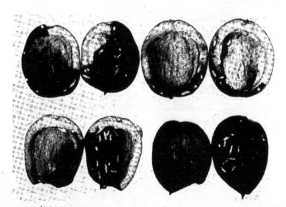

Fig. 464. Larvae of the walnut husk fly exposed in the husks. X 0.7. After Boyce (100).

Seasonal History:- The trend of the seasonal history of the walnut husk fly for the 4 years 1928-32 inclusive, can be seen from fig. 465. It can be seen that the flies began to emerge from July 5 to August 20, depending on the season, the peak of oviposition was from August 29 to September 5, and the peak of emergence of larvae from the nuts was from September 28 to October 4.

Injury:- The moist decay of the inner husk tissue, where the larvae have fed, causes a staining or blackening of the shell of the walnut, (fig. 466) probably

593

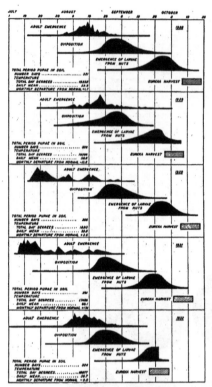

Fig. 465

Fig. 465 (left). Seasonal history of the walnut husk fly in the Chino-Pomona area (1928-1932). After Boyce (100).

Fig. 466 (lower left). Discoloration of walnut husks caused by husk fly. After Boyce (100).

Fig. 467 (lower right). Upper row, "sticktights" caused by walnut husk fly infestation; lower row, uninfested walnuts with husks split in normal manner. After Boyce (100).

Fig. 466

Fig. 467

caused by tannin released from the infested tissue. This stain cannot be removed by bleaching and results in the infested nuts being classed as culls. Usually they are cracked and the kernels are graded and marketed under various trade names, depending on their quality. Records have shown that the kernels in infested nuts suffer somewhat in appearance and quality. An additional loss results from the increased cost of harvesting the nuts due to the fact that the injured husk usually dries and adheres to the nut. (fig. 467)

Host Plants:- The walnut husk fly oviposits or attempts to oviposit on a large number of fruits and nuts, when confined with them, and a small percentage of the larvae could probably reach maturity on a few varieties of fruits. A few peaches have been found to be infested when growing in the vicinity of infested walnut trees. It is believed that the walnut husk fly is not likely to become economically important on crops other than the walnut.

Natural Enemies:- A mite, Pediculoides ventricosus New., can occasionally be found inside the egg cavity feeding on the eggs of the husk fly. These mites do not appear to become sufficiently abundant to be important as a control. Likewise true spiders, of which a number of species were identified, attack the adult flies, but are of minor importance. The nymphs and adults of the bug Triphleps insidiosus (Say) often feed on the eggs, inserting their beaks into the egg cavity. Another bug, Zelus renardi Koln., feeds on the adults, as well as the larvae of green lacewing fly, Chrysopa californica Coq. and 3 species of ants. No parasites of the husk fly were found in California.

Control:- The adults of this species are the stage most susceptible to control measures. Since for at least 10 days after emergence the females do not oviposit, the opportunity exists of poisoning them before the walnuts become infested. It was demonstrated by dissections that the adult husk flies were able to ingest particulate insecticide material. The most effective and practical poisons were found to be cryolite and barium fluosilicate. After satisfactory treatments were worked out in the early investigations, it became apparent by 1935 and 1936 that these same treatments were not resulting in as good control. Possibly the flies had undergone some change in their life history in adapting themselves to the subtropical climate of southern California, resulting in earlier emergence dates. It appeared also that the flies had increased their ability to penetrate the husks of the nuts in oviposition, and were more effectively ovipositing and developing in "resistant" varieties. (103)

Fortunately, in 1936 improvements were made in cryolite in the direction of smaller particle size. The pseudotracheae in the fleshy labella of the husk fly mouthparts are only about 6 microns in diameter, thus setting a definite limitation on the size of cryolite particles that can be ingested. This solid material may either be suspended in water from dew or fog or in the saliva emitted by the fly before feeding. Prior to 1937 only about 50% of the particles in commercial cryolite preparations were sufficiently small to pass through the husk fly mouthparts. In 1937 a preparation was available in which at least 90% of the particles were 5 microns or smaller in diameter. (104)

Boyce (104) recommended that a first treatment should be applied as a spray as soon as the flies begin to emerge from the soil. Emergence data are obtained by means of bait traps hung in the trees. The ingredients of the bait traps are as follows: casein, 7 oz. or a commercial grade of glycine, 3 oz.; commercial lye, 4 oz.; and water, 1 gallon. The spray consists of 4 lbs. of the finely divided cryolite per 100 gallons of water. Spreaders and stickers are not necessary. However, the spray should be applied thoroughly to all parts of the tree. Then a dust should be applied just prior to the peak of emergence of the flies (see fig. 465). The exact period for a given area can be obtained from the agricultural commissioners' offices. The recommended dust consists of 35% cryolite, 60% diluent and 5% petroleum oil. It was recommended that about 4 lbs. of dust should be applied per medium sized tree.

Although no entirely satisfactory treatment for the husk fly has been discovered, the use of cryolite is considered to be economically justifiable.

CHAPTER XXII
ALMOND PESTS

The almond appears to have originated in western Asia, but was probably introduced into Greece and North Africa in prehistoric times. Today it is grown in many of the warmer countries of the world, particularly in the Mediterranean basin. In North America it has been grown on a commercial scale mainly in California, where there were 111,169 acres in 1947.

The almond has a short chilling requirement, but nevertheless in southern California the yield is greatly reduced after the warmest winters because of slow and uneven opening of the buds. On the other hand, the almond blooms so early in the year that in the more temperate regions it is subject to damage from late frosts, and the weather may some years be too cool for insect activity at the period when insect pollination should take place. It may thus be considered to be a borderline crop between the subtropical and temperate climate crops and to have very exacting climatic requirements.

The principal pests of the almond are the several species of mites, although the peach tree borer, peach twig borer, and the San Jose scale are at times serious pests.

ROOT KNOT NEMATODE

Heterodera marioni (Cornu)

The root-knot nematode (see pp. 619 to 621), and possibly several other species, are of considerable importance as pests of fruit trees, including the almond, especially in light soils. The fact that so little discussion is heard about this matter is probably in large part owing to the fact that very little can be done about nematodes, once they are established in an orchard.

Older almond trees withstand a nematode infestation fairly well, but young trees are often killed. Peach root is usually not as satisfactory as almond for heavy production, but seedlings of several peach varieties, such as Shalil, Yunnan, and Bokhara, are highly resistant to nematode attack, and it might be desirable to use these resistant rootstocks in nematode-infested soils.

CLOVER MITE

Bryobia praetiosa Koch.

The clover mite, perhaps in California known more commonly as the almond mite or brown mite, is more injurious to the almond than to any other crop in this state.

Appearance:- The egg is spherical and deep red. The six legged larva has an almost globular bright red body. After the mite has fed for a short time, it changes to the brown or greenish color of the nymphal instars. The adults are the largest of the plant-infesting mites in California, being 0.75 mm long. They are rusty brown, olive green, or reddish in color with amber or orange legs. The dorsum of the body is flat. They are readily distinguished from other species of mites by their long front legs, which are as long as the body, and the 4 anterior lobes (fig. 468). The male is slightly smaller than the female and has a more pointed body posteriorly.

Life History:- Eggs are laid singly or in masses throughout the summer and fall on the slightly roughened bark of 2- or 3-year old wood, in crotches, and around the bases of and on the fruit spurs. In winter the eggs may be so abundant on the bark as to give it a bright red color. In northern regions only the eggs are found in winter, but in California, Arizona, New Mexico and the Southern States

596

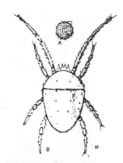

Fig. 468. Clover mite, Bryobia praetiosa. A, egg; B, adult. X 64. After Newcomer and Yothers (648).

Fig. 469. A six-spotted thrips, Scolothrips sexmaculatus, found feeding on mites on an almond tree. Photo by Roy J. Pence.

the adults may be found under the loose bark of the trees or on clover, malva, and other green host plants. The eggs hatch about the time the leaf and flower buds open. Three or four days are spent in each instar. There are 3 generations per year. Webs are not spun by this mite.

Injury:- The mites extract chlorophyll from the leaves, turning them pale yellow and causing them to fall prematurely. Sometimes the trees are almost completely defoliated by the end of August. The defoliation greatly weakens the tree and reduces the following year's crop. If the mites are controlled, the foliage may regain its green color.

Natural Enemies:- The clover mite, as well as the other mite species to be discussed later, are attacked by a large number of coccinelid beetles, syrphid fly and green lacewing fly larvae and predaceous mites, as previously discussed in connection with the citrus red mite. The six spotted thrips, Scolothrips sexmaculatus (Perg.) (fig. 469), may also be seen wherever mites are abundant on almonds.

Control:- Almond trees that are suffering from various causes are most apt to be severely attacked by the clover mite as well as other mite species. Plenty of irrigation may be all that a tree needs to ward off a serious infestation of mites, although heavy irrigation does not always insure the tree against damage.

Essig and Hoskins (288) list the control measures for the clover mite, in order of preference. In winter: (1) 4% tank mix or emulsive oil or 5% oil emulsion; (2) lime sulfur 1-10. In summer: (3) sulfur dust; (4) DN dusts; (5) tank mix or commercial oil spray at 1.5 to 2% plus a suitable spreader; (6) lime sulfur 1-50 plus 5 lbs. of wettable sulfur per 100 gallons of water.

PACIFIC MITE AND TWO-SPOTTED MITE

Tetranychus pacificus McGregor and T. bimaculatus Harvey

These two mites are similar in appearance, life history, and in the nature of their injury to the almond. Although they can be distinguished on the basis of certain structural characters, for convenience they will be considered as a single species in this discussion. In contrast with the clover mite, these species are profuse web-spinners, especially T. bimaculatus, (fig. 470). Large quantities of this webbing and its accumulation of dust, along with the whitish or yellowish discoloration of the foliage and excessive leafdrop, are the superficial characteristics indicating the presence of the above mite species. The mites are generally first present in appreciable number in June, and usually begin their attacks in the tops of the trees or the tips of the branches, spinning their web on both the upper and lower surfaces of the leaves. Their attacks do not begin as early as those of the clover mite, but the nature of the injury and its symptoms are similar to those caused by the latter species. The feeding of the mites on both the upper and lower leaf surfaces on the almond is an exception to the general rule on most fruit trees and other plants, on which the infestations are almost entirely confined to the lower leaf surfaces.

Life History:- In the almond orchard the Pacific and two-spotted mites spend the winter in the adult and immature stages on winter-growing plants or they may hibernate among leaves or in the soil. The barely visible globular eggs are found attached to the leaf or scattered throughout the webbing. The eggs are whitish as compared to the red color of the clover or European red mites. They hatch in from 3 to 6 days. Each mite may lay from 50 to over 100 eggs.

The rounded, colorless, almost transparent 6-legged larvae become greenish upon feeding a while. They do not move about much. After molting, the nymphs have 8 legs and wander about freely, especially as they approach maturity. The adults are yellowish green, usually with a large irregular dark spot or a cluster of small ones on each side of the upper surface (fig. 449). They average less than 0.5 mm long. The hibernating adults have a more reddish color. The males are somewhat smaller than the females and the abdomen narrows somewhat posteriorly.

Control:- As in the case of the clover mite, proper irrigation may in itself prevent a serious infestation by Pacific or two-spotted mites. If not, sulfur dusts may be used, in areas where they have been found to be effective. If sulfur dust is not effective, DN dust should be tried, or the trees may be sprayed with 1.5% light medium summer spray oil.

Fig. 470. Cross-section of a folded almond leaf, showing depth of webbing by two-spotted mites. Photo by Roy J. Pence.

EUROPEAN RED MITE

Paratetranychus pilosus (Canestrini and Fanzago)

This species is one of the important deciduous fruit pests in Europe and in the United States and Canada. It is so similar to the citrus red mite (p. 369) in appearance that the two were long considered to be the same species. McGregor (552) described the citrus red mite as a new species, Paratetranychus citri. It had previously been known as Tetranychus mytilaspidus Riley. In 1919 he described the anatomical differences between P. pilosus and P. citri (553), but since these differences could only be seen with the aid of oil-immersion lenses "supplemented by keen eyesight," it is understandable that not all entomologists accepted his diagnosis. In 1928, however, McGregor and Newcomer (566) published a paper which proved conclusively that the deciduous and citrus forms were distinct species. It was known that unfertilized females of either the deciduous or the citrus forms gave birth only to males. When citrus-form males were crossed with deciduous-form females, or vice versa, mating took place, but only male progeny resulted. This failure to interbreed was taken as further evidence that the two forms were distinct species. Neither form thrived on the hosts of the other. Certain color differences were found which did not change upon exchange of hosts. The eggs of each species have a stalk, but the citrus red mite attaches many guy fibrils to the top of the stalk, and these radiate to the substrate of the egg. The deciduous mite does not do this. Finally, it was shown that at the time of the investigation the habitats of the two species did not overlap. That this is not the case at present is evidenced by the fact that citrus red mite is a pest of citrus and the European red mite is a pest of apples (if the trees are treated with DDT) in the same orchard on the Los Angeles campus of the University of California at approximately 34° north latitude.

Description:- The description given for the citrus red mite (p. 369) will serve for the present species except that while the former is of a purplish red color, including the dorsal tubercles which give rise to the conspicuous white spines, the color of the European red mite may be described as being more or a brownish red and the dorsal tubercles are whitish. The male of the citrus red mite is dark red while the male of the European red mite is at first greenish, changing to brick-red or orange with age. The length of the adult European red mite is stated to be 0.32 to 0.37 mm and the length of the male 0.27 to 0.30 mm. (553)

ALMOND PESTS

Life History:- The winter is passed in the egg stage. The eggs may be found on the smaller branches and twigs of the trees, particularly on the lower sides or in the forks of the two limbs or other roughened places. The method of hatching and the appearance of the larval and nymphal stages are similar to those of the citrus red mite and need not be discussed here. In (fig. 471) is shown the egg, newly hatched larva, protonymph, deutonymph, and adult male and female.

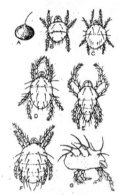

Fig. 471. European red mite. Paratetranychus pilosus. A, egg; B, larva; C, protonymph; D, deutonymph; E, adult male; F, adult female; G, adult female, lateral view. After Newcomer and Yothers (648).

Injury:- Similar to that of other mites infesting almond trees. This mite is not apt to be as abundant on almond trees as the previously discussed species.

Control:- As for clover mite. Since both the clover mite and the European red mite overwinter in the egg stage and the eggs are in exposed locations, dormant sprays, oil or lime sulfur, can be effectively used. In summer, sulfur dust, DN dusts or summer oil sprays are recommended.

Mite-Resistant Almond Trees:- The development of mite-resistant almond trees offers considerable promise as a means of combating these important pests. Several hybrid trees resistant to mites have already been produced in breeding experiments conducted cooperatively by the U. S. Department of Agriculture and the University of California.

PEACH TWIG BORER (50)

Anarsia lineatella Zeller

This is an Old World lepidopterous insect with a long history in this country. It is a pest of the peach, almond, apricot, nectarine, plum, and prune. In California, where the peach twig borer was first discovered about 60 years ago, it is found in greatest abundance in the upper San Joaquin and Sacramento Valleys.

Appearance. Egg : Pearly white, turning orange; reticulated; oval. Larva: when mature, averages 10 mm in length, and is reddish brown, with a black head. In sharp contrast to the general reddish brown color of the mature larva is the yellowish-white of the intersegmental membranes. Chrysalis: naked, smooth, dark brown, 6 mm long. Adult moth: about 8 mm long; steel gray, with white and dark scales; narrow, fringed wings with an expanse of 12 to 15 mm, held roof-like over the body when at rest. The palpi resemble horns (fig. 472). The males are somewhat smaller than the females.

Fig. 472. Adult peach twig borer, Anarsia lineatella. X 3. After Bailey (50).

Life History:- From September 1 to March 15 the larvae may be found in hibernation chambers or hibernaculae, the presence of which is indicated by small, reddish brown tubes or "chimneys" which extend upwards from the surface of the bark (fig. 473). These consist of bark fragments fastened together by silk, and are constructed by the larvae on the upper surfaces in the crotches of the younger branches. The chimney indicates the location of the silk-lined burrow beneath the bark. It is 2 or 3 times as long as the larvae within. With the approach of warm weather in the spring, the larvae burrow their way out of the cells. Since almond trees start growth earlier than the other trees such as the peach or plum, the hibernating twig borer larvae also become active earlier on this host. It has been noted that the clover mite likewise becomes active sooner on almond than on other trees.

Fig. 473 (left). The peach twig borer. A, chimneys of hibernaculae in the crotch of a peach tree; B, cross section of hibernacula, showing caterpillar. Fr. E. O. Essig: INSECTS OF WESTERN NORTH AMERICA. Copyright, 1926 by The Macmillan Company & used with their permission.

Fig. 474 (right). Wilting of the tips of almond shoots caused by peach twig borer larvae. After Bailey (50)

After leaving their hibernaculae, the larvae crawl to the base of a bud and feed, or they may crawl between the expanding leaves and eat their way down into the tips of the twigs, causing the characteristic wilting (fig. 474). Upon reaching the harder wood lower down in the twig, the larva bores its way out, leaving its exit hole on the side of the twig.

Bailey gives the duration of the various stages as follows: egg, 5-7 days; larva, 10-20 days; pupa, 6-7 days; adult longevity, 10 days (in the laboratory).

There are 4 generations per year. The second and third generation larvae feed on twigs or nuts. The fourth generation larvae do not feed, but construct the hibernaculae in which they will hibernate throughout the winter.

Pupation takes place in old pruning wounds, curled bark, in rough crevices on the main trunk of the tree, or in curled leaves. A silken web is spun over the place of concealment. Pupation requires from 7 to 20 days, depending on the season.

The first generation moths emerge between May 1 and May 15, the second generation between August 1 and August 10, and the third about the first week in September. The first brood of moths lay their eggs on the twigs, the second and third on the twigs or fruits, and the fourth appear to oviposit near the crotches on the young branches.

Injury:- The damage caused by twig borers consists of injury to buds and young twigs (fig. 474), and the injury to the nuts (fig. 475). While the shells of the almond are soft, the larvae can penetrate them and cause grooves in the kernels by their feeding. After the shells are hard, the larvae may bore only into the hulls. The larvae can damage a considerable percentage of nuts, but even if they damage only a small percentage, the nuts become unmarketable. (1011)

Fig. 475. Twig borer injury to green almonds early in June. X 0.5. After Bailey (50).

Natural Enemies:- A small hymenopterous larva, Hyperteles lividus Ashm. was observed one winter to have killed from 70 to 95% of the immature larvae in their hibernaculae in many orchards (245). After devouring all but the head capsule of the larva, the parasite bores its way out through a small hole at the base of the "chimney." Another important natural enemy is the grain or itch mite, Pediculoides ventricosus (Newport), which was observed to have destroyed as high as 95% of the overwintering larvae (586). Thirty other insect enemies of the peach twig borer, mainly hymenopterous parasites and all of minor importance, have been listed (50).

Control:- Dormant lime sulfur 1-10 applied under high pressure in the spring as the buds are opening has long been a standard control practice. However, a more satisfactory treatment consists of 3 lbs. of basic lead arsenate and 1/4 to 1/3 lb. of spreader to 100 gallons of water. To avoid the poisoning of bees, the lead arsenate spray should be applied after about two thirds of the blossoms have fallen. In cases of severe infestation, an additional arsenical spray may be necessary in May, when the first wilted shoots appear. The prunings and almond hulls should be burned so as to kill the overwintering larvae. (1011)

On almonds, 2 or 3% of an oil emulsion of 80 viscosity and 80 U.R. with 1-1/2 to 2 quarts to 100 gallons of 30% sodium dinitro-o-cresylate or 1 lb. to 100 of 20 to 30% wettable dinitro powder, can be used as a dormant spray to reduce the number of overwintering larvae. This combination should not be used on peaches because of adverse tree reaction. Wettable DDT powder appears to be somewhat superior to lead arsenate, but its use will probably increase the mite populations. Only future experience can determine whether improved control of peach twig borer by DDT will be sufficient to overcome the disadvantages attending the use of this material. (50)

PACIFIC PEACH TREE BORER

Sanninoidea exitiosa graefi (Hy. Edwards)

This is a serious pest in many parts of California. The damage is caused by the white larvae, about 1.5 inches long, which burrow under the bark near the surface of the soil, but may extend their burrows down to the bases of the main roots. If the larvae are not killed, the infested tree may die.

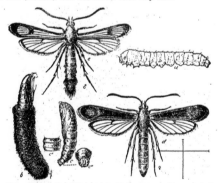

Fig. 476. Pacific peach tree borer, Sanninoidea exitiosa graefi. a, larva; b, cocoon; c, pupa; d, adult female; e, adult male. After Moulton, U.S.D.A.

Description and Life History:- The adult moths are day fliers. They are clearwinged moths, resembling wasps (fig. 476). The body color is black or steel blue with the body appendages and fringes of the wings black. The females have a wingspread of 30-34 mm and the males 25-30 mm. They lay their flat, oval eggs singly, scattered about on the trunks of the trees. Upon hatching from the eggs, the young larvae seek the crevices of the bark, feed a little, but migrate downward. After they are about a half inch long, they cease migrating and spend the remainder of their larval life in one burrow, usually below the soil surface. Pupation takes place in a case near the surface or at the surface of the burrow. The moths fly and lay eggs all summer and larvae of all sizes may be found at any period of the year even though there is but one generation per year (1015).

Control:- The larvae may be dug out of their burrows with a knife or they may be killed in their burrows with a wire probe. This work is usually done in the fall and spring. The wounds are covered with a good asphalt paint.

The Pacific peach tree borer can be effectively combatted with paradichloro-benzene (PDB). In California this is usually applied between the first of May and November. The soil temperature should be above 55° F and the soil moisture should not be excessive. The soil surface about the base of the tree is leveled, then from 3/4 to 1 ounce of the crystals is sprinkled around the tree in a continuous ring 2 inches wide, with the inner margin of the ring 2 to 4 inches from the bark. Then 2 to 4 inches of soil is placed over the PDB and the soil is packed well by tamping it with the shovel. (283)

RED HUMPED CATERPILLAR

Schizura concinna (Abbot & Smith)

These striking black-and-yellow-striped caterpillars with red head and red humps are not found as often on almonds as on some other trees, but they sometimes cause damage in late summer and early fall. They feed voraciously, and may defoliate a tree in a few days if not controlled. Entire colonies may be pruned out of the tree and destroyed, or they may be treated with lead arsenate spray or dust.

INDIAN MEAL MOTH

Plodia interpunctella (Hübner)

As far as known, the Indian meal moth does not attack almonds in the orchards, but feeds on the kernels of stored nuts. Damage is done by the larvae, which are pale in color and about 1/2 inch long. The fore wings of the adult moth have a pale basal portion and dark outer portion and have an expanse of 5/8 inch (fig. 503). These moths have been controlled by heat or fumigation treatments.

CHAPTER XXIII
PECAN PESTS

The pecan is stated to be perhaps the most important orchard species native to the United States (153). It is native to the majority of the Southern States and as far north as Kansas and Illinois. A large portion of the crop is still harvested from wild trees, or wild trees top-worked to the best varieties. Pecans have been extensively planted, however, in all the Gulf States, but especially in Georgia and Florida. Pecans are grown in California, but not on a large commercial scale; there were only 943 acres in 1947.

As the pecan industry developed in the southern United States, the insects and diseases became increasingly important. Among the insects, the most serious are the pecan nut case bearer, the hickory shuck worm, and the black pecan aphid. An extensive discussion of the biology and control of the pecan insects has been provided by Moznette et al (627) and the following is based largely on their paper, except that in some instances the results of recent experiments with the newer insecticides have been added.

A. INSECTS INJURING THE NUTS

SOUTHERN GREEN STINKBUG AND OTHER PLANT BUGS

Black pit and kernal spot of the pecan, caused by certain sucking insects, occur sporadically in restricted sections and frequently become serious to an individual grower. These maladies are caused by the feeding of the southern green stinkbug, Nezara viridula (L.) (fig. 477), and the leaf-footed bug, the latter being similar to the Western leaf-footed plant bug (fig. 292) previously discussed as a pest of citrus in California.

Fig. 477 (left). Southern green stinkbug infesting pecan nuts. After Moznette et al (627).

Fig. 478 (right). Spots on pecan kernals caused by the southern green stinkbug. After Moznette et al (627).

Black pit is a blackening of the inside of the pecan. Nuts with black pit drop prematurely. Sometimes brown or black spots or stains show up on the shuck, and after the nuts drop these soon spread over the entire surface. Punctures made by the bugs after the nuts have passed the "water stage" (about July 1) do not cause black pit, but instead the condition known as kernal spot, which consists of dark brown or black spots on the kernals (fig. 478). Kernal spot can not be detected until the shells have been removed.

SUBTROPICAL ENTOMOLOGY

Biology:- The southern green stinkbug is usually light green although it may be darker, with a pinkish or purplish tinge in winter. The adult has the shield-shaped form characteristic of the family Pentatomidae, and is about a half an inch long. The adults hibernate in winter, then lay clusters of 36 to 116 eggs in the under side of foliage from early April to the mid-November. There may be as many as 4 generations per year occurring on a wide variety of plants and on fruit trees. Only the adult bugs feed on pecan nuts. The seasonal history of the leaf-footed bug is similar to that of the stinkbug. It breeds on thistle, yucca, basketflower, jimsonweed, and cowpea.

Control:- Insecticides, hand picking, and trap crops have been found to be impractical in pecan orchards in controlling the bugs that cause black pit and kernal spot. Growers depend chiefly on the orchard sanitation. The soybean and certain species of Crotalaria serve as hosts for the injurious bugs and should not be used as covercrop or intercrop. The velvet bean can be safely used. If no covercrop or intercrop is planted, cultivation to keep down weeds during summer and fall is advisable. Sanitation in winter will reduce the number of hibernating adults.

SPITTLE BUG

Clastoptera obtusa (Say)

The presence of these bugs is indicated by a spittle-like, frothy substance on small pecan nuts and about the tender growing shoots during spring and early summer. The froth is produced by the nymphs of the spittlebugs and serves to protect them while they are engaged in sucking the sap from the tender buds or nuts.

Control:- This species does not cause serious injury, and it may be easily controlled with nicotine while the nymphs are under the "spittle" in early summer. If the trees are sprayed for pecan scab in the middle of May, nicotine sulfate at 1-1000, combined with the Bordeaux, will control spittlebugs.

PECAN NUT CASEBEARER

Acrobasis caryae (Grote)

In Texas the pecan nut casebearer often destroys from 1/3 to 3/4 of the pecan crop, while in Louisiana, northern Florida, Georgia, Alabama and Mississippi it annually damages from 5 to 40% of the crop. It is one of the most important insects attacking the pecan.

Description:- The egg is oval; 1/50 inch long; greenish white later changing to reddish. The larva (fig. 479, above) is about 1/2 inch long; general color a dirty olive green. The pupa is about 1/3 inch long; first with a decided olive-green cast, later brown and without conspicuous markings. The adult has approximately 5/8 inch of wing spread; color inconspicuous dark-gray with a ridge or tuft of long, dark scales extending across each fore wing near the middle (fig. 479, below).

Life History:- After passing the winter as partly grown larvae in small, tightly woven cases similar to those of the leaf casebearer the pecan nut casebearers feed for a short time on the buds to which the cocoons are attached. Later these same larvae bore into the young tender shoots and after feeding 2 weeks or more, they pupate within the tunneled shoots. The pupal period lasts 11 to 18 days. In southern Georgia the moths appear between May 7 and 24, about the time the nuts are setting. In from 2 to 4 days after emergence, oviposition begins, the eggs being placed singly on the outer end of the nut. Only 1 or 2 eggs are laid in a cluster of nuts. Three or 4 generations appear annually.

Injury:- The first-generation larvae appear in May and early June and bore into the recently set nuts. Characteristic masses of frass, held together by fine silken threads, form a silk-lined tube, possibly protecting the larvae while they are entering the nuts and in addition holding the nuts together and keeping them from falling before the larvae are mature (fig. 480). From 2 to 5 nuts may be destroyed by a single larva in the period of less than a month required for its development.

The second-generation larvae attack the nuts in the same way but cause less damage, since the nuts are larger and not so many are required to sustain the larvae. Still later generations cause little damage. For the most part they gnaw only through the shucks, which seems to not interfere with the development of the nuts.

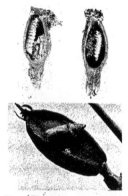

Control:- Good control of the pecan nut casebearer has been obtained by spraying twice with 13 fluid ounces of nicotine sulfate combined with 3 quarts of mineral-oil emulsion, or with 1 quart of fish oil, per 100 gallons of water. Often a single application has sufficed. Under the semiarid conditions in Texas, lead arsenate at 6 lbs. per 100 gallons in 2 applications has been found to be effective. Under humid conditions, however, lead arsenate is not safe. Timing is important so as to preclude damage from the first generation of larvae.

Fig. 479. Above, pecan nut casebearer larvae in small nuts; below, adult. X 2. After Moznette et al (627).

The pecan nut casebearer may be effectively and economically controlled with 1 application of DDT early in the season at the time the young nuts are beginning to turn brown at the tips. Four years of experimental work has led to the recommendation of 2 lbs. of 50% DDT wettable powder per 100 gallons water or Bordeaux mixture for pecan nut casebearer in the Southeastern States. A 1% summer oil emulsion spray or some other recommended material should be applied if a leaf scorch caused by red spiders appears in DDT-sprayed orchards. A second application 7 to 9 days after the first may be required when the red spider infestation is heavy. An adhesive or sticker does not increase the effectiveness of the DDT spray (692).

Fig. 480. Characteristic masses of frass on pecans infested with pecan nut casebearer. After Moznette et al (627).

HICKORY SHUCKWORM

Laspeyresia caryana (Fitch)

The hickory shuckworm is one of the most important pests of pecans and often causes serious damage wherever pecans are grown in the Southeastern States.

Description:- The name "shuckworm" refers to the white larvae, about 10 mm long (fig. 483, above). The inconspicuous adults are rarely noticed. They have a wingspread of about 15 mm, are smoky black in general coloration, but with some short streaks across the front margin of the fore wings.

Fig. 483. Above, hickory shuckworm larva in shuck of a nearly mature pecan nut, X 1; below, pupal skins protruding from shucks of pecan nut. After Moznette et al (627).

Life History:- In northern Florida and southern Georgia the first moths, arising from overwintering larvae in fallen pecan or hickory shucks, begin to appear about February 15, with maximum emergence usually during April. Not many pecans are infested before the middle of June because the majority of the spring brood of moths die before the nuts have set. The small, whitish, flattened eggs are laid on the shuck of the nut. The tiny larvae enter the green nuts, working more in the interior of the nuts than in the shucks, until the shells harden. The larva prepares a small silk-lined cocoon and cuts a circular hole in the shuck, to permit the adult to escape, before pupating. Prior to the emergence of the moth, the pupal skin is thrust a short distance through the circular cut (fig. 483, below).

In the northern pecan-growing districts the shuckworm has 1 or 2 generations per year, but farther south it may have as many as 5. The shuckworm populations may become very high in the last generations.

Control:- Sanitation, such as disposal of the pecan shucks containing the overwintering larvae, and plowing these shucks under in the early spring may aid in control. No satisfactory artificial control measures are reported although experiments are being made with new materials.

PECAN WEEVIL

Curculio caryae (Horn)

The pecan weevils or hickory-nut weevils, are widely distributed, but cause greatest damage to pecan in the Piedmont of North Carolina, South Carolina, Georgia and Alabama, and in parts of Mississippi, Louisiana, and Texas. The greatest losses are suffered by the Stuart, Schley, and Rome varieties. Other varieties, which form their kernals later, are not severely attacked. It is reported that pecan weevil damage is on the increase, and the experience of 1946 has shown that during a year of poor crops, the pecan weevil may destroy practically all the crop (626).

Description of Adult:- The adult weevil (fig. 481) is about 10 mm long, exclusive of its long beak, which is slightly longer than the body. The male is about the same size as the female, but its beak is only about half as long. The adult weevil is light brown or grayish. If the scales are rubbed off, the color becomes reddish brown.

Seasonal History:- As soon as the kernals develop the females drill holes through the shells of either hickory or pecan nuts and deposit 2 to 4 or more eggs in separate pockets, but through the same puncture. Full grown grubs can be found in about a month (fig. 482), emerging from the nuts late in September or early in

 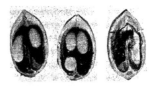

Fig. 481 (left). Pecan weevils: Left, male; right, female. X 2.0.
After Moznette et al (627).

Fig. 482 (right). Full grown larvae (grubs) of the pecan weevil.
After Moznette et al (627).

October. These grubs enter the soil and continue to do so until December or January. Cells are formed 1 to 10 inches below the surface, and pupation takes place in these cells. The adult weevil forces its way up through the ground the following summer. The heavy broods appear in alternate years.

Injury:- Injured nuts are often stained by juice coming from punctures made by the adults while the nuts are in the "water stage", which is in midsummer. Dark patches develop on the surface of the nuts and in a few days they drop from the trees. They are discolored and empty. The injury is similar to "black pit." The greater damage, however, is caused by the destruction of nuts in the fall by the grubs. The kernal is destroyed to such an extent that only a black powdered refuse remains. The emerging grubs make holes about 1/8 inch in diameter. Infestations are also characterized by the failure of the shucks to separate from the shell at the time normal nuts are opening.

Control:- Before oviposition begins, in August, the majority of weevils are in the lower limbs, and these can be conveniently jarred to remove the beetles. Later the adult weevils are found throughout the trees, and a correspondingly greater part of the tree must be jarred. A pole, padded with a piece of automobile tire or similar material is used for the jarring. The beetles fall onto large sheets, such as the regular harvesting sheets, and are then destroyed by crushing or placing in cans of kerosene. The jarring may be done during any time of the day and six jarrings, starting in late July or early August should suffice for good control of the weevils.

Two applications of lead arsenate spray in September were moderately effective in protecting pecan nuts from injury by the pecan weevil. Under certain conditions this treatment may result in a large increase of the aphid (Monellia sp.), but will cause a large reduction in the population of overwintering pecan nut casebearers (656). The relative ineffectiveness of stomach poisons is due to the fact that the weevils do not feed very much on the outer surface of the developing nuts. It has been demonstrated that 2 applications of DDT wettable powder at 6 pounds to 100 gallons can reduce the pecan weevil infestation to 1% in severely infested orchards. The last Bordeaux application is delayed to July 30, when it is combined with the first of the 2 DDT applications. The second application is made early in August. The hope is expressed that DDT applied for several seasons may eliminate a pecan weevil infestation from an orchard or at least reduce infestations to the point where spraying every year will not be necessary (626).

Susceptible varieties of pecan trees will be more heavily infested with pecan weevils, as well as certain other pecan pests, when growing near hickory trees, which also harbor these same pests.

SUBTROPICAL ENTOMOLOGY

B. INSECTS INJURING THE FOLIAGE AND SHOOTS

PECAN PHYLLOXERA

Phylloxera devastatrix Pergande

This and other species of phylloxera cause tumor-like swellings or galls to appear, occasionally in large numbers, on the leaves, leafstalks, nuts, and succulent shoots of the pecans (fig. 484). These may be found in all sections, but are most prevalent in Louisiana, Mississippi and Arkansas. The most susceptible varieties are Schley, Stuart, and Success.

Fig. 484. Galls of phylloxera on pecan twig.
After Moznette et al (627).

Seasonal History:- In the spring the young insects may be found on the unfolding buds, soon to be enveloped by galls. The latter are green or yellowish green, occasionally tinged with red. The mature galls are brown or black. After mature phylloxera lay eggs within the gall, they give rise to winged forms which are released when the gall splits open, usually by the last of May or first of June. The winter is passed in the egg stage in protected places on the branches.

Injury:- The infested twigs become malformed and weak, and may die; defoliation takes place, and nuts may fail to fill out properly or may drop. Entire limbs may be killed in severe infestations.

Control:- About 3/4 pint of nicotine sulfate to 4 lbs. of potash-fish oil soap, 2 qts. of lubricating-oil emulsion, or 2-1/2 gallons of lime-sulfur to 100 gallons of water is applied when the swelling buds show about a half inch of green. This spray mixture has given good control without injury to pecan trees.

BLACK PECAN APHID

Melanocallis caryaefoliae (Davis)

In many districts the black pecan aphid is one of the most destructive defoliators of pecan trees. The varieties suffering the most injury are the Schley, Alley, Van Denan, and Stuart, while the Curtis, Moore, and Moneymaker are usually not appreciably injured. Repeated applications of Bordeaux spray tend to increase the aphid infestations.

Life History:- The tiny, black, overwintering eggs hatch in the spring. The young aphids are pale green, but become darker with age. They feed on the opening buds and leaflets until they become adults, all of which are winged females. These

608

are about 1/16 inch long, with dark green bodies with a series of large black humps on the back and sides. This species reproduces parthenogenetically until fall. The aphids are scattered about -- not grouped together in colonies as aphids usually are. Both sides of the leaves are attacked, and hardened leaves in the inner branches are preferred. Each female gives birth to 75-100 eggs, and since there are about 15 generations per year, enormous aphid populations can be found by late summer.

Injury:- The first signs of black aphid infestation are the bright yellow spots on the leaves, which eventually become necrotic and turn brown in color. Many spots will cause the leaf to drop prematurely. Defoliation is apt to be first noted in the central portion of the tree. Although the symptoms of black aphid injury may be seen as early as May, extensive defoliation usually does not occur before August or September. Black aphid infestations prevent the proper filling of the nuts and reduce production the following year.

Control:- Nicotine sulfate (40% nicotine), at 1-400 with either Bordeaux or with summer oil emulsion containing 0.5% oil, is applied in June. Orchards regularly sprayed with Bordeaux should be watched especially carefully for aphids, for they are the most susceptible to infestation. The spray must be thoroughly applied to all parts of the tree to yield satisfactory control.

PECAN LEAF CASEBEARER

Acrobasis juglandis (Le Baron)

The pecan leaf casebearer is a serious pest of pecan only in northern Florida, the southern parts of Georgia, Alabama, Mississippi, Louisiana, and in Texas, but has been reported from other localities both on pecan and hickory.

Description:- The caterpillars are brown when young, but change to dark green. When full grown, they are a little over 1/2 inch long. The pupae are a deep, shiny mahogany brown. The moths have a white or dusky gray body, and the wings are a mixture of gray and brown in color and have an expanse of 2/3 inch.

Life History:- The majority of the caterpillars mature in May or early June. They pupate within their pupal "cases" (fig. 485, left). The moths appear from about May 15 to the first week in August. They deposit the eggs on the under side of leaves. They hatch a little after the middle of May until early in August. The larvae feed very sparingly on the under side of the leaves in the summer and early fall while building little winding cases which serve as protection. Shortly before the leaves drop in the fall the larvae leave their summer quarters and become

Fig. 485 (left). Left, larva of pecan leaf casebearer and its case; right, winter hibernaculae around a pecan bud. After Moznette (627).

Fig. 486 (right). Injury to pecan leaves from larvae of pecan leaf casebearer. After Moznette et al (627).

609

established around the buds in small oval cases called hibernaculae (fig. 485, right). There is but one generation per year.

Injury:- The greatest injury is caused early in the spring by the larvae which have emerged from their hibernaculae and feed on the unfolding buds and leaves (fig. 486). Sometimes these larvae keep pecan trees defoliated for weeks at this period of the year. When abundant, they may also feed on the pistillate blossoms and greatly reduce yield.

Control:- One thorough application of calcium arsenate early in July, has resulted in control of the pecan leaf casebearer according to investigations made in Georgia. The calcium arsenate should always be used with Bordeaux to prevent injury to the foliage or nuts. A 2-1-100 Bordeaux formula is sufficient as a safener, but should be increased to 6-2-100 if a fungicide is needed for scab-control.

FALL WEBWORM

Hyphantria cunea (Drury)

Late in the summer and in the fall, the fall webworm is perhaps the most commonly noticed insect in the pecan orchards. Trees which the larvae defoliate have nuts that are not well filled and yield little or no crop the following year.

Fig. 487. Larvae of fall webworm feeding within their webs. After Moznette et al (627).

Life History:- The larvae construct webs over the twigs and foliage within which they feed (fig. 487). When fully grown, the larvae are a little over an inch long and are covered with long white and black hairs. Fully grown larvae leave their webs and pupate in flimsy, hairy cocoons beneath rubbish on the ground under the loose flakes of bark, or just below the surface of loose soil. The first generation moths may appear by the latter part of June, but the majority not until later, in the summer. The moths are usually pure white (fig. 488), but may have a few black spots on both sides of the wings. The abdomen is often yellow with black spots. The second generation larvae feed during the summer and fall and give rise to the overwintering pupae.

Control:- Calcium arsenate or lead arsenate at 2 lbs. - 100 applied when the caterpillars are small will result in control. Bordeaux mixture should be added as a safener. A more practical measure, if the insects are not abundant, is to remove the webs with a long-handled tree pruner or a long bamboo pole with a hook.

WALNUT CATERPILLAR

Datana integerrima (Grote & Robinson)

Colonies of the walnut caterpillar (fig. 489) often defoliate pecan trees in the South. Small trees, and especially nursery stock, are sometimes completely defoliated, and large branches on older trees may be stripped of foliage.

Life History:- Winter is spent in the pupal stage. From April 15 to July 15 the moths emerge and lay their eggs (fig. 490) in masses of as many as several hundred on the underside of the leaflets. The eggs hatch in a little less than a week into brownish striped caterpillars. These feed at first only on the underside of the leaflets, but later devour all but the midribs and petioles. The larger

Fig. 488. Moth and egg mass of the fall webworm. X 1.5. After Moznette et al (627).

Fig. 489. Colony of walnut caterpillars on pecan twig. After Moznette et al (627).

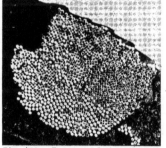

Fig. 490. Egg mass of the walnut caterpillar. X 2.0. After Moznette et al (627).

caterpillars, when preparing for molting, leave their feeding places and crawl to the trunk or larger limbs and shed their skins. These may adhere to the bark for weeks. After molting, the larvae ascend the tree and continue feeding. After feeding 25 days or more the larvae are full grown. They then crawl down and enter the soil. In a day or two they transform to brown pupae at a depth of a few inches. In the South there are two generations, the second feeding late in the summer and in the fall.

Control:- As for fall webworm. While molting on the tree trunks, the larvae may be destroyed by crushing or other means.

PECAN CIGAR CASEBEARER

Coleophora caryaefoliella (Clemens)

This pest, although usually of minor importance, sometimes seriously damages buds and foliage during the spring. It also feeds on the hickory and black walnut.

SUBTROPICAL ENTOMOLOGY

It may be found from Florida to the western border of Texas and as far north as New Hampshire, but does its greatest damage along the Gulf Coast.

<u>Life History</u>:- The partly grown larva hibernates in a case attached to the trunk or a large limb. These larvae feed on the buds as they open and later attack the foliage until about the middle of May, when they are full grown. There are several generations a year.

<u>Control</u>:- Calcium arsenate at 2 lbs. per 100 gallons, with Bordeaux as a safener, is the recommended treatment.

PECAN BUD MOTH
Gretchena bolliana (Slingerland)

This bud feeder causes a stunted growth as well as excessive branching in pecan nurseries or on young orchard trees.

<u>Life History</u>:- The hibernating adults are gray moths with blackish-brown patches and streaks and a wing expanse of a little over a half inch. As soon as the buds open, oviposition begins, first on the branches near the buds, but later on the leaves. After feeding about 25 days, the larvae pupate, pupation taking place in rolled-up leaves or infested buds, or occasionally under loose flakes of bark or the crown of the tree. There are several generations per year.

<u>Control</u>:- If the trees are kept vigorous in the nursery, the terminal buds unfold so rapidly that the larvae do not have time to cause serious damage. Calcium arsenate, with Bordeaux as a safener, will aid in keeping this pest in check. Four or 5 applications must be made at intervals of 3 or 4 weeks. This is no greater number of applications than is sometimes required for the control of nursery blight, when the Bordeaux must be used anyway.

PECAN CATOCALA
Catocala viduata (Guenée)

In the spring, pecan foliage is sometimes consumed by gray caterpillars measuring from 2.5 to 3 inches in length. When abundant, these caterpillars can strip the foliage, leaving only stems and petioles. The moths deposit eggs on the underside of flakes of bark in the fall and the winter is spent in the egg stage. The caterpillars feed throughout the spring and early summer and pupate in flimsy cocoons. The moths emerge from late June until fall.

<u>Control</u>:- When control measures are required, arsenical sprays applied while the caterpillars are still small are effective. These may in some cases be combined with Bordeaux mixture applied for scab control.

GREEN LEAF WORM
Isodyctium sp.

In some parts of Louisiana, small, green spiny larvae may be found at about the time the pistillate bloom appears. Their feeding causes a "shot hole" appearance of the foliage, the holes becoming progressively larger as the larvae develop.

<u>Life History</u>:- The adults, which are sawflies (Hymenoptera) are about 1/5 inch long. After emerging from the ground in April, they deposit small, pale-green eggs on tender leaflets. The newly hatched larvae are about 1/10 inch long but attain a length of 5/8 inch when full grown. They usually feed on the underside of leaves, but when resting, they are found along the outer margin of the leaves. The winter is spent by the larvae in earthern cocoons in the soil, 1 to 3 inches under the surface.

<u>Control</u>:- Spraying with arsenicals, as for other leaf-feeding insects, at the time of the first appearance of the pistillate bloom.

JUNE BEETLES
Phyllophaga spp.

The June beetles often seriously defoliate pecan trees in the spring, especially in the case of small trees growing adjacent to uncultivated land. As is the usual habit of June beetles, these insects feed at night and spend the day just beneath the surface of the ground. The brown, robust adults are 1/2 to 3/4 inch long.

612

They lay their eggs in the ground and the larvae, known as "white grubs," feed on the roots of grasses and other plants and require 2 years or more for their development.

Control:- On small trees beetles have been handpicked or shaken onto sheets on the ground at night and then destroyed. Lead arsenate applications have also been used in control.

The June beetles are seldom abundant in well cultivated orchards unless these are near uncultivated fields. The latter are not an important source of infestation, however, if they are plowed as often as once a year.

C. INSECTS INJURING THE TRUNK AND BRANCHES

TERMITES

Injury to the roots caused by the feeding of termites is sometimes sufficiently severe to kill nursery stock or small trees (fig. 491). Injury is most apt to occur if pecan trees are planted on recently cleared land containing stumps and dead roots, for termites usually live in dead wood and only incidentally feed on live tissue. Termites must live and feed in complete darkness. They attack only parts of trees which are underground or an inch or two above ground inside the trunk. By the time the trees show signs of injury it is usually too late to save them. Sometimes only the outer shell of the taproot remains by the time the tree shows outward signs of its mishap.

Control:- The stump and dead roots must be removed from recently cleared land before it is used for planting the nursery or orchard.

Fig. 491. Termite injury to roots of pecan nursery stock. After Moznette et al (627).

OBSCURE SCALE

Chrysomphalus obscurus (Comstock)

The obscure scale is the only scale insect of widespread importance among several species that attack the pecan. It also attacks oak, hickory, and certain other trees, and appears to be more important in bottomland than in hill-land orchards. Infestations tend to begin on the lower, inner portions of the tree and show a rather spotty distribution. The killing of branches 3 inches or less in diameter constitutes the main injury from this pest. Trees damaged by the obscure scale have fewer fruiting branches and twigs, and in their weakened condition are more subject to attack by borers and other insects.

Description:- These are armored scales (see fig. 451) whose grayish coloring closely simulates that of the bark on which they are found. The scale covering is about 1/8 inch long and 1/10 inch wide and tapers slightly posteriorly. The armor of the male is oval and only about half as large as that of the female.

Life History:- There is only one generation a year. The purplish eggs hatch in a few days after they are layed. The crawlers may settle under the armor of their parent or some old scale or an exposed limb or twig surface. Large numbers are seen from the middle or latter part of May to early August.

Control:- Lubricating-oil emulsion during the dormant period is an effective treatment, but may injure pecan trees, especially those low in vigor, even in concentrations as low as 3%. Therefore only 2% lubricating oil emulsion is recommended for trees in a weakened condition and 3% is recommended for the treatment of healthy trees.

PECAN CARPENTER WORM

Cossula magnifica (Strecker)

This lepidoperous larva mines in the sound wood of pecan, hickory, and oak, but fortunately it is rarely abundant. A pile of coarse reddish pellets may be seen at the bases of infested trees. The exit holes of the larvae in the limbs are circular and about the size of a lead pencil.

The moths lay eggs on small pecan branches in summer. The larvae mine successively larger branches until in the fall the half-grown larvae bore into the trunk. The full-grown larvae are about 1.25 inches long, have a thick pink body, and shiny head. They pupate late in the spring and the adults appear in May or June.

Control consists of injecting a medicine dropper full of carbon disulfide into the burrow and closing the hole at once.

COLEOPTEROUS BORERS

In general the same types of borers that attack the walnut also attack the pecan. Proper care of the trees and orchard sanitation are recommended for flat-headed apple tree borers, Chrysobothris femorata Olivier. Borers may be removed with a knife, the wound being subsequently painted with a prepared pruning compound or a mixture of 1 part of creosote and 3 parts of coal tar.

The red-shouldered shot-hole borer, Xylobiops basilare (Say), attacks only injured or dying trees, and infestations are characterized by small round emergence holes in the limbs and twigs. Control consists of keeping the trees healthy and reducing the sources of infestation near the orchard.

The twig girdler, Oncideres cingulatus (Say), attacks many kinds of deciduous trees, but in the South it is found mainly on pecan, hickory, and persimmon. When abundant, they damage trees by severing the branches and twigs, and these may sometimes be seen hanging in the trees and literally covering the ground. Nurseries often suffer from the girdling of terminal shoots. The severed branches can be gathered and burned, and thereby larvae which would develop in them are destroyed.

LEPIDOPTEROUS BORERS IN RECENTLY TOP-WORKED TREES

The American plum borer, Euzophera semifuneralis (Walker), is a pyralid moth occuring in various parts of the United States and attacking many hosts. This moth has a wing expanse of 18.25 mm. It is generally ochre-fuscous in color, with the fore wings a pale gray marked with reddish brown and black. The caterpillars when mature are 25 mm long and somewhat resemble those of codling moth. They enter the cracks, wounds, and sunburned areas of unhealthy trees.

The pecan borer, Conopia scitula (Harr.), belongs to the family Aegeridae, a family of clear-winged moths resembling certain species of wasps, and of which the best known species are the peach tree borers Sanninoidea exitiosa and S. exitiosa graefi.

Injury:- In Texas, the larvae of the above-mentioned moths often cause severe injury to recently top-worked pecan trees. They destroy grafts and patch buds by feeding on the cambium and the callus of the scion and stock. They usually enter graft wounds through cracks in the grafting wax. The larvae may girdle patch budded sprouts and the latter may then be blown from the tree by the wind (694).

Control:- Injury from the borers can be reduced by utilizing the inlay-bark graft method because it permits a close fit of scion and stock. Scions can best be protected from borer injury by covering all graft wounds with a grade of graft-ing wax that does not crack until after the graft has formed a good union.

Borer larvae in graft wounds can be killed by applying to the infested areas, with a paint brush, a solution of 1 pound of paradichlorobenzene to 2 quarts of oil.

In the top-working of pecan trees, care should be taken to avoid cutting back the branches any more than is necessary, for severe cutting back lowers the vital-ity of the trees and favors the development of the borers.

NATURAL ENEMIES OF PECAN INSECTS

Since the important pests of pecan are of American origin, the natural enemies have all long since had a chance to attain their maximum effectiveness. It is un-likely that present artificial control measures can be to any important extent be displaced by virtue of any possible advantage that could be taken of natural enemies. The egg parasite _Trichogramma minutum_ has been propagated in large numbers on the eggs of the Angoumois grain moth and liberated in pecan orchards, but little practical control of any pecan pests has been reported from this effort at biological control.

CHAPTER XXIV
FIG PESTS

On matters concerning the fig, we may now take advantage of the wealth of up to date information on practically all aspects of the history and culture of this interesting fruit crop in Ira J. Condit's recent book, "The Fig." We find that the cultivation of the fig tree probably originated in the fertile part of southern Arabia and gradually extended through western Asia. The fig became established in Greece and Italy very early, and subsequently throughout the entire Mediterranean region, where the six leading fig producing countries are now, in order of importance, Turkey, Greece, Italy, Algeria, Spain and Portugal. Fig trees are also extensively grown in Mexico, Chile and Argentina. The fig acreage in Texas has declined sharply from the maximum of 16,000 acres in 1928. California appears to be third or fourth in importance among the fig-producing districts of the world. In 1947 there were 37,078 acres, the principal counties being Fresno, Merced, Tulare, Madera, and Yolo.

The varieties of figs grown in California are, in order of their commercial importance, the Adriatic, Calimyrna (Lob Injir), Kadota (Dottato), Mission, Turkey, and Brunswick.

Caprification

According to Condit (185), "Botanically, the fruits of the fig are the one-seed ovaries which line the inner wall of the receptacle......The fig is unique among fruits in having an apical orifice, or ostiole, which connects the cavity of the receptacle with the exterior." It can be seen that the flowers are borne inside the receptacle. The flowers may be pistillate or staminate. In the case of the edible figs, no pollen is produced. Many varieties, known as the "common" figs, will set seedless fruits parthenocarpically if no pollen is brought to them.

An orchard of the Calimyrna variety of fig.
Courtesy Dr. Robert M. Warner.

The Calimyrna and other Smyrna-type figs, however, must be pollinized. Some fruits of the first crop may mature without pollination, but none of the second crop. Pollen must be supplied from an inedible kind of fig known as the caprifig. There are no native insects that can bring about pollination of figs. When the Lob Injir figs, now known in California as the Calimyrna, were first introduced into this state, they failed to produce second crop fruit.

It was known since the time of Aristotle that the presence of a small black wasp-like insect, now known to be the hymenopteron _Blastophaga psenes_ (L.), was required for some varieties of figs to cause them to remain on the trees until mature. It was not until the 18th century that it was known that the tiny insect inhabitant of the fig effects its pollination and its subsequent development to maturity. It is now known that blastophaga carries pollen from the caprifig to the Calimyrna fig and this process is known as caprification. This was at first generally considered to be a peasant superstition. Gustav Eisen, who contributed so much to our early information on caprification, wrote that when he announced his conclusions about caprification and of the necessity of importing blastophaga at a horticultural meeting held in Fresno in 1887, "I was hooted down and some of the mob whistled," (185). Nevertheless, blastophaga was successfully introduced into California in 1899 and subsequent studies of the habits and life history of this insect clarified the true nature of caprification (274).

The Profichi:- The profichi (fig. 492, above) are caprifig fruits that appear in the spring on wood of the previous season's growth, or sometimes on older wood, about the time the leaves are appearing, usually in late March or early April. These

figs contain pistillate flowers and, near the "eye" of the fig may be found the staminate flowers. The figs inhabited by blastophaga are dark green, firm, and plump; while those not inhabited are yellowish green, more or less ribbed, and spongy. The pollen matures about the time the insects leave the fruit, and upon leaving, they carry pollen with them on their bodies.

The Mammoni:- These figs appear singly or doubly in the axils of leaves on wood of the current season's growth a week or two before the profichi crop is mature. The blastophaga enter the mammoni figs from the profichi crop, and fertile seeds are commonly found in the mammoni because of pollination afforded by the activities of the blastophaga.

The Mammae:- The mammae (fig. 492, below) are the figs that are found on the trees during the winter; and in California are commonly called the "carry over" crop. The proper development of the mammae fig is usually dependent on the presence of blastophaga.

Fig. 492. Mammae figs (below) from which blastophagas are issuing and can be seen on surface of fruits. Profichi figs (above) are of sufficient size for blastophagas to enter and oviposit. After Condit (184).

The Blastophaga:- All known wild fig trees are inhabited by "inquilines" known as blastophagas, which are dependent on these wild figs for their existence. These insects can only oviposit in the "gall flowers"[1] which are specially adapted for oviposition. The styles of these flowers are not as long as those of the female flowers. The female flowers in the same fruit are not suitable for oviposition

Different kinds of wild figs have within them different species of Blastophaga, but the species in the caprifig is B. psenes. This insect is absolutely dependent on the caprifig for its survival and conversely the caprifig is dependent upon the blastophaga. The female B. psenes (fig. 493) averages 2.5 mm long, has a glossy black body and nearly veinless wings. The wingless, amber colored males have the abdomen tapered and prolonged into a curved tube which is much longer than the head and thorax combined (fig. 493). Only about 10 to 15% of the individuals are males. From 300 to 400 eggs are layed by a female. These are white, elliptical, and bear a short petiole. The legless, white larvae hibernate in the mammae crop of caprifigs and those mammoni which remain on the trees throughout the winter. They pupate in early spring and emerge as adults about the first of April.

[1] Condit and Flanders (187) regard the term "gall" as a misnomer in this connection and propose the term psenocarp.

The interesting habits of the blastophaga were worked out in detail by Eisen
(273). The male emerges first, crawls over
the flowers within the fig, finds one with
a female blastophaga within, and fertilizes
her through an opening he has gnawed
through the ovary of the flower, just beneath the style. The female emerges
through this same opening and leaves the
fig through the eye. In doing so she
covers her moist body with pollen from the
mass of staminate flowers located about the
eye. If in her subsequent flights she
happens to find a caprifig, she enters,
usually losing her wings among the bracts
of the eye as she forces her way through
this narrow aperture. These lost wings can
be seen wedged between the outer scales
"like feathers stuck under the band of a
hat." Their presence indicates that one or
more blastophagas have entered.

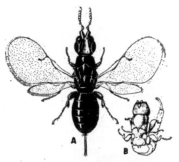

Fig. 493. The blastophaga, Blastophaga psenes. A, female, B, male.
Both X 23.5. Arranged from Condit
(184).

Inside the hollow synconium of the fig,
the female evidently feels her way about with her ovipositor. Upon alighting on
the stigma of a flower, she extends her ovipositor and thrusts it through the
center of the stigma and down the whole length of the style to the entrance to the
ovary of the flower. The egg is wedged between the nucleus of the ovary and its
surrounding integument. The ovipositor is then withdrawn and the lower part of
the stigma canal is filled by a filiform appendage of the egg. This process is repeated on other flowers. Pollen is thus carried to the stigmas of the flowers.
Usually the blastophaga female becomes so exhausted that she is unable to leave the
fig after her many ovipositions.

The developing larvae feed on
the ovary of the "gall flower."
The integument becomes large, hard,
brownish and glass-like in appearance. The "gall flowers" or
psenocarps would not develop, however, even if the blastophaga had
not developed within them.

If the blastophaga female
happens to enter a Calimyrna fig
instead of a caprifig she can not
oviposit because the styles of the
pistillate flowers are too long
(fig. 494) and her ovipositor can
not reach the ovary. Nevertheless,
in her wandering about the in-

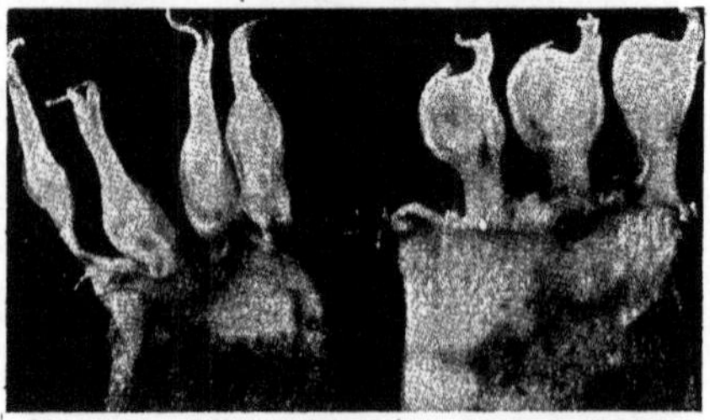

Fig. 494. Left, long-styled flowers of the
edible fig; right, short-styled gall flowers,
of the caprifig. After Condit (184).

terior of the synconium of the fig she unwittingly liberates pollen from her body
into the stigmas and effects the pollination of the flowers.

How Caprifigs Are Handled:- In California caprifigs have usually been planted
in rows or in groups among the Calimyrna trees, but there is a tendency toward separate plantings of caprifigs. This would facilitate endosepsis control.

The Calimyrna (Lob Injir) figs are in an ideal condition for caprification
when they are either light green and glossy or dark green and dull and about an
inch in diameter. If the figs have attained the dimensions of 1.5 to 2.0 inches
in diameter and have not by that time been pollinized, they begin to dry out and
eventually drop off if caprification is not provided for. One blastophaga per fig
suffices, and a count of 21 figs in which only one insect per fig had entered,
showed an average of 850 seeds per fig. (185).

618

FIG PESTS

Generally the figs are distributed at intervals of 4 days over a period of about 3 weeks. A 16-foot tree requires 2 baskets, each containing 2 figs, every 4 days, or a total of 16 figs per season. In one experiment the use of 1 caprifig to 18 square feet of bearing surface of the fig tree resulted in the largest average yields of passable Calimyrna figs, although one caprifig 14 to 22 square feet compared favorably with 1 to 18 (776). The beginning of the emergence of the males or the shedding of pollen by the stamens indicate when caprifigs are ready to pick. They may be held in cold storage at 36° to 40° F until ready to be placed in the orchard.

Endosepsis:- There may be a continuous infection of the caprifigs by the endo-sepsis fungus, _Fusarium moniliforme_ var. _fici_, the organism being carried from one crop to another, and hence to the commercial varieties, by the blastophaga. In-vestigations showed that when blastophagas used for caprification came from an orchard where endosepsis clean-up had been practiced, there was a much lower in-cidence of endosepsis than when the blastophagas were obtained from infected capri-figs (776). The California Fig Institute has briefly outlined measures to be used in reducing infection by endosepsis (915):

Picking

"Mammae figs should be picked before the first profichi in the orchard become receptive to the fig wasp. Early profichi figs if caprified before the clean-up will introduce disease into the Calimyrna crop.

Sorting

"All frost damaged, moldy or otherwise damaged fruit should be sorted out at once and burned or fumigated with methyl bromide to kill the wasps within. The large, more developed mammae figs should be sorted out for use before the others.

Storage

"The mammae figs can be stored for several weeks if kept cool. On the other hand, development of the wasps may be speeded up by subjecting the figs to a warm temperature.

Dipping

"Mammae figs should be split through the eye by making a shallow cut and then dividing them by hand to avoid injuring the galls. The figs should be submerged in a semesan solution for not less than 15 minutes. The semesan is mixed at the rate of one ounce (approximately 3 level tablespoons) in four gallons of water. A 14-ounce tin will make 56 gallons of the solution. The figs should be drained and then placed in the caprifig trees.

Redipping

"The dipping gives only temporary protection to the emerging wasps. After four days the semesan is exhausted by contact with the fig tissues and the fungus grows out to the surface again. Then the figs should be removed and redipped, or destroyed and freshly cut and dipped figs put in their place."

PESTS OF THE TREES AND GREEN FIGS

ROOT-KNOT NEMATODE

Heterodera marioni (Cornu)

The root-knot nematode has a world wide distribution and attacks many host plants. It is considered to be a pest primarily in light soils. In heavy soils it is more difficult for the larvae to travel in search of a host. However, severe infestations have been reported in heavy soils (900). The fig tree is very susceptible to nematode attack. This worm can penetrate the fibrous roots of the fig tree and cause the characteristic knot or gall to develop (fig. 495). An infestation may result in the death of many roots, whereupon new rootlets are produced by the tree to replace them. Based on their observations in Florida, Watson and Goff (928) state that fig trees infested with root-knot nematode "are very easily injured to such an extent as to be unprofitable and short-lived." They have found that in the majority of districts in Florida having sandy soils, figs are successful only when planted near a building under which their roots can penetrate, for they are then not so susceptible to root-knot. Figs are often successful, however, in the clay soils of northern Florida, where root-knot is not such a serious menace. In the San Joaquin Valley of California, many growers recognize the root-knot nematode as the most important obstacle to satisfactory production in their orchards.

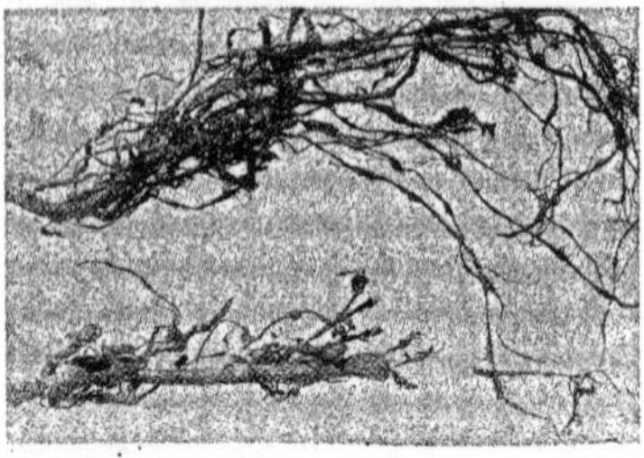

Fig. 495. Root-knot on fig roots. After Condit.

Description:- The mature female (fig. 496, D and F) is shiny white, gourd-shaped or pear-shaped and about 1/12 inch long. An egg mass as large as her own body is often found near or attached to her body. This egg mass may be entirely inside the root tissue or, if it has broken out into the soil, it appears as a tan, reddish, or brown spot on the outside of the gall. Individual eggs and larvae (fig. 496,A) are so small as to be ordinarily invisible to the unaided eye, but they are noticeable when present in large numbers. The young forms are long and slender and as they work their way through soil or swim through water they are cork-screw shaped. The males retain this slender form (fig. 496, E), but the females begin to assume their pear-shaped form as they approach maturity (fig. 496, B and C). The females usually lie just beneath the surface of the root, with the sharper head end toward the inside. As a result of their growth and the substances they give off, the roots they infest form galls, somewhat as a tumor or cancer is formed in the animal body.

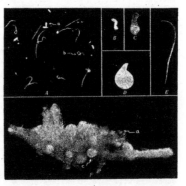

Fig. 496. Root-knot nematode, _Hederodera marioni_. A, eggs and larvae; a, egg, b, larva. B and C, partly grown females; D, mature female; E, mature male, F, gall from tobacco root, partly dissected: a, egg mass; c, female. After Tyler (900).

Life History:- The newly-hatched larvae may migrate to a new location on the same root on which they originated, or they may travel in the film of water that surrounds the soil particles. There they can survive for months, without, of course, undergoing further development until they find food. Many of them find and enter roots. As long as susceptible roots are available, the soil population of nematodes is always increasing despite the probable large numbers of individuals that fail to find further food.

The rootlet is usually attacked near the growing tip, and once established, the larvae continue feeding in the same spot, drawing their sustenance from large nectarial cells formed in the galls. The females become greatly swollen and extrude a yellowish jelly-like secretion in which the small white eggs are deposited. As many as 500 to 3,000 eggs are laid by a female, and fertilization of these eggs is not necessary to insure their hatching. Not many males are produced, and these are short-lived. The root-knot nematode has no true cyst stage, but dormancy may occur. (900)

Injury:- Whether or not it is caused by a secreted toxic substance, the presence of the nematode causes an irritation that stimulates abnormal growth of the plant cells. This causes a distortion of the sap-conduction vessels. The gnarled and broken vessels form into the knots which are the characteristic feature of injury. The infested area is the first part of the root to die, and the large galls are usually decayed to some extent. The infected rootlets are not able to transport water and nutrients in the normal manner. As a result, all the vital functions of the plant are adversely affected. (900)

Condit (185) reports the results of A. L. Taylor with fig seedlings grown in infested and nematode-free soil in Georgia: "After four months all of the infected seedlings but one were dead and this one died shortly after. All of the uninfected seedlings were living and had grown to an average height of 25.7 cm. The results of this experiment leave no doubt that the seedlings had neither resistance nor tolerance to root-knot". (870)

Control (185):- Nothing satisfactory in the way of chemical control has been developed for the treatment of the soil which will not adversely affect the cultivated plants growing in the soil at the time of treatment. Treatment of the soil before planting can be satisfactorily and economically accomplished, especially since the advent of two relatively new compounds which are being used on a large scale for this purpose, namely, DD (a mixture of 1, 2-dichloropropane and 1, 3-dichloropropene) and EDB (ethylene dibromide) (see pp. 229 and 230). Since it is entirely practical to destroy the nematodes in the nursery before cuttings are put in the soil, the grower may now at least start his orchard with nursery stock free of nematodes.

Some investigation has been made of the possibility of developing fig varieties resistant to root-knot nematode. In the University's experimental fig orchard in Riverside, California, not a single one of more than 120 fig varieties or of a large number of seedlings showed freedom from root-knot. There was evidence, however, that certain trees grow and thrive despite the nematode infestation, and it is suggested that such trees should be tested for their possible value as rootstalks. In Florida a nursery is grafting the common fig on stock of one of the two species of Ficus indigenous to that state. This is being sold as a stock resistant to root-knot.

Cultural measures that induce vigorous tree growth may be considered as control measures against the nematodes, for vigorous trees may grow and yield good crops despite a considerable infestation of nematodes. Crotalaria spectabilis is being utilized as a trap crop for nematodes, and this appears to be a promising development in nematode control.

Two nematode species other than the root-knot nematode, namely, the meadow nematode, Anguillulina pratensis, and the banana nematode, Pratylenchus, are known to be injurious to fig trees.

FIG MITE

Aceria ficus (Cotte)

The rust mite (fig rust mite) was first reported in California in 1922 (289) and has since been reported throughout the State, as well as from Oregon and Florida (189). It is probably world-wide in distribution.

Description and Habits:- This minute eriophyid is so small that it is invisible to the unaided eye. The adult female varies from 160 to 202 microns in length and the male is about 140 microns long. The eggs are laid in the bud, on the leaves, in the figs, or on the branches. The mites feed by sucking the plant juices, but do not cause galls like other eriophyids.

The fig mites possess an anal sucker composed of 2 symmetrical semicircular flaps (fig. 497). This can be attached to a surface and the mites can then raise themselves and move their bodies about in the air. This can be noticed whenever the mites are disturbed and it is believed that it is a reaction enabling them to attach themselves to passing objects and thus disseminate themselves.

In winter the mites may be found in terminal buds — from 10 or more to several hundred per bud. Sometimes the predaceous mite, Seilus pomi (Parrott) greatly reduces the fig mite population. The mites reach the peak of their infestation in June or July on the leaves. As many as a thousand mites may be found per leaf.

Another eriophyid, Rhyncaphytoptus ficifoliae Keifer occurs on the fig tree in California. It is not as numerous as Aceria ficus. It is easily distinguished from the latter species in that it is covered with a flocculent coating of white waxy material. With one exception, these mites were found only on the lower leaf surfaces of the Adriatic, Calimyrna, and Black Mission figs.

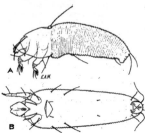

Fig. 497. Fig rust mite, Eriophyes fici. A, lateral view X 460; B, ventral view X 500. After McGregor.

Injury:- The fig mites are of importance mainly because they constitute the principle prey of the smut-and-mold-carrying predacious mites and thrips which follow them into the fig, and because by their feeding they scar the inner surfaces of the fig and cause dead material upon which smut and mold can develop. However, the grass thrips, Frankliniella occidentalis, which also can carry the smut and mold spores, may be abundant in the summer and enter a certain percentage of the figs. (52)

Fig mites may be present under the bracts of developing Kadota figs in September. Such figs are dark around the eye. The mites continue to live under the bracts as the fig develops and pushes past them, and their continuous feeding causes strips of scarred tissue on the outside of the fig. This blemish, if prominent, prevents figs from being used for canning. Considerable injury of this type can be found some years. Observations indicate that an increase in this particular type of injury may some years be correlated with the use of dormant oil spray for fig scale control.

A severe case of an uncommon type of injury was noted in a small fig planting on the Los Angeles Campus of the University of California in August, 1949. Hundreds of fig mites were found on terminal bud bracts, resulting in the dropping of the very small immature terminal leaves and the stunting of the twigs. Nearly all twig terminals were affected in this manner in the planting. Two per cent light medium oil spray was applied to the infested trees, but this, of course, did not reach the mites protected by the bud bracts.

Injury of the type described above has been recorded (185) but appears to be rare, at least as far as economically important damage is concerned.

Control:- Various sulfur, oil, or oil-nicotine formulations were found to result in some reduction of mites and thrips when applied as sprays. These may be applied (1) before the first crop figs are infested and the mites are on the leaves and (2) before the second crop figs are infested (52). Condit (185) recommends 2% medium oil spray for fig mites.

FIG PESTS

PACIFIC MITE

Tetranychus pacificus McGregor

The Pacific mite (fig. 430) attacks various deciduous trees and grapevines in California and Oregon, being probably the most common and the most important of the mites or "red spiders" (see p. 564). Among fig varieties the Kadota is preferred. The mites can cause defoliation and a weakening of the tree. This is usually first noticed on the upper part of the tree. Severe infestations are indicated by abundant webbing.

An important control measure is to keep the trees in as vigorous a condition as possible, for vigorous trees are not very susceptible to attack by Pacific mites. A summer oil emulsion can be used effectively against this species, care being taken to apply the spray before the infestation becomes acute, for after abundant webbing is seen it is usually too late to obtain satisfactory control (777). Some sulfur dusting has been done by plane, although sulfur dust is not very effective against this species.

CITRUS MEALYBUG (34)

Pseudococcus citri (Risso)

This mealybug (fig. 314) has long been a pest of fig trees in Louisiana where fig trees are grown mainly as home garden trees. Specimens were sent to G. F. Ferris in 1924 and were identified as citrus mealybug, Pseudococcus citri Risso. The mealybugs pass the winter in cracks under the bark and in leaves and other trash at the base of the tree. They become established on the new leaves upon the approach of warm weather in the spring, become abundant in April and May, and reach the peak of infestation about the time the figs begin to ripen. If not controlled, they may completely cover a tree, causing defoliation and reduction in size and quality of fruit and sometimes they are sufficiently abundant to kill the tree.

Control:- According to recommendations issued by the Department of Entomology of the Louisiana Agricultural Experiment Station, control is based on sanitation measures combined with (1) sprays of 3% oil-nicotine while the tree is dormant or summer oil emulsion or Loro, preferably combined with soap and nicotine in summer, (2) tanglefoot to prevent ants from ascending the tree and (3) sprinkling 5% DDT around the base of the tree every 2 or 3 weeks during the spring and summer to control ants. Ant control usually results in mealybug control.

MEDITERRANEAN FIG SCALE

Lepidosaphes ficus (Signoret)

This scale insect is thought to have been brought to this country from Algeria about 1905, but has now spread over the fig-growing districts of the San Joaquin Valley, California. Besides the usual methods of dissemination, such as the carrying of the crawlers by the wind and birds and insects, it is believed that caprification may aid in its spread, for the insect commonly overwinters in the caprifig (777).

Life History:- This species (fig. 498) resembles the oyster shell scale, Lepidosaphes ulmi, or the purple scale of citrus, L. beckii. The overwintering scales are dark brown, with a greasy appearing coat of wax, and are less than 1/8 inch long. They are found on one-and-two year old wood. They lay eggs from February to June. The first brood of crawlers usually hatches while the leaves of the fig tree are unfolding. The first generation matures from about June 15 to July 1 and the development of the second generation takes place approximately between June 25 and August 31. The scales that settle and develop on the leaves are lighter in color, smaller, and different in shape than those on the wood. There is then another third, or overwintering generation, or in some years there may be a partial fourth. These generations overlap considerably. The females for the most part settle on the leaves in the summer and on the twigs in the fall, the latter forming the overwintering brood. Some scales settle on the fruit.

Injury:- The greatest losses are caused by the scales that settle on the fruit. If the figs are canned, the portions beneath and surrounding the scales remain dark green. Such figs must be made into jam, and bring lower returns to the grower. Packers object to handling figs that are heavily infested with fig scale, claiming that they are light in weight and have a warty appearance.

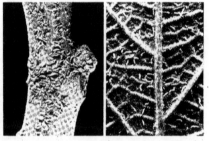

Fig. 498. Fig scale, Lepidosaphes ficus. Left, on twig; right, on underside of leaf. X 4. Photo by Roy J. Pence.

Control:- The following recommendations for the control of the Mediterranean fig scale were made by Stafford and Barnes (829).

Dormant Oil

"The fig scale is most conveniently controlled in the winter with dormant oil sprays.

"What to use:- Use an oil spray consisting of 4 gallons of heavy dormant spray oil tank-mixed with 4 ounces of blood albumin spreader to 100 gallons of water. Dormant emulsive or dormant soluble oils may be employed at the rate of 4 gallons to 100 gallons, or dormant oil emulsions may be used at the rate of 5 gallons per 100 gallons.

"When to spray:- Apply the spray during the dormant season, after the bark has been wet by winter rains and before the buds begin to swell in the spring. The best results can be secured by spraying late in the season, but there is danger that rains and wet ground will interfere if the work is delayed too long.

"How to apply sprays:- Use at least 300 pounds' pump pressure and spray guns that will drive the spray to all parts of the tree. Apply the spray from all sides, preferably walking around the tree while applying it. Every square inch of bark must be thoroughly wet with spray, since only those scales covered with spray will be killed. In orchards with heavy incrusted infestations careful spraying for at least two successive years will be necessary to secure control."

Summer Oil

"Preliminary experiments have indicated that if for some reason the fig scale has not been controlled during the dormant period, a spray properly applied about the middle of May will give good control.

"What to use:- Use a properly emulsified light-grade summer oil at the rate of 2 gallons to 100 gallons of water. The oil may be emulsified in the tank with blood albumin as recommended for the dormant oil, or commercial formulations may be used.

"When to spray:- The light-oil spray should be applied about the middle of May on Kadota figs in the Merced area. On adriatics in the Fresno area this spray should probably be applied a week earlier. By the middle of May there is considerable foliage on the trees, and complete coverage is very difficult.

"How to apply spray:- Use at least 400 pounds pressure and spray guns that will drive the spray to all parts of the tree. Apply the spray to all sides, walking around the tree while applying it. The leaves and bark must be thoroughly wet with spray, since, as with the dormant application, only scales covered with spray will be killed."

New Insecticides:- Preliminary tests have indicated that parathion is very effective in controlling fig scale, but further investigation must be made as to proper dosage and timing and the residue problem must be studied. The possibility of controlling mites and other pests with parathion, as well as the fig scale, en-hances its interest in connection with the control of fig pests. (828)

FIG BEETLE (652)

Cotinis texana Casey.

The fig beetle (fig. 499) appears to be native to Arizona along the Gila River and all its tributaries and along the upper tributaries of the Bill Williams River. It may be found at elevations ranging up to 6000 feet, but reaches its maximum number between 1000 and 3000 feet. This species is also found in the Mesilla Valley in New Mexico and is common in west Texas in the Rio Grande Valley. It is a serious pest of the soft skinned fruits, particularly figs, peaches and grapes, in Arizona. It has forced some growers to abandon the growing of these crops. Cotinis texana was found feeding on figs and grapes in Riverside, California, in September, 1934 (568), and it has since been found in a number of other localities in southern California.

 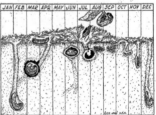

Fig. 499 (left). The fig beetle, Cotinis texana, feeding on a fig. X 1. Photo by Roy J. Pence.

Fig. 500 (right). The life cycle of the fig beetle. After Nichol (652).

Description. Egg:- Pearly white when first laid and 1.8 mm wide and 2.3 mm long. Greatly increases in size in 3 or 4 days by absorption of water.

Larva:- When full grown about 2 inches long, with yellow-brown chitinized head and legs. The legs are small and weak and are never used for locomotion. The larva moves on its back with an undulating motion of its entire body facilitated by transverse rows of stiff, short brown hairs on the dorsum of the thoracic region.

Pupa:- Yellow, in snugly fitting earthen pupal cases from 20 to 25 mm wide and 27 to 35 mm long.

Adult:- Varies in length from 0.75 to 1.3 inches. Prothorax and elytra velvety-green, except that each elytron has a marginal band of brownish-yellow. This band may be present or absent, or present in part on the prothorax. Head and clypeal horn shiny-green. The venter is brown to cuprous green and the legs are robust, shiny-green or brown.

Life History:- The life history of the fig beetle, diagrammatically depicted in fig. 500 A, was worked out by Nichol (652). The eggs are found 2.5 to 5 inches be-low the surface of the ground. The laying of from 50 to 211 eggs by different fe-males has been observed. The egg swells after oviposition and neatly fills the earthern cell prepared by the thrust and removal of the ovipositor. At soil temperatures of 80° to 85° F the eggs hatch in 7 to 8 days. The egg-laying period extends from early August to late October.

The larvae return to their permanent burrows after feeding on organic matter on the ground surface. Their tunnels may be from 4 to 12 inches deep at first, but third instar larvae live in tunnels from 12 to 24 inches beneath the surface. An enlarged chamber is excavated at the bottom of each burrow. Here the larva or "grub" rests between feeding periods. The larvae are always found on their backs (fig. 500). They feed only under cover of litter, where they are protected from predators.

Pupation begins in May after the larva has constructed the pupal case 2 to 5 inches below the surface, often along the line of its old tunnel.

The emergence of the adults usually takes place from the first week to the third week in July. They cannot dig their way to the surface when the soil is dry, but a rain of 0.25 inch or more will bring a wave of adults to the surface.

Injury:- The adult beetles will feed on a wide variety of fruits, including the cactus fruit, which was probably the original host, but in Arizona this beetle is economically important only on figs, peaches, and grapes. They will start feeding where the skin has been broken away, as by birds, or, if forced to make the original puncture, they do so with their clypeal horn, for their mouthparts are weak and unsuitable for tearing the skin. The very disagreeable odor, especially of the females, and the copious and greasy excrement, ruins fruits even slightly injured by the beetles.

Control:- These beetles commonly breed under manure piles and haystack bottoms. It is believed that they may have been introduced into California in manure. Thus the cleaning of such breeding places in February, March, and April is recommended as a control measure. Removal of manure and other organic matter starves the larvae and causes the soil to harden, imprisoning many adults. Flooding of the infested soil will also result in the destruction of many insects because neither the eggs nor the youngest of the first stage larvae can survive in soil saturated with water 48 hours or longer.

No satisfactory control of the adults has been worked out. Since they do not feed on the epidermis of the fruit, stomach poison is of no avail. It is likely, however, that some of the recently developed organic contact insecticides of high residual value might afford satisfactory control.

Other Species of Cotinis:- A related species, the green June beetle, Cotinis nitida L. attacks a wide variety of fruits, including figs, peaches, grapes, plums, apricots, prunes, nectarines, pears, apples, and berries, the first three named being the most frequently attacked. This species occurs east of the Mississippi River and south of latitude 29°. C. nitida is similar in appearance to C. texana, but it is somewhat smaller and the yellowish brown margins of the elytra are continued forward on the thorax. (218)

THREE-LINED FIG-TREE BORER

Ptychodes vittatus Fabricius

This cerambycid beetle has caused damage to fig trees in Louisiana and Texas. It is a gray-brown beetle with 3 scalloped white stripes extending almost the full length of the insect. The female is 1 inch long by 1/4 inch wide and the male is somewhat smaller. The antennae are more than twice the length of the body. The larva is a white legless grub 2 inches long when full grown.

The adults first appear in March and lay eggs throughout the summer, depositing them on the bark of the trunk and larger branches. From 130 to 200 eggs are deposited per female. After feeding near the surface for 1 to 3 weeks, the larvae

FIG PESTS

usually burrow more deeply into the wood, mining for 2 or 3 months or longer. Then
they pupate and emerge as fully developed adults.

Except when excessively numerous, the three-lined fig-tree borers will not
attack healthy fig trees. A tree having portions of the bark broken, a split trunk,
broken branches, or untreated wounds made by the pruning saw, will often be infested by several hundred borers, while only a few feet away perfectly sound trees
may be found uninfested. (373)

Control:- The avoidance of conditions mentioned as favoring the attack of
beetles will of course reduce damage. Heavily infested trees should be cut down
and burned. Highly prized individual trees may be protected by covering the trunk
and larger limbs with wire netting and in addition larvae may be removed by digging
them out with a knife. The larger larvae, which have become deeply imbedded in the
wood, can be killed by applying 2 or 3 drops of carbon disulfide into the tunnels
and immediately plugging the hole with mud. The resulting wounds can be treated
with a paint, as, for example a mixture of 5 parts of coal tar with 1 part of
either creosote or crude carbolic acid. (34)

PESTS OF FULLY RIPENED, FALLEN, OR PARTIALLY DRIED FRUIT

DRIED FRUIT BEETLE (777)

Carpophilus hemipterus (Linnaeus)

Among the insects damaging ripening and partially dried figs before they are
dried and stored in boxes, the dried fruit-beetle (fig. 501) is the most important.
This insect is a cosmopolitan species which was recorded
as a serious pest of figs in California as early as 1907.
The presence of the beetles in the fruit is, of course,
objectionable, and, in addition, it is believed that it
is the principal agency for the carrying of smut and
souring organisms into the fruit. No souring occurs in
fruits grown on experimental trees kept free from insects by protective netting.

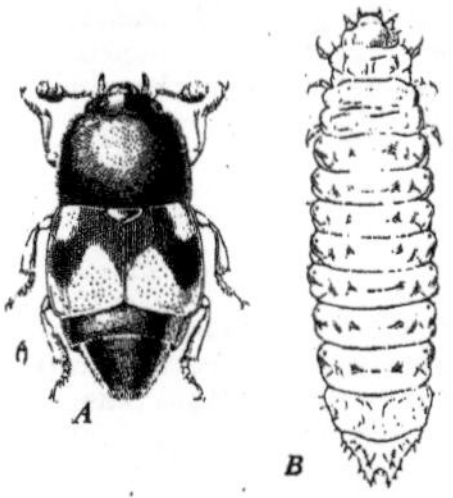

Description. Larva:- The newly hatched larvae are
0.25 inch long when full grown, white or yellowish, with
the head and posterior tip of the body amber brown, with
the body clothed with spine-like hairs, and with 2 large
tubercles at the posterior end of the abdomen and 2
smaller ones just anterior to these. All larval stages
are very active and disappear quickly when disturbed.

Fig. 501. Dried-fruit
beetle, Carpophilus hem-
ipterus. A, adult; B,
larva. Both X 17. After
Linsley and Michelbacker
(537).

Pupa:- Robust, oval, somewhat spiny, about 1/8 inch
long, white or pale yellow, becoming darker when nearly
mature.

Adult:- One-eighth inch long, oval, dull or shiny-
black, with 2 conspicuous amber-brown spots at posterior tips of the elytra and 2
smaller, more obscure spots of the same color at the lateral margins of the wing
covers, near their bases; antennae and legs reddish or amber; the surface of the
body finely punctate, each pit giving rise to a hair. The elytra are very short
and do not reach to the tip of the abdomen, the 2 posterior abdominal segments being exposed. The antennae are knobbed. (280)

Life History:- Incubation of eggs requires from 1 to 7 days, with an average
of 2.2; the period required for larval development varies from 6 to 14 days, average 10; the pupal stage varies from 5 to 11 days, average 6.8; and the life cycle
may vary from a minimum of 15 days in summer to several months in winter. A generation may develop between the ripening of the first crop of Adriatic figs and the
maturing of the second crop. During the winter the adults occur in decaying
melons, cull figs, moist raisins, etc. Sometimes larvae may be found, especially

627

SUBTROPICAL ENTOMOLOGY

in such favorable locations as in moist cull figs. The pupae generally overwinter in the soil.

Control:- Damage from the dried-fruit beetle can be reduced by frequent picking of figs in the orchard. Figs used for drying, and even for the fresh fruit market, are allowed to drop to the ground as they ripen and are picked up and washed. The beetle enters the figs on the ground and the shorter the period that the figs are exposed to the beetles the fewer will be the number infested.

Sometimes the dried fruit beetle infests figs in the trays in the "dry yard." They pupate in the soil under the trays. It has been recently discovered that treatment of the soil with benzene hexachloride (1.5% gamma isomer) at 100 lbs. per acre will result in the complete control of the beetle. The material may be disked into the soil. It appears to retain its effectiveness for 30 to 60 days or longer, depending on soil moisture and temperature. The benzene hexachloride should not be allowed to contact the fruit during drying. (193)

A DARKLING GROUND BEETLE (777)

Blapstinus fulginosus Casey

This dull black tenebrionid beetle is about 6 mm long and can be found under or among the clods of the soil. The larvae presumably feed on vegetable materials in the soil. They are called "false wireworms" because of their resemblance to the true wireworms, which are larvae of another family of beetles.

The adult darkling ground beetles may be found in the fig orchards throughout the year, but are most numerous in July and August, when, if they are abundant, they feed on and under fallen figs, and may eventually almost completely consume the figs. Some injury caused by the beetle girdling suckers at or near the surface of the ground has also been reported.

VINEGAR FLIES (777)

Drosophila sp.

Vinegar or pumace flies are often serious pests of figs in the orchard when souring is in progress. Split or souring figs may contain large numbers of these flies. The larvae may be encountered in ripening or partially dried figs, where they might be mistaken for larvae of the dried-fruit beetle except for the fact that they are legless, eyeless and have the typical pointed anterior section of the fly maggot. The flies, when they are abundant, appear to "overflow" from sour figs to sound figs. The Adriatic variety of fig is especially susceptible to vinegar fly infestation. Ordinarily these flies are considered only as a nuisance, and their infestation of sound figs is one of the few examples of their being of economic importance on growing crops.

Removal of breeding places, such as piles of fruit pits or decaying cull fruit, is recommended as a means of reducing the population of vinegar flies as well as that of the dried-fruit beetle.

PESTS OF WELL-DRIED, HARVESTED FIGS

THE DRIED FRUIT MITE (777)

Carpoglyphus passularum Hering

This is a very small light-colored mite, 0.40 to 0.50 mm long, which may sometimes be found infesting dried fruit. It is of European origin. Besides feeding on dried fruit, it may be found on cured hams and the pollen paste of bees. Sometimes these mites occur in great numbers in dried fruit, imparting a somewhat disagreeable odor.

FIG PESTS

INDIAN-MEAL MOTH (777)

Plodia interpunctella Hübner

The larvae of this cosmopolitan pest feed on all kinds of dried fruits, nuts, flour, cereals, corn meal, chickpeas, beans, peanuts, and chocolate. In common with other moth pests of dried fruits, the adults take no food. They have a wing expanse of 0.5 to 0.75 inch. They can be quite readily distinguished from other dried fruit moths by the wing coloration pattern. The fore wings are a reddish brown on the outer two-thirds, while in the center is a copper-colored band. The basal third of the fore wing is dark gray. The wings are folded horizontally along the body when the moth is at rest (fig. 502, right).

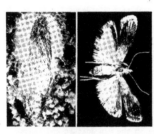

Fig. 502. Indian-meal moth, Plodia interpunctella. Left, pupa inside a dried jujube, X 3.5; right, adult, X 6.5. Photo by Roy J. Pence.

Life History:- The females tend to remain on or near the dried figs and other dried fruits while the males fly about more actively. These moths are nocturnal and fly readily, with a zig-zag flight, when disturbed. The moths live from 30 days in hot weather to 203 days in the cool weather of spring or fall.

The minute white eggs are laid singly or in small clusters, usually attached to the figs. The incubation period is from 2 to 4 days in summer to 6 to 14 days in cooler weather.

The full grown larvae range from 3/8 to 3/4 inch in length. They are dirty-white, yellowish, pinkish, or greenish, frequently with traces of 2 or more of these colors.

The newly hatched larvae usually begin feeding inside the fig, often among the flowers lining the cavity. By the time they are a week old they have constructed a cocoon-like structure of pellets of excrement webbed together. When full-grown, they frequently migrate out of the fig to spin cocoons. They are then much in evidence. The larval stage may require 21 days in summer, but it is very variable in duration.

The winter is passed by the larvae, which continue to feed and grow in heated buildings, but may spend 5 to 8 months in this stage in unheated buildings in the San Joaquin Valley.

The light brown pupae change to a darker brown with age. They are 4/16 to 7/16 inch long are usually inclosed in silken cocoons (fig. 502, left). The pupal period ranges from 4 to 8 days in the hot weather of summer to 30 days in cool weather.

There may be 4 or 5 generations of Indian-meal moth per year.

THE FIG MOTH (777)

Ephestia cautella Walker

The fig moth is of about the same size as the Indian-meal moth but is more slender and not so prominently marked. The fore wings of the majority of individuals are usually a mottled gray but may in some be strongly suffused with fawn-colored scales. The quick-darting manner of flight and rapid vibration of the wings is characteristic of the fig moth. The larvae of the fig moth are similar in habits to those of the Indian-meal moth.

Life History:- The eggs are at first whitish, but turn brownish before hatching. They are loosely attached to the fruit. They hatch in from 4 to 8 days. The larvae feed inside the fig and conspicuous webbing indicates that their growth is completed. They are on the average about a half inch long, of a dirty-white color, and are tinged with brown or purple, with several rows of prominent dots along the back. They usually leave the figs for pupation, and it is during this period that they are most frequently seen. The pupal cocoons are often spun in the cracks of dried-fruit boxes, and sometimes inside figs, near the eye. The average developmental period of moths emerging from mid-July to mid-August was found to be 81.9 days.

RAISIN MOTH

Ephestia figulilella Gregson

The biology of this dried fruit pest has already been discussed in Chapter 20.

Figs are likely to become heavily infested with raisin moths after the fruit has fallen to the ground. Prompt gathering of the fruit reduces the degree of infestation. The partly dried figs may be fumigated before they are spread on trays in the sun. It is recommended that waste figs of the early first crop should not be plowed under but should either be raked out of shaded locations so that the insects may be killed by the heat or be picked up and fed to stock.

THE DRIED FRUIT MOTH

Vitula serratilineella Ragonot

This insect was first reported as a pest in California in 1903. The species has been found in stored figs, raisins, and prunes in the San Joaquin and Santa Clara Valleys. The adults are mottled-gray moths which are apt to be found in dark corners of storage bins or between rows of stacked boxes. They are nearly twice as large as the Indian-meal moth and fig moth and the larvae are also larger. The larvae have the same habit as that possessed by the Indian-meal moth of migrating from the fruit in which they are feeding to pupate. From 62 to 115 days are required to complete the life cycle, making possible 3 generations per year in unheated dried-fruit storage houses in the San Joaquin Valley.

Parasites of the Larvae of Fig-Infesting Moths

Most important among the parasites attacking fig-infesting moths is the hymenopteron Microbracon hebetor (Say) (Habrobracon juglandis (Ash.)). This braconid species is 2 to 3 mm long, black with yellowish markings. It attacks the Lepidoptera of stored grains and dried fruits and occasionally the bee moth in hives (282). In one experiment, the life cycle of this parasite averaged 14.9 days (777).

SAW-TOOTHED GRAIN BEETLE (777)

Oryzaephilus surinamensis (Linnaeus)

This beetle is one of the most important pests attacking a wide variety of stored food products. It is able to reproduce rapidly and the adults may live remarkably long lives - over 3 years in some cases (44), but usually about a year.

Description:- This beetle (fig. 503) is about 1.8 inch long, slender, flattened, and chocolate-brown. The margins of the thorax possess tooth-like projections and on each side of the middle of the thorax is a shallow depression. The adults have well-developed wings, but do not appear to use them. The larvae are a little less than 1/8 inch long when fully grown. They are slender, pale, with a brownish head, and often with pale-brownish bands on the body segments. The pupae are at first pale, but later become darker, and have a row of marginal spines.

Fig. 503 (at right). Saw-toothed grain beetle. X 20. Photo by Roy J. Pence.

Life History:- In figs, the saw-toothed grain beetle lays its eggs inside the dried fruit, among the seeds. Oviposition begins as soon as 5 days after emergence of the adults and from 45 to 285 eggs may be laid per individual. The eggs hatch in from 3 to 17 days. The larval stage lasts from 2 to 7 weeks. The pupae are found in a rude pupal cell or naked (44). In one experiment the life cycle of the saw-toothed grain beetle averaged 51.6 days at mean temperatures of from 75.5 to 86.8°F. An average period of 7.9 days was required for the incubation of the eggs, 37.4 days for the larval period, and 6.3 days for pupation.

CONTROL OF INSECTS ATTACKING DRIED FIGS (777)

Fumigation and Storage in the Orchard

The control of insects attacking dried figs has been the subject of considerable investigation on the part of the U.S.D.A. Dried Fruit Insects Laboratory at Fresno, California. All varieties of dried figs may be expected to be ruined by insect attack unless they are either marketed promptly or fumigated and stored in a tight enclosure. Realizing this, some growers have constructed suitable equipment for fumigating and storing their crop. Providing the temperature is not below 60° F, properly conducted fumigation in a tight inclosure destroys all insects and their eggs. The fumigated fruit may then be protected against reinfestation by storage in an insect proof bin. It is recommended that the fumigation be repeated about once in 3 weeks as an additional safeguard. Sulfur houses are sometimes used as fumigation chambers or storage bins, usually with some changes to make the door gas-tight. Methyl bromide is now the most commonly used gas for the fumigation of stored products, mainly because of its great penetrativity.

Storage temperature below 40° to 45° F prevents the development of insects, but does not kill them. It has been found that 32° F and 55% relative humidity provide the best condition for storage of dried fruits. A temperature of 32° F was better than 40° F or higher in preventing browning, in retaining sulfur dioxide, ascorbic acid, and carotene, as well as in controlling the insects (60).

A U.S.D.A. "Leaflet" describes means which may be used to prevent insect damage to dried fruits prepared at home. This Leaflet contains a popular discussion of exclusion, use of high temperatures, cold storage, fumigation, storage containers, and sanitation as these measures may be conveniently utilized to control home-dried fruit insects on a small scale. (774)

Precautions

Linsley and Michelbacher (537) have given the following advice in the use of fumigants in enclosures.

"When a building or storage chamber can be isolated or completely evacuated the following fumigants can be used: hydrocyanic acid, chloropicrin, methyl bromide, or carbon disulfide. If buildings or chambers cannot be completely isolated and where parts of the building are occupied one of the following should be used: mixture of ethylene oxide and carbon dioxide; mixture of ethylene dichloride and carbon tetrachloride; mixture of propylene dichloride and carbon tetrachloride; or carbon tetrachloride." All these fumigants are poisonous to man, however, and should be used with extreme caution.

CHAPTER XXV
OLIVE PESTS

The olive is indigenous to Asia Minor and the African continent and is now grown extensively in the Mediterranean Basin. In the United States, olive growing is confined almost entirely to California, where in 1947 there were 28,307 acres. The leading olive-producing counties were, in order of importance, Tulare, Butte, Tehama, Sacramento, and Los Angeles. In California the principal pests are several species of scale insects. The olive fly, <u>Dacus oleae</u> Rossi (fig. 504), which is such a serious pest in the Mediterranean countries as to preclude the possibility of the development of a ripe olive industry such as we have in California, has not as yet become established in this state.

NEMATODES

According to Condit (186) there are two species of nematodes prevalent on the roots of olive trees in California, the lesion nematode, <u>Pratylenchus musicola</u>, and the root-knot or garden nematode, <u>Heterodera marioni</u>. The former attacks the bark portion of the root and does not kill the whole cambium or cause a rotting of the roots. Its work is indicated by longitudinal, black cracks in the bark of larger roots. Dr. Condit doubts that this species seriously impairs the health and productiveness of olive trees. The second species has been

Fig. 504. Olive fly, Dacus oleae. X 7. After Balachowsky and Mesnil (55).

discussed in connection with fig pests (p. 619). It is especially prevalent in sandy soils and may seriously weaken young olive trees, but older trees can be kept in good production despite the root-knot nematode if proper cultural methods are employed. Neither of these nematodes can be effectively controlled in an orchard.

BLACK SCALE

<u>Saissetia oleae</u> Bernard

The black scale is the most widely distributed of the olive insects and also causes the greatest damage. When abundant, it so weakens the tree that considerable defoliation takes place and the crop for the following season is reduced. The honey dew excreted by the scale supports the growth of sooty mold fungus, causing much difficulty in the harvesting, handling, and processing of the fruit.

The life cycle of the scale on olives is similar to its life cycle on citrus in the interior districts and has already been discussed (pp. 428-430).

Control (828):- Three methods of control are available: fumigation, oil spray, and DDT-kerosene.

Fumigation. From about mid-July on through the fall and winter and up to March 1, olive trees may be fumigated with HCN gas, using a 16 cc schedule on the Mission variety and 18 cc on others. To avoid injury fumigation should be discontinued sufficiently early so that the tents will be removed from fumigated trees 3 to 4 hours before sunrise, for if fumigation is continued later, the gas may not entirely leave the foliage by daybreak and as a result injury to the tips of twigs and a browning of the fruit stems may result. If the weather is cloudy or cool, the tents need not be removed until about 2 hours before sunrise.

Oil spray. In the summer after nearly all the eggs have hatched, oil spray can be satisfactorily used to control black scale. This may be in late July or early August in the southern San Joaquin Valley. Light medium oil at 1-2/3 to 2% is recommended. In the Sacramento and northern San Joaquin Valleys the hatch may continue throughout the summer and fall, in which case oil sprays should be applied beginning the 10th to 20th of August. Because of the possible injury to the trees, oil sprays should not be applied in fall or early winter, and likewise they should not be applied when the temperature is above 90°F. The effectiveness of the spray depends on how thoroughly it is applied.

DDT-kerosene. Since DDT cannot be applied on the fruit without leaving a residue and penetrating through the skin of the fruit, it can only be used after harvest. Young black scale can be controlled if 1-1/2 lbs. of 50% DDT wettable powder and 2 gallons of kerosene per 100 gallons of water are applied soon after the harvest. It is important that the spray be applied before the scales have reached the "rubber" stage, because the DDT-kerosene spray is then no longer effective. Often by the time the olives are harvested the black scale is already too large for effective results with DDT-kerosene.

OLIVE SCALE

Parlatoria oleae Colvée

The olive scale became noticeably abundant in an olive orchard near Fresno, California, in 1931, and by 1934 it had become so abundant that the crop was completely ruined for canning purposes (1018). It is now a serious pest over a wide area in southern San Joaquin Valley and occasional infestations have called for eradication measures in Los Angeles, San Joaquin, Sacramento, Placer, Sutter, and Yuba Counties. The latter infestations have been caused by the transportation of infested plants from the main infested areas. Eradication is very difficult because the olive scale is very resistant to existing control measures. Its eradication is further impeded by the fact that it can develop on over 200 host plants. This insect also occurs in Arizona, in the Mediterranean Basin, and in China.

Description:- The females are ovate circular and from 1 to 1.5 mm in average diameter. The scale covering is a dirty gray with a black exuvium (fig. 505. The males are much smaller and are more elongate than the females (fig. 505). The body beneath the scale covering varies from light reddish to dark purple in both sexes. Taxonomic characters of the pygidium of the female are shown in fig. 505.

Life History:- The winter is spent by the adult females. These lay eggs in late March which hatch in early May. The insect passes through the typical diaspine life history. In late July the eggs of the second generation start hatching. There are 2 generations per year.

Injury:- The scales infest the twigs and leaves in enormous numbers (fig. 506) yet heavily infested trees may show no apparent decrease in yield. In this respect the olive scale differs from the black scale. The greatest economic loss from the olive scale is therefore caused by the effect of these insects on the fruit. Infestation of the fruit in the spring may cause it to be severely deformed. Dark spots occur on the immature fruit (fig. 507) and while these become

Fig. 505. Olive scale, <u>Parlatoria oleae</u>. Above: left, male; right, female, X 18. Below, pygidium of <u>Parlatoria oleae</u>. Arranged after Ferris (308).

Fig. 506 (left). The olive parlatoria scale on twig and foliage of olive. Photo by Roy J. Pence.

Fig. 507 (right). Immature olives with dark spots caused by the olive scale. After Stafford (827).

less pronounced later in the summer, they again become more conspicuous as the fruit approaches maturity. The fruit is thus spoiled for pickling purposes. In addition, the oil content of the fruit may be reduced as much as 25% in heavy infestations.

Control:- This insect may be controlled by tent fumigation. Fumigation is especially effective on the Manzanillo variety of olives, for they will withstand a 20 cc HCN schedule before harvest and 22 cc after harvest. In the fumigation of the Mission variety, however, the dosage schedule should not exceed 18 cc. As explained in the discussion of black scale fumigation, the tents should be removed from the trees 3 or 4 hours before sunrise if the day is to be clear and warm and well before sunrise if the weather is to be cloudy or cool. (828)

Oil Spray:- A medium grade summer or foliage type of oil spray at 3% of actual oil may be applied in January, February, or the first week in March. A further prolongation of the period of treatment may result in fruit bud injury and reduction in crop. From mid-June to mid-July a 2% medium summer oil spray may be used, although where black scale is also a problem, the treatment may be delayed until about the first of August. It is more difficult to control olive scale than black scale, and thoroughness of application is imperative. Sometimes, when the olive scale infestation is very heavy, a winter spray followed by a summer spray may be required for adequate control.

Oil sprays sometimes result in adverse effects on trees and fruit. Adverse tree reaction is most apt to occur if the sprayed trees are suffering from insufficient soil moisture or if hot weather or strong winds follow the treatment. Injury appears to occur more frequently in the more southern olive sections, as in Tulare County or southern California. Summer-sprayed olives have more prominent and darker pores than unsprayed fruit and may be less suitable for the "green-ripe" process of pickling. In the "Spanish green" process of pickling the sprayed olives show an oil mottling, but this almost entirely disappears by the time the process is completed. The appearance of oil-sprayed olives made into "dark ripe" pickles is entirely satisfactory.

Parathion appears to be very promising in the control of the olive scale, but has not yet been recommended.

OLEANDER SCALE
Aspidiotus hederae (Vallot)

Although the olive grower may confuse this species with the olive scale, if a sufficient number of specimens are compared it will be seen that the armor of the

OLIVE PESTS

oleander scale is generally quite circular as compared to the somewhat ovate armor
of the olive scale. The armor is 1 to 2 mm in diameter, flat, gray, and possesses
a yellow or light brown central or subcentral exuvia. The armor of the male is
smaller and more elongate. The armor of the female can be lifted at any stage of
development and the yellowish body of the insect is revealed. This is a cosmopolitan species and has many host plants.

Life History:- The majority of individuals found in winter are adult females.
The young are born alive in April and when these are mature they give rise to a
second generation in July and August.

Injury:- The majority of oleander scales are on the leaves in the lower inside part of the tree. They do not appear to cause noticeable injury to the tree
even when relatively abundant. Economic injury is caused when these scales settle
on the fruit, for that portion of the olive beneath and surrounding the body of a
scale is delayed in maturing and remains green in conspicuous contrast to the dark
color of the remainder of the olive. If the scales settle in the olives early in
the season they cause a pitting and deformation of the fruit. The affected olives
must be culled out because they do not make good pickles. As in the case of the
olive scale, the oil content of the fruit may be reduced as much as 25% if the infestations are heavy.

Control:- As for olive scale.

OTHER OLIVE PESTS

Two beetles are occasional pests of olive trees in California (281). One of
these is the branch and twig borer, Polycaon confertus Lec. (fig. 381), a small
brown and black beetle from 1/8 to 1/4 inch long, which also attacks many other
fruit trees. Injury occurs when branches break off at the holes made by these
beetles at the base of a bud or in the fork of a small branch. Infested branches
should be pruned off and burned. The olive bark beetle, Luperisinus californicus
Swaine is a small robust beetle about 1/8 inch long, black, with whitish scales
forming a typical pattern (fig. 508). The "shot hole" appearance on the bark,

indicating the location of the exit holes, as
well as the size and appearance of the beetles,
might lead one to confuse it with the common
shot-hole borer, Scolytus rugulosus, of
deciduous fruit trees. The olive bark beetle
prefers sickly trees, but may "work over" onto
healthy trees. Keeping the trees in a thrifty
condition and pruning off and burning infested
branches and twigs are recommended as control
measures.

Fig. 508. Olive bark beetle, Lep-
erisinus californicus, on an olive
twig. X 8. Photo by Roy J. Pence.

The California red scale, Aonidiella
aurantii, occasionally attacks olive trees.
HCN fumigation is recommended as a control
measure on the rare occasions when control may
be necessary.

A tiny eriophyid mite, Oxypleurites maxwelli Keifer, first noticed in 1938, is
quite widely distributed in olive trees in California, feeding on terminal buds, in
blossoms, and on upper leaf surfaces, especially on young, tender leaves. The mite
population declines when the foliage hardens, but mites can be found on the leaves
at all times of the year. They apparently cause no damage (186).

CHAPTER XXVI
AVOCADO PESTS

The cultivated species of the avocado are native to Mexico and Central and South America. Besides growing both in the wild and cultivated state in these regions of their origin, they are now cultivated in California, Florida, the West Indies, and in Algeria, Australia, Canary Islands, southern France, Hawaiian Islands, Madagascar, Madeira, Mauritius, New Zealand, Israel, Philippine Islands, Polynesia, South Africa, and southern Spain. California has close to 18,000 acres of avocados and it is estimated that in Florida there are about 7,000 acres. Considerable interest in avocado growing is reported in the lower Rio Grande Valley of Texas, where there are now several large experimental plantings.

Depending on the variety, the avocado may be grown in all three of the climatic zones of southern California, the coastal, the intermediate and the inland (see map, p. 352). The intermediate zone is at present the most important producing area, probably mainly because it is suitable for the growing of the Fuerte, the most popular avocado variety.

The commercial growing of avocados in California was not seriously considered until about 1910. As is usually the case when a fruit industry is newly established in a region, a period of relative freedom from serious pests was enjoyed. Certain native species became established on the avocado in California, however, and others were introduced. At present the avocado can be said to have its share of insect pests, although fortunately these pests are more easily controlled than those attacking certain other subtropical fruits. Likewise the natural enemies of avocado pests seem to be more effective in keeping the pest populations down to sub-economic proportions than they are on the other important subtropical fruit crops. In the words of D. F. Palmer, Agricultural Commissioner for San Diego County, Nature, assisted by the Division of Biological Control of the Citrus Experiment Station, "has been kind to the avocado grower" (674).

In California, the greenhouse thrips is at present the pest of greatest economic importance. It is especially serious as a pest in San Diego County, where over half of California's acreage of avocados is located. The avocado brown mite is more widely distributed, but is easily and inexpensively controlled with sulfur dust. The latania scale can be found on nearly any tree throughout all the avocado-growing areas, but within the last decade its rather effective control by natural enemies has lessened its importance as a pest as compared to what appears to have been its status in former years. Mealybugs, once serious pests, are now well controlled by predators and parasites, but can damage the scions of newly grafted trees if not promptly controlled artificially.

In Florida, the dictyospermum scale appears to be the most important avocado pest, especially in the south. In the north and central parts of the state the pyriform scale is the most important scale pest. In Florida not only the greenhouse thrips, but also the red-banded thrips, is a pest of avocados. These two species cause damage similar to that of the greenhouse thrips in California. The avocado red mite is a pest during the winter season in Florida while in California the closely related avocado brown mite reaches its greatest population density in the summer and fall.

Several very serious pests of avocados occur in Mexico which have not as yet become established in California or Florida.

EUROPEAN BROWN SNAIL

Helix aspersa (Müller)

The brown snail has been discussed as a pest of citrus (p. 359). On avocados, the snails feed on the blossoms and very young fruit, causing scars. During the

636

late winter and early spring months, which is the period when the snails are the most active, they can be controlled by poison bran mash scattered under the trees as recommended for citrus, about a pound per average-size tree. In California, ducks and geese are often allowed to run loose in avocado orchards to keep the snails under control by eating them.

AVOCADO RED MITE

Paratetranychus yothersi McGregor

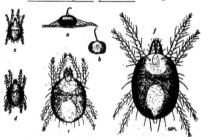

Fig. 509. The avocado red mite of Florida, Paratetranychus yothersi McG. a, b, Egg; c, larva; d, protonymph; e, deutonymph; f, adult. After Moznette (625).

This mite (fig. 509) is a common pest of the avocado in Florida, where its injury attracted the attention of growers as early as 1909. Its attacks are most severe on the tender West Indian avocado varieties, which are common in Florida. It causes severe defoliation and weakening of the tree. Besides attacking the avocado, this mite attacks the mango and has been recorded from camphor and eucalyptus in Florida and from elm, oak, and pecan in South Carolina.

Description (624). Egg:- The egg is globose, smoky amber, and bears a stalk at its apex. Occasional eggs have guy fibrils radiating to the leaf. The eggs are deposited singly and are found at first along the midrib, but later scattered quite generally over the entire leaf.

Larva:- The globular, light yellow larvae possess 6 legs and conspicuous carmine eyes. They become darker after feeding awhile.

Protonymph:- Another pair of legs is gained in this stage, these appearing behind the legs of the larva. The body color darkens still further.

Deutonymph:- This stage becomes larger and more elongate than the protonymph, resembling more the adult.

Adult:- The adult female is oval, rusty red and averages 0.30 mm in length. The body, including the legs, is covered with conspicuous bristles. The male is smaller than the female, averaging 0.22 mm in length and its abdomen narrows posteriorly. The eyes are red and are more conspicuous than in the female.

Life History:- Unfertilized females can lay eggs, but these give rise only to males. There is no webbing except for the occasional fibrils attached to the eggs.

Incubation of the eggs requires from 7 to 11 days. The average larval period is 2.58 days, the protonymph requires 2.8 days, and the deutonymph 2.84 days. The average life cycle is 14.2 days. There is a heavy shedding of foliage in March and April, resulting in a great reduction in avocado red mite population. Spring and summer rains also keep down the population by preventing the mites from establishing themselves on the foliage. They are most abundant in late fall and winter.

Injury:- The avocado red mite feeds on the upper leaf surface, extracting the chlorophyll and causing myriads of tiny white spots to form. The infested area, however, gradually becomes reddish in color, especially adjacent to the midrib. If the mites are sufficiently abundant, the infested leaves may fall.

Predators:- Moznette (625) considers that the predators keep down the avocado red mite populations to some extent. He lists Scolothrips sexmaculatus Perg., Chrysopa lateralis Guer., Scymnus utilis Horn, S. kinzeli Casey, and Leptothrips mali Hinds as being the most important of these predators.

Control:- Sulfur dust is effective and practical, and should be used when no other pest problem is involved. A combination of avocado red mite and other pests might be controlled by adding some other material to the sulfur, by using lime sulfur, or, if the dictyospermum scale is present, an oil spray would be effective for both mites and scale.

AVOCADO BROWN MITE

Paratetranychus coiti McGregor

A mite has been known to infest avocado trees in California for 20 years. It appears to have started in the Carlsbad region in San Diego County and to have spread rapidly from there. It now occurs throughout the coastal avocado areas. This species was believed to be the same as the avocado red mite in Florida until McGregor, in 1941, showed that on the basis of a rather striking difference in the male genitalia, the California mite should be considered as a new species. He named it after Dr. J. E. Coit, who called his attention to the mite in 1929. This species is not known to occur outside of California.

Fig. 510 (at left). Avocado brown mite, Paratetranychus coiti McG. 1, Adult female; 2, egg; 3, tarsal appendages of foreleg of male (lateral view); 4, tarsal appendages of female (lateral view); 5, tarsal appendages of female (ventral view). Adapted from McGregor (560).

Fig. 511 (at right). Avocado brown mite, Paratetranychus coiti. Left, adult female; center, stalked eggs; right, adult male. X 11. Photo by Roy J. Pence.

Description:- The stalked egg resembles that of P. yothersi. Likewise the immature stages range from pale to increasingly darker hues with increasing age. The adult female (figs. 510 and 511) is about 0.40 mm long, and broadly ovate, being about a third longer than wide. Twenty-six strong pale bristles are located dorsally, but do not arise from tubercles as is the case with the closely related citrus red mite. The cephalothorax is pinkish and the lateral area of the abdomen, and sometimes the median area, is occupied by many blotches of purplish or blackish brown. On older individuals the abdomen may be quite solidly blackish brown. The forelegs and palpi are rusty pink, with the other legs pale.

The male (fig. 511) is smaller than the female, averaging about 0.30 mm in length. The body is narrow and tapers to a point behind. The color is paler than that of the female.

Life History:- Observations made in the University experimental orchard at Los Angeles indicate that the duration of the various stages does not differ significantly, under similar temperature conditions, from that of the preceeding species. In the summer there may be two complete generations within a month. Under laboratory conditions at a constant temperature of 77° F. only 7 days are required to complete a generation, but at a constant temperature of 91.4° F. the mites died in all stages, including the egg (560).

The drop of old leaves at the blooming period in the spring causes a heavy loss of overwintering mites, but unlike the conditions in Florida, where the summer rains keep down the mite population on the avocado, the population may build up rapidly in California and reach its highest levels in summer and fall.

Injury:- The mites are found on the upper sides of the leaves, at first congregated along the midrib, then along the smaller veins, or even entirely over the upper leaf surfaces in heavy infestations. The area along the midrib, and finally along the smaller veins, becomes brownish. In addition to the typical discoloration of the leaf, an avocado brown mite infestation is characterized by the myriads of whitish hatched eggs and cast skins of the mites.

The only indications of webbing by the avocado red mite in Florida are said to be the fibrils attached to the eggs (624). This is not true of the avocado brown mite in California. The webbing is very delicate and may be invisible to the unaided eye, but examination with binoculars will reveal considerable webbing, especially in the region of the midrib and over depressed areas. The webbing may be described as light, however, in comparison with that of the majority of spider mite species.

The destruction of chlorophyll no doubt reduces the value of the leaf to the tree. The green color returns if the mites are controlled. In severe infestations there may be some defoliation. It appears that this species does not cause as much damage as one might expect when compared to the severe damage done to citrus by the related Paratetranychus citri when it is present in similar numbers. Some growers, at least in recent years, have discontinued treatments for the avocado brown mite.

Predators:- The predators normally found attacking the citrus red mite (pp. 380 to 382) also attack the avocado brown mite.

Control:- In recent years control measures have usually not been required in California except where DDT has been used for the control of the greenhouse thrips in which case the thrips and mite treatments can be combined. The avocado brown mite is especially well controlled with sulfur, which is most economically applied as a dust. From 1/4 to 1/2 lb. of sulfur is required per tree, depending on the size of the tree. Sulfur is only effective at temperatures above 70° F. Wettable sulfur at 2 lbs. per 100 gallons is also used in brown mite control, but severe injury to the foliage and fruit may result in areas removed from the coast, such as Vista, at temperatures around 85° F.

SIX SPOTTED MITE

Tetranychus sexmaculatus Riley

What appears to have been the first discovery of the six spotted mite (fig. 265) on avocado was made by Oldham and Thorne in August, 1946 (664). A number of infestations have since been discovered. The mites are located only on the undersides of the leaves, as in the case of citrus, but the characteristic depressions formed by the mites on citrus leaves are absent on avocado foliage. However, a characteristically discolored area is formed along the midrib and larger veins where the colonies occur (fig. 512).

Fig. 512. Lower surface of an avocado leaf, showing nature of injury from six-spotted mite. Photo by Roy J. Pence.

BROAD MITE

Hemitarsonemus latus (Banks)

This mite (fig. 513) has been found attacking the tips of avocado and citrus seedlings in greenhouses, causing a characteristic crinkling and dwarfing of terminal foliage (fig. 514). It is found on many varieties of plants throughout the country in greenhouses and is controlled with sulfur dust.

AVOCADO BUD MITE

Epitrimerus myersi Keifer

These tiny eriophyid mites (fig. 515) were first found on avocado in 1938. They appear to be quite widely distributed throughout the avocado-growing area in California. They can be most readily found under the "buttons" of the fruits (fig. 516) where they are attached to the stems, and the writer has found as many as 186 in one such location. Despite their abundance at times, there is no evidence as yet that the mites cause appreciable injury of any kind.

GREENHOUSE THRIPS

Heliothrips haemorrhoidalis (Bouché)

This is almost a cosmopolitan pest in tropical and subtropical regions and, in greenhouses, in temperate regions. In California and Florida it occurs outdoors on avocado and citrus and many other hosts. In California the greenhouse thrips has been known to occur on avocados for many years, but was considered as one of the minor pests. Within the last decade, however, it has rapidly increased the severity of its attacks. Except during years following winters of exceptionally low temperatures, the greenhouse thrips populations may become so great, if uncontrolled, that from 50 to 90% of the fruit may become scarred, with a consequent loss in its market value.

640

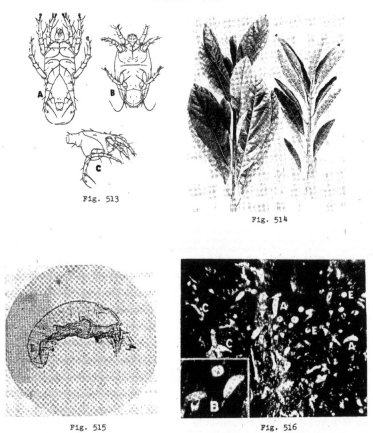

Fig. 513

Fig. 514

Fig. 515

Fig. 516

Fig. 513. Broad mite, Hemitarsonemus latus. A, male (ventral view); B, female (ventral view); C, male (side view). After Hirst (429).

Fig. 514. Injury in greenhouses from the broad mite. Left, normal foliage of an avocado terminal; right, dwarfed and crinkled foliage of an infested terminal. Photo by Roy J. Pence.

Fig. 515. Epitrimerus myersi greatly enlarged. Photo by Roy J. Pence.

Fig. 516. A bud mite, Epitrimerus myersi, on exposed stem end of an avocado: adults (A), eggs (E) and cast skins (C). X 16. B, 2 adults and an egg. X 30. Photo by Roy J. Pence.

Hosts and Varietal Susceptibility:- Other subtropical fruits attacked are the mango, sapote, cherimoya and guava. Among ornamental plants, the carissa, rose, arbutus, viburnum, laurestinus, statis, mandevilla, fuchsia, eugenia, hibbertia, myrtle, azalea, euonymous, cypress, eucalyptus and mesymbryanthemum are especially severely attacked and should be removed if they are growing adjacent to avocado plantings, so as to be eliminated as sources of infestation. Among the avocado varieties, the Mexican seedlings, such as Northrop and Puebla, are especially severely attacked and, although they are of no commercial value, they may serve as sources of infestation for commercial varieties. They should either be removed or treated at the first sign of thrips.

Among the least susceptible varieties are the Anaheim and Nabal. The Fuerte and Dickinson are also relatively resistant when compared with such highly susceptible varieties as Itzamna, Hass, Carlsbad, Benik, Queen, Panchoy and Milly-C.

Description. Egg:- The eggs (fig. 517, A) are about 0.3 mm long when laid, and are white and reniform (kidney-shaped). They are inserted singly into the leaf tissue beneath the epidermis of either the upper or lower leaf surface or the fruit (fig. 518). They continue to increase in size and become considerably swollen and distorted near the end of the incubation period. This gradual increase in size causes a corresponding swelling of the leaf cuticle and the "egg humps" (fig. 519) denoting the locations of the eggs are then readily seen with the aid of a hand lens, although when the eggs are first laid there is no outward evidence of their locations.

Nymph:- The newly hatched nymph (fig. 517, B) is about 2.5 mm long and has a small, tapered abdomen which distends after feeding and becomes slightly yellowish. Before molting, the first-instar nymph attains the length of about 0.75 mm. It is yellowish white in color and has red eyes, as do all the immature stages. The second instar nymph varies from a length of 0.75 mm after the first molt to about 1.0 mm just before pupation. It is yellowish white, like the older individuals of the first instar.

A peculiar characteristic of the nymphs is their habit of carrying a globule of liquid faeces on the tip of the last abdominal segment, where it is supported by six anal setae. This is shown in fig. 519. This faecal liquid is first reddish, then becomes black. The globule of liquid increases in size until it falls off, then another globule begins to form.

Prepupa:- The prepupa (fig. 517, E) is a millimeter or more in length and yellowish white. The antennae are somewhat shorter than those of the second-instar nymph. Two pairs of short wing pads are present. The mouthparts of the prepupa, as well as those of the following pupal stage, are non-functional and there is no feeding.

Pupa:- The pupa is slightly larger than the prepupa. It is also at first yellowish white, with the yellowish hue becoming more intense with age. The pupa may be distinguished from the prepupa by the fact that the antennae are bent backward over the head, the wing pads are much longer and the eyes are much larger (fig. 517, F). Just before molting, the fringe of hairs that constitute the characteristic feature of the wings of the Thysanoptera are folded forward.

Adult:- Upon immerging from the pupal skin, the adult is whitish throughout. The wing tassels are folded forward, but soon unfold into their functional position. Within an hour the head and thorax become black and a few hours later the abdomen also. Thus the adult female a few hours after emergence is black, with whitish legs, antennae and wings (fig. 517, I). The length of the body is about 1.25 mm. The body is markedly reticulated, especially on the head and thorax. The antennae are 8-segmented. The eyes are large and conspicuous, and have large, irregularly arranged facets. The fore wings are broad at the base and have 2 longitudinal veins.

Only one instance has been found in the literature in which a male is claimed to have been found. J. C. Crawford states that he found a male, together with a

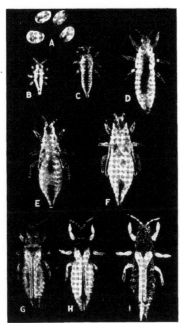

Fig. 517

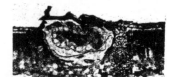

Fig. 518.

Fig. 517. Photomicrographs of greenhouse thrips, Heliothrips haemorrhoidalis. A, eggs in late stage of development; B, newly emerged 1st instar
nymph; C, fully developed 1st instar
nymph; D, 2nd instar nymph, E, prepupa;
F, pupa; G, newly emerged adult; H,
adult 1 hour old; I, adult several days
old. Photo by Roy J. Pence.

Fig. 519

Fig. 518. Cross-sections of greenhouse thrips eggs inserted beneath the epidermis
of an avocado leaf in the typical manner. X 80. Sectioned by Edith Bernstein;
photo by Roy J. Pence.

Fig. 519. Larvae, prepupa and adults of the greenhouse thrips on carissa. Note
egg humps (arrows), some with exit holes of egg parasites. X 12. Photo by
Roy J. Pence.

single female, on a eugenia fruit from Santa Marta, Colombia, collected at the port of New York. A description of the male, together with a diagram of the apical portion of the ninth abdominal segment, was published (204).

Life History:- As previously stated, the parthenogenetic female lays her eggs singly under the epidermis of the leaf or fruit (fig. 518). She first makes an incision with the strong serrated ovipositor (fig. 520) located near the tip of the abdomen, then thrusts the egg into the incision. The eggs are large, for the size of the insect, and dissections have never revealed more than 3 fully developed eggs in the abdomen at a time. No more than 1 or 2 eggs are laid per day. The adult lives a long time, however, and Rivnay (744) states that she may lay as many as 60 eggs.

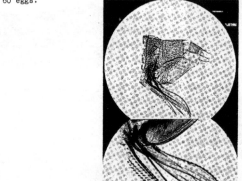

Fig. 520. Ovipositor of the greenhouse thrips: Above, attached to posterior portion of abdomen, (X 110); below, ovipositor further enlarged, (X 400). Photo by Roy J. Pence.

In studies made by the writer on a tree of the Northrop variety in the experimental orchard at the University of California, Los Angeles, 5 adult thrips confined to avocado leaves were examined nearly every day from June 26, 1947 to the day of their death. The 5 individuals were selected from numerous individuals on heavily-infested avocado leaves because they showed by their appearance that they were not more than a few hours old at the beginning of the experiment. The thrips were placed in celluloid cages stoppered with corks into which screens were fitted to allow normal temperature and humidity within the cage. A slit was cut into each cage so that it could be clamped over the edge of a leaf (fig. 521). Of the 5 thrips in the experiment, one died in 20 days and layed only 3 eggs. The performance of the other thrips is shown in table 12 (see fig. 522 for temperature data).

Boyce and Mabry (114) found that the majority of adults lived from 45 to 55 days, but that no single individual laid more than 25 eggs in studies made on Valencia orange leaves and fruit in an insectary at temperatures ranging from approximately 67° F at night to 86° F during the day and at a relative humidity of 49%.

Fig. 521. Cages for confining greenhouse thrips to an avocado leaf (above) or a small portion of the leaf (below) in life history studies. X 0.4.

Table 12

Longevity, number of eggs laid, and number of eggs hatched, for 4 caged greenhouse thrips on avocado leaves under outdoor conditions.

Cage Number	No. Days Thrips alive	No. Eggs Laid	No. Eggs Per Day*	Number Hatched	Per Cent Hatched
1	62	37	0.71	22	59.4
2	49	31	0.80	16	51.6
3	64	42	0.78	12	28.5
4	78	43	0.63	20	46.5
Average	63.2	38.2	0.73	17.5	46.5

*The preoviposition period for each thrips was considered to be 10 days. See fig. 522 for temperatures.

Life history studies made indoors by the writer, at temperatures that averaged close to 23° C (73.4° F), revealed that the preoviposition period was 6.3 days; egg stage 21.4 days; first nymphal instar 6.8 days; second nymphal instar 6.4 days, prepupal stage 1.3 days and pupal stage 3.9 days. The average life cycle, from egg to egg, was found to be 46.1 days under the conditions of the experiment.

The life history and seasonal history of the greenhouse thrips, based on observations of 5 individuals in each of 3 cages on a Northrop tree in the experimental orchard at the University of California, Los Angeles, and on a tree of the Benik variety in Carlsbad, California, is shown in fig. 522, p. 646. Based on the records of the most rapidly developing individuals in each generation, there were on the University Campus nearly 5 generations of greenhouse thrips from November 19, 1946, to the same date in 1947. In Carlsbad, during the same period, there were nearly 6 generations. Five or six generations can thus be said to be the greatest possible number of generations per year under the conditions of the experiment. It was noticed that some individuals were still ovipositing when their

645

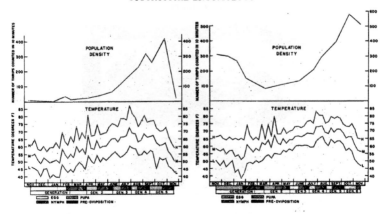

Fig. 522. The seasonal life history and the seasonal fluctuation in population density of the greenhouse thrips at U. C. L. A. (left) and Carlsbad (right).

first-born-daughters had reached maturity. This, coupled with considerable variation in the duration of the life cycle of different individuals, results in a great overlapping of generations. All stages and instars may be found feeding together in a single colony (fig. 519).

In fig. 522 it will be noted that along with the life history experiment, and on the same trees, a study of the seasonal fluctuation of the thrips population was made. The curves shown in the figure are based on counts of the number of thrips of all stages that could be found in 10 minutes. In making the counts an effort was made to locate the most heavily infested leaves in the four quadrants of the tree. It will be noted from fig. 522 that at U. C. L. A. the population of active stages was practically nil during the winter months. The thrips overwintered mainly in the egg stage. These began to hatch on February 15. On the other hand the active stages were rather abundant during the winter in Carlsbad. Reduction in numbers was caused more by the loss of old infested leaves during the blossoming period than by cold weather.

It will be noted from the temperature chart in fig. 522 that there were 11 weeks at U. C. L. A. when the mean minimum temperature was below 45° F, while at Carlsbad there were only 4 weeks with mean minimum temperatures that low. The following winter (1947-48) was even colder, and at U. C. L. A. thrips never appeared during the year 1948 on the tree from which the data shown in fig. 522 were taken, while on the experimental tree at Carlsbad they did not begin to appear until October.

The greenhouse thrips is a sluggish insect. The nymphs and adults occasionally move about slowly and may sometimes travel a distance of several feet in 10 or 12 hours in search of suitable feeding location. The prepupa moves very little and the pupa apparently not at all. The adult has never been seen in flight outdoors, and apparently cannot be provoked to fly, but in the laboratory the writer observed one flying from an infested leaf toward the light of a microscope lamp - a distance of possibly 2 feet.

Injury:- The mouthparts of thrips (fig. 523) are asymmetrical, that is, the left mandible is strongly developed and functional while the right mandible is

646

vestigial or lacking. The maxillae are both present. The greenhouse thrips feeds by puncturing the epidermal cells with its mandible. With a forward and backward movement of the head, similar to the action of a pick ax, the mandible is thrust into the epidermis. Borden (93) believes the finer stylets of the maxillae may be used in lancing the softer tissues. The plant juices being released, the mouth cone is placed over the puncture and the juices are sucked into the mouth, Thrips mouthparts are commonly called "rasping-sucking," but there is no evidence from microscopical study (fig. 524) that the thrips rasp or tear the tissue. They merely puncture the epidermis, suck out the contents of a cell, and move on to another. The apparently "scarred" tissue which appears in old infestations, and is especially abundant on the fruit, appears to be a belated reaction of the injured tissue to the feeding of the thrips.

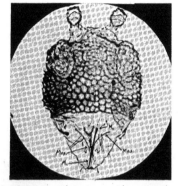

Fig. 523. Head of greenhouse thrips cleared so as to reveal the structure of the mouthparts. M, mandible; Max., maxilla. Photo by Roy J. Pence.

The thrips do not attack the young leaves, nor do they attack fruit before they are the size of a hen's egg. If eggs are laid in leaves that are too young, the rapid proliferation of the leaf tissue surrounding the eggs causes them to become greatly distorted or even crushed.

Thrips injury on the foliage begins to become apparent, usually sometime in June, in the form of small, whitish, silvery, or ashy-gray patches on the upper leaf surfaces where the thrips are found in greatest numbers. The greatest numbers of thrips are found inside and on the north sides of the tree, away from the direct rays of the sun.

The whitish discoloration of foliage and fruit (figs. 524, 525) caused by the earlier infestations changes to a brownish discoloration later in the season. The epidermis of both leaves and fruits becomes thickened, hardened and cracked (fig. 525, below), and the characteristic black specks of thrips excrement become noticeable on the infested parts. It is not likely that much damage is done to the tree by extraction of sap and chlorophyll. The commercial damage consists mainly in the reduced value of the fruit owing to cullage of the discolored, cracked, and scarred fruit. The loss may

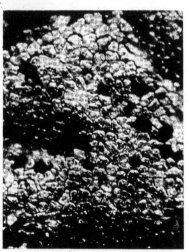

Fig. 524. Whitish appearance of leaf cuticle caused by sucking out of chlorophyl by the thrips. The few dark cells seen in the photograph have not been pierced. Greatly enlarged. Photo by Roy J. Pence.

Fig. 525. Thrips-infested avocados. Above: left, uninfested fruit of
Itzamna variety; right, fruit turned white by thrips feeding, but not yet
turned brownish and cracked. Below: a Fuerte avocado with leathery,
brown, cracked skin caused by prolonged infestation. Photo by Roy J. Pence.

amount to 50 to 90% of the value of the crop some years and with some of the more
susceptible varieties.

Natural Enemies:- What appears to be the most important natural enemy of the
greenhouse thrips is a hymenopterous egg parasite, Megaphragma mymaripenne Timb.
(fig. 526). At times over half of the "egg blisters" that may be observed on
avocado leaves and fruit have the exit holes of this parasite, indicating that this
species may be of some importance in keeping down the thrips population.

The only predator which has been actually observed by the writer feeding on
greenhouse thrips is a predatory thrips which has been identified as Franklinothrips vespiformis (Crawf.) (fig. 527). This species is in the Aeolothripidae, a
family containing several species which spin cocoons in which to pupate. This is
one of a number of biological characteristics of certain species of Thysanoptera
and Hemiptera which verge closely onto those of the more specialized higher orders
with complete metamorphosis.

The adult is 1.6 mm long, generally dark brown except for the first two and
part of the third abdominal segment, which are white, as is also a basal and a
median area on the first pair of wings. The second pair of wings is colorless.
The abdomen is constricted at its base, giving the insect a wasp-like appearance.
The larva has a white head and thorax and a red abdomen.

Franklinothrips vespiformis was found abundantly on Pike sapote trees in the
University orchard in Los Angeles and has often been observed feeding on thrips in
these trees. In the same orchard, however, this species has never been found on

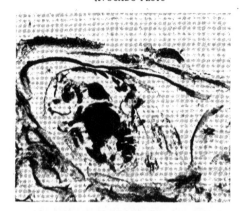

Fig. 526. The parasite Megaphragma mymaripenne
inside the egg of a greenhouse thrips in situ.
The head, thorax and abdomen of the thrips can be
distinguished. X 230. Microtome section prepared by Edith Bernstein; photo by Roy J. Pence.

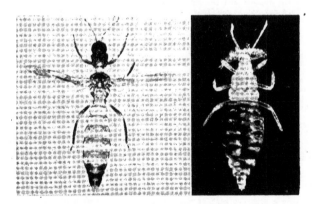

Fig. 527. Franklinothrips vespiformis, a predator on greenhouse
thrips. Left, adult; right, nymph feeding on a greenhouse thrips.
X 40. Photo by Roy J. Pence.

thrips-infested avocado trees, nor has it ever been found on avocado trees in any locality in California. Moulton (623), however, states that the species is found on avocados and citrus in Florida, Texas, Cuba, Nicaragua, and Brazil, and E. A. McGregor found it on citrus in the Imperial Valley in 1926, the first record of the species in California.

Another predatory thrips, the black hunter, Leptothrips mali (Fitch), has been found sparingly under circumstances which indicate it may prey on greenhouse thrips. This thrips is 1 to 1.7 mm long and dark brown to black, but with white wings.

Control:- Before the advent of DDT as an agricultural insecticide, 3/4 pint of standard pyrethrum extract and 1/2 gallon of light-medium oil per 100 gallons was generally recommended as a spray. Preliminary experiments showed that a dust of 5% DDT and 50-75% sulfur at 1/2 to 1 lb. per tree resulted in a great reduction in the thrips population and undoubtedly had considerable residual effect, yet in cases of heavy infestation some damage to the fruit was noted late in the fall. The present recommendation is to spray with 2 lbs. of 50% wettable DDT powder to 100 gallons. Since DDT increases the mite infestation, a suitable miticide should be added to the spray. Wettable sulfur at 2 lbs. to 100 gallons has been used in past years, but one instance of severe damage to foliage and fruit was observed at Vista when such a spray was applied at a temperature of 84° F. One pound of wettable sulfur to 100 gallons has been found to be effective in controlling avocado brown mite, and is recommended as an addition to the DDT in the inland areas. Neotran or DN-111 at 1 lb. - 100 may also be added to the DDT spray but are more expensive than sulfur. One per cent light medium oil, added to the DDT, has controlled mites and in addition makes possible the destruction of the thrips eggs buried beneath the epidermis of the leaves and fruit. Thus an ovicidal effect is obtained in addition to the destruction of the active stages and the prolonged residual effect obtained with DDT alone.

When DDT powder is used without oil, it is recommended that adjuvants be added to increase the deposit and the adhesiveness of the insecticide. Without such an adjuvant there may not be sufficient DDT deposited to control the thrips for an entire season, especially on the Hass and Itzamna varieties.

Avocado trees characteristically grow close to the ground and spread out and interlace their branches. This growth habit prevents the use of motorized equipment within an orchard of mature trees. If power sprayers are to be used, two or three hundred feet of hose must be pulled into the orchard from the road or from such "picking drives" as may be provided to facilitate the hauling of the fruit from the orchard. Obviously, spray towers are not used, but this does not appear to seriously reduce the effectiveness of the treatment, for thrips avoid the outer foliage at the tops of the trees, seeking the cooler foliage farther removed from the direct rays of the sun.

Many analyses for the amount of DDT deposited by sprays and dusts have been made by F. A. Gunther. These analyses have shown that the amount of DDT either deposited on the surface of the avocado or penetrating beneath the peel have been far below the provisional tolerance of 7 parts per million allowed by the Federal Food and Drug Administration on certain crops.

New Insecticides:- Three of the newer insecticides have been found to be more toxic than DDT to the active stages of the greenhouse thrips, namely, the gamma isomer of benzene hexachloride, parathion, and tetraethylpyrophosphate (TEPP). However, since a complete kill of thrips can be obtained with DDT at the recommended dosage, and it has a good residual effect, there appears to be no reason at present for changing from DDT to other insecticides. Any of the above materials in combination with 1/2 to 1 per cent light medium spray oil result in the destruction of the eggs as well as the active stages of the greenhouse thrips.

AVOCADO PESTS

RED-BANDED THRIPS (757)

Selenothrips rubrocinctus (Giard)

This is an important enemy of cacao in the West Indies whence it gained a foothold in Florida and now attacks avocados and mangos, along with the greenhouse thrips. The most noticeable characteristic of this species, as the common name implies, is a bright red band across the body of all stages and instars (fig. 528).

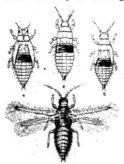

Fig. 528. The red-banded thrips, Seleno-thrips rubrocinctus. After Russell (757).

The adult female is about the size of the greenhouse thrips, dark brown or black in color, except the aforementioned reddish band covering the first 3 segments of the abdomen and the anal segment. The wings are dark. The male is much smaller and is rarely seen.

The life history of the red-banded thrips is similar to that of the greenhouse thrips and it causes similar injury. The heavy summer rains in Florida are said to cause the destruction of large numbers of this species.

Control:- As for greenhouse thrips.

AVOCADO BLOSSOM THRIPS (625)

Frankliniella cephalicus Crawford

This is another avocado pest in Florida. It is about 1 mm long and pale-yellow. It was first collected in the mountains near Guadalajara, Mexico, on a native acacia-like plant, and its mode of entry into Florida is not known. The thrips lay eggs in the stems bearing the flower cluster as well as in the petioles of the individual flowers. They frequently cause a considerable shedding of the bloom, but the principal injury is due to their feeding on the stamens and other flower parts. They have been controlled with nicotine sulfate and soap solution.

AVOCADO LACEBUG (625)

Acysta perseae Heidemann

The lacebugs are small, flat bugs with a reticulated surface that is often lace-like in appearance. The avocado lacebug (fig. 529, p. 652) is about 2 mm long, blackish-brown with yellowish white legs, and with iridescent wings. It is often found in large numbers on avocado trees in the dry winter months in Florida. It usually occurs in colonies on the undersides of the leaves and deposits its eggs in clusters. Yellowish areas, even as seen through the foliage from the upper side of the leaf, indicate the location of a colony of lacebugs.

651

Fig. 529.   Avocado lacebug. X 21.   After Moznette (625).

Control:- Nicotine sulfate at 1-900, with 2-4 lbs. of fish-oil soap to 100 gallons of spray, is recommended for control of the avocado lacebugs. This may be added to a lime-sulfur solution in a combined spray for the avocado red mites and the lacebugs.

AVOCADO WHITEFLY (625)

Trialeurodes floridensis (Quaintance)

This species is much smaller than those attacking citrus, but similar in habits. The adults are less than 1 mm long, with pale yellow bodies and white wings. It is a native insect found on avocados in Florida wherever they are grown. It also has been collected from papaya, banana, guava, and annona. The avocado whitefly is a pest both in the orchard and the nursery. It is responsible for the development of sooty mold fungus and the affected fruit must be cleaned before it is packed.

Seasonal History:- The avocado whitefly spends the winter in the pupal stage. The adults are found in great numbers on new growth in the spring, and the females mate and deposit about 100 eggs on the lower surfaces of young leaves in March. The pearly white eggs are usually placed in circles. The young are semitransparent, and yellowish. There are 3 nymphal instars requiring 5, 5-6, and 7 days, respectively, then a pupal stage of 15 to 30 days in summer and from 3 to 6 months in fall and winter. There are 3 generations and sometimes a partial fourth.

Control:- An oil emulsion is recommended in the fall at the time the foliage begins to harden, to be repeated some time in the spring after the fruit has set. The fall spray should be at 1.5% and the spring spray at 1.25%. This program will also result in the control of the dictyospermum scale.

MIRIDS (1008a)

Rhinacloa subpallicornis Knight and Lygus fasciatus Reuter

In recent years there have been occasional severe infestations of plant bugs known as mirids in avocado orchards in Florida. These are small, soft-bodied insects, not over 1/4 inch in length. They attack the opening buds and cause the young fruit to drop or to become malformed. Leaves are attacked while still in the bud. The punctured tissue becomes necrotic, so that when the leaves expand they contain many holes.

Control:- It is important to treat promptly when the infestation is first observed, using nicotine sulfate, pyrethrum or rotenone at high concentrations. Control measures are aimed at the nymphs, for the adults are hard to kill.

AVOCADO PESTS

HARLEQUIN BUG

Murgantia histrionica (Hahn)

This is a black bug with bright red markings on the dorsum. It varies from 8 to 11 mm in length. The cylindrical eggs simulate tiny white barrels with black hoops. It is a serious garden pest throughout the country. Occasionally this species attacks avocado trees and may cause serious damage. It may be controlled by dusting with toxaphene, DDT, or benzene hexachloride.

FALSE CHINCH BUG

Nysius ericae (Shilling)

This is a small, light or dark gray bug, 3 to 4 mm long (fig. 293). The pale gray nymphs, with reddish brown abdomen, swarm from dry grasslands into adjacent cultivated areas, attacking nearly any green plant. There are from 4 to 7 generations per year. On a number of occasions young avocado trees have been attacked and severely injured. For control, DDT or benzene hexachloride dusts have been used with fair success in recent years, but one experiment with 5% chlordan dust in a heavily-infested peach orchard indicates that this material may prove to be superior to other insecticides used to date (see p. 396).

LONG-TAILED MEALYBUG

Pseudococcus adonidum (Linnaeus)

The long-tailed mealybug is the most important of 5 species of mealybugs which may be found on avocado trees, the others being P. gahani, P. maritimus, P. citri, and P. colemani. The biology of the long-tailed mealybug was discussed on page 421.

The long-tailed mealybug was a serious pest of the avocado in San Diego County prior to 1942. Especially in the vicinity of Carlsbad, avocado orchards were heavily infested with this insect, and the sooty mold fungus which accompanies such infestations was much in evidence. At that time there was evidence of only a small amount of parasitization by the tiny hymenopterous parasite Coccophagus gurneyi. Cryptolaemus beetles which were liberated to combat the long-tailed mealybug were not able to control this pest.

In the spring of 1941, the San Diego County Department of Agriculture liberated two other hymenopterous parasites, Tetracnemus peregrinus and Anarhopus sydneyensis, in avocado orchards in the Carlsbad areas. These were collected in other areas in which they had been introduced by the Division of Biological Control of the Citrus Experiment Station. Two years later no commercial damage to avocado trees was found in orchards previously heavily infested with mealybugs.

In recent years much grafting has been done in the coastal areas on varieties which have been found to be commercially unsuitable in the localities in which they were planted. On grafted trees, long-tailed mealybugs are still an important problem. The scions are covered with paper bags to keep the direct sunlight off the tender new foliage. The shade afforded by these bags makes it possible for the mealybugs to attack this foliage. Unless they are controlled, the mealybugs usually kill the scion (fig.530).

Fig. 530. Long-tailed mealybugs infesting a tender terminal of a newly-grafted avocado tree. X 4.5. Photo by Roy Pence.

Predators and parasites are not able to control the mealybugs before they destroy the scions. The paper bags do not appear to encourage the mealybugs any more than to afford shade, for they are just as abundant on the scions of trees shaded by parasols (experimentally) as they are on scions covered by bags. In fact, every year in the spring, tender terminal sprouts in shady portions of avocado trees are attacked by long-tailed mealybugs, but these infestations have not been considered to be of practical importance. A similar infestation on the scion of a grafted tree, however, results in its death.

Fig. 531. Newly grafted avocado stump, showing the two scions. Photo by Roy J. Pence.

Control:- In grafting, the trunk or several larger limbs of the tree are cut off and usually 2 scions are placed in clefts at either side of the sawed-off area (fig. 531). After grafting is done, the top of the trunk or limb and the sides, for about 6 inches from the top, should be treated with DDT. The following materials have been used with success: 10% DDT dust, a spray of 50% wettable DDT powder at about a third of an ounce to a gallon of water, and a slurry consisting of 1 ounce of 50% wettable DDT powder to 100 ml of water. The latter can be painted onto the area to be protected, and since it leaves a greater amount of DDT on the tree, it should be the most effective method of application. It may also prove to be the most convenient. However, all three methods of application of DDT have resulted in good protection of the scions until the new growth is about a foot long, after which it no longer requires protection.

When the graft cleft is cracked, it is the practice to reseal the cleft, and any insecticide applied before the second application of sealing substance is thereby covered over. In a number of experiments, it was only on trees on which the graft clefts were resealed after the application of the insecticide that any more than an occasional mealybug was found. The grower should be notified if and when the graft clefts are to be resealed and the insecticide should be reapplied to the affected trees.

DDT spray for greenhouse thrips results in a reduction of the mealybug population for at least a half year after treatment, as compared with untreated checks. It is not likely, therefore that the general use of DDT spray will aggravate the mealybug problem by the destruction of natural enemies.

LATANIA SCALE

Hemiberlesia lataniae (Signoret)

The latania scale inhabits many regions of the world on a large number of hosts. In California, however, it was practically unknown until a survey made by state and county officials in 1928 revealed that it was the principal scale pest of avocado. Another more thorough survey was made by the State Bureau of Entomology and Plant Quarantine in 1930. Of 27,868 trees inspected, representing practically all the avocado properties in the state, 12,245 or 43.9%, were infested with one or more species of scale insects. Of these, 85.9% were infested with latania scale, 17.6% with oleander scale, Aspidiotus hederae (Vallot), 9.3% with dictyospermum scale, Chrysomphalus dictyospermi (Morg.), 6.2% with greedy scale, Aspidiotus camelliae Sign., and 5.1% with California red scale, Aonidiella aurantii (Mask.). In addition, black scale was found in 41 orchards, soft (brown) scale in 30, hemispherical scale in 5, European fruit lecanium in 3, and barnacle scale in 1 (573). The per cent of trees infested with dictyospermum scale in 1930 is not representative of the relative importance of the pest today, but was occasioned by the fact that the pest was at the time abundant on avocado trees in the Whittier residential area, a situation that no longer obtains. In commercial acreage, it was found on only 0.8% of the trees examined.

Description:- The armor or scale covering of the adult (fig. 532) is removable at all times. It is circular, 1.5 to 2 mm in diameter, and rather strongly convex. The armor can easily be confused with that of the oleander or greedy scales, being of the same gray color. The exuviae of the latania scale are central or subventral and generally yellow or light brown in color.

Fig. 532 (at left). A small avocado with latania scale. Note pits caused by the scales. X 2. Photo by Roy J. Pence.

Fig. 533 (at right). Pygidium of latania scale, Hemiberlesia lataniae. a, Circumgenital pores; b, median lobes; c, outer pair of lobes. After McKenzie (569).

By comparing the cleared and stained pygidium of the latania scale (fig. 533) with those of the oleander and greedy scales (fig. 363), it can be readily distinguished from the latter species. The pygidium of the latania scale has only one pair of lobes, in that respect being similar to the greedy scale, nevertheless a microscopic examination will show definite differences. Circumgenital gland openings do not occur on the pygidium of the greedy scale, but are present on the pygidium of the latania scale (569).

Life History and Description:- Upon turning over the scales, one may often find the yellow eggs, or the sulfur-yellow crawlers which hatch therefrom, within a few hours. The latter usually settle near the parent within a half a day after hatching, and like other armored scales, they start secreting the wax which forms the scale covering. In about 2 weeks the insect undergoes its first molt, the molting process requiring 2 or 3 days. In 16 to 19 days after the first molt, the second molt occurs, and the insect enters the adult stage. Males have been recorded by several authors, but have not been found in California. Isolated females lay eggs, proving that they can reproduce parthenogenetically. In 26 to 30 days after the second molt, crawlers appear, and in summer it has been found that the life cycle of the latania scale is 56 to 65 days in coastal San Diego County (569).

Injury:- The scale is usually most abundant on the branches or twigs, but may appear on the leaves and fruit as the infestation increases. The smaller twigs may be killed. The fruit is degraded or culled because of the presence of the scales on the peel, although the quality of the fruit is not affected. On the Fuerte variety and possibly other thin-skinned varieties, the rostralis of the scale appears to cause an irritation in the flesh, as indicated by nodules adhering to the inside of the peel when it is removed (fig. 534, p. 656). Corresponding depressions occur in the flesh of the ripe fruit. The Anaheim appears to be the most severely attacked of all varieties.

In former years the latania scale was a much more important problem than it is today. In one packing house the fruit was washed in a special solution and vigorously brushed to remove the scales, and a considerable acreage was fumigated or sprayed. At present there are apparently no treatments being made for latania scale, although a small percentage of the fruit is infested in some orchards. The improved situation with respect to the latania scale seems to be due to the effective work of predators and parasites.

655

Natural Enemies:- It appears that the most important coccinellid predator of latania scale is the twice-stabbed lady beetle, Chilocoris stigma (Say) (figs. 535 and 536). This beetle is 4 - 5 mm long, nearly hemispherical, and shining black except for a red spot slightly anterior to the middle of each elytron. The cylindrical, orange eggs are laid singly or in small groups in the crevices of the bark. The larvae are 5 - 7 mm long and black except for a yellow transverse band. They possess many long, branched spines.

Generally a large percentage of adult latania scales have holes eaten into the armor (fig. 535) which are made by the twice-stabbed lady beetles. One is impressed by the great destruction wrought by a relatively few beetles and can only imagine how effective these predators might be if they were reared and distributed in large numbers in the orchards where infestations of economic importance occur.

In addition to their good work as predators, the beetles carry about, generally beneath their elytra (wing covers), the "travelling form" (fig. 537, left) of a parasitic mite, Hemisarcoptes malus Banks. These mites have a remarkable ability for clinging to the beetles, even being dragged about by the legs of the latter, and sometimes with only their long anal hairs as a means of attachment. Upon reaching the scales, the "travelling form" drops off and attempts to crawl beneath the armor of the latania scale. It can only enter after the newly-born scale crawlers have started to emerge and have raised the waxy margin of the scale sufficiently to allow for their exit. After getting under the scale armor, the mite transforms to a form of greatly different appearance (fig. 538), feeds on the under side of the scale and lays large numbers of remarkably large eggs (fig. 538).

These Hemisarcoptes mites have been becoming increasingly abundant in recent years and probably destroy many scales. They belong to the family Canestrinidae, which appears to be related both to the Sarcoptidae and the Tyroglyphidae. The family comprises only a few species, all parasitic on insects. The majority of species occur on beetles and some species are found under the elytra. H. malus has been found on Florida red scale in Florida (596). A predaceous mite, Cheletomimus berlesei (Oudemans), commonly feeds on latania scale crawlers (fig. 537, right).

A small, yellowish chalcid parasite, Aphytis diaspidis (Howard) (fig. 539), attacks the latania scale, the larva feeding externally, but beneath the armor, on the body of the scale (fig. 224). This parasite does not seem to be important.

Fig. 534 (at left). Latania scale on Fuerte avocado. Note nodules in inner surface of peel (left) and corresponding pits in the flesh (right). X 2/3. After Mackie (389).

Fig. 535 (at right). Twice-stabbed lady beetle, Chilocoris stigma, feeding on latania scale. Note holes in the scales caused by the feeding of the beetle. X 8. Photo by Roy J. Pence.

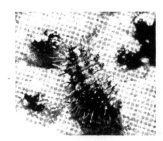

Fig. 536.

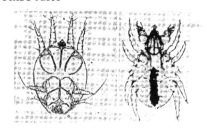

Fig. 537

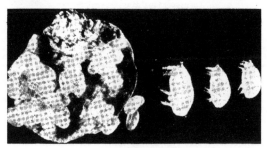

Fig. 538

Fig. 539

Fig. 536. Larva of the twice-stabbed lady beetle feeding on latania scales. X 13.
Photo by Roy J. Pence.

Fig. 537. Left, the "travelling form" of Hemisarcoptes malus; right, Cheletomimus
berlesei a mite found with H. malus under a latania scale. Greatly enlarged. Photo by Roy J. Pence.

Fig. 538. A form of Hemisarcoptes malus. This form is always found under the
bodies of the latania scale. Left, eggs; right, 3 mites; all found
under the body of the scale shown in the figure. Photo by Roy J. Pence.

Fig. 539. A hymenopterous parasite, Aphytis diaspidis, that has been recovered
from latania scale in California. Drawing by H. Compere.

Control:- HCN fumigation was at one time employed in controlling the latania scale (569). Light tents made of "350 sheeting" were pulled over the trees by means of long bamboo poles. The HCN gas was usually generated by sprinkling calcium cyanide dust on the ground under the tent. The dust was placed in tin cans, with the bottom perforated, attached to long handles so that the operator could shake the material out of the can without having to enter the tent. The number of "units" was calculated as explained before in the discussion on citrus fumigation (p. 242), and 2 ounces of calcium cyanide were used for each unit of volume. The gas was generated upon exposure of the calcium cyanide to the moisture in the atmosphere. Liquid HCN was also sometimes used.

Light medium spray oil, emulsives at 1-3/4% and emulsions at 2%, can also be used in the control of latania scale.

DICTYOSPERMUM SCALE

Chrysomphalus dictyospermi Morgan

As previously stated, a State survey in 1930 showed the dictyospermum scale (see p. 469) to be quite abundant on avocado trees in residential properties in Whittier, California, at that time, but very scarce in commercial orchards. After the known infestations of appreciable importance were controlled, this insect never again built-up in noticeable infestations on avocado in California. In Florida, however, it is the principle scale insect in the nursery as well as in the orchard (625). It attacks the twigs and branches, and, in the case of severe infestations, the foliage also. The branches at the base of a tree are the most severely attacked. The foliage-bearing twigs and branches may become so weakened that the entire tree eventually becomes weak and very sparsely foliaged. Severely infested branches and twigs become roughened and crack, allowing for entrance of destructive fungi.

Control:- Oil emulsions at 1.5% are recommended during the dormant period of the tree, from December 15 to February 1, using 2 applications within a 3-weeks period (625).

CALIFORNIA RED SCALE

Aonidiella aurantii (Maskell)

The California red scale (see p. 443), if it is present at all, will usually be found on only occasional trees in the orchard, generally in proximity to some more favored host, such as citrus. Sometimes a tree may be severely attacked (fig. 540) while neighboring trees are entirely free of the insects. An exception to the generally spotty distribution of the red scale in an avocado orchard was a rather heavy infestation in a 15-acre orchard of the Fuerte and Ryan varieties at Monrovia, California, which was sprayed with 1-2/3% light medium oil, with good results, on January 25 - 26, 1946.

PYRIFORM SCALE

Protopulvinaria pyriformis Cockerell (625)

This is another Florida species which not only extracts the sap from the foliage, but excretes large quantities of honeydew which supports the sooty mold fungus. As the name implies, it is a pear-shaped scale, about 3 mm long, and secreting from its margin curled waxy filaments. The winter is spent by the female in a half-grown condition. These mature in the spring and lay eggs. The males emerge from scales which have no cottony secretions. There are several overlapping generations per year, infesting only the foliage. Control as for dictyospermum scale.

TEA SCALE

Parlatoria theae Cockerell
The tea scale (fig. 541) was found infesting avocado trees at Point Loma, San Diego County, in 1948 and was eradicated by the San Diego County Commissioner's

Office in March of that year. This species probably
had been introduced from the Orient, where it is a
pest of the tea plant.

SOFT SCALES

Among the soft scales attacking the avocado are
the soft (brown) scale, Coccus hesperidum L, hemispherical scale, Saissetia hemisphaerica (Targ.),
black scale, S. oleae (Bern.), and the European fruit
lecanium, Lecanium corni Bouché. These are all quite
effectively controlled by parasites, and only rarely
do they increase to more than a localized infestation
in an occasional tree in the orchard. If control appears to be necessary, these insects can be controlled
by means of oil spray, providing the treatment is
timed to coincide with the occurrence of the young,
vulnerable stages or if repeated applications of spray
are made to insure the control of the progeny of those
insects which may have been resistant to the spray at
the time of treatment.

OMNIVOROUS LOOPER (569)

Sabulodes caberata Guenée

As the common name implies, this insect feeds on
a large number of plants. The term "looper" derives
from the crawling habits which the insect shares with
other species of the family Geometridae. The members
of this family are called loopers, measuring worms,
spanworms, or geometers, because of their looping
method of locomotion. Prolegs are absent from the
middle of the body and there are only
2 or 3 pairs at the posterior end,
so the larvae must pull these up
close to the true legs, thus looping
their bodies, then the prolegs are
firmly attached to the leaf and the
fore end of the body is thrust forward.

Fig. 540. California
red scale on an avocado
leaf. X 0.50.

Evidence of the presence of
this native insect can be seen on
almost any avocado tree in California, yet strangely enough, it seldom
becomes of economic importance.
However, in some orchards occasional
trees may be severely attacked. Although the omnivorous looper is distributed throughout California, it
has not been recorded in any other
western state.

Description and Life History:-
The eggs are ovoid, about 1.5 mm
long, and largest at the anterior
end. They are deposited in clusters
on the undersides of the leaves,
each cluster consisting of from 3 to
80 eggs resting horizontally on the
leaf surface. The eggs are at first
metallic green, but in 2 days turn

Fig. 541. Parlatoria theae Ckll. The
large individuals are females; the small,
elongate ones are males. X 20. Photo by
Roy J. Pence.

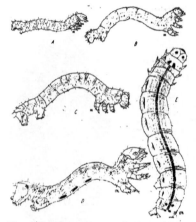

Fig. 542. The omnivorous looper, showing the 5 instars. After McKenzie (569).

to a chocolate brown. They hatch in 8 or 9 days.

The first instar larva (fig.542,A) is at first pale yellow and is the same length as the egg. It eats only the epidermis of the upper surface of the leaves, leaving a characteristic brownish membrance. All the other larval instars eat all the way through the leaf. Both types of injury are shown on the leaf in fig. 543. Looper larvae may eat holes into the leaf away from the margin, but are more apt to eat their way in from the margin of the leaf.

The remaining instars are a yellow to pale green or pink, with yellow, brown, or green stripes on the sides and back, besides a number of black markings.

The second instar larva is 1/4 inch long and has 2 dark bands along the sides of the body. The third-instar larva is about 1/2 inch long and has 4 dark bands along the side of the body. The fourth-instar larva is a little over 3/4 inch long and has 2 black dots on the front of the head; otherwise it is practically identical in appearance to the previous instar. The fifth and last instar of the larval stage is 1-1/2 to 2 inches long. The greatest damage is done by this instar. One individual may eat an avocado leaf in a single day.

The pupa may be found webbed between 2 leaves (fig. 544, above) or inside a leaf which has been folded and webbed together by the full-grown larva for the purpose of pupation (fig. 26). It is at first a pearly white (fig. 544, below), but becomes dark brown as the time for the emergence of the adult approaches. It is about 1-1/4 inches in length.

The adult moth (fig. 545, below) is dull brown or yellow above and nearly white beneath. Two irregular darker transverse median bands cross the upper surface. It has a wing expanse of about 2 inches. These attractive moths are seldom seen during the day because they cling to the undersides of the leaves with their wings spread as shown in fig. 545, and fly about only at night. They lay from 200 to 300 eggs.

The life cycle of the omnivorous looper requires about 1-1/2 months in summer in coastal San Diego County, and it appears that there may be 5 or 6 generations per year.

Injury:- The injury from omnivorous looper is nearly entirely confined to the foliage (fig. 543). If a sufficient percentage of the foliage is removed, the tree is of course weakened and the following year's crop is reduced. This degree of injury seldom occurs. Until recently it was believed that the omnivorous looper seldom feeds on the fruit. In July, 1949, however, many first-and second-instar omnivorous looper larvae were found feeding on young Fuerte avocado fruits near Vista, California (fig. 543a). This appears to be the first record of appreciable damage to the fruit from these insects.[1] About half the fruits, which were at the time from 1 to 1-1/2 inches in length, were injured by the larvae throughout a 3-acre orchard.

[1] A similar injury to young avocado fruits, also first noted in July, 1949, is caused by adult chrysomellid beetles, Diachus auratus Fab. These tiny beetles are 1.5 to 2 mm long, metallic bronze in color, and are common on willow. Essig (282) has found them injuring the tender tips of prune trees in California. The injury to avocados was called to the writer's attention by Mr. Howard B. Sheldon in his Fuerte avocado orchard near Santa Paula, California.

Fig. 543   Fig. 543a

Fig. 544   Fig. 545

Fig. 543. Injury to avocado foliage by the omnivorous looper. Note larva on edge of upper leaf. X 0.5. Photo by Roy J. Pence.

Fig. 543a. Above, shallow grooves eaten into the peel of young avocados by early instars of the omnivorous looper; below, deep pits eaten out by later instars. X 0.7.

Fig. 544: Above, two avocado leaves webbed together to serve as a protection during pupation of omnivorous looper. X 13. Below, pupa of omnivorous looper in early period of development; it darkens with age. X 2. Photo by Roy J. Pence.

Fig. 545. Above, larva of omnivorous looper feeding on a leaf, X 2. Below, adult moth with wings in normal position when at rest, X 1.8. Photo by Roy J. Pence.

Natural Enemies:- The most important parasite of the omnivorous looper is the hymenopteron Habrobracon xanthonotus (Ashm.). This insect stings the caterpillar near the head and paralyzes it, then oviposits on various parts of the victim. The eggs hatch in 1 or 2 days, the larval period lasts 4 to 6 days, and then the parasite pupates beneath the body of the caterpillar in a flimsy white cocoon. From 8 to 10 days are spent in the pupal stage. McKenzie (569) records that in 1932 at least 70% of the omnivorous loopers were parasitized by H. xanthonotus, yet serious injury from omnivorous loopers occurred that year.

Several other parasites have been found attacking the omnivorous looper, including the egg parasite Trichogramma minutum Riley.

A fungus disease similar to the wilt disease of the silkworm attacks the third, fourth and fifth instars of the larval stage. McKenzie (569) found that all but 2 out of 88 caterpillars used in an experiment were killed by this fungus and it is an important factor in keeping the looper population down to its usual numbers.

Control:- Until the advent of DDT, lead arsenate spray was recommended as a control measure. The formula was 4 lbs. acid lead arsenate, and 6 ounces blood albumin spreader to 100 gallons of water.

It has been noted that in the considerable acreage that has been sprayed with DDT for greenhouse thrips, the omnivorous looper has not become a pest. About an hour after spraying, the larvae begin to fall to the ground in a paralyzed condition. Collected and brought to the laboratory, they die within 10 to 12 hours. If spraying needs to be done for omnivorous looper alone, it is likely that 2 lbs. of wettable DDT powder to 100 gallons, plus suitable "wetter-sticker" or "deposit-builder" would be effective and would result in the control of greenhouse thrips, long-tailed mealybugs, amorbia, June beetles, fuller rose weevils and false chinch bugs. If from 1 to 2 pounds of wettable sulfur is added, the avocado brown mite can also be controlled.

In the control of the aforementioned infestation of omnivorous loopers attacking avocado fruits, 1 lb of 50% DDT wettable powder was used to 100 gallons. This resulted in control of the larvae on the fruits, but a few remained alive on the undersides of some leaves. It is difficult to wet the undersides of avocado leaves, and the addition of an adhesive to the spray material results in a greater quantity of insecticide sticking to the leaf surface at the moment of contact of the spray with the foliage.

Fig. 546. Egg mass of the amorbia. Note form of larvae as seen through egg shells. Highly magnified. Photo by Roy J. Pence.

AMORBIA (569)

Amorbia essigana Busck

The amorbia is a tortricid (leaf roller) moth which was first recorded as occurring in California by E. O. Essig in 1922. It has been found in the coastal counties of southern California and seems to prefer the avocado as a host.

Life History and Description:- The amorbia lays its eggs (fig. 546) on the upper surface of the avocado leaf, along the midrib. The greenish eggs are laid in flat masses of from 5 to 100, and from 400 to 500 eggs may be laid by a single moth. These hatch in from 13 to 15 days.

Although there may be as many as 7 larval instars, pupation may take place after the fifth or sixth instar. The first-instar larva is small and yellowish green. The larvae of the remaining instars are similar in appearance, except for size, the seventh-instar larva being 3/4 inch to 1-1/8 inch long (fig. 547). Larvae of all instars usually spend the day hidden between two leaves that have been webbed together, in this habit resembling the larvae of the omnivorous loopers. The larvae wriggle violently and fall to the ground if the leaves are pulled apart. The duration of the larval stage, including 7 instars, was found to be 63.6 days under outdoor conditions in late summer at Encinitas, California.

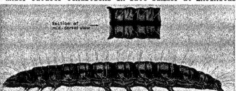

Fig. 547. Last instar larva of the amorbia. X 3.0. After C. M. Dammers from McKenzie (569).

The pupa (fig. 548) is from 1/2 to 3/4 inch long. At first it is pale green, later changing to chocolate brown. As in the case of the omnivorous looper, the larvae web two leaves together to conceal themselves during pupation. The pupal stage requires an average of 17 days in summer.

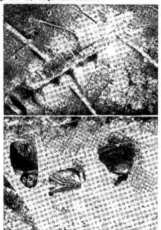

Fig. 548. Above: the amorbia webs itself in between two leaves prior to pupation. One leaf has been withdrawn. Below: pupa with one leaf withdrawn. X 2. Photo by Roy J. Pence.

The adult (fig. 549, p. 664) has a wing expanse of 1 inch, being only a little over a half the size of the omnivorous looper, and the fore wings are reddish brown. The moths when at rest have the bell-shaped outline so often seen in the tortricids. The apical point of the fore wings are usually somewhat notched. The adults are nocturnal and rest on the undersides of the leaves during the day. The life cycle of the amorbia requires about 2 months in summer at Encinitas. There appear to be 4 or 5 generations per year.

Fig. 549 (at left). Amorbia with wings folded in repose. X 5.5. Photo by Roy J. Pence.

Fig. 550 (at right). Injury of the amorbia as it appears on mature fruit. Photo by Roy J. Pence.

Injury:- On the leaves the amorbia causes a skeletonizing similar to that of the omnivorous looper. Although it is found more often on the leaves than on the fruit, the fact that it feeds on the fruit makes it of special economic importance when it is abundant. Fortunately it is seldom as abundant as the omnivorous looper, and usually one has difficulty finding it in an avocado orchard.

The young fruits are the ones most apt to be attacked, and the scars caused by the feeding become enlarged and more conspicuous as the fruit matures (fig. 550).

In August, 1949, amorbia larvae together with larvae of the orange tortrix, Argyrotaenia citrana (Fernald) were found to be destroying the buds of newly budded avocado trees. The larvae fed on the bark adjoining the inserted buds, on the buds themselves, and on any growth which may have developed from the buds. The larvae were identified by Mr. E. L. Atkins, who also reared the moths from captured larvae.

Parasites:- Tachinid parasites, tentatively identified by H. T. Reinhard as Phorocera erecta Coquillett, often destroy a high percentage of the larvae. The egg parasite Trichogramma minutum Riley appears to be of small value.

Control:- As for omnivorous looper.

AVOCADO LEAF ROLLER (625)

Gracilaria perseae Busck

This is a very small gray moth, about a fourth of an inch in length, which deposits its eggs singly on the new growth of the avocado in Florida. The larvae feed on the undersides of the leaves, and in the course of their feeding they roll the leaves back from the tip. The full-grown larvae construct silken cocoons in the folds of the leaf and pupate within these folds. Lead arsenate at 2 lbs.-100 has been effective in controlling these pests.

JUNE BEETLES

Serica sp. and Coenonycha testacea (Cazier)

Of the June beetles attacking avocados, the species Serica fimbriata Lec. and S. alternata Lec. are the most familiar to the layman. They are found in practically any locality and attack many trees and other crops besides the avocado. These species are rather large and robust, S. fimbriata being 1/2 inch and S. alternata

3/8 inch long. The former is a smooth or velvety brown, with faintly striated elytra (fig. 551, above) while the latter is of a uniform shiny brown color.

June beetles are most injurious in young orchards planted near uncultivated land. They fly in from their breeding places in untilled fields and brush land and eat the foliage on the trees at night. They may completely defoliate hundreds of trees in a single orchard. During the day they burrow into the soil to a depth of from 1/2 inch to 2 inches and reappear the following night to resume their feeding.

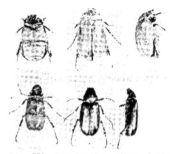

Coenonycha testacea Cazier (fig. 551, below) occurs over a wide area in California on certain species of native vegetation, but was first discovered on

Fig. 551. June beetles. Above, Serica fimbriata; below, Coenonycha testacea. X 2. Photo by Roy J. Pence.

avocado trees in an orchard 2 miles south of Fallbrook, California, in February, 1946. Here the beetle did severe damage, stripping the foliage from young trees and completely destroying a small avocado nursery by feeding on the buds of newly budded seedlings. The beetle reappeared in the same orchard the following year, and also in neighboring orchards, and has been steadily increasing the range of its feeding on avocados in the Fallbrook area since 1946.

Coenonycha testacea belongs to the same family (Scarabaeidae) as the Serica beetles, but is smaller, and is distinctly narrower. This beetle measures about 8 mm in length and 3.5 mm in width and approaches a rectangular shape in contrast to the broadly oval shape of the Serica beetles. C. testacea is shiny brown in color. This species first begins to feed on avocado foliage very early, about the first of February, and is found in appreciable numbers for only about a month. The Serica beetles appear about 3 months later, but the period of their activity is much longer.

Control:- All 3 species discussed above are readily controlled with 5% DDT dust. Control can be effected by applying the dust either to the foliage or to the ground beneath the tree. In either case the beetles get enough DDT on their tarsi in the course of their diurnal activity to result in their death. Probably the application of the dust both on the tree and on the ground would insure the best results.

BLOSSOM ANOMALA (625)

Anomala undulata Melsheimer

In Florida, another scarabaeid beetle, known as the blossom anomala, may visit the avocado blossoms in large numbers and cause serious injury. The adult varies from 1/4 to 5/16 inch in length. It has a black thorax with a yellowish border and the elytra are yellowish brown, sometimes merging on black.

Sometimes the spikes are completely stripped of blossoms by the blossom anomala, and at times they are completely cut off by this beetle. The beetles feed at night and spend the day in the soil much as do the June beetles, to which the anomala is closely related.

For control, the blossom spikes were in former years sprayed with lead arsenate, but it is likely that DDT or some other of the more recent organic insecticides would now prove to be a more effective remedy.

665

SUBTROPICAL ENTOMOLOGY

FULLER ROSE BEETLE

Pantomorus godmani (Crotch)

The Fuller rose beetle (fig. 379) is about 3/8 inch long and is a uniform pale brown. It feeds on the younger foliage of avocado trees and may at times do some damage to the younger trees. The foliage has the same ragged appearance as infested citrus foliage (fig. 380). The smooth elliptical, pale yellow eggs are laid in masses under the bark, on the ground near the tree, or near the bases of smaller plants. The legless white larvae live on the roots of various plants and pupate in the soil. The adults are unable to fly, so it is possible to prevent their ascent by a tanglefoot or cotton band if the lower branches do not make contact with the ground. However, DDT or cryolite dust provide a simpler method of combating the pest. The vegetable weevil, Listroderes obliquus Klug, which also occasionally feeds on avocado foliage, may likewise be controlled by means of these dusts.

AMBROSIA BEETLES

Occasionally avocado trees which may appear to be somewhat weakened, and yet yield good crops and are commercially desirable, are attacked by small beetles which cause the trunk and larger limbs to be riddled with small holes, about 0.5 mm

in diameter. Specimens sent to Mr. Peter Ting for identification were found to be small, cylindrical, brownish to black ambrosia beetles, Xyleborus xylographus (Say) and Monarthrum sp., the former (fig. 552) being the more numerous of the two species. These beetles infest many native trees in California. They often extend their brood chambers deep into the heartwood of the trees they attack and in these chambers they culture ambrosia fungi which serve as food for the larvae. In vigorous trees the flow of sap drowns out the eggs and larvae. These beetles are no threat to healthy trees but may accelerate the death of weakened trees.

Fig. 552. Ambrosia beetles,
Xyleborus xylographus (Say).
X 9. Photo by Roy J. Pence.

The infested trunk of a large avocado tree was cut into three sections and each section split in two halves. To the bark of one half of each of the three sections, a 5% solution of DDT in kerosene was applied by means of a paint brush. The untreated halves served as untreated checks. No activity of the beetles, as indicated by "saw-dust" dropping from the holes, or by gum exudations, was noted after the application of DDT-kerosene, although such signs of activity were still abundant on the untreated trunk sections over a year after the experiment was made.

The branch and twig borer, Polycaon confertus Lec., and other species of beetles occasionally cause damage by boring into the branches of avocado trees. These have previously been discussed in connection with other fruit crops.

AVOCADO TREE GIRDLER (1010a)

Heilipus squamosus Le Conte

This curculionid beetle, closely related to the avocado seed weevils of Mexico and Central America, was first recognized as a pest in southern Florida in October, 1947, although it is believed to have been in the region for at least 10 years before its discovery. Among 512 trees in a young avocado planting, between 8 and 10 per cent of the trees were lost because of the attacks of this insect.

Description and Habits:- The avocado tree girdler is about 1/2 inch long and is black with irregular white areas on its wing covers (fig. 552a). It deposits its eggs on the bark at ground level. The pale yellowish larvae become 1/2 to 5/8 inch in length when full grown (fig. 552a). There appears to be 1 generation per year. The adults are most abundant in April, May and June.

666

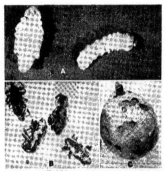

Fig. 552a. The avocado girdler, **Heilipus squamosus**. A, pupa and larva; B, adults,
C, healed-over scars made by the avocado girdler while the fruit was young. After
Wolfenbarger (1010a).

Injury:- Masses of frass at the bases of the trees indicate the presence of
the larvae boring within the trunk. In small trees the larvae are nearly always
found within 6 inches of the ground level, and only in large trees with 12-to 24-
inch bases are they found up as high as 2 feet. The smaller trees are the more
seriously injured. Yellowish leaves, which fall prematurely, are symptoms of injury which have been observed during February and March. Often only a part of the
tree shows these symptoms. The adult beetles sometimes feed on young avocado fruits,
eating away portions about 1/6 to 1/2 inch in diameter and up to 1/10 inch deep.
The fruit usually remains on the tree and scar tissue grows over the wounds. The
adults also feed on terminal buds and other parts of the tender terminal twigs.
This weevil is now considered to be the most important avocado pest of southern
Florida.

Control:- The larvae of the avocado tree girdler may be cut out of the trunk
with a knife. Some of the newer insecticides are being tested against this pest.

SEED WEEVILS (569)

Seed weevils are among the most important
pests of the avocado from Mexico to Panama, but no
destructive species has been found in the United
States. In California a species of weevil known
as the broad-nosed grain weevil, Cauliphilus
latinasus Say, (fig. 553), may be found on avocado
seeds, but only after they have fallen to the
ground. It bores through the rotting flesh of the
avocado and enters the seed. It is of interest to
us only from the standpoint that it should not be
confused with the dangerous seed weevils of Mexico, Central America, and Panama.

Fig. 553. Broad-nosed grain
weevil, Cauliphilus latinasus.
X 8. Photo by Roy J. Pence.

The broad-nosed grain weevil is about 3 mm long. It resembles the well-known
granary weevil, Sitophilus granarius (L.), having the typical prolonged snout of
the weevils, but it is darker brown in color.

FIRE ANT

Solenopsis geminata (Fabricius)

The fire ant (see p. 494) has girdled and killed young avocado trees in California and probably in other regions. Some believe the ants injure the bark in

667

order to bring about a flow of sap, upon which they feed. The avocado tree "bleeds" copiously when injured. In California a 5% chlordan dust has been successfully used in the control of this pest.

RATS

One often finds partly devoured avocados hanging in trees and others in the same condition on the ground. This damage is usually done by rats, and may at times reach serious proportions. Although space has not permitted a discussion of rodents and other vertebrates as pests of the various subtropical fruits, an exception will be made of the rats attacking avocado fruits because the problem pertains specifically to the avocado and written information on the subject is almost entirely lacking.

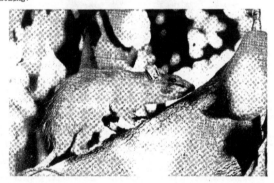

Fig. 554. The roof rat in an avocado tree. X 0.4. Photo by Roy J. Pence.

The principal species involved are the roof rat, Rattus rattus alexandrinus, also known as the Alexandrine or gray rat (fig. 554), and the black rat, Rattus rattus rattus, but especially the former. The black rat resembles the roof rat except for its almost solidly black color. It is found only near salt water, such as at seaports and adjacent towns, but may be found in some coastal avocado orchards. The Norway rat, Rattus norvegicus, is a larger species with a shorter tail. It is relatively unimportant as an avocado pest. The native wood rats, also known as "pack rats" or "trade rats," especially Neotoma fuscipes, will eat avocados, but a more serious damage results from their feeding on the bark, which sometimes results in a complete girdling and death of branches.

Control:- Trapping is a very effective control measure, using almost any human food for bait. The ordinary spring snap trap, such as used for mice, except that it is larger, appears to be the most effective. Contrary to popular opinion, nothing is gained by wearing gloves when setting traps for rats or mice or by boiling, washing, or smoking the traps to remove the human odor. Some success will be had if the traps are set and baited at once, but trapping is even more effective if the traps are placed and baited, but left unset for 3 to 5 nights before being baited and set. By this time the rats have overcome their usual fear of a new object. In avocado orchards rats have to be trapped near or under woodpiles, near the trunks of trees where rats have been feeding on the fruit, and, in fact, in practically any location. When placed on the ground, the base of the trap may be countersunk in the soil with the trigger projecting above the surface. For especially wary rats, the traps may be covered with fine soil or sawdust, just so the action of the trigger or spring is not adversely affected. Some rats will avoid exposed traps, but can be caught if the trap is shielded by a board or box, leaving a space for the rat to enter.

AVOCADO PESTS

If rats are numerous, it may be expedient to first poison as many as possible, then trap the few which would not be poisoned. The chief poisons at present are red squill, zinc phosphide, and ANTU. Thallium sulfate and 1080 are effective, but are not recommended for general public use because of the great danger to humans. Indeed, except for red squill, all rat poisons are dangerous to humans as well as to pets and other domestic animals, and greatest care should be used to conspicuously label the stocks of poisoned bait with the word "POISON" and to lock them up when they are not in use. Care should also be exercized in the placing of the baits and in the disposal of unused baits and of poisoned rodents. A thorough discussion of trapping and poisoning methods may be found in California Agricultural Extension Circular No. 142, "Control of rats and mice" (852).

AVOCADO PESTS IN MEXICO (756, 890)

The avocado, most generally called aquacate in Mexico, is indigenous to that country and has been consumed by Mexicans since time immemorial. The species Persea drymifolia is well known in the highlands of the center and south of the Republic and P. americana is common in the tropical regions, where the fruit is called pahua. Another species, less frequently encountered, is P. schiedeana, known by the name chinini or coyó, which can be found in Veracruz State and a few other regions. There are few cultivated orchards, and very little has been done in the way of selection, improvement, and standardization of varieties, yet the trees growing wild or as dooryard trees probably contribute a considerable tonnage of this nutritious fruit to the Mexican dietary.

As might be expected, many species of native insects adapted themselves to the avocado in Mexico, and some of these are very serious pests. It becomes apparent to anyone familiar with the avocado pest situation in Mexico that the avocado is not inherently resistant to insect attack, as growers in some of the avocado-growing sections of California are prone to believe. A brief discussion of the avocado pests of Mexico may be appropriate here because of the possibility of their introduction into California or Florida despite the existing quarantines.

AVOCADO SEED WEEVILS

Heilipus lauri Boheman and others.

Heilipus lauri, known in Mexico as barrenador del hueso, picudo, or polilla, is 12 to 15 mm long, of a dark brown color with two incomplete transverse dorsal bands of yellow on the elytra (fig. 555, A). The typically grub-like legless larva is 10 to 15 mm long and of a dirty white color. There is one generation per year and the winter is spent in the adult stage. The elongate eggs, 1 to 2 mm long, are deposited in May, June or July under the epidermis of the developing fruit, making in the latter a "half moon" puncture. The larva makes its way through the pulp to the seed, where it feeds and spends its larval and pupal existence, although, in

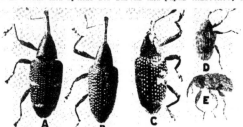

Fig. 555. Foreign avocado seed weevils. A, Heilipus lauri; B, H. pittieri; C, H. perseae; D and E, Conotrachelus perseae. X 3.5. From Barber (59) and Dietz and Barber (238).

may offer some point of entry, as through a puncture or wound:

It is recommended that the infested and fallen fruits be collected and buried to a depth of at least 1 meter. Community action is required for best results.

Heilipus lauri is the avocado seed weevil that is responsible for the quarantine against Mexican avocados (see p. 52).

In Costa Rica occurs a weevil that is either another form of Heilipus lauri or another species, H. pittieri Barber (fig. 555, B). This weevil causes injury to avocados similar to that of H. lauri. In the Canal Zone another species of weevil, H. perseae Barber (fig. 555, C), feeds on the seed, pulp, skin, young twigs and leaves of the avocado. This species is 11 to 15.5 mm long. When first emerging, the adults are reddish with 6 prominent yellow spots. The reddish color eventually becomes blackish. (238) These weevils are also controlled by orchard sanitation. The fallen fruits and seeds should be gathered and burned or buried. In Quatemala is still another destructive weevil with habits similar to those of H. lauri. The scientific name of this weevil is Conotrachelus perseae Barber (fig. 555, D and E). It is a shiny almost black beetle the elytra of which are moderately clothed with a mixture of rose-red, pale brownish, and white hairs. It averages about 6 mm in length (59). In Mexico a species of Cossonus, also attacks the seed, reducing it to a powder.

COPTURUS

Copturus sp.

In a number of localities in Mexico, especially in some avocado orchards near Atlixco, Puebla, the larvae of a curculionid beetle Copturus sp. (fig. 556, A), have caused great destruction of avocado trees by boring into the branches and twigs, often causing them to break at the point where they are most extensively mined. Some trees which have been infested for a number of years are progressively reduced in size as the outer branches die-back or break off (fig. 557). This insect is called barrenador de las ramas. It is 4 to 5 mm long, grayish in color with a broad transverse band of white on the elytra. It is deeply punctate and rather densely covered with hairs.

A couple of specimens which the writer reared out of infested twigs taken near Atlixco were sent to L. L. Buchanan at the U. S. National Museum, and he informs us that this species is probably undescribed. It is different than a "Copturus"[1] perseae Gunther, which attacks avocados in Costa Rica or Copturomimus perseae Hustache in Colombia. The latter (fig. 556, B) is a severe pest of avocado trees, causing damage similar to that of Copturus sp. in Mexico. It has been controlled with a spray having 3% of "Geserol A-5", a preparation containing 5% DDT (584). A 50% wettable DDT powder would at present, of course, be the appropriate material to use, and at a much reduced concentration. The spray was repeated 2 or 3 times at monthly intervals. A 5-gallon sprayer was carried up and down ladders to reach all parts of the tree. The beetles are susceptible to DDT because they do considerable crawling about over the surfaces of the twigs and branches. Possibly Copturus sp. in Mexico could also be controlled with repeated DDT sprays.

LEAF GALL

Trioza anceps Tuthill

Perhaps the most spectacular injury caused by any avocado pest in Mexico is caused by a psyllid (fig. 556, D) which deposits its tiny, oval, yellow eggs in punctures made in the epidermis of the upper leaf surface. Upon hatching out, the

---

[1] This species is now believed to belong in the genus Copturomimus, thus making C. perseae Hustache a homonym, according to information received from C. F. W. Muesebeck.

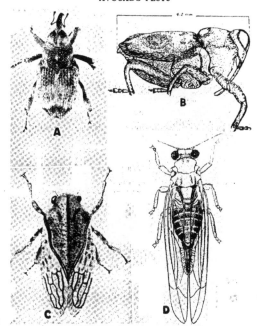

Fig. 556. A, a weevil, Copturus sp.; B, a weevil, Copturomimus perseae; C, a tree-hopper, Metcalfiella monogramma; D, a psyllid, Trioza anceps. A, C, D are avocado pests in Mexico; B in Colombia. A, original; B, After Mariño Moreno (584); C, after Ruiz Valencia (756); D, after Tinoco Corona (890).

Fig. 557. Avocado tree in Atlixco, Mexico, with peripheral branches killed by Copturus beetles.

nymphs produce a secretion which causes a rapid proliferation of cells followed by the development of an elongate gall (fig. 558), which may become 6 to 8 mm long and 2 to 4 mm wide. These are called urnas or urnitas by the Mexicans. The galls are at first green, but later become reddish at the terminal end. The trees often become covered with thousands of these galls, resulting in a bizarre appearance. Inside the galls the nymphs feed and reach the adult stage. The adult emerges on the lower side of the leaf through a circular hole made at the base of the gall.

Fig. 558. Galls of the psyllid Trioza anceps on avocado leaves X 0.5. Courtesy of Señor Jacques de Choulot, Atlixco, Mexico.

The avocado tree can support a large number of psyllids with no apparent injury, but very heavy infestations result in defoliation, and some observers believe that even a moderate infestation will result in reduced production of fruit.

AVOCADO TREEHOPPER (138)

Metcalfiella monogramma Germar

The avocado treehopper is probably indigenous to Mexico, and is known as the periquito del aguacate or by other terms commonly used for treehoppers, such as salton or torito. The adults (fig. 556, C) vary in color from a burnt red of the young adult to the brown or straw-color of the older individuals, and some have a green prothorax. The females are 9 to 12 mm long and 4.25 to 6 mm wide and the males are 8 to 10 mm long and 3.75 to 5.25 mm wide.

This insect attacks only avocados. In severe infestations such a great number of eggs are laid that the egg punctures alone cause the twigs to die. However, the principal damage ordinarily is from the extraction of sap by the nymphs. There appear to be no important natural enemies. There have been no effective control measures in the past, but some of the more recent synthetic organic insecticides might be effective because of their prolonged residual effect.

AVOCADO LACEBUG

Acysta perseae Heideman

This bug has been previously described as a pest of the avocado in Florida (p. 651). In Mexico, oil spray, lime sulfur, and nicotine sulfate have been used as control measures. Oil emulsions and Bordeaux mixture have given good results owing to the fact that the latter is a good repellent, acting as such throughout the entire dry season.

OTHER SPECIES

The caterpillars of certain papilionid, sphingid and saturnid moths are injurious at times. The larvae of buprestid and cerambycid beetles and even termites are apt to attack old or sickly trees. Lacebugs, (Tingidae), the red-banded thrips, leafhoppers, mealybugs, the black pyriform scale and leaf miners sometimes infest avocado trees in Mexico, but are seldom serious.

CHAPTER XXVII

Like the grape, the fig and the olive, the date was an important fruit, and, indeed, an important staple food for many people in the hot deserts south and east of the Mediterranean, since earliest recorded history. It is in this section of the world that it is grown to the greatest extent even today. A great amount of heat is necessary to ripen the fruit, yet temperatures as low as 5°F. are tolerated. Low humidity is also desirable. In this country the date is grown commercially mainly in the hottest deserts of California and Arizona, although a few plantings occur in Texas. In California there are 5,066 acres.

Since the eradication of the date palm scale, Parlatoria blanchardi (fig. 559)(see p. 53), certain species of nitidulid beetles and the date mite may be considered to be the most important pests of the date in California. In comparison with the majority of subtropical fruit crops, the date palm is comparatively free of pests.

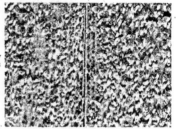

Fig. 559. Date palm scale, eradicated in a joint Federal and State campaign terminating in 1936. X 2.5. Photo by Roy J. Pence.

DATE MITE

Paratetranychus simplex Banks

The female (fig. 560) is 0.3 mm long, usually flesh to pale amber in color, with a few olive-amber markings tending to be greenish in individuals feeding on grasses. There are 26 pale dorsal body setae. These do not arise from tubercles. The egg is spherical, without stalk, and pale amber in color. The eggs are deposited on the surface of the dates and among the fibrils of webbing. The male is 0.224 mm long, narrower and more wedge-shaped than the female. (559)

Paratetranychus simplex is synonymous with P. heterohychus Ewing. It was once also considered the same as the Asio-African species which is such a serious pest of dates in the Old World, but after a critical comparison of specimens, McGregor found that Old World individuals, obtained from Iraq and Algeria, differ with respect to details of the male genitalia and the tarsal appendages of Leg I of the male. Consequently, the Old World mites were described by McGregor as a new species, Paratetranychus afrasiaticus. Paratetranychus viridis (Banks), which has been found on pecan in Texas and South Carolina, has been stated

Fig. 560. Date mite, Paratetranychus simplex. After McGregor (559).

by some authors to occur on dates in California, but E. A. McGregor states[1] that he has no record of this species infesting dates or occurring in California. The females are identical to those of P. simplex, but the two species can be differentiated on the basis of differences in the male genitalia.

[1]Correspondence of August 11, 1948.

673

SUBTROPICAL ENTOMOLOGY

A mite commonly occurring on grasses throughout southern California and Arizona resembles Paratetranychus simplex so closely that the two species cannot be distinguished with a hand lens, but it nevertheless has taxonomic characters that place it in a distinct species. McGregor named this species P. stickneyi and suggested the common name "grass mite."

The date mite can cause serious injury to dates in California by scarring, distorting and discoloring the fruit. Treatment is often required, and fortunately the mites are easily controlled by means of sulfur dust. This is applied by means of orchard dusters to which appropriate lengths of very light aluminum tubing are attached. Fig. 561 shows how effectively the sulfur dust controls the mites. The

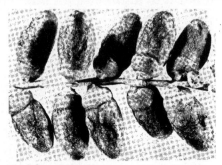

Fig. 561. Injury to dates caused by the date mite. Note freedom from injury at base of each date where growth took place after treatment with sulfur dust. X 1. Photo by Roy J. Pence.

date increases in size by expanding at its base. As can be seen in fig. 561, the fruit had suffered severe injury before the sulfur dusting. After the dust treatment, the new tissue developing at the base of each date was free of mites and also free of blemish.

The Old World date mite, Paratetranychus afrasiaticus Mc G., was found to be a serious pest in all parts of Mesopotamia, where the mites were spinning webs over the clusters of dates. Dust covers the webs and the dates beneath never ripen properly. A species of wasp was believed to be a factor in the spread of the mites. It was observed that no serious damage was caused by the date mite except when trees were growing far from water or in "ill-kept unweeded gardens." (135)

RED DATE SCALE

Phoenicococcus marlatti Cockerell

The red date scale infests the commercial date palm, Phoenix dactylifera, as well as two ornamental species, P. canariensis and P. reclinata. It occurs in the Mediterranean region and in Arizona and California.

The adult female of the red date scale is 1.3 mm long, oval, light pink to deep red, and is partly enveloped by masses of cottony filaments upon which it rests. These insects undergo the usual coccid life history except for the fact that the adult males, which are smaller and more elongate than the females, are wingless and have no mouthparts. (617)

This is a species that is able to survive the heat and dryness of the desert because of its ability to find a suitable ecological niche in the date palm. The majority of the scales are found behind the fibre, deep down on the leaf bases and

DATE PESTS

fruit stems, where it is humid. The insect soon succumbs if it is exposed to the dryness and heat of the desert. This scale also infests the roots of the palm.

NITIDULID BEETLES

Within relatively recent years, four species of nitidulid beetles have become prominent as date pests in the Coachella Valley, California, feeding on ripened dates. The biology and control of these beetles have been investigated by Barnes and Lindgren (62) in a cooperative project between the U.S.D.A. and the University of California. More recently Lindgren and Vincent have added additional information.[1]

Description:- The species involved are Urophorus humeralis (Fabricius), Carpophilus hemipterus (Linnaeus), C. dimidiatus (Fabricius), and Haptoncus luteolus (Erickson) (fig. 562). Urophorus humeralis is black and is the largest of the 4 species, being 4 to 5 mm long. The dried fruit beetle, Carpophilus hemipterus, is also black, but with buff-colored spots on the elytra. The elytra are so short that they leave the 2 posterior abdominal segments exposed. This species is 3 to 4 mm long and somewhat flattened. The corn sap beetle, C. dimidiatus, is pitchy black to dull yellow brown, with the elytra always somewhat paler. It is 2 to 3 mm long. Haptoncus luteolus is the smallest of the 4 species, being about 2 mm long. It is yellowish brown, but considerably lighter in color than the preceding species.

Fig. 562. Nitidulid beetles attacking dates. Left to right: Urophorus humeralis, Carpophilus hemipterus, C. dimidiatus, and Haptoncus luteolus. X7. Photo by Roy J. Pence.

Carpophilus dimidiatus and C. hemipterus comprised about 99% of the beetle population on dates in 1947 and 1948.

Life Histories:- The date gardens are infested throughout the year. Both larvae and adults were found feeding on waste dates from November through May. Eggs are laid in the fruit. The full-grown larvae enter the soil, pupate, and emerge throughout the winter and spring. The June drops of the new crop of dates become available as food at about the time the waste dates of the old-crop dates become useless as food for larvae and adults. The new-crop fruit continues to be attacked throughout the remainder of the season, thus the life cycle is repeated many times a year.

The mean monthly temperature during July and August at Indio, in the Coachella Valley, is 91.4° F. At 90° F temperature, Carpophilus hemipterus can develop from egg to adult in 12.4 days and C. dimidiatus in 14.7 days. At 65° F, this period is increased to 42.2 days for C. hemipterus and 49.2 days for C. dimidiatus.

Control:- Sprays of benzene hexachloride, chlordan, parathion and Compounds 497 and 118 result in a high initial kill of the nitidulid beetles and also leave a residue that is effective against these pests for several weeks to a month after application. It is believed on the basis of preliminary tests that insecticidal dusts containing the above compounds, applied to the bunches of dates 10 days to 2 weeks before the fruit is picked, might result in the control of the beetles.

The nitidulid beetles breed in fruit and truck-crop waste throughout the Valley besides breeding in dates within the gardens. It is believed that the cleaning up of breeding places should greatly reduce the beetle population.

Dates that fall on damp soil provide the ideal condition for the breeding of nitidulid beetles. If the irrigation water is not allowed to wet that portion of

---

[1]Lindgren, D. L. and L. E. Vincent. 1949. Investigations on the control of date insects and mites. Unpublished manuscript on file at the Citrus Experiment Station, Riverside, California.

the soil directly under the palms, where the majority of the dates fall, the latter will dry out and provide a less favorable environment for the development of the beetles.

DRIED FRUIT MOTHS

The striped larvae of the raisin moth, Ephestia figulilella Greg. and the yellowish-white larvae of the Indian meal moth, Plodia interpunctella (Hbn.), have been observed in dates in the gardens when the picking of the fruit has been unduly delayed. Frequent picking of the fruit, followed by prompt fumigation, will preclude the possibility of damage from these pests.

Fumigation of Dates in Storage

Methyl bromide has replaced HCN as a fumigant for the control of insect pests infesting stored dates (mainly the dried fruit beetle, Carpophilus hemipterus). Because of the great penetrating properties of this gas, vacuum fumigation is no longer necessary. One pound of methyl bromide is used per 1,000 cu. ft. of space, including the load, in the practically gas-tight chambers in which such fumigations take place. The dates are exposed to the gas for 24 hours at a temperature above 60°F and below 90°F. During the injection of the gas, and for at least 15 minutes thereafter, a fan at least 10 inches in diameter is in operation. Immediately following completion of the treatment, and before being handled by workers, the fumigated dates should be properly aired. Safety precautions should be strictly observed in fumigating with methyl bromide. (41)

Analyses made by the State Department of Agriculture showed that fumigation of dates with methyl bromide according to the above schedule left no appreciable amount of bromine residue and did not impart a disagreeable flavor to Deglet Noor dates. If used as recommended, methyl bromide fumigation should not be deleterious to the fruit or to the consumer. (41)

CHAPTER XXVIII

PESTS OF MINOR SUBTROPICAL FRUITS

THE POMEGRANATE

In California there are 663 acres of pomegranates, nearly all in Tulare and Fresno counties. In the latter county there is one orchard containing 200 acres of pomegranates. Many are grown as border trees, especially around vineyards. The pomegranate is also grown in home plantings in the Gulf States.

Brevipalpus lewisi McGregor

Beginning about 1943, a russetting, checking and scabbing of the rind of pomegranates was noticed in Tulare county. This type of injury became increasingly more in evidence until by 1947 it caused the culling of from 50% to 90% of the fruit in some orchards. The matter was called to the attention of the University in October, 1947, and the injury was found to be caused by a pseudoleptid mite which was determined by E. A. McGregor to be a new species which he named Brevipalpus lewisi (565). This species was observed doing considerable damage to the fruit of a small untreated Lisbon lemon orchard near Porterville in 1942 and 1943 (see p. 378), and was considered at the time to belong to the genus Tenuipalpus (521). The injury to lemons (fig. 273) closely resembles that found on pomegranates.

Description and Habits:- The adult mite averages 0.244 mm long and is "carrot to salmon red" in color. A dark area occurs in about the center of the mite. The proximal end of the femur is strongly constricted. During the winter the adults may be found grouped under flakes of loose bark (fig. 563). These overwintering mites, as well as mites reared in the laboratory, are a pale salmon color. In July or August, depending on the season, the mites may be seen on the fruit, at first near the stem end, then progressively further downward.

Injury:- The mites cause a brownish discoloration which extends progressively further from the stem end of the fruit as the season advances and may eventually envelop the entire fruit. This is followed by a cracking of the rind and the formation of scab tissue between the cracks. After the cracks are formed, the mites and their cast skins are found in these cracks.

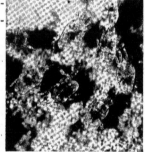

Fig. 563. Adults of Brevipalpus lewisi overwintering under a flake of bark of a pomegranate tree. X 22. Photo by Roy J. Pence.

A small percentage of the fruits will have a cracked and discolored surface whether mites are present or not. An inexperienced observer might mistake this type of blemish for mite injury, but the two types of blemish can be readily distinguished. In fig. 564, the two upper fruits are mite-infested. The checked area radiates out from the stem end and will be found at varying distances from the stem, depending on the age of the infestation. The light brown discolored area extends beyond the checked area and grades off in severity without being sharply delimited. The lower two fruits in fig. 564 have a blemish that is occasionally found on pomegranates, but is not caused by the mites. The checked and dark brown discolored areas coincide and are sharply defined. There is no tendency for the blemished area to be concentrated at the stem end of the fruit.

Control:- In experiments made in 1948, it was shown that all the commonly used mite treatments will control the brevipalpus mite, but since sulfur dust is highly

677

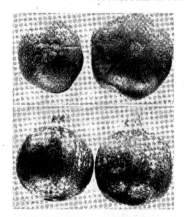

Fig. 564 (left). Above, injury to pomegranates caused by Brevipalpus lewisi; below, a checking and discoloration of the peel of, the pomegranate not associated with Brevipalpus infestations. Photo by Roy J. Pence.

Fig. 565 (right). The twig girdler. a, adult; b, larval mines; c, girdling by adult. After Moznette.

effective as well as being inexpensive, it is recommended as a control measure. In 1948, experimental and commercial control treatments made in June and July, with 1/2 lb. of sulfur dust per tree, resulted in excellent control, while untreated check trees were severely infested, with a high percentage of blemished fruit.

Western Leaf-Footed Plant Bug

The Western leaf-footed plant bug, Leptoglossus zonatus (Dallas) (fig. 292), has caused damage to pomegranate trees, as well as nearby citrus trees, in the Imperial Valley, California. On citrus, the removal of the bugs by jarring the trees, as well as tent-fumigation, have been employed in their control.

The citrus thrips (Scirtothrips citri), greenhouse thrips (Heliothrips haemorrhoidalis), melon aphid (Aphis gossypii), citricola scale (Coccus pseudomagnoliarum) and black scale (Saissetia oleae), also attack the pomegranate, but have seldom, if ever, been serious pests.

THE PERSIMMON

The Oriental persimmon, long an important fruit in China and Japan, is also cultivated in California, where there are 1,213 acres. There are also a few commercial plantings in Texas. The American persimmon grows wild from the Gulf of Mexico northward nearly to Iowa and to Connecticut. The wild fruit is highly prized in sections where it grows, and a few varieties have been propagated from wild trees bearing especially large fruit of superior flavor.

The persimmon is remarkably free of pests. In California there are none of economic importance, although the long-tailed mealybug and the latania scale may sometimes be seen on persimmon trees, especially under the sepals of the fruits. The soft scale, Coccus hesperidum, and the black scale, Saissetia oleae, also occasionally infest the persimmon.

In the Eastern States, where the persimmon is indigenous, it has been stated that the twig girdler, Oncideres cingulata (Say) (fig. 565), is the most important pest (335). The larva gnaws out a tunnel beneath the bark, encircling the twig and

678

causing it to break. It is recommended that the broken twigs should be picked off the ground in June or early July and burned, thus destroying the insects inside them. The flat-headed borer, Chrysobothris femorata Fab., is also an occasional pest of persimmon trees, especially young trees unprotected from sunburn. A moth, the persimmon root borer, Sannina uroceriformis Walk., occurs in the South Atlantic states. The larva may bore 17 to 22 inches below the surface of the ground or in the trunk above ground (724).

MANGO

The mango is considered to be a tropical rather than a subtropical fruit, yet Florida, with its occasional frosts and freezes, has introduced numerous modern varieties, a few of which are now proving to be a source of many promising seedlings (856). There is an experimental 2-1/2 acre planting of mangos at Leucadia, San Diego County, California and scattered trees throughout the State. Another planting of 20 acres is contemplated in San Diego County.

G. F. Moznette, in 1922, prepared a bulletin on "Insects of the Mango in Florida and How to Combat Them" in which he discussed the blossom anomola, Anomala undulata Mels., which attacks the blossom spike; the avocado red mite, Paratetranychus yothersi McG., which attacks the mango during the dry winter months; the red banded thrips, Selenothrips rubrocinctus (Giard); the mango shield scale, Coccus acuminatus Sign.;[1] the tessellated scale, Eucalymnatus tessellatus Sign.; the Florida red scale, Chrysomphalus aonidum L., the mango scale, Leucaspis indica Marlatt, and the Florida wax scale, Ceroplastes floridensis Comst.

In California the writer has observed only the long-tailed mealybug, Pseudococcus adonidum (L.), as a pest of the mango.

Among the above species, the blossom anomola, the avocado red mite, the red banded thrips, the Florida red scale, and the Florida wax scale have been discussed elsewhere.

Mango Shield Scale, Coccus acuminatus Signoret

The mango shield scale, a soft-bodied scale, and thus a honeydew-producing species, occurs in Florida wherever the mango is growing. It may be found infesting the lower leaf surfaces. The adult female is thin and flat, bluntly pointed in front, yellowish green, irregularly marked with black. Like the avocado, the mango sheds its older leaves in the spring, and many scales must perish at this time, but a sufficient number of crawlers have transferred to new growth to continue the infestation on the tree. There appear to be three overlapping generations per year in southern Florida.

This species has been satisfactorily controlled by means of oil spray.

Tessellated Scale, Eucalymnatus tessellatus Signoret

The tessellated scale is another soft-bodied scale often attacking mangos in Florida. It is oval, dark brown, with a mosaic appearance on the upper surface. The seasonal history and nature of the injury are similar to that of the mango shield scale. Oil sprays are recommended for control.

Mango Scale, Leucaspis indica Marlatt

This species is a pest of the mango in India, and was probably introduced into Florida from that country. A sac-like, waxy, white covering loosely adheres to the small, narrow convex body of this diaspine scale. When not abundant, it may be seen in cracks or under loose bark, but when it is found in large numbers it occurs on the branches and trunk of the tree and the general appearance of the infestation

---

[1] A "mango shield scale" is referred to by Sturrock (856) as being the most common of several scale insects attacking the mango in the West Indies. According to him, this species is Coccus mangiferae Green. Sturrock discusses the pests and diseases of the mango from a world standpoint.

then resembles somewhat that of the chaff scale. This species is found only on the mango in Florida. It can be controlled by oil spray.

Since the publication of the aforementioned bulletin by Moznette, other pests have begun to attack the mango in Florida (755). The broad mite, Hemitarsonemus latus (Banks) (fig. 513), has been observed in a mango nursery. Its symptoms, including a crinkling of the terminal leaves, rolling of the leaf margins, glazing of the leaf surfaces, and defoliation, are similar to those the writer has observed in attacks of the same species attacking citrus and avocados in greenhouses in California (see fig. 514). Dusting sulfur or lime sulfur sprays are recommended for control.

A small brownish ambrosia beetle, Hypocryphalus mangiferae (Stebbing), about 1/16 inch long, burrows into the branches and trunks of the trees, carrying fungi with it which cause a staining, decay, and serious damage to the infested branches. In preliminary tests, DDT spray has caused a reduction in the number of beetles.

The red banded thrips, Selenothrips rubrocinctus Giard, has been increasing in numbers in recent years. Nicotine sulfate at 1-800 or rotenone extractive sprays will control the thrips, providing they are wet by the spray. Two pounds of 10% gamma benzene hexachloride to 100 gallons of water may give satisfactory control, and 2 lbs. of 50% DDT or 50% chlordan to 100 gallons, plus 1% oil emulsion, are also effective (755). Scale insects have increased, in some cases, following the use of the chlorinated hydrocarbons.

PAPAYA

The papaya is a tropical fruit indigenous to the American tropics, possibly southern Mexico, but now grown in tropical regions throughout the world. There have been at times as many as several hundred acres of papayas in subtropical southern Florida.

Insect Pests in Florida (1008)

The papaya fruit fly, Toxotrypana curvicauda Gerst., was introduced into Florida about 1905. It is about a half inch long and has an ovipositor as long as the body proper. This long ovipositor can penetrate the entire distance through the flesh of the fruit. The larvae usually feed on the seeds and the lining of the seed cavity, then eat their way out of the fruit and drop to the ground for pupation in the soil Sanitation, consisting of picking up and destroying all dropped fruits, or fruits on the tree which turn yellow prematurely, or small fruits giving indications of being infested, is recommended for control. A poison bait spray, as used for the Mediterranean fruit fly, is also used in the control of the papaya fruit fly.

The papaya webworm, Homolapalpia dalera, has caused much damage to papaya plantings. This caterpillar forms a small web under which it eats its way into the growing fruit or its stem. Pyrethrum or rotenone sprays have given fair control when applied with enough pressure to remove the web. Lead arsenate, 1 lb. to 50 gallons, with at least 4 lbs. of hydrated lime as a safener, may also be used.

The root-knot nematode, Heterodera marioni (Cornu), frequently attacks the roots of the papaya plants, especially in soil previously planted to preferred hosts of the nematode. Vigorous plants are able to resist the attack fairly well. Infested land can be freed from nematodes by planting for a season with immune cover crops. Newly-cleared land will usually be free from nematodes.

In California the papaya is grown commercially in glasshouses at Leucadia. The only insect pest has been the long-tailed mealybug (fig. 566) which has been controlled by applying DDT or chlordan sprays to the stems of the plants and on the ground around the stems.

Fig. 566 (left). A California glasshouse papaya infested with long-tailed mealybug. X 0.5. Photo by Roy J. Pence.

Fig. 567 (right). Brevipalpus inornatus. After McGregor (565).

Pests of the Papaya in Hawaii (435)

In Hawaii the most important pests of the papaya are the mites. The species which has been known as the common red spider, Tetranychus telarius L., but which may prove to be T. althaeae von Hanst, causes injury characteristic of this species on other plants. Infested foliage becomes a yellowish-green. Brevipalpus inornatus (Banks) (Tenuipalpus bioculatus McG.) (fig. 567) causes calloused, grayish, scaly, or cracked areas on the fruit, which may be sunken below the normal level of the skin. The broad mite, Hemitarsonemus latus (Banks) (fig. 513), causes the infested leaves to be reduced in size and to become yellowish green to yellow. The lamina of the leaves become distorted, crinkled and leathery. Wettable sulfur sprays are used for control.

Aphids and thrips sometimes cause damage to the papaya plants in the first few weeks after they are placed in the field. The aphid species involved are Myzus persicae Sulz, Macrosiphum gei Koch and Aphis gossypii Glover. The onion thrips, Thrips tabaci Lind., may cause a serious distortion of the foliage of young plants. Aphids and thrips may be controlled with nicotine sulfate at 1 - 400 to which soap has been added (6 lbs. per 100 gallons) and this may be combined with the wettable sulfur spray for mites.

PESTS OF MISCELLANEOUS FRUITS

The Queensland or Macadamia nut may become infested with greenhouse thrips, Heliothrips haemorrhoidalis Bouché (fig. 568, left). A row of these trees in the orchard at the Los Angeles Campus of the University of California were severely infested with these pests in 1947, some of the trees being nearly entirely defoliated. A 10% DDT dust resulted in complete control, with no reappearance of the thrips. The sapote is also susceptible to infestation by greenhouse thrips (fig. 568, right), and this species may cause injury to the fruit of the cherimoya.

The loquat may at times bear a heavy infestation of the latania scale, Hemiberlesia lataniae Sign., but as in the case of the avocado, this insect does not cause as much damage as might be expected from the numbers of scale sometimes occurring on the tree. It may be controlled with oil spray. The ilicis mite, Paratetranychus ilicis McG. (fig. 569), causes a brownish discoloration along the midrib of the upper surface of the loquat leaf. Severe infestations cause a discoloration of the entire leaf. This mite can be readily controlled with DN dusts, Neotran or oil spray. The ilicis mite is darker than the majority of mites. Its color ranges from ferruginous to reddish brown, but with large irregular black areas on the dorsum. The dorsal bristles are prominent and arise from prominent tubercles. The eyes are carmine and rather conspicuous.

In the experimental orchard of the University of California, Los Angeles, some of the cherimoya trees in 1946-47 were rather severely infested with the snowy tree

Fig. 568. Injury from greenhouse thrips to the foliage of macadamia nut (left) and fruit of sapote (right). Photo by Roy J. Pence.

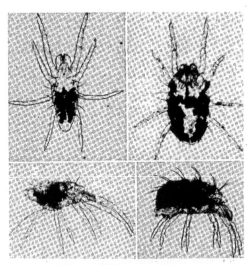

Fig. 569. Paratetranychus ilicis. Left, males; right, females. Greatly enlarged. Photo by Roy J. Pence.

682

cricket, Oecanthus niveus (De Geer)(fig. 570). This is a very pale green species about 14 mm long. An infestation is first indicated by the small round holes in the twigs and branches, similar to those made by the "shot hole borer" and other bark beetles. If one of these holes is examined closely, the minute white projections of the "cap" of the long pale yellow egg may be seen. The whitish nymphs feed on the flowers, young fruit, and foliage of fruit trees and sometimes cause damage by injuring the blossoms and scarring the fruit. Twigs bearing many egg punctures may be easily broken.

The snowy tree cricket also occasionally attacks persimmon trees, on which it is a carrier of a pathogenic fungus attacking the twigs.

Fig. 570. A snowy tree cricket Oecanthus niveus, on a cherimoya twig. X 2. Photo by Roy J. Pence.

683

LITERATURE CITED

1. Abbott, W. S. 1925. A method of computing the effectiveness of an insecticide. Jour. Econ. Ent. 18:265-7.
2. Adams, J. A. and E. H. Wheeler. 1946. Rate of development of milky disease in Japanese beetle populations. Jour. Econ. Ent. 39:248-54.
3. Adams, N. K. 1941. The physics and chemistry of surfaces. Third Ed. Oxford Univ. Press.
4. Afzal Husain, M. 1945. The citrus aleyrodidae (Homoptera) in the Punjab and their control. Mem. Ent. Soc. India No. 1, 41 pp, 3 pls.
5. Alexander, C. C., C. C.Cassill, F. P. Dean, and E. J. Newcomber. 1944. The use of oleic acid and aluminum sulfate to increase deposits of nicotine bentonite. Jour. Econ. Ent. 37:610-17.
6. Allen, T. C. 1947· Suppression of insect damage by means of plant hormones. Jour. Econ. Ent. 40:814-17.
7. _____and J. W. Brooks. 1940. The effect of alkaline dust diluents on toxicity of rotenone bearing roots as shown by tests with houseflies. Jour. Agr. Res. 60:839-45.
8. _____, R. J. Dicke, and H. H. Harris. 1944. Sabadilla, Schoenocaulon spp., with reference to its toxicity to houseflies. Jour. Econ. Ent. 37:400-407.
9. Allison, J. R. 1931. Quantity of oil retained by citrus foliage after spraying. Calif. Citrog. 16:481.
10. Allman, S. L. 1939. The Queensland fruit fly. Observations on breeding and development. New South Wales Agr. Gaz. 50:499-501, 547-549.
11. _____1942. Foliage poison spray for Queensland fruit fly. New South Wales Agr. Gaz. 53:186-87.
12. Anonymous. 1935. The fruit tree leaf roller (Tortrix postvittana). New South Wales Agr. Gaz. 46:640.
13. _____1936. A report on Sciobius granosus (citrus snout beetle) with reference to its occurrence in the Muden Valley and recommendations for its control. Citrus Grower No. 44, pp. 23-25, 27-35, 37-40.
14. _____1936. Grape vine mites. New South Wales Agr. Gaz. 47:457.
15. _____1938. Insect pests and their control. New South Wales Agr. Gaz. 49:391-95.
16. _____1939. Mite control on potted plants. Florist's Review. 83(2148):16.
17. _____1939. Orange cultivation and production in Spain. Hadar 12(1):9-12.
18. _____1940. Citrus cultivation in U.S.S.R. Hadar 13:251.
19. _____1942. Insect pests. New South Wales Agr. Gaz. 53:36-40.
20. _____1943. Campana de erradicación de la mosca de la fruta en la ciudad de Arica y valles de Azapa, Codpa y Timar. Bol. Dep. Sanid. veg. 2(2):175-80.
21. _____1943. Report on the marketing of citrus fruits in India. Govt. Ind. Marketing Ser. 43.
22. _____1944. Insect pests. New South Wales Agr. Gaz. 55(9):402-406.
23. _____1945. Control of red scale. Citrus News 21(10):154-55, 159.
24. _____1946. DDT and other insecticides and repellants developed for the armed forces. U.S.D.A. Misc. Publ. No. 606. 71 pp, 27 figs.
25. _____1946. The dependence of agriculture on the beekeeping industry - a review. U.S.D.A. Bur. Ent. & Plant Quar. E-584.
26. _____1946. Report on the citrus industry survey, 1945. Bur. Agr. Econ., Dept. Comm. & Agr., Commonwealth of Australia. Bul. No. 1.
27. _____1947. Agricultural Statistics 1947. U.S. Dept. Agr. See specifically pp. 178-183.
28. _____1947. Bromo analog of DDT. Agr. Chem. 2(6):59.
29. _____1947. Plagas determinantes de los cítricos. Dir. Gral. de Agric. (Mexico).
30. _____1947. The bronze orange bug (Rhoecocorus sulciventris). New South Wales Agr. Gaz. 58:197-99.

LITERATURE CITED

31. Anonymous. 1947. Warning on use of hexaethyl tetraphosphate. U.S.D.A., Bur. Ent. & Pl. Quar. Mineog. Apr. 3, 1947.
32. _____1948. Aircraft for spraying and dusting. U.S.D.A., Bur. Ent. & Pl. Quar. EC-2
33. _____1948. Sulphur and sulphuric acid. Chem. Engin. 55:104-5, 128.
34. _____1948. Fig. insects. La. Agr. Exp. Sta. Insect Pest Cont. Ser. Leaflet No. 17. Revised Jan. 1948.
35. _____1949. AIFA adopts new name, broader membership basis. To serve entire agricultural pest control industry. AIF News 7(3):1-3.
36. Armitage, H. M. 1945. Twenty-sixth annual report of the Bureau of Entomology and Plant Quarantine. Calif. St. Dept. Agr. Bul. 34:157-212.
37. _____1946. The grapeleaf skeletonizer in California. The Blue Anchor 23(2): 6, 25-32.
38. _____1946. Twenty-seventh annual report of the Bureau of Entomology and Plant Quarantine. Calif. St. Dept. Agr. Bul. 35:185-229.
39. _____1947. Twenty-eighth annual report of the Bureau of Entomology and Plant Quarantine. Calif. St. Dept. Agr. Bul. 36:141-201.
40. _____1949. The oriental fruit fly. Calif. Citrog. 34:236, 252.
41. _____and J. B. Steinweden. 1945. Fumigation of California dates with methyl bromide. Calif. St. Dept. Agr. Bul. 34:101-107.
42. Autuori, M. 1941. Contribuicao para o conhecimento de sauva (Atta spp.- Hymenoptera-Formicidae). I. Evolucao do sauveiro. (Atta sexdens rubropilosa Forel, 1908). Arg. Inst. Biol. 12:197-228, 4 pls. 4 figs., 10 refs. Sao Paulo, 1941. Abst. Rev. Appl. Entom. Ser. A. 31:478.
43. Ayoutantis, A. 1940. Scale insects observed on citrus in the island of Crete. Int. Rev. Agric. 31(1) 1-2 (Rome).
44. Back, E. A. and R. T. Cotton. 1926. Biology of the saw-toothed grain beetle, Oryzaephilus surinamensis Linne. Jour. Agr. Res. 33:435-52.
45. _____and R. T. Cotton. 1935. Industrial fumigation against insects. U.S.D.A. Circ. 369.
46. Bacon, O. G., A. E. Michelbacher, and W. H. Hart. 1948. Catalina cherry moth attacks walnuts. Diamond Walnut News 30(5):10.
47. _____, A. E. Michelbacher, and G. L. Smith. 1945. Control of Catalina cherry moth on walnuts. Calif. Agr. Exp. Sta. Circ. 365, pp. 85-87.
48. Baier, W. E. 1932. Brief report on effect of oil sprays on quality of Navels. Calif. Fruit Growers Exch., Office of Field Dept., Pest Cont. Circ. No. 114.
49. Bailey, S. F. 1942. The grape or vine thrips, Drepanothrips reuteri. Jour. Econ. Ent. 35:382-86.
50. _____1948. The peach twig borer. Calif. Agr. Exp. Sta. Bul. 708.
51. Baker, A. C., W. E. Stone, C. C. Plummer, and M. McPhail. 1944. A review of studies on the Mexican fruit fly and related Mexican species. U.S. Bur. Ent. & Pl. Quar. Misc. Publ. No. 531, 155 pp., 82 figs., 8 tables.
52. Baker, E. W. 1938. Fig mites and insects and their control. Pacific Rural Press 135:300.
53. _____1939. The fig mite, Eriophyes ficus Cotte, and other mites of the fig tree, Ficus carica Linn. Calif. St. Dept. Agr. Bul. 28:266-75.
54. _____1945. Studies on the Mexican fruitfly known as Anastrepha fraterculus. Jour. Econ. Ent. 38:95-100.
55. Balachowsky, A. and L. Mesnil. 1935. Les Insectes nuisibles aux plantes cultivées. Busson, Paris.
56. Balock, J. W. and D. F. Starr. 1945. Mortality of the Mexican fruitfly in mangoes treated by the vapor-heat process. Jour. Econ. Ent. 38:646-51.
57. Bancroft, W. D. 1912-13. The theory of emulsification. Jour. Phys. Chem. 16:177-233, 345-72, 475-512, 739-57; 17:501-19.
58. Banks, N. 1915. The Acarina or mites. U.S.D.A. Office of Secretary Rpt. 108.
59. Barber, H. S. 1919. Avocado seed weevils. Proc. Ent. Soc. Wash. 21:53-60. 1 pl.
60. Barger, W. R., W. T. Pentzer, and C. K. Fisher. 1948. Low temperature storage retains quality of dried fruit. Food Industries, March, 1948.
61. Barnes, D. F. 1944. Notes on the life history and other factors affecting control of the grape leaf folder. U.S.D.A. Bur. Ent. & Pl. Quar. E-616.
62. _____and D. L. Lindgren. 1947. Progress of work on beetle infestation in dates. Date Growers' Inst., 24th Ann. Rpt. pp. 3-4.

LITERATURE CITED

Barnhart, C. S. 1943. A pattern of thrips feeding recorded. Jour. Econ. Ent. 36:476-7.

Barrett, R. E. 1932. An annotated list of the insects and arachnids affecting the various species of walnuts or members of the genus Juglans Linn. Univ. of Calif. Publ. in Entom. 5:275-309.

Bartholomew, E. T. 1936. The effects of oil sprays on plant tissues. Third Western Shade Tree Conference Proc. 3:53-61.

——and E. C. Raby. 1935. Photronic photoelectric turbidimeter for determining hydrocyanic acid in solutions. Indus. & Engin. Chem., analyt. ed. 7:68-9.

——, W. B. Sinclair and B. E. Janes. 1939. Factors affecting the recovery of hydrocyanic acid from fumigated citrus tissues. Hilgardia 12:473-95.

——, W. B. Sinclair and D. L. Lindgren. 1942. Measurements on hydrocyanic acid absorbed by citrus tissues during fumigation. Hilgardia 14:373-408.

——, W. B. Sinclair and E. C. Raby. 1934. Granulation (crystalization) of Valencia oranges. Calif. Citrog. 19:88-89, 106, 108, 3 figs.

Basinger, A. J. 1923. The eradication campaign against the white snail (Helix pisana) at La Jolla, California. Calif. St. Dept. Agr. Mo. Bul. 12(1-2):7-11.

——1924. A supposedly beneficial insect discovered to be a citrus pest. Jour. Econ. Ent. 17:637-39.

——1931. The European brown snail in California. Calif. Agr. Exp. Sta. Bul. 515.

——1931. Field key for the determination of some of the common mealybugs infesting nursery stock in California. Calif. St. Dept. Agr. Mo. Bul. 20:189-93.

——1936. Notes on the orange worms Argyrotaenia (Tortrix) citrana Fern. and Platynota stultana Wlsm. Jour. Econ. Ent. 29:131-34.

——1938. The orange tortrix, Argyrotaenia citrana. Hilgardia 11:635-69.

——and A. M. Boyce. 1936. Orange worms in California and their control. Jour. Econ. Ent. 29:161-68.

Bastionsen, O. and O. Hassel. 1947. A new isomer of hexachlorocyclohexane with zero dipole moment. (In English). Acta Chem. Scand. 1:683.

Batchelor, L. D., O. L. Braucher, and E. F. Serr. 1945. Walnut production in California. Calif. Agr. Exp. Sta. Circ. 364.

——and H. J. Webber. 1948. The Citrus Industry. Vol. II. Univ. Calif. Press, Berkeley, Calif.

——and M. H. Kimball. 1947. Walnut growers turn squirrel catchers. Diamond Walnut News 19(3):4-6.

Bedford, E. C. G. 1943. The biology and economic importance of the South African citrus thrips, Scirtothrips aurantii (Faure). Publ. Univ. Pretoria 2(Nat. Sci.) no. 7, 68 pp., 8 pls., 7 figs.

——1943. The biology of the citrus thrips. Farming in South Africa 18(205):175-80.

Beran, F. 1948. Die Frostspritzung, eine Möglichkeit zur Erhöhung der Wirksamkeit ölhältiger Winterspritzmittel. Pflanzenschutz Berichte 11-12:161-75.

Berger, E. W. 1910. Whitefly control. Fla. Agr. Exp. Sta. Bull. 103.

——1921. Natural enemies of scale insects and whiteflies in Florida. Florida State Plant Bd. Quart. Bul. 5:141-154, 10 figs.

——1944. Red aschersonia for control of whiteflies. Citrus Ind. 25(4): 16-17, 22.

Bergold, G. 1947. Die Isolierung des Polyeder-Virus und die Natur der Polyeder. Zeitsch. für Naturforsch. 2b (3-4):122-43, 51 figs., 12 tab.

——1948. Bündelförmige Ordnung von Polyederviren. Zeitsch. für Naturforsch. 3b (1-2):25-26 (13 fig.).

Bishopp, F. G. 1946. The insecticide situation. Jour. Econ. Ent. 39:449-59.

Bliss, C. I. 1935. The calculation of the dosage-mortality curve. Ann. Appl. Biol. 22:134-67.

——, B. M. Broadbent, and S. A. Watson. 1931. The life history of the California red scale Chrysomphalus aurantii Maskell: progress report. Jour. Econ. Ent. 24:1222-29.

Bodenheimer, F. S. and H. Z. Klein. 1934. On some moths injurious to citrus trees in Palestine. Hadar 7(1):8-10.

687

LITERATURE CITED

93. Borden, A. D. 1915. The mouthparts of the Thysanoptera and the relation of thrips to the non-setting of certain fruits and seeds. Jour. Econ. Ent. 8:354-60.

94. _____1934. The tank-mixture method for dormant oil spraying of deciduous fruit trees in California. Calif. Agr. Exp. Sta. Bul. 579.

95. _____1948. Low volume spraying. Calif. Agr. 2(6):5, 16.

96. _____1948. The orange tortrix -- pest of citrus becoming of economic importance on deciduous fruit. Calif. Agr. 2(8):13.

97. Bousquet, E. W., P. L. Salzberg, and H. F. Dietz. 1935. New contact insecticides from fatty alcohols. Ind. Eng. Chem. 27:1342-44.

98. Bowen, C. V. and W. F. Barthel. 1943. Nornicotine in commercial nicotine sulfate solutions. Jour. Econ. Ent. 36:627.

99. Box, H. E. 1941. Citrus moth investigations. Report on investigations carried out from December, 1939, to August, 1941. Asuansi, Colonial Development Fund, 1941. Abst. Rev. Appl. Ent. A 30:505-6.

100. Boyce, A. M. 1934. Bionomics of the walnut husk fly, Rhagoletis completa. Hilgardia 8(11):363-579.

101. _____1935. The codling moth in Persian walnuts. Jour. Econ. Ent. 28:864-73.

102. _____1936. The citrus red mite (spider) problem in southern California. Calif. Citrog. 21:418, 438-41.

103. _____1937. The increasing importance of the walnut husk fly. Diamond Walnut News 19(2):6-7.

104. _____1937. How to control the walnut husk fly in 1937. Diamond Walnut News 19(4):8-9.

105. _____1938. Dinitro-o-cyclohexylphenol in the control of the citrus red mite. Jour. Econ. Ent. 31:781.

106. _____1948. Insects and mites and their control. The Citrus Industry, Univ. of Calif. Press. (Chapter 14 by A. M. Boyce).

107. _____and B. R. Bartlett. 1939. Red spiders on walnuts and their control. Diamond Walnut News 21(4):4-5.

108. _____and B. R. Bartlett. 1941. Lures for the walnut husk fly. Jour. Econ. Ent. 34(2):318.

109. _____and W. H. Ewart. 1946. Regarding the uses of DDT for control of citrus thrips and citricola scale in central California during the season of 1946. Univ. of Calif. Citrus Exp. Sta. Div. of Ent. News Letter No. 25.

110. _____and J. F. Kagy. 1941. Control of the citrus red mite. Calif. Citrog. 26:154, 171-73.

111. _____, J. F. Kagy, C. O. Persing, and J. W. Hansen. 1939. Studies with dinitro-o-cyclohexylphenol. Jour. Econ. Ent. 32:432-50.

112. _____and R. B. Korsmeier. 1941. The citrus bud mite, Eriophyes sheldoni Ewing. Jour. Econ. Ent. 34:745-56.

113. _____, R. B. Korsmeier, and C. O. Persing. 1942. The citrus bud mite and its control. Calif. Citrog. 27:134-25, 134-38, 140-41.

114. _____and Janet Mabry. 1937. The greenhouse thrips on oranges. Calif. Citrog. 23:19-20, 28.

115. _____and J. C. Ortega. 1947. Control of orange worms. Calif. Citrog. 32:318-19.

116. _____and C. O. Persing. 1939. Tartar emetic in the control of citrus thrips on lemons. Jour. Econ. Ent. 32:153.

117. _____and D. T. Prendergast. 1936. Dinitro-orthocyclohexylphenol offers promise in control of citrus red mite. Jour. Econ. Ent. 29:218-19.

118. _____, D. T. Prendergast, J. F. Kagy, and J. W. Hansen. 1939. Dinitro-o-cyclohexylphenol in the control of mites on citrus and Persian walnuts. Jour. Econ. Ent. 32:450-67.

119. _____and W. R. Stanton. 1933. Application - an important factor in codling moth control. Diamond Walnut News 15(2):3-4.

119a. Bodenheimer, F. S. 1935. A visit to the citrus district of southern Turkey April 1934. Hadar 8:10-14.

120. Boyden, B. L. 1941. Eradication of the parlatoria date scale in the United States. U.S.D.A., Bur. Ent. & Pl. Quar. Misc. Publ. No. 433.

121. Branigan, E. J. 1916. A satisfactory method of rearing mealybugs for use in parasite work. Calif. St. Com. Hort. Mo. Bul. 5:304-6.

122. Brescia, F. and I. B. Wilson. 1947. Larvicidal treatment of large areas by ground dispersal of DDT aerosols. Jour. Econ. Ent. 40:309-13.

LITERATURE CITED

123. Brescia, F. and I. B. Wilson. 1947. Treatment of native villages with aerosol generator. Jour. Econ. Ent. 40:313-16.
124. _____and I. B. Wilson. 1947. Aerosol generator as used for sand fly control. Jour. Econ. Ent. 40:316-19.
125. Brett, C. H. and W. C. Rhoades. 1948. Grasshopper control with parathion, benzene hexachloride, chlorinated camphene, and chlordane. Jour. Econ. Ent. 41:16-18.
126. Brinley, F. J. and R. H. Baker. 1927. Some factors effecting the toxicity of hydrocyanic acid for insects. Biol. Bull. 53:201-207.
127. Bronson, T. E. and S. A. Hall. 1946. Hexaethyl tetraphosphate. Agr. Chem. 1(7):19.
128. Bruner, S. C. and L. Ochoa. 1946. Metodo de azufre y arsénico para combatir la bibijagua. Estac. Exp. Agr., Cuba, Circ. No. 86.
129. _____and Valdes Barry, F. 1946. La bibijagua. Estac. Exp. Agr., Cuba, Circ. No. 87.
130. Brunton, J. G. 1947. Cryolite as an insecticide. Agr. Chem. 2:27-28, 64.
131. Buffa, P. 1898. Contributo allo studio anatomico della Heliothrips haemorrhoidalis Fabr. Revista di Patologia vegetale 7:94-108, 129-36.
132. Burger, O. F. and A. F. Swain. 1918. Observations on a fungus enemy of the walnut aphis in southern California. Jour. Econ. Ent. 11:278-89.
133. Busck, A. 1917. The pink bollworm: Pectinophora gossypiella. Jour. Agr. Res. 9:343-369. See specifically pages 362-66.
134. Busvine, J. R. and S. Barnes. 1947. Observations on mortality among insects exposed to dry insecticidal films. Bull. Ent. Res. 38:81-90.
135. Buxton, P. A. 1920. Insect pests of the date palm in Mesopotamia and elsewhere. Bull. Ent. Res. 11:287-303.
136. Byars, L. P. 1921. Notes on the citrus nematode. Phytopath. 11:90-94.
137. Byrd, N. J., R. E. Blair, and H. C. Phillips. 1948. California fruit and nut crop acreage estimates as of 1947. Calif. St. Dept. Agr. Bul. 37:37-64.
138. Camacho, A. A. 1944. El periquito del aguacate. Fitófilo 3(3):3-54.
139. Camacho, A. O. 1944. Aplicación de emulsión de aceite en contra de las escamas de los cítricos. Fitófilo (Mexico) 3(3):36-41.
140. Camp, A. F. 1945. Citrus industry of Florida. Part I. Citrus growing in Florida pp. 102-3. Fla. Dept. Agr. in cooperation with Univ. of Fla.
141. Campau, E. J. and H. F. Wilson. 1944. Dispersants for rotenone. A study of the effect of dispersants on the toxicity of rotenone dusts. Soap 20:117, 119, 121.
142. Campbell, F. L. and F. R. Moulton. 1943. Laboratory procedures in studies of the chemical control of insects. Publ. Amer. Assoc. Adv. Sci. No. 20.
143. _____, W. N. Sullivan, L. E. Smith, and H. L. Haller. 1934. Insecticidal tests of synthetic organic compounds - chiefly tests of sulfur compounds against culicine mosquito larvae. Jour. Econ. Ent. 27:1176-85.
144. Campbell, R. E. 1914. A new coccid infesting citrus trees in California (Hemip.) Entom. News 25:222-4.
145. Campbell, R. W. 1948. Recent developments in spray equipment. Kansas St. Hort. Soc. Biennial Rpt. 49:85-90.
146. Carlisle, P. J. 1933. Manufacture, handling, and use of hydrocyanic acid. Ind. Eng. Chem. 25:959-64.
147. Carman, G. E. 1948. Provisional suggestions for the use of DDT to control red scale on citrus. Univ. Calif. Citrus Exp. Sta., Div. of Ent. News Letter No. 36.
148. Carter, R. H. 1930. Solubilities of fluosilicates in water. Ind. Eng. Chem. 22:886-87.
149. _____1931. The incompatibility of lime with fluosilicates. Jour. Econ. Ent. 24:263-68.
150. _____and R. L. Busbey. 1939. The use of fluorine compounds as insecticides, a review with annotated bibliography. U.S.D.A. Bur. Ent. & Pl. Quar. E-466.
151. Carter, W. 1943. A promising new soil amendment and disinfectant. Science 97(2521):383-84.
152. Chamberlin, J. C. 1941. Entomological nomenclature and literature. Edwards Bros. Inc., Ann Arbor, Mich.

LITERATURE CITED

153. Chandler, W. H. 1942. Deciduous orchards. Lea & Febiger, Philadelphia, Pa.
154. Chapman, P. J., A. W. Avens and G. W. Pearce. 1944. San Jose scale control experiments. Jour. Econ. Ent. 37:305-6.
155. _____, J. G. Pearce and A. W. Avens. 1943. Relation of composition to the efficiency of foliage or summer type petroleum fractions. Jour. Econ. Ent. 36:241-47.
156. Chapman, R. N. 1931. Animal ecology. McGraw-Hill Book Co., New York, London.
156a. Chen, S. H. 1940. Two new Dacinae from Szechwan. Sinensia, 11(1-2):131-36.
157. Cherian, M. C. 1942. Our present position with regard to the control of fruit pests. Madras Agr. Jour. 30(1):14-17.
158. Chisholm, R. D. 1948. A review of DDT formulations. U.S.D.A. Bul. E-742.
159. _____, L. Koblitsky, W. C. Fest, and E. D. Burgess. 1947. Insecticide duster. U.S.D.A. Bur. Ent. & Pl. Quar. ET-237.
160. Clark, S. W. 1931. The control of fire ants in the lower Rio Grande Valley. Texas Agr. Exp. Sta. Bul. No. 435.
161. _____and W. H. Friend. 1932. California red scale and its control in the lower Rio Grande Valley of Texas. Texas Agr. Exp. Sta. Bul. 455.
162. Clausen, C. P. 1915. Mealybugs of citrus trees. Calif. Agr. Exp. Sta. Bul. 258.
163. _____1927. The citrus insects of Japan. U.S.D.A. Tech. Bul. 15.
164. _____1931. Insects injurious to agriculture in Japan, U.S.D.A. Bul. 168.
165. _____1933. The citrus insects of tropical Asia. U.S.D.A. Cir. 266.
166. _____1936. Insect parasitism and biological control. Ann. Ent. Soc. Amer. 29:201-23.
167. _____1940. Entomophagous insects. McGraw-Hill Book Company, Inc., New York and London.
168. _____and P. A. Berry. 1932. The citrus blackfly in Asia, and the importation of its natural enemies into tropical America. U.S.D.A. Tech. Bul. 320.
169. Clermont, Philippe de. 1854. Note sur la preparation de quelques ethers. (Paris) Acad. des Sci. Compt. Rend. 39:338-341.
170. Cobb, N. A. 1913. Notes on Monnochus and Tylenchulus. Jour. Wash. Acad. Sci. 3:288-89.
171. _____1914. Citrus root nematode. Jour. Agr. Res. 2:217-230.
172. Cook, S. A. 1931. The webbing spider of citrus trees. Jour. Agr., Victoria 29:82-85.
173. Cockerell, T. D. A. 1899. Some insect pests of the Salt River Valley and the remedies for them. Ariz. Agr. Exp. Sta. Bul. No. 32.
174. Colby, G. E. 1907. Determination of kerosene in kerosene emulsion by the centrifugal method. U.S.D.A. Bur. Chem. Bul. 105, p. 165.
175. Cole, F. R. 1925. The natural enemies of the citrus aphid, Aphis spiraecola (Patch). Jour. Econ. Ent. 18:219-23.
176. _____1941. Red scale on rampage. (Digest of a discussion by F. R. Cole). Calif. Citrog. 27:61.
178. Compere, H. 1926. New coccid-inhabiting parasites (Encyrtidae, Hymenoptera) from Japan and California. Univ. Calif. Publ. in Entom. 4:33-50. See specifically pp. 44-50.
179. _____1929. Description of a new species of Coccophagus recently introduced into California. Univ. Calif. Publ. in Entom. 5:1-3.
180. _____1940. Parasites of the black scale, Saissetia oleae, in Africa. Hilgardia 13:387-425.
181. _____and S. E. Flanders. 1934. Anarhopus sydneyensis Timb., an encyrtid parasite of Pseudococcus longispinus (Targ.) recently introduced into California from Australia. Jour. Econ. Ent. 27:966-73.
182. _____and H. S. Smith. 1927. Notes on the life history of two oriental chalcidoid parasites of Chrysomphalus. Univ. Calif. Publ. in Entom. 4:63-73.
183. _____and H. S. Smith. 1932. The control of the citrophilus mealybug, Pseudococcus gahani, by Australian parasites. Hilgardia 6:585-618.
184. Condit, I. J. 1920. Caprifigs and caprification. Calif. Agr. Exp. Sta. Bul. 319.
185. _____and W. T. Horne. 1933. A mosaic of the fig in California. Phytopath. 27(11):887-96.

LITERATURE CITED

186. Condit, I. J. 1947. Olive culture in California. Calif. Agr. Ext. Service Cir. 135.
187. _____ and S. E. Flanders. 1945. "Gall-flower" of the fig, a misnomer. Science 102:128-30.
188. _____, W. E. Hoffman, and H. C. Wang. 1935. Observations on the culture of oranges near Swatow, China. Lingnan Agr. Jour. 1(2):175-248. (Chinese, with English summary).
189. _____ and W. T. Horne. 1933. A mosaic of the fig. Phytopath 23:7.
190. Cook, F. C. and N. E. McIndoo. 1923. Chemical, physical, and insecticidal properties of arsenicals. U.S.D.A. Dept. Bul. 1147.
191. Cooley, C. E. 1947. International air commerce and plant quarantine. Jour. Econ. Ent. 40:129-32.
192. Coquillett, D. W. 1890. The use of hydrocyanic acid gas for the destruction of the red scale. Insect Life 2:202-207.
193. Corbett, W. H. 1948. Dried fruit beetle controlled in dry yard. Fig Leaf 1(6):6.
194. Cotterell, G. S. 1940. Citrus fruit-piercing moths - summary of information and progress. Pap. 3rd W. Afr. Agr. Conf. Nigeria June 1938 Sect. Gold Coast 1:11-24, 5 pls. Abstr. Rev. Appl. Ent. 29:106-7.
195. Cottier, W. 1938. Citrus pests: red mite and thrips. New Zealand Jour. Agric. 57:138-39.
196. Cotton, R. T. 1941. Insect pests of stored grain and grain products. Burgess Publishing Co., Minneapolis, Minn.
197. _____, A. I. Balzer, and H. D. Young. 1944. The possible utility of DDT for insect-proofing paper bags. Jour. Econ. Ent. 37:140.
198. _____ and R. C. Roark. 1928. Ethylene oxide as a fumigant. Ind. Eng. Chem. 20:805.
199. Courtney, W. D., E. P. Breakey and L. L. Stitt. 1947. Hot-water tanks for treating bulbs and other plant materials. Wash. St. Col. Agr. Exp. Sta. Pop. Bul. No. 184.
200. Cox, A. J. 1941. Preliminary draft of compatibility data regarding economic poisons used for plant protection. Calif. State Dept. Agr. Special Publ. No. 184, 81 pp.
201. _____ 1943. Terminology of insecticides, fungicides and other economic poisons. Jour. Econ. Ent. 36:813-21.
202. Crane, D. B. 1933. Rotenone- a new parasiticide. Cornell Vet. 23:15-31.
203. Crawford, D. L. 1927. Investigations of the Mexican fruit fly (Anastrepha ludens Loew) in Mexico. Calif. State Dept. Agr. Mo. Bul. 16:422-35.
204. Crawford, J. C. 1940. The male of Heliothrips haemorrhoidalis (Bouché) (Thysanoptera). Proc. Ent. Soc. Wash. 42:91-2.
205. Creighton, J. T. 1945. Platypus compositus attacking citrus. Jour. Econ. Ent. 38:706.
206. Cressman, A. W. 1933. Biology and control of Chrysomphalus dictyospermi (Morg.). Jour. Econ. Ent. 26:698-706.
207. _____ 1935. Pests. U.S.D.A. Farmers' Bul. 1031:23-27.
208. _____ 1941. Susceptibility of resistant and non-resistant strains of the California red scale to sprays of oil and cube resins. Jour. Econ. Ent. 34:859.
209. _____ 1943. Effectiveness against the California red scale of cube resins and nicotine in petroleum spray oil. Jour. Agr. Res. 67:17-26.
210. _____ and B. M. Broadbent. 1944. Changes in California red scale populations following spray of oils with and without derris resins. Jour. Econ. Ent. 37:809-813.
211. _____ and L. H. Dawsey. 1936. The comparative insecticidal efficiency against the camphor scale of spray oils with different unsulphonatable residues. Jour. Agr. Res. 52(11):865-78.
212. _____ and H. K. Plank. 1935. The camphor scale. U.S.D.A. Bur. Ent. & Pl. Quar. Cir. 365.
213. Daasch, L. W. 1947. Infrared spectroscopic analysis of five isomers of 1,2,3,4,5,6-hexachlorocyclohexane. Ind. Eng. Chem. 19:779-85.
214. Davidson, W. M. 1914. Walnut aphides in California. U.S.D.A. Bul. 100.
215. _____ 1916. Economic Syrphidae in California. Jour. Econ. Ent. 9:454-57.
216. _____ and R. L. Nougaret. 1921. The grape phylloxera in California. U.S.D.A. Bul. 903.

LITERATURE CITED

217. Davis, A. C. and H. D. Young. 1934. Sulfur fumigation of mushroom houses. Jour. Econ. Ent. 27:518-25.
218. Davis, J. J. and P. Lugenbill. 1921. The green June beetle or fig eater. N. C. Agr. Exp. Sta. Bul. 242:1-35.
219. Dean, M. L. 1926. The Western Plant Quarantine Board, its purposes and organization. Jour. Econ. Ent. 19:388-94.
220. DeBach, P. 1943. The importance of host-feeding by adult parasites in the reduction of host populations. Jour. Econ. Ent. 36:647-58.
221. ———1947. Predators, DDT, and citrus red mite populations. Jour. Econ. Ent. 40:598-9.
221a. ———1949. The establishment of the Chinese race of Comperiella bifasciata on Aonidiella aurantii in southern California. Jour. Econ. Ent. 41:985.
222. ———and C. A. Fleschner. 1947. Ladybirds, lacewings, parasites tested as long-tailed mealybug controls in California citrus. California Agriculture 1 (13):1, 3.
223. ———, C. A. Fleschner and E. J. Dietrick. 1948. Natural control of the California red scale. Calif. Citrog. 34:6, 38-39.
224. ———and H. S. Smith. 1941. The effect of host density on the rate of reproduction of entomophagous parasites. Jour. Econ. Ent. 34:741-45.
225. ———and H. S. Smith. 1947. Effects of parasite population density on rate of change of host and parasite populations. Ecology 28:290-98.
226. Delgado de Garay, Alfonso. 1946. Cooperación entre los Estados Unidos de Norteamerica y la República de México, para el control del gusano rosado del algodonero. Jour. Econ. Ent. 39:95-98.
227. de Ong, E. R. 1924. The preparation and use of colloidal sulfur as a control for red spider. Jour. Econ. Ent. 17:533-8.
228. ———1931. Present trend in oil sprays. Jour. Econ. Ent. 24:978-85.
229. ———1948. Chemistry and uses of insecticides. Reinhold Publ. Co., New York.
230. ———1949. Soil fumigation: summary of chemicals and their uses. Calif. Fruit and Grape Grower. 3(1):7-8.
231. ———, H. Knight, and J. C. Chamberlin. 1927. A preliminary study of petroleum oil as an insecticide for citrus trees. Hilgardia 2:251-384.
232. Deonier, C. C. and H. A. Jones. 1946. TDE, 1,1-dichloro-2, 2-bis(p-chlorophenyl) ethane, as an anophelene larvicide. Science 103:13-14.
233. de Stefani, T. 1913. Helix pisana and the damage it is capable of doing to citrus trees. Nuovi Annali di Agricoltura Siciliaro, Anno II, N. S. Fasc. I, 1913, Palermo. Translated from the Italian by H. S. Smith.
234. Dethier, V. G. 1947. Chemical insect attractants and repellents. 289 pp: The Blakiston Co., Philadelphia, Toronto.
235. Dickson, R. C. 1941. Inheritance of resistance to hydrocyanic acid fumigation in the California red scale. Hilgardia 13:515-22.
236. ———and D. L. Lindgren. 1947. The California red scale. Calif. Citrog. 32:524, 542-44.
237. ———and E. Sanders. 1945. Factors inducing diapause in the Oriental Fruit moth. Jour. Econ. Ent. 38:605-6.
238. Dietz, H. F. and H. S. Barber. 1920. A new avocado weevil from the Canal Zone. Jour. Agr. Res. 20:111-15.
239. ———and J. Zetek. 1920. The black fly of citrus and other subtropical plants. U.S.D.A. Bur. of Ent. Bul. No. 885.
240. Donohoe, H. C., P. Simmons, D. F. Barnes, C. H. Kaloostian, and C. K. Fisher. 1943. Preventing damage to commercial dried fruits by the raisin moth. U.S.D.A. Leaflet No. 236.
241. Dover, M. W. 1947. Testing insecticide residues. Soap and Sanitary Chemicals. 23:139, 141, 143, 193.
242. Dowden, P. B., D. Whittam, H. K. Townes, and N. Hotchkiss. 1945. DDT applied against certain forest insects in 1944, particularly with aerial equipment. U.S.D.A. Bur. Ent. & Pl. Quar. E-663.
243. Dozier, A. L. 1924. Insect pests and diseases of the satsuma orange. Gulf Coast Citrus Exchange Educ. Bul. No. 1.
243a. Du Charme, E. P. 1949. Resistance of Poncirus trifoliata rootstock to nematode infestation in Argentina. Citrus Ind. 30(5):16-17.
244. du Noüy, P. Lecomte. 1919. A new apparatus for measuring surface tension. Jour. Gen. Phys. 1:521.

692

LITERATURE CITED

245. Duruz, W. P. 1923. The peach twig borer (Anarsia lineatella Zeller). Calif.
     Agr. Exp. Sta. Bul. 355.
246. Ebeling W. 1931. Method for determination of the efficiency of sprays and
     HCN gas used in the control of the red scale. Calif. St. Dept. Agr.
     Mo. Bul. 20:669-72.
247. _____1932. Experiments with oil sprays used in the control of the California
     red scale, Chrysomphalus aurantii (Mask.) (Homoptera-Coccidae), on lemons.
     Jour. Econ. Ent. 25:1007-1012.
248. _____1933. Variation in the population density of the California red scale,
     Aonidiella aurantii Mask., in a hilly lemon grove. Jour. Econ. Ent.
     26:851-54.
249. _____1933. Some differences in habits and structure between citrus thrips
     and flower thrips. Calif. St. Dept. Agr. Mo. Bul. 22:381-89.
250. _____1935. A new scolytid beetle found in the bark of lemon trees (Coleoptera,
     Scolytidae). Pan-Pac. Ent. 11:21-23.
251. _____1936. Effect of oil spray on California red scale at various stages of
     development. Hilgardia 10:95-125.
252. _____1936. Effect of spray materials on citrus trees. Calif. Citrog.
     21:370, 409.
253. _____1939. The grape bud beetle, Glyptoscelis squamulata Crotch. Calif.
     St. Dept. Agr. Bul. 28:459-65.
254. _____1939. The penetration of spray liquids into plant tissue. Verh. VII
     Internatl. Kong. für Entom., Berlin, 1938, 4:2966-82.
255. _____1939. The rôle of surface tension and contact angle in the performance
     of spray liquids. Hilgardia 11:665-98.
256. _____1940. A method for determination of oil-spray residue on citrus foli-
     age. Jour. Econ. Ent. 33:900-904.
257. _____1940. Toxicants and solids added to spray oil in control of California
     red scale. Jour. Econ. Ent. 33:92-102.
258. _____1944. The citrus industry and its pest problems in Mexico. Calif.
     Citrog. 29:278.
259. _____1945. Citrus in South America. Calif. Citrog. 30(11):338-39.
260. _____1945. DDT and rotenone used to control the California red scale. Jour.
     Econ. Ent. 38:556-63.
261. _____1945. DDT for control of the grape bud beetle. Jour. Econ. Ent.
     38:600.
262. _____1945. El cultivo de los cítricos en Mexico, Colombia, el Perú, y Chile.
     La Hacienda 40(12):600-603.
263. _____1945. El problema de los insectos en los árboles cítricos del Perú.
     Publication of the Department of Agricultural Information, Ministerio
     de Agricultura, Lima, Peru.
264. _____1945. Problemas relacionadas con las pestes que afectan a los citrus
     y otras plantas subtropicales en Chile. Agricultura Técnica (Chile)
     5:197-212.
265. _____1945. Properties of petroleum oils in relation to toxicity to potato
     tuber moth larvae. Jour. Econ. Ent. 38:26-34.
266. _____1946. Problemas entomológicas de los citrus en Colombia. Agricultura
     Tropical 2(2):9-10 and 2(3):41-44.
267. _____1946. Results with DDT in control of the grape bud beetle in Coachella
     Valley in 1946. Univ. Calif. Citrus Exp. Sta., Div. of Ent. News Letter
     No. 27.
268. _____1947. DDT preparations to control certain scale insects on citrus.
     Jour. Econ. Ent. 40:619-32.
269. _____, F. A. Gunther, J. P. LaDue, and J. J. Ortega. 1944. Addition of ex-
     tractives of rotenone-bearing plants to spray oils. Hilgardia 15:675-701.
270. _____and L. J. Klotz. 1936. The relation of pest control treatment to water
     rot of navel oranges. Calif. St. Dept. Agr. Mo. Bul. 25:360-68.
271. _____, L. J. Klotz, and E. R. Parker. 1938. Experiments on Navel water spot.
     Calif. Citrog. 23:410, 434-36.
272. Eckert, J. K. and A. Mallis. 1937. Ants and their control in California.
     Calif. Agr. Exp. Sta. Circ. 342.
273. Eisen, G. 1896. Biological studies on figs, caprifigs and caprification.
     Calif. Acad. Sci. Proc. 5:897-1003.

LITERATURE CITED

274. Eisen, G. 1901. The fig. U.S.D.A. Div. Pom. Bul. 9:1-317.
275. Ellisor, L. O. 1936. A toxicological investigation of nicotine on the goldfish and the cockroach. Iowa State Col. Jour. Sci. 11:51-53.
276. _____and C. R. Blair. 1940. Effect of temperature upon the toxicity of stomach poisons. Jour. Econ. Ent. 33:760-62.
277. _____and G. F. Turnipseed. 1940. Control of the major pests of the satsuma orange in South Alabama. Ala. Agr. Exp. Sta. Bul. 248.
278. Essig, E. O. 1913. The branch and twig borer. Calif. State Dept. Agr. Mo. Bul. 2:587.
279. _____1914. The mealybugs of California. Calif. St. Dept. Agr. Mo. Bul. 3:97-143.
280. _____1915. The dried-fruit beetle. Jour. Econ. Ent. 8:396-400.
281. _____1917. The olive insects of California. Calif. Agr. Exp. Sta. Bul. 283.
282. _____1926. Insects of Western North America. The Macmillan Co., New York.
283. _____1926. Paradichlorobenzene as a soil fumigant. Calif. Agr. Exp. Sta. Bul. 411.
284. _____1930. Plant quarantine. Calif. State Dept. Agr. Mo. Bul. 19:477-82.
285. _____1931. A history of entomology. The Macmillan Co., New York.
286. _____1933. Insects and agriculture. Jour. Econ. Ent. 26:869-72.
287. _____1942. College entomology. The Macmillan Co., New York.
287a. _____1949. Aphids in relation to quick decline and tristezia of citrus. Pan-Pacific Entom. 25:13-23.
288. _____and W. M. Hoskins. 1944. Insects and other pests attacking agricultural crops. Calif. Agr. Exp. Sta. Circ. 87.
289. _____and E. H. Smith, 1922. Two interesting new blister mites. Calif. St. Dept. Agr. Mo. Bul. 11:63.
290. Evans, A. C. and H. Martin. 1935. The incorporation of direct with protective insecticides and fungicides I. The laboratory evaluation of water-soluble wetting agents as constituents of combined washes. Jour. Pomol. and Hort. Sci. 13:261-292.
291. Everett, P. 1945. Scale insects and their control. New Zealand Jour. Agr. 70:85-86.
292. Ewart, W. H. 1946. DDT for citricola scale and citrus thrips in central California. Calif. Citrog. 31:393-94.
293. _____1949. The present status of the use of new insecticides on citrus in central California Part I. Citrus thrips and citricola scale. Univ. Calif. Citrus Exp. Sta. News Letter No. 39.
294. _____and P. DeBach. 1947. Important factors to consider in the use of DDT for the control of citrus thrips and citricola scale in central California in 1947. Univ. Calif. Citrus Exp. Sta. Ent. News Letter No. 30.
295. _____and P. DeBach. 1947. DDT for control of citrus thrips and citricola scale. Calif. Citrog. 32:242-45.
296. _____and P. DeBach. 1948. Suggestions concerning the use of DDT for the control of citrus thrips and citricola scale in central California in 1948, and a brief account of the new insecticide parathion. Univ. Calif. Citrus Exp. Sta., Div. Entom. News Letter No. 33. 6 pp. (Mimeogr.)
297. Ewing, H. E. 1939. A revision of the mites of the subfamily Tarsoneminae of North America, the West Indies, and the Hawaiian Islands. U.S.D.A. Tech. Bul. 653.
298. Eyer, J. R. 1945. Solid baits for codling moth. Jour. Econ. Ent. 38:344-46.
299. Fahey, J. E. 1941. A study of clays used in the preparation of tank-mix nicotine bentonite sprays.
300. Fawcett, H. S. 1908. Fungi parasitic upon Aleyrodes citri. Special studies I, Univ. Florida, 41 pp., 19 figs., 8 pls.
301. _____1912. Stem-end rot and scale insects. Fla. Agr. Exp. Sta. Press Bul. 195.
302. _____1927. The recent decay of Navel oranges. Calif. Citrog. 12:219.
303. _____, L. J. Klotz, and A. R. C. Haas. 1933. Water spot and water rot of citrus fruits. Calif. Citrog. 18:165, 175.
304. Fay, R. W., E. L. Cole, and A. J. Buckner. 1947. Comparative residual effectiveness of organic insecticides against house flies and malaria mosquitoes. Jour. Econ. Ent. 40:635-40.

LITÉRATURE CITED

Fenton, F. A. and G. A. Bieberdorf. 1948. DNO289 and DN-Dry Mix No. 1 show promise in dormant sprays for pecan nut casebearer control. Down to Earth (Dow Chem. Co.) 4(3):10.

Ferris, G. F. 1918. The California species of mealybugs. Stanford Univ. Publ. Univ. Series.

_____1928. The principles of systematic entomology. Stanford University Press.

_____1937-1942. Atlas of the scale insects of North America. Series I to IV. Stanford Univ. Press.

Finkle, P., H. C. Draper, and J. H. Hildebrand. 1923. The theory of emulsification. Jour. Amer. Chem. Soc. 45:2780-88.

Finney, D. J. 1947. Probit analysis: A statistical treatment of the sigmoid response curve. The Macmillan Co., New York.

Finney, G. L., S. E. Flanders and H. S. Smith. 1944. The potato tuber moth as a host for mass production of *Macrocentrus ancylivorus*. Jour. Econ. Ent. 37:61-64.

_____, S. E. Flanders and H. S. Smith. 1947. Mass culture of *Macrocentrus ancylivorus* and its host, the potato tuber moth. Hilgardia 17:437-83.

Fisher, R. A. 1941. Statistical methods for research workers. 8th ed. Oliver and Boyd, London.

_____and W. A. Mackenzie. 1923. Studies in crop variation II. The manurial response of different potato varieties. Jour. Agr. Sci. 13:311.

Fisk, F. W. and H. H. Shepard. 1938. Laboratory studies of methyl bromide as an insect fumigant. Jour. Econ. Ent. 31:79-84.

Fiske, W. F. 1910. Superparasitism: an important factor in the natural control of insects. Jour. Econ. Ent. 3:88-97.

Flanders, S. E. 1925. Longevity of the adult codling moth. Jour. Econ. Ent. 18:752-3.

_____1935. Host influence on the prolificacy and size of *Trichogramma*. Pan Pacific Ent. 11:175-77.

_____1940. Biological control of the long-tailed mealybug, *Pseudococcus longispinus*. Jour. Econ. Ent. 33:754-59.

_____1940. Environmental resistance to the establishment of parasitic Hymenoptera. Ann. Ent. Soc. Amer. 33:245-53.

_____1941. Dust as an inhibiting factor in the reproduction of insects. Jour. Econ. Ent. 34:470-71.

_____1941. Peculiar habits of beneficial insects. Calif. Citrog. 26:285,306.

_____1942. *Metaphycus helvolus*, an encyrtid parasite of the black scale. Jour. Econ. Ent. 35:690-98.

_____1942. Biological observations on the citricola scale and its parasites. Jour. Econ. Ent. 35:830-3.

_____1942. Oosorption and ovulation in relation to oviposition in the parasitic Hymenoptera. Ann. Ent. Soc. Amer. 35:251-66.

_____1943. The susceptibility of parasitic Hymenoptera to sulfur. Jour. Econ. Ent. 36:469.

_____1944. Control of the long-tailed mealybug on avocados by hymenopterous parasites. Jour. Econ. Ent. 37:308.

_____1944. Diapause in the parasitic Hymenoptera. Jour. Econ. Ent. 37:408-11.

_____1945. Coincident infestations of *Aonidiella citrina* and *Coccus hesperidum*, a result of ant activity. Jour. Econ. Ent. 38:711-12.

_____1945. The surplus male. A problem of biological control. Calif. Citrog. 30:267-68, 275.

_____1947. Elements of host discovery exemplified by parasitic Hymenoptera. Ecology 28:299-309.

_____1948. Biological control of yellow scale. Calif. Citrog. 34:56, 76-77.

_____1949. Using black scale as a "foster host." Calif. Citrog. 34:222-24.

Fleming, W. E. 1930. Treatment of soil to destroy the Japanese beetle. Jour. Econ. Ent. 23:502-508.

Fletcher, W. F. 1915. The native persimmon. U.S.D.A. Farmers' Bul. 685.

Fleury, A. C. 1937. Bureau of Plant Quarantine. Calif. St. Dept. Agr. Bul. 26:540-62.

Fonseca, J. 1933. Insectos e acarinos productores de manchas e lesos nos fructos citricos. Sec. da Agr. Ind. e Com. do Estado de São Paulo.

LITERATURE CITED

338. Foote, F. J. and K. D. Gowans. 1947. Citrus nematode. Calif. Citrog. 32:522-23, 540-41.

339. Ford, W. K. 1933. Some observations on the bionomics of the false codling moth, *Argyroploce leucotreta* Meyr., in Southern Rhodesia. British So. Africa Co. Mazoe Citrus Exper. Sta. Ann. Rep. 1933:9-34.

340. Foster, S. W. 1910. On the nut feeding habits of the codling moth. U.S.D.A. Bur. Ent. 80(V):67-70.

341. Fraenkel, G. and K. M. Rudall. 1947. The structure of insect cuticles. Proc. Royal Soc. London, Series B, 134:111-43.

342. Frazier, N. W. 1948. Field tests on some pests of grapes. Calif. Agr. 2(4):13-16.

343. _____ and E. M. Stafford. 1947. Control of the leafhopper *Erythroneura elegantula* on grape with DDT and other insecticides. Jour. Econ. Ent. 40:487-95.

344. Frear, D. E. H. 1942. Chemistry of insecticides and fungicides. D. Van Nostrand Co., New York, London.

345. _____ 1948. Chemistry of insecticides, fungicides, and herbicides. Second Edition. D. Van Nostrand, Inc., New York, London.

346. _____ and E. J. Seiferle. 1947. Chemical structure and insecticidal efficiency. Jour. Econ. Ent. 40:736-41.

347. Fredrick, J. M. 1943. Some preliminary investigations of the green scale, *Coccus viridis* (Green), in South Florida. Fla. Entom. 26:13-15. 78 pp., 16 figs., 3 pls.

348. Freeborn, S. B. 1931. Citrus scale distribution in the Mediterranean basin. Jour. Econ. Ent. 24:1025-31.

349. French, O. C. 1942. Spraying equipment for pest control. Calif. Agr. Exp. Sta. Bul. 666.

350. _____ 1947. Utilizing airplanes for pest control. Farm Implement News 68 (6):60, 112, 114, 116.

351. _____ 1947. The use of the airplane for pest control. The Rice Journal 50(3):20, 21-26.

352. Friend, W. H. 1947. Control of citrus pests in the lower Rio Grande Valley. Prog. Rpt. Texas Agr. Exp. Sta. April 19, 1947. 2 pp.

353. Frost, S. W. 1942. General entomology. McGraw-Hill Book Co., Inc. New York and London.

354. Garcia, C. E. 1939. The citrus rind borer and its control. Philippine Jour. Agr. 10 (1):89-92.

355. Gause, G. F. 1934. The struggle for existence. The Williams and Wilkins Company, Baltimore.

356. Gentner, L. G., H. E. Morrison and W. B. Rasmussen. 1948. Aerosol generator applications of DDT for codling moth control. Jour. Econ. Ent. 41:67-69.

357. Geoffroy, E. 1895. Contribution à l'étude du Robinia Nicou Aublet au point de vue botanique, chimique et physiologique. Ann. Inst. Colon. Marseille 3:1-86.

358. Glaser, R. W. and J. W. Chapman. 1913. The wilt disease of gypsy moth caterpillars. Jour. Econ. Ent. 6:479-88.

359. Glasgow, R. D. 1948. Application of insecticides: old and new methods. Ind. Eng. Chem. 40:675-79.

360. _____ and D. L. Collins. 1946. The thermal aerosol for generator for large scale application of DDT and other insecticides. Jour. Econ. Ent. 39:227-35.

361. Gleissner, B. D., F. Wilcoxin and E. H. Glass. 1947. O, o-diethyl o-p-nitrophenyl thionophosphate, a promising new insecticide. Agr. Chem. 2 (10):61.

362. Gnadinger, C. B. 1936. Pyrethrum flowers. McLaughlin Gormley King Co., Minneapolis, Minn.

363. Goff, C. C. and A. N. Tissot. 1932. The melon aphid, *Aphis gossypii* Glover. Fla. Agr. Exp. Sta. Bul. 252.

364. Golding, F. V. 1945. Fruit-piercing lepidoptera in Nigeria. Bul. Ent. Res. 36(2):181-84.

365. Gomez Clemente, F. 1943. Cochinillas que atacan a los ágrios en la región de Levante. Bol. de Pat veg. (Spain) 12:299-328.

696

LITERATURE CITED

365a. Gomeza Ozamiz, J. M. 1945. Ion, Revista Española de Quimica Aplicada 5:745-50. (See also Rev. Appl. Entom., Ser. A, 36(6):199, June, 1948).

366. González-Regueral y Bailly, F. 1932. Las Cochinillas de los ágrios y su tratamiento. Estac. Fitop. Agric. Levante (Spain).

367. Gooden, E. L. and C. M. Smith. 1935. Dimorphism of rotenone. Jour. Amer. Chem. Soc. 57:2616-18.

368. Goodhue, L. D. 1942. Insecticidal aerosol production. Spraying solutions in liquified gases. Indus & Engin. Chem., Indus. Ed. 34:1456-9.

369. _____1946. Aerosols and their application. Jour. Econ. Ent. 39:506-9.

370. _____and F. F. Smith. 1945. DDT in aerosol form to control insects on vegetables. Jour. Econ. Ent. 38:179-182.

371. _____and W. N. Sullivan. 1942. The preparation of insecticidal aerosols by the use of liquified gases. U.S.D.A. Bur. Ent. & Pl. Quar. ET-190.

372. Gossard, H. A. 1901. The cottony cushion scale. Fla. Agr. Exp. Sta. Bul.56.

373. Gould, H. P. 1919. Fig growing in the South Atlantic and Gulf States. U.S.D.A. Farmers' Bul. 1031. Portion on insect pests by J. R. Horton, pp. 28-35.

374. Gough, H. C. 1945. A review of the literature on soil insecticides. Imp. Inst. of Ent. 161 pp.

375. Goulden, C. H. 1939. Methods of statistical analysis. Wiley, New York.

376. Graham, C. 1948. Control of grasshoppers in apple and peach orchards. Jour. Econ. Ent. 41:111.

377. Graham, H. A. and A. Nicholson. 1947. Farm equipment sprouts wings. Country Gentlemen 117 (1):19, 80-84.

378. Gray, G. P. 1915. New fumigating machines. Calif. St. Comm. Hort. Mo. Bul. 4:68-80.

379. _____and E. R. de Ong. 1926. California petroleum insecticides. Laboratory and field test. Ind. Eng. Chem. 18:175-80.

380. _____and A. F. Kirkpatrick. 1929. Resistant scale insects in certain localities to hydrocyanic acid fumigation. Jour. Econ. Ent. 15:400-404.

381. _____and A. F. Kirkpatrick. 1929. Resistant scale investigations. Calif. Citrog. 14 (8):308, 336 (9):364, 380-81.

381a. _____and A. F. Kirkpatrick. 1929. The protective stupefaction of certain scale insects by hydrocyanic acid vapor. Jour. Econ. Ent. 22:878-92.

381b. _____and A. F. Kirkpatrick. 1929. Resistance of the black scale, Saissetia oleae Bern., to hydrocyanic acid fumigation. Jour. Econ. Ent. 22:893-97.

382. Green, E. E. 1909. Coccidae of Ceylon. p. 262.

383. Griffiths, J. T. Jr. and J. R. King. 1948. Comparative compatibility and residual toxicity of organic insecticides as based on grasshopper control. Jour. Econ. Ent. 41:389-92.

384. _____and C. R. Stearns. 1947. A further account of the effects of DDT when used on citrus trees in Florida. Fla. Entom. 30:1-8.

385. _____and W. L. Thompson. 1947. The grasshopper menace in Florida for 1947. Cit. Ind. (8):3,4, 18.

386. _____and W. L. Thompson. 1947. The use of DDT on citrus trees in Florida. Jour. Econ. Ent. 40:386-388.

387. Groves, A. B. 1942. The elemental sulfur fungicides. Va. Agr. Exp. Sta. Tech. Bul. 82.

388. Gunn, D. 1921. The false codling moth (Argyroploce leucotreta Meyn). Union of South Africa Dept. Agr. Sci. Bul. No. 21.

389. Gunther, F. A. 1945. Quantitative estimation of DDT and DDT spray or dust deposits. Ind. Eng. Chem. 17:149.

390. _____1946. Chlorination of benzene. Chem. and Ind. 44:399.

391. _____1948. Sample manipulation and apparatus useful in estimating surface and penetration residues of DDT in studies with leaves and fruits. Hilgardia 18:297-316.

392. _____and R. C. Blinn. 1947. Alkaline degradation of benzene hexachloride. Jour. Amer. Chem. Soc. 69:1215.

393. _____and J. P. LaDue. 1944. Determination of oil deposit on citrus leaves by the steam-distillation method. Jour. Econ. Ent. 37:52-6.

394. _____, D. L. Lindgren, M. I. Elliott, and J. P. LaDue. 1946. Persistence of certain DDT deposits under field conditions. Jour. Econ. Ent. 39:624-27.

697

LITERATURE CITED

395. Gunther, F. A. and F. M. Turrell. 1945. The location and state of rotenone in the root of <u>Derris elliptica</u>. Jour. Agr. Res. 71:61-79.

396. Haag, H. B. 1931. Toxicological studies of <u>Derris elliptica</u> and its constituents. I. Rotenone. Jour. Assoc. Off. Agr. Chem. 17:336-39.

397. Haas, A. R. C. 1934. Relation between the chemical composition of citrus scale insects and their resistance to hydrocyanic acid and fumigation. Jour. Agr. Res. 49:477-92.

398. _____and H. J. Quayle. 1935. Copper content of citrus leaves and fruit in relation to exanthema and fumigation injury. Hilgardia 9:143-77.

399. Hall, S. A. and M. Jacobson. 1948. Chemical assay of tetraethyl pyrophosphate. Agr. Chem. 3(7):30-31.

400. Hall, W. J. 1930. The South African citrus thrips in Southern Rhodesia. British South Africa Co. Publ. 1. 56 pp. 8 pls.

401. _____and W. K. Ford. 1933. Notes on some citrus insects of Southern Rhodesia. The British So. Africa Co. Mazoe Citrus Exp. Sta. Publ. No.2a.

402. Haller, H. L. and C. V. Bowen. 1947. Basic facts about benzene hexachloride Agr. Chem. 2 (1):15-17.

403. Hamilton, H. W. 1947. Economic poison laws. Soap and Sanitary Chemicals. 23(11):114-15.

404. Hammer, O. H. and J. W. Davidson. 1948. DN-289 enjoys broad range of usefulness. Down to Earth (Dow Chem. Co.):4(3):10.

405. Hansberry, T. R. and C. H. Richardson. 1935. A design for testing technique in codling moth spray experiments. Iowa State Co. Jour. Sci. 10:27-35.

406. Hansen, J. W. 1947. Hexaethyl tetraphosphate. Jour. Econ. Ent. 40:600.

407. Hardison, A. C. 1941. Controlling scale cooperatively. Calif. Citrog. 26:283, 308-309.

408. Hardman, N. F. and R. Craig. 1941. A physiological basis for differential resistance of the two races of red scale to HCN. Science 94 (2434):187.

409. Harris, J. S. 1947. Tetraethyl pyrosphosphate. Agr. Chem. 2(10):27-29, 65-66.

410. Hartzell, A. and F. Wilcoxon. 1932. Some factors affecting the efficiency of contact insecticides. II. Chemical and toxicological studies of pyrethrum. Contrib. Boyce Thompson Inst. 4:107-17.

411. _____and F. Wilcoxon. 1934. Organic thiocyanogen compounds as insecticides. Contr. Boyce Thompson Inst. for Pl. Res. 6:269-77.

412. Hazleton, L. W. and E. Godfrey. 1948. Progress report on parathion toxicity. Insect. Dept., Agr. Chem. Div., American Cyanamid Co.

413. Hawes, Ina L. and Rose Eisenberg. 1947. Bibliography on aviation and economic entomology. U.S.D.A. Bibl. Bul. No. 8, April, 1947.

414. Hawkins, J. H. 1946. Effect of calcic and magnesic diluents of calcium arsenate on bean yields. Jour. Econ. Ent. 39:145-48.

415. Hayes, W. B. 1945. Fruit growing in India. Kitabistan, Allahabad. 283 pp. Ilus.

416. Hayward, K. J. 1941. Las cochinillas de los cítricos Tucumanos y su control. Estac. Exper. Agric. de Tucumán Bol. 32.

417. _____1944. Los Moscas de last frutas en Tucumán. Circ. Estac. Exp. Agric. Tucumán No. 126.

418. _____1944. La cochinilla blanca de los cítricos (<u>Unaspis citri</u> Comstock) en Tucumán. Circ. Estac. Exp. Agric. Tucumán No. 124.

419. Headlee, T. J., J. M. Ginsburg and R. S. Filmer. 1930. Some substitutes for arsenic in codling moth control. Jour. Econ. Ent. 23:45-53.

420. Hely, P. C. 1944. The white louse scale (<u>Prontaspis citri</u>). A pest of coastal Citrus trees. New South Wales Agr. Gaz. 55:283-85.

421. _____1938. The bronze orange bug. A pest in citrus orchards. New South Wales, Agr. Gaz. 49:378-80.

422. _____1945. The citrus green tree hopper (<u>Caedicia strenua</u> Wlk.) New South Wales Agr. Gaz. 56:166-68, 178.

423. Henderson, C. E. and H. V. McBurnie. 1943. Sampling technique for determining population of the citrus red mite and its predators. U.S.D.A. Circ. 671:1-11.

424. Henderson, Y. and H. W. Haggard. 1943. Noxious gasses and the principles of respiration influencing their action. Reinhold Publ. Corp., New York.

425. Hensill, G. S. 1947. Insecticide application by vapo-diffusion. Agr. Chem. 2(11):21-23.

698

LITERATURE CITED

426. Hensill, G. S. and W. M. Hoskins. 1935. Factors concerned in the deposit of sprays. I. The effect of different concentrations of wetting agents. Jour. Econ. Ent. 28:942-50.

426a. Hesse, A. J. 1940. A new species of Triphleps (Hemiptera-Heteroptera, Anthocoridae) predaceous on the citrus thrips (Scirtothrips aurantii Faure) in the Transvaal. Jour. Ent. Soc. So. Africa 3:66-71.

427. Hine, C. H. 1948. Conference on health aspects of agricultural chemicals, July 12, 1948. Mimeo. Copy from Bureau of Adult Health. State Dept. of Public Health; Berkeley, Calif.

428. Hinman, E. J. and F. T. Cowan. 1947. New Insecticides in grasshopper control. U.S.D.A. Bur. Ent. and Pl. Quar. E-722.

429. Hirst, S. 1921. On some new parasitic mites. Zool. Soc. London, Proc. 1921:769-802.

430. Hockenyos, C. L. and B. L. Patton. 1945. Petroleum oils used by pest control operators. Soap and Sanitary Chemicals 21(3):109.

431. Hodgson, R. W. 1947. Fruit size problems of California and Florida. Calif. Citrog. 32:322, 344-46.

432. Hoek, H. van den. 1934? Report on the citrus industry of Brazil. So. Afr. Coop. Citrus Exch.

433. Hoffman, W. E. 1926. A stink bug injurious to citrus in South China. Pan-Pacific Sci. Cong. Proc. 2:2030-38.

434. _____1934. Tree borers and their control in Kwangtung. Lingnan Agric. Jour. 1(1):37-59.

435. Holdaway, F. G. 1941. Papaya production in the Hawaiian Islands. Part IV. Insect pests of the papaya and their control. Hawaii Agr. Exp. Sta. Bul. 87.

436. Holland, E. A. 1938. Suicide by ingestion of derris root sp. in New Zealand. Trans. Roy. Soc. Trop. Med. & Hyg. 32:293-94.

437. Holloway, J. K. 1948. Biological control of Klamath weed--Progress report. Jour. Econ. Ent. 41:56-57.

438. _____, C. F. Henderson and H. V. McBurnie. 1942. Population increase of citrus red mite associated with the use of sprays containing inert granular residues. Jour. Econ. Ent. 35:348-350.

439. _____and Huffaker, C. B. 1949. Klamath weed beetles. California Agriculture 3(2):3, 10.

440. _____and T. R. Young, Jr. 1943. The influence of fungicidal sprays on entomogenuous fungi and the purple scale in Florida. Jour. Econ. Ent. 36:453-7.

441. Horton, J. R. 1915. Control of citrus thrips in California and Arizona. U.S.D.A. Farmers' Bulletin 674.

442. _____1922. A swallow-tail butterfly injurious to California orange trees. Calif. State Dept. Agr. Mo. Bul. 11:377-87.

443. _____and C. E. Pemberton. 1915. Katydids injurious to oranges in California. U.S.D.A. Bul. 256.

444. Hoskins, W. M. 1941. Some recent advances in the chemistry and physics of spray oils and emulsions. Jour. Econ. Ent. 34:791-98.

445. _____, A. M. Boyce and J. F. Lamiman. 1938. The use of selenium in sprays for the control of mites on citrus and grapes. Hilgardia 12:115-75.

446. _____and A. S. Harrison. 1934. The buffering power of the contents of the ventriculus of the honeybee and its effect upon the toxicity of arsenic. Jour. Econ. Ent. 27:924-42.

447. Howard, L. O. 1891. The methods of pupation among the Chalcidae. U.S. Div. Ent. Insect Life 4:193-6.

448. _____1930. A history of applied entomology. The Smithsonian Institution, Washington, D. C. 564 pp.

449. _____and W. F. Fiske. 1911. Importation into the United States of the parasites of the gypsy and brown-tail moths. U.S.D.A. Bur. Ent. Bul. 91:1-312.

450. Howard, N. F., C. A. Weigel, C. M. Smith, and L. F. Steiner. 1943. Insecticides and equipment for controlling insects on fruits and vegetables. U.S.D.A. Misc. Publ. 526, 41 pp.

451. Hrenoff, A. K. and C. D. Leake. 1939. Toxicity studies on 2-4-dinitro-6-cyclohexylphenol, a new insecticide. Univ. Calif. Publ. Pharmacol. 1:151-60.

LITERATURE CITED

452. Hsing, Yun Fan. 1947. The toxicity of organic arsenicals. Jour. Econ. Ent. 40:883-95.
453. Hu, C. C. 1931. Citrus culture in China. Calif. Citrog. 16:502, 43-4.
454. Hubbard, H. G. 1885. Insects affecting the orange. U.S.D.A. Div. Ent. 83 pp.
455. Hurst, H. 1940. Permeability of insect cuticle. Nature 145:462-63.
456. _____1945. Enzyme activity as a factor in insect physiology and toxicology. Nature 156:194-98.
457. Hutson, R., J. M. Merritt, and F. Parmelee. 1938. Fixed nicotines and lead arsenate. Jour. Econ. Ent. 31:374-82.
458. Ingle, L. 1947. Toxicity of chlordane to white rats. Jour. Econ. Ent. 40:264-268.
459. Irons, F. 1947. Application equipment. Agr. Chem. 2 (5):23-25, 69.
460. Isely, D. and F. D. Miner. 1946. Sulfur as an insecticide. Jour. Econ. Ent. 39:93-94.
461. Ivy, E. E., C. R. Parencia, Jr., and K. P. Ewing. 1947. A chlorinated camphene for control of cotton insects. Jour. Econ. Ent. 40:513-17.
461a. Jackson, S. N. 1949. Grapevine hoplia control. California Fruit and Grape Grower 3(4):13.
462. Jacob, H. E. 1947. Grape growing in California. Calif. Agr. Ext. Cir. 116.
463. Jeppson, L. R. 1947. Field studies with new chemicals for control of citrus mites. Calif. Citrog. 32:331.
464. _____1948. Citrus red mite control. Calif. Citrog. 33:138, 168.
465. Jones, E. P. 1936. Investigations on the cotton boll worm, Heliothis obsoleta Fabr. British So. Africa Co. Mazoe Citrus Exp. Sta. Rep. 1934:21-82.
466. Jones, H. A., L. C. McAlister, Jr., R. C. Bushland, and E. F. Knipling. 1945. DDT impregnation of underwear for control of body lice. Jour. Econ. Ent. 38:217-23.
467. Jones, L. S. and D. T. Prendergast. 1938. Methods of obtaining an index to density of field populations of citrus red mites. Jour. Econ. Ent. 30:934-40.
468. Jones, P. R. and J. R. Horton. 1911. The orange thrips. U.S.D.A. Bur. Ent. Bul. No. 99, 16 pp.
469. Jones, R. M. 1933. Reducing inflammability of fumigants with carbon dioxide Ind. & Eng. Chem. 25:394-96.
470. Jones, S. C. 1937. Lime sulfur to control forest tent caterpillar. Jour. Econ. Ent. 30:678.
471. Jordan, Karl. 1888. Anatomie und Biologie der Physapoda. Zeit. für wiss. Zool. 47:541-620.
472. Kagy, J. F. 1936. Toxicity of some nitro-phenols as stomach poisons for several species of insects. Jour. Econ. Ent. 29:397-405.
473. _____and C. H. Richardson. 1936. Ovicidal and scalicidal properties of solutions of dinitro-o-cyclohexylphenol in petroleum oil. Jour. Econ. Ent. 29:52-61.
474. Kahan, A. R. 1941. Preliminary observations on the use of nicotine sulfate against citrus psylla. Indian Science Congress Proc. 28th p. 192.
475. Kalichevsky, V. A. 1943. The amazing petroleum industry. Reinhold Publ. Corp. New York.
476. Kannan, K. K. 1928. The large citrus borer of South India. Mysore Dept. Agr. Ent. Ser. Bul. 8.
477. Kauer, K. C., R. B. DuVall, and F. N. Alquist. 1946. Abstracts of papers 110th meeting of the Amer. Chem. Soc. Sept. 9-13, 1946.
478. Kearns, C. W., L. Ingle and R. L. Metcalf. 1945. A new chlorinated hydro-carbon insecticide. Jour. Econ. Ent. 38:661-68.
478a. _____, C. J. Weinman and G. C. Decker. 1949. Insecticidal properties of some new chlorinated organic compounds. Jour. Econ. Ent. 42:127-34.
479. Keifer, H. H. 1938. Eriophyid studies. Calif. State Dept. Agr. Bul. 27:181-206.
480. _____1947. Annual report of the Bureau of Entomology and Plant Quarantine: Systematic Entomology, by H. H. Keifer. St. Dept. Agr. Bu. 36:168-73.
481. Kelley, R. H. 1947. Chemical application by helicopter. Agr. Chem. 2(7):36-39.

LITERATURE CITED

482. Kendall, T. A. 1942. The program of the Los Angeles County Insectary on mealybug control. Exch. Pest Cont. Circ. 87:330.

483. Khan, A. W. 1942. Citrus aleurodidae (white flies of citrus) in the Punjab and their control. Punjab Fruit Jour. 6:1218-9.

484. Kido, G. S. and T. C. Allen. 1947. Colloidal DDT: its use in insecticide sprays. Agr. Chem. 2(6):21-23, 67.

485. King, H. L. and D. E. H. Frear. 1944. Relation of chemical constitution of some n-heterocyclic compounds to toxicity as fumigants. Jour. Econ. Ent. 37:629-33.

486. _____and D. E. H. Frear. 1944. Relation of chemical compositions of some n-heterocyclic compounds to toxicity against Tribolium confusum. Jour. Econ. Ent. 37:637-40.

487. _____, D. E. H. Frear, and L. E. Dills. 1944. Relation of chemical constitution of some n-heterocyclic compounds to toxicity to Aphis rumicis. Jour. Econ. Ent. 37:637-40.

488. Klein, H. Z. 1939. The present position of citrus insect pests and their control. Hadar 12:157.

489. _____1947. The rust mite of citrus trees in Palestine in 1946. Hassadeh 27(7):355-6.

490. _____and J. Perzelan. 1940. A contribution to the study of Pseudococcus comstocki in Palestine. Hadar 13:107-10.

491. Klotz, L. J. and D. L. Lindgren. 1944. Types of fumigation injury on citrus. Calif. Citrog. 29:244-45.

492. Knapp, H. B. and E. C. Auchter. 1941. Growing trees and small fruits. John Wiley & Sons. See specifically pp. 304-55.

493. Knight, H. 1925. Factors affecting efficiency in fumigation with hydrocyanic acid. Hilgardia 1:49-51.

494. _____1932. Some notes on scale resistance and population density. Pomona College Jour. Ent. and Zool. 24(1):1.

495. _____1942. Some observations on oil deposit. Jour. Econ. Ent. 35:330-32.

496. _____, J. C. Chamberlin, and C. D. Samuels. 1929. On some limiting factors in the use of saturated petroleum oils as insecticides. Plant Physiology 4:299-321.

497. _____and C. R. Cleveland. 1934. Recent developments in oil sprays. Jour. Econ. Ent. 27:269-89.

498. Knowlton, G. F. and M. J. Janes. 1933. Lizards as predators of the beet leafhopper. Jour. Econ. Ent. 26:1011-16.

499. Kurosawa, M. 1938. Notes on three unrecorded thrips from greenhouse in Nippon. Kontyu 12:121-29 (in Japanese).

500. LaFollete, J. R. 1945. Compatibility of materials used as sprays and dusts on certain citrus trees in California. Exch. Pest Cont. Circ. Subject Series No. 3, October, 1945.

501. LaForge, F. B., and H. L. Haller. 1932. Rotenone XIX. The nature of the alkali soluble hydrogenation products of rotenone and its derivatives and their bearing on the structure of rotenone. Jour. Amer. Chem. Soc. 54:810-18.

502. _____, H. L. Haller and L. E. Smith. 1933. The determination of the structure of rotenone. Chem. Rev. 12:181-213.

503. Lamiman, J. F. 1933. Control of the grape leafhopper in California. Calif. Agric. Ext. Circ. 72.

504. _____1935. The Pacific mite, Tetranychus pacificus McG., in California Jour. Econ. Ent. 28:900-3.

505. Lange, H. W. 1936. The biology of the orange tortrix, Eulia (Argyrotaenia) Citrana Fern. Calif. St. Dept. Agr. Mo. Bul. 25:283-85.

506. _____1944. The western grape leaf skeletonizer Harrisina brillians in California. Calif. St. Dept. Agr. Bul. 33:98-104.

507. _____1946. The use of two new soil fumigants, D-D and EDB, for wireworm control in California. Calif. Agr. Exp. Sta. Offset leaflet, February 1946.

508. Langford, G. S. 1947. Entoma -- a directory of insect and plant pest control. Seventh Edition, 1947. Published by the Eastern Branch of the American Association of Economic Entomologists. G. S. Langford, Editor.

LITERATURE CITED

509. Langford, G. S., M. H. Muma and E. N. Cory. 1945. DDT as an automatic kill-
     ing agent in Japanese beetle traps. Jour. Econ. Ent. 38:199-204.
510. Langhorst, H. J. 1947. Hydrocyanic acid gas fumigation. Agr. Chem. 2(4):
     30-73, 69.
511. Latta, R. 1932. The vapor-heat treatment as applied to the control of
     narcissus pests. Jour. Econ. Ent. 25:1020-25.
512. Läuger, P., H. Martin, and P. Müller. 1944. Über Konstitution und toxische
     Wirkung von natürlichen und neuen synthetischen insectentotenden Stoffen.
     Helv. Chem. Acta. 27:892-928.
513. _____, R. Pulver, C. Montiegel, R. Wiesmann, and H. Wild. 1946· Mechanism
     · of intoxication of DDT insecticides in insects and warm-blooded animals.
     Geigy Co., Inc.
514. Leary, J. C., W. I. Fishbein, and L. C. Salter. 1946. DDT and the insect
     problem. McGraw-Hill Book Co., Inc. New York, London.
515. Lee, H. A. and J. P. Martin. 1927. The development of more effective dust
     fungicides by adding oxidizing agents to sulfur. Science 66:178.
516. Lehman, A. J. 1948. The pharmacology of the newer insecticides. Mimeo.
     Circ. of Fed. Food & Drug Adm., Div. of Pharmac.
517. Lemmon, A. B. 1946. Bureau of chemistry. Calif. St. Dept. Agr. Bul.
     35:264-282.
518. _____1946. Bureau of Chemistry. Calif. St. Dept. Agr. Bul. 36:243-58.
519. Lepage, H. S. 1943. A "escama vermelha" dos Citrus em Sao Paulo. Agr.
     Inst. Biol. (1943):311-330.
520. Lewis, H. C. 1932. Citrus dusting equipment. Calif. St. Dept. Agr. Mo.
     Bul. 21:324-39.
521. _____1944. Injury to citrus by tenuipalpus mites. Calif. Citrog. 29:87.
522. _____1948. Three years' use of DDT on citrus in Central California. Calif.
     Citrog. 33:546-47.
523. Lieu, K. O. V. 1945. The study of wood borers of China. I. Biology and
     · control of the citrus-root-cerambycids, Melanauster chinensis, Forster
     (Coleoptera). Fla. Ent. 27 (4):62-101.
524. Lindgren, D. L. 1935. The respiration of insects in relation to the heat-
     ing and fumigation of grain. Minn. Agr. Exp. Sta. Tech. Bul. 109:1-32.
525. _____1938. The stupefaction of red scale, Aonidiella aurantii, by hydro-
     cyanic acid. Hilgardia 11:213-225.
526. _____1941. Factors influencing the results of fumigation of the California
     ·red scale. Hilgardia 13:491-511.
527. _____and R. C. Dickson. 1942. Spray-fumigation experiments on California
     red scale. Jour. Econ. Ent. 35:827-45.
528. _____and R. C. Dickson. 1945. Repeated fumigation with HCN and the develop-
     ment of resistance in the California red scale. Jour. Econ. Ent.
     · 38:296-99.
529. _____and P. D. Gerhart. 1947. The response of California red scale to
     fumigation with ethylene dibromide and ethylene dibromide-HCN. Jour.
     Econ. Ent. 40:680-82.
530. _____, P. D. Gerhart and L. E. Vincent. 1947. Nylon fumigating tents.
     Calif. Citrog. 32:142.
531. _____, J. P. LaDue and R. C. Dickson. 1945. Laboratory studies with rote-
     none oil in sprays to control the California red scale. Jour. Econ.
     Ent. 567-72.
532. _____, J. P. LaDue and Doris Dow. 1944. Preliminary studies with DDT in
     control of the California red scale. Calif. Citrog. 29:180-181.
533. _____, J. P. LaDue and L. S. Harris. 1946. Preliminary field experiments ·
     on the control of purple scale with DDT incorporated in petroleum oil.
     Calif. St. Dept. Agr. Mo. Bul. 35:100-102.
534. _____and W. B. Sinclair. 1944. Relation of mortality to amounts of hydro-
     cyanic acid recovered from fumigated resistant and non-resistant citrus
     scale insects. Hilgardia 16:303-315.
535. _____and W. B. Sinclair. 1945. Sorption of HCN by insect pupae. Jour.
     Econ. Ent. 38:617.
536. Lindquist, A. W. and C. C. Deonier. 1946. Toxicity and repellency of cer-
     tain organic compounds to larvae of Lucilia sericata. Jour. Econ. Ent.
     39:589-97.

LITERATURE CITED

536a. Lindquist, A. W., A. H. Madden, H. G. Wilson, and E. F. Knipling. 1945. DDT as a residual-type treatment for control of houseflies. Jour. Econ. Ent. 38:257-61.

537. Linsley, E. G. and A. E. Michelbacher. 1943. Insects affecting stored food products. Calif. Agr. Exp. Sta. Bul. 676.

538. Lipkin, U. R. and S. S. Kurtz, Jr. 1941. Temperature coefficient of density and refractive index for hydrocarbons in the liquid state. Ind. Chem. Anal. Ed. 13:291-95.

539. Lockwood, S. 1931. An economic survey of the navel orange worm, Myelois venipars Dyar, in Arizona. Calif. St. Dept. Agr. Mo. Bul. 20:655-60.

540. Lodeman, E. G. 1896. The spraying of plants. (Edited by L. H. Bailey, 1913). Macmillan Co., New York.

541. Losada, S. B. 1945. Que son las cochinillas de los citrus y como se combaten. Ministerio de la Economia Nacional (Colombia) Bol. No. 3.

542. Ludvik, G. F. and G. C. Decker. 1947. Toxicity of certain esters of phosphorus acids to aphids. Jour. Econ. Ent. 40:97-100.

543. Lyle, C. 1947. Achievements and possibilities in pest eradication. Jour. Econ. Ent. 40:1-8.

544. McCallan, S. E. A. and F. Wilcoxon. 1931. The fungicidal action of sulfur. II. The production of hydrogen sulphide by sulphured leaves and spores and its toxicity to spores. Contrib. Boyce Thompson Inst. 3:13-38.

545. McClure, F. J. 1933. A review of fluorine and its physiological effects. Physiol. Rev. 13:277-300.

546. McCutcheon, J. W. 1946. Synthetic detergents vs. soap. Soap and Sanitary Chemicals 22(9):37-39, 141, 143.

547. McDonnell, D. C. 1947. New Insecticide Act. Soap and Sanitary Chemicals 23(8):114-117, 152A.

548. _____and J. J. T. Graham. 1917. The decomposition of dilead arsenate by water. Jour. Amer. Chem. Soc. 39:1912-18.

549. _____and C. M. Smith. 1916. The arsenates of lead. Jour. Amer. Chem. Soc. 38:2027-38.

550. _____and C. M. Smith. 1916. The arsenates of lead. II. Equilibrium in the system PbO, AS_2O_5, H_2O. Jour. Amer. Chem. Soc. 38:2366-69.

551. _____and C. M. Smith. 1917. The arsenates of lead. III. Basic arsenates. Jour. Amer. Chem. Soc. 39:937-43.

552. McGregor, E. A. 1916. The citrus red mite named and described for the first time. Ann. Ent. Soc. Amer. 9:284-88.

553. _____1919. The red spiders of America and a few European species likely to be introduced. U.S. Natl. Mus. Proc. 56:641-79, pls. 76-81.

554. _____1928. The citrus thrips and its control. Calif. Citrog. 14:2, 26-27.

555. _____1932. The ubiquitous mite, a new species on citrus. Proc. Ent. Soc. Wash. 34:60-64.

556. _____1934. A new spinning mite on citrus at Yuma, Arizona. Proc. Ent. Soc. Wash. 34:256-59.

557. _____1934. The relationship of fineness of sulfur particles to effectiveness against the citrus thrips in Central California. Jour. Econ. Ent. 27:543-46.

558. _____1935. The Texas citrus mite, a new species. Proc. Ent. Soc. Wash. 37:161-65.

559. _____1939. The specific identity of the American date mite; description of two new species of Paratetranychus. Proc. Ent. Soc. Wash. 41:247-56.

560. _____1941. The avocado mite of California, a new species. Proc. Ent. Soc. Wash. 43:85-8.

561. _____1942. Recently discovered mite on citrus. Calif. Citrog. 27:270.

562. _____1942. The taxonomic status of the so-called "common red spider." Proc. Ent. Soc. Wash. 44:26-29.

563. _____1943. A new spider mite on citrus in southern California. Proc. Ent. Soc. Wash. 45:127-29.

564. _____1944. The citrus thrips, measures for its control, and their effect on other citrus pests. U.S.D.A. cir. No. 708.

565. _____1949. Nearctic mites of the family Pseudoleptidae. So. Calif. Acad. Sci. Memoirs 3(2):1-45, 10 figs., 10 pls.

703

LITERATURE CITED

566. McGregor, E. A. and E. J. Newcomer. 1928. Taxonomic status of the deciduous-fruit Paratetranychus with reference to the citrus mite (P. citri). Jour. Agr. Res. 36:157-81.

567. McIndoo, N. E. 1943. Present status and future trends of nicotine as an insecticide. Jour. Econ. Ent. 36:473-75.

568. McKenzie, H. L. 1934. The green fruit or peach beetle, Cotinus texana Casey, in California. Jour. Econ. Ent. 27:1110.

569. _____ 1935. Biology and control of avocado insects and mites. Calif. Agr. Exp. Sta. Bul. 592.

570. _____ 1937. Morphological differences distinguishing California red scale, yellow scale and related species. Calif. Publ. in Entom. 6:323-36.

571. _____ 1938. The genus Aonidiella (Homoptera:Coccoidea:Diaspididae). Microentomology 3 (1):1-36, fig. 1-16.

572. _____ 1942. Two new species related to red scale (Homoptera:Coccoidea:Diaspididae). Calif. St. Dept. Agr. Bul. 31:141-56; figs. 1-2.

573. Mackie, D. B. 1931. A report on the coccids infesting avocados in California. Calif. State Dept. Agr. Mo. Bul. 20:419-41.

574. _____ and W. B. Carter. 1937. Methyl bromide as a fumigant. A preliminary report. Calif. St. Dept. Agr. Mo. Bul. 26:153-62.

575. _____ and W. B. Carter. 1942. Fruit and vegetable treatments. A loose-leaf booklet. 7 pp. Calif. Dept. Agr.

576. _____ and W. B. Carter. 1942. Stored products treatments. A loose-leaf booklet 25 pp. Calif. Dept. Agr.

577. _____, J. B. Steinweden, W. B. Carter and S. S. Smith. 1942. Nursery stock treatments, new series. A loose-leaf booklet. 100 pp. Calif. Dept. Agr.

578. McClintock, J. A. and W. B. Fisher. 1945. Spray chemicals and application equipment. Greenlee Co., Chicago.

579. Maehler, K. L. 1948. The oriental fruit fly in Guam. Jour. Econ. Ent. 41:991-92.

580. Mail, G. A. 1948. New chemical weapon spells doom for moths. Pest Control and Sanitation 3:11-13.

581. Mallis, A. 1945. Handbook of pest control. MacNair-Dorland Company.

582. Marcovitch, S. 1935. Experimental evidence on the value of strip farming as a method of the natural control of injurious insects with special reference to plant lice. Jour. Econ. Ent. 28:62-70.

583. _____ and W. W. Stanley. 1929. Cryolite and barium fluosilicate: their use as insecticides. Tenn. Agr. Exp. Sta. Bul. 140.

584. Mariño Moreno, E. 1947. Copturomimus perseae Hustache. Nueva especie entomológica, grave plaga del aguacate en Colombia. Revista del Facultad Nacional de Agronomia (Colombia). 7(26):167-247.

585. Markwood, L. N. 1936. Nicotine peat - a new insoluble nicotine insecticide. Ind. Eng. Chem. 28:561-63.

586. Marlatt, C. L. 1898. The peach twig-borer. U.S.D.A. Farmers' Bul. 80:1-5.

587. _____ 1929. Recent developments in federal plant quarantine activities. Jour. Econ. Ent. 22:461-68.

588. Marloth, R. H. and F. J. Stofberg. 1939. The effect of lead arsenate and copper carbonate sprays on the quality of oranges. J. Pomol. 16:329-45.

589. Marshall, J. 1937. Inverted spray mixtures and their development with reference to codling moth control. Wash. Agr. Exp. Sta. Bul. 350.

590. _____ 1948. Oil spray investigation in British Columbia. Jour. Econ. Ent. 41:592-95.

591. Martin, H. 1936. The scientific principles of plant protection. See specifically page 301 et. seq. Edward Arnold & Co., London.

592. _____ 1946. Insecticides: Chemical constitution and toxicity. Jour. Soc. Chem. Ind. 65:402-5.

592a. _____ and R. L. Wain. 1943. Insecticidal action of DDT. Nature 154:512-13.

593. Mason, A. C. and R. D. Chisholm. 1945. Ethylene dibromide as a fumigant for the Japanese beetle. Jour. Econ. Ent. 38:717-18.

594. Mason, A. F. 1936. Spraying, dusting, and fumigating plants. Macmillan Co., New York.

595. Matheson, R. 1944. Entomology for introductory courses. Comstock Publ. Co., Ithaca, New York.

LITERATURE CITED

Mathis, W. 1947. Biology of the Florida red scale in Florida. The Florida Entomologist 29:14-35.

May, A. W. S. 1947. Larger horned citrus bug control with DDT. Queensland Agr. Jour. 65:186-87.

Mentzer, R. L., F. C. Daigh and W. A. Connell. 1941. Agents for increasing the toxicity of pyrethrum to mosquito larvae and pupae. Jour. Econ. Ent. 34:182-86.

Metcalf, C. L. and W. P. Flint. 1932. Fundamentals of insect life. McGraw-Hill Book Co., New York.

_____and W. P. Flint. 1939. Destructive and useful insects. Second Edition. McGraw-Hill, New York.

_____and G. S. Hockenyos. 1930. The nature and formation of scale insect shells. III. State Acad. Sci. Trans. 22:166-84.

Metcalf, R. L. 1947. Relative toxicities of isomeric hexachlorocylohexanes and related materials to thrips. Jour. Econ. Ent. 40:522-25.

_____1948. Some insecticidal properties of fluorine analogues of DDT. Jour. Econ. Ent. 41:416-21.

Metcalf, Z. P. 1940. How many insects are there in the world? Ent. News 51:219-22.

Michelbacher, A. E., G. F. Macleod and R. F. Smith. 1943. Control of diabrotica, or western spotted cucumber beetle, in deciduous fruit orchards. Calif. Agr. Exp. Sta. Bul. 681.

_____, W. W. Middlekauff and D. Davis. 1949. Codling moth investigations in northern California and recommendations for 1949. Calif. Agr. 3(3):3-4.

Middlekauff, W. W., A. E. Michelbacher, and E. Wegenek. 1949. Walnut aphid control. Calif. Agr. 3(4):6, 14.

Miles, W. R. and L. H. Beck. 1947. Infrared absorption in field studies of olfaction in bees. Science 106:512.

Miller, R. L. 1929. The biology and control of the green citrus aphid, Aphis spiraecola Patch. Fla. Agr. Exp. Sta. Bul. 203.

_____, I. P. Bassett, and W. W. Yothers. 1933. Effect of lead arsenate insecticides on orange trees in Florida. U.S.D.A. Tech. Bul. 350.

Monte, O. 1941. Dois percevejos prejudiciais as laranjas. Biológico 7(7):187-91. São Paulo.

Moore, W. 1933. Studies of the "resistant" California red scale, Aonidiella aurantii Mask in California. Jour. Econ. Ent. 21:65-77.

_____and J. J. Willaman. 1917. Studies in greenhouse fumigation with hydrocyanic acid:physiological effects on the plant. Jour. Agr. Res. 11:319-38.

Morozov, S. F. 1935. The penetration of contact insecticides. I. Methods of investigation and general properties of the cuticle with regard to its permeability. (In Russian). Zashch. Rast. ot Vrd. 6:38-58. (Rev. Appl. Ent. Series A 24:587-88).

Morrill, A. W. and E. A. Back. 1911. Whiteflies injurious to citrus in Florida. U.S.D.A. Bur. Ent. Bul. 92.

_____and E. A. Back. 1912. Natural control of whiteflies in Florida. U.S. Bur. Ent. Bul. 102.

Morrison, H. 1921. Red date palm scale. Jour. Agr. Res. 21:669-76.

Morse, F. W. 1887. The use of gases against scale insects. Calif. Agr. Exp. Sta. Bul. 71.

Mote, D. C. 1922. A new orange pest in Arizona. Calif. St. Dept. Agr. Mo. Bul. 11:628-33.

Motz, F. A. 1942. The fruit industry of Argentina. U.S.D.A., Office of Foreign Agricultural Relations, Foreign Agr. Rept. No.1.

_____1942. The fruit industry of Brazil. U.S.D.A., Office of Foreign Agricultural Relations, Agr. Rpt. No. 2.

_____and L. D. Mallory. 1944. The fruit industry of Mexico. U.S.D.A. For. Agr. Rpt. No. 9.

Moulton, D. 1932. The Thysanoptera of South America. Rev. de Entomologia (Brazil) 2(4):451-84.

Moznette, G. F. 1922. The red spider on the avocado. U.S.D.A. Bul. 1035.

_____1922. The avocado: its insect enemies and how to combat them. U.S.D.A. Farmers' Bul. 1261.

705

LITERATURE CITED

626. Moznette, G. F. 1947. DDT for pecan weevil in the Southeastern States. U.S.D.A. Bur. Ent. and Pl. Quar. E-723.

627. _____, C. B. Nickels, W. C. Pierce, T. L. Bissell, J. B. Demaree, J. R. Cole, H. E. Parson, and J. R. Large. 1940. Insects and diseases of the pecan and their control. U.S.D.A. Farmers' Bul. 1829.

628. Muesebeck, C. F. W. 1946. Common names of insects approved by the American Association of Economic Entomologists. Jour. Econ. Ent. 39:427-48.

629. Muller, Hermann. 1873. Probosces capable of sucking the nectar of An- agraecum sesquipedale. Nature 8:223.

630. Munger, F. 1942. A method for rearing citrus thrips in the laboratory. Jour. Econ. Ent. 35:373-75.

631. Murphy, D. F. 1933. Insecticidal activity of aliphatic thiocyanates. II. Mealybugs. Indus. & Eng. Chem. 25:638-9.

632. _____1936. Insecticidal activity of aliphatic thiocyanates. III. Red spiders and mites. Jour. Econ. Ent. 29:606-611.

633. _____and C. H. Peet. 1932. Insecticidal activity of aliphatic thiocyanates. I. Aphis. Jour. Econ. Ent. 25:123-29.

634. Myburg, A. C. 1948. A study of a new method of controlling the false codling moth. Citrus Grower No. 170:1-3, No. 171:5-6, 12.

635. Myers, J. G. 1935. Second report on an investigation into the biological control of West Indian insect pests. Bul. Ent. Res. 26:181-252.

636. Nagai, Ki. 1902. Researches on the poisonous principles of Rohten I. Jour. Tokyo Chem. Soc. 23:744-77.

637. National Plant Board. 1932. Report of secretary of National Plant Board. Jour. Econ. Ent. 25:486-88.

638. Naude, T. J. 1947. Insect pests and their control. Farming in South Africa 22(251):245-59.

639. Needham, J. G., J. R. Traver and Yin-Chi Hsu. 1935. The biology of Mayflies. Comstock Publ. Co., Ithaca, New York.

640. Nel, R. G. 1933. A comparison of Aonidiella aurantii and Aonidiella citrina, including a study of the internal anatomy of the latter. Hilgardia 7:417-66.

641. Nelson, H. D. 1943. The position of the rostralis of the California red scale feeding on lemons. Jour. Econ. Ent. 36:750-51.

642. Nelson, R. H. 1936. Observations on the life history of Platynota stultana Wlsm. on greenhouse rose. Jour. Econ. Ent. 29:306-12.

643. Newcomb, D. A. 1947. Oil spray vs. fumigation. Calif. Citrog. 32:210-12.

644. Newcomb, R. V. 1947. "Speed spraying" an improved method of applying orchard sprays. Agr. Chem. 2(3):26-27, 63.

645. Newcomer, E. J. and R. H. Carter. 1933. Casein ammonia, a practical emulsifying agent for the preparation of oil emulsions by orchardists. Jour. Econ. Ent. 26:880-887.

646. _____and F. D. Dean. 1946. Effect of xanthone, DDT, and other insecticides on the Pacific mite. Jour. Econ. Ent. 39:783-86.

647. _____, F. D. Dean, and F. W. Carlson. 1946. Effect of DDT, xanthone, and nicotine bentonite on the wooly apply aphid. Jour. Econ. Ent. 39:674-76.

648. _____and M. A. Yothers. 1929. Biology of the European red mite in the Pacific Northwest. U.S.D.A. Tech. Bul. 89.

649. Newell, W. and A. C. Brown. 1939. Eradication of the citrus backfly in Key West, Fla. Jour. Econ. Ent. 32:680-82.

650. Newhall, A. G. 1947. Comparison of some volatile soil fumigants. Agr. Chem. 2(8):30-31, 63.

651. Newstead, R. 1900. Monograph of the Coccidae of the British Isles. Ray Society of London.

652. Nichols, A. A. 1935. A study of the fig beetle, Cotinis texana Casey. Ariz. Agr. Exp. Sta. Tech. Bul. 55.

653. Nicholson, A. J. 1933. The balance of animal populations. Jour. Animal Ecology (Supplement):132-178.

654. Nicholson, H. J. 1939. Indirect effects of spray practice on pest population. Verh. VII Internat'l. Kong. für Entom., Berlin, 1938, 33:3022-8.

655. _____and V. A. Bailey. 1935. The balance of animal populations. Part I. Proc. Zool. Soc. London, Part III, pp. 551-98.

656. Nickels, C. B. and Pierce, W. C. 1946. Effect of lead arsenate sprays on the pecan weevil and other pecan insects. Jour. Econ. Ent. 39:792-94.

LITERATURE CITED

657. Norton, L. B. 1940. Distribution of nicotine between water and petroleum oils. Ind. Eng. Chem. 32:241-44.

658. _____and M. A. Yothers. 1929. Sterility in the San Jose Scale. Jour. Econ. Ent. 22:821-22.

659. O'Kane, W. C. 1923. Chemistry in the control of plant enemies. Ind. & Eng. Chem. 15:911.

660. _____1947. Report of the Crop Protection Institute. Jour. Econ. Ent. 40:158-159.

661. _____1947. Results with benzene hexachloride. Jour. Econ. Ent. 40:133-34.

662. _____, G. L. Walter, and H. G. Guy. 1933. Studies of contact insecticides. VI. New Hamp. Agr. Exp. Sta. Tech. Bul. 54.

663. _____, W. A. Westgate, L. C. Glover and P. R. Lowry. 1930. Surface tension, surface activity, and wetting ability as factors in the performance of contact insecticides. Studies of contact insecticides. I. New Hamp. Agr. Exp. Sta. Tech. Bul. 39.

664. Oldham, H. T. and F. T. Thorne. 1947. Six-spotted mite on avocado. Jour. Econ. Ent. 40:279.

665. Omer-Cooper, J. 1940. Remarks on the false codling moth. Rev. Appl. Ent. 29:227-28.

666. Ortega, J. C. 1948. DDT controls codling moth, if used correctly. Diamond Walnut News 30(2):6-8.

667. _____1948. Codling moth on walnuts. Calif. Agr. 2(7):4, 14 and Diamond Walnut News 30 (2):6-8.

668. Osborn, H. 1937. Fragments of Entomological History. Published by the author. Columbus, Ohio.

669. Osburn, M. R. 1942. Cost of control measures for the citrus rust mite. Citrus Ind. 23(9):3, 7, 11.

670. _____and W. Mathis. 1946. Effect of cultivation on Florida red scale populations. Jour. Econ. Ent. 39:571-74.

671. _____and H. Spencer. 1943. Experiments in the control of citrus rust mite. Citrus Ind. 24(6):3, 6-7, 13-15, 18.

672. _____and N. Stahler. 1946. Use of DDT to control the little fire ant. Citrus Ind. 27 (6):3, 7, 11.

673. Pagden, H. T. 1931. Two citrus fruit borers. Malay States Dept. Agr. (Scientific Series) No. 7, pp. 1-29.

674. Palmer, D. F. 1944. Avocados and their insect allies. Calif. Citrog. 29:162-63.

675. _____1945. Control of greenhouse thrips on oranges. Exch. Pest Cont. Circ. No. 129, p. 501.

676. Parker, E. R., M. B. Rounds and C. B. Cree. 1943. Orchard practices in relation to yield and quality of Valencia oranges. Part II. Calif. Citrog. 28:260, 268-69.

677. Parker, W. B. 1933. Vapo-dust - a development in scientific pest control. Jour. Econ. Ent. 26:718-20.

678. Parker, W. L. and J. H. Beacher. 1947. Toxaphene: a chlorinated hydrocarbon with insecticidal properties. Del. Agr. Exp. Sta. Bul. No. 264, Tech. No. 36.

679. Patch, E. M. 1915. Pink and green aphid of potato. Maine Agr. Exp. Sta. Bul. 242.

680. _____1938. Without benefit of insects. Bul. Brooklyn Ent. Soc. 33:1-9.

681. Pearce, G. W. and P. J. Chapman. 1942. Blood albumin for use as an emulsifier. New York St. Agr. Exp. Sta. Bul. 698. See specifically pp. 15-16.

682. _____, P. J. Chapman, and A. W. Avens. 1942. The efficiency of dormant type oils in relation to their composition. Jour. Econ. Ent. 35:211-20.

683. Pence, R. J. 1946. DDT found effective against oak moth, but injurious to camellias. Pacific Coast Nurseryman 4(12):5.

684. Penny, D. C. 1921. The results of using certain oil sprays for the control of the fruit tree leaf-roller in the Pajaro Valley, California. Jour. Econ. Ent. 14:428-33.

685. Pepper, B. B. and L. A. Carruth. 1945. A new plant insecticide for the control of the European corn borer. Jour. Econ. Ent. 38:59-66.

686. Perry, Josephine. 1946. The petroleum industry. Longmans, Green & Co., New York, Toronto.

LITERATURE CITED

687. Persing, C. O. 1935. A discussion of various oils in spray combinations
     with lead arsenate, cryolite and barium fluosilicate. Jour. Econ. Ent.
     28:933-40.
688. _____1943. Chemical control of snails in citrus groves. U.C. Citrus Exp.
     Sta. Div. of Ent. News Letter No. 22, Dec., 1943.
689. _____1945. Snail baits. Calif. Citrog. 30:94-95.
690. _____and A. M. Boyce. 1941. Suggestions for equipping a modern citrus
     duster for the application of tartar emetic-sugar solutions in the con-
     trol of citrus thrips. Univ. of Calif., Citrus Exp. Sta. Div. of Ent.
     News Letter No. 17.
691. _____, A. M. Boyce and C. S. Barnhart. 1943. Present status of citrus
     thrips control. Calif. Citrog. 28:142-165.
692. Phillips, A. M. 1948. Tests with DDT for control of the pecan nut case-
     bearer in the southeastern states. U.S.D.A. Bur. Ent. & Pl. Quar.
     E-746.
693. Phillips, V. T. 1946. The biology and identification of trypetid larvae
     (Diptera:Trypetidae). Memoirs Amer. Ent. Soc. No. 12. 151 pp., 16 pls.
694. Pierce, W. C. and C. B. Nickels. 1941. Control of borers on recently top-
     worked pecan trees. Jour. Econ. Ent. 34:522-26.
695. Pierce, W. D. 1917. A manual of dangerous insects likely to be introduced
     into the United States through importations. U.S.D.A. 256 pp., 107 fig.,
     49 pls.
696. Pierpont, R. L. 1945. Development of a terpene thiocyano ester (Thanite)
     as a fig spray concentrate. Del. Agr. Exp. Sta. Bul. 253.
697. Pilat, M. 1935. Histological researches into the action of insecticides on
     the intestinal tube of insects. Bul. Ent. Res. 26:165-72.
697a. Pinto da Fonseca, J. 1934. Relacão das principais pragas observadas nos
     anos de 1931, 1932 e 1933, nas plantas de maior cultivo no estado de
     S. Páulo. Arch. Inst. Biol. 5:263-89.
698. Planes, S. 1944. La "rona" de los frutos cítricos. Bol. Pat. Veg. Ent.
     Agric. 13:47-54. Also Rev. Appl. Entom. 35:149.
699. Plank, H. K. and A. W. Cressman. 1934. Some predatory habits of the orange
     bagworm Platoeceticus gloverii (Packard). Calif. State Dept. Agr.
     Mo. Bul. 23:207-9.
700. Platt, R. F. 1948. Red scale eradication in eastern Riverside County.
     Exch. Pest Cont. Circ. 159:620-21.
701. Plummer, C. C. 1946. Laboratory studies with DDT against the Mexican fruit-
     fly. U.S.D.A. Bur. Ent. & Quar. E-700.
702. _____and Baker, E. W. 1946. The effect of sublethal dosages of tartar
     emetic on the Mexican fruitfly. Jour. Econ. Ent. 39:775-80.
703. Pope, W. T. 1934. Citrus culture in Hawaii. Haw. Agr. Exp. Sta. Bul. 71.
704. Potts, S. F. 1946. Particle or droplet size of insecticides and its rela-
     tion to application, distribution and deposit. Jour. Econ. Ent.
     39:716-20.
705. Pratt, F. S., A. F. Swain and D. N. Eldred. 1931. A study of fumigation
     problems: "protective stupefaction," its application and limitations.
     Jour. Econ. Ent. 24:1041-63.
706. Quayle, H. J. 1907. Insects injurious to the vine in California. Calif.
     Agr. Exp. Sta. Bul. 192:119-22.
707. _____1908. The grape leafhopper. Calif. Agr. Exp. Sta. Bul. 198.
708. _____1911. The red or orange scale. Calif. Agr. Exp. Sta. Bul. 222.
709. _____1911. The black scale. Calif. Agr. Exp. Sta. Bul. 223.
710. _____1912. The purple scale. Calif. Agr. Exp. Sta. Bul. 226.
711. _____1912. Red spiders and mites of citrus trees. Calif. Agr. Exp. Sta.
     Bul. 239.
712. _____1914. Citrus fruit insects in Mediterranean countries. U.S.D.A. Bul.
     134.
713. _____1915. The citricola scale. Calif. Agr. Exp. Sta. Bul. 255.
713a. _____1916 Are scales becoming resistant to fumigation? Calif. Univ. Jour.
     Agr. 3:333-34, 358.
714. _____1916. Dispersion of scale insects by the wind. Jour. Econ. Ent.
     9:486-93.
715. _____1917. Some comparisons of Coccus citricola and C. hesperidum. Jour.
     Econ. Ent. 10:373-376.

LITERATURE CITED

716. Quayle, H. J. 1918. The Mediterranean and other fruit flies. Calif. Agr. Exp. Sta. Cir. 315.
717. _____1919. Daylight fumigation with hydrocyanic acid. Calif. Citrog. 4:292.
718. _____1922. Resistance of certain scale insects in certain localities to hydrocyanic acid fumigation. Jour. Econ. Ent. 15:400-404.
718a. _____1926. The codling moth in walnuts. Calif. Agr. Exp. Sta. Bul. 402.
719. _____1928. Fumigation with calcium cyanide dust, Hilgardia 8:207-32.
720. _____1929. Citrus fruit insects in Mediterranean countries. U.S.D.A. Bul. 134.
721. _____1929. Western leaf-footed plant bug. Calif. Citrog. 14:125.
722. _____1933. Bordeaux spraying and fumigation injury. Calif. Citrog. 18: 166-84.
723. _____1934. Another species of thrips on citrus fruits. Jour. Econ. Ent. 28:1100.
724. _____1938. Insects of citrus and other subtropical fruits. Comstock Publ. Co. 583 pp., 377 figs.
725. _____1938. The development of resistance to hydrocyanic acid in certain scale insects. Hilgardia 11:183-210.
726. _____1939. Advisability of treating picking boxes to prevent red scale spread. Calif. Citrog. 24:111.
727. _____1940. Time interval in double fumigation. Calif. Citrog. 26:4.
728. _____1941. Control of citrus insects and mites. Calif. Agr. Ext. Service Circ. 123.
729. _____1942. A physiological difference in the two races of red scale and its relation to tolerance to HCN. Jour. Econ. Ent. 35:813-16.
730. _____1943. The increase in resistance in insects to insecticides. Jour. Econ. Ent. 36:493-500.
731. _____and W. Ebeling. 1934. Spray-fumigation treatment for resistant red scale on lemons. Calif. Agr. Exp. Sta. Bul. 583.
732. _____and P. W. Rohrbaugh. 1934. Temperature and humidity in relation to HCN fumigation for the red scale. Jour. Econ. Ent. 27:1083-95.
733. Rakshpal, R. 1945. Citrus fruit sucking moths and their control. Indian Fmg. 6(10):441-43. (Delhi).
734. Reed, A. H. 1931. To review founding of citrus industry during La Fiesta de Los Angeles. Calif. Citrog. 16:507, 532.
734a. Reiniger, C. H. 1942. Contribucao ao estudo dos possiveis insetos vetores de virus dos "citri" no Brasil. Biol. Esc. Nac. Agron. 2:225-45.
735. Rhodes, W. W. 1947. Beer cans for aerosol dispensing. Soap and Sanitary Chem. 23(10):122.
736. Richards, A. G. 1947. The organization of arthropod cuticle: a modified interpretation. Science 105:170-171.
737. Richardson, C. H. 1945. Rate of penetration of nicotine into the cockroach from solutions of various hydrogen ion concentrations. Jour. Econ. Ent. 38:710-11.
738. _____and H. H. Shepard. 1930. The effect of hydrogen-ion concentration on the toxicity of nicotine, pyridine, and methylpyrrolidine to mosquito larvae. Jour. Agr. Res. 41:337-48.
739. _____and C. R. Smith. 1926. Toxicity of dipyridyls and certain other organic compounds as contact insecticides. Jour. Agr. Res. 33:597-609.
740. Richter, P. O. and R. K. Colfee. 1937. Nicotine in oil. A promising insecticide for horticultural purposes. Jour. Econ. Ent. 30:166-74.
741. Ripper, W. E. 1944. Biological control as a supplement to chemical control of insect pests. Nature 153:448-452.
742. _____1948. The development of a helicopter spraying machine. Bul. Ent. Res. 39:1-12, 8 pls.
743. _____, R. M. Greenslade, J. Heaths and H. Barker. 1948. New formulation of DDT, with selective properties. Nature 161:484-85.
744. Rivnay, E. 1934. Biology of the greenhouse thrips Heliothrips haemorrhoidalis Bouché in Palestine. Hadar 7:241-46.
745. _____1939. Studies on the biology and control of Pseudococcus comstocki in Palestine. Hadar 13:107-10.
746. Roark, R. C. 1927. Importance of patent literature to economic entomologists. Jour. Econ. Ent. 20:640-41.

LITERATURE CITED

747. Roark, R. C. 1930. Pyrethrum and soap, a chemically incompatible mixture. Jour. Econ. Ent. 23:460-62.
748. _____1937. An insect that breathes through its nose. Jour. Econ. Ent. 30:522-27.
749. _____1942. Definition of aerosol. Jour. Econ. Ent. 35:105-6.
750. _____1944. "Incompatibility" of insecticides. Jour. Econ. Ent. 37:302.
751. Roberti, D. 1946. A serious attack of red spider (Tetranychus telarius) on Citrus along the Sorrento coast. Int. Bul. Plant Prot. 20(3):4 pp. (cf. Rev. Appl. Ent. Ser. A 35(pt.9):287).
752. Rohrbaugh, P. W. 1934. Penetration and accumulation of petroleum spray oils in the leaves, twigs, and fruit of citrus trees. Plant Phys. 9:699-730.
753. Roney, J. N. and H. C. Lewis. 1948. Citrus thrips in Arizona. Calif. Citrog. 33:302.
754. Roux, W. K. 1938. Report on research work done on the control of the boll worm on citrus, 1935 to 1937. Citrus grower, No. 59, pp. 3, 5-7.
755. Ruehle, G. D. and D. O. Wolfenbarger. 1948. Diseases and pests of the mango in Florida. Univ. of Fla. Subtrop. Exp. Sta. Mimeo. Rpt. No. 13.
756. Ruiz Valencia, G. 1912. Cultivo y explotación del aguacate. Est. Agric. Cent. (Mexico) Bol. No. 71.
757. Russel, H. M. 1912. The red-banded thrips. U.S.D.A. Bul. 99.
758. Russell, L. M. 1948. The North American species of whiteflies of the genus Trialeurodes. U.S.D.A. Misc. Publ. No. 635.
759. Ryan, H. J. 1937. Los Angeles County Agricultural Commissioner's Office. The Tax Digest 15 (6):186-188, 208-212. Published by the California Taxpayers' Association.
760. Sachanen, A. N. 1945. The chemical constituents of petroleum. Reinhold Publ. Co., New York.
761. Sanchez Colin, S. 1942. Cultivo del limonero y industrilización del limón. Sec. de Agric. y Fom. (Mexico).
761a. _____1944. Interpretación de un viaje de estúdios a las zonas citrícolas de Nueva Leon y Veracruz. Fitófilo (Mexico) 3(5):47-52.
762. San Juan, J. M. 1924. Prays citri Mill, a rind insect pest of Philippine oranges. The Philippine Agriculturist 12:339-48.
763. Schmidt, C. T. 1947. Dispersions of fumigants through soil. Jour. Econ. Ent. 40:829-37.
764. Schräder, G. 1942. Hexaalkyl and hexacycloalkyl esters of tetraphosphoric acid. German Patent No. 720,577, issued April 9, 1942; assigned to I. G. Farbinindustrie A. G. Abstract in Chem. Abs. 37:2388 (1943).
765. _____1943. Hexaesters of tetraphosphoric acid. U. S. Patent 2,336,302, issued Dec. 7, 1943; vested in the Alien Property Custodian.
766. Schweig, C. and A. Grünberg. 1936. The problem of the black scale (Chrysomphalus ficus Ashm.) in Palestine Bul. Ent. Res. 27:677-717.
767. Scott, F. M. and K. C. Baker. 1947. Anatomy of Washington Navel orange rind in relation to water spot. Bot. Gaz. 108:459-75.
768. Seifriz, W. 1925. Studies in emulsions I. Types of hydrocarbon oil emulsions. Jour. Phys. Chem. 29:587-95.
769. Shafer, G. D. 1911. How insecticides kill I. Mich. Agr. Exp. Sta. Tech. Bul. 11, 65 pp.
770. Shepard, H. H. 1939. Chemistry and toxicology of insecticides. Burgess Publ. Co., Minneapolis, Minnesota.
771. _____and R. H. Carter. 1933. The relative toxicities of some fluorine compounds as stomach insecticides. Jour. Econ. Ent. 26:913.
772. _____, D. L. Lindgren and E. L. Thomas. 1937. The relative toxicity of insect fumigants. Minn. Agr. Exp. Sta. Tech. Bul. 120.
773. Silvestri, F. 1929. Preliminary report on the citrus scale-insects of China. Fourth Intern. Cong. Ent., Ithaca, N.Y. 1928. 2:897-904.
774. Simmons, P. 1943. Preventing insect damage in home-dried fruits, U.S.D.A. Leaflet No. 235.
775. _____and C. K. Fisher. 1945. Ethyl formate and isopropyl formate as fumigants for packages of dried fruits. Jour. Econ. Ent. 38:715.
776. _____, C. K. Fisher, I. J. Condit, A. N. Hanson, and J. G. Tyler. 1947. Caprification of Calimyrna figs. Calif. Sta. Dept. Agr. Bul. 36:115-21.

LITERATURE CITED

777. Simmons, P., W. R. Reed, and E. A. McGregor. 1931. Fig insects in California. U.S.D.A., Bur. Ent. & Plant Quar. Circ. No. 157, 71 pp.

778. Sinclair, W. B., E. T. Bartholomew and W. Ebeling. 1941. Comparative effects of oil spray and hydrocyanic acid fumigation on the composition of orange fruits. Jour. Econ. Ent. 34:821-29.

779. _____and R. C. Ramsey. 1944. The picric acid method for determining minute amounts of hydrocyanic acid in fumigated insects. Hilgardia 16:291-300.

780. Slade, R. E. 1945. The gamma isomer of hexachlorocyclohexane (Gammexane). Chem. and Ind. 40:314-19.

781. Slawson, H. H. 1948. Experiment station digest. Agr. Chem, 3(10):43.

782. Smit, B. 1947. Fruit-sucking moths. Farming in South Africa 22(258): 758-60.

783. _____1947. The control of fruit flies. Farming in South Africa 22(258): 721-30.

784. Smith, C. F. 1940. Toxicity of some nitrogenous bases to eggs of Lygaeus kalmii. Jour. Econ. Ent. 33:724-27.

785. Smith, C. R. 1935. Occurrence of anabasine in Nicotiana glauca. Jour. Amer. Chem. Soc. 57:959.

786. Smith, E. H. and G. W. Pearce. 1948. The mode of action of petroleum oils as ovicides. Jour. Econ. Ent. 41:173-80.

787. Smith, F. F., R. A. Fulton and P. Brierley. 1947. Use of DDT and H.E.T.P. as aerosols in greenhouses. Part I. Agr. Chem. 2(12):28-31, 61.

788. _____, P. H. Lung, and R. A. Fulton. 1948. Parathion in aerosols for control of pests on greenhouse ornamentals. U.S.D.A., Bur. Ent. & Pl. Quar. E-759.

789. Smith, H. S. 1912. The chalcidoid genus Perilampus and its relation to the problem of parasite introduction. U.S. Bur. Ent. Tech. Ser. 19, 4:33-69.

790. _____1916. An attempt to redefine the host relationships exhibited by entomophagous insects. Jour. Econ. Ent. 9:477-86.

791. _____1917. The life-history and successful introduction into the United States of the Sicilian mealybug parasite. Jour. Econ. Ent. 10:262-68.

792. _____1929. Multiple parasitism: its relation to the biological control of insect pests. Bul. Ent. Res. 20:141-49.

793. _____1934. University of California to continue search for red scale parasites. Calif. Citrog. 19:263, 280-82.

794. _____1935. The rôle of biotic factors in the determination of population densities. Jour. Econ. Ent. 28:873-98.

795. _____1939. Insect populations in relation to biological control. Ecol. Monog. 9:311-320.

796. _____1941. Status of biological control of scale pests. Calif. Citrog. 26:58-76, 77.

797. _____1942. Biological control of the black scale. Calif. Citrog. 27:266, 290.

798. _____1945. Cost of producing Macrocentrus by the potato-tuber-worm method. Jour. Econ. Ent. 38:316-19.

799. _____and H. M. Armitage. 1931. The biological control of mealybugs attacking citrus. Calif. Agr. Exp. Sta. Bul. 509.

800. _____and A. J. Basinger. 1947. History of biological control in California. Calif. Cultivator 94:720, 729.

801. _____and H. Compere. 1916. Observations on the Lestophonus, a dipterous parasite of the cottony cushion scale. Calif. Sta. Dept. Agr. Mo. Bul. 5:384-390.

802. _____and H. Compere. 1928. Preliminary report on the insect parasites of the black scale, Saissetia oleae (Bern.). Univ. of Calif. Publ. in Entom. 4:231-334.

803. _____and H. Compere. 1928. The introduction of new insect enemies of the citrophilus mealybug from Australia. Jour. Econ. Ent. 21:664-69.

804. _____and H. Compere. 1931. An imported parasite attacks the yellow scale. Calif. Citrog. 16:328.

805. _____(Chairman), E. O. Essig, H. S. Fawcett, G. M. Peterson, H. J. Quayle, R. E. Smith, and H. R. Tolley. 1933. The efficacy and economic effects of plant quarantines in California. Calif. Agr. Exp. Sta. Bull. 553.

LITERATURE CITED

806. Smith, H. S. and S. E. Flanders. 1947. Biological control of insect and
      weed pests. Syllabus for course in biological control at the University
      of California (mimeographed).
807. _____and S. E. Flanders. 1948. Search for parasites of red scale reoriented.
      Calif. Citrog. 34:4, 17-18, 20.
808. Smith, L. M. 1932. A shot hole borer. Calif. Agr. Ent. Ser. Circ. 64.
809. _____1945. Control of the shot-hole borer. Univ. of Calif. Div. of Ent.
      (Berkeley) News Letter (May, 1945).
810. _____1948. Red spiders on grapes. Calif. Agr. 2(4):2-5, 12.
811. _____and E. M. Stafford. 1948. The bud mite and the erineum mite of grapes.
      Hilgardia 18:317-34.
812. Smith, M. I., H. Bauer, E. F. Stohlman and R. D. Lillier. 1946. The phar-
      macologic action of certain analogues and derivatives of DDT. Jour.
      Pharm. and Exp. Therapeutics 88:359-365.
813. Smith, M. R. 1936. Distribution of the Argentine ant in the United States
      and suggestions for its control or eradication. U.S.D.A. Circ. 387.
814. Smith, R. E. 1921. The preparation of nicotine dust as an insecticide.
      Calif. Agr. Exp. Sta. Bul. 336.
815. _____and J. P. Martin. 1923. A self-mixing dusting machine for applying
      dry insecticides and fungicides. Calif. Agr. Exp. Sta. Bul. 357.
816. Smith, R. H. 1926. The efficacy of lead arsenate in controlling the codling
      moth. Hilgardia 1:403-53.
817. _____1932. Observations of the reactions of citrus trees in California to
      oil spray. Citrus Leaves 12(1):6-8, 25; 12(2):5-6.
818. _____1932. The tank mixture method of using oil spray. Calif. Agr. Exp.
      Sta. Bul. 527.
819. _____1944. Bionomics and control of the nigra scale, Saissetia nigra.
      Hilgardia 16:225-88. 21 figs.
820. _____,H. U. Meyer, and C. O. Persing. 1934. Nicotine vapor in codling moth
      control. Jour. Econ. Ent. 27:1192-95.
821. Smyth, H. F. and H. F. Smyth, Jr. 1932. Relative toxicity of some fluorine
      and arsenical insecticides. Ind. Eng. Chem. 24:229-32.
822. Snedecor, G. W. 1937. Statistical methods applied to experiments in agri-
      culture and biology. 3rd ed. Iowa State Col. Press.
823. Snodgrass, R. E. 1935. Principles of insect morphology. McGraw-Hill Book
      Co., New York and London.
824. Sokoloff, V. P. and L. J. Klotz. 1942. Mortality of the red scale on
      citrus through infection with a spore-forming bacterium. Phytopath.
      32:187-98.
825. Sontakay, K. R. 1943. Lemon butterfly and its control. Indian Farming
      4:456-7.
825a. _____1945. The bark-eating borer of orange. Indian Farming 6:74-5.
826. Spencer, H. and M. R. Osburn. 1948, Randomized-block arrangement for in-
      secticide experiments on citrus trees. Fla. Entom. 31:1-7.
827. Stafford, E. M. 1945. Control of scale insect pests of olives in the
      Sacramento and San Joaquin Valleys. Calif. Agr. Exp. Sta. Lithoprint:
      Oct., 3 pp.
828. _____1948. Scale insect pests of olives. Calif. Fruit and Grape Growers
      2(6):22-23.
829. _____and D. F. Barnes. 1947. Control of fig scale. U.C. Div. Ent. and
      Parisit. News Letter, January, 1947.
830. _____and D. F. Barnes. 1948. Biology of the fig scale in California.
      Hilgardia 18:567-98.
831. _____and D. F. Barnes. 1948. Fig scale. Calif. Agr. 2(4):9-12..
832. _____and N. W. Frazier. 1946. 1947 suggestions for grape leafhopper con-
      trol with DDT. Univ. Calif. Div. Ent. and Parisit. (Berkeley) News
      Letter, Oct. 1946.
833. _____, N. W. Frazier, G. L. Smith, and L. M. Smith. 1946. Grape insects.
      Univ. Calif. Div. Entom. & Parisit. (Berkeley). News Letter, File
      10.1.
834. Stammers, F. M. G. and F. G. S. Whitfield. 1947. The toxicity of DDT to
      man and animals. Bul. Ent. Res. 38:1-73.

Annual Meeting, N. J. Mosq. Exterm. Assoc., 1946.

837. _____, W. L. Parker, D. MacCreary, and W. A. Connell. 1947. A chlorinated bicyclic terpene used to control certain fruit and vegetable insects. Jour. Econ. Ent. 40:79-83.

838. Steiner, L. F., and R. F. Sazama. 1938. Experiments with tank-mix nicotine-bentonite-soybean oil for codling moth control. Jour. Econ. Ent. 31:366-74.

839. Steinhaus, E. A. 1945. Insect pathology and biological control. Jour. Econ. Ent. 38:591-96.

840. _____1946. Insect microbiology. Comstock Publ. Co., Ithaca, N.Y.

841. _____1948. Polyhedrosis ("Wilt Disease") of the alfalfa caterpillar. Jour. Econ. Ent. 41:859-65.

842. _____1947. A new disease of the variegated cutworm, Peridroma margaritosa (Haw.) Science. 106:323.

842a. _____, K. M. Hughes and Harriette Blochwasser. 1949. Demonstration of the granulosis virus of the variegated cutworm. Jour. Bact. 57:219-224.

843. Steinweden, J. B. 1929. Bases for the generic classification of the coccoid family Coccidae. Ent. Soc. Amer. Ann. 22:197-245.

844. _____1945. Methyl bromide fumigation. Calif. St. Dept. Agr. Mo. Bul. 34:4-13.

845. Stewart, W. S. 1948. Instructions for the use of 2, 4-D sprays for control of preharvest drop of citrus fruit. Univ. Calif. Citrus Exp. Sta. Plant Phys. News Letter No. 3.

846. _____and W. Ebeling. 1946. Preliminary results with the use of 2, 4-dichlorophenoxyacetic acid as a spray-oil amendment. Bot. Gaz. 108:286-94.

847. _____and C. Gammon. 1947. Fog application of 2, 4-D to wild grape and other plants. Amer. Jour. Bot. 34:492-96.

848. _____and L. A. Riehl. 1948. Addition of 2, 4-D to oil sprays used for citrus pest control. Univ. of Calif. Citrus Exp. Sta. Plant Phys. News Letter No. 4.

849. _____and L. A. Riehl. 1948. Addition of 2, 4-D to oil sprays. Calif. Citrog. 33:456-58.

850. Stickney, F. 1924. Date palm insects. Date Grower's Inst., First Ann. Rept. pp. 16-17.

852. Storer, T. I. 1948. Control of rats and mice. Calif. Agr. Ext. Ser. Circ. 142.

853. Strand, A. L. 1930. Measuring the toxicity of insect fumigants. Ind. Eng. Chem. 2:4-8.

854. Streeter, L. R. and W. H. Rankin. 1930. The fineness of ground sulfur sold for dusting and spraying. New York Agr. Exp. Sta. Tech. Bul. 160.

855. Strickland, E. H. 1928. Can birds hold injurious insects in check? Scientific Monthly 26:48-56.

856. Sturrock, D. 1944. Notes on the mango. Published by Stuart Daily News Inc., Stuart, Fla.

857. Suit, R. F. 1948. Spreading decline of citrus in Florida. Citrus Ind. 29 (2):12, 14-17.

858. Sullivan, W. N., L. D. Goodhue, and J. H. Fales. 1940. Insecticide dispersion. A new method of dispersing pyrethrum and rotenone in the air. Soap and Sanit. Chem. 16(6):121, 123, 125.

859. Summerville, W. A. T. 1931. The larger horned citrus bug. Queensland Agr. Jour. 36:543-83, 2 figs., 7 pls.

860. Susainathan, B. 1924. The fruit moth problem in the northern Circars. Agr. Jour. Ind. 19:402-4.

861. Swain, A. F. 1919. A synopsis of the aphididae of California. Univ of Calif. Publ. in Entom. Tech. Bul. 3(1):1-221.

862. _____and J. K. Primm. 1942. Mildew prevention on fumigation tents. Calif. Citrog. 27:148, 184.

713

LITERATURE CITED

863. Sweetman, H. L. 1936. The biological control of insects. Comstock Publ. Company Inc., Ithaca, N. Y.

864. Swingle, H. S., and O. I. Snapp. 1931. Petroleum oils and oil emulsions as insecticides, and their use against the San Jose scale on peach trees in the south. U.S.D.A. Tech. Bul. 253:1-48.

865. Synerholm, M. E., A. Hartzell and V. Cullmann. 1947. Pyrethrin extenders. Contrib. Boyce Thompson Inst. 15:35-45.

866. Tattersfield, F. 1932. The loss of toxicity of pyrethrum dusts on exposure to air and light. Jour. Agr. Sci. 2:396-417.

867. _____and J. T. Martin. 1935. The problem of the evaluation of rotenone - containing plants. I. _Derris elliptica_ and _Derris malaccensis_. Ann. Appl. Biol. 22:578-605.

868. _____, C. Potter, and Mrs. E. M. Gillham. 1947. The effect of medium on the toxicity of DDT to aphids. Bul. Ent. Res. 37:497-502.

869. Taubmann, G. 1930. Untersuchungen über die Wirkungen organischer Rhodanide. Arch. Exp. Path. u. Pharmakol. 150:257-84.

870. Taylor, A. L. 1943. The effects of root-knot on fig seedlings. 1. Dis. Rep. 27:224.

871. Thomas, E. E. 1913. A preliminary report of a nematode observed on citrus roots and its possible relation with the mottled appearance of citrus trees. Calif. Agr. Exp. Sta. Circ. 85.

872. _____1923. The citrus nematode, _Tylenchulus semipenetrans_. Calif. Agr. Exp. Sta. Tech. Paper. No. 2.

873. Thompson, M. A. 1937 Summary of state and territorial plant quarantines affecting interstate shipments. U.S.D.A. Misc. Pub. No. 80. 132 pp.

874. Thompson, W. L. 1935. Lime-sulfur sprays for the combined control of purple scale and rust mites. Fla. Agr. Exp. Sta. Bul. 282.

875. _____1939. Cultural practices and their influence upon citrus pests. Jour. Econ. Ent. 32:782-8.

876. _____1939. Notes on _Chaetoanaphothrips orchidii_ (Moulton) found attacking citrus fruit in Florida. Fla. Entom. 22(4):65-69.

877. _____1940. Thrips attacking citrus fruits. The Citrus Industry. October, 1940, pp. 5, 8-9, 12-13, 17.

878. _____1940. Plant bugs in citrus groves. The Citrus Industry. Oct. 1940, pp. 16-17.

879. _____1940. Notes on dictyospermum scale infesting citrus. Citrus Ind. 21(11):5, 9.

880. _____1942. The effect of magnesium deficiency on infestations of purple scale on citrus. Jour. Econ. Ent. 35:351-54.

881. _____1942. Some problems of control of scale insects on citrus. Citrus Ind. 23(6):6-7, 14-15, 18-19.

882. _____1945. Control of the shot-hole borers in citrus trees. Citrus Ind. 26(13):3, 21.

883. _____1946. Preventative sprays for mite control on citrus. Citrus Ind. 27:14, 16-17, 20-21.

884. _____and J. T. Griffiths. 1947. New insecticides and their application on citrus. Proc. Fla. St. Hort. Soc., Sixtieth Meet., pp. 86-90.

885. _____and J. W. Sites. 1945. Relationship of solids and ratio to timing of oil sprays on citrus. Citrus Ind. 26(5):6-8, 14.

886. Thurston, H. W., Jr., and D. E. H. Frear. 1939. The importance of standard-ized procedures in diluting liquid lime sulfur. Phytopath. 29:993-95.

887. Timberlake, P. H. 1913. Preliminary report on the parasites of _Coccus hesperidum_ in California. Jour. Econ. Ent. 6:293-303.

888. _____1923. Descriptions of new chalcid-flies from Hawaii and Mexico (Hymen-optera). Proc. Haw. Ent. Soc. 5:395-417.

889. _____1929. Three new species of the hymenopterous family Encyrtidae from New South Wales. Univ. Calif. Pubs. Ent. 4:5-18.

890. Tinoco Corona, L. 1945. Algunos insectos enemigos del aguacate en las zonas productoras de Queretaro, Guanajuato y Tamaulipas. Fitófilo (Mexico) 4(6):352-63.

891. Tissot, A. N. 1926. Some observations on the life history of the citrus aphid (_Aphis spiraecola_ Patch). Fla. Entom. 10(2):26-27.

LITERATURE CITED

Todd, J. N. 1938. Effective duration of toxicity to the Mexican bean beetle of derris deposits on foliage. Jour. Econ. Ent. 31:478-79.

Tubbs, D. W. 1947. Citrus white fly in Fullerton area. Exch. Pest Cont. Circ. 147:474.

Tucker, R. P. 1930. Some notes on the soluble sulfurs. Calif. St. Dept. Agr. Mo. Bul. 19:422-29.

_____1936. Oil sprays. Chemical properties of petroleum unsaturates causing injury to foliage. Ind. Eng. Chem. 28:458-61.

Turner, N. 1943. Reversals in the order of effectiveness of insecticides. Jour. Econ. Ent. 36:725-28.

Turrell, F. M. 1948. Chemical and physical factors in the sulfur burn on citrus. Scalacs 3:285.

_____and M. Chervenak. 1949. Metabolism of radioactive elemental sulfur applied to lemons as an insecticide. Bot. Gaz. (in press).

Tutin, F. 1928. Investigations on tar distillates and other spray liquids. Part I. Ann. Rept. Agri. and Hort. Res. Sta., Long Ashton. pp. 81-90.

Tyler, J. 1944. The root knot nematode. Calif. Agr. Exp. Sta. Circ. 330:1-33. Rev. ed. 1944.

Ullyett, G. C. 1946. Red scale on citrus and its control by natural factors. Fmg. in S. Afr. 1946 repr. no. 25, 5 pp. Also Rev. Appl. Ent. 36:172.

Urbahns, T. D. 1920. Grasshopper control in citrus orchards. Calif. Citrog. 5:252, 271.

U. S. Dept. of Agr. 1947. Yearbook of Agriculture. See specifically pp. 648-50.

Vaile, R. S. 1924. A survey of orchard practices in the citrus industry of southern California. Calif. Agr. Exp. Sta. Bul. 374.

van den Hoek, M. 1922. Report on the citrus industry of Brazil. South Afr. Coop. Citrus Exch., Ltd.

Van der Linden, T. 1912. Über die Benzol-hexachloride und ihen Zerfall in Trichlor-benzole. Ber. 45:231-47.

van de Merwe, C. P. 1941. The citrus psylla (Spanioza erytreae, del G.). Sci. Bull. Dept. Agric. S. Afr. No. 233, 12 pp., 1 fig.

Van der Meulen, P. A. 1930. A note on the relation between insecticidal action and the physical properties of soap solutions. Jour. Econ. Ent. 23:1011-12.

Van Volkenberg, H. H. 1935. Biological control of an insect pest by a toad. Science 82:278-9.

Vlugter, J. C., H. I. Waterman and H. A. van Westen. 1935. Improved methods of examining mineral oils, especially the high boiling components. J. Inst. Petr. Techn. 21:661-700. II. 700-8.

Voitenko, M. P. 1937. Heliothrips haemorrhoidalis Bouché and its control (in Russian). Plant Prot. No. 14, pp. 112-13. Also Rev. Appl. Entom.

LITERATURE CITED

921. Watson, J. R. 1941. Fall clean-up for white flies and scale insects. Citrus Ind. 22(10):9, 13.
922. _____1944. A Cuban pest of citrus, *Cryptocephalus marginicollis* Suffr: Florida Ent. 27:13, 18.
923. _____1944. Control of citrus aphids. Citrus Ind. 25(3):14-15.
924. _____1944. Control of rust mites. Citrus Ind. 25(7):15, 18.
925. _____and E. W. Berger. 1932. Citrus insects and their control. Fla. Agr. Ent. Service Bul. 67.
926. _____and E. W. Berger. 1937. Citrus insects and their control. Fla. Agr. Ent. Bul. 88,
927. _____and H. E. Bratley. 1944. The effect of mulches on the root-knot nematode. Fla. Agr. Exp. Sta. Rpt. for year ending June 30, 1944-49.
928. _____and C. C. Goff. 1937. Control of root-knot in Florida. Fla. Agr. Exp. Sta. Bul. 311:3-22.
929. Webber, H. J. and L. D. Batchelor. 1943. The Citrus Industry. Vol. I. Univ. Calif. Press. Berkeley, Calif.
930. Weber, A. L., H. C. McLean, H. C. Driggers, and W. J. O'Neill. 1937. Influence of different materials on coverage and adhesiveness of sprays and their effect on residue removal from apples. N. J. Agr. Exp. Sta. Bul. 627.
931. Weber, G. A. 1930. The Plant Quarantine and Control Administration: its history, activities and organization. viii + 198 p. The Brookings Institution, Washington, D. C.
932. _____1930. The Bureau of Entomology: its history, activities and organization, vii + 177 p. The Brookings Institution, Washington, D. C.
933. Weddell, J. A. 1944. Fruit-sucking moths. Queensland Agric. Jour. 59(2): 89-92, 3 figs.
934. Weed, A. 1940. Pyrethrum deterioration. Soap 16:101-3.
935. Weigel, C. A. 1927. Hot-water bulb sterilizers. Jour. Econ. Ent. 37:605-9.
936. Weir, H. M., W. T. Houghton, and T. M. Majewski. 1930. The acid treatment of lubricating distillates. Ind. & Eng. Chem. 22:1293-1300.
937. Weldon, G. P. 1947. Evolution of California's Department of Agriculture. Calif. Cult. 94:744.
938. Wene, G. P. and W. A. Rawlins. 1945. Compatibility of cryolite and copper fungicides. Jour. Econ. Ent. 38:655-7.
939. _____1947. The fog aerosol machine to control vegetable insects. Jour. Econ. Ent. 40:675-79.
940. West, T. F., and G. A. Campbell. 1946. DDT the synthetic insecticide. Chapman & Hall Ltd. London.
941. Weyrauch, W. 1942. La hormiga "coquis." Bol. Mus. Hist. Nat. "Javier Prado." 6(21):193-201, 5 figs.
942. Wheeler, E. W. 1923. Some braconids parasitic on aphids and their life history. Ann. Ent. Soc. Amer. 16:1-29.
943. Wheeler, Wm M. 1907. The fungus-growing ants of North America. An. Mus. Nat. Hist. 23:669-807, 31 figs., 80 refs.
944. White, F. A. 1947. Notes on citrus nematodes. Calif. Citrog. 32:312-13.
945. White, R. T. and S. R. Dutky. 1942. Cooperative distribution of organisms causing milky disease of Japanese beetle grubs. Jour. Econ. Ent. 35:679-82.
946. Whitcomb, W. D. 1935. Naphthalene as a greenhouse fumigant. Mass. Agr. Exp. Sta. Bul. 326.
947. Wigglesworth, V. B. 1933. The physiology of the cuticle and of ecdysis in *Rhodnius prolixus*. Quart. J. Micr. Sci., 76 pp. 269-318.
948. _____1939. The principles of insect physiology. E. P. Dutton and Co., Inc., New York. (See specifically p. 344).
949. Wilcoxon, F. and S. E. A. McCallen. 1930. The fungicidal action of sulfur: I. The alleged rôle of pentathoinic acid. Phytopath. 20:391-417.
950. Wildermuth, V. L. 1916. California green lacewing fly. Jour. Agric. Res. 6:515-25.
951. Wilkes, B. G. and J. N. Wickert. 1937. Synthetic aliphatic penetrants Ind. and Eng. Chem. 29:1234-39
952. Wille, J. E. 1943. Entomología agrícola del Perú. Agr. Exp. Sta., La Molina, Peru.

LITERATURE CITED

953. Wirtschafter, Z. T. and E. D. Schwartz. 1939. Acute ethylene dichloride Poisoning. Jour. Ind. Hgy. and Tox. 21:126-31.

954. Woglum, R. S. 1909. Fumigation investigations in Calif. U.S.D.A. Bur. Ent. Bul. 79.

955. _____1913. Report on a trip to India and the Orient in search of the natural enemies of the citrus white fly. U.S.D.A. Bur. Ent. Bul. 120.

956. _____1920. A recently discovered citrus pest, Platynota tinctana (Walk.) in California. Calif. State Dept. Agr. Mo. Bul. 9:341-43.

957. _____1920. Fumigation of certain plants with hydrocyanic acid: conditions influencing injury. U.S.D.A. Bul. 907, pp. 1-43.

958. _____1921. Control of the Argentine ant in California Citrus orchards. U.S.D.A. Bul. 965.

959. _____1923. Bordeaux and fumigation injury. Calif. Citrog. 18:166-84.

960. _____1925. Observations on insects developing immunity to insecticides. Jour. Econ. Ent. 18:593-97.

961. _____1932. Report on the influence of oil sprays on Navel fruit loss in eastern Los Angeles County. Calif. Fruit Growers Exchange, Office of Field Dept., Pest Control Circ. No. 114.

962. _____1935. Army worms or cutworms - central California. Exch. Pest. Cont. Circ. 5:14.

963. _____1936. Collapse of leaf tissue. Exch. Pest Cont. Circ. 23:64.

964. _____1937. The effect of cold weather on scale insects and spider. Exch. Pest Cont. Circ. 26:71.

965. _____1938. Fruit spotting by leafhoppers in central California. Exch. Pest Cont. Circ. 40:128.

966. _____1939. Grasshoppers in central California. Exch. Pest. Cont. Circ. 53:182.

967. _____1939. Citrus insect conditions in central California; grasshoppers. Exch. Pest Cont. Circ. 54:184.

968. _____1939. Effect of the recent hot spell on citrus pests. Exch. Pest. Cont. Circ. 58:201.

969. _____1939. Fuller's rose beetle active this autumn. Exch. Pest Cont. Circ. 60:212.

970. _____1940. False chinch bug on small citrus trees. Exch. Pest. Cont. Circ. 61:215.

971. _____1940. Purple scale. Exch. Pest Cont. Circ. 68:247.

972. _____1940. The use of oil spray during December. Exch. Pest. Cont. Circ. 72:263.

973. _____1941. Mealybugs have shown some increase. Exch. Pest Cont. Circ. 84:317.

974. _____1942. The Argentine ant. Calif. Citrog. 27:155.

975. _____1942. "Silvering" of interior Valencias by citrus thrips. Exch. Pest. Cont. Circ. 92:354.

976. _____1943. Present condition of important insect pests: mealybug. Exch. Pest. Cont. Circ. 102:393.

977. _____1943. A new type cable-operated tower. Exch. Pest Cont. Circ. 102: 393-94.

978. _____1943. Further analysis of Navel fruit drop from oil spray. Exch. Pest Cont. Circ. 108:415-17.

979. _____1943. Tent pullers - the latest advance in fumigation. Calif. Citrog. 28(8):143, 158-59.

980. _____1944. Boom sprayers in citrus orchards. Exch. Pest Cont. Circ. Subject Series No. 1. pp. 1-8.

981. _____1944. Tent pulling apparatus in California citrus groves. Exch. Pest Cont. Circ. Subject Series No. 2, pp. 9-18.

982. _____1945. Bud mite control on lemons. Exch. Pest. Cont. Circ. 124:480-81.

983. _____1945. Damage to lime trees from oil sprays. Exch. Pest Cont. Circ. 129:502.

984. _____1946. The Arizona navel orange worm established in California. Exch. Pest Cont. Circ. 138:538.

985. _____1946. Recent spread of the citrus rust mite. Exch. Pest Cont. Circ. 141:549-50.

986. _____1946. Bud mite control. Exch. Pest Cont. Circ. 142:553.

LITERATURE CITED

987. Woglum, R. S. 1946; Fumigation on nights of freezing weather is hazardous. Exch. Pest Cont. Circ. 143:555.
988. _____1946. A rotary motion boom sprayer. Exch. Pest Cont. Circ. 143:558.
989. _____1947. Citrus aphis control. Exch. Pest. Cont. Circ. 147:512.
990. _____1947. The fruit tree leaf roller on citrus in central California. 148:577-78.
991. _____1947. Brown rot, septoria, mottle-leaf and leafhopper control in central California. Exch. Pest Cont. Circ. 154:601.
992. _____1947. Control of citrus rust mite; particularly on oranges. Exch. Pest Cont. Circ. 155:603-4.
993. _____1947. Brown rot control. Exch. Pest Cont. Circ. 155:604.
994. _____1947. The citrus black fly campaign in Sonora gets underway. Exch. Pest Cont. Circ. 156:609-10.
995. _____1948. Red scale eradication in central and northern California. Exch. Pest. Cont. Circ. 157:612-13.
996. _____1948. Spring oil spraying of oranges. Exch. Pest. Cont. Circ. 158:618.
997. _____1948. The California - Arizona citrus growers campaign against the citrus black fly on the west coast of Mexico. Exch. Pest Cont. Ciro. Spec. Rpt. No. 5.
998. _____1948. Black fly in Mexico now being handled by the U. S. Bureau of Entomology. Exch. Pest Cont. Circ. 162:634.
999. _____1948. The purple scale and the Glover's scale: how they may be separated. Exch. Pest Cont. Circ. 168:656.
1000. _____1948. A willow pest that occasionally injures oranges. Exch. Pest. Cont. Circ. 168:658.
1001. _____and A. D. Borden. 1921. Control of the Argentine ant in California citrus orchards. U.S.D.A. Bul. 965.
1002. _____and H. C. Lewis. 1947. Nurse rigs speed citrus spraying. Calif. Citrog. 32:285, 306-7.
1003. _____and H. C. Lewis. 1947. Leaf roller attacks citrus trees. Calif. Citrog. 32:308-9.
1004. _____and M. B. Rounds. 1920. Daylight orchard fumigation. Jour. Econ. Ent. 13:476-85.
1005. _____and H. S. Smith. 1947. The citrus black fly. Calif. Citrog. 32: 412-15.
1006. Woke, P. A. 1940. Effects of some ingested insecticides on the midgut wall of the southern armyworm larva. Jour. Agr. Res. 61:321-29.
1007. Wolcott, G. N. 1933. Economic Entomology of the West Indies. 450 pp. Ent. Soc. of Puerto Rico.
1008. Wolfe, H. S. and S. J. Lynch. 1940. Papaya culture in Florida. Fla. Agr. Exp. Sta. Bul. 350.
1008a. _____, L. R. Toy and A. L. Stahl. 1946. Avocado production in Florida, (Revised by H. S. Wolfe). Fla. Agr. Exp. Sta. Bul. 129. See specifically pp. 85-91.
1009. Wolfenbarger, D. O. 1946. Observations on the airplane for application of sprays and dusts. Jour. Econ. Ent. 39:503-5.
1010. _____1947. Tests of some newer insecticides for control of subtropical fruit and truck crop pests. Fla. Entom. 29:37-44.
1010a. _____1948. Heilipus squamosus Lec., A new enemy of the avocado. Yearbook of the California Avocado Society (1948): 98-102.
1011. Wood, M. N. 1947. Almond culture in California. Calif. Exp. Sta. Circ. 103.
1012. Woodhill, A. R. and S. L. Allman. 1927. The dickey rice weevil. New South Wales Agr. Gaz. 38:791-99
1013. Woodstock, W. H. 1946. Method of producing neutral esters of molecularly dehydrated phosphoric acids. U. S. Patent 2,402,703, issued June 25, 1946; assigned to Victor Chemical Works.
1014. Woodworth, C. W. 1902. The red spiders of citrus trees. Calif. Agr. Exp. Sta. Bul. 145.
1015. _____1902. The California peach-tree borer. Calif. Agr. Exp. Sta. Bul. 143.
1016. _____1903. Fumigation dosage. California Agr. Exp. Sta. Bul. 152.
1017. Worthy, W. and H. Wilcomb. 1941. Red scale grading in field and packing house. Calif. Citrog. 26:252, 272.

LITERATURE CITED

1018. Wymore, F. H. 1935. The olive scale *Parlatoria oleae* Colv. Jour. Econ. Ent. 28:552.

1019. Yeomans, A. H. 1948. Field-model aerosol machines. U.S.D.A. Bur. Ent. & Plant Quar. ET-258.

1020. Yothers, W. W. and A. C. Mason. 1930. The citrus rust mite and its control. U.S.D.A. Tech. Bul. No. 176.

1021. Young, H. D. 1916. The generation of hydrocyanic acid gas in fumigation by portable machines. Calif. Agr. Exp. Sta. Circ. 139.

1022. Yust, H. R. 1943. Productivity of the California red scale on lemon fruits. Jour. Econ. Ent. 36:868-72.

1023. _____ and R. L. Busbey. 1942. A comparison of the susceptibility of the so-called resistant and nonresistant strains of California red scale to methyl bromide. Jour. Econ. Ent. 35:343-45.

1024. _____, R. L. Busbey and H. D. Nelson. 1942. Reaction of resistant and non resistant strains of California red scale to fumigation with HCN. Jour. Econ. Ent. 35:816-20.

1025. _____, H. D. Nelson, and R. L. Busbey. 1943. The influence of repeated fumigation with HCN on the susceptibility of the California red scale. Jour. Econ. Ent. 36:872-74.

1026. _____, H. D. Nelson, and R. A. Fulton. 1946. Effect of oil sprays on re-sistance of California red scale to HCN fumigation. Jour. Econ. Ent. 39:822-23.

1027. _____, H. D. Nelson, and R. L. Busbey. 1943. Fumigation of yellow scale with HCN under laboratory conditions. Jour. Econ. Ent. 35:825-26.

1028. _____, I. R. Welborn, Jr., and H. D. Nelson. 1948. Nylon tents for fumi-gation of citrus. Calif. Citrog. 33:382, 405-6.

1029. Zacher, F. 1935. Oele, Fetter und Seifen im Vorratsschutz. Mitt. Ges. Vorratsschutz 11(1):6-9.

1030. Zeidler, O. 1874. Verbindungen von Chloral mit Brom-und Chlorbenzol. Ber. 7:1180-81.

1031. Zimmerman, O. T. and I. Lavine. 1946. DDT killer of killers. Industrial Research Service, Dover, New Hampshire 180 pp., 24 figs.

1032. Zimmerman, P. W. and A. Hartzell. 1947. Hexaethyl tetraphosphate and tetraethyl pyrophosphate: I. Their effects on plants. II. Their toxicities to insects and mites. Contr. Boyce Thomp. Inst. 15:11-19.

Lightning Source UK Ltd.
Milton Keynes UK
UKHW021845271118
333060UK00011B/854/P